0 200 400 600 800 km
INDIA
TROPIC OF CANCER
Mandalay
MYANMAR
Akyab
Hanoi
Haiphong
LAOS
Gulf of Tonkin
Chiang Mai
Vientiane
Vinh
Yangon
Pathein
Mawlamyine
Hue
Da Nang
THAILAND
Coco Channel
Dawei
Bangkok
CAMBODIA
VIETNAM
South China Sea
Phnom Penh
Andaman Sea
Gulf of Thailand
Ho Chi Minh City
Long Xuyen
Spratly Islands
Phuket
Songkhla
Great Channel
George Town
Kota Kinabalu
Banda Aceh
Ipoh
MALAYSIA
Kepulauan Natuna
BRUNEI
Bandar Seri Begawan
Strait of Malacca
Kuala Lumpur
Medan
Melaka
MALAYSIA
Pulau Simeulue
Singapore
SINGAPORE
Borneo
Kuching
Pulau Nias
Sumatra
EQUATOR
Pontianak
Samarinda
Makassar Strait
Padang
Pulau Bangka
Pulau Siberut
INDONESIA
Kepulauan Mentawai
Palembang
Billiton
Banjarmasin
INDIAN OCEAN
Java Sea
N
Tanjungkarang-Telukbetung
Selat Sunda
Jakarta
Semarang
Madura
Surabaya
Bandung
Java
Lombok
Bali
Denpasar
Christmas Island

A FIELD GUIDE TO THE
REPTILES OF SOUTH-EAST ASIA

A FIELD GUIDE TO THE
REPTILES OF SOUTH-EAST ASIA

INDRANEIL DAS

MYANMAR, THAILAND, LAOS, CAMBODIA,
VIETNAM, PENINSULAR MALAYSIA, SINGAPORE,
SUMATRA, BORNEO, JAVA, BALI

Illustrated by Robin Budden, Sandra Doyle, Rachel Ivanyi, Szabolcs Kokay,
Jonathan Latimer, Denys Ovenden, Lyn Wells

BLOOMSBURY
LONDON · NEW DELHI · NEW YORK · SYDNEY

Bloomsbury Natural History
An imprint of Bloomsbury Publishing Plc

50 Bedford Square	1385 Broadway
London	New York
WC1B 3DP	NY 10018
UK	USA

www.bloomsbury.com

BLOOMSBURY and the Diana logo are trademarks of Bloomsbury Publishing Plc

First published in 2010 by New Holland Publishers (UK) Ltd
This edition published in 2015 by Bloomsbury Publishing Plc

www.bloomsbury.com

Copyright © 2010 Indraneil Das

Plate illustrations by:
Robin Budden 1–5, 10–14, 38–40
Sandra Doyle 49–58
Rachel Ivanyi 28–37
Szabolcs Kokay 41–48, 72–74
Jonathan Latimer 59–70
Denys Ovenden 6–9, 71
Lyn Wells 15–27
Illustrations copyright © 2010 Bloomsbury Publishing Plc

Marilyn Dunn has asserted her right under the Copyright, Designs and Patents Act, 1988, to be identified as Author of this work.

All rights reserved. No part of this publication may be reproduced or transmitted in any form or by any means, electronic or mechanical, including photocopying, recording, or any information storage or retrieval system, without prior permission in writing from the publishers.

No responsibility for loss caused to any individual or organization acting on or refraining from action as a result of the material in this publication can be accepted by Bloomsbury or the author.

British Library Cataloguing-in-Publication Data
A catalogue record for this book is available from the British Library
Library of Congress Cataloguing-in-Publication data has been applied for.

ISBN: HB: 978-1-4729-2057-7
ePDF: 978-1-4729-2059-1
ePub: 978-1-4729-2058-4

2 4 6 8 10 9 7 5 3 1

Printed in China by C & C Offset Co. Ltd.

Bloomsbury Publishing Plc makes every effort to ensure that the papers used in the manufacture of our books are natural, recyclable products made from wood grown in well-managed forests. Our manufacturing processes conform to the environmental regulations of the country of origin.

To find out more about our authors and books visit www.bloomsbury.com. Here you will find extracts, author interviews, details of forthcoming events and the option to sign up for our newsletters.

CONTENTS

Acknowledgements	6
Abbreviations, Conventions and Symbols	6
Introduction	7
Defining Reptiles	7
Morphological Specializations	7
Identification of Reptiles	8
Reptile Ecology	11
Region Covered	13
The Environment	15
Reptile Conservation	15
Book Organization	15
Management of Snakebite	16
PLATES	**18**
SPECIES ACCOUNTS	**166**
CROCODILES	**166**
Crocodylidae Crocodiles	166
Gavialiidae Gharials	167
TURTLES	**167**
Geoemydidae Asian Hardshelled Turtles	167
Cheloniidae Sea Turtles	173
Dermochelyidae Leatherback Sea Turtles	174
Emydidae New World Hardshelled Turtles	174
Platysternidae Asian Big-headed Turtles	174
Testudinidae Tortoises	175
Trionychidae Softshell Turtles	176
LIZARDS	**178**
Agamidae Agamid Lizards	178
Anguidae Glass Lizards	195
Dibamidae Worm Lizards	196
Eublepharidae Eyelid Geckos	198
Gekkonidae Geckos	199
Lacertidae 'Typical' Lizards	225
Lanthanotidae Earless Monitor	226
Leiolepidae Butterfly Lizards	226
Scincidae Skinks	228
Shinisauridae Crocodile Lizards	254
Varanidae Monitor Lizards	255
SNAKES	**256**
Acrochordidae Wart Snakes	256
Anomochilidae Giant Blind Snakes	257
Pythonidae Pythons	257
Colubridae 'Typical' Snakes	259
Viperidae Vipers & Pit Vipers	302
Cylindrophiidae Pipe Snakes	312
Elapidae Cobras, Kraits, Coral Snakes & Sea Snakes	313
Homalopsidae Puff-faced Water Snakes	324
Natricidae Water Snakes	330
Pareatidae Slug-eating Snakes	344
Pseudoxenodontidae False Cobras	346
Typhlopidae Blind Snakes	348
Xenodermatidae Strange-skinned Snakes	351
Xenopeltidae Sunbeam Snakes	352
Xenophidiidae Spine-jawed Snakes	353
Glossary of Technical Terms	354
Selected Bibliography	356
Internet Resources	367
Index	369

ACKNOWLEDGEMENTS

This work is dedicated to the memory of six early explorers and scholars of South-East Asian herpetology: Heinrich Boie (1794–1827), Heinrich Kuhl (1797–1821), Henri Mouhot (1826–1861), Malcolm Arthur Smith (1875–1958), William Theobald (1829–1908) and Frank Wall (1868–1950).

I owe a debt to my editor, Krystyna Mayer, for putting this project together, for her editorial acumen and patience.

I am deeply grateful to illustrators Robin Budden, Sandra Doyle, Rachel Ivanyi, Szabolcs Kokay, Jonathan Latimer, Denys Ovenden and Lyn Wells, whose artistic skills are evident from the finished product. Many of the illustrations were prepared from rather limited sources of information, such as text descriptions, photographs and museum specimens. Others in New Holland who helped the project take off include Ken Scriven, for his input and encouragement, James Parry, who originally conceived the project, and commissioning editor Simon Papps.

For information/images/translations/companionship during field trips, I am grateful to Kraig Adler, Natalia Ananjeva, Harry Andrews, E. Nicolas Arnold, Kurt Auffenberg, the late Walter Auffenberg, Mark Auliya, Christopher Austin, Raul Bain, David Barker, Aaron Bauer, David Bickford, Vladimir Bobrov, Wolfgang Böhme, Timothy Brophy, Rafe Brown, John Cadle, Ashok Captain, Chan Chew Lun, Lawan Chanhome, Joseph Charles, Michael Cota, Merrel Jack Cox, Jennifer Daltry, Ilya Darevsky, Patrick David, Stuart Davies, Geoff Davison, Maxmilian Dehling, Arvin C. Diesmos, Cheong-Hoong Diong, Julian Dring, David S. Edwards, Linda Ford, Jack Frazier, Uwe Fritz, Maren Gaulke, Genevieve Gee, Frank Glaw, David Gower, Ulmar Grafe, Allen Greer, Jesse Grismer, Lee Grismer, Wolfgang Grossmann, Andreas Gumprecht, Rainer Günther, Alexander Haas, Jacob Hallermann, Amir Hamidy, James Hanken, Harold Heatwole, Ronald Heyer, Tsutomu Hikida, Marinus Hoogmoed, Rick Hudson, Jaafar Ibrahim, Ivan Ineich, Robert Inger, Djoko Iskandar, John Iverson, Karen Jensen, Jiang Jian-Pe, Ulrich Joger, Vladimir Kharin, Arnold Kluge, André Koch, Gunther Köhler, Gerald Kuchling, Umi Laela, Maklarin Lakim, Edgar Lehr, Tzi-Ming Leong, Alan Leviton, Harvey Lillywhite, Lim Boo Liat, Kelvin Kok Peng Lim, Colin McCarthy, William McCord, Jimmy McGuire, Steven Mahony, Anita Malhotra, Edmond Malnate, Ulrich Manthey, Joseph Martinez, Masafumi Matsui, the late Sherman Minton, Akira Mori, G. Mumpuni, John Murphy, Jonathan Murray, the late Jarujin Nabhitabhata, Ngo Van Tri, the late Wirot Nutphand, Peter Kee Lin Ng, Samhan bin Nyawa, Cord Offermann, Nikolai Orlov, Mark O'Shea, Hidetoshi Ota, Steve Platt, Peter Praschag, Peter C. H. Pritchard, Ding Qi Rao, Arne Rasmussen, Jens Rasmussen, Anders Rhodin, Herbert Rösler, José Rosado, Charles A. Ross, John Rudge, Klaus-Dieter Schulz, Saibal Sengupta, Rick Shine, Gerold Schipper, Irwan Sidik, the late Joseph Slowinski, Hobart Smith, Bryan Stuart, Robert Stuebing, Jeet Sukumaran, Tan Heok Hui, Tog Tan, Kumthorn Thirakhupt, Frank Tillack, Michihisa Toriba, Truong Nguyen, the late Garth Underwood, Jens Vindum, Peter Paul van Dijk, Johan van Rooijen, Miguel Vences, Gernot Vogel, Harold Voris, Van Wallach, David Warrell, Romulus Whitaker, Antony Whitten, Guin Wogan, Perry Wood, Wolfgang Wüster, Norsham Yaakob, Paul Yambun, Yuichirou Yasukawa, Darren J. Yeo, Yong Hoi Sen, Pui Min Yong, Timothy Youmans, Er-Mi Zhao, Zhou Ting, Thomas Ziegler, Nikolay Zinenko and George Zug.

I am grateful to Andrew Alek Tuen, Director, Institute of Biodiversity and Environmental Conservation, Universiti Malaysia Sarawak, and my other colleagues at this institute. I am thankful to my wife, Genevieve V.A. Gee, for the unenviable chore of reading the proofs, as well as for looking after the household during the time spent writing and researching for this book. Finally, my thanks to my friend, Patrick David, Musée National d'Histoire Naturelle, Paris, for his comments on the manuscript before its final journey to the press.

Abbreviations and Conventions

ASEAN Association of South East Asian Nations
asl above sea level
ca circa
IUCN International Union for Conservation of Nature and Natural Resources
nr near
SCL straight carapace length (of turtles)
SVL snout-vent length (of most lizards)
TL total length (of snakes and anguid lizards)

INTRODUCTION

This book provides a convenient means for both non-specialists and professional herpetologists working in the field to identify all valid species of reptile currently known in South-East Asia. In this work, South-East Asia refers to all political entities on the Asian mainland east of India and south of China, and bordered by the Sahul Plate to the south (see page 13 for further details).

Given the intense sampling now taking place in many countries in the region, and from habitats that were until recently inaccessible, new species discoveries, as well as rediscoveries of species hitherto known only from textual descriptions, are now routine. New methods for differentiating cryptic species, especially molecular techniques, have also added significantly to the arsenal for species prospectors. A work of this nature is thus likely to become rapidly outdated, and one hopes it can be revised at periodic intervals, incorporating all new information that becomes available.

Not all species covered here can be reliably recognized in the field: for a number of groups, particularly within the squamates (lizards and snakes), reliable identification can be carried out only after careful scale counts, in addition to examination of dentitional and hemipeneal structures, which is challenging to all except those with access to a microscope or at least a good hand lens.

DEFINING REPTILES

Unlike mammals and birds, which are natural units (or share a common ancestry), reptiles are not easy to define. Contemporary classifications based on both bones and molecular data show that modern birds are 'nested' well inside the group. Exclusion of these feathered cousins therefore renders 'reptiles' incomplete. Reptiles belong to a natural group called 'amniotes', which is characterized by the production of water-tight eggs that are relatively better buffered from the external environment than the eggs of other vertebrates, such as fish and frogs. Amniotes comprise most of the land vertebrates, including mammals, birds, turtles, the tuatara, lizards, snakes and crocodiles. Within this group, with a membership of 20,000 species, are the classes Synapsida, which includes mammals, and Sauropsida, which embraces all other groups (including reptiles and birds). In the latter group, the turtles have been argued to be either the sister group to the clade Sauria, or a sister group to the superorder Lepidosauria. Recent molecular evidence also shows a close relationship between birds and crocodilians. Adding extinct amniotes contributes further complexity to these relationships.

'Reptiles' as defined in this work follows Gauthier et al. (1988) in including 'the most recent common ancestor of extant turtles and saurians, and all of its descendants plus, following conventional usage in the herpetological literature, crocodilians'.

MORPHOLOGICAL SPECIALIZATIONS

While the monophyly (descent from a common ancestor) of the group, as traditionally defined, is itself in question, the use of the word 'reptiles' in this work refers to crocodiles, turtles, snakes and lizards, which do share a number of morphological features. These include: a. a tabular (bone on the posterolateral corner of the skull table) that is small when present; b. a suborbital foramen (hole under the eyeball); c. a supraoccipital anterior crista (prominent ridge above the orbit) and d. a narrow supraoccipital plate (median bone at the back of the skull, which borders the foramen magnum).

CROCODILES

Crocodiles (order Crocodylia) and their relatives today comprise just three living families with about two-dozen species, although a rich assemblage existed in former times. They are some of the largest of living reptiles. They have an enlarged head with powerful jaws equipped with numerous conical and pointed teeth; a broad and posteriorly flattened skull; external ear openings; shortened forelimbs and a long to extremely long elongated snout. A 'third eyelid', the nictitating membrane, covers the eyes while they are underwater. Other external characteristics shared by crocodiles and their close kin, alligators, include rather tough scaled skin, webbed feet and a laterally compressed, muscular tail bearing distinct ridges, the last two features being associated with a life in water and swimming.

Internal characteristics shared by the group include a skull with temporal fenestrae (bilaterally symmetrical holes in the temporal bones); well-developed neural spines (spinous processes of the vertebral column); a well-developed secondary palate and a four-chambered heart.

TURTLES AND TORTOISES

The order Chelonii (Chelonia or Testudines of some authors) includes turtles, tortoises and terrapins, which are arguably the most easily recognizable group within the reptiles. They are characterized by external features such as a lack of teeth (in all living species), and most famously by a shell located above (the carapace) and underneath (the plastron) the body. The carapace and plastron are associated with protection of the head, limbs, tail and internal organs, and derive from the fusion of the vertebral column with the ribs, sternum, and pectoral and pelvic girdles. Significantly, the girdles are located inside the ribs. A majority of turtles have added another element to this nearly impregnable fortress – a layer of keratinized scutes or scales, further reinforcing the shell.

Only a few families of living turtles have done away with scutes altogether, and have a bony shell enveloped by leathery skin. Shell scutes of juvenile Leatherback Turtles (*Dermochelys coriacea*) are lost

IDENTIFICATION OF REPTILES

In many South-East Asian reptiles, body size, shape, form, colour and patterning are diagnostic. In lizards and snakes, scale patterns and counts on the head and body are often key to identification.

MEASUREMENTS OF SIZE

Turtles, Terrapins and Tortoises
There are several ways of assessing the size of turtles, terrapins and tortoises. The size of the species in this book is reported as the measurement along the midline of the straight length of the carapace (SCL).

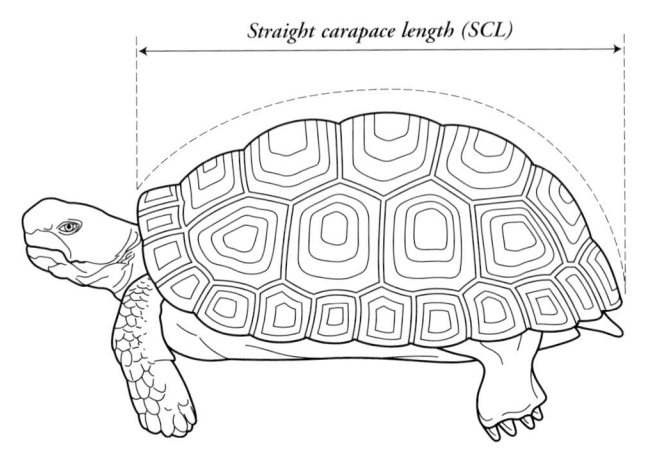

Straight carapace length (SCL)

Lizards
For all lizards apart from glass snakes, size is determined by measuring the body length from snout to vent (SVL), rather than the total length. Many lizards may lose and regenerate their tail, so knowledge of their total length is of little advantage for identification purposes.

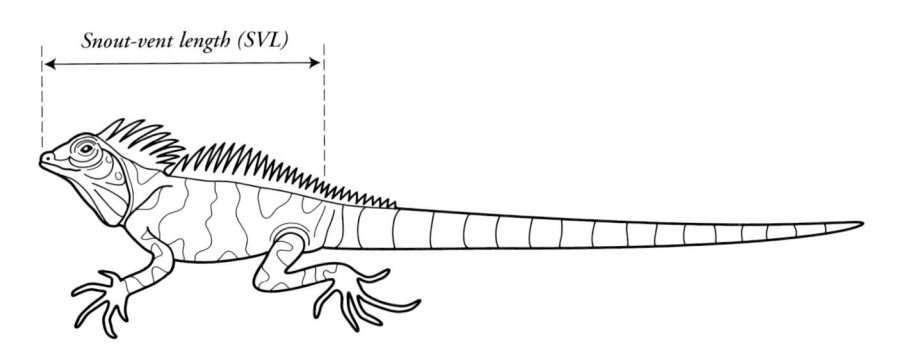

Snout-vent length (SVL)

Snakes and Glass Snakes
In this book, the size of snakes and glass snakes (a group of mostly limbless lizards) is reported as total length (TL), measured from the tip of the snout to the tip of the tail.

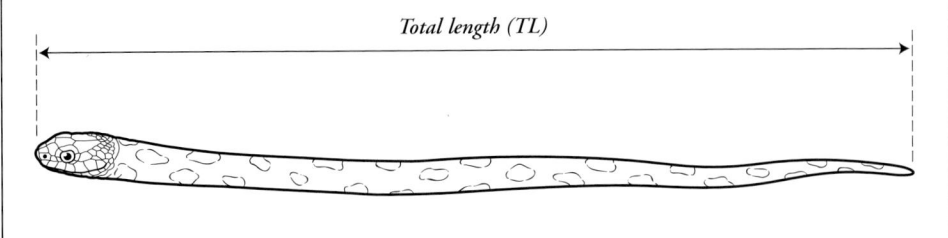

Total length (TL)

MORPHOLOGICAL TRAITS AND SCALE TYPES

Turtles, Terrapins and Tortoises

The scales (or scutes) on the hard, bony upper shell (carapace) and lower shell (plastron) of hardshelled tortoises and turtles are arranged in a specific manner and greatly facilitate identification.

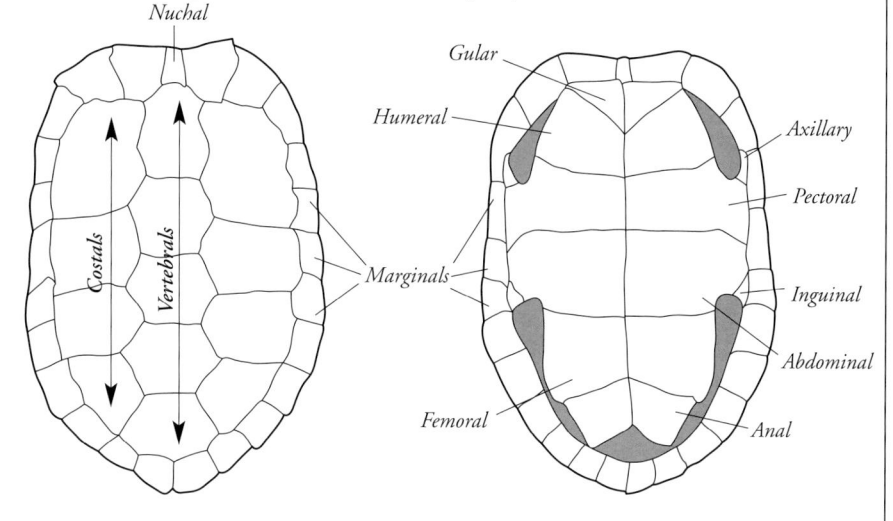

Dorsal view, showing carapace

Ventral view, showing plastron

Head Scales of Lizards

Scale position and size are useful pointers in the identification of lizards (and snakes, see page 10). The positions and names of the scales are more or less typical in most squamates.

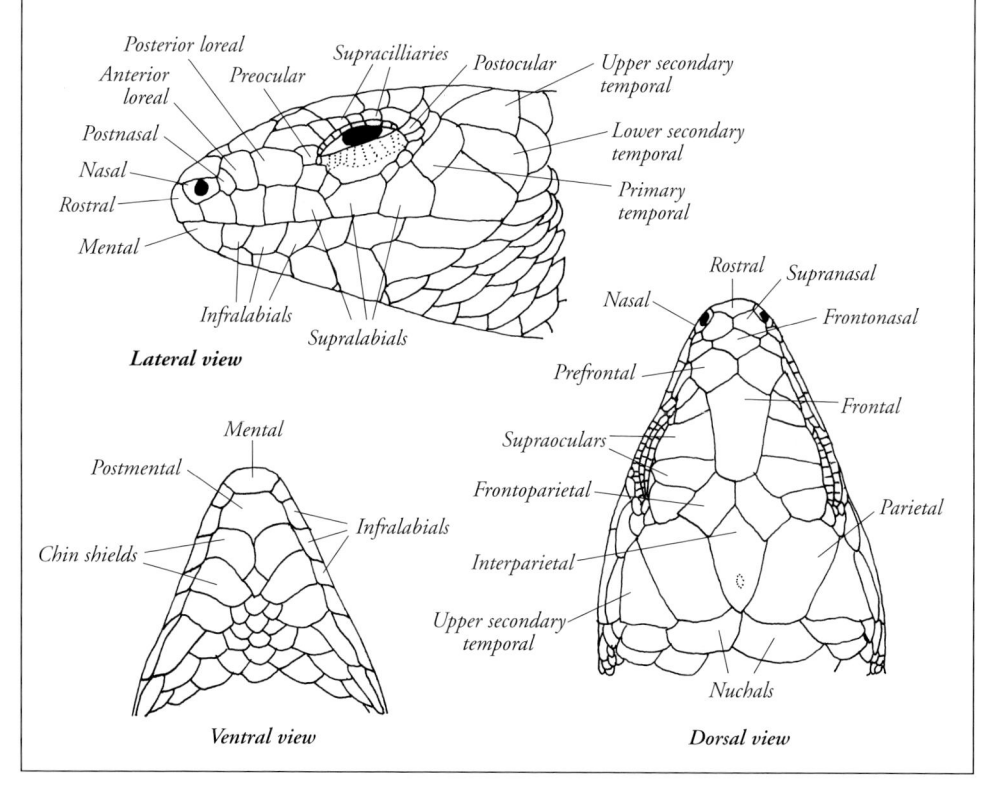

Lateral view

Ventral view

Dorsal view

Body Scale Counts of Snakes

Counts of the number of scale rows on a snake's body are useful for identifying species. A body scale count is made midway between the head and the cloaca, where the number of rows is the highest. The ventral scale row is not counted.

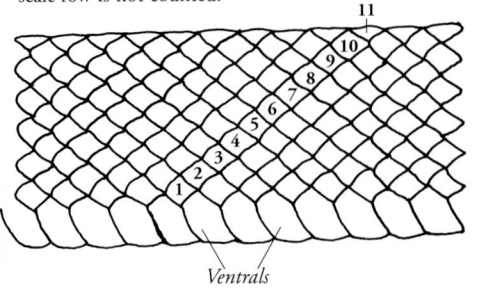

Tail Scale Counts of Snakes

The 'subcaudal scale count' is the number of scales, or pairs of scales (depending on the species) under the tail. The count is made from the first scale or pair of scales below the anal, to the scale just in front of the terminal scale at the tail-tip.

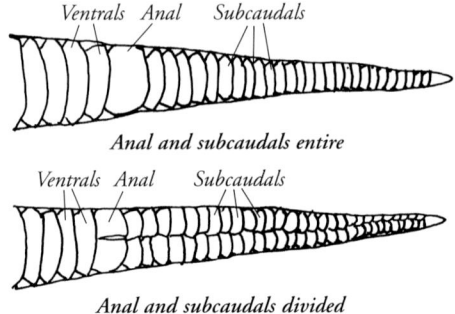

Anal and subcaudals entire

Anal and subcaudals divided

Head Scales of Snakes

The scales on the heads of lizards can be identified by their shape and relative position.

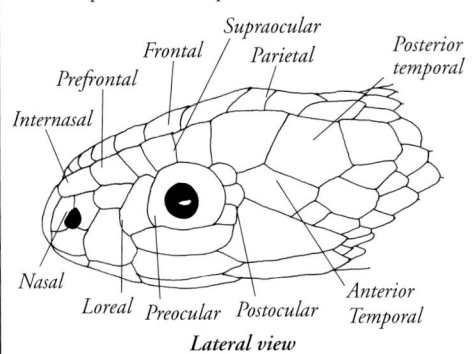

Lateral view

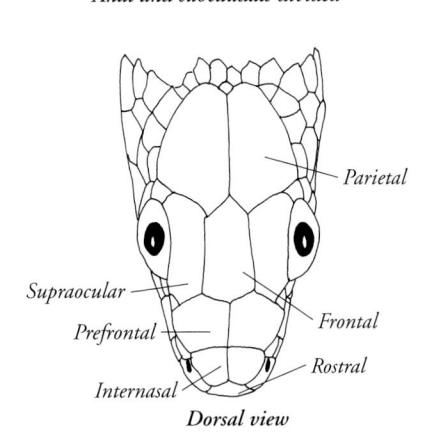

Dorsal view

Dentitional Types in Snakes

Snakes can be placed in one of four groups, according to the form of teeth they have, which is associated with the method used to capture and kill prey. Snake teeth can grow back when lost, and snakes may have several sets of teeth throughout their life – this is necessary because their teeth are often lost during feeding. In **aglyphous snakes**, which are not venomous, each tooth is more or less the same shape and size. **Opisthoglyphous snakes** are similar to aglyphous snakes, but possess weak venom, which is injected by a pair of backwards pointing, enlarged teeth at the back of the maxillae (outer front upper jawbones). The snake must bite the prey, move it to the back of its mouth in order to penetrate it and allow the venom to seep into it along grooves in the fangs. In **proteroglyphous snakes** a substantially enlarged 'fang' in front of the oral cavity points downwards and folds around a venom channel, forming a hollow needle that is injected into prey. **Solenoglyphous snakes** have the most sophisticated venom delivery method, unique to the vipers. The fangs are long and typically folded back. Vipers can open their mouths to almost 180 degrees, and the fangs swing into position to allow them to penetrate deep into prey.

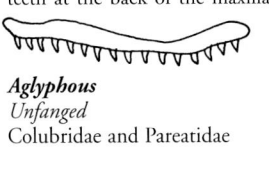

Aglyphous
Unfanged
Colubridae and Pareatidae

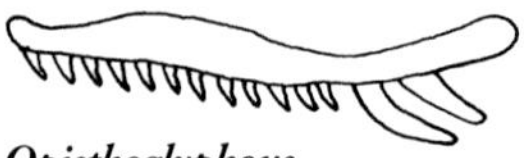

Opisthoglyphous
Rear-fanged
Homalopsidae, Natricidae and Pseudoxenodontidae

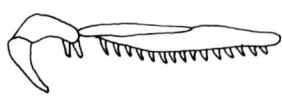

Proteroglyphous
Immovable anterior fanged
Elapidae

Solenoglyphous
Moveable anterior fanged
Viperidae

with growth, being replaced by smooth skin in adults. Aquatic turtles tend to have webbed feet and streamlined shells, while tortoises predictably possess fingers and toes free of a web, and more rounded or elevated shells. In several highly aquatic freshwater turtles, such as the river terrapins *Batagur affinis* and *B. baska*, the snout is upturned and the nostrils are placed relatively high, permitting respiration with the rest of the body submerged.

Internal characteristics associated with the group include an akinetic (with non-moving components) skull; a complete or emarginated temporal region, lacking temporal fenestrae; reduced dermal roof elements; no parietal foramen (opening in the midline of the skull roof containing a sensory organ) and a shoulder girdle that is internal to the ribs and shell.

LIZARDS AND SNAKES

The order Squamata comprises the final grouping within the reptiles, and contains two groups – the lizards and snakes. Although these groups are externally easy to tell apart (except for the several unrelated groups of limbless lizards, and snakes with vestiges of hind limbs, mentioned below), they are united in possessing the following external characteristics: a relatively slender body covered with scales; frequently, a parietal foramen (a light-sensitive organ); paired hemipenes and numerous sharp teeth. Increased capacity of movement by their skulls and mandibles (compared to that of turtles and crocodiles) permits members of the Squamata to ingest relatively large prey.

Internal characteristics uniting members of the group include a single temporal arch (which has been lost in the gekkotans and snakes); a moveable quadrate; a single temporal fenestra and median cranial elements, including premaxillae, frontals and parietals, which are frequently fused, and show the loss of the temporal arch (the bridge-like extension of the jugal and squamosal bones).

The fundamental differences between the two groups are ecological rather than morphological – lizards are primarily predators of invertebrates, while snakes tend to consume vertebrates. Nonetheless, the numerous exceptions to the rule tend to complicate matters, and several large-growing lizards are also known to consume snakes (in addition to other, sometimes large vertebrates).

A number of lizard families have secondarily lost their limbs (or have degenerated hind limbs), most famously within the Anguidae, Dibamidae and Scincidae families. Meanwhile in members of the Pythonidae and Boidae snake families, vestiges of hind limbs exist in the form of cloacal spurs.

Snakes have a spectacle or brille covering their eyelids, giving them the characteristic 'unblinking' stare, while lizards tend to have moveable eyelids (exceptions being members of the Gekkonidae family). Snakes also lack palpebrals (enlarged scales forming the upper eyelid), a parietal eye and foramen (the 'eye' being a sensory structure opening through the top of the skull), a tympanum, a pectoral girdle and forelimbs. A number of characteristics in snakes, such as a long, thin forked tongue, recurved teeth and male combat, are shared with one group of lizards – the monitors – suggesting a close relationship.

REPTILE ECOLOGY

USE OF HABITAT

The greatest species richness of reptiles occurs in tropical rather than temperate regions of the world, due in part to the complexity of available habitats and microhabitats, greater diversity of prey types, and prevailing climatic conditions (including aseasonality, permitting year-round activity), in addition to other factors (such as geological history).

Reptiles inhabit a broad range of habitats in South-East Asia, including grasslands, freshwaters and peat swamps, dry deciduous forests, lowland dipterocarp forests, karst-dominated forests, montane forests, sea coasts and coral reefs. Within a region, more species and larger communities of reptiles are encountered in lowland forests than in the highlands or in other habitats. In moist deciduous forest settings, for instance, species space themselves out, presumably to reduce competition. Within such communities there may be terrestrial and fossorial forms, in addition to arboreal and aquatic ones.

In many natural reptile communities, habitat specialization is evident. Thus, rock-dwelling geckos equipped with specialized scansors-scales on their fingers and toes may scuttle over karst limestone regions, while arboreal snakes ascend apparently effortlessly to the forest canopy, aided by keeled ventral scales and binocular vision. In a similar setting, fossorial species may seem to effortlessly disappear into loose soil, into which they dig with their sharp snouts, resistance to their passage being reduced through the development of highly polished scales that make them appear iridescent. Adaptations in aquatic snakes include the dorsal (as opposed to lateral) placement of the nostrils and eyes, a streamlined body and, in the case of sea snakes, the flattening of the tail-tip, enabling it to function somewhat like a paddle.

TIME OF ACTIVITY

Time of activity is another dimension ecologically separating species, and two obvious categories here are diurnal (day-active) species, which have rounded pupils, and nocturnal (night-active) species, with elliptical pupils. Breaking down this division somewhat are species that are crepuscular, active during the low light associated with dawn or dusk, these tending to possess elliptical pupils.

Timing of activity may remain unchanged throughout the year in some species, or fluctuate seasonally. Certain species may display a bimodal activity pattern, depending on ambient temperature regimes or the activity of specific prey species. Other species, such as those that inhabit the tropical rainforest floor, are most active under conditions of reduced light by day.

THE SENSES

For such a diverse group of organisms, generalizations on sensory biology tend to be difficult. Within the groups, sight ranges from good in the case of the binocular vision of certain tree snakes (genus *Ahaetulla* and others), critical for judging distances, to non-existent, as in several species of burrowing blind snake (genus *Typhlops*) that lack externally visible eyes.

Hearing is similarly reduced in the reptiles, and snakes have been widely considered 'deaf', although recent experimental data shows that the relatively vocal King Cobra (*Ophiophagus hannah*) may well be capable of perceiving some sound in the form of air-borne vibrations.

Olfactory senses are better developed throughout the group, and may be critical for finding food and detecting the odour of conspecifics, particularly during the breeding season. Pit vipers also possess enlarged pits, located between the nostrils and eyes, and pythons labial pits, whose function is to detect the warmth of endothermic prey in darkness.

Finally, mention needs to be made of the Jacobson's organ, situated on the palate of snakes and some lizards, where the two-tined tongue is applied after sampling the external environment to sense its chemical nature, including the possible presence of prey and predatory species.

SOCIAL RELATIONSHIPS

Reptiles tend to be solitary. Social behaviour is the exception rather than the norm, and reptiles aggregate only seasonally, for reproduction – for example during the 'arribadas' or mass-nesting of the Olive Ridley Sea Turtle (*Lepidochelys olivacea*) – or for hibernation in communal sites (as in snakes in temperate regions). Lack of appropriate exposed sites may also force turtles to bask in large numbers on banks and on logs. Parental care is relatively rare within reptiles, and the most famous examples occur among the crocodilian species, which remain near nests and with their emergent young for a few days, much like the birds to which they are distantly related.

ANTI-PREDATOR RESPONSES

Defensive behaviour enhances survival among living organisms, and a variety of strategies is employed by reptiles. In order to stay out of sight of their enemies, a majority of reptiles have colours and patterns that can be described as 'surface mimicry' or crypsis, being hardly discernable from their surroundings. Others display warning coloration to advertise their bite (such as the venomous *Calliophis* or coral snakes). Many use flight as an effective way of rapidly escaping from a threatening situation. Still others feign death, becoming immobile to the touch or manipulation by a potential predator, turning into a ball of coils or expanding parts of their body, all with the intention of confusing the would-be predator.

Venomous snakes, especially vipers and cobras, may hiss, expelling inhaled air in an act of displeasure, while warning of a lethal bite. Cobras may additionally raise their head and spread a hood, a behaviour mimicked by a few groups of unrelated and non-venomous snakes. Some species, while non-venomous, adopt colours of unrelated venomous species, thereby gaining the advantage of protection in being mistaken for a dangerous snake.

A few groups of lizards, most famously the geckos and skinks, can readily detach their tail (caudal autotomy) at signs of danger, at a predetermined fracture plane within the caudal vertebrae. The lost tail is gradually replaced, but may differ in shape and size, and the bone is replaced by a cartilaginous rod. Caudal autotomy (but apparently without regrowth) is also known in snakes of the families Natricidae (genus *Amphiesma*) and Colubridae (genus *Sibynophis*), and occurs between the caudal vertebrae.

HEAT CONTROL IN THE BODY

A majority of the world's living reptiles display ectothermy, whereby the body temperature depends primarily on the absorption of heat energy from the environment. This is fundamentally different from the condition in birds and mammals, which employ endothermy, where the body temperature is dependent primarily on heat produced via the metabolism, and the dissipation of heat to the environment. However, the body temperature of reptiles is not controlled by a single environmental temperature, and individual reptiles are capable of actively regulating their body temperature through the selection of substrates or microhabitats that show a range of temperatures. Typically, an increase in body temperature is achieved by basking, through obtaining heat from the sun. Other ways to gain or loose heat include absorbing heat from the substratum, panting or shunting heat away from extremities exposed to the cold. Ectotherms thus need to allocate a lower proportion of the energy derived from food to maintaining an optimal body temperature, in relation to endotherms.

DIETS AND FORAGING

Few reptiles are herbivores, and those that are tend to be large in body size compared to their carnivorous relatives. One explanation for this is that, being large, reptiles cannot harvest enough animal prey food, and are thus facultatively herbivorous. The juvenile stages of these species (including iguanians, sea turtles and freshwater turtles) display carnivorous habits.

Carnivorous reptiles hunt in two distinct ways. Sit-and-wait species wait in ambush for their prey, detecting it visually. They employ relatively low levels of activity and appear to have poorly developed chemosensory systems. Examples of these species include agamid lizards and geckos. On the other hand, widely foraging species such as monitor lizards actively hunt for prey, using more developed chemosensory systems. These species show high activity levels and appear to harvest more prey calories per unit time.

Dietary specializations are relatively rare in reptiles compared to other vertebrate groups. The most widespread example is a diet of ants in flying lizards (agamid lizards of the genus *Draco*). Other specialized diet types include earthworms (colubrid snakes of the genera *Blythia, Calamaria* and *Plagiopholis*), freshwater and land snails (geoemydid turtles of the genus *Malayemys* and members of the snake family Pareatidae), softshelled crabs (homalopsid snakes of the genus *Fordonia*) and fish (crocodilians of the Gavialiidae family and the Hydrophiidae sea snakes). Some snakes have further specialization: those of the genus *Psammodynastes* consume heavily scaled vertebrates (including fish, skinks and other snakes), and one sea snake, the Beaded Sea Snake (*Aipysurus eydouxii*), eats only fish eggs.

REPRODUCTIVE DIVERSITY

A majority of reptiles reproduce sexually and produce eggs that hatch outside the mother's body. Notable exceptions include the presence of a single sex in some species, represented in the region by several geckos and the ubiquitous Brahminy Blind Snake (*Ramphotyphlops braminus*), which are all-female species. Within reptiles that produce eggs (oviparous), clutch size and offspring size are typically related to other aspects of the ecology of the species. While larger females can produce larger eggs and/or larger clutches, other considerations may be important. These include the number of times a female can breed in a season, and whether additional parental investments, such as care of the eggs and/or neonates, are required. A few species employ ovoviviparity or egg-retention within the body of the female – a form of parental care. All species of crocodilians, turtles, terrapins and tortoises are oviparous.

Foraging strategies, described earlier, are known to influence reproduction. For instance, the clutch size of widely foraging species tends to be smaller than that of sit-and-wait ones, as members of the former group cannot weigh themselves down while pursuing prey. Clutch size itself is subject to considerable variation, even within orders. Within turtles, it may range from a single, relatively large egg, as in the Spiny Hill Turtle (*Heosemys spinosa*), to more than 150 eggs in the marine turtles. Among squamates, clutch size may range from a single egg in Sauter's Keelback (*Amphiesma sauteri*), to 124 eggs in the Reticulated Python (*Broghammerus reticulatus*).

REGION COVERED

The region currently referred to as South-East Asia was variously referred to as 'Indes Orientalis' or 'Eastern India' in medieval Europe, and in even earlier times as 'Land below the Wind and Golden Khersonese' (literally, 'land of gold') by Arab and Indian seafarers. Interest in the region grew from the 1500s, fuelled primarily by the desire for spices and timber. Centralized societies existed here well before this time in north-eastern Thailand (between 1400 and 1000 BC), and these had extensive trading connections with societies in China and India. Subsequently, other important political centres developed, including Co Loa in Vietnam and Fu-Nan in Cambodia. Following the retreat of the European colonial powers around the middle of the 1900s, several nation states emerged, which formed the Association of Southeast Asian Nations (ASEAN). The members of this primarily economic block comprise Myanmar, Thailand, Cambodia, Laos, Vietnam, Malaysia, Singapore, Indonesia, Brunei Darussalam, the Philippines and Timor Lesté.

In this work, South-East Asia refers to all political entities on the Asian mainland east of India and south of China, and bordered by the Sahul Plate to the south. (See also map on front endpapers, and map on back endpapers for topography.) Thus, the eastern islands of Indonesia, including New Guinea, have been omitted, as have the archipelagic nations of Timor Lesté and the Philippines. Brief introductions to the countries whose reptile fauna is dealt with in this work are given below.

Brunei

Negara Brunei Darussalam, total land area 5,270sq km, is located on the north-western coast of Borneo, and is divided into two parts, both wedged into the Malaysian State of Sarawak. To its north is the South China Sea, and it has a coastline of 161km. The capital is Bandar Seri Begawan. The climate is aseasonal, tropical. The coastal plains on the eastern sector rise to mountains further south; on the western sector beyond the coastal plains lie extensive lowland and hill dipterocarp forests. The highest mountain is Bukit Pagon (1,850m asl). The human population in 2008 was 381,371. Environmental problems relevant to biodiversity include logging on Brunei's borders and smoke from forest fires, resulting in haze. Important centres for protection of biological diversity include the Temburong National Park and the Pulau Selurong Nature Reserve.

Cambodia

The Kingdom of Cambodia, total land area 176,520sq km, is part of the region popularly referred to as 'Indo-China', and borders the Gulf of Thailand in the south. Its political boundaries include Thailand to the west and north-west, Vietnam in the east and Laos to the north-east. The coastline is 443km in length. The capital is Phnom Penh. Subtropical monsoon forests characterize Cambodia, and there is a distinct dry season. The topography is essentially low and flat, with mountains in the northern and south-western parts of the country. Major wetlands include the Mekong River and the great lake of Tonle Sap. The highest mountain is Phnom Aural (1,810m asl). The human population in 2008 was 14,241,640. Issues relevant to the protection of biological diversity include loss of forest cover as a result of the removal of timber, and mining for gems. Important areas for biological diversity conservation include the Kirirom National Park, Phnom Bokor National Park and Botum–Sakor National Park.

Indonesia

The Republic of Indonesia, total land area 1,826,440sq km, straddles the archipelagos between the Indian Ocean and the Pacific Ocean. Indonesia shares just three of the 17,508 islands with other political entities – Borneo with Malaysia and Brunei, Timor with Timor Leste, and New Guinea with Papua New Guinea. It has a coastline of 54,716km. The capital is Jakarta. The climate of the islands of Indonesia is mostly aseasonal and tropical, with the highlands experiencing a cooler climate. The coastal plains give rise to tall mountains, chiefly on the larger islands. The highest mountain is Puncak Jaya (5,030m asl) on New Guinea. The human population in 2008 was 237,512,352. Major environmental problems include logging for timber, the clearing of land for shifting cultivation and industrial pollution. Areas gazetted for protection of biological diversity within the Greater Sundas include Meru Betiri National Park, Kerinchi Seblat National Park and Bentuang Karimun National Park.

Laos
The Lao People's Democratic Republic, total land area 230,800sq km, is another country of Indo-China, bordering Vietnam on the west, and is north-east of Thailand and north of Cambodia. It has a short border with China in the north, and is landlocked. The capital is Vientiane. The climate is subtropical with distinct dry and wet seasons in the north, more tropical in the south. Laos has large tracts of primary forests, and the land itself is mostly rugged with some plains. The Mekong River forms part of the western boundary, and the highest peak is Phou Bia (2,817m asl). The human population in 2008 was 6,677,534. Important issues in biodiversity conservation include deforestation and the presence of unexploded ordnance. Protected areas include the Nam Et and Phou Loei National Biodiversity Conservation Area, Phu Luang National Biodiversity Conservation Area and Khammouane Limestone National Biodiversity Conservation Area.

Malaysia
Malaysia, total land area 328,550sq km, is bisected into two parts: on the South-East Asian mainland it forms the long peninsula jutting into the South China Sea, and on the island of Borneo it occupies the twin states of Sarawak and Sabah. On the mainland it is bounded by Thailand to its north; on Borneo it is bounded by Kalimantan (Indonesia); the independent Sultanate of Brunei is wedged within the state of Sarawak. Malaysia has a coastline of 4,675km. The capital is Kuala Lumpur. The climate is primarily aseasonal, tropical, the northern states of Peninsular Malaysia showing a hint of the dry season. The land itself includes coastal plains, rising to mountains in the interior. The highest mountain is Gunung Kinabalu on Borneo (4,100m asl). The human population in 2008 was 25,274,132. Factors that threaten biodiversity include loss of forest cover as a result of commercial logging, clearance of land for shifting cultivation and perhaps also hunting. Important areas for biodiversity include Taman Negara, Bako National Park, Lambir Hills National Park and Gunung Kinabalu Park.

Myanmar
The Union of Myanmar ('Burma' up to 1989), total land area 657,740sq km, is the most north-western country in South-East Asia. It borders the Andaman Sea and Bay of Bengal to the east, sharing borders on the north and north-east with Bangladesh, India and China. To its west is a narrow contact with Laos, and in the south-east it contacts western Thailand. Its coastline is 1,930km in length. The capital is Yangon. The climate can be described as tropical monsoon, the lowlands of central Myanmar being ringed on all sides by gentle to steep mountains, including the highest mountain in South-East Asia (Hkakabo Razi, 5,881m asl). The human population in 2008 was 47,758,180. The key environmental problem related to biological diversity is loss of forest cover as a result of timber harvesting. Important protected areas include Kathapa National Park, Popa Mountain Park and Chhatin Wildlife Sanctuary.

Singapore
The Republic of Singapore, total land area 682.7sq km, is an island state located at the southern tip of Peninsular Malaysia. It has a coastline of 193km. The local climate of this city-state is aseasonal, tropical. While most of the island comprises relatively flat areas, the central plateau is gently undulating. The highest point is Bukit Timah (166m asl). The human population in 2008 was 4,608,167. Threats to local biodiversity include fragmentation and industrial pollution. Important protected areas include Bukit Timah Nature Reserve, Pasir Ris Park and Sungei Buloh Wetland Reserve.

Thailand
The Kingdom of Thailand (formerly Siam), total land area 511,770sq km, borders the Andaman Sea and Gulf of Thailand in the west and south, Myanmar to the north-east and east, northern Peninsular Malaysia in the deep south, and the Indo-Chinese countries of Laos and Cambodia on the north-east, east and south-east. The coastline is 3,219km in length. The capital is Bangkok. The climate is subtropical with a distinct dry season in the north, more aseasonal in the southern peninsula. Apart from the dry central plain and low-lying areas of the south, much of the country is mountainous, especially along the borders. The highest mountain is Doi Inthanon (2,576m asl). The human population in 2008 was 65,493,296. Factors that threaten local biodiversity include deforestation, and hunting and gathering of wildlife. Important areas for biological diversity include Khao Yai National Park, Doi Inthanon National Park and Phu Kao-Phu Phan Kam National Park.

Vietnam
The Socialist Republic of Vietnam is the largest country in Indo-China, with a total land area of 325,360sq km, and borders the Gulf of Thailand, the Gulf of Tonkin and the South China Sea. To the north it is bounded by China, in the east by Laos, and west of southern Vietnam is Cambodia. The coastline is

3,444km in length. The capital is Hanoi. The climate ranges from subtropical monsoon in the north, with distinct dry and wet seasons, to aseasonal tropical in the south. The landscape consists of extensive low-lying areas, including deltas, as well as mountains in the northern, north-western and central regions. The highest mountain is Fan Si Pan, currently known as Phang Si Pang (3,144m asl). The human population in 2008 was 86,116,560. Major factors that threaten biological diversity include agricultural practices such as slash-and-burn agriculture, and also logging. Important areas for biodiversity include Phong Nha-Ke Bang National Park, Vu Quang National Park and Cat Ba Island National Park.

THE ENVIRONMENT

The richness of plant and animal life in the region referred to as South-East Asia is attributed to two factors – geological history and present-day climate – and a number of reptile lineages are found nowhere else in the world. Several groups, especially geckos and skinks, reach their greatest diversification in the region, having terrestrial, arboreal, fossorial and even aquatic forms.

Historically, nearly all of South-East Asia was covered with humid tropical forests, although some regions, including parts of central Myanmar and Sumatra, were naturally dry and supported savannah-type vegetation. Others, such as the Annamite (Truong Son) Mountains and north-western Borneo, remained unaffected by the glacials, and thus harbour ancient lineages of animals and plants.

Starting in the 19th century, the logger and his chain-saw has been the most destructive of all to arrive in the region. Vast swathes of once-forested land have been cleared from South-East Asia, and the pace of forest-clearing stops locally only when all of the commercially valuable trees have been removed. The effects of this practice on the land's biodiversity, not to mention the soil, water and human well-being, have been nothing short of catastrophic.

REPTILE CONSERVATION

Reptile exploitation in South-East Asia ranges from the occasional capture of a gigantic python or sea turtle by indigenous groups for subsistence, to large-scale removal of turtles, lizards and snakes for the food and traditional medicine trade. Hunting pressure can be intense, and is matched only by the acceleration of habitat loss, particularly in the last 30 years. Several local species of turtle and at least two crocodilians are now seriously threatened, and the status of many more species is unclear due to lack of scientific data on identities, populations and distribution.

Few reptile species are deemed charismatic by the general public, and few are scientifically managed. Indeed, conservation and management priorities are generally set by non-biological criteria such as economic value and appeal. Nonetheless, a number of species are known to be of importance to ecosystems and local food webs, and have been identified as key to helping maintain such functions. Additionally, reptiles can aid the dispersal of plants, function as beneficial scavengers, facilitating the release of nutrients locked up in dead tissue, and help to control agricultural pests including rats and mice, and locusts and other insect pests. Unfortunately, reptiles are on the menu of other predatory vertebrates and invertebrates, including humans, several life-saving drugs are derived from snake venoms, and innumerable folk remedies and indigenous systems of medicine are based on various reptile body parts.

A number of actions can be recommended to augment the conservation of reptiles regionally. Identification of species in need of protection is an obvious first step. While progress has been substantial for turtles and crocodiles, the conservation status of most lizard and snake species remains unknown. Efforts need to be made to produce inventories of areas with intact forest cover, especially global and regional hotspots, in order to understand species' distribution and habitat association. Also needed are life-history studies on local reptile assemblages that will provide scientific information on aspects of their ecology and behaviour, which is essential for their conservation and management.

Habitat protection, especially the conservation of lowland rainforests (a majority of local reptiles are restricted to these forests), is fundamental in the conservation of reptile biodiversity. Also in need of protection are specialized forest types such as mangrove and peat swamps, montane forests and limestone karst areas, each of which harbours unique assemblages of reptile species. Efforts also must be made to connect existing protected areas in order to create potentially larger areas under natural forest cover, thus increasing the viability of populations within them.

Trade in wildlife is the earliest occupation of humans, and while a small degree of sustainable use (in non-consumptive forms, such as ecotourism) can sometimes be encouraged, large-scale organized trade, targeting selected species for overseas markets, should be strictly monitored, controlled and – in the vast majority of cases – stopped altogether.

Finally, the socio-cultural value of reptiles across the many cultures of South-East Asia is a much-neglected area of study. Important insights can be gained from such studies that can add to our knowledge of reptile biology, while at the same time enhancing conservation practice.

BOOK ORGANIZATION

This work covers all currently valid species known from the region. The cut-off date for inclusion was 31 December 2008. A majority of the species are illustrated. For each species, the following are supplied:

1. **Common English name** (coining a new name in a few cases where no common name existed).

2. **Current scientific name.**

3. **Symbols** where relevant next to the common name. These denote a species that may pose a danger to humans. The symbols are as follows:

💡 MILDLY VENOMOUS SNAKE Bites from small snakes may cause slight envenomation in humans, while bites from adults of some snake species can cause mortality in humans.

☠ DEADLY VENOMOUS SNAKE Bites from these snakes may cause mortality in humans.

🐊 LARGE-GROWING REPTILE Adults of these species may be dangerous to humans.

4. **Maximum length attained** Straight carapace length (SCL) in turtles, terrapins and tortoises; total length (TL) in crocodilians, snakes and anguid lizards; and snout-vent length (SVL) in all other lizards. See also diagrams on page 8. Size measurements of more than 200 millimetres are provided in metres; measurements below 200 millimetres are given in millimetres.

MANAGEMENT OF SNAKEBITE

Snakes are most likely to bite humans when they have been startled, provoked and/or cornered. Make sure you familiarize yourself with the basic procedure for snakebite treatment before travelling to areas with venomous snakes.

Several well-known dangerously venomous snakes occur in South-East Asia. Among them are members of two families – the Elapidae (comprising cobras, kraits, coral snakes and sea snakes) and the Viperidae (consisting of the vipers, including pit vipers).

Venom from the Elapidae affects the nerves, hence 'neurotoxic' venom. It blocks the conduction of nerve impulses to muscles. Symptoms of venomous snakebite include loss of muscle control, which is manifested by drooping eyelids, a loss of muscle tone in the facial features and paralysis of the diaphragm, resulting in an inability to breathe.

Venom from the Viperidae affects the circulatory system and is called 'haematoxic' venom. It damages the walls of the blood vessels. Symptoms of snakebite in this case include severe local pain and swelling, non-clotting of the blood and kidney failure.

Bites from some of the larger keelback snakes (including members of the genera *Rhabdophis* and *Macropisthodon*) may also be serious enough to warrant medical treatment for envenomation.

Antivenom should be administered only by qualified physicians, when signs of local or systemic envenomation are evident. Ideally, the antivenom used should be from the same species and from the same geographical area. A hospital in the tropics should have staff knowledgeable about local snakes of medicinal importance, and the possible symptoms of their bites. They should also have access to appropriate antivenom, and epinephrine to treat anaphylaxis. Other basic facilities for treatment of venomous snakebite include a system for assisted breathing (especially for serious cases of neurotoxic envenomation) and treatment for acute renal failure (for bites from vipers, especially *Daboia russelii*).

If someone you know suffers a snakebite, bear in mind the following:

• For bites from species belonging to the families Elapidae and Viperidae, arrange to transport the person to a hospital immediately.

• A person bitten by a snake needs to be calmed and reassured, and kept immobile as much as possible, as movement can increase the systemic absorption of venom.

• In the case of a bite by an elapid species, apply a pressure bandage (as firm as you would apply to a sprained ankle) to contain the venom and prevent it from spreading. Bandage as much of the limb as possible. Apply a splint to keep the limb immobile. Do not remove the bandage before professional help is available.

• An accurate description of the snake responsible for the bite greatly aids treatment. However, do not endanger yourself or others by trying to capture the snake for identification purposes.

• Elevate the bite site. This results in reduced blood flow to and from the bite site, slowing the spread of the venom.

• Do not apply a tourniquet unless you are a doctor. Oxygen starvation of a limb can cause more damage than the envenomation, and the procedure is rarely lifesaving.

• Do not allow the patient to consume alcohol, which elevates metabolism and promotes vasodilatation (widens the blood vessels), causing a more rapid onset of symptoms. Also do not administer stimulants or pain medication.

• Do not cut open the bite site, or try to suck out the venom.

Apart from the dangers posed by venomous snakes, large-growing members of several families (especially pythons) can give a painful, crushing bite, causing severe lacerations that require stitches.

5. **Morphological characters used in identification** (see also diagrams, pages 9–10).

6. **An account of coloration in life.**

7. **Description of diagnostic characteristics of subspecies** that occur in the region, where applicable.

8. **Notes on habits and behaviour,** including habitat associations, elevational range, unique habits, diet and reproduction, where known. Elevational range is given in metres above sea level (m asl). In lizard and snake families that are known to contain both oviparous and oviviparous species, this information is given within the species accounts (where known). Where all species within a family are either oviparous or oviviparous, this is stated in the introduction to the family only.

9. **Distributional range** within South-East Asia (country-wise) and notes on occurrence in extralimital areas. Localities are arranged in geographical order, starting from Myanmar in the north, to the islands of Indonesia in the south. For each country, if the taxon is widespread, no specific localities are provided. If it is represented by a few sites (typically less than five) locally, they are listed.

10. **Conservation status**, according to the 2008 IUCN Red List of Threatened Species, Red List Categories, Version 3.1 (2001) (http://www.iucnredlist.org/). The IUCN threat categories are as follows:

EXTINCT (EX)
A taxon is Extinct when there is no reasonable doubt that the last individual has died. A taxon is presumed Extinct when exhaustive surveys in known and/or expected habitat, at appropriate times (diurnal, seasonal, annual), throughout its historic range have failed to record an individual. Surveys should be over a time frame appropriate to the taxon's life cycle and life form.

EXTINCT IN THE WILD (EW)
A taxon is Extinct in the Wild when it is known only to survive in cultivation, in captivity or as a naturalized population (or populations) well outside the past range. A taxon is presumed Extinct in the Wild when exhaustive surveys in known and/or expected habitat, at appropriate times (diurnal, seasonal, annual), throughout its historic range have failed to record an individual. Surveys should be over a time frame appropriate to the taxon's life cycle and life form.

CRITICALLY ENDANGERED (CR)
A taxon is Critically Endangered when the best available evidence indicates that it meets any of the criteria A to E for Critically Endangered, and it is therefore considered to be facing an extremely high risk of extinction in the wild.

ENDANGERED (EN)
A taxon is Endangered when the best available evidence indicates that it meets any of the criteria A to E for Endangered, and it is therefore considered to be facing a very high risk of extinction in the wild.

VULNERABLE (VU)
A taxon is Vulnerable when the best available evidence indicates that it meets any of the criteria A to E for Vulnerable, and it is therefore considered to be facing a high risk of extinction in the wild.

NEAR THREATENED (NT)
A taxon is Near Threatened when it has been evaluated against the criteria but does not qualify for Critically Endangered, Endangered or Vulnerable category now, but is close to qualifying for or is likely to qualify for a threatened category in the near future.

LEAST CONCERN (LC)
A taxon is Least Concern when it has been evaluated against the criteria and does not qualify for Critically Endangered, Endangered, Vulnerable or Near Threatened. Widespread and abundant taxa are included in this category.

DATA DEFICIENT (DD)
A taxon is Data Deficient when there is inadequate information to make a direct, or indirect, assessment of its risk of extinction based on its range and/or population status. A taxon in this category may be well studied, and its biology well known, but appropriate data on abundance and/or range are lacking. Data Deficient is therefore not a category of threat. Listing of taxa in this category indicates that more data is required and acknowledges the possibility that future research will reveal that threatened classification is appropriate.

NOT EVALUATED (NE)
A taxon is Not Evaluated when it has not yet been evaluated against the criteria.

LOWER RISK is no longer used by the IUCN in the evaluation of taxa, but persists in the Red List for taxa evaluated before 2001, when Version 3.1 of the Red List was first used, and not re-evaluated since. Before 2001, Near Threatened and Conservation Dependent were both subcategories of the category Lower Risk. Had the category been assigned with the same data today, the species would be designated Near Threatened in either case.

Plate 1. CROCODYLIDAE & GAVIALIIDAE

1. MARSH CROCODILE *Crocodylus palustris*, p. 166.
TL 5m Snout relatively broad and heavy; no longitudinal ridges anterior to eyes; no postoccipital scutes; adults grey to brown, usually without dark bands.

2. SALTWATER CROCODILE *Crocodylus porosus*, p. 166.
TL 6.2m Snout relatively broad and heavy; longitudinal ridges anterior to eyes; postoccipital scutes present; adults black spotted or blotched on pale yellow, grey or greyish-olive background.

3. SIAMESE CROCODILE *Crocodylus siamensis*, p. 166.
TL 3.5m Snout relatively broad; longitudinal ridges anterior to eyes; postoccipital scutes present; dorsum pale yellow, grey or olive-grey with black spots or blotches.

4. GANGES GHARIAL *Gavialis gangeticus*, p. 167.
TL 7m Snout slender, parallel-sided; neck armour continuous with back armour; dorsum olive to tan with dark blotches or bands on body and tail.
(4a) Female Snout-tip lacks swelling.
(4b) Male Snout-tip with distinctive knob.

5. MALAYAN FALSE GHARIAL *Tomistoma schlegelii*, p. 167.
TL 5.5m Snout slender, tapering gradually, tip rounded; neck armour continuous with back armour; dorsum brown with black spots and bands; tail with broad black bands.

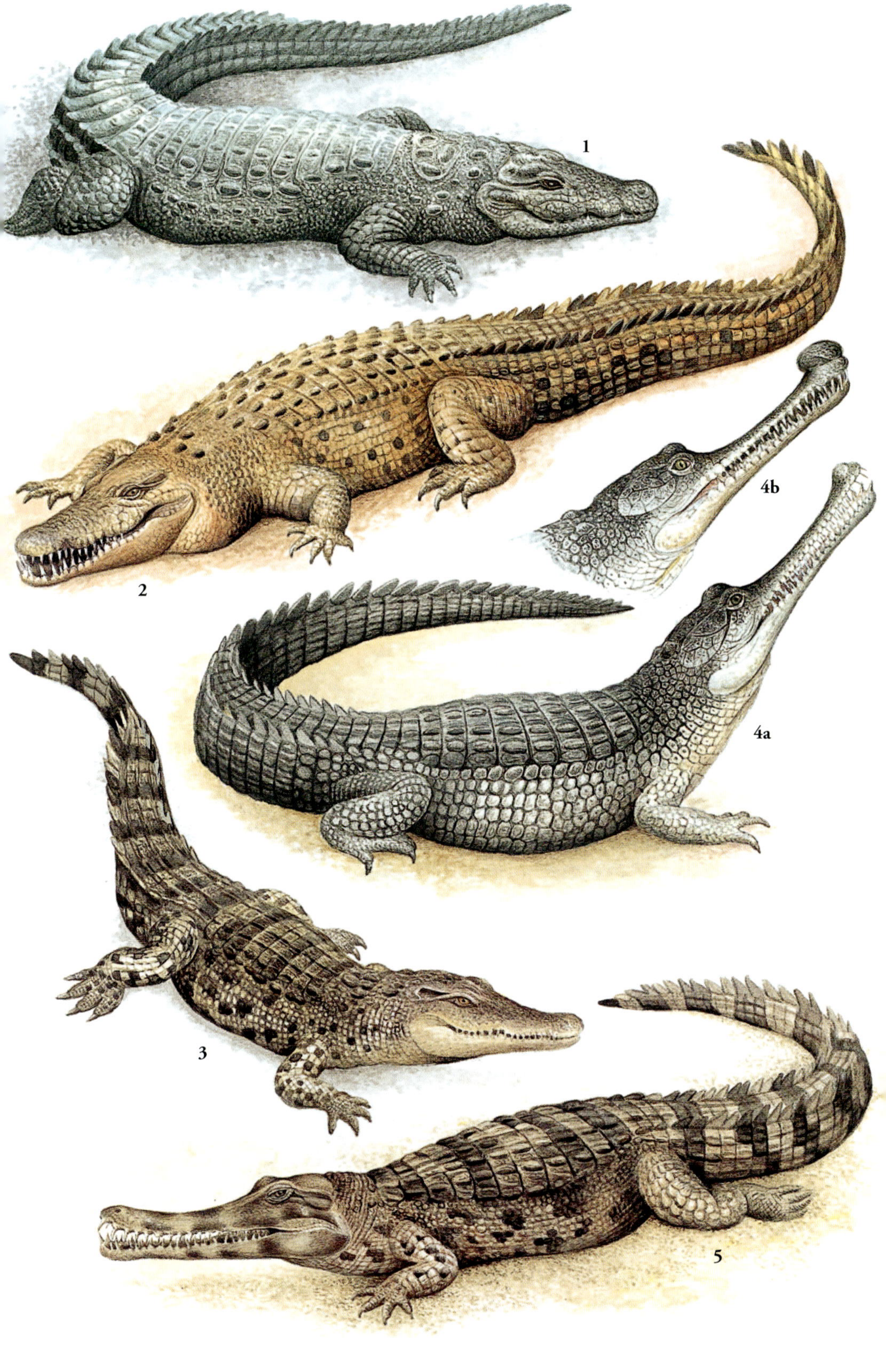
1
2
4b
4a
3
5

Plate 2. GEOEMYDIDAE

1. SOUTHERN RIVER TERRAPIN *Batagur affinis*, p. 167.
SCL 600mm Carapace domed, heavily buttressed; head small with upturned snout; 4 claws on forelimb; carapace olive-grey or brown.
(**1a**) **Female** Head brown.
(**1b**) **Male** Head black.

2. PAINTED TERRAPIN *Batagur borneoensis*, p. 168.
SCL 600mm Carapace oval, flattened; forelimbs with 5 claws; carapace light brown or olive with 3 longitudinal black stripes.
(**2a**) **Female** Forehead grey.
(**2b**) **Male** Forehead white with red stripe between eyes.

3. BURMESE PAINTED TERRAPIN *Batagur trivittata*, p. 168.
SCL 580mm Carapace low with distinct vertebral keel; head with pointed, slightly upturned snout; 5 claws on forelimb; carapace olive-green with 3 longitudinal black stripes; forehead turns crimson with lozenge-shaped black area in males; carapace dark brown, forehead and neck greenish-olive with black crown patch in females (illustrated).

4. INDO-CHINESE BOX TURTLE *Cuora galbinifrons*, p. 169.
SCL 190mm Carapace high-domed and smooth; plastron with transverse hinge behind pectoral and abdominal scutes; face with narrow dark stripes.
(**4a**) **Carapace** Narrow yellow or cream vertebral stripe.
(**4b**) **Plastron** Yellow or brownish-yellow with dark blotches.

5. MALAYAN BOX TURTLE *Cuora amboinensis*, p. 168.
SCL 216mm Carapace high-domed and smooth; plastron with transverse hinge behind pectoral and abdominal scutes; face with longitudinal yellow stripes.
(**5a**) **Carapace** Olive, brown or nearly black.
(**5b**) **Plastron** Yellow or cream with single black blotch on each scute.

6. KEELED BOX TURTLE *Cuora mouhotii*, p. 169.
SCL 180mm Carapace elongated, tricarinate, distinctly flat-topped; marginals serrated posteriorly; weak transverse plastral hinge in adult females.
(**6a**) **Carapace** Dark or light brown.
(**6b**) **Plastron** Yellow or light brown with dark brown blotches on each scute.

7. THREE-STRIPED BOX TURTLE *Cuora trifasciata*, p. 169.
SCL 230mm Carapace elevated, arched and elongated; plastron with transverse hinge behind pectoral and abdominal scutes; head with black postorbital stripe enclosing brown or olive triangle behind eye.
(**7a**) **Carapace** Light brown with 3 longitudinal black stripes; scutes with thin radiating black pattern.
(**7b**) **Plastron** Yellow; scutes black with yellow border.

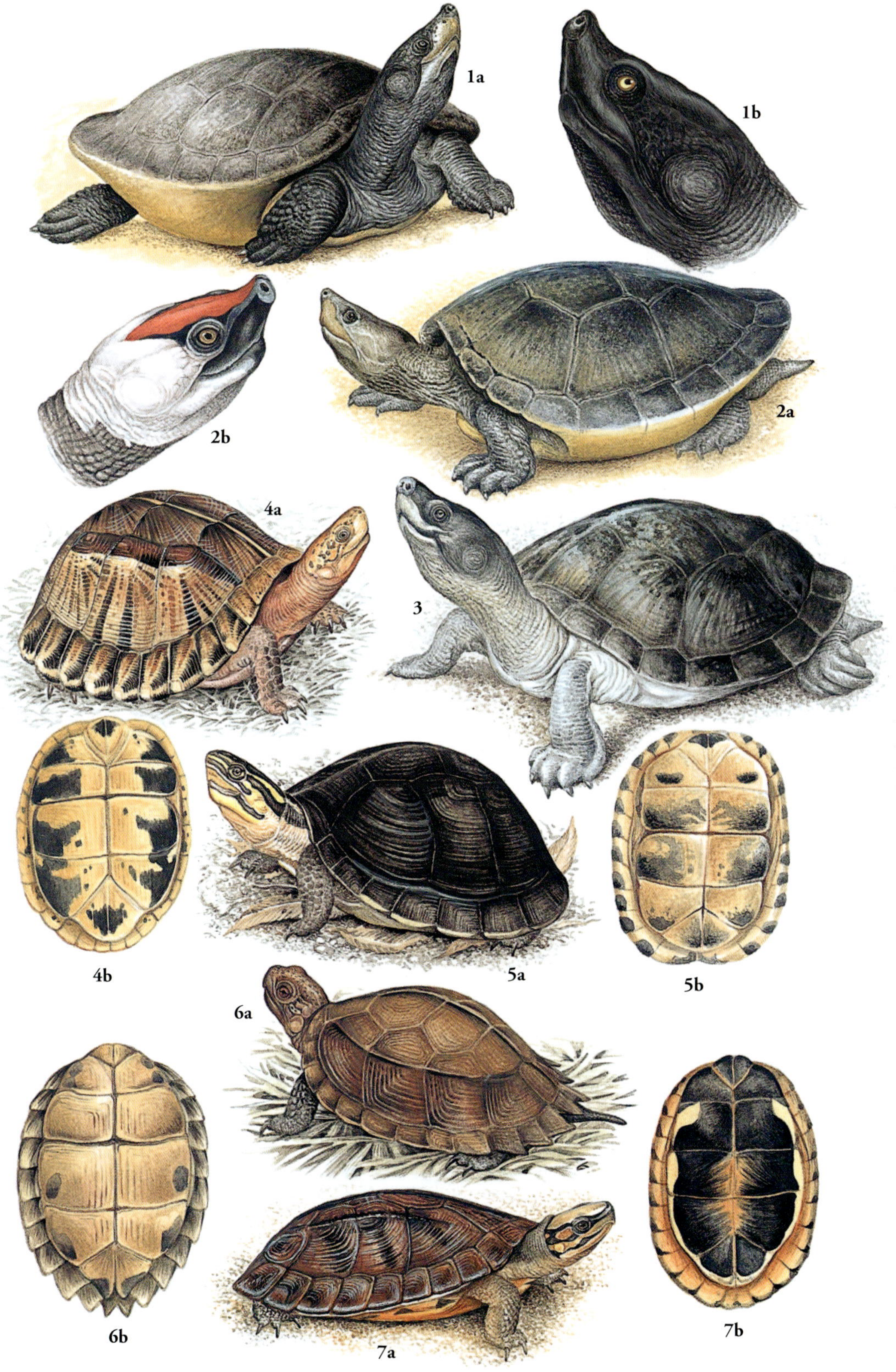

1a
1b
2b
2a
4a
3
4b
5a
5b
6a
6b
7b
7a

Plate 3. GEOEMYDIDAE

1. BLACK-BRIDGED LEAF TURTLE *Cyclemys atripons,* p. 169.
SCL 236mm Carapace ovoid, elongated, depressed, tricarinate; forehead with enlarged scales; femoral midseam shorter or equal to anal midseam; anal notch narrow to wide.
(1a) Carapace Chestnut-brown, unpatterned or with fine radiating black lines; head and neck striped; forehead spotted or speckled with black.
(1b) Plastron Hinge in adults; femoral midseam shorter or equal to anal midseam; anal notch narrow to wide; yellow with radiating dark pattern.

2. COMMON LEAF TURTLE *Cyclemys dentata,* p. 169.
SCL 210mm Carapace oval, depressed, tricarinate; forehead with enlarged scales; femoral midseam shorter than anal midseam; anal notch narrow, acute-angled.
(2a) Carapace Dark brown, unpatterned or with fine black lines; forehead dark speckled; temples and neck with dark stripes.
(2b) Plastron Yellow, unpatterned or with radiating dark pattern; plastron of juveniles mottled with black.

3. GRAY LEAF TURTLE *Cyclemys fusca,* p. 170.
SCL 242mm Carapace oval, depressed, tricarinate; forehead with enlarged scales; femoral midseam greater or equal to anal midseam; anal notch wide.
(3a) Carapace Dark brown, sometimes with radiating dark lines; forehead greenish-yellow, lighter than temporal region; neck dark, lacking stripes.
(3b) Plastron Dark brown to black, sometimes with radiating dark lines.

4. OLDHAM'S LEAF TURTLE *Cyclemys oldhamii,* p. 170.
SCL 254mm Carapace rectangular, depressed, tricarinate; forehead with enlarged scales; femoral midseam greater or equal to anal midseam; anal notch wide.
(4a) Carapace Dark or light brown; forehead speckled; neck with or without stripes;
(4b) Plastron Dark brown, black or yellow, sometimes with radiating dark lines; juvenile plastron brown or yellow, with large central plastral figure and ocellated pattern along submarginal seams.

5. VIETNAMESE LEAF TURTLE *Cyclemys pulchristriata,* p. 170.
SCL 227mm Carapace oval, elongated, depressed, tricarinate; forehead with enlarged scales; femoral midseam shorter or equal to anal midseam; anal notch narrow to wide.
(5a) Carapace Chestnut-brown or pale brown with wide radiating dark lines or thick black speckling; forehead with dark speckling; temples and neck with distinct stripes.
(5b) Plastron Yellow with short, thick radiating dark lines; juvenile plastron with large dark brown specks and ocellated pattern along submarginal seams.

6. BLACK-BREASTED LEAF TURTLE *Geoemyda spengleri,* p. 170.
SCL 125mm Carapace depressed, tricarinate; anterior marginals serrated; eyes large; upper jaw hooked; outer faces of forelimbs with enlarged scales.
(6a) Carapace Dark reddish-orange, orange-yellow, olive or light brown with black lines or wedges over keels; forehead brown with yellow postocular stripe; tympanic region and throat yellow-spotted.
(6b) Plastron Dark brown, edges yellow.

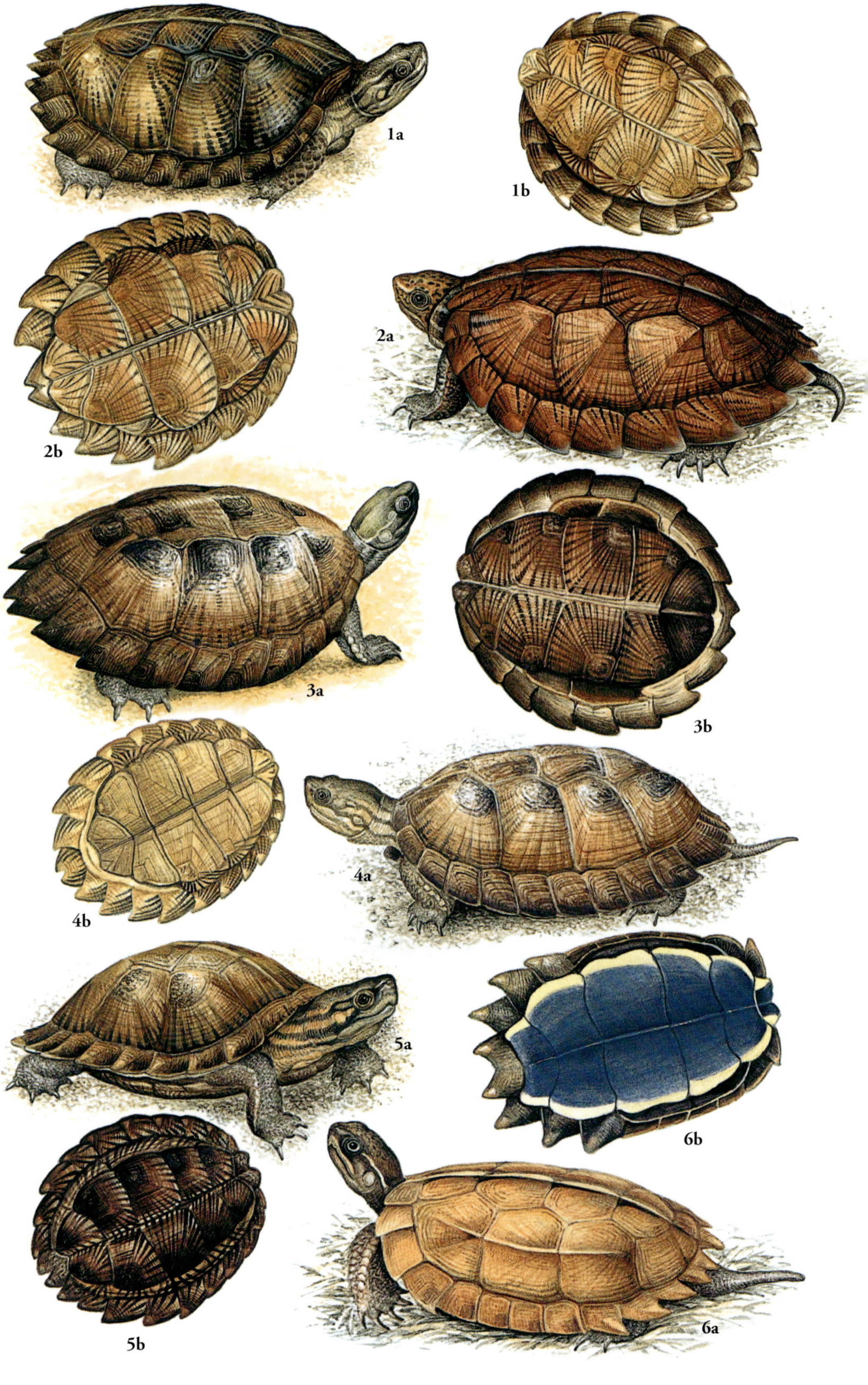
1a
1b
2b
2a
3a
3b
4b
4a
5a
6b
5b
6a

Plate 4. GEOEMYDIDAE

1. YELLOW-HEADED TEMPLE TURTLE *Heosemys annandalii*, p. 170.
SCL 506mm Carapace elongated, raised, flat-topped in adults; shell outline rounded; posterior marginals serrated.
(**1a**) **Carapace** Black with orange lines near marginals and on vertebral keel in juveniles.
(**1b**) **Plastron** Pale orange with grey vermiculations, turning pale grey and eventually to black.

2. ARAKAN FOREST TURTLE *Heosemys depressa*, p. 170.
SCL 242mm Carapace depressed; vertebral region flattened with obtuse keel; posterior marginals serrated; toes with rudiments of webs.
(**2a**) **Carapace** Light brown, sometimes with dark brown mottling.
(**2b**) **Plastron** Yellowish-brown with black blotches or radiating lines on each scute.

3. GIANT ASIAN POND TURTLE *Heosemys grandis*, p. 171.
SCL 480mm Carapace tricarinate, elevated in adults, depressed in juveniles; lateral keel does not extend across fourth costal.
(**3a**) **Carapace** Dark brown, lighter around marginals; keels yellowish-orange or brown.
(**3b**) **Plastron** With radiating pattern emerging from darker blotch in each scute in juveniles.

4. SPINY HILL TURTLE *Heosemys spinosa*, p. 171.
SCL 220mm Carapace oval, arched in adults; strong vertebral keel; marginals of adults smooth; those of juveniles with greatly expanded marginals bearing distinct spines.
(**4a**) **Carapace** Brown.
(**4b**) **Plastron** Brown, scutes with radiating lines.

5. MALAYAN SNAIL-EATING TURTLE *Malayemys macrocephala*, p. 171.
SCL 300mm Carapace tricarinate; head large, especially in old females.
(**5a**) **Carapace** Dark brown with black areoli and yellow rim.
(**5b**) **Plastron** Yellow, each scute with large black blotch.

6. MEKONG SNAIL-EATING TURTLE *Malayemys subtrijuga*, p. 171.
SCL 210mm Carapace tricarinate; head large, especially in old females.
(**6a**) **Carapace** Dark brown with black areoli and yellow rim.
(**6b**) **Plastron** Yellow, each scute with large black blotch.

7. VIETNAMESE POND TURTLE *Mauremys annamensis*, p. 171.
SCL 170mm Carapace oval, low, tricarinate in juveniles; lateral keels indistinct in adults; posterior marginals weakly serrated.
(**7a**) **Carapace** Carapace dark grey-brown.
(**7b**) **Plastron** Yellow or yellowish-orange, each scute with large black blotch.

8. ASIAN YELLOW POND TURTLE *Mauremys mutica*, p. 171.
SCL 194mm Carapace oval, depressed, tricarinate; vertebral keel low; nuchal, small axillary and inguinals present.
(**8a**) **Carapace** Yellow-brown or brown; marginals dark.
(**8b**) **Plastron** Yellow, each scute with square or rectangular dark blotch.

9. CHINESE STRIPE-NECKED TURTLE *Mauremys sinensis*, p. 171.
SCL 240mm Carapace oval, moderately flattened, not depressed; juvenile carapace tricarinate, keels low and discontinuous, keels disappear in adults; Vertebrals IV and V wider than long.
(**9a**) **Carapace** Reddish-brown to greyish-brown.
(**9b**) **Plastron** Yellow, each scute with large blackish-brown blotch.

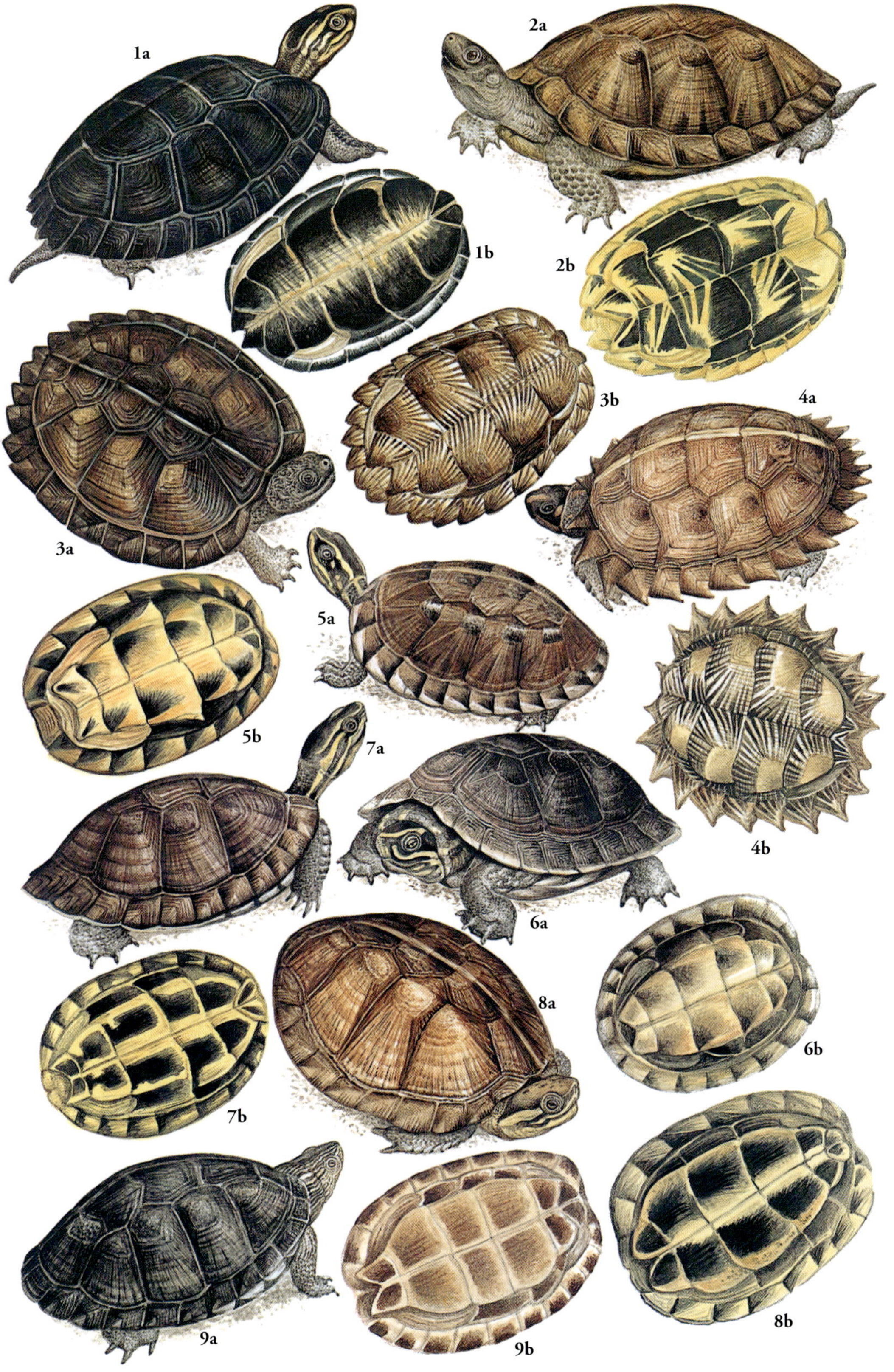
1a
2a
1b
2b
3b
4a
3a
5a
5b
7a
4b
6a
8a
6b
7b
9a
9b
8b

Plate 5. GEOEMYDIDAE

1. INDIAN BLACK TURTLE *Melanochelys trijuga*, p. 172.
SCL 280mm Carapace elongated, high in adults, depressed in juveniles; tricarinate; short snout.
(1a) Carapace Brown or blackish-grey.
(1b) Plastron Dark with pale yellow border (except in old individuals).

2. BURMESE EYED TURTLE *Morenia ocellata*, p. 172.
SCL 155mm Carapace domed, smooth-shelled; vertebral keel low (juveniles) or absent (adults); posterior marginals unserrated; head small with pointed snout.
(2a) Carapace Greenish-brown or olive; vertebrals and costals with large yellow ocelli with dark brown centre.
(2b) Plastron Unpatterned yellow.

3. MALAYAN FLAT-SHELLED TURTLE, *Notochelys platynota*, p. 172.
SCL 400mm Carapace flat with low, interrupted vertebral keel; 6–7 vertebrals; weak plastral hinge.
(3a) Carapace Olive, yellowish-brown or brick-red (adults); bright green (hatchlings).
(3b) Plastron Yellowish-orange, each scute with black blotch.

4. MALAYAN GIANT TURTLE *Orlitia borneensis*, p. 172.
SCL 800mm Carapace narrow, humped (juveniles) or smooth (adults); head large; band-like scales on outer faces of forelimbs.
(4a) Carapace Unpatterned black, brown or grey.
(4b) Plastron Yellowish-orange or brown, sometimes with dark flecks.

5. BEALE'S FOUR-EYED TURTLE *Sacalia bealei*, p. 172.
SCL 143mm Carapace elongated, unicarinate, slightly depressed; posterior marginals not serrated; vertebral keel low; plastral buttresses weak.
(5a) Carapace Yellowish-brown to chocolate-brown; anterior margin with numerous black or dark brown speckles.
(5b) Male plastron Yellow to light olive, sometimes with dark vermiculations.
(5c) Female plastron Yellow to light olive, sometimes with dark blotches.

6. FOUR-EYED TURTLE *Sacalia quadriocellata*, p. 172.
SCL 140mm Carapace elongated, unicarinate, slightly depressed; posterior marginals not serrated; vertebral keel low; plastral buttresses weak.
(6a) Carapace Dark brown with radiating dark lines; anterior region lacks dark speckling, or speckling much reduced.
(6b) Male plastron Yellow with solid black patches.
(6c) Female plastron Cream with dark vermiculations and stippling.

7. BLACK MARSH TURTLE *Siebenrockiella crassicollis*, p. 173.
SCL 200mm Carapace oval with serrated posterior marginals; vertebral region flattened, unicarinate (adults) or tricarinate (juveniles).
(7a) Carapace Dark grey or nearly black.
(7b) Plastron Pale grey with large dark areas in each scute.

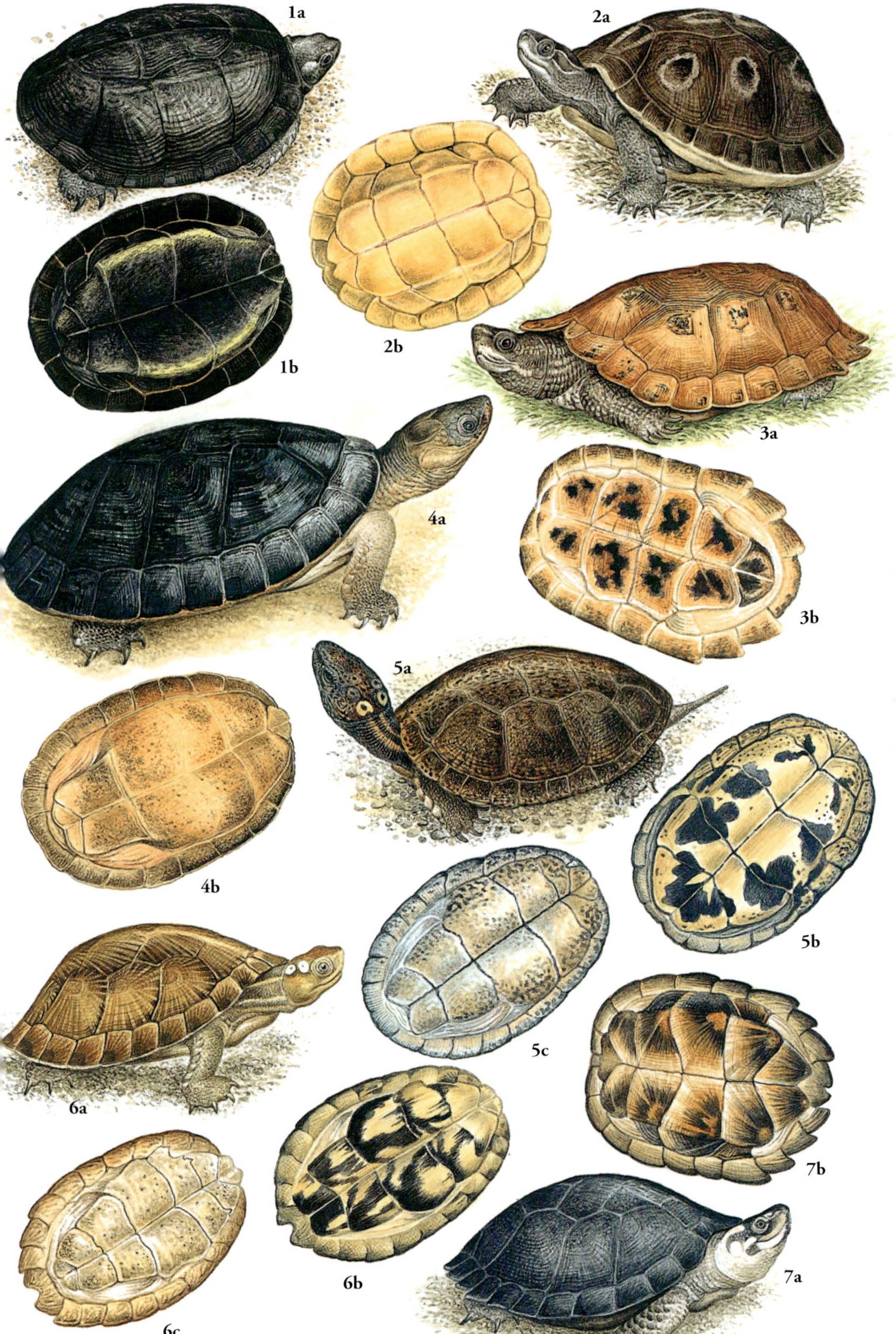
1a
2a
2b
1b
3a
4a
3b
5a
4b
5b
5c
6a
7b
6b
7a
6c

Plate 6. CHELONIIDAE & DERMOCHELYIDAE

1. LOGGERHEAD SEA TURTLE, *Caretta caretta*, p. 173.
SCL 1,200mm Carapace elongated with tapering end; costals 5 pairs, infralabial scutes lack pores; 13 marginal scutes.
(**1a**) **Carapace** Reddish-brown.
(**1b**) **Plastron** Yellowish-brown or yellowish-orange.
(**1c**) **Hatchling** Lateral keels.

2. GREEN TURTLE *Chelonia mydas*, p. 173.
SCL 1,400mm Carapace heart-shaped; paired prefrontals; scutes non-overlapping; upper jaw without hook.
(**2a**) **Carapace** Olive or brown with radiating dark pattern.
(**2b**) **Plastron** Pale yellow.
(**2c**) **Hatchling** No lateral keels.

3. HAWKSBILL SEA TURTLE *Eretmochelys imbricata*, p. 173.
SCL 1,000mm Carapace heart-shaped; 2 pairs of prefrontals; scutes overlapping; upper jaw relatively narrow, elongated.
(**3a**) **Carapace** Olive-brown; juveniles with darker blotches.
(**3b**) **Plastron** Pale yellow (adults) or dark grey (juveniles).
(**3c**) **Hatchling** No lateral keels.

4. OLIVE RIDLEY SEA TURTLE *Lepidochelys olivacea*, p. 173.
SCL 80mm Carapace heart-shaped; scutes juxtaposed; 5–9 pairs of costals; inframarginals with pores; upper jaw hooked.
(**4a**) **Carapace** Olive-green or greyish-olive.
(**4b**) **Plastron** Greenish-yellow (adults) or cream (juveniles).
(**4c**) **Hatchling** Lateral keels.

5. LEATHERBACK SEA TURTLE *Dermochelys coriacea*, p. 174.
SCL 2.5m Carapace elongated, tapering; 7 ridges on carapace and 5 on plastron; shell covered with skin; limbs paddle-like and clawless.
(**5a**) **Carapace** Black or blackish-blue with pale flecks.
(**5b**) **Plastron** Pale grey or pinkish-grey.
(**5c**) **Hatchling** Carapace covered with tiny scales.

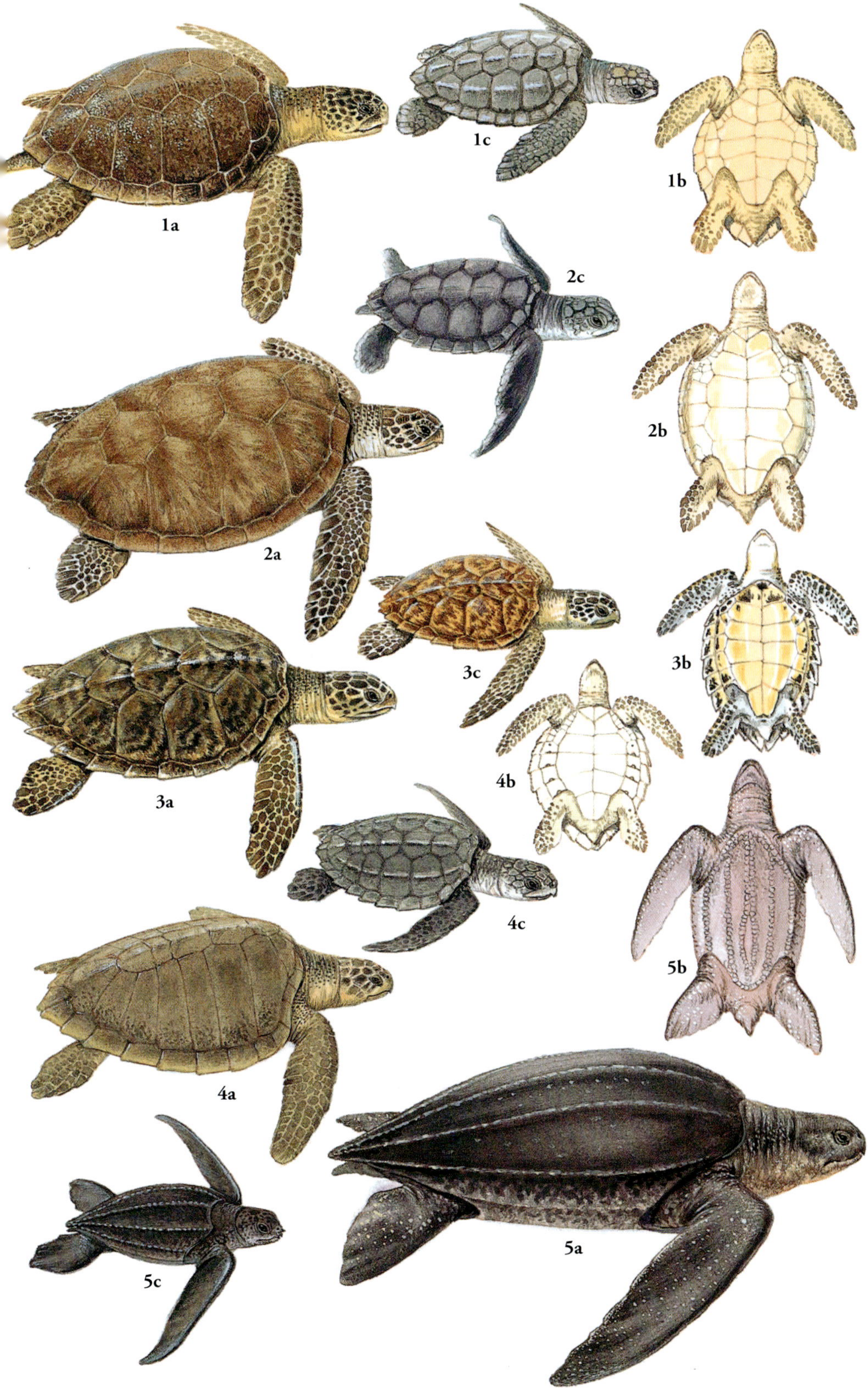
1a
1c
1b
2c
2b
2a
3c
3b
4b
3a
4c
5b
4a
5a
5c

Plate 7. EMYDIDAE, PLATYSTERNIDAE & TESTUDINIDAE

1. RED-EARED SLIDER *Trachemys scripta*, p. 174.
SCL 280mm Shell rounded with nearly smooth outline; plastron lacks hinge; carapace green with yellow lines, turning darker with growth; orange or red patch on temples.
(1a) Adult Carapace olive-grey; males with elongated claws on fingers; plastron with dark speckles.
(1b) Juvenile Carapace green with narrow yellow lines; plastron with rounded patches.
(1c) Hatchling plastron Dark ocelli on scutes.
(1d) Adult plastron Yellow with solid dark mark on each scute.

2. BIG-HEADED TURTLE *Platysternon megacephalum*, p. 174.
SCL 200mm Head large and covered with undivided scales; jaws hooked; throat with flattened rounded tubercles.
(2a) Adult Shell oval, depressed; tail covered with squarish scales; lines on forehead.
(2b) Head subspecies *P. m. peguense* Black-bordered pale postorbital stripe.
(2c) Plastron subspecies *P. m. peguense* Dark plastral seams.
(2d) Head subspecies *P. m. shiui* Heavy speckling of yellow, orange or pink.
(2e) Plastron subspecies *P. m. shiui* Speckling of yellow, orange or pink on shell.
(2f) Juvenile Shell flattened; tail long, slender.

3. BURMESE STAR TORTOISE *Geochelone platynota*, p. 175.
SCL 260mm Shell convex; no nuchal scute.
(3a) Carapace Black or dark brown, vertebral and costal scutes with yellow centres and radiating lines.
(3b) Plastron Yellow with scutes with dark patches.

4. ELONGATED TORTOISE *Indotestudo elongata*. p. 175.
SCL 330mm Carapace domed; flattened dorsally with arching sides; plastron elongated with deep notch posteriorly; nuchal scute narrow.
(4a) Carapace Yellowish-brown or olive with scattered dark blotches.
(4b) Plastron Yellowish-brown or olive with scattered dark blotches.

5. ASIAN GIANT TORTOISE *Manouria emys*, p. 175.
SCL 500mm (*M. e. emys*); 580mm (*M. e. phayrei*) Shell low; vertebral region depressed; distinct growth rings on carapace scutes; posterior marginals weakly serrated; outer surfaces of forelimbs bear large scales; pair of tuberculate scales on thighs.
(5a) Carapace subspecies *M. e. emys* Medium brown.
(5b) Carapace subspecies *M. e. phayrei* Blackish-brown.
(5c) Plastron subspecies *M. e. emys* Pectoral scutes small and separated.
(5d) Plastron subspecies *M. e. phayrei* Pectoral scutes large and fused.

6. IMPRESSED TORTOISE *Manouria impressa*, p. 175.
SCL 302mm Shell low; vertebrals and costals concave; posterior marginals strongly serrated; outer surfaces of forelimbs bear large scales; single tuberculate scale on each thigh.
(6a) Carapace Scutes translucent brown to yellowish-orange with streaks.
(6b) Plastron Yellowish-brown with radiating dark lines.

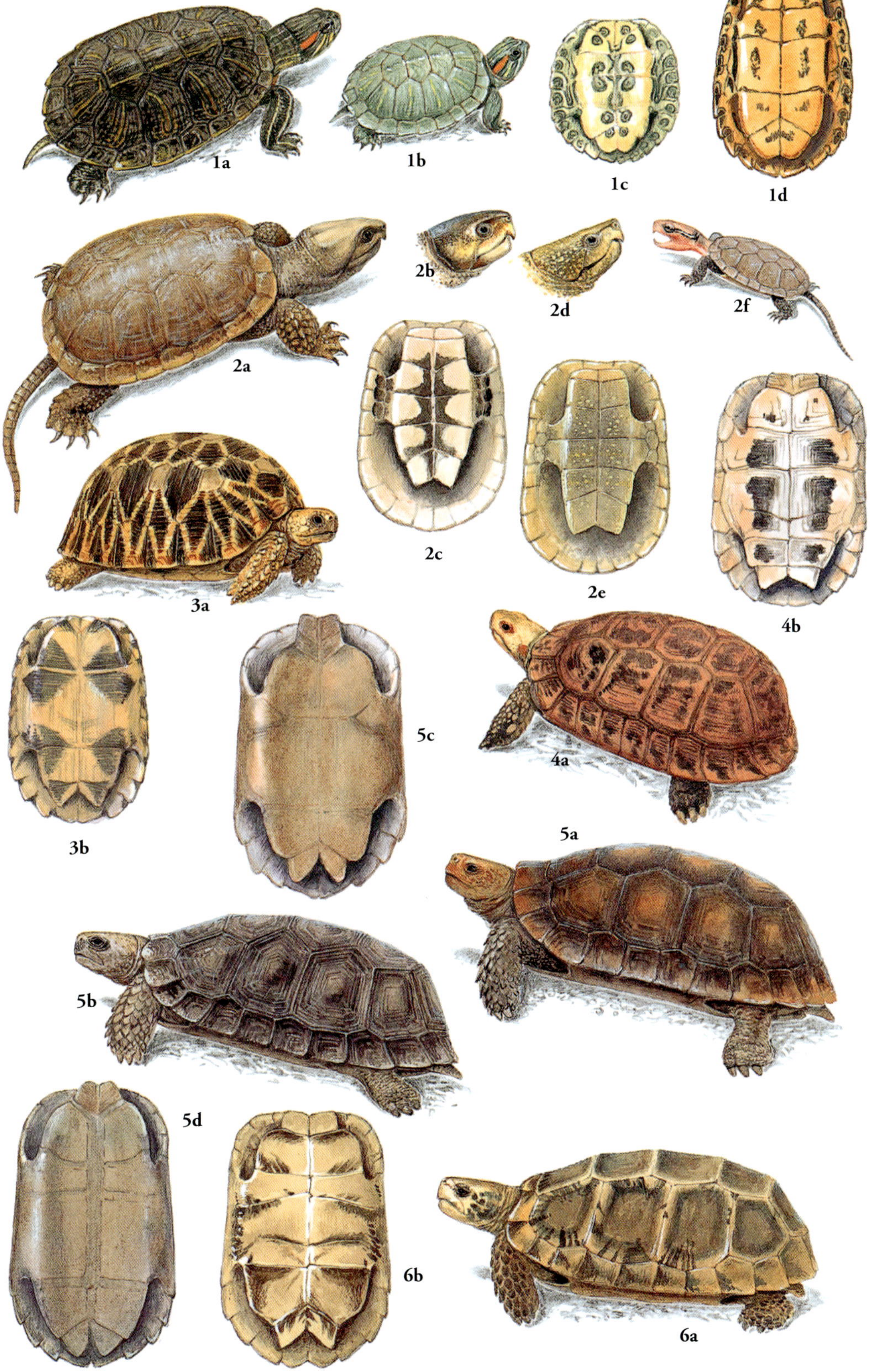
1a
1b
1c
1d
2a
2b
2d
2f
2c
2e
4b
3a
3b
5c
4a
5a
5b
5d
6b
6a

Plate 8. TRIONYCHIDAE

1. ASIAN SOFTSHELL TURTLE *Amyda cartilaginea*, p. 176.
SCL 750mm Carapace with rounded sides; head narrower; tubercles at anterior carapace margin.
(1a) Carapace Greenish-grey or olive, sometimes with yellow-bordered black spots or radiating streaks.
(1b) Plastron Cream (males); grey (females).

2. MAINLAND NARROW-HEADED SOFTSHELL TURTLE *Chitra chitra chitra*, p. 176.
SCL 1,220mm Carapace rim smoothly joins cartilagineous part to skin of neck; head small with tiny proboscis; indistinct dark speckling and ocelli on chin; broad costal markings.
(2a) Carapace Brown or yellowish-brown with dark-edged, bright yellow, greenish-yellow or tan stripes, including an inverted chevron-like marking on anterior of carapace and neck.
(2b) Plastron Cream or pale pink.
(2c) Head No X-shaped figure between eyes.

3. JAVANESE NARROW-HEADED SOFTSHELL TURTLE *Chitra chitra javanensis*, p. 176.
SCL 1,220mm Carapace rim smoothly joins cartilagineous part to skin of neck; head small with tiny proboscis; ocelli in eye region absent; bold black speckling and ocelli on chin.
(3a) Carapace Dark brown; midline and lateral vertebral carapacial stripes missing; distinct bell-shaped mark on carapace anterior.
(3b) Head X-shaped figure between eyes.

4. BURMESE NARROW-HEADED SOFTSHELL TURTLE *Chitra vandijki*, p. 176.
SCL 412mm Carapace rim smoothly joins cartilagineous part to skin of neck; head small with tiny proboscis; eyes situated close to snout-tip.
(4a) Carapace Chocolate-brown or olive-green; no dark vertebral stripe.
(4b) Plastron White or pale pink.
(4c) Head Transverse, light dark-bordered bar connecting eyes, and 1–2 pairs of light dark-bordered ocelli posterior to transverse bar between or behind eyes.

5. MALAYAN SOFTSHELL TURTLE *Dogania subplana*, p. 176.
SCL 350mm Carapace with straight sides; head large, bearing down-turned snout; adults with carapace hinge.
(5a) Carapace Dark olive or brown with dark median stripe.
(5b) Plastron Cream or grey.
(5c) Head Dark postocular stripe.
(5d) Juvenile Carapace dark yellow or olive with 2–3 pairs of black-centred eye-like spots.

6. INDIAN FLAPSHELL TURTLE *Lissemys punctata*, p. 176.
SCL 370mm Carapace oval, domed; plastron with 7 callosities; paired flaps cover hind limbs when retracted; entoplastral callosity small in adults.
(6a) Carapace Olive-green with dark yellow blotches.
(6b) Plastron Cream or pale yellow.
(6c) Head Yellow spots.

7. BURMESE FLAPSHELL TURTLE *Lissemys scutata*, p. 177.
SCL 370mm Carapace oval, domed; plastron with 7 callosities; paired flaps cover hind limbs when retracted; entoplastral callosity large in adults.
(7a) Carapace Brownish-olive, sometimes with fine black spots or reticulations.
(7b) Plastron Yellow.
(7c) Head Dark postocular stripe bounding a paler one.

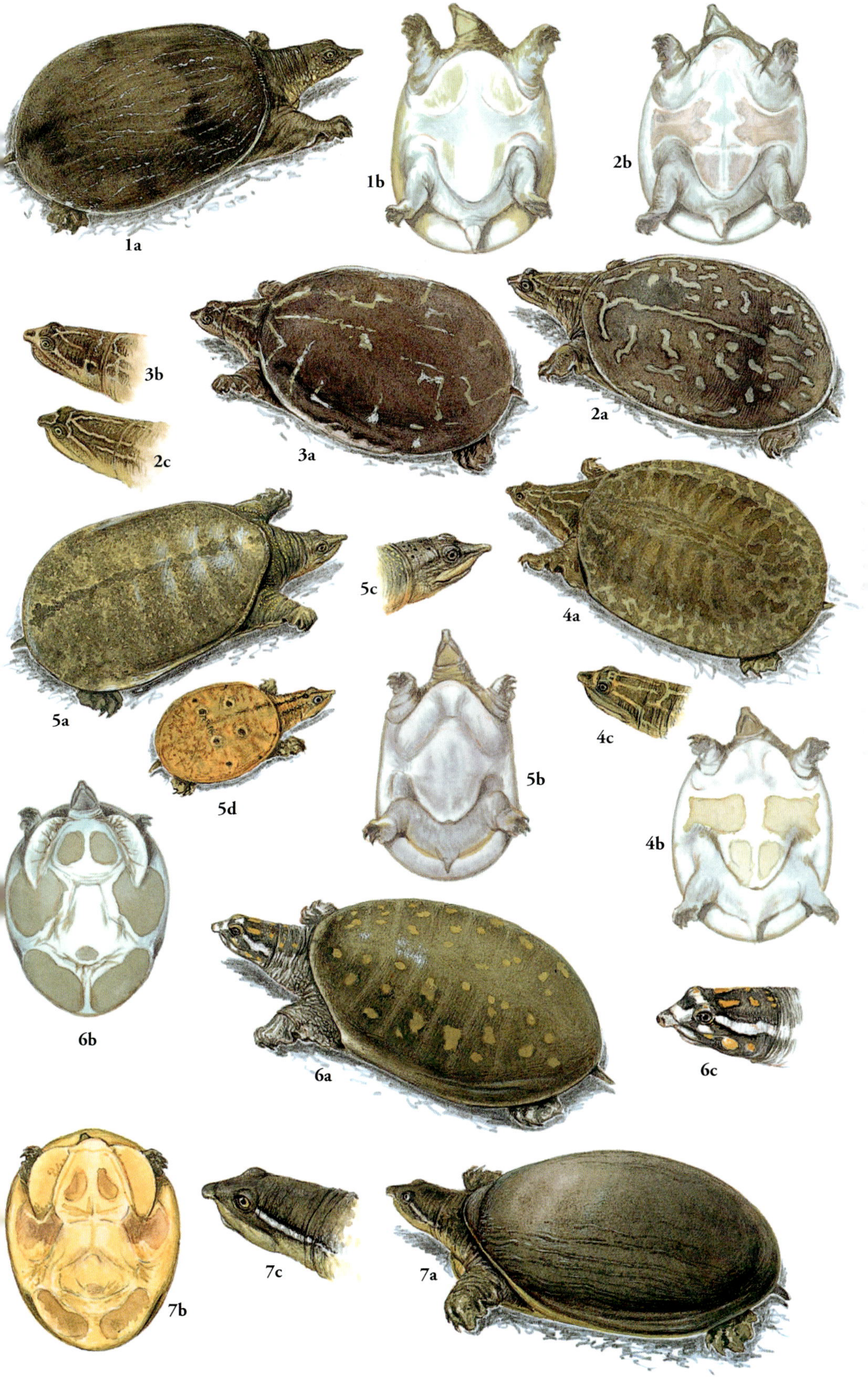
1a
1b
2b
3b
2c
3a
2a
5c
4a
5a
5d
5b
4c
4b
6b
6a
6c
7c
7a
7b

Plate 9. TRIONYCHIDAE

1. BURMESE SOFTSHELL TURTLE *Nilssonia formosus*, p. 177.
SCL 650mm Carapace rounded, smooth (adults) or with longitudinal rows of tubercles (juveniles); series of enlarged blunt tubercles above neck.
(**1a**) **Carapace** Olive-grey to brown with dark reticulations (adults); 4 dark-centred, light-bordered occelli (juveniles).
(**1b**) **Plastron** Unpatterned cream.

2. WATTLE-NECKED SOFTSHELL TURTLE *Palea steindachneri*, p. 177.
SCL 426mm Carapace oval; anterior carapace tuberculate.
(**2a**) **Carapace** Unpatterned olive, brown or grey.
(**2b**) **Plastron** Cream or yellow.
(**2c**) **Hatchling** Anterior carapace bears raised tubercles.

3. ASIAN GIANT SOFTSHELL TURTLE *Pelochelys cantorii*, p. 177.
SCL 1,500mm Carapace flattened; head short; eyes close to tip of snout; proboscis extremely short and rounded.
(**3a**) **Carapace** Olive or brown, sometimes spotted or streaked with lighter or darker shades with lighter outer edge.
(**3b**) **Plastron** Unpatterned white, sometimes with pink blotches.

4. CHINESE SOFTSHELL TURTLE *Pelodiscus sinensis*, p. 177.
SCL 250mm Carapace oval, longer than wide, smooth (adults) or with longitudinal rows of low tubercles (juveniles).
(**4a**) **Carapace** Unpatterned olive to greyish-green (adults); light-bordered spots (juveniles).
(**4b**) **Plastron** Unpatterned cream, grey or yellow.
(**4c**) **Hatchling** Plastron pinkish-red with large black blotches.

5. INDO-CHINESE GIANT SOFTSHELL TURTLE *Rafetus swinhoei*, p. 177.
SCL 600mm Carapace oblong; 2 plastral callosities.
(**5a**) **Carapace** Olive-green with numerous yellow spots encircled by yellow dots, sometimes forming stripes, especially in juveniles; pattern lost in adults.
(**5b**) **Plastron** Unpatterned pale olive-grey.

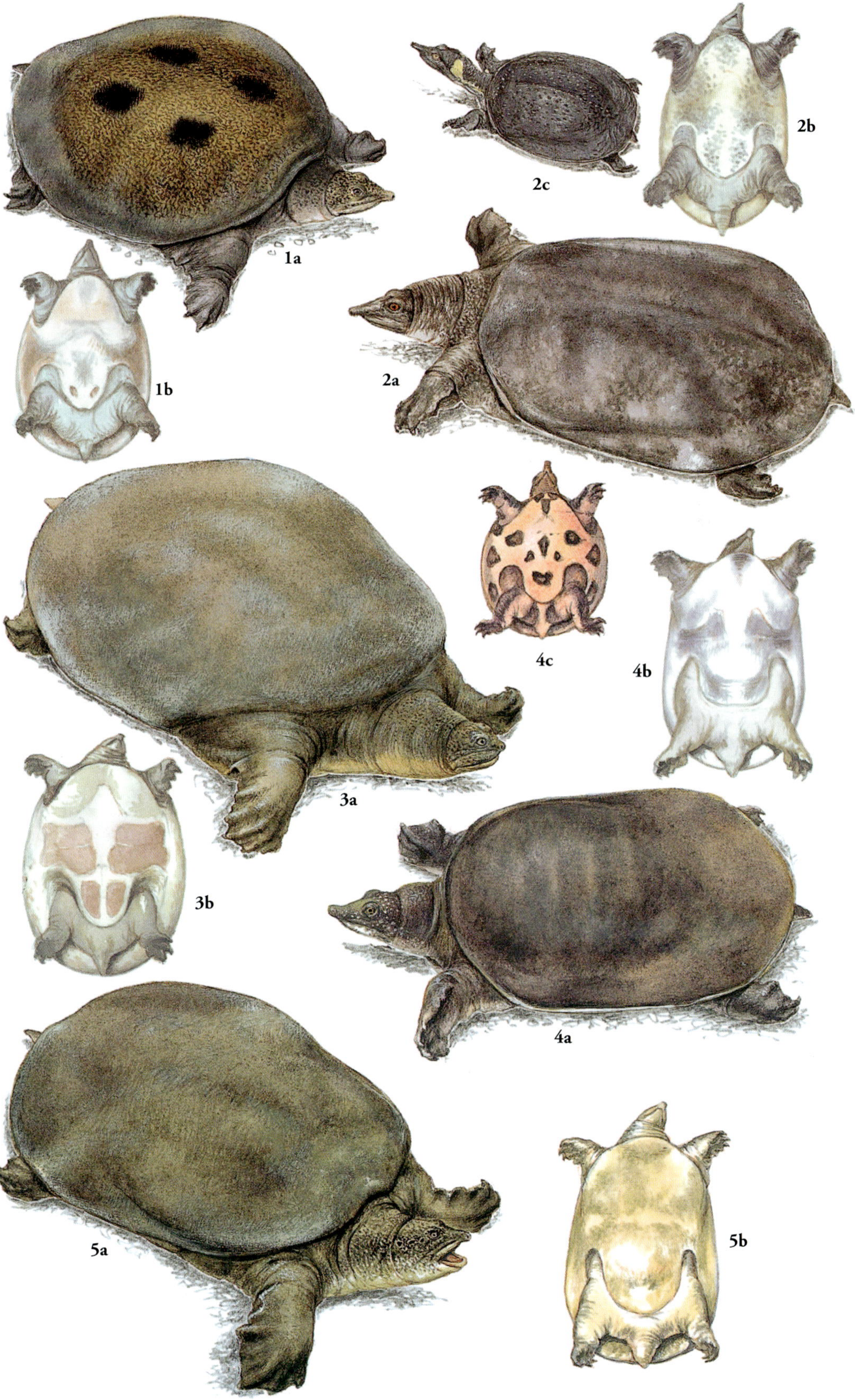
1a
1b
2a
2b
2c
3a
3b
4a
4b
4c
5a
5b

Plate 10. AGAMIDAE

1. GREATER SPINY LIZARD *Acanthosaura armata*, p. 178.
SVL 140mm Superciliary and occipital spines reach level of nuchal crest; nuchal crest separated or joined to dorsal crest; dorsum grey, brown to nearly black, with darker marbling; diamond-shaped mark on axilla; triangular patch covers eye.
(**1a**) **Male** Head and dorsum brighter.
(**1b**) **Female** Head and dorsum paler.

2. INDO-CHINESE SPINY LIZARD *Acanthosaura capra*, p. 178.
SVL 137.9mm Occipital spine between tympanum and nuchal crest; nuchal crest comprises broad lanciform scales; dorsal and nuchal crests separate, comprising short scales; dorsum green or olive with black spots and yellow spots encircled with black.
(**2a**) **Male** Head green, gular pouch bluish-green.
(**2b**) **Female** Head greenish-yellow; gular pouch yellow with green streaks.

3. CROWNED SPINY LIZARD *Acanthosaura coronata*, p. 178.
SVL 137.5mm Occipital spine between tympanum and nuchal crest; nuchal and dorsal crest continuous, low, comprising triangular scales.
(**3a**) **Male** Head and dorsum light green with grey-brown mottling; dorsum with dark cross-bars.
(**3b**) **Female** Head and dorsum brownish-red; dorsum with dark cross-bars.

4. MASKED SPINY LIZARD *Acanthosaura crucigera*, p. 179.
SVL 140mm Nuchal crest tall, separated from low dorsal crest; spines of nuchal and dorsal crests broad basally; dorsum greenish-yellow with indistinct dark network of enclosing pale yellow spots; dark facial mask.
(**4a**) **Male** Gular pouch pale brown.
(**4b**) **Female** Gular pouch dark.

5. SCALE-BELLIED SPINY LIZARD *Acanthosaura lepidogaster*, p. 179.
SVL 111mm Nuchal crest comprises 6 conical scales; dorsal crest low, composed of subtriangular scales; dorsum bright green changeable to dark brown; indistinct diamond-shaped black mark on nape.
(**5a**) **Male** Head and dorsum brighter.
(**5b**) **Female** Head and dorsum paler.

6. NATALIA'S SPINY LIZARD *Acanthosaura nataliae*, p. 179.
SVL 158mm No occipital spine above tympanum; nuchal and dorsal crests distinct, comprising lanceolate scales pointing posteriorly, separated from each other.
(**6a**) **Male** Dorsum yellowish-brown changeable to red, brown or yellow; head, nuchal and dorsal crests and limbs red; black mask covers orbit and tympanum; dark brown postocular stripe.
(**6b**) **Female** Head and dorsum emerald-green.

7. LONG-SNOUTED SHRUB LIZARD *Aphaniotis acutirostris*, p. 179.
SVL 72mm Snout acute, longer than eye diameter; projecting convex scale above rostral; dorsum brown with darker variegation.

8. BROWN SHRUB LIZARD *Aphaniotis fusca*, p. 179.
SVL 67mm Snout rounded, not longer than diameter of eye; dorsum dark brown or brownish-olive.

9. ORNATE SHRUB LIZARD *Aphaniotis ornata*, p. 180.
SVL 57mm Snout rounded; dorsum medium brown to brownish-red; small yellow spot on each side of eyelid.
(**9a**) **Male** Snout-tip with fleshy conical appendage; low nuchal crest comprising erect scales.
(**9b**) **Female** Snout-tip without fleshy appendage; no nuchal crest.

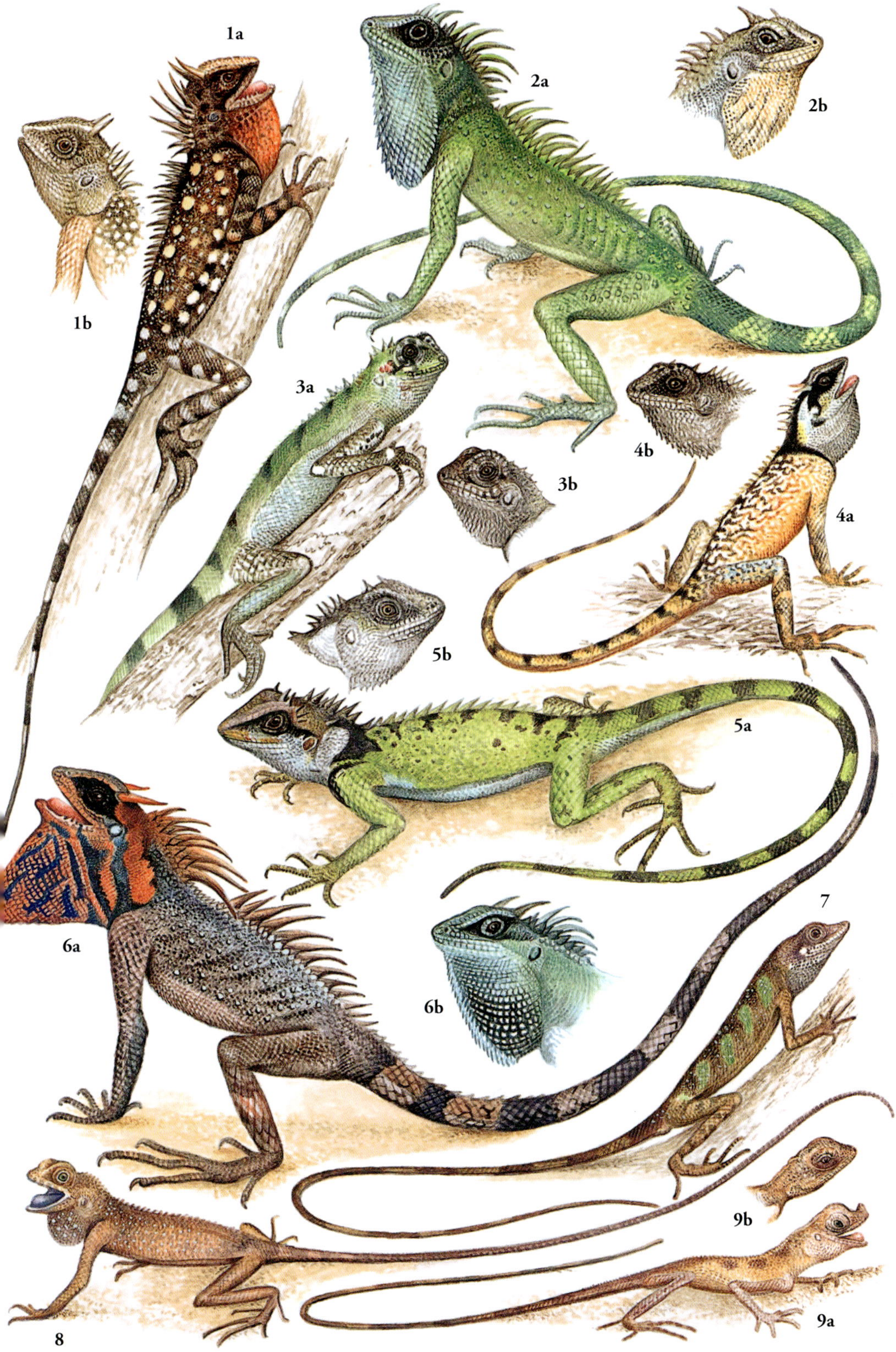
1a
1b
2a
2b
3a
3b
4a
4b
5a
5b
6a
6b
7
8
9a
9b

Plate 11. AGAMIDAE

1. CRESTED GREEN LIZARD *Bronchocela cristatella*, p. 180.
SVL 130mm Nuchal crest small, erect, flattened, continuous with dorsal crest, which is a serrated ridge not extending to tail; dorsum bright green, sometimes with white or light blue spots or bars.

2. HAYEK'S FOREST LIZARD *Bronchocela hayeki*, p. 180.
SVL 120mm Nuchal crest comprises crescentic scales, joined to low dorsal crest of slightly enlarged scales; dorsum pale green with scattered white patches; head pale yellow, cream, tan or green.

3. MANED FOREST LIZARD *Bronchocela jubata*, p. 180.
SVL 140mm Nuchal crest large with falciform scales, directed posteriorly; dorsal crest lower, extending to tail; dorsum green changeable to brown or black, with yellow or red spots or vertical bars.

4. CHANTHABUN FOREST LIZARD *Bronchocela smaragdina*, p. 181.
SVL 113mm Nuchal crest low, composed of small erect scales; no dorsal crest; dorsum emerald-green; white or yellow stripe along lower flanks that extends to tail-base.

5. VIETNAMESE FOREST LIZARD *Bronchocela vietnamensis*, p. 181.
SVL 122mm Nuchal crest comprises 6–12 small erect scales; no dorsal crest; dorsum green; yellow band on belly to inguinal region and posterior side of femur and lateral side of tail.

6. COLLARED FOREST LIZARD *Calotes chincollium*, p. 181.
SVL 142.9mm Nuchal crest composed of erect compressed scales; dorsal crest follows without gap; dorsum grey with 4 irregular saddle-like brown markings.
(6a) Male Dark brown patch over eyes and tympanic region; dorsal saddle-like markings light and faded posteriorly.
(6b) Female Dark brown postocular patch to tympanic region; dorsal saddle-like markings distinct.

7. FOREST CRESTED LIZARD *Calotes emma*, p. 181.
SVL 115mm Long spine above eye and 2 above tympanum; nuchal and dorsal crests continuous; dorsum olive-brown with dark brown dorsal bars or transverse spots; light dorsolateral stripe.
(7a) Male Distinct transverse dark bands on dorsum; sometimes a pale dorsolateral stripe; interstitial skin purple or black.
(7b) Female Dorsolateral stripe indistinct or absent; interstitial skin not purple or black.

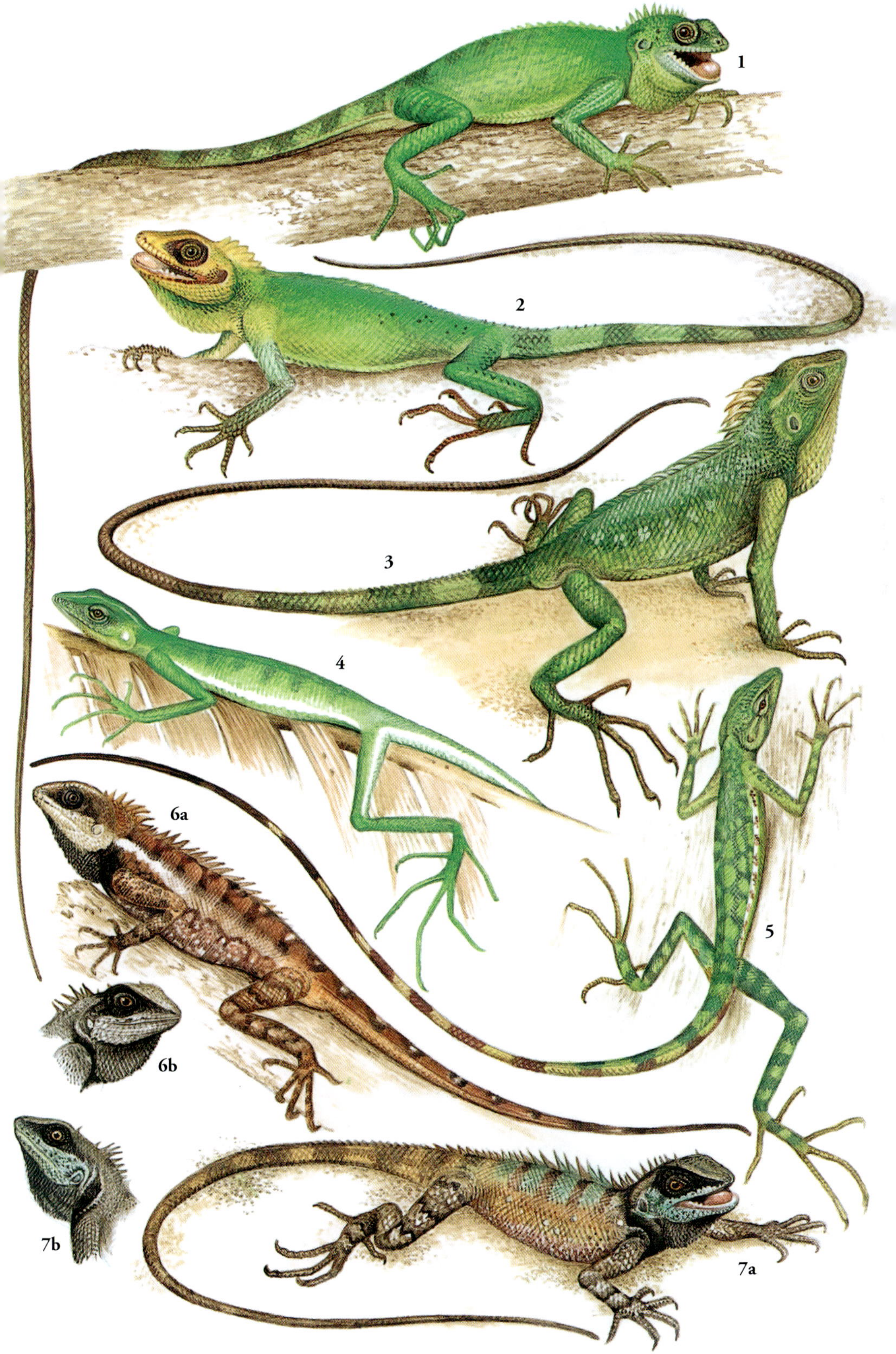
1
2
3
4
5
6a
6b
7a
7b

Plate 12. AGAMIDAE

1. HTUNWIN'S FOREST LIZARD *Calotes htunwini*, p. 181.
SVL 91.4mm Pair or cluster of supratympanic spines; scales comprising dorsal crest small; dorsum beige to light tan; paired dark brown, cream-centred nuchal spots; throat faded orange.
(**1a**) **Male** SVL to 91.4mm; forearm band always absent.
(**1b**) **Female** SVL to 84.3mm; forearm band present in half the population.

2. AYEYARWADY FOREST LIZARD *Calotes irawadi*, p. 182.
SVL 106.8mm Pair or clusters of spines above tympanum; dorsum bronze-brown; pair of dark brown nuchal spots on posterolateral edge of interparietal; faded palmate striping on throat.
(**2a**) **Male** SVL to 106.8mm; nuchal spot present in half the population.
(**2b**) **Female** SVL to 77.9mm; nuchal spot always present.

3. JERDON'S FOREST LIZARD *Calotes jerdoni*, p. 182.
SVL 100mm Dorsal crest present, nuchal crest weak; dorsum bright green, sometimes with pair of black-edged brown bands, and yellow, orange or brown blotches.

4. BLUE FOREST LIZARD *Calotes mystaceus*, p. 182.
SVL 140mm Dorsal and nuchal crests continuous; 2–3 spines behind eye; dorsum greyish-brown, turning bright blue to turquoise, with 3–5 large dark spots on flanks, during breeding season.

5. GARDEN LIZARD *Calotes versicolor*, p. 183.
SVL 95mm Two separated spines above tympanum; no antehumeral fold.
(**5a**) **Male** Head orange or bright red; black patch on throat in breeding males, fading to dull grey.
(**5b**) **Female** Yellow or greyish-olive.

6. KINABALU FOREST LIZARD *Complicitus nigrigularis*, p. 183.
SVL 75.5mm Middorsal crest slightly large, strongly keeled; anterior 9 scales of crest elongated and compressed; dorsum dark brownish-grey with 2 broad transverse white bands on trunk; white subocular spots; gular pouch black with pair of white spots near distal end.

7. BOULENGER'S TREE LIZARD *Dendragama boulengeri*, p. 183.
SVL 73mm Bony ridge covered with large scales extends from above tympanum to orbit; short dorsal crest comprising 13–14 scales; nuchal crest comprises 6–7 erect scales; dorsum bluish-green or olive-grey with transverse dark bands; oval dark patch on sides of neck, edged by narrow yellow stripe.

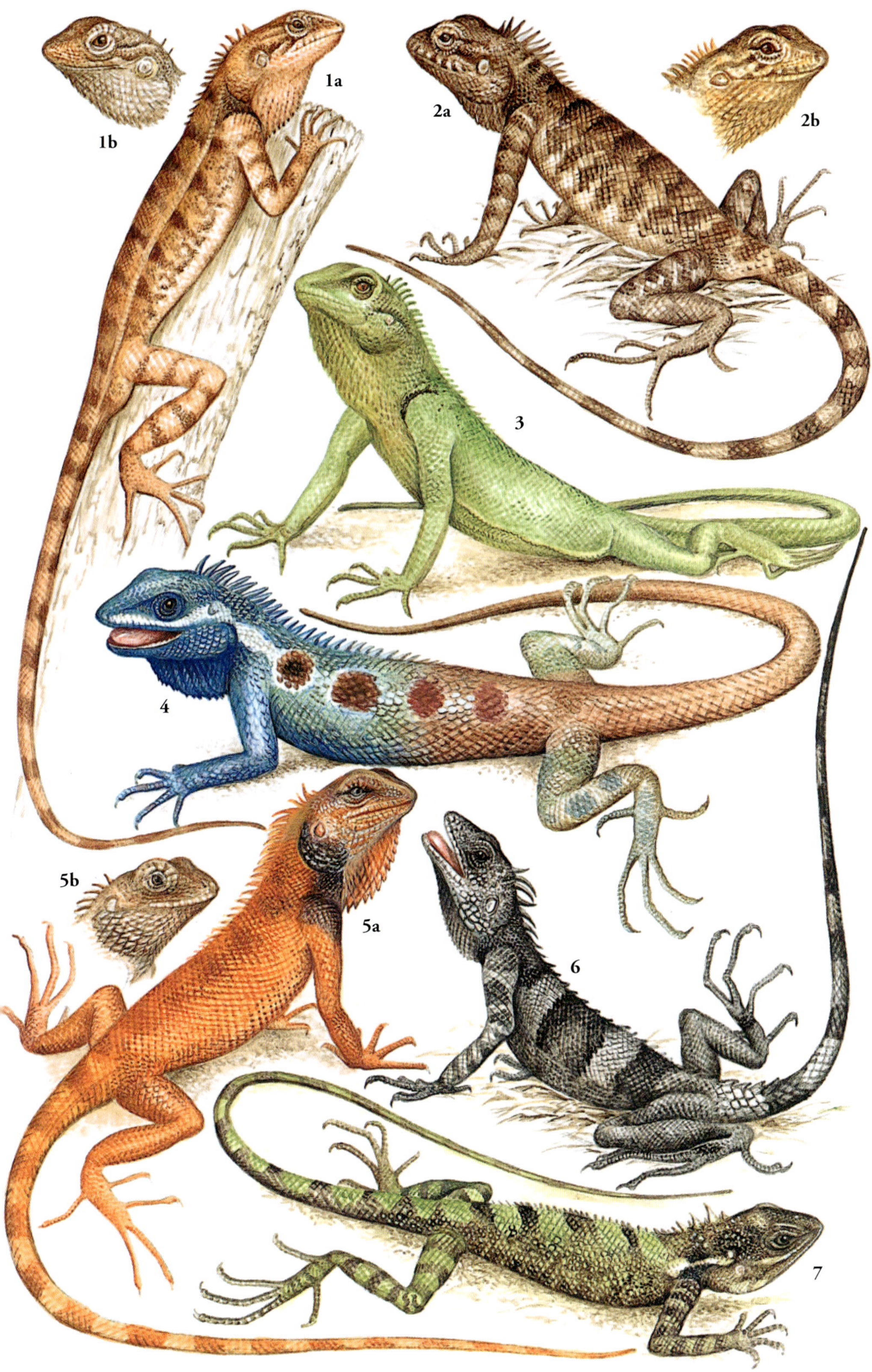
1a
1b
2a
2b
3
4
5a
5b
6
7

Plate 13. AGAMIDAE

1. BLANFORD'S FLYING LIZARD *Draco blanfordii*, p. 183.
SVL 130mm Nostrils directed dorsally; no thorn-like scale on supraciliary edge; patagial ribs typically 5, exceptionally, 6; dorsum bluish-grey or brownish-grey; patagia of males with longitudinal white lines, of females with black cross-bars; patagium with 5 transverse dark bands with pale spots.

2. HORNED FLYING LIZARD *Draco cornutus*, p. 183.
SVL 88.9mm Nostrils oriented laterally; distinct thorn-like scale over eye; patagial ribs 6; dorsum bright green to greenish-brown in males; tan or light brown in females; patagium reddish-orange with dark spots or bands; gular pouch bright yellow.

4. CRESTED FLYING LIZARD *Draco cristatellus*, p. 183.
SVL 90mm Nostrils oriented laterally or slightly obliquely; patagial ribs 5–6; dorsum brownish-grey with transverse dark bars or spots; patagium dark brown with lighter lines.

4. FRINGED FLYING LIZARD *Draco fimbriatus*, p. 184.
SVL 132mm Nostrils oriented laterally; spinous projection over eye; patagial ribs 5; dorsum and patagium greyish-brown with grey and pale green markings.

5. BEAUTIFUL FLYING LIZARD *Draco formosus*, p. 184.
SVL 101mm Nostrils directed dorsally; no thorn-like scale on supraciliary edge; patagial ribs 5; dorsum bluish-grey or brownish-grey; patagium olive-grey dorsally with distinct or obscure greyish-black or brown cross-bars or spots.

6. RED-BEARDED FLYING LIZARD *Draco haematopogon*, p. 184.
SVL 94mm Nostrils oriented dorsally; no thorn-like scale above eye; patagial ribs 5; dorsum dark olive or brownish-grey with indistinct lighter and darker spots; patagium black with yellow spots.

7. INDO-CHINESE FLYING LIZARD *Draco indochinensis*, p. 184.
SVL 107.8mm Nostrils dorsal, directed upwards; no thorn-like scale on superciliary edge; patagial ribs 5; dorsum brownish-grey with small dark spots; patagium light brown to dark brown on edges, with 6 transverse light-edged bars.

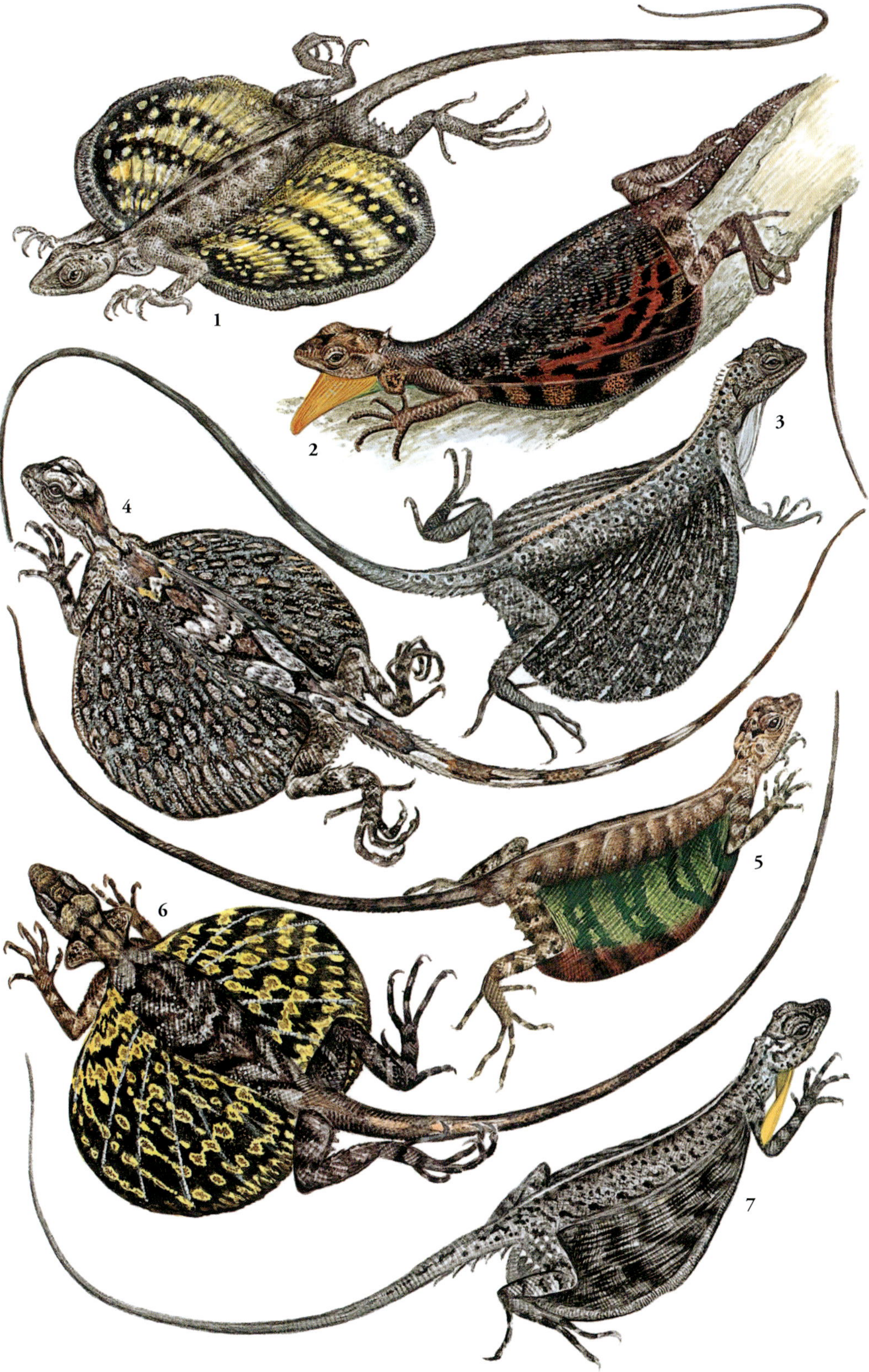
1
2
3
4
5
6
7

Plate 14. AGAMIDAE

1. ORANGE-WINGED FLYING LIZARD *Draco maculatus*, p. 184.
SVL 82mm Nostrils lateral, directed outwards; patagial ribs 5; dorsum pale blue or olive-grey, male patagium yellow, sometimes with orange outer edge, and narrow longitudinal dark stripes dorsally; that of females (illustrated) grey with narrow white lines and large yellow blotches.

2. LARGE FLYING LIZARD *Draco maximus*, p. 185.
SVL 140mm Nostrils dorsal; no spinous projection above eye; patagial ribs 6; dorsum green or greyish-olive with brownish-olive pattern of bands; patagium red or yellowish-orange with black bands showing discontinuous olive-brown lines.

3. BLACK-BEARDED FLYING LIZARD *Draco melanopogon*, p. 185.
SVL 93mm Nostrils dorsal; no spinous projections above eye; patagial ribs 5; dorsum olive or green with brownish-grey bands or diamond-shaped spots; patagium black, sometimes with scattered yellow-orange spots.

4. OBSCURE FLYING LIZARD *Draco obscurus*, p. 185.
SVL 113mm Nostrils oriented upwards; patagial ribs 5; dorsum grey-brown with scattered dark brown flecks; 5 indistinct butterfly-shaped blue marks across vertebral region; patagium yellowish-olive or yellowish-orange with 5–6 broad dark bands enclosing yellowish-olive or tan-grey areas.

5. FIVE-BANDED FLYING LIZARD *Draco quinquefasciatus*, p. 185.
SVL 110mm Nostrils dorsal; patagial ribs 6; dorsum bright green in males, brownish-olive in females (illustrated), with dark speckling; patagium olive-brown, olive-yellow or orange-red above, with 5 dark brown or black cross-bars.

6. COMMON FLYING LIZARD *Draco sumatranus*, p. 185.
SVL 85mm Nostrils lateral; patagial ribs typically 6; dorsum yellowish-brown or light brown with dark brown blotches; gular pouch bright yellow with black dots at base.

7. NARROW-LINED FLYING LIZARD *Draco taeniopterus*, p. 185.
SVL 83mm Nostrils dorsal; patagial ribs 5; dorsum pale grey or greenish-brown; patagium pale grey or greenish-yellow with 5 irregular black bands.

8. JAVANESE FLYING LIZARD *Draco volans*, p. 186.
SVL 96mm Nostrils lateral; patagial ribs typically 6; dorsum yellowish-tan or dark brown; distinct black edge to pale yellow-brown or yellowish-tan patagium in males.

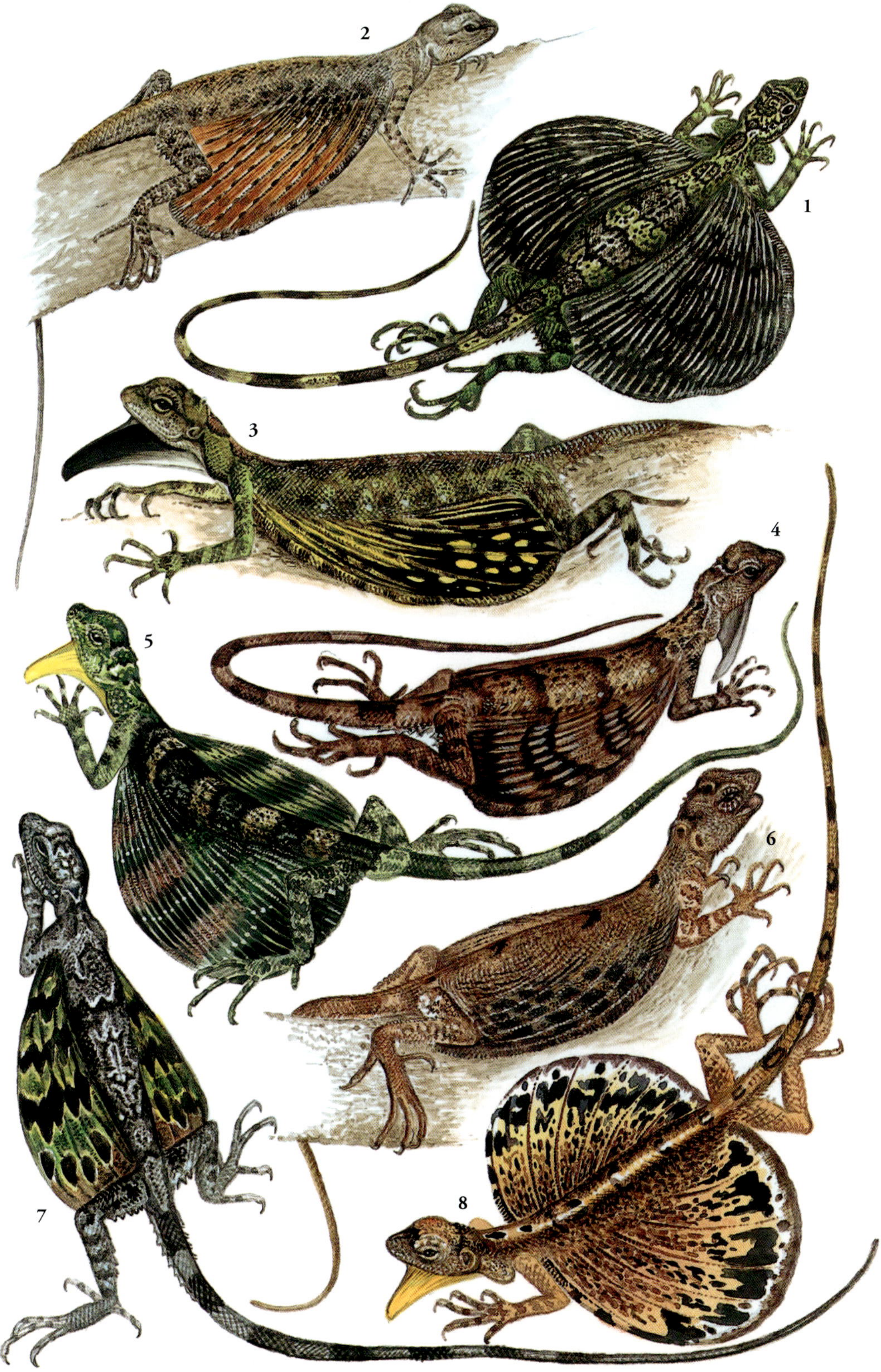
2
1
3
4
5
6
7
8

Plate 15. AGAMIDAE

1. ABBOTT'S ANGLE-HEADED LIZARD *Gonocephalus abbotti*, p. 186.
SVL 143mm Dorsal crest lower than nuchal crest; superciliary border raised; dorsum green; radiating dark lines around eyes; iris bright red.

2. BLUE-NECKED ANGLE-HEADED LIZARD *Gonocephalus bellii*, p. 186.
SVL 150mm Dorsal crest comprises tall scales anteriorly; dorsum greyish-brown in males, reddish-brown in females; gular pouch pale green at base, with tip indigo blue in males; gular region with pink flecks in females.

3. BEYSCHLAG'S ANGLE-HEADED LIZARD *Gonocephalus beyschlagi*, p. 186.
SVL 126mm Nuchal and dorsal crests continuous, slightly notched at nape, on dorsum, sail-like and interconnected; dorsum green with yellow streak on nape; iris brown; flanks with large yellow spots.

4. BORNEAN ANGLE-HEADED LIZARD *Gonocephalus bornensis*, p. 186.
SVL 136mm Nuchal and dorsal crests continuous, comprising elongated scales; dorsum green with dark bands; flanks with oval light spots; gular pouch pale with dark stripes.
(4a) Adult Nuchal crest high.
(4b) Female Nuchal crest low.

5. TIOMAN ANGLE-HEADED LIZARD *Gonocephalus chamaeleontinus*, p. 186.
SVL 170mm Raised and curved eyebrows; dorsal crest with triangular to lanceolate scales; dorsum green with yellow spots, cross-bars and reticulated pattern; iris brownish-red; gular pouch with blue stripes.

6. MARQUIS DORIA'S ANGLE-HEADED LIZARD *Gonocephalus doriae*, p. 186.
SVL 163mm Dorsal crest as high as nuchal crest; dorsum green with wavy grey pattern or with large areas of orange; iris pink; gular pouch yellow, orange or grey with 6 bluish-grey stripes.

7. GIANT ANGLE-HEADED LIZARD *Gonocephalus grandis*, p. 187.
SVL 160mm Nuchal and dorsal crests separate in males; dorsal crest absent but nuchal sail present in females; dorsum green to grey; flanks with yellow spots in males, brownish-green in females.
(7a) Male Dorsal crest.
(7b) Female No dorsal crest; cheek unpatterned.
(7c) Juvenile male Low dorsal crest; cheek with striped pattern.

8. KLOSS'S ANGLE-HEADED LIZARD *Gonocephalus klossi*, p. 187.
SVL 165mm Nuchal scales lance-like; nuchal and dorsal crests separate; dorsum olive-brown with pale brown flecks; iris powder-blue; head with greyish-black mask; gular pouch cream.

9. KUHL'S ANGLE-HEADED LIZARD *Gonocephalus kuhlii*, p. 187.
SVL 100mm Nuchal crest low; dorsum greenish-olive with pale flecks on head; iris brown; cream band on shoulder; sometimes vertical red or yellow bands on dorsum.

10. SIKULIKAP ANGLE-HEADED LIZARD *Gonocephalus lacunosus*, p. 187.
SVL 145mm Dorsal crest separate from nuchal crest; dorsum green or grey; enlarged flank scales greenish-olive or yellow; iris dark blue; head and gular pouch with turquoise markings.

11. BLUE-EYED ANGLE-HEADED LIZARD *Gonocephalus liogaster*, p. 187.
SVL 140mm Nuchal and dorsal crests continuous, comprising lanceolate scales; dorsum brown or green with reticulated dark flank pattern in males, with yellow cross-bars in females; iris bright blue.
(11a) Male Dorsal crest high; skin surrounding eye reddish-orange.
(11b) Female Dorsal crest low; skin surrounding eye brown.

12. LARGE-SCALED ANGLE-HEADED LIZARD *Gonocephalus megalepis*, p. 188.
SVL 140mm Nuchal and dorsal crests separate; dorsum green with indistinct brown bands; iris indigo blue; some rounded scales on body bright yellow; pinkish-cream stripe along body; gular pouch pale green.

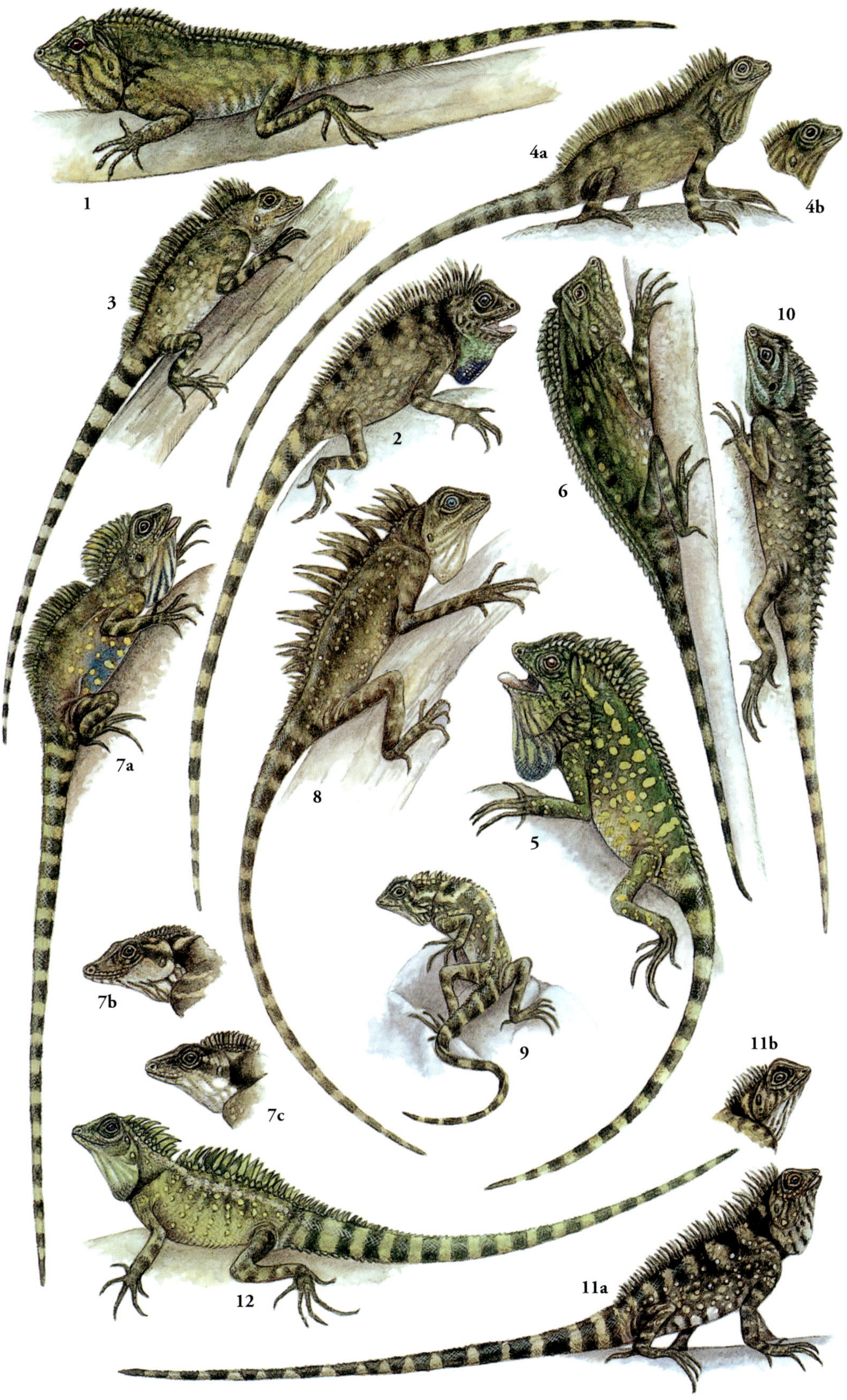

1
4a
4b
3
2
6
10
7a
8
5
7b
9
11b
7c
12
11a

Plate 16. AGAMIDAE

1. MJÖBERG'S ANGLE-HEADED LIZARD *Gonocephalus mjobergi*, p. 188.
SVL 88mm Single large flat scale below tympanum; gular pouch small, its edge feebly serrated and covered with small scales; nuchal crest present; dorsal crest a small ridge; dorsum pale green changeable to brownish-grey, with narrow reticulated grey pattern that encloses yellow spots on lower flanks.

2. ROBINSON'S ANGLE-HEADED LIZARD *Gonocephalus robinsoni*, p. 188.
SVL 152mm Nuchal and dorsal crests continuous, decreasing in height caudally; gular pouch large; flanks with scattered enlarged scales; dorsum green with oblique dark cross-bars and yellow spots forming pale bars; labials white with dark bars on sutures; black postocular streak extends to tympanum.
(2a) Female Gular pouch yellow.
(2b) Male Gular pouch pink.

3. BECCARI'S HORNED MOUNTAIN LIZARD *Harpesaurus beccarii*, p. 188.
SVL 86mm Dorsum with longitudinal series of smooth rhomboidal scales; long anterior rostral appendage and short posterior one; dorsal crest comprises 13 scales; dorsum bright green; 2 oblique white stripes on head and neck.

4. BORNEAN HORNED LIZARD *Harpesaurus borneensis*, p. 188.
SVL 59mm Cylindrical rostral appendage surrounded by 4 enlarged scales; nuchal crest developed; dorsal and caudal crests in males; gular pouch present; dorsum reddish-brown with black spots in 7 oblique series and yellow-brown bands.

5. MODIGLIANI'S HORNED LIZARD *Harpesaurus modigliani*, p. 189.
SVL 83mm Rostral horn present; gular pouch present; high nuchal crest or spinous scales; less spinose scales from nape to mid-tail; dorsum azure-blue; supralabials cream; gular pouch brown with light patches; rostral horn grey.

6. THREE-BANDED HORNED LIZARD *Harpesaurus tricinctus*, p. 189.
SVL 64mm Rostral horn present; gular scales tubercular; dorsal crest a low serrated ridge; dorsum brown with 3 broad, transverse yellow bands on body, anterior narrowest on shoulder, others descend to venter.

7. KINABALU CRESTED DRAGON *Hypsicalotes kinabaluensis*, p. 189.
SVL 145mm Nuchal and dorsal crests separate, continuing to tail; gular pouch distinct in males; dorsal scales heterogenous; dorsum green with chocolate-brown and black spots forming bands; gular pouch pale red with black and white stripes in anterior margin.

8. CHAPA MOUNTAIN LIZARD *Japalura chapaensis*, p. 189.
SVL 59.6mm Nuchal crest present; dorsum pale greyish-tan with 5 transverse dark brown bars; pale dorsolateral stripe.

9. CRESTLESS MOUNTAIN LIZARD *Japalura planidorsa*, p. 190.
SVL 53mm Occipital region with numerous spinous tubercles; temporal spines conical; ridge of enlarged keeled scales; dorsum yellowish-brown; series of dark streaks across back; orange or red gular region.

10. SWINHOE'S MOUNTAIN LIZARD *Japalura swinhonis*, p. 190.
SVL 82.7mm Nuchal crest present; dorsal crest low, extending to part of tail; dorsum brownish-grey with light dorsolateral stripe; broad and angular dark cross-bars in intervening areas.

11. YUNNAN MOUNTAIN LIZARD *Japalura yunnanensis*, p. 190.
SVL 75mm Dorsum covered with small keeled scales, intermixed with larger and strongly keeled scales; low nuchal crest comprising separated triangular spines; dorsal crest a serrated ridge; dorsum olive-green with dark green markings and pale cross-bars; pale dorsolateral stripe.

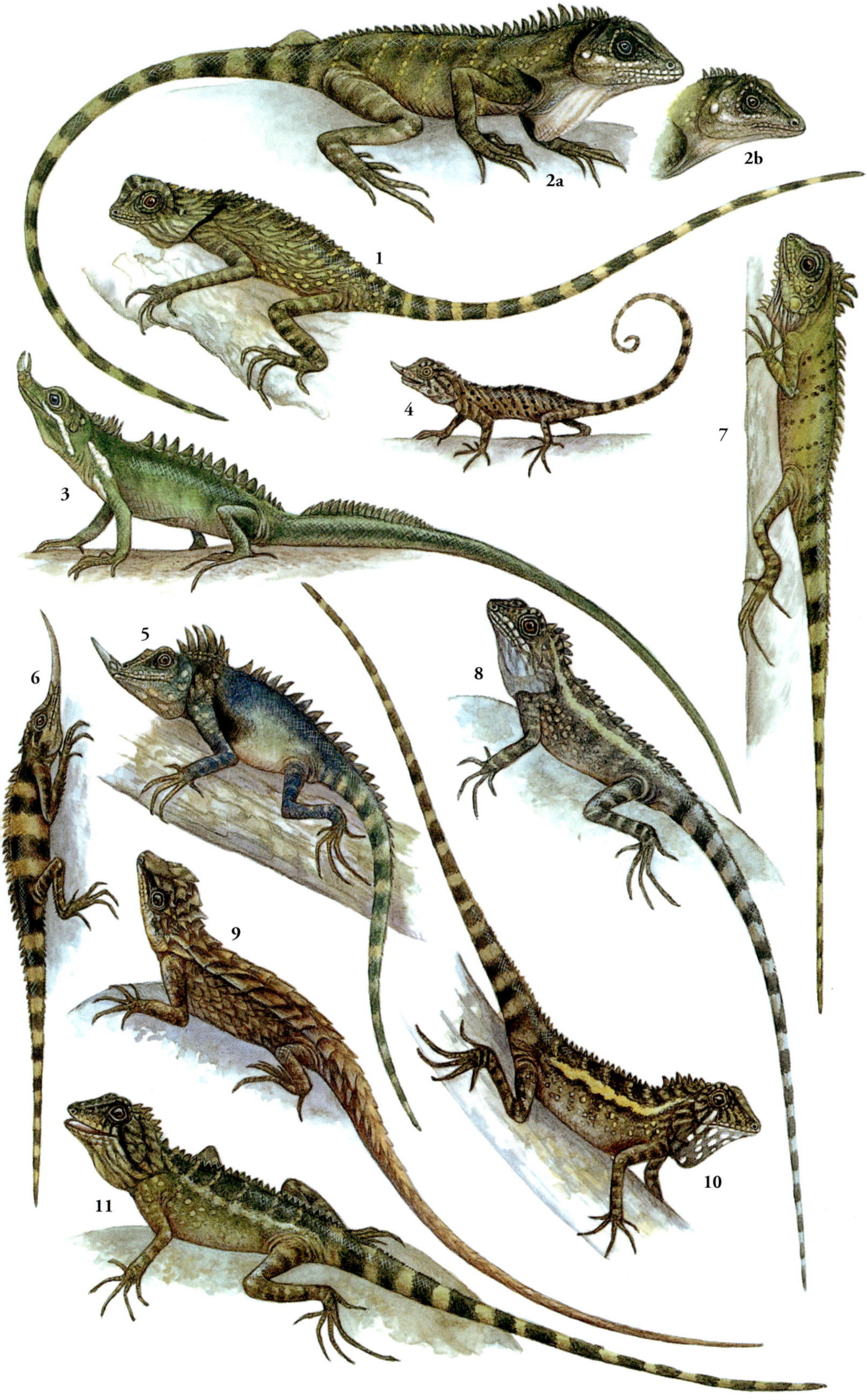

2a
2b
1
4
7
3
5
6
8
9
10
11

Plate 17. AGAMIDAE

1. LUDEKING'S CRESTED LIZARD *Lophocalotes ludekingi*, p. 190.
SVL 92mm Tympanum large; head large, expanded below tympanum; males bright green or dark brown with large white patch behind tympanum and indistinct transverse olive bands on body; females lack light vertebral band.

2. PHU WUA LIZARD *Mantheyus phuwuanensis*, p. 191.
SVL 90mm Gular pouch surrounded by U-shaped fold; femoral glands present; limbs and digits elongated; dorsum dark brown or olive-green with green spots; flanks brown or bluish with black speckles and pale orange stripes.

3. BORNEAN SHRUB LIZARD *Phoxophrys borneensis*, p. 191.
SVL 66mm No spine above eye; nasal in contact with supralabial; continuous row of infraorbitals; gular scales sharply keeled; dorsum brown to greyish-brown; 2 dark interorbital bars, anterior one narrower.

4. LARGE-HEADED SHRUB LIZARD *Phoxophrys cephalus*, p. 191.
SVL 84mm No spine above eye; nuchal crest comprises 7–8 thick conical scales; nasal in contact with supralabials; 2 continuous rows of infraorbitals; dorsum pale green with wavy dark green or greyish-green bands; 2 enlarged cream spines on sides of lower jaw.

5. BLACK-LIPPED SHRUB LIZARD *Phoxophrys nigrilabris*, p. 191.
SVL 58mm No spine above eye; nuchal crest comprises 6–12 compressed scales; nasal separated from supralabials; gular scales distinctly keeled; 4 rows of large scales near tail-base; dorsum brown, green or olive with or without transverse blue bands.

6. SPINY-HEADED SHRUB LIZARD *Phoxophrys spiniceps*, p. 191.
SVL 60.3mm Enlarged spine above eye; nasal in contact with second supralabial; infraorbitals in single continuous row; gular scales keeled; dorsum greenish-grey with brown bars and pale stripes.

7. INDO-CHINESE WATER DRAGON *Physignathus cocincinus*, p. 192.
SVL 250mm Nuchal and dorsal crests continuous, separated from caudal crest; no gular pouch; distinct gular fold; dorsum bright green changeable to brownish-green; flanks with 3–5 narrow bluish-green stripes in juveniles; in adults, stripes obscure or absent.
(7a) Adult Nuchal crest; dark postocular stripe.
(7b) Juvenile No nuchal crest; no postocular stripe.

8. SHORT-FOOTED LONG-HEADED LIZARD *Pseudocalotes brevipes*, p. 192.
SVL 77.5mm Nuchal crest comprising 6–7 erect compressed scales; enlarged spinose scale above tympanum; low denticulate dorsal crest; dorsum brown; cheeks sometimes bluish-grey; yellow and orange stripes on gular pouch.

9. YELLOW-THROATED LONG-HEADED LIZARD *Pseudocalotes flavigula*, p. 192.
SVL 72mm Nuchal crest composed of 6 small spines; dorsal crest a low ridge; large lateral scales; dorsum light olive-green with saddle-like brown patches; flanks spotted with white.

10. FLOWER'S LONG-HEADED LIZARD *Pseudocalotes floweri*, p. 192.
SVL 98mm Nuchal crest with 8 erect spines; snout length twice orbit diameter; 6 scales between nasal and supraorbital; dorsum brown; enlarged dorsal scales blue; spines light brown.

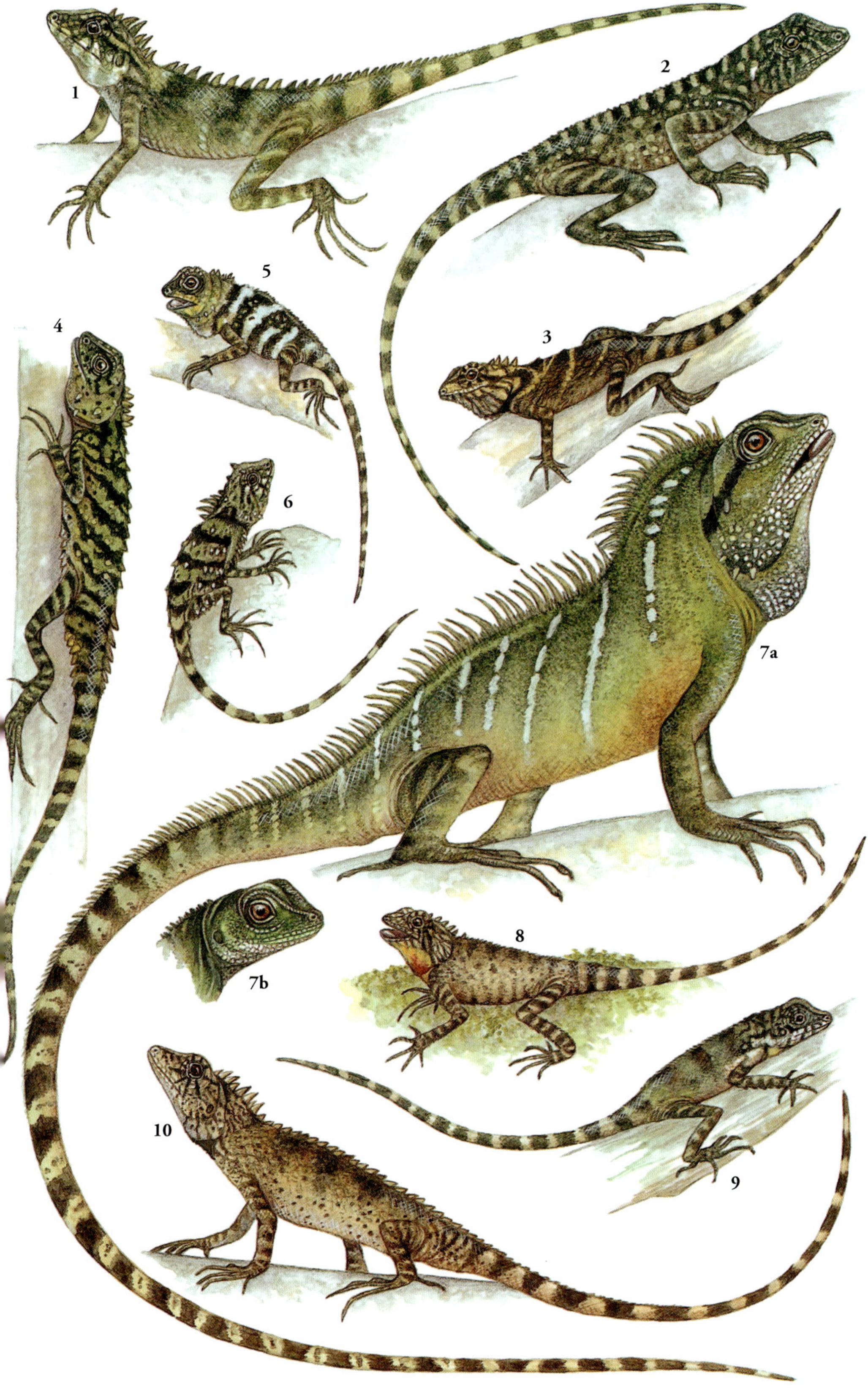
1
2
5
3
4
6
7a
7b
8
10
9

Plate 18. AGAMIDAE

1. KAKHIEN HILLS LONG-HEADED LIZARD *Pseudocalotes kakhienensis*, p. 193.
SVL 125mm Nuchal crest composed of 7–9 separated spines; no gular pouch; dorsum pale greenish-olive with light and dark brown variegation; labials with dark bars on sutures; dark postocular stripe to tympanum.

2. BUKIT LARUT LONG-HEADED LIZARD *Pseudocalotes larutensis*, p. 193.
SVL 77.3mm Nuchal crest comprises 5 erect compressed scales; dorsum pale tan-yellow with greenish cast; 3 broad dark bands on dorsum.

3. BORNEAN LONG-HEADED LIZARD *Pseudocalotes sarawacensis*, p. 193.
SVL 82mm Nuchal crest comprises 7 tall flexible scales; prefrontal scales slightly large, forming inverted V; dorsum dark brown with indistinct green areas; nuchal crest and throat with orange-red wash.

4. JAVANESE LONG-HEADED LIZARD *Pseudocalotes tympanistriga*, p. 194.
SVL 80mm Nuchal crest denticulated; forehead scales strongly keeled; gular pouch small; dorsum grass-green with or without olive-brown cross-bars.

5. KON TUM LONG-HEADED LIZARD *Pseudocophotis kontumensis*, p. 194.
SVL 87.8mm Nuchal crest comprises 6 erect scales; low dorsal crest; tympanum concealed; no gular pouch; dorsum brown with green areas on head; light green stripe below eye extends to neck.

6. MOUNT VICTORIA FAN-THROATED LIZARD *Ptyctolaemus collicristatus*, p. 194.
SVL 91.3mm No dorsal crest; faint gular folds; males with high nuchal crest comprising large, flattened and triangular scales; dorsum pale greyish-brown with mottling or dark brown saddles; flanks greyish-brown anteriorly turning yellowish-brown posteriorly, with brown reticulations; gular pouch bright yellow medially.

7. GREEN FAN-THROATED LIZARD *Ptyctolaemus gularis*, p. 194.
SVL 80mm No dorsal crest; 3 longitudinal folds on each side of throat; gular pouch present; low nuchal crest comprising small triangular scales; dorsum olive-brown; fold on back deep blue; 5 broad transverse bands on body; green dorsolateral band sometimes on anterior flanks.

8. SUMATRAN HORNED LIZARD *Thaumatorhynchus brooksi*, p. 194.
SVL 60mm Cylindrical rostral appendage; dorsal and nuchal crests present, not continuous; gular pouch small; dorsum dark green; flanks with cream stripes; lobes of nuchal crest blue.

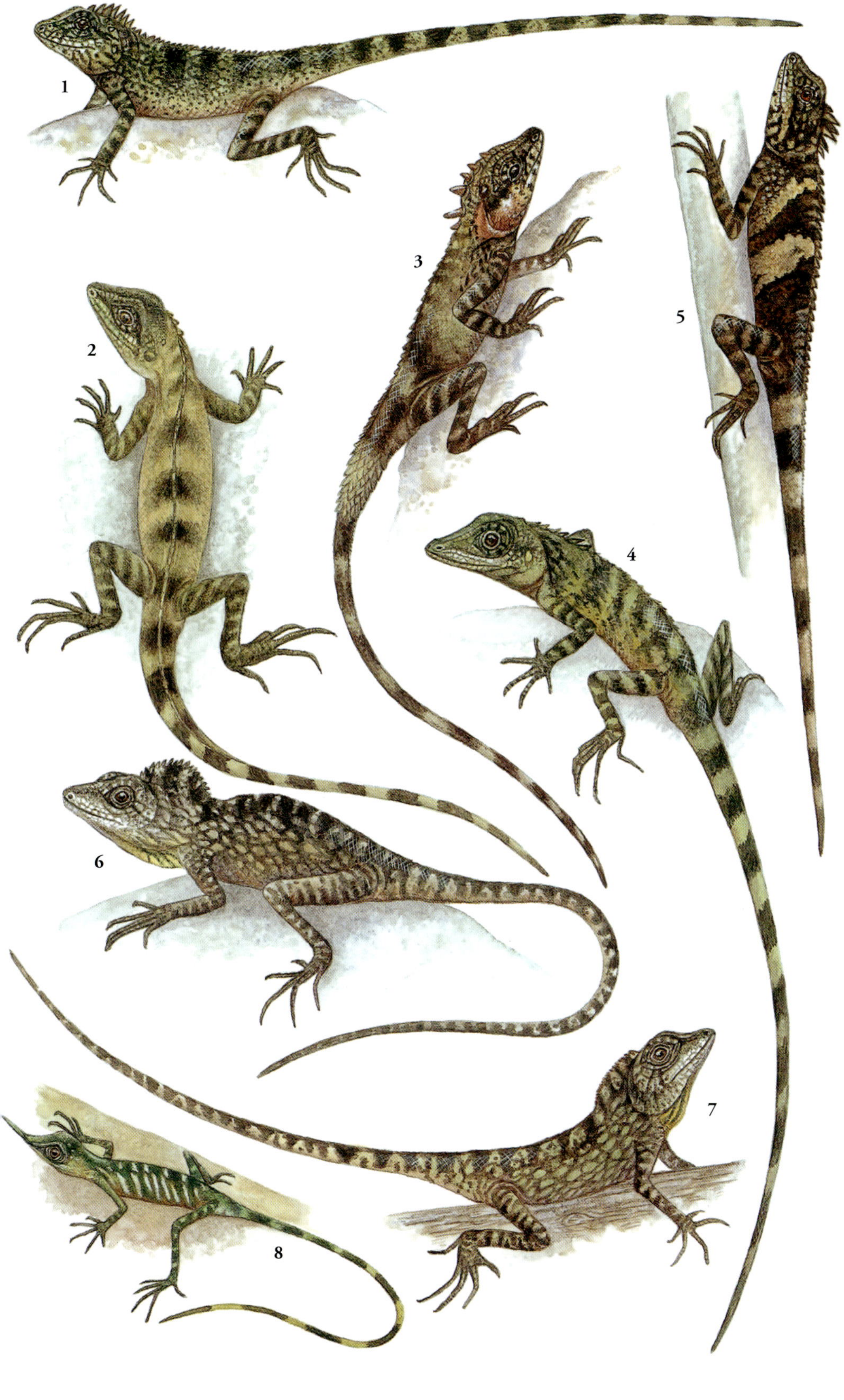
1
2
3
4
5
6
7
8

Plate 19. ANGUIDAE & DIBAMIDAE

1. BORNEAN GLASS SNAKE *Ophisaurus buettikoferi*, p. 195.
TL 375mm Frontonasal narrower than frontal width; auricular opening as large as nostril; transverse rows of dorsals 98–105; dorsum brown with dark lateral band; small blue interparietal spot; anterior dorsum with irregular transverse series of blue spots.

2. INDIAN GLASS SNAKE *Ophisaurus gracilis*, p. 195.
TL 589.5mm Frontonasal narrower than frontal width; auricular opening as large as nostril; transverse rows of dorsals 88–94; dorsum reddish-brown, light brown or yellow; transverse rows of black-edged blue spots on forepart of dorsum.

3. HART'S GLASS SNAKE *Ophisaurus harti*, p. 195.
TL 596mm Frontonasal narrower than frontal width; auricular opening minute, smaller than nostril; 2 scales in line between rostral and prefrontal; transverse rows of dorsals 94–100; 2 scales in line between nostril and prefrontal; dorsum brown or pale olive with transverse blue bars, or unpatterned.

4. SOKOLOV'S GLASS SNAKE *Ophisaurus sokolovi*, p. 195.
TL 176mm Frontonasal subequal to frontal; auricular opening three times nostril size; transverse rows of dorsals 88–92; dorsum brownish-beige with longitudinal rows of dark spots anteriorly; olive cross-bars across body.

5. BOO LIAT'S WORM LIZARD *Dibamus booliati*, p. 196.
SVL 102.7mm Postocular single; no rostral suture; interparietal posteriorly bordered by 4 nuchal scales; dorsum brownish-red, scales dark-edged; nuchal band cream; snout-tip and throat pinkish-white.

6. DE ZWAAN'S WORM LIZARD *Dibamus dezwaani*, p. 196.
SVL 123.1mm Postoculars 2; frontonasal entire; incomplete rostral sutures; labial and nasal sutures complete; dorsum purplish-brown, scales edged with dark brown; no nuchal or body bands; supralabials cream.

7. GREER'S WORM LIZARD *Dibamus greeri*, p. 197.
SVL 82mm Incomplete medial-rostral nasal suture; single postocular; supralabials 2; posteromedial edge of infralabials bordered by long narrow scales; dorsum dark brown, sometimes with 3 bright blue rings, each up to 9 scales wide, 2 on body and 1 on tail.

8. INGER'S WORM LIZARD *Dibamus ingeri*, p. 197.
SVL 96mm Nasal and labial sutures complete, extending from ocular to nostril; frontonasal divided; postoculars 2; Supralabial I bordering ocular; dorsum unpatterned brown, except posterior half of body, with pale linear blotches; snout-tip and supralabials yellowish-cream; wide cream nuchal band.

9. WHITE-TAILED WORM LIZARD *Dibamus leucurus*, p. 197.
SVL 136mm Single postocular; nasal suture complete; rostral sutures incomplete; 3 small scales bound infralabials; dorsum medium brown, each scale with darker edge; snout-tip and tail-tip pale brown; no pale nuchal or body bands.

10. TIOMAN WORM LIZARD *Dibamus tiomanensis*, p. 198.
SVL 92.5mm Rostral sutures incomplete; nasal and labial sutures complete; single postocular; dorsum dark brown dorsally and ventrally; snout and jaws light brown.

11. VORIS'S WORM LIZARD *Dibamus vorisi*, p. 198.
SVL 89.2mm Rostral suture incomplete; nasal suture complete; no labial suture; postoculars 2; dorsum unpatterned brown; snout-tip, sides of head and supralabials yellowish-cream; no nuchal band; pale brown body band.

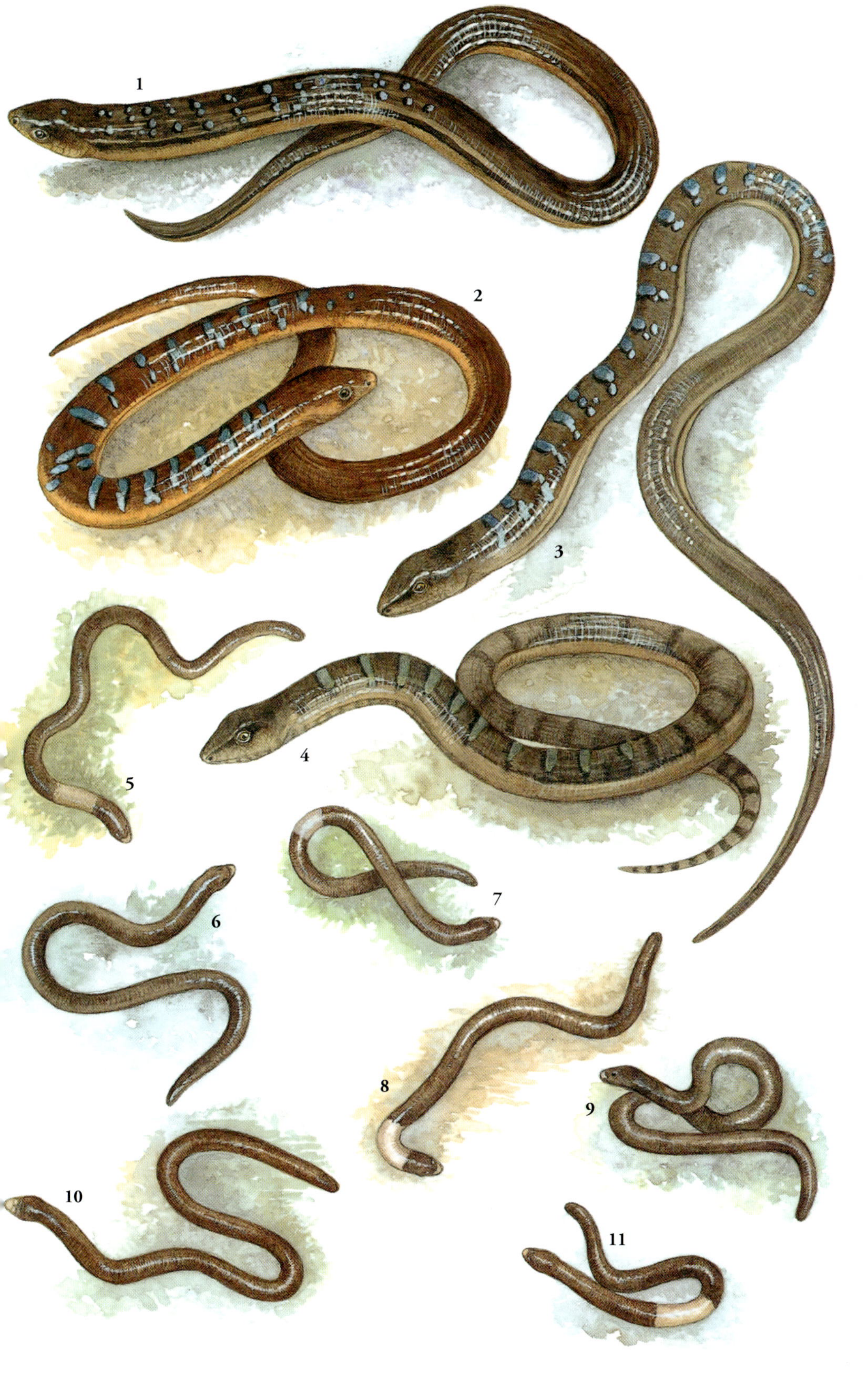
1
2
3
4
5
6
7
8
9
10
11

Plate 20. EUBLEPHARIDAE & GEKKONIDAE

1. CAT GECKO *Aeluroscalabotes felinus*, p. 198.
SVL 122mm Eyelids fleshy; tail capable of being curled laterally; transverse enlarged lamellae restricted to bases of digits; dorsum tan-brown, sometimes with orange or white spots and dark or pale vertebral stripe on body and tail.

2. VIETNAM LEOPARD GECKO *Goniurosaurus araneus*, p. 198.
SVL 124mm Eyelids fleshy; claws within 4 scales; dorsum dull yellow-grey; bands nearly immaculate and lacking mottling, comprising wide dorsal bands between limb insertions; nuchal loop posteriorly protracted.

3. LICHTENFELDER'S LEOPARD GECKO *Goniurosaurus lichtenfelderi*, p. 199.
SVL 83mm Eyelids fleshy; 24 paravertebral tubercles; dorsum dark or light brown; 2 dorsal body bands, black-edged with wide border, between insertions of limbs in adults; juveniles (illustrated) with numerous stripes; nuchal loop rounded posteriorly.

4. CHINESE LEOPARD GECKO *Goniurosaurus luii*, p. 199.
SVL 119mm Eyelids fleshy; dorsals granular; internasal scales present; no rows of enlarged tubercles on dorsum; dorsum grey-brown to cream with dark and light mottling; nuchal loop wide and not tapered backwards.

5. PENANG DAY GECKO *Cnemaspis affinis*, p. 200.
SVL 50.8mm Dorsal tubercles keeled; paravertebral tubercles 20–24; dorsum dark grey to brown; flanks with light markings; dark shoulder patch encloses white or yellow ocellus; yellow or white postscapular spot.

6. BAUER'S DAY GECKO *Cnemaspis baueri*, p. 200.
SVL 34.9mm Dorsum with enlarged tubercles in linear series; median subcaudals enlarged; dorsum dark brownish-olive with greyish-black spots on vertebral region; dorsal tubercles lack white tips.

7. CHANTHABURI DAY GECKO *Cnemaspis chanthaburiensis*, p. 201.
SVL 41mm Dorsal surface with subequal granular scales, lacking tubercles; smooth ventral and subcaudal scales; dorsum chocolate-brown or greyish-olive; white paravertebral markings from nape to tail; markings on nape and between shoulders oval and coalesced at midline, or separated by a pale stripe.

8. TITI WANGSA DAY GECKO *Cnemaspis flavolineatus*, p. 202.
SVL 41.2mm Dorsal tubercles keeled; median subcaudal row not enlarged; dorsum yellowish- or olive-brown with a distinct yellow vertebral stripe; no dark spots on neck or white markings on flanks.

9. KENDALL'S DAY GECKO *Cnemaspis kendallii*, p. 202.
SVL 80mm Dorsum with large and small scattered scales; median subcaudals tricarinate; dorsum pale brown with oblong dark brown spots forming 7 interrupted bands.

10. KUMPOL'S DAY GECKO *Cnemaspis kumpoli*, p. 202.
SVL 52mm Median subcaudals enlarged; dorsum olive-grey with 4 transverse blackish-grey bars; dark postocular stripe meets at back of head, edged by olive-grey line.

11. LIM'S DAY GECKO *Cnemaspis limi*, p. 202.
SVL 88.2mm Dorsum with enlarged smooth tubercles; dorsum dark green with oval-elongated black paravertebral spots enclosed within pale bands; green flecks on dorsum and in areas between dark stripes on sides of neck.

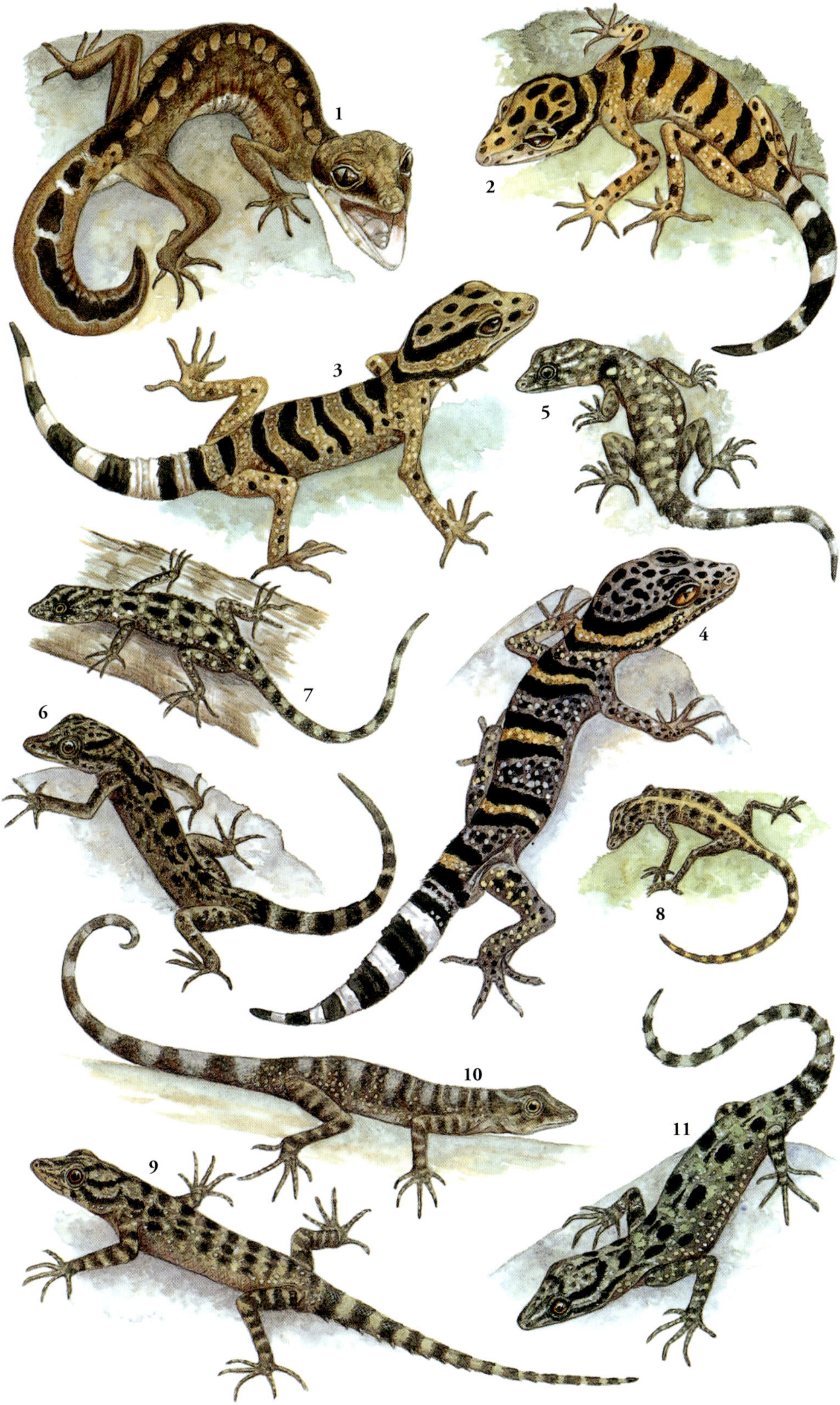

1
2
3
5
4
7
6
8
10
11
9

Plate 21. GEKKONIDAE

1. GADING DAY GECKO *Cnemaspis nigridia*, p. 203.
SVL 69.8mm Dorsal surface with small granular scales and scattered large keeled tubercles; males lack precloacal or femoral pores, or precloacal groove; dorsum brownish-olive with black blotches; 2 pairs of elongated dark brown spots on nape and axilla; single elongated spot on vertebral region.

2. PEMANGGIL DAY GECKO *Cnemaspis pemanggilensis*, p. 203.
SVL 76mm Dorsal surface with multi-ridged tubercles; no precloacal and femoral pores; median subcaudal scales enlarged and keeled; dorsum unpatterned black to grey with darker spots and stripes; dark bifurcating stripe on snout; dorsal stripe extends to nape, forming tripartite band, middle stripe to nape forming second band, ventral-most stripe extends to forelimb base.

3. PHUKET DAY GECKO *Cnemaspis phuketensis*, p. 204.
SVL 29.1mm Dorsal surface with small keeled scales and scattered spinous tubercles on paravertebral region; spines on flanks; no precloacal and femoral pores; median subcaudals not enlarged; dorsum olive with sinuous dark greyish-brown markings on nape and body.

4. THAI DAY GECKO *Cnemaspis siamensis*, p. 204.
SVL 42mm Dorsal surface with tubercles in 12–20 longitudinal rows; pores occasionally absent; median subcaudals enlarged; dorsum olive-grey with pale mottling; dark brown blotches may form cross-bars; 2 pale spots at back of head, separated by brown spot.

5. ANGLED BENT-TOED GECKO *Cyrtodactylus angularis*, p. 205.
SVL 95mm Dorsal surface with small scales intermixed with larger conical tubercles; subcaudals enlarged; preanofemoral scales 25; 6 precloacals in angular series and femorals slightly enlarged and continuous with precloacals in males; dorsum grey-brown to mid-brown with 2 series of large angular spots connected mesially; forehead with indistinct angular spots.

6. AYEYERWADY BENT-TOED GECKO *Cyrtodactylus ayeyarwadyensis*, p. 205.
SVL 78mm Dorsal surface with tubercles in 22–24 rows, keeled, oblong; 10–28 precloacal pores in single series or with scattered gaps of 1 poreless scale in males; subcaudals without enlarged midventral plates; dorsum mid-brown with 10 transverse rows of rectangular dark brown patches comprising rows of paired paravertebral marks and pair of lateral marks.

7. BADEN BENT-TOED GECKO *Cyrtodactylus badenensis*, p. 206.
SVL 74.1mm Dorsal surface with small scales and larger conical tubercles in 6 rows; femorals enlarged; no femoral pores; transversely enlarged subcaudals; dorsum dark chocolate-brown with 4 white cross-bars; cone-shaped marks arranged in series of 3–5.

8. KINABALU BENT-TOED GECKO *Cyrtodactylus baluensis*, p. 206.
SVL 86mm Dorsal surface with small scales and larger tubercles in 21–24 rows; no precloacal groove; precloacal pores 9–10; femoral pores 6–9; sharp boundary of scale size between ventral scales; dorsum brown to yellowish-brown with irregular dark spots or cross-bars.

9. SHORT-TOED BENT-TOED GECKO *Cyrtodactylus brevidactylus*, p. 206.
SVL 88mm Dorsal surface with enlarged keeled tubercles in 27 rows; subcaudals not enlarged; precloacal pores 8; no femoral pores; dorsum pale greyish-brown to white with 3–4 large dark blotches, edged with narrow dark brown.

10. SHORT-FINGERED BENT-TOED GECKO *Cyrtodactylus brevipalmatus*, p. 206.
SVL 72mm Dorsal surface with tubercles in 14–18 rows; 9 enlarged precloacal pores; 6–7 enlarged femoral pores in males; subcaudals enlarged; dorsum brown with small dark spots not arranged in linear series.

11. CAO VAN SUNG'S BENT-TOED GECKO *Cyrtodactylus caovansungi*, p. 207.
SVL 94mm Dorsal surface with rounded-triangular tubercles not arranged in longitudinal rows; 9 precloacal pores; 8 pairs of enlarged femoral scales; dorsum brown with 4 transverse bands with irregular edges; irregular small dark spots in middle part of light bands.

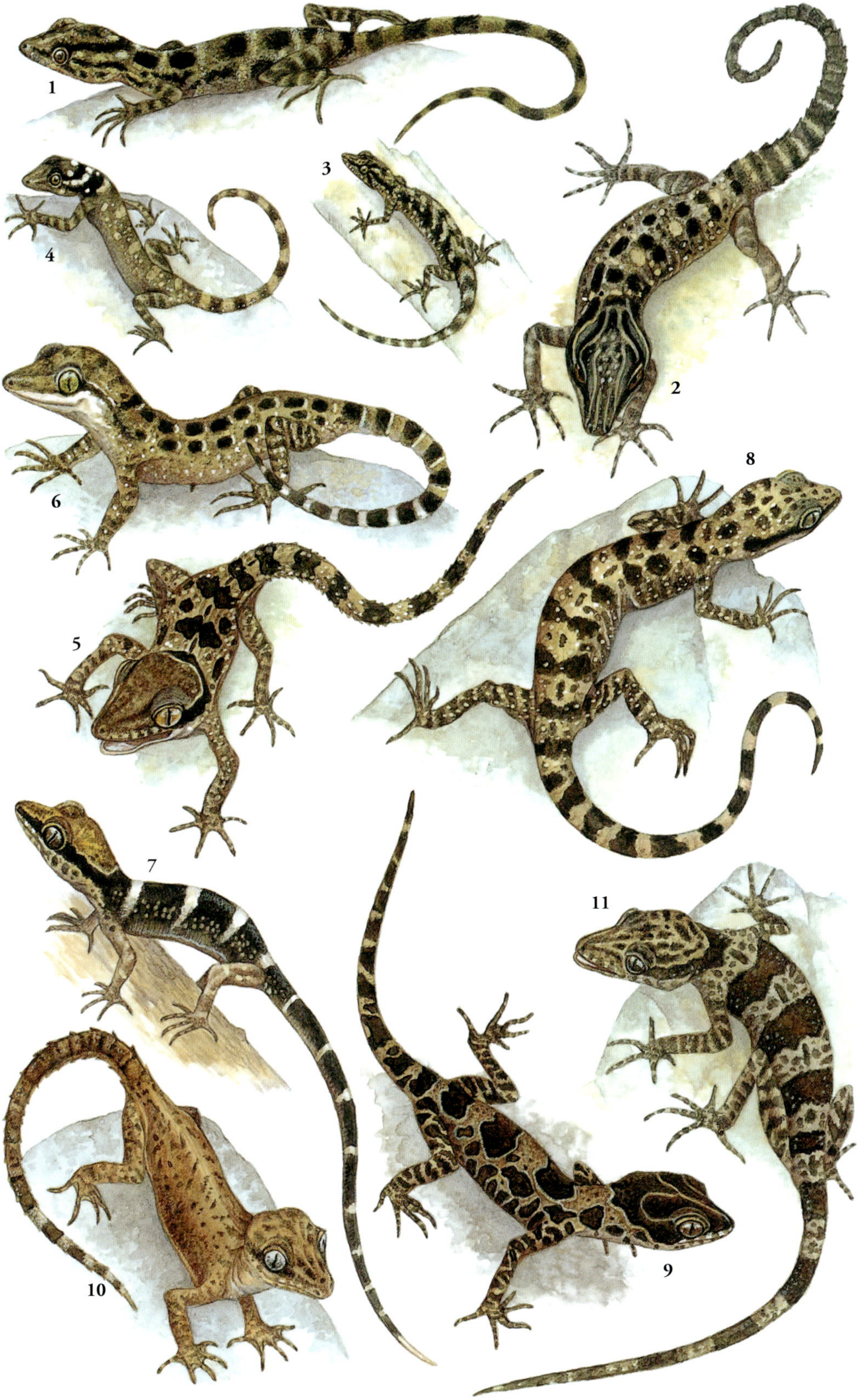
1
2
3
4
5
6
7
8
9
10
11

Plate 22. GEKKONIDAE

1. NIAH BENT-TOED GECKO *Cyrtodactylus cavernicolus*, p. 207.
SVL 80.8mm Dorsal surface with granular scales interspersed with 20–22 rows of trihedral or conical tubercles; no precloacal and femoral scales; dorsum brown with dark-edged brown cross-bars, changeable to brown-black.

2. CHANHOME'S BENT-TOED GECKO *Cyrtodactylus chanhomeae*, p. 207.
SVL 79mm Dorsal surface with 16–18 rows of keeled tubercles; 32–34 pore-bearing preanofemoral scales in males; dorsum pale brown with 3 cream-edged dark bands; brown nuchal loop.

3. CHANQUANG BENT-TOED GECKO *Cyrtodactylus chauquangensis*, p. 207.
SVL 99.3mm Dorsal surface with homogeneous and rounded scales, not arranged in linear series; precloacal pores present; no femoral pores and femoral scales; dorsum chocolate-brown with 5 blackish-brown bands; dark postocular stripe meets at nuchal region.

4. SHAN STATE BENT-TOED GECKO *Cyrtodactylus chrysopylos*, p. 207.
SVL 79mm Dorsal surface with keeled tubercles in 16 rows; no precloacal groove and femoral pores; 10 precloacal pores; dorsum mottled purplish-brown with 6 chocolate-brown and pale orange bands.

5. PETERS'S BENT-TOED GECKO *Cyrtodactylus consobrinus*, p. 208.
SVL 125mm Dorsal surface with tubercles in 18–20 irregular rows; 9–14 preanal scales forming narrow angular series.
(5a) Adult Head and dorsum dark chocolate-brown; body with 4–8 transverse white or yellow bands; dark-edged intervening areas.
(5b) Juvenile Head and dorsum blackish-grey to jet-black, with lemon-yellow bands narrower than dark intervening areas.

6. DARK-COLLARED BENT-TOED GECKO *Cyrtodactylus cryptus*, p. 208.
SVL 90.8mm Dorsal surface with granular scales and rounded conical tubercles in 20 rows; forehead tuberculate; 9–11 precloacal pores in angular series; dorsum brownish-olive with 3–4 transverse buff-margined, violet-brown bands; forehead with brown blotches.

7. GUNUNG LAWIT BENT-TOED GECKO *Cyrtodactylus elok*, p. 208.
SVL 67.5mm Dorsal surface with tubercles in 6–10 rows, separated by 4–9 granules; no femoral pores in males; no precloacal groove; dorsum dark brown; broad yellowish-orange to silver-brown middorsal band, crossed by 5–7 dark brown bars.

8. TAMARIND BENT-TOED GECKO *Cyrtodactylus fumosus*, p. 209.
SVL 75mm Dorsal surface with granular scales and scattered, rounded, smooth or weakly keeled tubercles; 42–52 preanofemoral pores; dorsum greyish- or pinkish-brown with transverse blackish-brown bars.

9. GANS'S BENT-TOED GECKO *Cyrtodactylus gansi*, p. 209.
SVL 63mm Dorsal surface with rounded conical tubercles in 20–25 rows; 16–29 large precloacal pores in angled series; dorsum light to mid-brown with transverse dark markings forming 7 cross-bars.

10. INGER'S BENT-TOED GECKO *Cyrtodactylus ingeri*, p. 210.
SVL 80.2mm Dorsal surface with large tubercles in 17 irregular rows; 7–9 precloacal pores form angular series; dorsum grey or yellowish-brown with 5–6 diamond-shaped, paired brown paravertebral blotches.

11. NAM NAO BENT-TOED GECKO *Cyrtodactylus interdigitalis*, p. 210.
SVL 80mm Dorsal surface with smooth or weakly keeled tubercles in 18–22 rows; precloacal pores 14; dorsum reddish-brown with dark bands and sometimes 2 longitudinal dark stripes with pale edges.

12. CARDAMOM MOUNTAINS BENT-TOED GECKO *Cyrtodactylus intermedius*, p. 210.
SVL 85mm Dorsal surface with granular scales and conical tubercles; 8–10 precloacal pores; dorsum greyish-brown or dark brown with 4–5 yellow-edged dark brown bands, including nuchal loop.

13. JARUJIN'S BENT-TOED GECKO *Cyrtodactylus jarujini*, p. 211.
SVL 90mm Dorsal surface with granular scales, intermixed with tubercles in 18–20 rows; preanofemoral pores 26 pairs; dorsum mid-brown with dark brown blotches, fused to form bands.

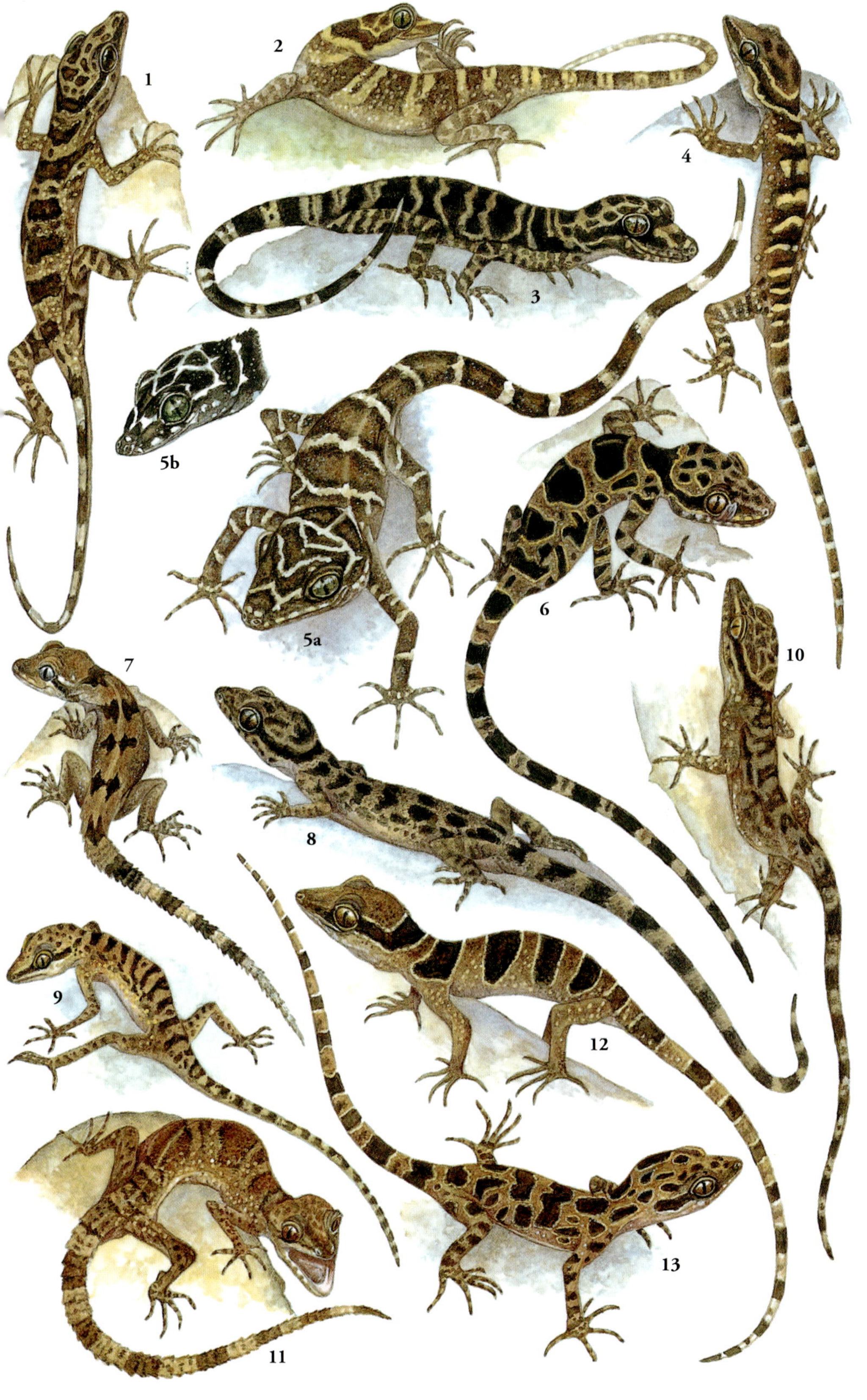

1
2
3
4
5a
5b
6
7
8
9
10
11
12
13

Plate 23. GEKKONIDAE

1. KHASI HILLS BENT-TOED GECKO *Cyrtodactylus khasiensis*, p. 211.
SVL 85mm Dorsal surface with small granular scales intermixed with larger keeled tubercles; lateral fold of enlarged scales; 12–14 precloacal pores; dorsum dark greyish-brown with dark spots.

2. MALAYAN BENT-TOED GECKO *Cyrtodactylus malayanus*, p. 211.
SVL 117mm Dorsal surface with keeled tubercles in 18–20 rows; precloacal pores indistinct in males; subcaudals large; dorsum chestnut-brown with narrow light cross-bars; series of isolated dark spots along vertebral region.

3. CLOUDED BENT-TOED GECKO *Cyrtodactylus marmoratus*, p. 211.
SVL 74.4mm Dorsal surface with granular scales intermixed with rounded, weakly keeled tubercles; precloacal pores 16; femoral pores 3–10; dorsum light brown with dark brown spots or cross-bars.

4. MATSUI'S BENT-TOED GECKO *Cyrtodactylus matsuii*, p. 212.
SVL 105mm Dorsal surface with large tuberculate scales arranged in 18 irregular rows; 7–8 precloacal pores forming angular series; dorsum yellowish-brown or pale brown with irregular dark cross-bars; forehead with small dark spots.

5. OLDHAM'S BENT-TOED GECKO *Cyrtodactylus oldhami*, p. 212.
SVL 65mm Dorsal surface with granular scales, with scattered tubercles; precloacal pores 4 pairs in angular series and enlarged femoral scales in males; dorsum brown with elongated, dark-edged cream spots, arranged in 4 longitudinal series.

6. BUTTERFLY BENT-TOED GECKO *Cyrtodactylus papilionoides*, p. 212.
SVL 93mm Dorsal surface with 12–14 rows of tubercles; continuous series of 29–33 preanofemoral pores; dorsum beige-brown to mid-brown with 4 butterfly-shaped dark patches edged with yellow and cream.

7. PEGU BENT-TOED GECKO *Cyrtodactylus peguensis*, p. 213.
SVL 85mm Dorsal surface with keeled tubercles; no precloacal groove. Subspecies *C. p. peguensis* (illustrated) shows irregular blackish-brown marks edged with cream on dorsum. Subspecies *C. p. zebraicus* (not illustrated) shows 8 transverse stripes.

8. PHONG NHA KE BANG BENT-TOED GECKO *Cyrtodactylus phongnhakebangensis*, p. 213.
SVL 96.3mm Dorsal surface with 11–20 longitudinal rows of tubercles; preanofemoral pores 32–42; dorsum grey with 4 brownish-grey serrated bands with light yellow edges.

9. FALSE LINED BENT-TOED GECKO *Cyrtodactylus pseudoquadrivirgatus*, p. 213.
SVL 84mm Dorsal surface with tubercles in 16–24 rows; precloacal pores in angular series 5–9; dorsum light brown with dark brown mottling, stripes or bands.

10. GROOVED BENT-TOED GECKO *Cyrtodactylus pubisulcus*, p. 213.
SVL 77mm Dorsal surface with tubercles in 17–22 rows; precloacal pores 3–5 pairs; dorsum grey or brown with dark cross-bars or blotches, sometimes arranged in longitudinal series.

11. BEAUTIFUL BENT-TOED GECKO *Cyrtodactylus pulchellus*, p. 213.
SVL 115mm Dorsal surface with small scales intermixed with larger rounded scales; longitudinal groove in pubic region of males; 4 precloacal pores; 15–20 femoral pores; dorsum yellowish-brown with 4 dark brown cross-bars edged with yellow or cream.

12. FOUR-STRIPED BENT-TOED GECKO *Cyrtodactylus quadrivirgatus*, p. 214.
SVL 71mm Dorsal surface with tubercles in regular longitudinal rows; precloacal pores; dorsum grey, fawn or dark brown with 4 longitudinal black lines separated by lighter areas.

13. RUSSELL'S BENT-TOED GECKO *Cyrtodactylus russelli*, p. 214.
SVL 116mm Dorsal surface with conical to keeled tubercles in 22 rows; 15 precloacal pores; 16–19 pairs of femoral pores; dorsum mid-brown to dark brown with elongated dark paravertebral blotches and longitudinal bands on dorsolateral margins.

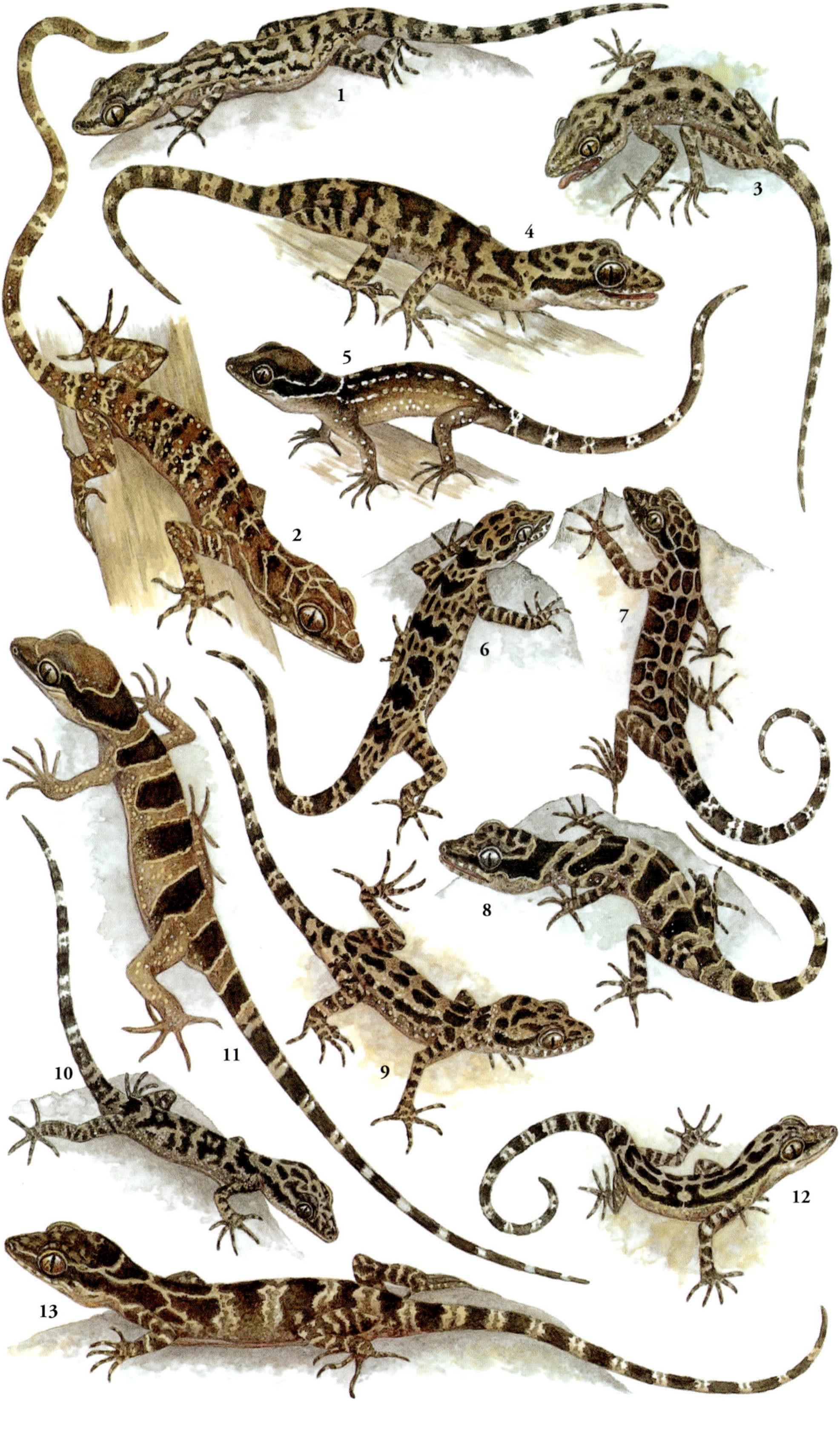
1
3
4
5
2
6
7
8
11
9
10
12
13

Plate 24. GEKKONIDAE

1. PENINSULAR MALAYSIAN BENT-TOED GECKO *Cyrtodactylus semenanjungensis*, p. 214.
SVL 69mm Dorsal surface with conical keeled tubercles; no enlarged femoral scales or pores; dorsum grey with dark mottling; 7 irregular, wide dark bands; ends of anterior-most body band join with dark lateral stripes on neck to connect with posterior portion of dark nuchal loop.

2. SERIBUAT BENT-TOED GECKO *Cyrtodactylus seribuatensis*, p. 214.
SVL 75mm Dorsal surface with 27–35 rows of tubercles distributed from back of head to tail; preanofemoral pores 40–44; dorsum light grey with narrow dark bands; head and labials with fine dark spots.

3. SLOWINSKI'S BENT-TOED GECKO *Cyrtodactylus slowinskii*, p. 214.
SVL 108mm Dorsal surface with 10–22 rows of flattened, weakly keeled tubercles; femoral pores 11; subcaudals with enlarged scales; dorsum mid-brown with 6–8 paired chocolate-brown patches, bordered with dark brown and edged with narrow yellowish-cream edges.

4. SUMONTHA'S BENT-TOED GECKO *Cyrtodactylus sumonthai*, p. 215.
SVL 70.66mm Dorsal surface with 12 rows of tubercles; 2 precloacal pores; dorsum yellowish-cream; 4 pale brown bands edged with dark brown, distinct dorsally and fading on flanks.

5. SWORDER'S BENT-TOED GECKO *Cyrtodactylus sworderi*, p. 215.
SVL 80mm Dorsal surface with large keeled tubercles; no femoral pores; 8–9 precloacal pores; dorsum dark brown; yellow spots on nape, meeting at midline and continuing as vertebral stripe or spots, to tail-base; indistinct paravertebral yellow spots.

6. THIRAKHUPT'S BENT-TOED GECKO *Cyrtodactylus thirakhupti*, p. 215.
SVL 80mm Dorsal surface with 14 rows of keeled tubercles; no precloacal or femoral pores; dorsum brown with 6 transverse grey bands edged with dark brown, first one across nape forming collar.

7. STRIPED BENT-TOED GECKO *Cyrtodactylus tigroides*, p. 215.
SVL 83.2mm Dorsal surface with 13 rows of keeled tubercles; 8–9 precloacal pores; 5–7 pairs of femoral pores, dorsum mid-brown; yellowish-cream body bands with slightly paler markings, each edged with dark brown borders; pale band across nape, 4 across trunk.

8. PULAU TIOMAN BENT-TOED GECKO *Cyrtodactylus tiomanensis*, p. 216.
SVL 86mm Dorsal surface with keeled tubercles; preanofemoral pores 19; dorsum chocolate-brown; dark brown temporal patch; body with 4 olive-yellow bands edged with greyish-brown narrow bands.

9. VARIEGATED BENT-TOED GECKO *Cyrtodactylus variegatus*, p. 216.
SVL 71mm Dorsal surface with granular scales intermixed with larger tuberculate scales; preanofemoral pores 32; dorsum grey spotted and marbled with black, edged with white.

10. WAKES'S BENT-TOED GECKO *Cyrtodactylus wakeorum*, p. 216.
SVL 64mm Dorsal surface with smooth scales, and oval or rounded keeled tubercles in 24 rows; 12 precloacal pores; no femoral pores; dorsum mid-brown with 5 chocolate-brown bands on body, posteriorly edged with narrow cream border.

11. YOSHI'S BENT-TOED GECKO *Cyrtodactylus yoshii*, p. 216.
SVL 96mm Dorsal surface with small scales and tubercles in 17 rows; no femoral pores; precloacal pores 8–12; dorsum grey with 5 V-shaped dark cross-bars between nape and inguinal region; dark brown V extends from posterior corner of orbit to nape.

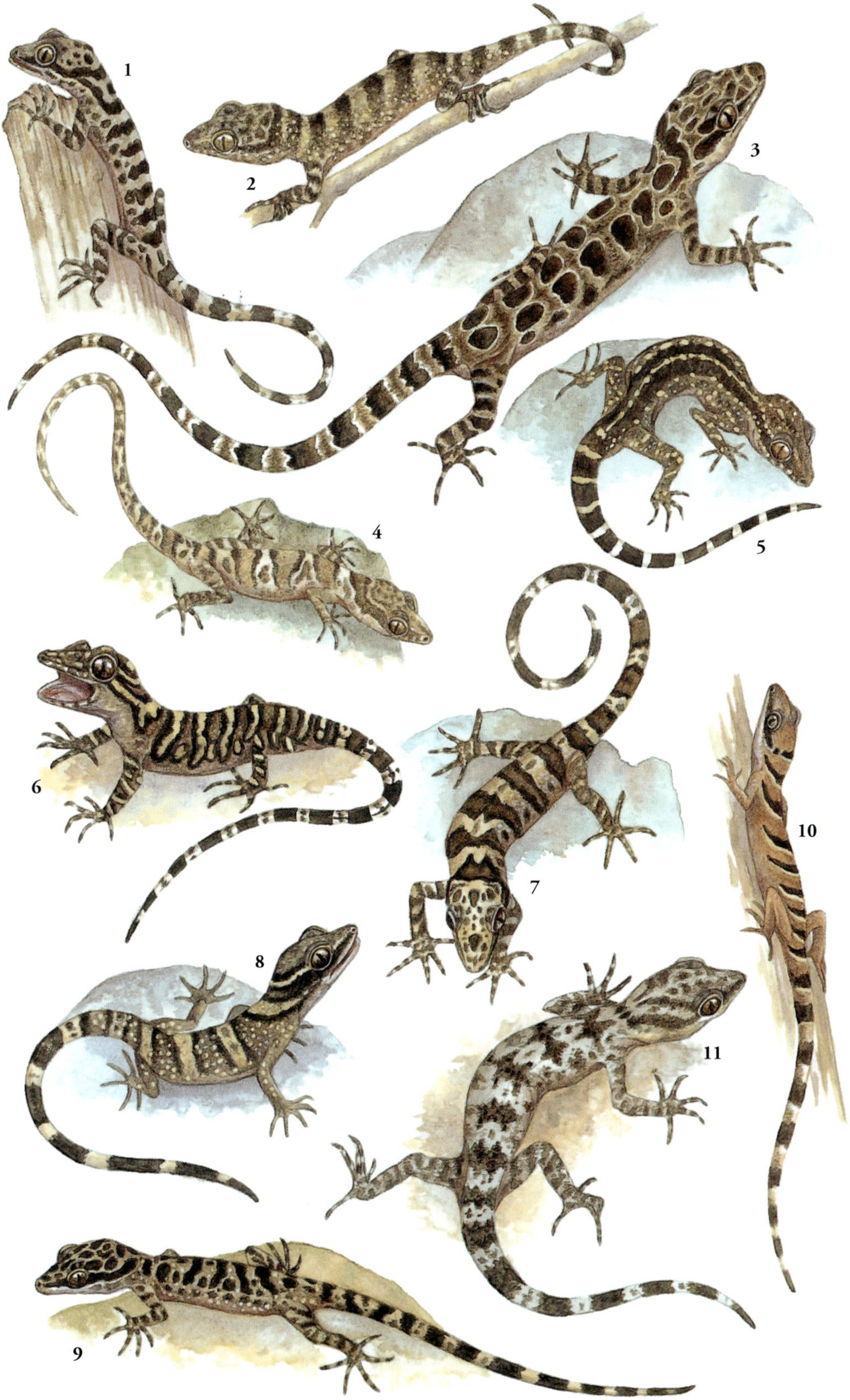

1
2
3
4
5
6
7
8
9
10
11

Plate 25. GEKKONIDAE

1. ORANGE-TAILED GROUND GECKO *Dixonius hangseesom*, p. 216.
SVL 42.12mm Dorsal surface with irregular, imbricating and scattered tubercles in 12–14 rows; no femoral pores; dorsum beige, grey to yellowish-tan with dark cross-bands and reticulations; tail bright orange with indistinct dark bands.

2. DARK-SIDED GROUND GECKO *Dixonius melanostictus*, p. 217.
SVL 50mm Dorsal surface with enlarged trihedral or keeled tubercles, median rows of which are smaller than other rows, in 10–11 rows; 9 precloacal pores; dorsum yellowish-brown or lavender-grey; dorsal spots absent or present in 4 longitudinal series.

3. SPOTTED GROUND GECKO *Dixonius siamensis*, p. 217.
SVL 57mm Dorsal surface with granular scales separating tubercles; flanks with tuberculate scales; preanal pores 8; dorsum lavender-brown to grey with an indistinct series of 17 pink, buff or yellow dots.

4. VIETNAMESE GROUND GECKO *Dixonius vietnamensis*, p. 217.
SVL 46.3mm Dorsal surface with keeled scales, small on vertebral region and larger laterally; precloacal pores 6; dorsum olive-grey with brownish-olive blotches; dark stripe from rostrum, nearly meeting 2 broken bands at back of head.

5. FEHLMANN'S FOUR-CLAWED GECKO *Gehyra fehlmanni*, p. 217.
SVL 51mm Dorsal surface with enlarged scales on axilla, smaller scales on back, juxtaposed, flattened; 22 preanofemoral pores in an arched series; dorsum pale brown with scattered pale and black flecks, and dark bands.

6. KANCHANABURI FOUR-CLAWED GECKO *Gehyra lacerata*, p. 218.
SVL 55mm Dorsal surface with small scales, lacking tubercles; 17–20 precloacal pores; dorsum grey with scattered large dark grey spots and double row of pale rounded spots.

7. COMMON FOUR-CLAWED GECKO *Gehyra mutilata*, p. 218.
SVL 64mm Dorsal surface with smooth granular scales; 25–41 preanofemoral pores; dorsum pale, nearly translucent grey to pinkish-grey, typically with pale vertebral area; indistinct white band along face.

8. BADEN'S GECKO *Gekko badenii*, p. 218.
SVL 76.5mm Dorsal surface with granular scales, with tubercles in 12–17 rows; 14–18 precloacal pores; no femoral pores; dorsum brown; forehead with light spots; 4–8 narrow transverse light bands with dark edges on vertebral region.

9. TOKAY GECKO *Gekko gecko*, p. 218.
SVL 185mm Dorsal surface with granular scales interspersed with subconical tubercles in 12 longitudinal rows; 10–24 precloacal pores; femoral pores present; dorsum slaty- or bluish-grey with red or orange spots, sometimes also 7–8 pale bands or spots.

10. GROSSMANN'S GECKO *Gekko grossmanni*, p. 218.
SVL 90mm Dorsal surface with small granular scales, lacking tubercles; 12–14 precloacal pores; no femoral pores; dorsum yellowish-grey, changeable under stress to greyish-black, with several rows of cream spots and flecks, and blue and grey marbling.

11. JAPANESE GIANT GECKO *Gekko japonicus*, p. 219.
SVL 74mm Dorsal surface with small imbricate scales; 6–9 precloacal pores; dorsum grey with 6 indented dark cross-bands that are medially interrupted by light flecks.

12. WARTY HOUSE GECKO *Gekko monarchus*, p. 219.
SVL 102mm Dorsal surface with large tuberculate scales in 16–17 longitudinal rows; preanofemoral pores 23–42; dorsum greyish-brown with dark brown blotches arranged in 7–9 pairs.

13. PALMATED GECKO *Gekko palmatus*, p. 219.
SVL 79mm Dorsal surface with tubercles in 12 rows; 24–27 precloacal pores; dorsum greyish-tan; paired rounded or oval-elongated spots on occipital and nuchal region, followed by 4–6 middorsal spots; light middorsal stripe.

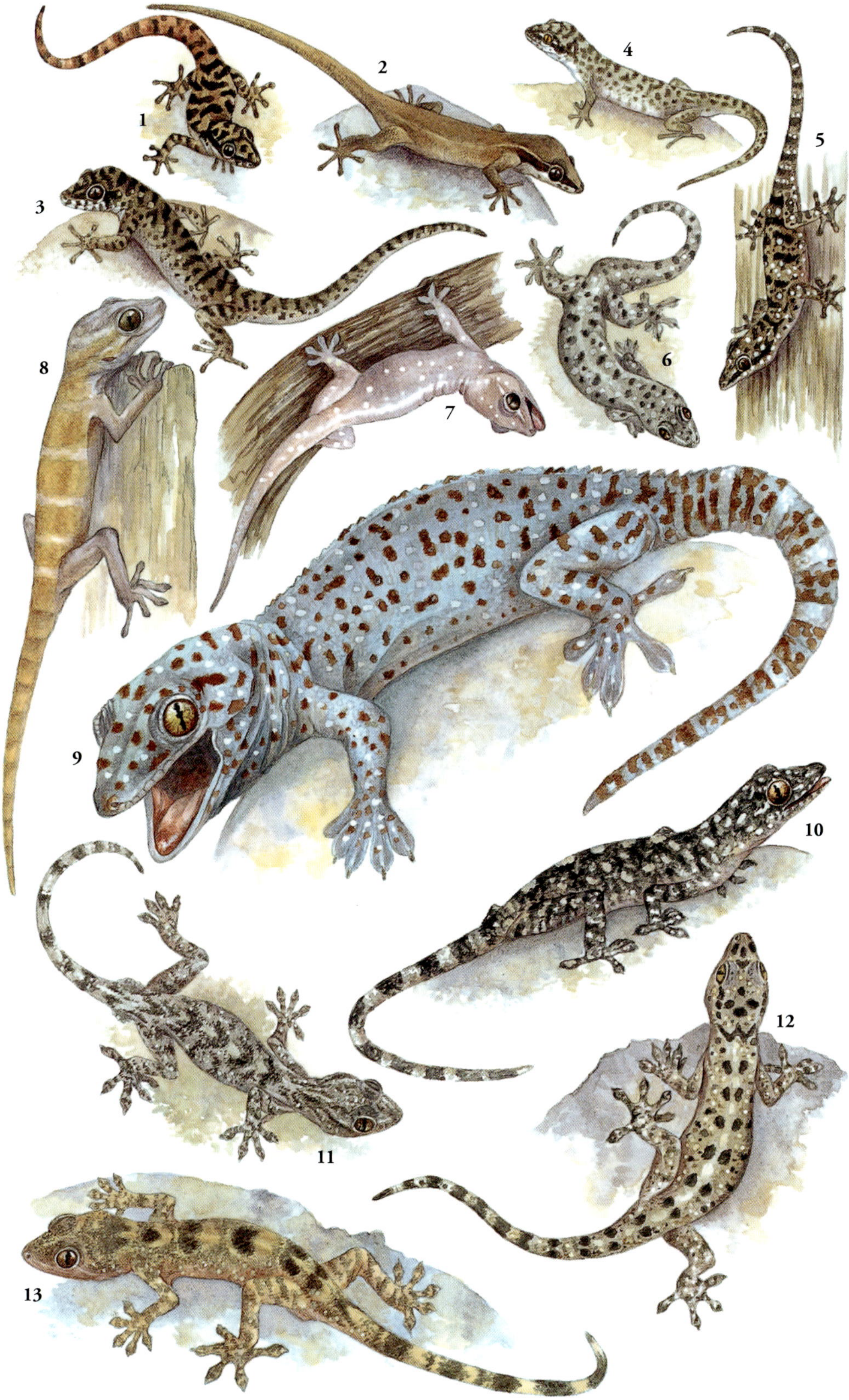
1
2
3
4
5
6
7
8
9
10
11
12
13

Plate 26. GEKKONIDAE

1. SANDSTONE GECKO *Gekko petricolus*, p. 219.
SVL 101mm Dorsal surface with non-enlarged tubercles; precloacal pores 9–10; no femoral pores; dorsum yellow; forehead lavender-grey; numerous rounded cream spots on head and body.

2. SEVEN-SPOTTED BENT-TOED GECKO *Gekko scientiadventura*, p. 220.
SVL 73mm Dorsal surface with small scales, lacking tubercles; precloacal pores 5–8; dorsum yellow to brown, typically with 7 large light spots, sometimes expanded to lateral narrow wavy bands.

3. SIAMESE GECKO *Gekko siamensis*, p. 220.
SVL 141mm Dorsal surface with tubercles in series of 16–19; precloacal pores 11–12; no femoral pores; dorsum purplish-grey; occipital region white-spotted; dark nuchal loop; V-shaped dark mark from auricular opening and extending beyond axilla; white-spotted dark bands on body.

4. SMITH'S GIANT GECKO *Gekko smithii*, p. 220.
SVL 191mm Dorsal surface with scattered tubercles; 11–16 precloacal pores; dorsum greyish-brown with transverse series of white spots.

5. ULIKOVSKI'S GECKO *Gekko ulikovskii*, p. 220.
SVL 140mm Dorsal surface with granular scales and evenly scattered tubercles, not arranged in linear series; 10–15 precloacal pores; no femoral pores; dorsum yellow or yellowish-green with up to 8 narrow light bands.

6. BOWRING'S HOUSE GECKO *Hemidactylus bowringii*, p. 220.
SVL 51mm Dorsal surface with small granular scales; 18–27 preanofemoral pores; dorsum light brown, mid-brown or tan with dark brown smudges, and occasionally white spots.

7. BROOKE'S HOUSE GECKO *Hemidactylus brookii*, p. 220.
SVL 65mm Dorsal surface with small granular scales and tubercles in 16–20 rows; 7–12 preanofemoral pores; dorsum dark brown to light grey with dark spots usually arranged in groups.

8. FRILLY FOREST GECKO *Hemidactylus craspedotus*, p. 221.
SVL 62mm Dorsal surface with scattered tubercles; precloacal and femoral pores; dorsal surface greyish-brown with 2 rows of rectangular dark spots.

9. ASIAN HOUSE GECKO *Hemidactylus frenatus*, p. 221.
SVL 67mm Dorsal surface with small scales; 23–36 preanofemoral pores; dorsum greyish-brown or dusky brown, sometimes with darker markings.

10. GARNOT'S HOUSE GECKO *Hemidactylus garnotii*, p. 221.
SVL 65mm Dorsal surface with small scales; 14–19 enlarged femoral scales; no preanofemoral pores; dorsum brownish-grey to yellowish-tan with 5 longitudinal rows of cream spots from nape to inguinal region, or longitudinal brown stripes.

11. FRILLY HOUSE GECKO *Hemidactylus platyurus*, p. 221.
SVL 69mm Dorsal surface with small scales as well as granules; 34–36 femoral pores; dorsum light grey to mid-brown with darker variegations or elongated dark brown spots.

12. STEJNEGER'S HOUSE GECKO *Hemidactylus stejnegeri*, p. 222.
SVL 59.6mm Dorsal surface with small scales, lacking tubercles; precloacal and femoral pores indistinct within slightly enlarged preanal and femoral scales; dorsum greyish-tan with indistinct cream spots on body and limbs.

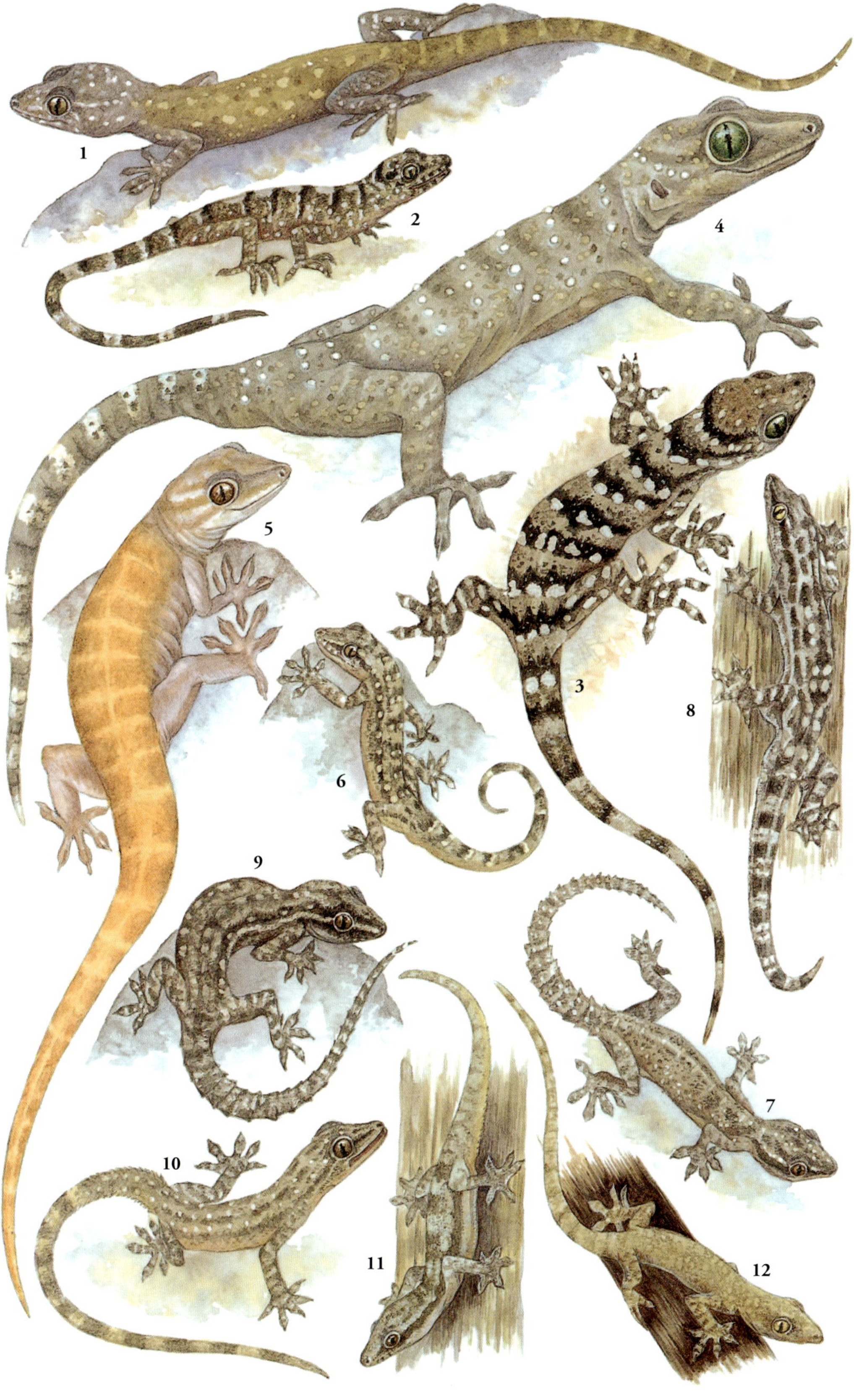

1
2
4
5
3
8
6
9
7
10
11
12

Plate 27. GEKKONIDAE

1. HARTERT'S WORM GECKO *Hemiphyllodactylus harterti*, p. 222.
SVL 45mm Dorsal surface with granular scales in 30–36 rows, lacking tubercles; 26–42 preanofemorals; 2–3 postcloacal tubercles; dorsum greyish-brown with irregular brown pattern; dark lateral stripe from head.

2. COMMON WORM GECKO *Hemiphyllodactylus typus*, p. 223.
SVL 60mm Dorsal surface with granular scales; 10–12 precloacal pores; 8–10 femoral pores; dorsum dark brown or yellowish-brown with dark and pale brown spots, blotches or specks, sometimes forming longitudinal stripes.

3. COMMON MOURNING GECKO *Lepidodactylus lugubris*, p. 223.
SVL 49mm Dorsal surface with granular scales, lacking tubercles; preanofemoral scales 25–31; dorsum cinnamon-brown, pale brown or greyish-brown; dark stripe along face; cross-bars on body and tail W-shaped.

4. KINABALU MOURNING GECKO *Lepidodactylus ranauensis*, p. 223.
SVL 47.7mm Dorsal surface with small scales, lacking large tubercles; 35–37 preanofemoral pores; dorsum greyish-brown; forehead greyish-brown or with slightly reddish-brown area; paired triangular dark markings on dorsolateral tail-base.

5. BROWN'S CAMOUFLAGE GECKO *Luperosaurus browni*, p. 223.
SVL 66.4mm Dorsal surface with rounded, flattened or slightly convex tubercles; preanofemoral pores 28–32; dorsum light grey with minute black spots on head, body and limbs; 5 dark broken chevrons on middorsum.

6. CROCKER RANGE CAMOUFLAGE GECKO *Luperosaurus sorok*, p. 224.
SVL 34.7mm Dorsal surface with rounded, convex and granular scales; no femoral and precloacal pores in females; dorsum pale grey with dark grey double chevrons fused middorsally on forehead and body, dark areas edged with black.

7. YASUMA'S CAMOUFLAGE GECKO *Luperosaurus yasumai*, p. 224.
SVL 38.9mm Dorsal surface with conical or spinose tubercles; femoral and precloacal pores; dorsum brownish-tan; numerous pale, cloudy markings on head, body and tail; 2 distinct rounded ivory spots on middorsum.

8. HORSFIELD'S PARACHUTE GECKO *Ptychozoon horsfieldii*, p. 224.
SVL 80mm Dorsal surface with small scales, lacking tubercles; femoral pores 8–11; precloacal pores 10–11; dorsum grey or brown with black mottling; dark band from eye to beyond tympanum; butterfly-shaped dark mark on axilla; trunk with 3 other wavy bands.

9. KUHL'S PARACHUTE GECKO *Ptychozoon kuhli*, p. 224.
SVL 107.8mm Dorsal surface with scattered granules, tubercles in 2–6 straight rows; 18–28 precloacal pores; dorsum grey or reddish-brown with 4–5 wavy, transverse dark brown bands; dark brown line from eye to first dorsal band.

10. SMOOTH PARACHUTE GECKO *Ptychozoon lionotum*, p. 224.
SVL 95mm Dorsal surface with smooth scales, lacking scattered granules or tubercles; 16–25 precloacal pores; dorsum greyish-brown with transverse dark brown bands; dark brown line from eye to first dorsal band.

11. KINABALU PARACHUTE GECKO *Ptychozoon rhacophorus*, p. 224.
SVL 75mm Dorsal surface spinose or with scattered thorny tubercles; 17 precloacal pores; dorsum brownish-green, sometimes with dark mottling and indistinct wavy bands.

12. THREE-BANDED PARACHUTE GECKO *Ptychozoon trinotaterra*, p. 225.
SVL 71.3mm Dorsal surface with or without midvertebral row of enlarged tubercles; 19–21 preanofemoral pores; dorsum light grey with 3 transverse dark bands; intervening areas with dark mottling; forehead dark brown.

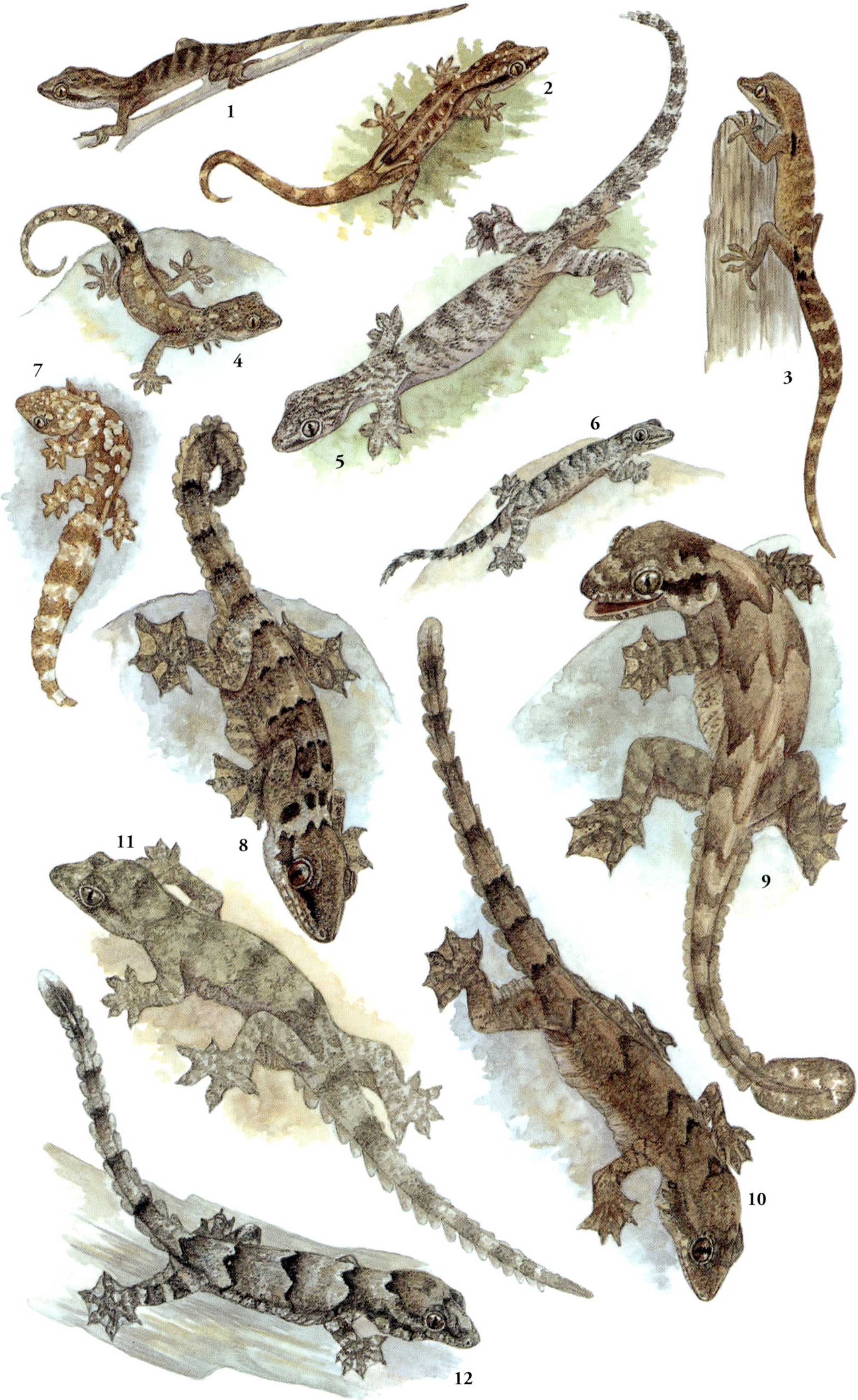

1
2
3
4
5
6
7
8
9
10
11
12

Plate 28. LACERTIDAE, LANTHANOTIDAE & LEIOLEPIDAE

1. HAN'S GRASS LIZARD *Takydromus hani*, p. 225.
SVL 79mm Head short, less than twice as long as wide; femoral pores 6–8; dorsum green; flanks paler; thin black stripe across eye to tympanum.

2. KÜHNE'S GRASS LIZARD *Takydromus kuehnei*, p. 225.
SVL 60mm Head short, as long as wide; femoral pores 4–5; dorsum olive-brown; limbs and tail yellowish-olive; flanks dark olive-brown; pale line edged with dark brown across nostril and eye extends to axilla.

3. LONG-TAILED GRASS LIZARD *Takydromus sexlineatus*, p. 225.
SVL 65mm Head at least twice as long as wide; 1–2 femoral pores; dorsum with coffee-brown vertebral stripe, yellow or cream stripe from orbit to flanks, paravertebral stripe to beyond tail-base; dorsolateral stripe dark brown, unpatterned; flanks sometimes greenish-yellow in anterior first third.

4. WOLTER'S GRASS LIZARD *Takydromus wolteri*, p. 226.
SVL 58.8mm Head short, less than twice broad; single femoral pore; dorsum olive with light dorsolateral streak and dark olive lateral band; dark-edged postocular streak to inguinal region.

5. BORNEAN EARLESS MONITOR *Lanthanotus borneensis*, p. 226.
SVL 200mm Body slender, elongated; limbs short; forehead covered with small granular scales; eye reduced; lower eyelid with clear window; tongue forked; 6 parallel rows of large tuberculate scales; dorsum unpatterned brownish-orange or with dark vertebral stripe.

6. BELL'S BUTTERFLY LIZARD *Leiolepis belliana*, p. 226.
SVL 170mm Body robust, depressed; scales across undersurface of tibia 7–12; femoral pores 13–20 pairs; midventrals 10–18, as broad as 3–4 dorsals; lamellae under Toe IV 32–41; dorsum greyish-olive or blackish-brown with black-edged pale yellow spots and 3 longitudinal stripes; flanks bluish-black with 7–9 broad vertical orange bars.

7. BÖHME'S BUTTERFLY LIZARD *Leiolepis boehmei*, p. 227.
SVL 126mm Body robust, depressed; enlarged scales across lower part of tibia 12–14; femoral pores 15–19; dorsum blackish-olive with 2 pale grey uninterrupted lateral stripes; series of 6–7 oval grey spots on dorsum; flanks with indistinct oblique light stripes.

8. SPOTTED BUTTERFLY LIZARD *Leiolepis guttata*, p. 227.
SVL 184mm Body robust, depressed; scales across undersurface of tibia 12–14; femoral pores 19–22; dorsum pale greyish-olive with pink spots and 3 pale dorsolateral stripes; flanks bluish-black with 7 broad vertical white bars.

9. EYED BUTTERFLY LIZARD *Leiolepis ocellata*, p. 227.
SVL 156mm Body robust, depressed; dorsum greyish-olive or blackish-grey with black-edged pale yellow spots and 3 longitudinal stripes that extend to pelvic region; dorsum covered with oval light ocelli whose dark edges form vertical bars; flanks orange with vertical black bands.

10. REEVES'S BUTTERFLY LIZARD *Leiolepis reevesii*, p. 227.
SVL 151mm Body robust, depressed; midventrals 13–18; dorsum buff brown-grey with spot markings producing broad reticulations and ocelli, with light centres and grey edges; flanks orange with vertical black bars.

11. RED-BANDED BUTTERFLY LIZARD *Leiolepis rubritaeniata*, p. 228.
SVL 134mm Body robust, depressed; femoral pores 15–16; scales across undersurface of tibia 7–12; dorsum greyish-olive or blackish with oval bluish-grey spots with black network; females with pale ocelli in network; irregular orange dorsolateral stripes above limbs; flanks with 8 black bars in males, unicolored in females.

12. TRIPLOID BUTTERFLY LIZARD *Leiolepis triploida*, p. 228.
SVL 148mm Body robust, depressed; femoral pores 17–21; midventrals 17–22; dorsum olive-brown with 3 dull-bordered dorsal stripes from axilla to inguinal regions, enclosing 2 rows of oval eye-like areas, created by yellow spot within grey area, within reticulated pattern; flanks darker with 5 yellowish-cream cross-bars.

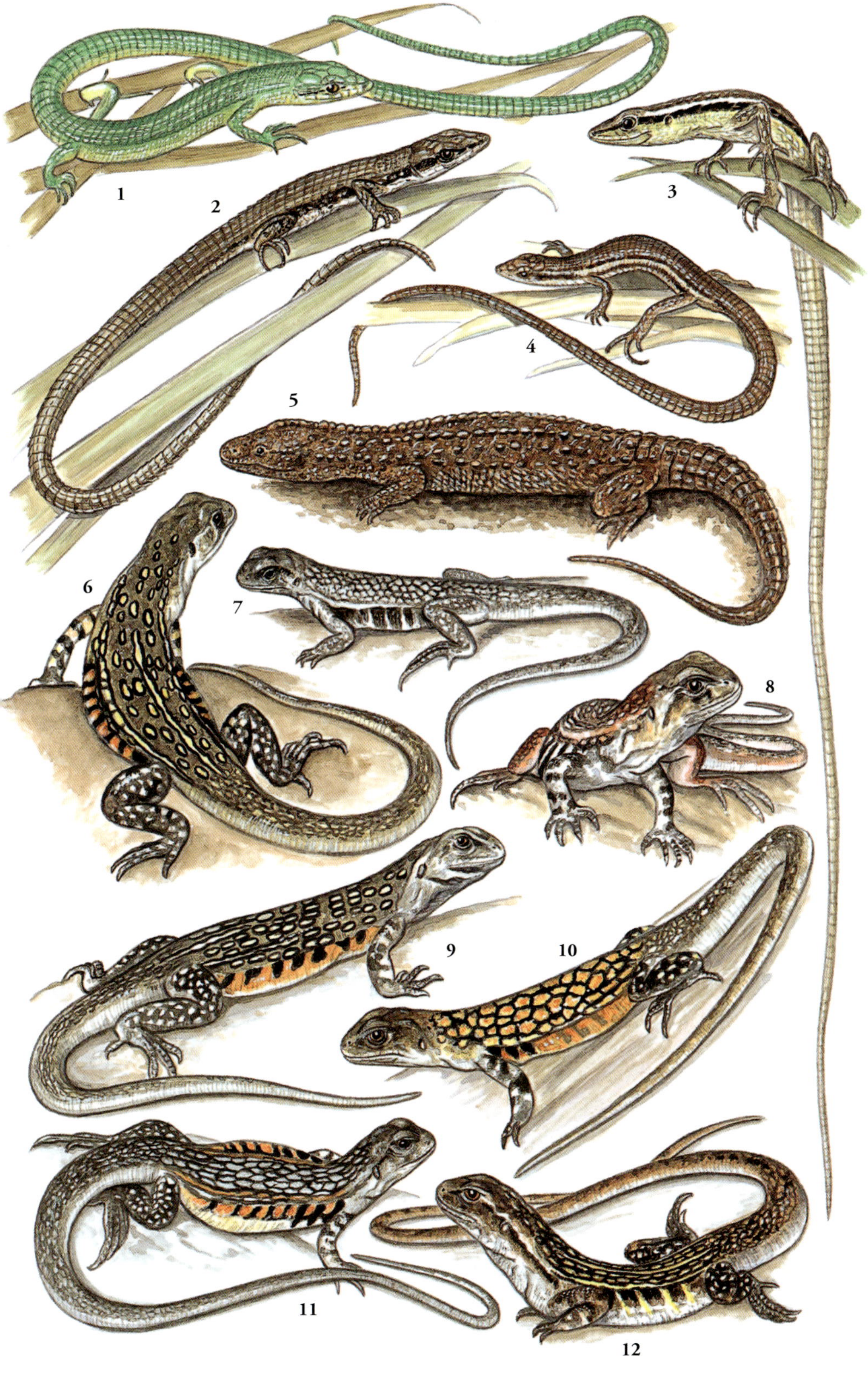
1
2
3
4
5
6
7
8
9
10
11
12

Plate 29. SCINCIDAE

1. STRIPED BORNEAN TREE SKINK *Apterygodon vittatum*, p. 228.
SVL 96mm Dorsals keeled; prefrontals separated from each other; anterior dorsum black; rest of dorsum brownish-grey with dark and light spots; light cream or yellow stripe from snout-tip to back of head; pale lateral stripe.

2. CHINESE SLENDER SKINK *Ateuchosaurus chinensis*, p. 228.
SVL 83.8mm Snout short; no parietals; tympanum sunk; dorsum greyish- or reddish-brown, scales with dark central spot; flanks pale, uniform or spotted with black and white.

3. BORNEAN LIMBLESS SKINK *Brachymeles apus*, p. 229.
SVL 131mm Limbless; lower eyelid with single large scale; prefrontals separated; tail blunt-tipped; dorsum reddish-brown, darkening posteriorly; snout-tip lighter; ocular scales darker.

4. BALINESE SNAKE-EYED SKINK *Cryptoblepharus balinensis*, p. 229.
SVL 43mm Snout acute; orbit surrounded by granular scales; lower eyelid coalesced with rudimentary upper eyelid; dorsum olive with 4 black areas, separated by 3 light olive lines and edged ventrally by olive on flanks.

5. BEACH SNAKE-EYED SKINK *Cryptoblepharus cursor*, p. 229.
SVL 43mm Snout acute; orbit surrounded by granular scales; lower eyelid coalesced with rudimentary upper eyelid; dorsum light greenish-olive with 2 narrow dark lines not covering vertebral scale rows meeting at tail-base; silvery-white lateral stripe, edged with white-spotted dark stripe.

6. BLUE-TAILED SNAKE-EYED SKINK *Cryptoblepharus renschi*, p. 229.
SVL 50mm Snout acute; lower eyelid coalesced with rudimentary upper eyelid; dorsum black to brownish-black with 5 bluish-white or yellow bands; median stripe and paravertebral bands.

7. GREY TREE SKINK *Dasia grisea*, p. 230.
SVL 130mm Snout elongated; 3 strong keels on dorsal scales; prefrontals in broad contact; supranasals in broad contact; paravertebrals not widened; dorsum light or dark brown with 8–14 narrow dark rings.

8. OLIVE TREE SKINK *Dasia olivacea*, p. 230.
SVL 115mm Snout elongated; dorsum yellowish-orange or greenish-brown, sometimes black-spotted or with 13–16 transverse dark bands.

9. HALF-BANDED TREE SKINK *Dasia semicincta*, p. 230.
SVL 130mm Snout elongated; vertebrals 59–63, not enlarged; 3 weak keels on dorsals; prefrontals typically separated; supranasals in broad contact; dorsum grey-black with 5–8 dark rings.

10. MANGROVE SKINK *Emoia atrocostata*, p. 230.
SVL 97.5mm Snout tapering; prefrontals narrowly in contact or separate; interparietal distinct, narrow; dorsum greyish-olive flecked with dark brownish-grey; dark lateral stripe from sides of head to beyond base of tail.

11. COMMON BLUE-TAILED SKINK *Emoia caeruleocauda*, p. 230.
SVL 65mm Snout short, tapering; frontonasal broader than long; interparietal distinct; pair of enlarged nuchals; dorsum with dark vertebral stripe in males, yellow in females.

12. BLUE-TAILED SKINK *Emoia cyanura*, p. 231.
SVL 61mm Snout short, tapering; prefrontals separated; dorsum brownish-black or dark brown with pale vertebral and dorsolateral stripes, fading posteriorly to blue or brown; pale midlateral stripe.

13. INDO-PACIFIC MOLE SKINK *Eugongylus rufescens*, p. 231.
SVL 143mm Dorsals smooth (adults) or weakly keeled (juveniles); tail nearly as thick as body; dorsum glossy brown, unpatterned or with transverse dark bars.

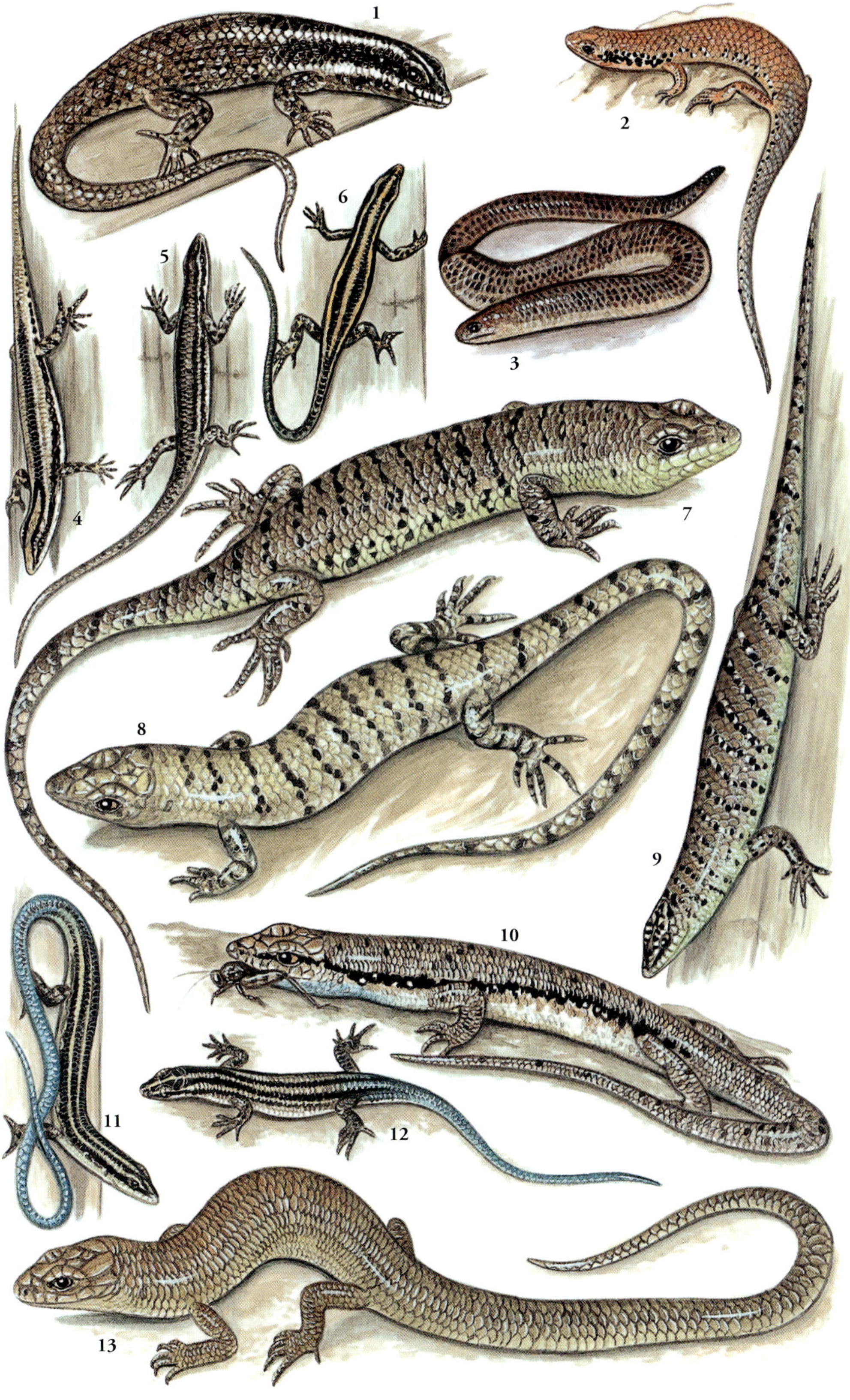
1
2
3
4
5
6
7
8
9
10
11
12
13

Plate 30. SCINCIDAE

1. STRIPED GROUND SKINK *Eutropis dissimilis*, p. 231.
SVL 150mm Snout short; auricular opening oval with 3–4 lobules; lower eyelid with clear window; scales with 3 keels; dorsum dark brown or pale olive with yellow stripes, edged with black dots, which may be joined to form lines; sides of body with small white spots.

2. LONG-TAILED GROUND SKINK *Eutropis longicaudata*, p. 232.
SVL 140mm Snout short; auricular opening small with or without lobules anteriorly; lower eyelid scaly; frontoparietals in broad contact; middorsals slightly keeled; dorsum mid-brown or reddish-brown, sometimes with 7 broken longitudinal black stripes; dark grey band on flanks from eye to beyond inguinal region.

3. LARGE-EYED GROUND SKINK *Eutropis macrophthalma*, p. 232.
SVL 108mm Snout acute; lower eyelid scaly; auricular opening large with 2–3 pointed lobules on anterior edge; dorsals with 3 keels; preanals slightly enlarged; dorsum iridescent greenish-brown; 2 dorsal rows of 5 black-spotted scales; black lateral stripe; flanks grey to bluish-grey.

4. LITTLE GROUND SKINK *Eutropis macularia*, p. 232.
SVL 75mm Snout short; limbs well developed; lower eyelid scaly; auricular opening small with several indistinct lobules anteriorly; dorsal scales with 5–9 keels; dorsum bronze-brown, unicoloured or spotted; flanks darker, spotted with white.

5. COMMON SUN SKINK *Eutropis multifasciata*, p. 233.
SVL 137mm Snout short; lower eyelid scaly; auricular opening small with small pointed lobules; dorsal scales with 3 (rarely 5) keels; dorsum bronze-brown with yellow or red stripe along flanks; series of white spots or streaks along flanks.

6. FOUR-KEELED GROUND SKINK *Eutropis quadricarinata*, p. 233.
SVL 50mm Snout short; lower eyelid scaly; auricular opening rounded, lacking lobules; dorsal scales with 4 distinct keels; dorsum olive-brown, unpatterned or with small black spots arranged longitudinally; flanks sometimes with dark lines.

7. BLACK-BANDED GROUND SKINK *Eutropis rudis*, p. 233.
SVL 120mm Snout short; forehead scales at posterior rugose; dorsal scales with 3 strong keels; dorsum olive-brown with light-edged dark brown line along sides of head and body; flanks white-spotted; throat of adult males crimson, that of females unpatterned cream.

8. RED-THROATED GROUND SKINK *Eutropis rugifera*, p. 233.
SVL 65mm Snout short; lower eyelid scaly; auricular opening small with small pointed lobules; dorsal scales with 5 (rarely 7) distinct keels; dorsum dark brown or bronze-brown, lighter on flanks, with 5–7 longitudinal greenish-cream stripes, sometimes broken into spots.

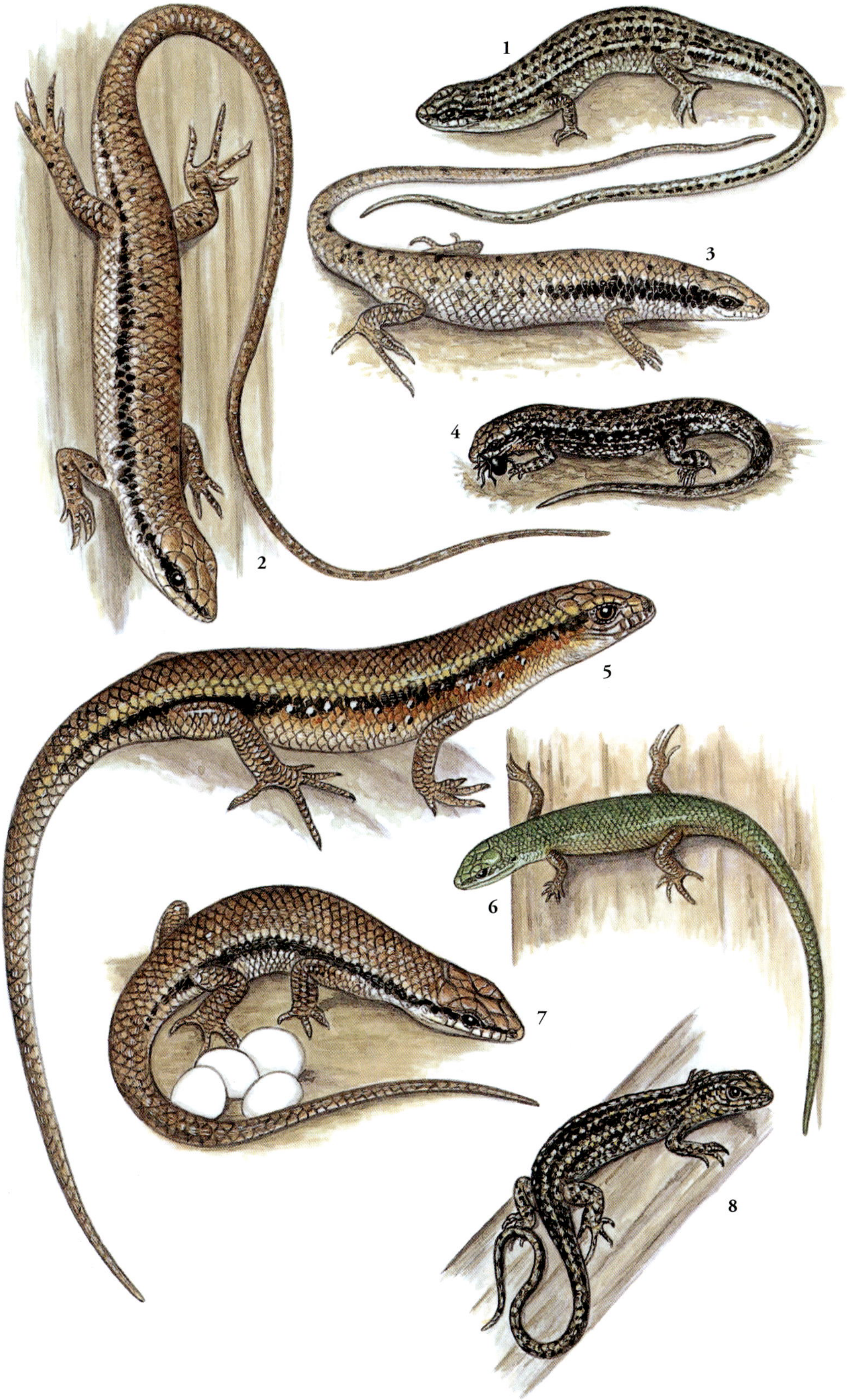
1
2
3
4
5
6
7
8

Plate 31. SCINCIDAE

1. COUNT GYLDENSTOLPE'S LIMBLESS SKINK *Isopachys gyldenstolpei*, p. 234.
SVL 220mm Limbless; nasals in contact; frontal and frontonasal fused; dorsum dark grey-brown; yellow paravertebral stripes; narrow vertebral stripe to tail-base; dark dorsal stripe with serrated edges caudally.

2. VYNER'S TREE SKINK *Lamprolepis vyneri*, p. 235.
SVL 66mm Snout obtuse; lower eyelid scaly; supranasals not in contact; dorsals smooth; dorsum with 4 longitudinal stripes consisting of series of black dorsal scales with central olive-grey area; flanks with black-edged brown scales; forehead olive-grey; scales edged with black.

3. BUKIT LARUT LIMBLESS SKINK *Larutia larutensis*, p. 235.
SVL 191mm Limbs reduced; second pair of chin shields separated by 2 gulars; first pair of chin shields contact only 1 infralabial; dorsum blue-black or brown; single light nuchal band.

4. SERIBUAT LIMBLESS SKINK *Larutia seribuatensis*, p. 235.
SVL 115mm Limbs reduced with 2 claws on hands and feet; well-defined digits with 4 subdigital lamellae; dorsum chocolate-brown; paired light-coloured dorsolateral stripes; 3 yellow nuchal bands.

5. SUMATRAN LIMBLESS SKINK *Larutia sumatrensis*, p. 236.
SVL 176mm Limbs reduced with 2 claws on hands and feet; dorsum brownish-grey; no nuchal bands.

6. THREE-LINED LIMBLESS SKINK *Larutia trifasciata*, p. 236.
SVL 250mm Limbs reduced; second pair of chin shields separated by 2 gulars; well-defined digits with 4 subdigital lamellae; dorsum blue-black or brown; 3 light nuchal bands, first complete with smooth borders, contacting posterior parietals; second and third close behind, incomplete.

7. BORNEAN STRIPED SKINK *Lipinia inexpectata*, p. 236.
SVL 40.6mm Body slender, elongated; snout acute; no auricular opening; lower eyelid with clear spectacle; dorsum tan brown with dark grey-brown stripes; paired paravertebral stripes; paired dorsal stripes from temporals to inguinal region.

8. SARAWAK STRIPED SKINK *Lipinia nitens*, p. 237.
SVL 33.6mm Body slender, elongated; snout pointed; no auricular opening; limbs reduced; dorsum metallic green; flanks spotted with black and green; pale yellow vertebral stripe with jagged-edged black lines.

9. MALAYAN STRIPED SKINK *Lipinia surda*, p. 237.
SVL 50mm Body slender, elongated; auricular opening indicated by depression; dorsum dark brown, lacking vertebral or paravertebral stripes.

10. COMMON STRIPED SKINK *Lipinia vittigera*, p. 237.
SVL 43mm Body slender, elongated; snout elongated and acute; lower eyelid with transparent disc; auricular opening small; dorsum brownish-black with yellow vertebral stripe; flanks with dark and pale spots.

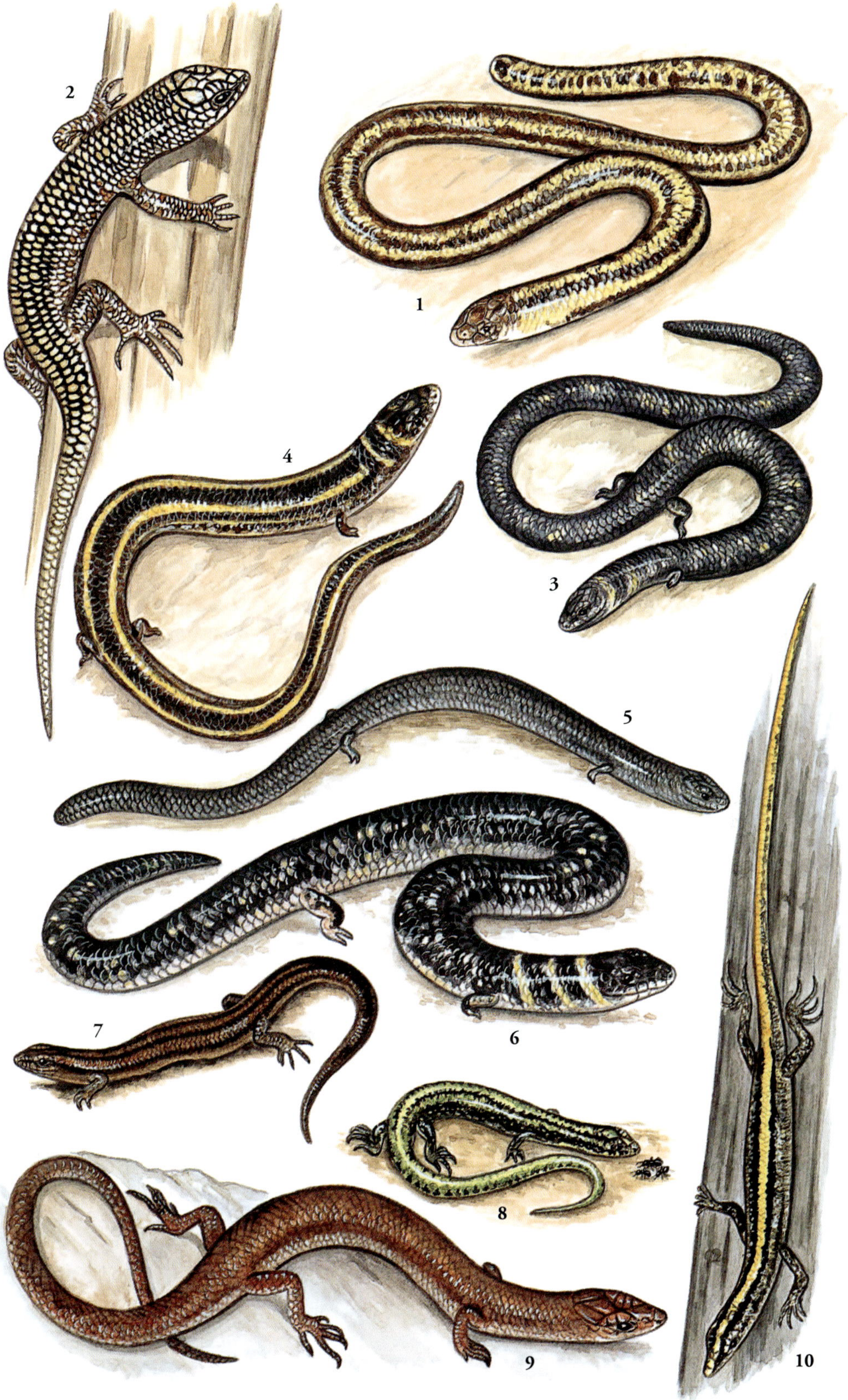

2
1
4
3
5
10
7
6
8
9

Plate 32. SCINCIDAE

1. WHITE-SPOTTED SUPPLE SKINK *Lygosoma albopunctatum*, p. 238.
SVL 65mm Body elongated; lower eyelid scaly; auricular opening rounded; scales smooth or feebly keeled; dorsum brown to reddish-brown; scales dark spotted, forming longitudinal series.

2. BAMPFYLDE'S GIANT SKINK *Lygosoma bampfyldei*, p. 238.
SVL 142.1mm Body robust; lower eyelid scaly; 3 large auricular lobules; scales smooth; dorsum pale grey-brown up to lower flanks; brownish-black band extends ventrally; cream band anteriorly from backs of eyes; area around eye with dark patches.

3. BÖHME'S SUPPLE SKINK *Lygosoma boehmei*, p. 238.
SVL 86mm Body robust; dorsal scales with pseudokeels; scaly lower eyelid; dorsum reddish-brown or brownish-black; anterior supralabials and infralabials with greyish-black-edged sutures.

4. BOWRING'S SUPPLE SKINK *Lygosoma bowringii*, p. 238.
SVL 58mm Body slender, elongated; lower eyelid scaly; auricular opening rounded; dorsals smooth or weakly keeled; dorsum bronze-brown; flanks with dark band bearing white and black spots, forming longitudinal lines in juveniles.

5. ANNAM SUPPLE SKINK *Lygosoma corpulentum*, p. 239.
SVL 170mm Body robust; eyelids scaly; scales smooth; auricular openings lack lobules; dorsum light yellowish- or chocolate-brown, densely mottled with dark brown on back and flanks.

6. HAROLD YOUNG'S SUPPLE SKINK *Lygosoma haroldyoungi*, p. 239.
SVL 141.3mm Body robust, elongated; limbs reduced; head broad; scales smooth; dorsum yellowish-brown with dark brown spots, which are small anteriorly, larger posteriorly; forehead dark brown; 27 dark brown bands on body, sometimes forming reticulated pattern.

7. HERBERT'S SUPPLE SKINK *Lygosoma herberti*, p. 239.
SVL 67mm Body slender; lower eyelid scaly; scales with 5 keels; dorsum bronze-brown; dark postocular stripe reaches flanks; flanks, sides of neck and tail with pale spots.

8. KORAT SUPPLE SKINK *Lygosoma koratense*, p. 240.
SVL 106mm Body robust, elongated; lower eyelid scaly; auricular opening small; scales smooth; dorsum reddish-brown, dorsal scale-tip black; flanks greenish-yellow, lateral scale-tip black; head scales with central spots and dark edges.

9. LINED SUPPLE SKINK *Lygosoma lineolatum*, p. 240.
SVL 63mm Body slender; lower eyelid with clear window; dorsum brown with dark dots, forming dorsolateral line; sides of neck and anterior flanks sometimes with white spots.

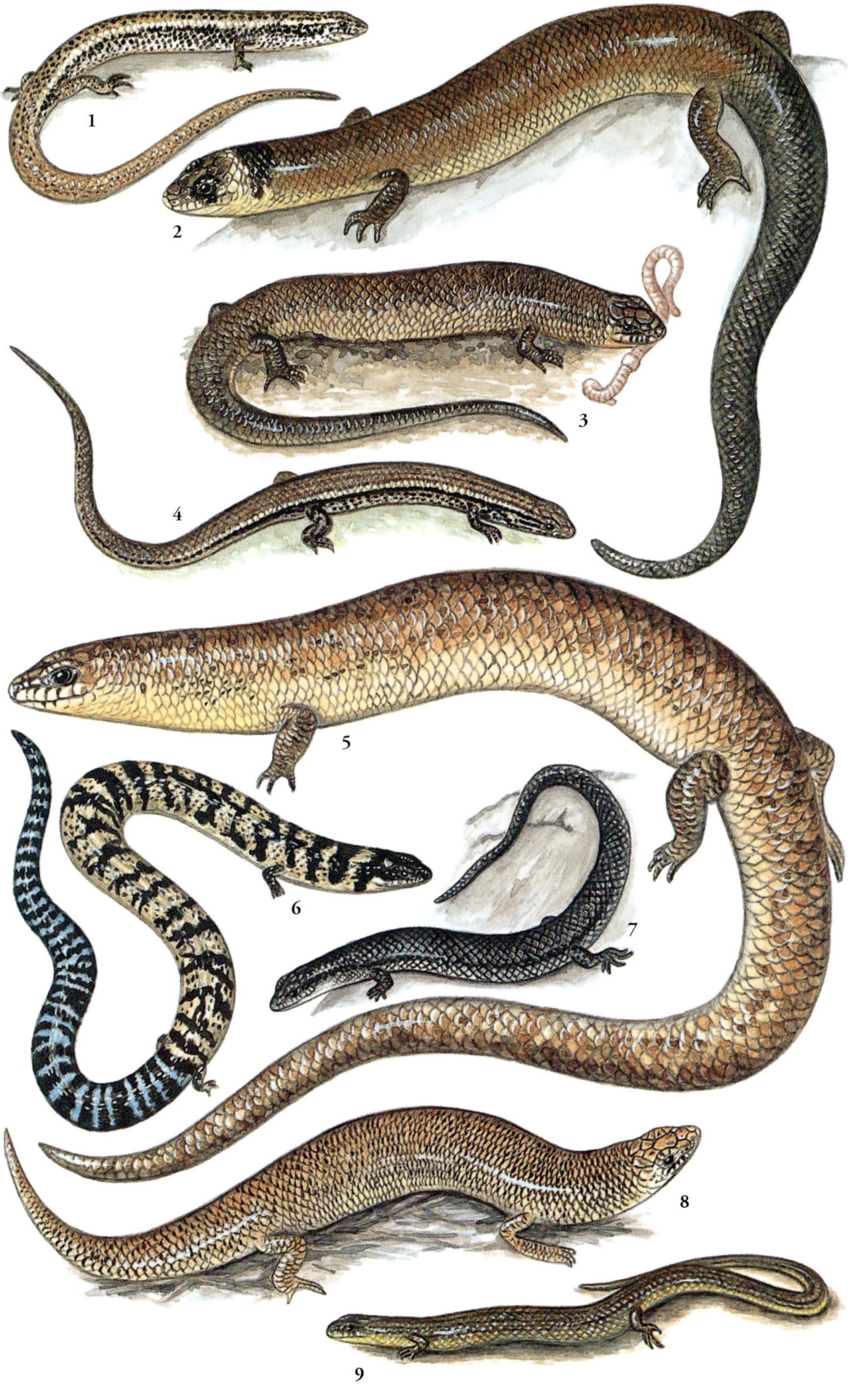
1
2
3
4
5
6
7
8
9

Plate 33. SCINCIDAE

1. SPOTTED SUPPLE SKINK *Lygosoma punctatum*, p. 240.
SVL 85mm Body slender, elongated; lower eyelid with undivided transparent disk; auricular opening rounded; scales smooth; dorsum bronze-brown with 4–6 rows of black spots, lateral ones more distinct, forming 4–6 longitudinal stripes on juvenile dorsum; broad cream stripe along body.

2. SHORT-LIMBED SUPPLE SKINK *Lygosoma quadrupes*, p. 240.
SVL 96mm Body extremely elongated; auricular opening narrow; scales smooth; dorsum yellowish-brown with dark lines bordering edges of all dorsals; forehead and supralabials darker than rest of dorsum.

3. FALSE STRIPED SKINK *Paralipinia rara*, p. 240.
SVL 45mm Body slender; prefrontals in broad contact; lower eyelid with clear window; scales smooth; double row of basal subdigital lamellae; dorsum and flanks golden-brown with scattered black spots arranged in 2 linear rows; dark lateral band.

4. CHINESE BLUE-TAILED SKINK *Plestiodon chinensis*, p. 240.
SVL 134mm Body robust; auricular opening with 3–4 small lobules; dorsum olive or brownish-olive with scattered red or orange blotches on flanks.

5. FOUR-LINED BLUE-TAILED SKINK *Plestiodon quadrilineatus*, p. 241.
SVL 73mm Body slender; 2 median dorsals larger than scales on flanks; dorsum dark grey-brown; greenish-white or cream stripe along flanks from snout-tip to tail; light lateral line from labials crosses level of lower auricular opening and continues to inguinal region; head yellowish-brown; labials red.

6. TAMDAO BLUE-TAILED SKINK *Plestiodon tamdaoensis*, p. 241.
SVL 129mm Body slender; auricular opening large, vertical, with small lobules anteriorly; dorsum black or blackish-brown; narrow pale dorsolateral stripes; broad dark stripe from nostril to flanks, edged ventrally by narrow pale stripe; tail bright blue.

7. BLACK-SPOTTED GROUND SKINK *Scincella melanosticta*, p. 242.
SVL 57.4mm Body robust; lower eyelid with clear window; no auricular lobules; dorsum olive-, bronze- or golden-brown; large brown or black spots concentrated along vertebral region; dark brown or black stripe starts from nostril, narrow on head, widening on flanks, where it is broken up by cream spots.

8. REEVES'S GROUND SKINK *Scincella reevesii*, p. 242.
SVL 57.4mm Body slender, elongated; lower eyelid with clear window; no auricular lobules; dorsum bronze-brown; black spots on dorsum concentrated in vertebral region; dark brown stripe narrow on head, widening on flanks, where it is broken up by cream spots.

9. MOUNT VICTORIA GROUND SKINK *Scincella victoriana*, p. 243.
SVL 76.7mm Body robust; dorsal keels distinct; dorsal and lateral scales subequal; dorsum mid-brown; brown spots sometimes form longitudinal stripes on back; thick dark postocular stripe extends along flanks and continues along tail; several cream spots within dark stripe.

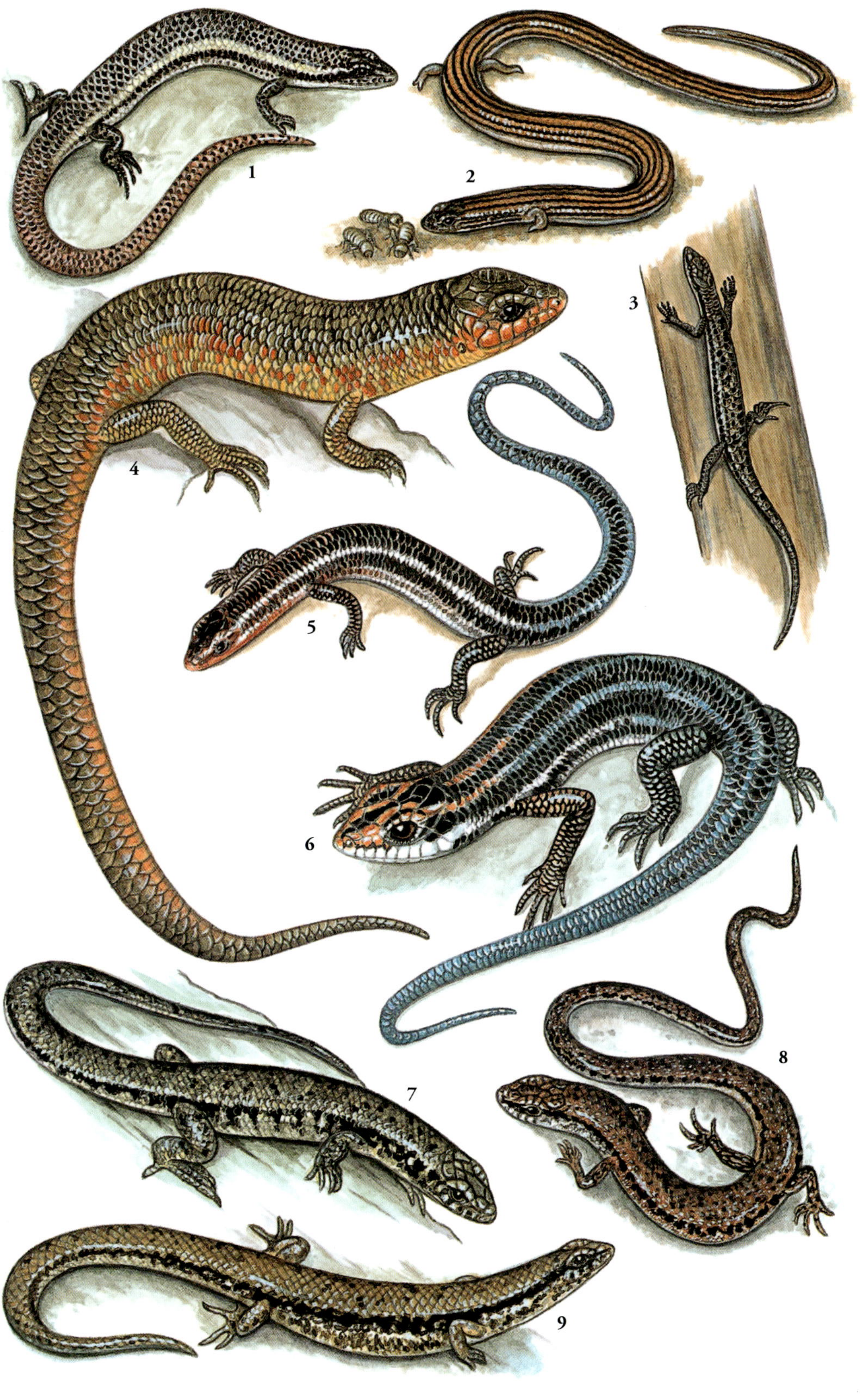
1
2
3
4
5
6
7
8
9

Plate 34. SCINCIDAE

1. OAK FOREST SKINK *Sphenomorphus aesculeticola*, p. 243.
SVL 43mm Body slender; limbs short; dorsum brown; many scales dark spotted, forming dark lines or chequered pattern; dark lateral band.

2. EARLESS FOREST SKINK *Sphenomorphus cryptotis*, p. 244.
SVL 82mm Body slender; limbs moderate; auricular opening concealed; dorsum dark brown with golden sheen; indistinct row of irregular dark middorsal spots; indistinct dorsolateral stripe, edged dorsally and ventrally with 2 rows of cream ocellae or comma-shaped spots.

3. BLUE-THROATED FOREST SKINK *Sphenomorphus cyanolaemus*, p. 244.
SVL 60mm Body slender; limbs relatively long; auricular opening lacking lobules; dorsum bronze- or olive-brown with two rows of yellow-brown spots; dark grey-brown dorsolateral stripe; head, throat and pectoral regions in adult males deep indigo-blue, in females (illustrated) light blue.

4. GRASSHOPPER-EATING FOREST SKINK *Sphenomorphus devorator*, p. 245.
SVL 58mm Body slender; limbs moderate; auricular opening large, deeply sunk; dorsum silver-grey; dark vertebral stripe disappears on tail; dark brown lateral stripe from nostril and from posterior corner of orbit to tail; labials with numerous black spots; forehead with irregular black spots.

5. INDIAN FOREST SKINK *Sphenomorphus indicus*, p. 246.
SVL 97mm Body slender; limbs moderate; tympanum deeply sunk; dorsum brown, unpatterned or with dark brown spots forming longitudinal lines; dark greyish-black stripe on flanks extends from eye to tail.

6. ISHAK'S FOREST SKINK *Sphenomorphus ishaki*, p. 246.
SVL 41mm Body slender; limbs small; dorsum dark brown or greyish-brown; scale centres lighter, producing spotted appearance; postocular stripe yellow, extending to axilla and becoming indistinct thereafter, edged ventrally by thick dark brown stripe; forehead with light spots.

7. GUNUNG KINABALU FOREST SKINK *Sphenomorphus kinabaluensis*, p. 246.
SVL 58mm Body slender; limbs long; dorsum light to dark brown; longitudinal rows of dark brown to yellow spots, and sometimes dark brown speckles; black dorsolateral stripe with small yellowish flecks.

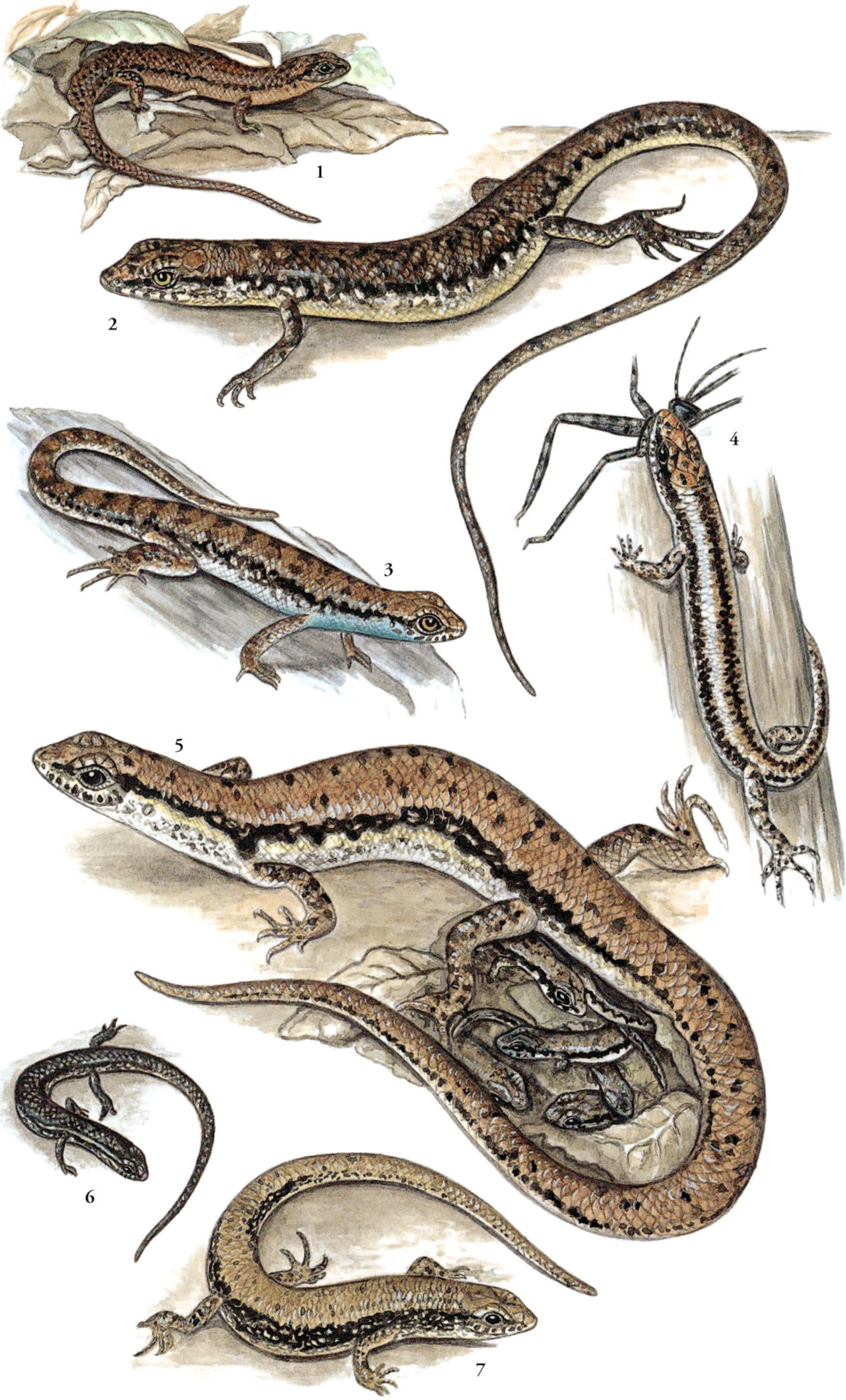
1
2
3
4
5
6
7

Plate 35. SCINCIDAE

1. SPOTTED FOREST SKINK *Sphenomorphus maculatus*, p. 247.
SVL 65mm Body slender; tympanum on surface, not deeply sunk; auricular opening lacks lobules; dorsum bronze-brown or brownish-pink, unpatterned or with dark green spots; 2 dark median series of spots; dark lateral band on flanks, spotted with white.

2. DWARF FOREST SKINK *Sphenomorphus mimicus*, p. 247.
SVL 36mm Body slender; lower eyelid scaly; dorsum brown with indistinct dark brown spots, especially around shoulders; sides of neck and flanks with fine brown flecks.

3. MANY-SCALED FOREST SKINK *Sphenomorphus multisquamatus*, p. 247.
SVL 69mm Body robust; no auricular lobules; dorsum dark greyish-brown with squarish black spots; labials without black bars; males usually with blue throat and sides of neck.

4. GUNUNG MURUD FOREST SKINK *Sphenomorphus murudensis*, p. 247.
SVL 50.4mm Body slender; lower eyelid scaly; no auricular lobules; dorsum dark brown, sometimes with black spots; dark band on flanks.

5. BUKIT LARUT FOREST SKINK *Sphenomorphus praesignis*, p. 248.
SVL 110mm Body robust; auricular opening lacks lobules; dorsum brownish-grey or chestnut-brown with black mottling; large black patches on sides of neck and flanks.

6. SABAH FOREST SKINK *Sphenomorphus sabanus*, p. 248.
SVL 58mm Body robust; no auricular lobules; dorsum pinkish-brown or olive-yellow with indistinct light spots; forehead orange-brown, scales with dark grey areas; no dark dorsolateral band; sides of neck and flanks of males ringed with orange; males usually with orange flush on flanks.

7. SELANGOR FOREST SKINK *Sphenomorphus scotophilus*, p. 249.
SVL 50mm Body slender; lower eyelid scaly; auricular opening lacks lobules; dorsum dark brown with brownish-black and yellow spots; dorsolateral series of rounded cream spots; supralabials and infralabials white with black spots.

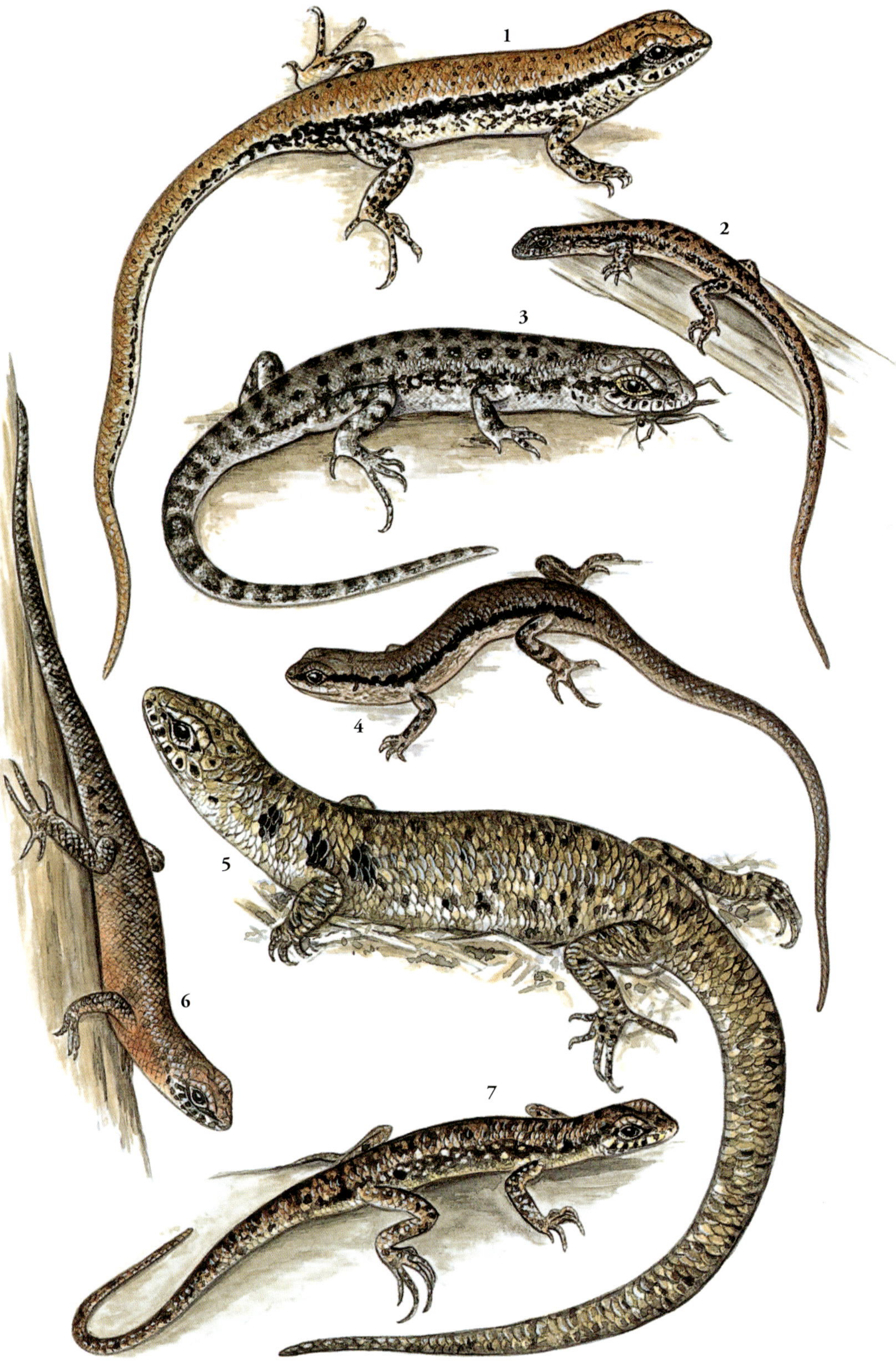
1
2
3
4
5
6
7

Plate 36. SCINCIDAE

1. STARRED FOREST SKINK *Sphenomorphus stellatus*, p. 249.
SVL 80mm Body slender; auricular opening lacks lobules; dorsum yellowish-, greenish- or bronze-brown with small star-like white areas; dark vertebral and lateral stripes in some populations; throat with 7–8 distinct longitudinal dark bands between scale rows.

2. TEMMINCK'S FOREST SKINK *Sphenomorphus temminckii*, p. 250.
SVL 56mm Body slender; lower eyelid scaly; auricular opening smaller than orbit; dorsum grey-brown or mid-brown with dark brown lateral stripe; sides of head with large pale and dark areas.

3. PALE FOREST SKINK *Sphenomorphus tersus*, p. 250.
SVL 92mm Body slender; lower eyelid scaly; auricular opening large, lacking lobules; dorsum dark brown, unpatterned or with indistinct dark brownish-black and black spots and variegations, spots arranged in longitudinal series; flanks lighter.

4. BAVI WATER SKINK *Tropidophorus baviensis*, p. 251.
SVL 91mm Body robust, dorsoventrally depressed; forehead scales smooth; dorsals keeled, especially scales on flanks, which are spinous; dorsum dark brown with cream spots and blotches, forming broken bands; flanks dark brown with lighter markings; forehead dark brown.

5. BECCARI'S WATER SKINK *Tropidophorus beccarii*, p. 251.
SVL 98mm Body robust in adults, slender in juveniles, rounded; dorsals smooth; dorsum dark brown or reddish-brown with dark brown blotches and cross-bars; sides of head and flanks with light spots; broad dark stripe on flanks, with vertical light bars or wedge-shaped marks.

6. BERDMORE'S WATER SKINK *Tropidophorus berdmorei*, p. 251.
SVL 97mm Body robust, rounded; forehead scales smooth; dorsals and laterals smooth or obtusely keeled; dorsum brown with transverse pale orange and dark marks.

7. BROOKE'S WATER SKINK *Tropidophorus brookei*, p. 251.
SVL 101mm Body robust in adults, slender in juveniles, rounded; dorsals keeled, forming 8 longitudinal ridges on dorsum; dorsum dark brown with darker spots and blotches, forming indistinct transverse bands; black spot on sides of neck; flanks with dark and white spots.

8. INDO-CHINESE WATER SKINK *Tropidophorus cocincinensis*, p. 252.
SVL 80mm Body robust, rounded; dorsals keeled in juveniles, smooth in adults; dorsum light brown or greyish-brown with indistinct darker markings; flanks brownish-black with small white spots or orange flecks.

9. LAOTIAN WATER SKINK *Tropidophorus laotus*, p. 252.
SVL 75mm Body robust, rounded; forehead scales smooth; dorsal and lateral scales smooth; dorsum dark brown, sometimes with light, V-shaped black-edged bars; flanks with small white spots.

10. SPINY-TAILED WATER SKINK *Tropidophorus hangnam*, p. 252.
SVL 78.2mm Body robust, dorsoventrally flattened; forehead scales smooth; dorsals weakly keeled; tail with spinous keels laterally; dorsum blackish-brown with transverse yellow bars; forehead reddish-brown.

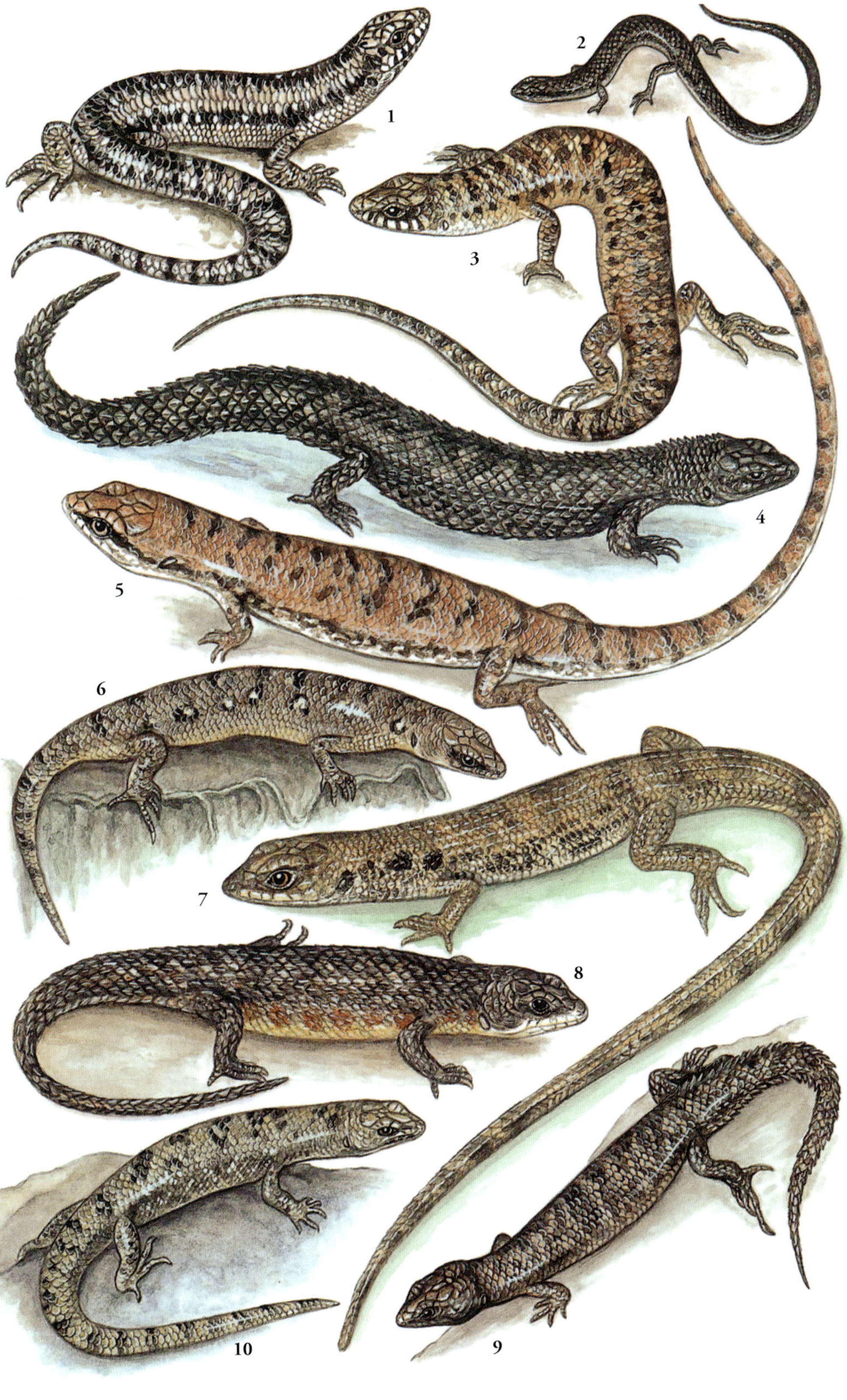
1
2
3
4
5
6
7
8
9
10

Plate 37. SCINCIDAE, SHINISAURIDAE & VARANIDAE

1. MOCQUARD'S WATER SKINK *Tropidophorus mocquardii*, p. 253.
SVL 95mm Body slender, rounded; forehead scales smooth; dorsals smooth; dorsum brown with transverse dark bands; flanks with white spots.

2. MURPHY'S WATER SKINK *Tropidophorus murphyi*, p. 253.
SVL 85.1mm Body robust, dorsoventrally depressed; forehead scales smooth; paravertebral scale rows smooth or weakly keeled; dorsum dark brown with transverse pale brown bands, 3 on head and 7 on body.

3. NOGGE'S WATER SKINK *Tropidophorus noggei*, p. 253.
SVL 110.2mm Body dorsolaterally depressed; tympanum superficial, ovoid; enlarged dorsal scales; dorsum dark brown, scales edged with light brown; 3 pale bands on neck, 6–9 pale brown bands on dorsum.

4. PERPLEXING WATER SKINK *Tropidophorus perplexus*, p. 254.
SVL 73mm Body slender, rounded; dorsal and lateral scales large and strongly keeled; scales on forehead rugose; dorsum rich brown with paler narrow cross-bars.

5. CHINESE WATER SKINK *Tropidophorus sinicus*, p. 254.
SVL 65mm Body robust, rounded; forehead scales strongly striated; dorsals with sharp keels; dorsum dark brown with 10 rusty-brown bars on body and tail; supralabials and infralabials cream with black bars.

6. CHINESE CROCODILE LIZARD *Shinisaurus crocodilurus*, p. 255.
SVL 300mm Body robust; head short; enlarged collar scales 10–12; dorsum grey or reddish-brown; red, orange and yellow blotches on head, throat and flanks of males; radiating dark lines from orbit.

7. BENGAL MONITOR LIZARD *Varanus bengalensis*, p. 255.
SVL 1,740mm Body slender; snout somewhat elongated; nostrils nearer eye than snout-tip; juveniles pale olive or dark grey dorsally, with yellow bands; adults unicoloured, or pattern less distinct.

8. DUMÉRIL'S MONITOR LIZARD *Varanus dumerilii*, p. 255.
SVL 1,500mm Body robust; head small, flattened; snout short and broad; nostrils elongated; nuchal scales oval, flat, smooth or feebly keeled; dorsum brownish-yellow or tan with dark temporal streak.
(8a) Adult Forehead tan or yellow.
(8b) Juvenile Forehead orange or yellowish-orange.

9. SOUTH-EAST ASIAN MONITOR LIZARD *Varanus nebulosus*, p. 255.
SVL 1,200mm Body slender; snout somewhat elongated; nostrils nearer eyes than snout-tip; nostril an oblique slit; juveniles dark grey dorsally, with yellow bands; adults dark grey.

10. ROUGH-NECKED MONITOR LIZARD *Varanus rudicollis*, p. 256.
SVL 1,460mm Body slender; snout relatively long; nuchal scales strongly keeled; dorsum blackish-grey in adults, with yellow tinge on neck and foreparts of body; neck with 3 black stripes.

11. WATER MONITOR LIZARD *Varanus salvator*, p. 256.
SVL 800mm Body robust in adults, relatively slender in juveniles; nostril rounded or oval, twice as far from orbit as from snout-tip; juveniles dark dorsally, yellow spotted; dorsum darkens with growth.

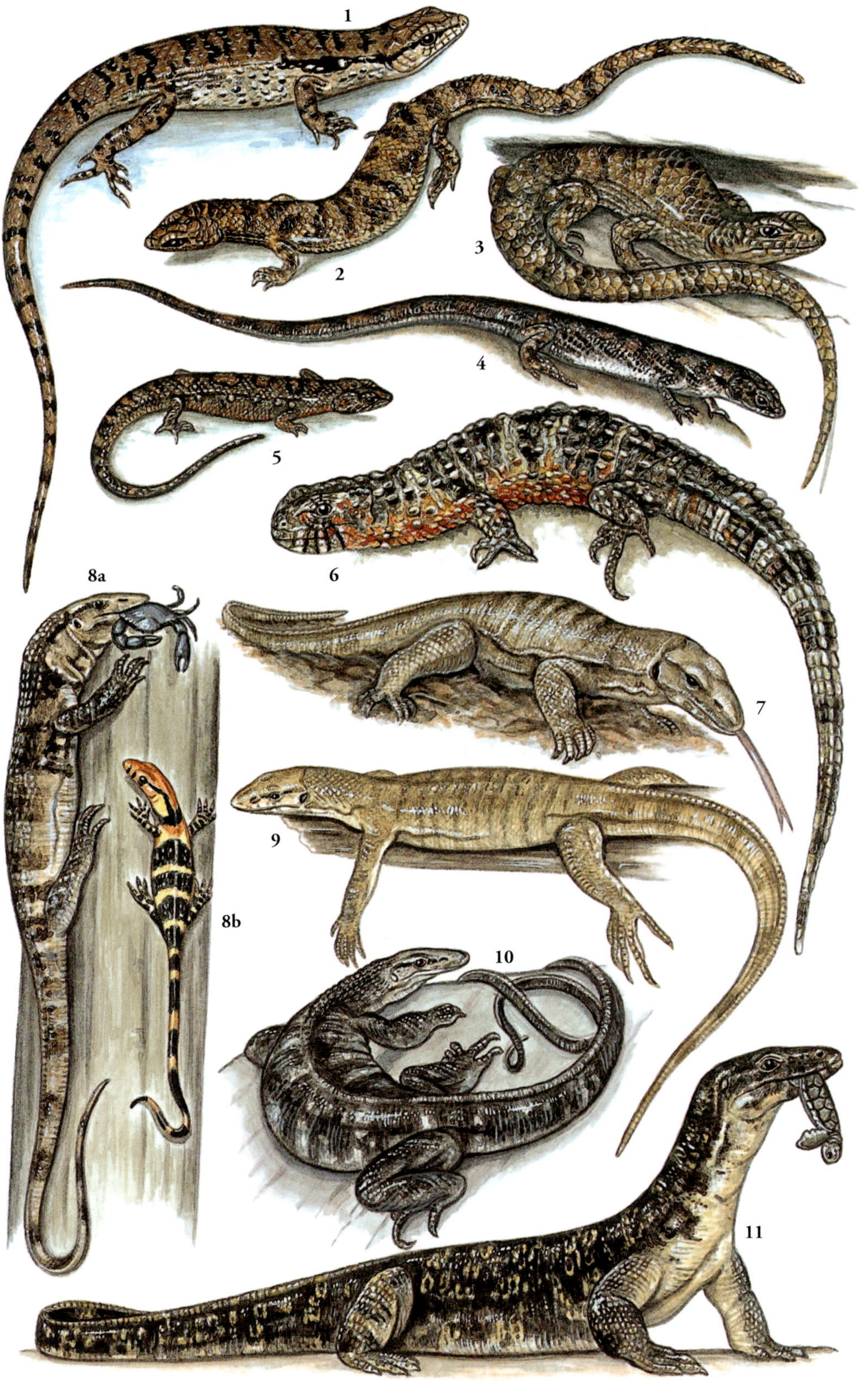
1
2
3
4
5
6
8a
7
9
8b
10
11

Plate 38. ACROCHORDIDAE, ANOMOCHILIDAE & PYTHONIDAE

1. WART SNAKE *Acrochordus granulatus*, p. 256.
TL 1,000mm Body stout, compressed, supralabials 8–11; midbody scale rows 100; dorsum olive, blue or blackish-grey with transverse cream bands, sometimes absent in adults.

2. ELEPHANT TRUNK SNAKE *Acrochordus javanicus*, p. 256.
TL 2m Body extremely stout, slightly compressed; supralabials 22–36; midbody scale rows 120–150; dorsum greyish-black; 2 diffuse longitudinal stripes and elongated dark blotches on flanks.

3. MALAYAN GIANT BLIND SNAKE *Anomochilus leonardi*, p. 257.
TL 228mm Body stout, rounded; azygous parietofrontal; midbody scale rows 17 or 19; ventral scales 214–252; subcaudals 6–7; dorsum glossy black or purplish-brown with oval yellow spots; yellow bar covers most of frontal and part of preoculars.

4. KINABALU GIANT BLIND SNAKE *Anomochilus monticola*, p. 257.
TL 521.2mm Body stout, rounded; azygous parietofrontal; midbody scale rows 19; ventrals 258–261; subcaudals 7–8; dorsum blue-black; no light line along flanks and large pale blotches on either side of vertebral; transverse yellow bar across snout; series of isolated pale yellow scales on flanks.

5. SUMATRAN GIANT BLIND SNAKE *Anomochilus weberi*, p. 257.
TL 230mm Body stout, rounded; paired parietofrontals; midbody scale rows 19; ventrals 242–248; subcaudals 6–8; dorsum black; pale stripe along flanks; large pale blotches on either side of vertebral.

6. RETICULATED PYTHON *Broghammerus reticulatus*, p. 258.
TL 9.83m Body elongated and slender; 2–3 anterior and 5–6 posterior infralabials with pits; infralabial pits better defined than supralabial pits and set in longitudinal groove; dorsum yellow or brown with rhomboidal dark markings; black median line runs from snout to nape.

7. BORNEAN SHORT PYTHON *Python breitensteini*, p. 258.
TL >2m Body short, robust; rostral with deep pit on each side; 2 supralabials with pits; anterior and posterior infralabials with weak pits; supralabial pits better defined than infralabial pits; dorsum pale yellow or tan with dark subrectangular blotches; scattered pale spots on vertebral region; flanks with smaller dark-edged grey spots or wavy bands.

8. BRONGERSMA'S SHORT PYTHON *Python brongersmai*, p. 258.
TL 2.6m Body short, robust; naso-preocular groove, comprising series of granular scales; anterior pair of parietals in broad contact in median suture; supralabial pits better defined than infralabial pits; tail short; midventrals 53–61; ventrals 160–178; subcaudals 24–36; dorsum reddish-brown, grey or brown; narrow dark stripe on middle of forehead, with vertebral spots sometimes coalescing to form elongated blotches or stripes; lower anterior flanks pale with longitudinal dark blotches.

9. SUMATRAN SHORT PYTHON *Python curtus*, p. 259.
TL >2m Body short, robust; naso-preocular groove, comprising series of granular scales; anterior pair of parietals in broad contact in median suture; supralabial pits better defined than infralabial pits; dorsum brownish-grey with longitudinal dark subrectangular blotches; flanks with longitudinal series of large blotches with black edges; sides of snout with dark stripes.

10. INDIAN ROCK PYTHON *Python molurus*, p. 259.
TL 7.6m Body thick, cylindrical; supralabials separated from orbit by row of subocular scales; sensory pits in rostral, first 2 supralabials and 14–18th infralabials; supralabial pits better defined than infralabial pits; dorsum dark brown or yellowish-grey with squarish, irregular chocolate-grey patches edged with black on dorsal surface and on flanks; dorsal and lateral spots dark; dark grey subocular stripe.

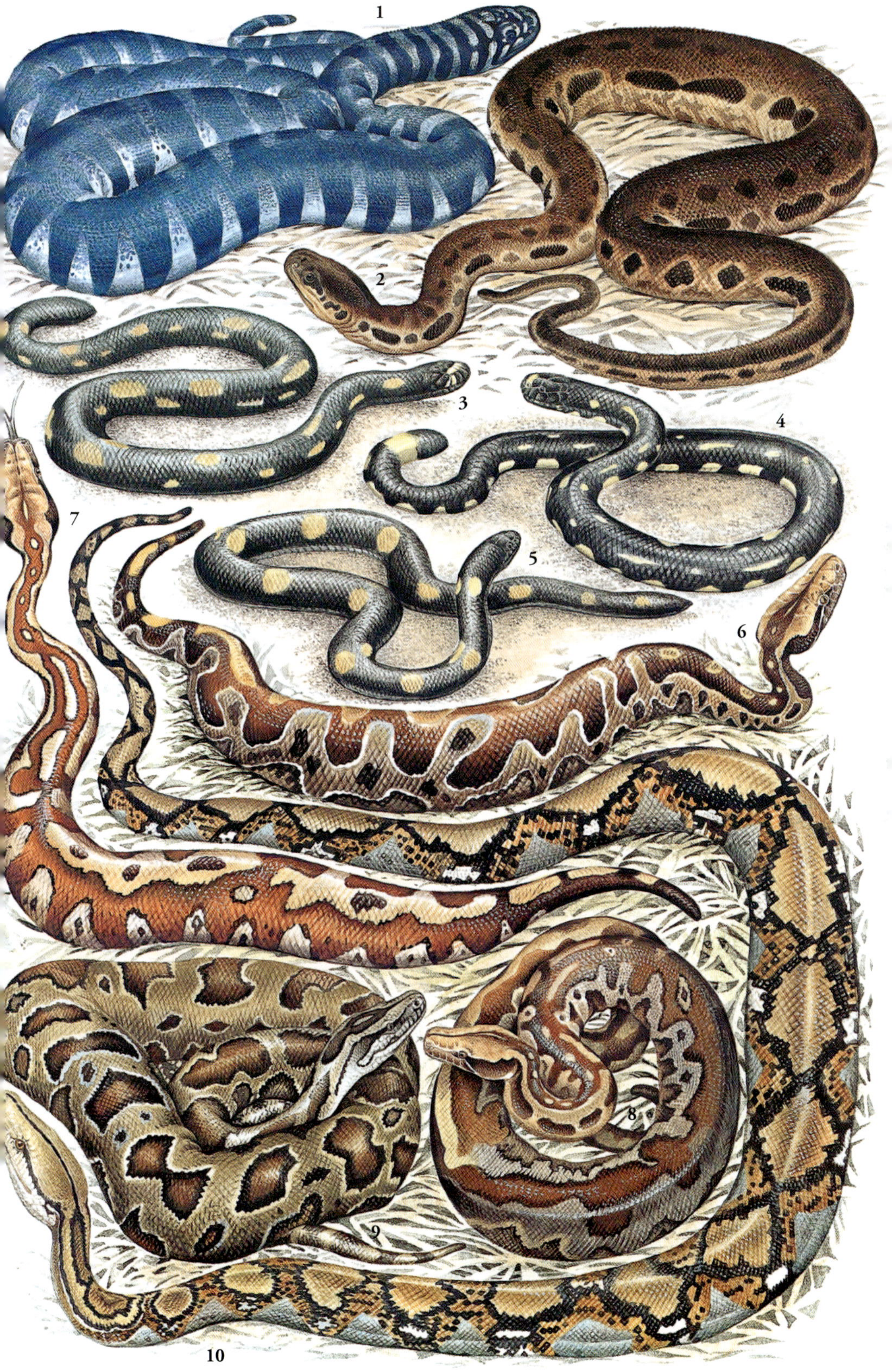
1
2
3
4
5
6
7
8
9
10

Plate 39. COLUBRIDAE

1. SPECKLE-HEADED VINE SNAKE *Ahaetulla fasciolata*, p. 259.
TL 1,690mm Snout long, ending in curled rostral scale; vertebrals enlarged; anal single; dorsum light brown, grey or pinkish-tan with numerous narrow, oblique dark bands on anterior of body; forehead with elongated or curved dark markings.

2. RIVER VINE SNAKE *Ahaetulla fronticincta*, p. 259.
TL 980mm Snout long; anal divided; dorsum bright green or brownish-yellow; skin between scales black and white, forming oblique lines; forehead with or without black spots.

3. MALAYAN VINE SNAKE *Ahaetulla mycterizans*, p. 260.
TL 920mm Snout elongated, with groove; anal entire; dorsum bright green, greyish-green or brown; in green morph, venter white with paired longitudinal green lines, and sometimes green line along middle.

4. INDIAN VINE SNAKE *Ahaetulla nasuta*, p. 260.
TL 2m Snout long with rostral appendange; groove in front of eyes; anal divided; dorsum bright green with longitudinal yellow line along outer margin of ventrals.

5. ORIENTAL VINE SNAKE *Ahaetulla prasina*, p. 260.
TL 1,970mm Snout elongated, with groove; anal divided; dorsum green, brown, yellow, dark grey or golden-yellow speckled with black; yellow stripe along lower flanks.

6. IRIDESCENT SNAKE *Blythia reticulata*, p. 260.
TL 514mm Head indistinct from neck; dorsals smooth, glossy; tail short; dorsum olive to dark, iridescent, giving it a bluish appearance; scales sometimes light-specked or light-bordered; juveniles with yellow collar; gap on dark vertebral line.

7. BENGKULU CAT SNAKE *Boiga bengkuluensis*, p. 260.
TL 1,673mm Head large, distinct from neck; vertebral scale rows enlarged; dorsum greenish-brown with 41 wide transverse cross-bars; interspaces with irregular dark green spots that extend ventrally.

8. BOURRET'S CAT SNAKE *Boiga bourreti*, p. 261.
TL 1,155mm Head large, distinct from neck; dorsum light greyish-brown with reddish-brown to black collar, 2 V-shaped reddish-brown to black bands that are sometimes fused, and a chequered pattern of light and dark blotches on top and sides of body.

9. GREEN CAT SNAKE *Boiga cyanea*, p. 261.
TL 1,870mm Head large, distinct from neck; dorsum emerald-green in adults, reddish-brown or olive with yellowish-green forehead in juveniles (illustrated); interstitial skin black; gular region sky-blue.

10. DOG-TOOTHED CAT SNAKE *Boiga cynodon*, p. 261.
TL 2.8m Head distinct from neck; vertebrals distinctly enlarged; dorsum brownish-tan or yellowish-brown with dark brown or reddish-brown bands, darkening posteriorly; head sometimes more yellow than rest of body; dark postocular stripe.

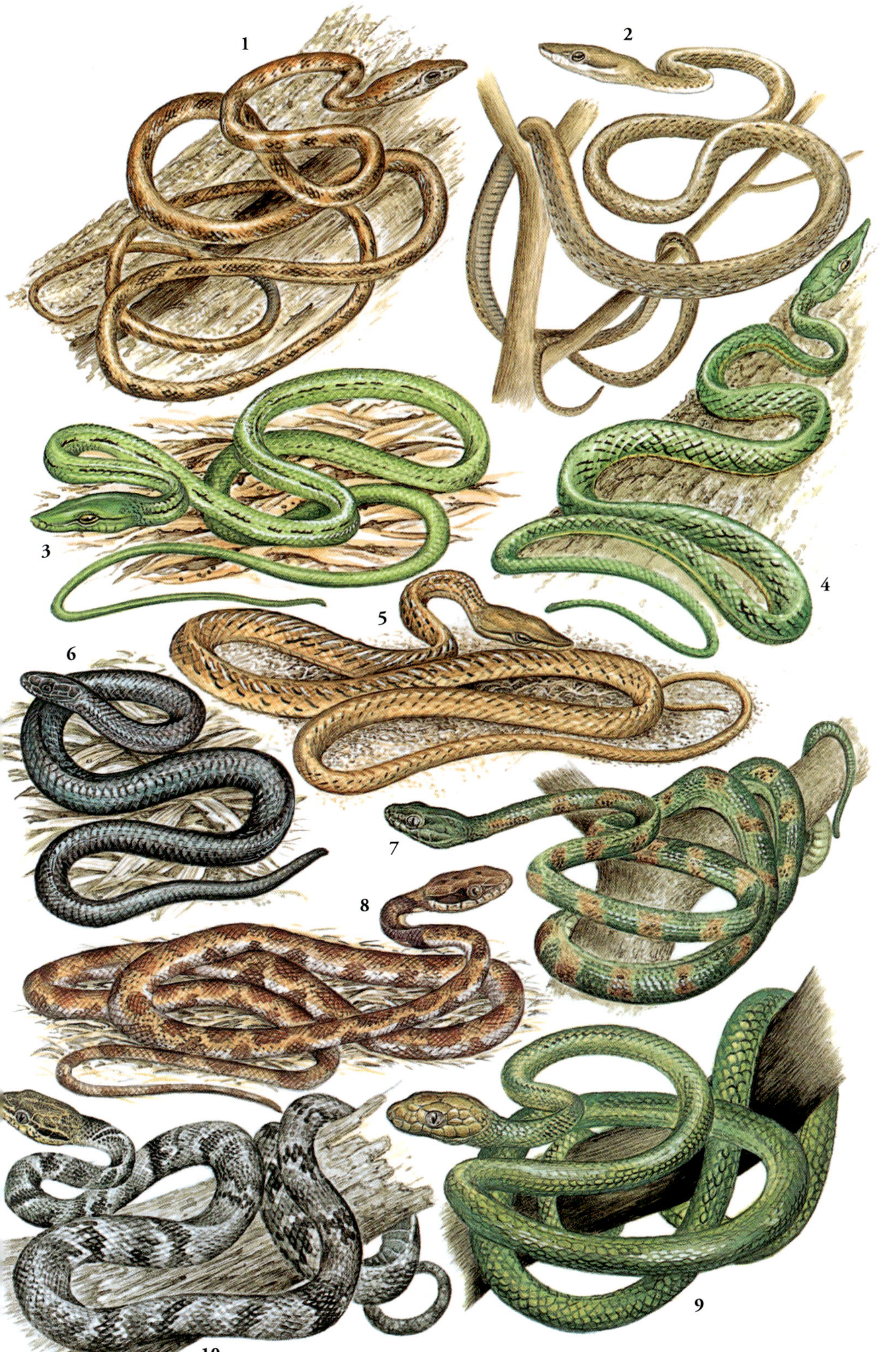
1
2
3
4
5
6
7
8
9
10

Plate 40. COLUBRIDAE

1. MANGROVE CAT SNAKE *Boiga dendrophila*, p. 261.
TL 2.5m Head distinct from neck; snout short, rounded; dorsum black with yellow rings around body.

2. WHITE-SPOTTED CAT SNAKE *Boiga drapiezii*, p. 262.
TL 2.1m Head distinct from neck; dorsum olive-grey to reddish-brown; vertebral region with paired pink spots anteriorly, sometimes fused to form bands; pink or cream spots on flanks.

3. EASTERN CAT SNAKE *Boiga gokool*, p. 262.
TL 1,200mm Head large, distinct from neck; dorsum yellowish-brown; series of vertical T- or Y-shaped marks; forehead with large, arrowhead-shaped brown mark, edged with black.

4. GUANGXI CAT SNAKE *Boiga guangxiensis*, p. 262.
TL ca 2m Head large, distinct from neck; dorsum olive or brownish-tan, turning dark grey posteriorly, with irregular black cross-bars; bright red spots between cross-bars on anterior body; forehead greyish-black.

5. JASPER CAT SNAKE *Boiga jaspidea*, p. 262.
TL 1,500mm Head large, distinct from neck; dorsum brown, reddish-brown or grey-brown with paired row of dark spots or bars on flanks; greyish-red vertebral stripe.

6. SQUARE-HEADED CAT SNAKE *Boiga kraepelini*, p. 262.
TL 1,520mm Head large, distinct from neck; dorsum brownish-grey with 57 dark cross-bars on flanks; lateral spots faint; forehead unpatterned brown.

7. MANY-SPOTTED CAT SNAKE *Boiga multomaculata*, p. 263.
TL 1,870mm Head large, distinct from neck; dorsum grey-brown with black postocular stripe; 2 brown lines edged with black from snout to back of head; series of irregular brown blotches on dorsum.

8. BLACK-HEADED CAT SNAKE *Boiga nigriceps*, p. 263.
TL 2m Head large, distinct from neck; dorsum straw-, olive- or reddish-brown; forehead often darker; labials cream or yellow.

9. TAWNY CAT SNAKE *Boiga ochracea*, p. 263.
TL 1,100mm Head large, distinct from neck; dorsum reddish-brown, ochre or coral red, unpatterned or with poorly defined transverse dark bars; dark subocular streak to angle of mouth; labials yellow or cream.

10. BANDED GREEN CAT SNAKE *Boiga saengsomi*, p. 263.
TL 2.1m Head large, distinct from neck; dorsum yellowish-green with narrow yellow areas; interstitial skin black; forehead olive; supralabials yellow.

11. THAI CAT SNAKE *Boiga siamensis*, p. 264.
TL 1,700mm Head distinct from neck; dorsum light brown with 87–98 V-shaped dark brown bands, more distinct anteriorly; posteriorly, bands bar-shaped with little posterior extension; flanks with dark and light alternating spots; forehead mid-brown; dark streak from posterior margin of eye to beyond last supralabial.

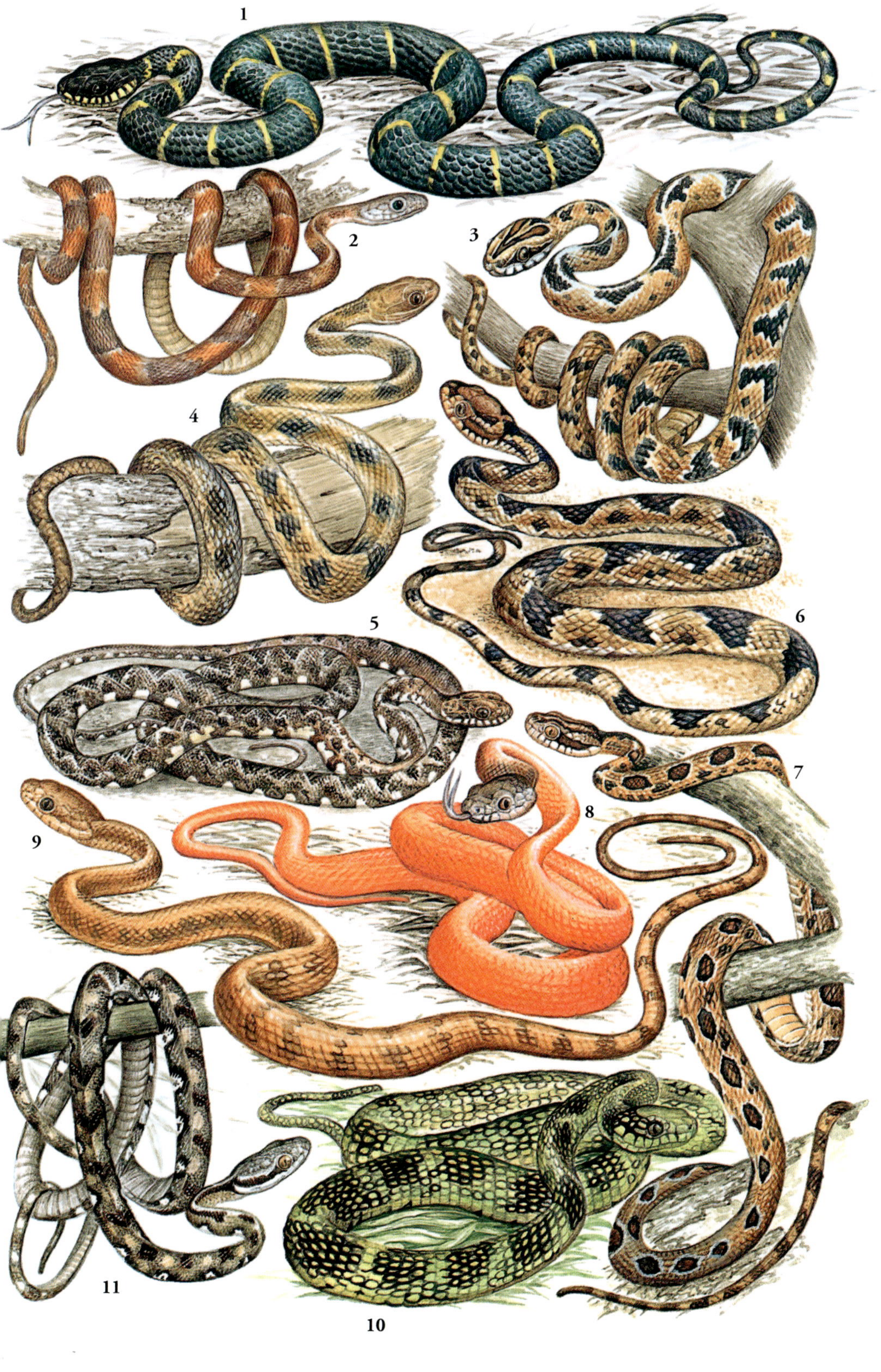
1
2
3
4
5
6
7
8
9
10
11

Plate 41. COLUBRIDAE

1. BICOLOURED REED SNAKE *Calamaria bicolor*, p. 265.
TL 450mm Body slender, cylindrical; head short, indistinct from neck; loreal absent; 3 infralabials contact anterior chin shields; dorsum blue-black or dark brown, unpatterned or with dark cross-bands; forehead dark brown, sometimes with 2 oblique dark bands crossing yellow labials.

2. BORNEAN REED SNAKE *Calamaria borneensis*, p. 265.
TL 374mm Body slender, cylindrical; head short, indistinct from neck; loreal absent; 2 infralabials contact anterior chin shields; dorsum greyish-brown; scattered dark spots or stripes on scales on middorsum; first scale row yellow, sometimes with dark cross-bars.

3. EVERETT'S REED SNAKE *Calamaria everetti*, p. 266.
TL 330mm Body slender; head short, indistinct from neck; loreal absent; 3 infralabials contact anterior chin shields; dorsum brown; scales with or without dark network; longitudinal stripe formed by dark areas on scales anteriorly; pale brown or yellow vertical bar at rear of head, and yellow nuchal collar.

4. GRABOWSKY'S REED SNAKE *Calamaria grabowskyi*, p. 266.
TL 470mm Body slender, cylindrical; head short, slightly distinct from neck; loreal absent; 3 infralabials contact anterior chin shields; dorsum dark brown; each scale with dark network; scattered dark brown or yellow spots on back; labials yellow; lateral bands composed of elongated dark spots.

5. LINED REED SNAKE *Calamaria griswoldi*, p. 267.
TL 490mm Body slender, cylindrical; head short, indistinct from neck; loreal absent; 3 infralabials contact anterior chin shields; dorsum dark brown with blackish-brown and yellow stripes; head dark brown; lower portions of supralabials yellow; oblique pale bar from parietals to gular region.

6. HILLENIUS'S REED SNAKE *Calamaria hilleniusi*, p. 267.
TL 370mm Body slender, cylindrical; head short, indistinct from neck; loreal absent; 3 infralabials contact anterior chin shields; dorsum brown; forehead brown; dark pigments ending in oblique line from upper edge of second supralabial to lower third or fifth; supralabials yellow.

7. INGER'S REED SNAKE *Calamaria ingeri*, p. 267.
TL 177mm Body slender, cylindrical; head short, indistinct from neck; loreal absent; 3 infralabials contact anterior chin shields; dorsum dark brown with intermittent light scale spots; 26 incomplete light transverse bands on body and tail.

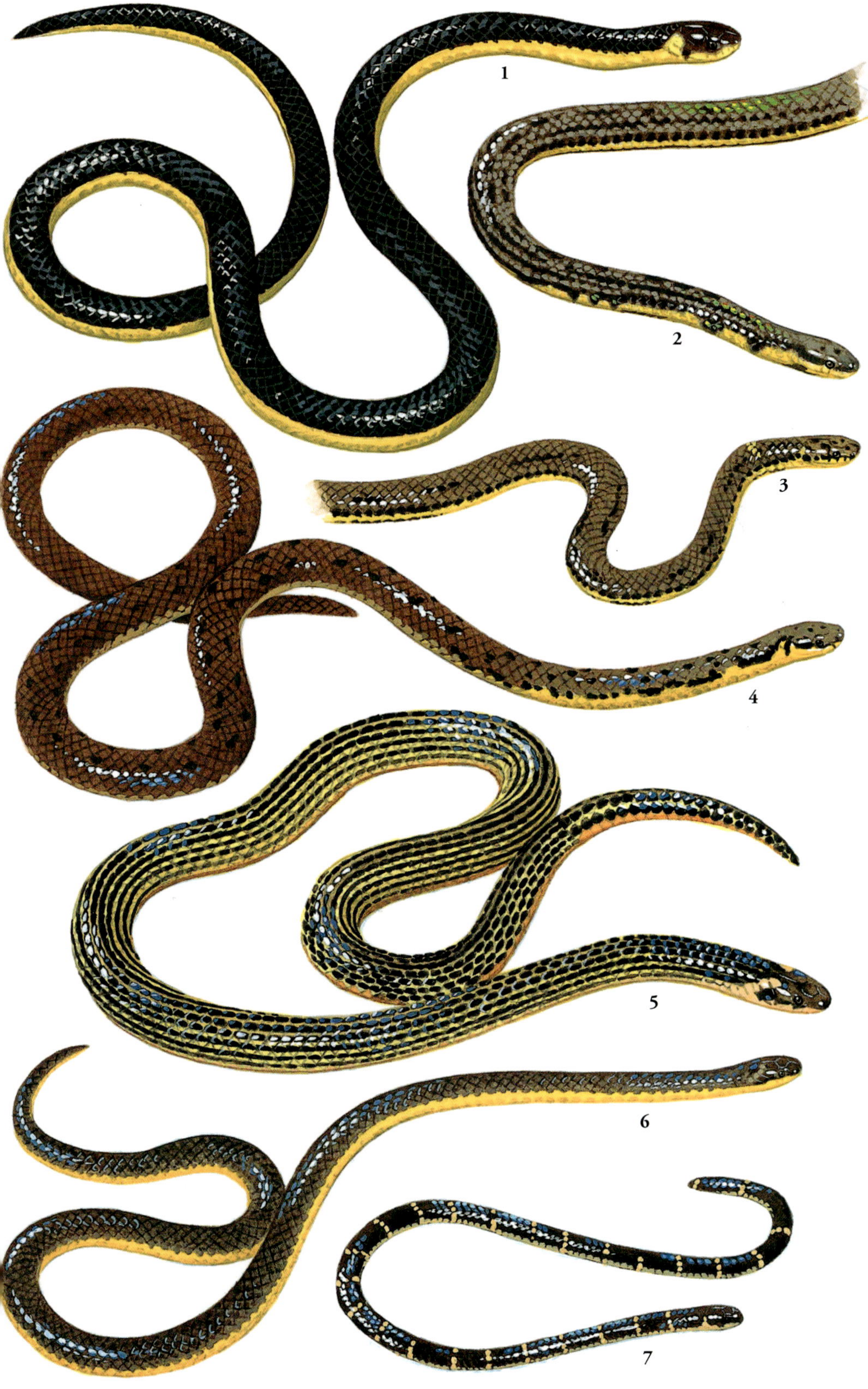
1
2
3
4
5
6
7

Plate 42. COLUBRIDAE

1. COLLARED REED SNAKE *Calamaria leucogaster*, p. 268.
TL 223mm Body slender, cylindrical; head short, slightly distinct from neck; 3 infralabials contact anterior chin shields; dorsum bright orange-red, olive or brown with longitudinal dark stripes; black collar; half ring around tail-base

2. LINNAEUS'S REED SNAKE *Calamaria linnaei*, p. 268.
TL 400mm Body slender, cylindrical; head short, indistinct from neck; 3 infralabials contact anterior chin shields; dorsum black to pale brown; dorsal scales light with dark network; sometimes isolated small scales on dorsum; light chevron behind forehead; venter orange, yellow or cream with dark pigments.

3. VARIABLE REED SNAKE *Calamaria lumbricoidea*, p. 268.
TL 640mm Body moderately robust, cylindrical; head short, indistinct from neck; 3 infralabials contact anterior chin shields.
(3a) Adult Dark brown head and dorsum.
(3b) Juvenile Red or pink head; rest of dorsum black with yellow or white bands.

4. BROWN REED SNAKE *Calamaria pavimentata*, p. 270.
TL 490mm Body slender, cylindrical; head short, indistinct from neck; 3 infralabials contact anterior chin shields; dorsum with narrow longitudinal dark stripes; solid black collar.

5. RED-HEADED REED SNAKE *Calamaria schlegeli*, p. 270.
TL 450mm Body slender, cylindrical; head short, indistinct from neck; 3 infralabials contact anterior chin shields; dorsum dark brown or black; head red, orange or dark brown.

6. SCHMIDT'S REED SNAKE *Calamaria schmidti*, p. 270.
TL 280mm Body slender, cylindrical; head short, indistinct from neck; 3 infralabials contact anterior chin shields; dorsum unpatterned blackish-grey with green and blue iridescence; scales pale margined.

7. NORTHERN REED SNAKE *Calamaria septentrionalis*, p. 270.
TL 450mm Body slender, cylindrical; head short, indistinct from neck; 3 infralabials contact anterior chin shields; dorsum dark brown or black; scales with small pale dots, forming network; yellow nuchal loop.

8. SHORT-TAILED REED SNAKE *Calamaria virgulata*, p. 271.
TL 370mm Body slender, cylindrical; head short, indistinct from neck; 3 infralabials contact anterior chin shields; dorsum dark brown; dorsal scales with light network; yellow nuchal collar sometimes present; longitudinal dark stripes along body sometimes present.

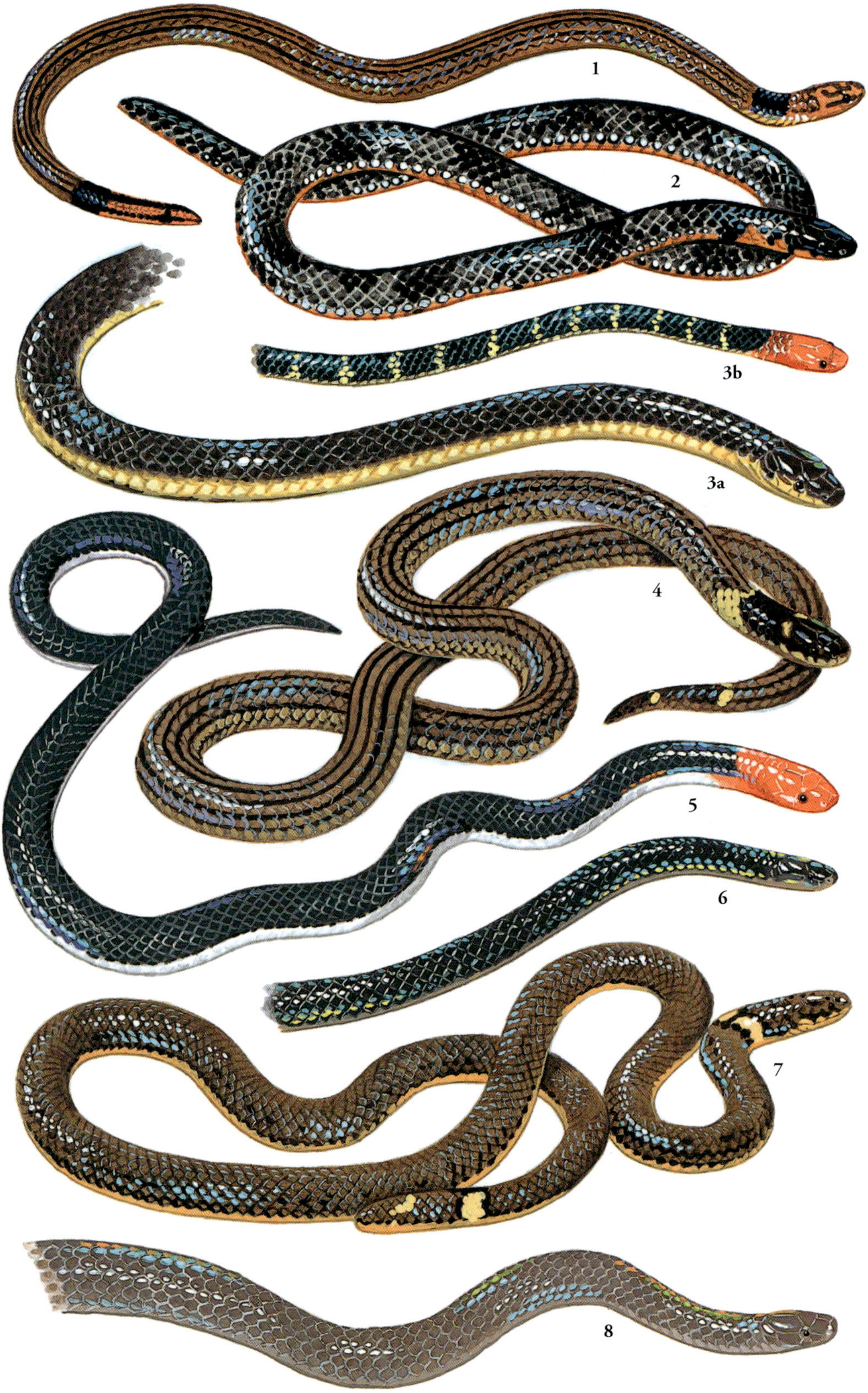
1
2
3b
3a
4
5
6
7
8

Plate 43. COLUBRIDAE

1. ORNATE FLYING SNAKE *Chrysopelea ornata*, p. 272.
TL 1,400mm Body slender; head depressed; ventrals with pronounced keels laterally.
(1a) Body Dorsum greenish-yellow or pale green; head black dorsally, with yellow and black cross-bars.
(1b) Flying snake When making glides, body is flattened like a ribbon.

2. GARDEN FLYING SNAKE *Chrysopelea paradisi*, p. 272.
TL 1,500mm Body slender; head depressed; ventrals with pronounced keels laterally; dorsum black; scale centres with green spot; vertebral sometimes with row of 3–4 pink or red spots; forehead with yellow bands.

3. TWIN-BARRED FLYING SNAKE *Chrysopelea pelias*, p. 272.
TL 740mm Body slender; head depressed; ventrals with pronounced keels laterally; dorsum red or orange with yellow or cream cross-bars, edged with black bars; forehead with 3 red cross-bars.

4. ENGGANO RAT SNAKE *Coelognathus enganensis*, p. 272.
TL 1,200mm Body slender; snout relatively long; dorsum pale brown; nape sometimes with indistinct black blotches and stripes; faint dark postocular stripe; dorsum either patternless or with rows of dark blotches on anterior; pale vertebral stripe sometimes present; posterior body patternless.

5. PHILIPPINE TRINKET SNAKE *Coelognathus erythrurus*, p. 273.
TL 1,670mm Body slender; snout long; dorsum brown to olive; narrow dark postocular stripe to angle of jaws; posterior third of body darker.

6. YELLOW-STRIPED TRINKET SNAKE *Coelognathus flavolineatus*, p. 273.
TL 1,800mm Body slender; snout long; dorsum brownish-grey or brownish-olive; dark postocular stripe and another one along nape; pale vertebral stripe sometimes present; several short dark stripes or elongated blotches on dorsum and flanks.

7. COPPER-HEAD TRINKET SNAKE *Coelognathus radiatus*, p. 273.
TL 2.3m Body slender; snout relatively long; dorsum greyish-brown or yellowish-brown with 4 black stripes along anterior of body; cream stripe between 2 upper stripes; lower stripes narrower and broken up; head coppery-brown with 3 radiating black lines from eyes.

8. INDONESIAN TRINKET SNAKE *Coelognathus subradiatus*, p. 273.
TL 2m Body slender; snout relatively long; dorsum yellowish-brown to olive with 4 narrow longitudinal stripes or black saddle-like blotches, or a pattern of both stripes and blotches; forehead yellow to dark brown with dark postocular stripe; supralabials yellowish-white.

9. GREATER GREEN SNAKE *Cyclophiops major*, p. 274.
TL 1,200mm Body moderately robust, cylindrical; snout elongated; dorsum bright green; scattered black spots in juveniles.

10. MANY-BANDED GREEN SNAKE *Cyclophiops multicinctus*, p. 274.
TL 1,070mm Body moderately robust, cylindrical; snout acute; dorsum green anteriorly, grey posteriorly; numerous cream cross-bars edged with black, on posterior body, sometimes fused on vertebral, or may alternate.

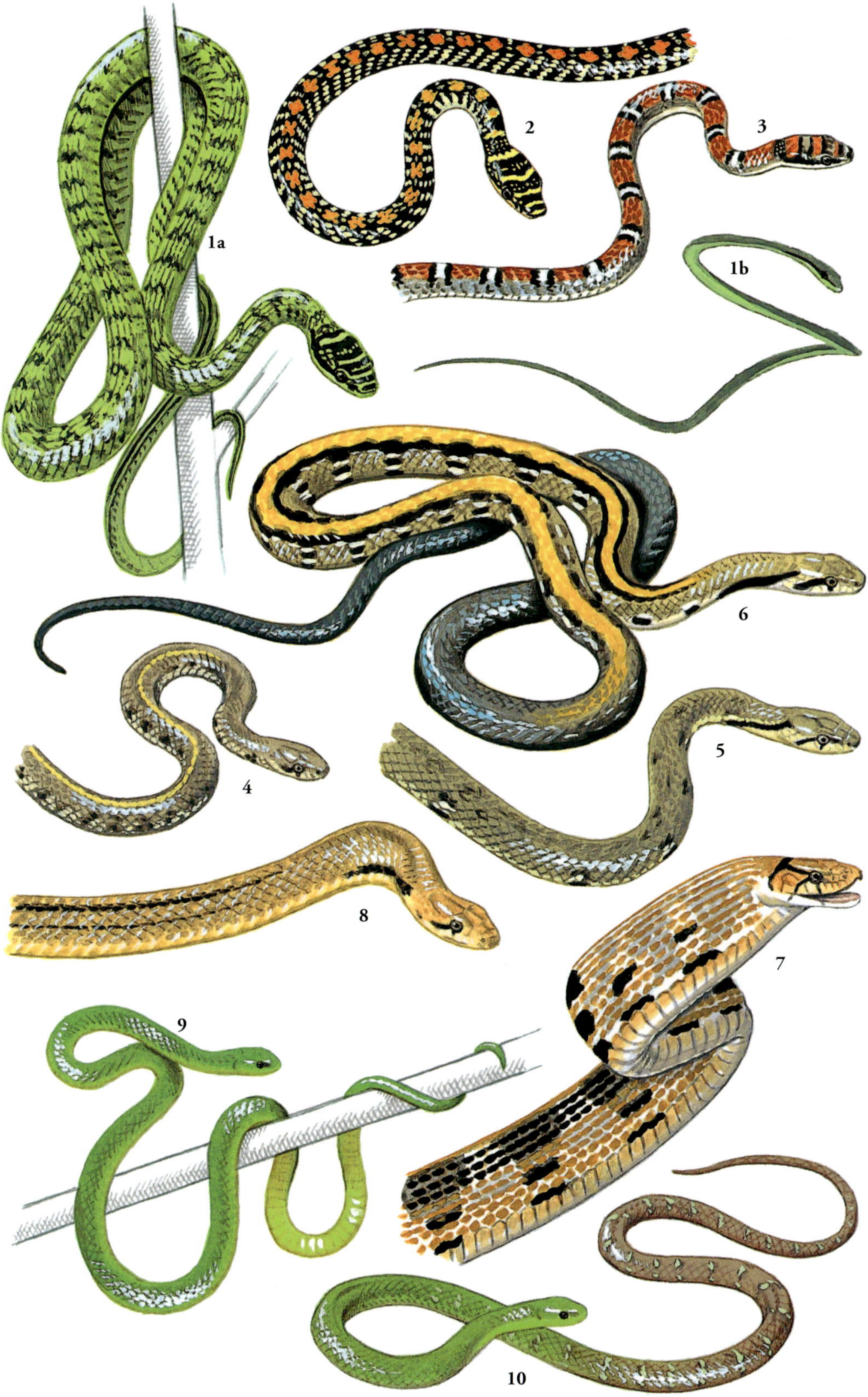

1a
2
3
1b
6
4
5
8
7
9
10

Plate 44. COLUBRIDAE

1. STRIPE-TAILED BRONZEBACK TREE SNAKE *Dendrelaphis caudolineatus*, p. 274.
TL 1,520mm Body slender; eye large; dorsum reddish-brown or olive-brown; pale green stripe on lower flanks, edged dorsally by narrow black stripe, and ventrally by broad black stripe.

2. BLUE BRONZEBACK TREE SNAKE *Dendrelaphis cyanochloris*, p. 275.
TL 1,430mm Body slender; eye small; dorsum olive; scales edged with blue; broad black temporal stripe to beyond neck, breaking up into spots; all dorsals except first row with black anterior and lower edges.

3. BEAUTIFUL BRONZEBACK TREE SNAKE *Dendrelaphis formosus*, p. 275.
TL 1,470mm Body slender; eye large; pupil rounded; dorsum reddish-brown; black postocular stripe from rostral extends to neck; 3 dark lateral stripes (exposed only in individuals in 'display') on posterior third of body; dorsals dark-edged.

4. KOPSTEIN'S BRONZEBACK TREE SNAKE *Dendrelaphis kopsteini*, p. 275.
TL 1,425mm Body slender; eye moderate; dorsum bronze-brown; black postocular stripe across lower half of temporal region to jaw end; vertebrals with black posterior margin; interstitial skin anteriorly brick-red; dorsum weakly banded.

5. NGANSON BRONZEBACK TREE SNAKE *Dendrelaphis ngansonensis*, p. 276.
TL 1,500mm Body robust; eye moderate; dorsum and forehead bronze-brown; broad black postocular stripe extends to neck, becoming bronze-brown; neck has black-grey lateral scales with blue edges.

6. PAINTED BRONZEBACK TREE SNAKE *Dendrelaphis pictus*, p. 276.
TL 1,250mm Body slender; eye large; dorsum bronze-brown or brownish-olive; yellow or cream ventrolateral stripe, edged with black; forehead brown with black postocular stripe covering over half temporal region and extending to neck; blue or greenish-blue patch on neck displayed when excited.

7. STRIATED BRONZEBACK TREE SNAKE *Dendrelaphis striatus*, p. 276.
TL 1,020mm Body slender; eye moderate; dorsum bronze-brown; labials yellow; dark stripe between nostril to orbit; dark postocular stripe covers temporals; neck yellow when inflated; oblique black bars laterally on body; interstital skin blue.

8. MOUNTAIN BRONZEBACK TREE SNAKE *Dendrelaphis subocularis*, p. 276.
TL 880mm Body slender; eye small; dorsum bronze-brown; scales with black edges; lower flanks beyond scale row 2 bright cream, olive or greenish-white; dark postocular stripe extends to sides of neck.

9. INDIAN BRONZEBACK TREE SNAKE *Dendrelaphis tristis*, p. 276.
TL 1,500mm Body slender; eye large; dorsum unpatterned purplish- or bronzy-brown; vertebral scales on neck and forebody yellow; buff flank stripe; light blue on neck between scales displayed when excited.

10. YELLOW LARGE-TOOTHED SNAKE *Dinodon flavozonatum*, p. 277.
TL 1,440mm Body slender; eye moderate; dorsum black with 85–95 narrow yellow cross-bars that bifurcate on flanks, enclosing dark spots.

11. VIETNAMESE LARGE-TOOTHED SNAKE *Dinodon meridionale*, p. 277.
TL 1,950mm Body slender; eye large; dorsum yellow with 97 black bands on body; scales on forehead with large grey areas with black edges.

12. RED LARGE-TOOTHED SNAKE *Dinodon rufozonatum*, p. 277.
TL 1,350mm Body slender; eye moderate; dorsum coral-red with 61–65 black or dark brown bands on body; flanks with dark blotches or indistinct vertical bars, alternating with cross-bars of dorsum.

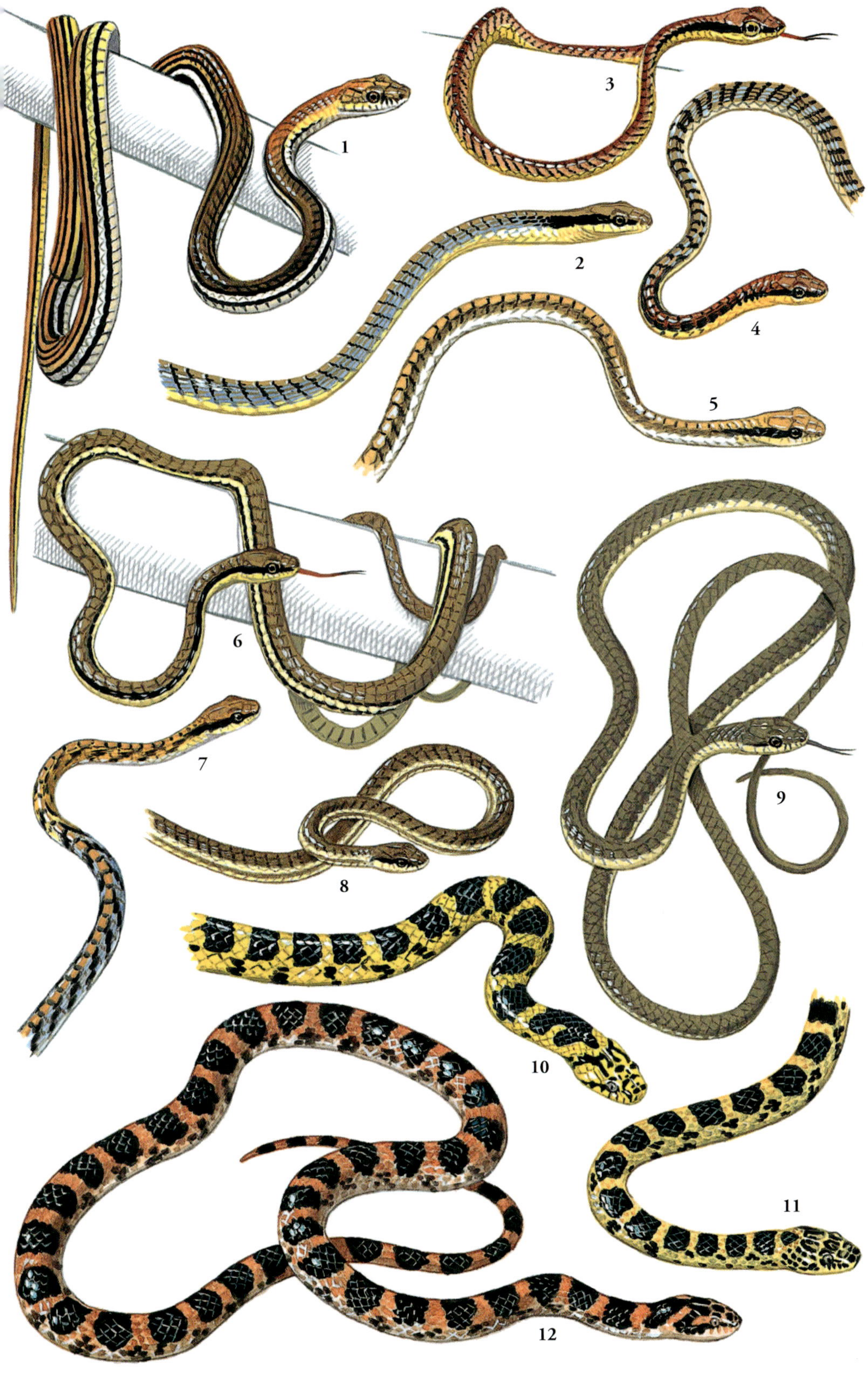
1
2
3
4
5
6
7
8
9
10
11
12

Plate 45. COLUBRIDAE

1. DAVISON'S BRIDLED SNAKE *Dryocalamus davisonii*, p. 278.
TL 920mm Body slender, compressed; head depressed, distinct from neck; dorsals smooth; dorsum black with 29–31 irregular pale green or white cross-bars that expand on flanks.

2. HALF-BANDED BRIDLED SNAKE *Dryocalamus subannulatus*, p. 278.
TL 600mm Body slender, compressed; head depressed, distinct from neck; dorsals smooth; dorsum tan or light brown with large transverse brown spots; on flanks, spots smaller.

3. THREE-BANDED BRIDLED SNAKE *Dryocalamus tristrigatus*, p. 278.
TL 650mm Body slender, compressed; head depressed, distinct from neck; dorsals smooth; dorsum dark brown with 3 yellow stripes; forehead shields with white edge.

4. KEEL-BELLIED WHIP SNAKE *Dryophiops rubescens*, p. 278.
TL 750mm Body slender, compressed; head distinct from neck; dorsals smooth; dorsum reddish-brown with small dark and pale spots; forehead with dark streaks; dark postocular streak.

5. KEELED RAT SNAKE *Elaphe carinata*, p. 279.
TL 2.4m Body robust; snout elongated; head indistinct from neck; dorsals keeled; prefrontals in broad contact with supraoculars.
(5a) Body Yellow, olive-clay or brown on anterior half; posterior half with 32 blackish-grey cross-bars separated by cream patches; posterior dorsals edged with black.
(5b) Head Yellow, olive-clay or brown; Supralabials IV–V contact orbit.

6. DARK-GREY GROUND SNAKE *Elapoides fuscus*, p. 279.
TL 500mm Body slender; head indistinct from neck; dorsals keeled; dorsum iridescent black, dark brown or reddish-brown, unpatterned or with yellow or red spots or flecks; posterior part black and iridescent.

7. MANDARIN TRINKET SNAKE *Euprepiophis mandarinus*, p. 279.
TL 1,700mm Body robust; snout short; head slightly distinct from neck; dorsals smooth.
(7a) Body Dorsum grey to greyish-brown, dorsal scales with brownish-red centres; dorsum and tail with rounded yellow blotches edged with black and yellow.
(7b) Forehead Yellow with V-shaped dark pattern; Supralabials III–IV or IV–V contact orbit.

8. ORANGE-BELLIED SNAKE *Gongylosoma baliodeirum*, p. 280.
TL 450mm Body slender; head slightly wider than neck; dorsals smooth; dorsum dark brown to reddish-brown with paired rows of cream spots; upper labials edged with dark grey.

9. STRIPED GROUND SNAKE *Gongylosoma longicauda*, p. 280.
TL 500mm Body slender; head wider than neck; dorsals smooth; dorsum brownish-red or black with a yellow or cream chevron at back of head and 5 orange or yellow stripes on dorsum.

10. PULAU TIOMAN GROUND SNAKE *Gongylosoma mukutense*, p. 280.
TL 429mm (subadult; adults unknown) Body slender; head wider than neck; dorsals smooth; dorsum red anteriorly, fading to brown-grey posteriorly; forehead brick-red; nuchal band confluent with vertebral stripe; white postocular patch; sometimes 5 incomplete white stripes anteriorly.

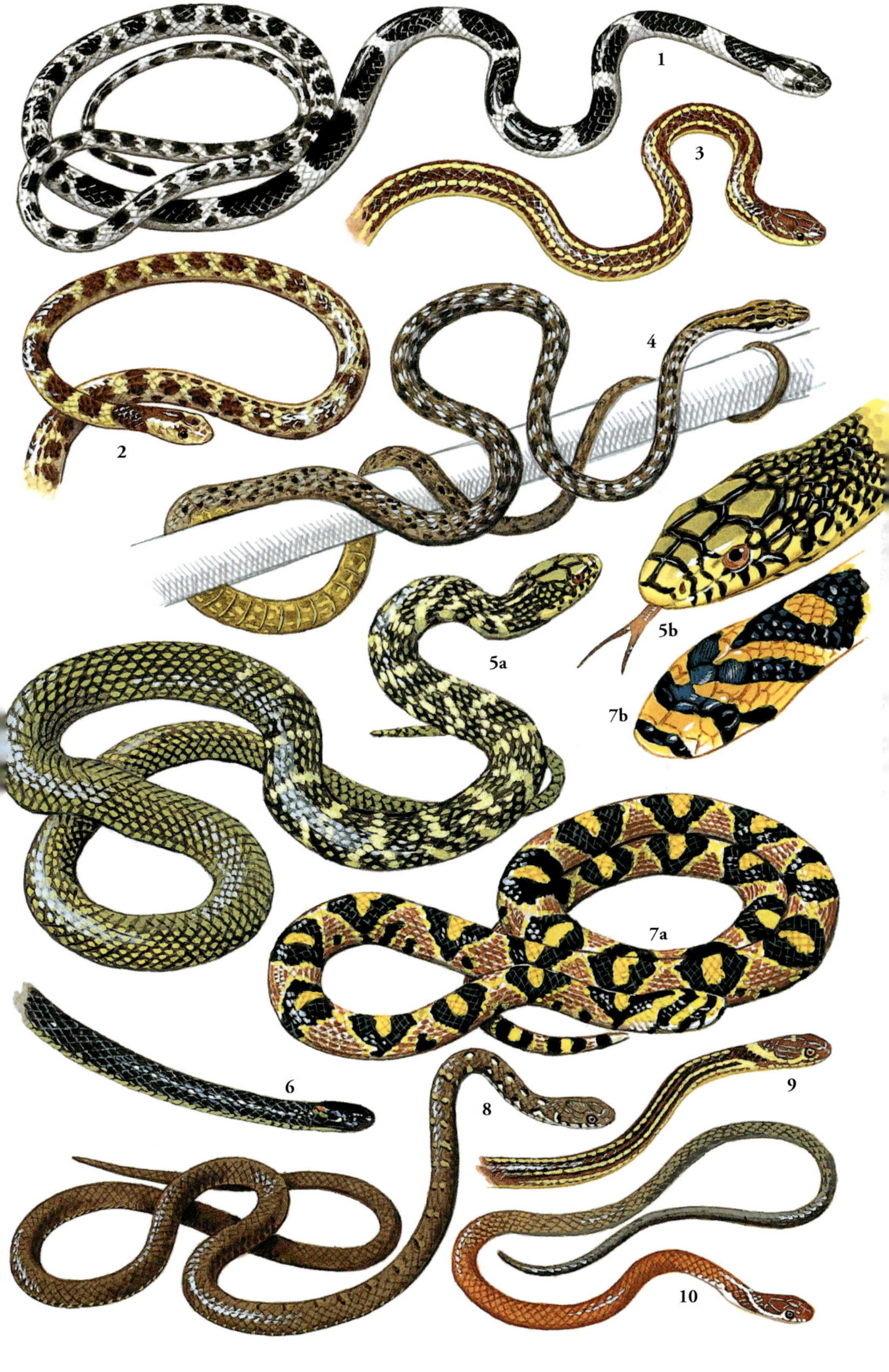

1
3
2
4
5b
5a
7b
7a
6
9
8
10

Plate 46. COLUBRIDAE

1. ROYAL TREE SNAKE *Gonyophis margaritatus*, p. 280.
TL 2m Body robust, elongated, compressed; head distinct from neck; dorsals smooth; dorsum bright green to greyish-olive; scales edged with black; several yellow or orange-red bands along posterior body; black postocular stripe.

2. RED-TAILED RACER *Gonyosoma oxycephalum*, p. 280.
TL 2.4m Body slender, elongated, compressed; head distinct from neck; dorsals smooth or weakly keeled; dorsum emerald-green or light green; black stripe along sides of olive-yellow head; juveniles olive-brown with narrow white bars towards posterior; tail yellowish-brown or reddish-orange.

3. STRIPE-NECKED SNAKE *Liopeltis frenata*, p. 281.
TL 760mm Body slender and cylindrical; head distinct from neck; dorsals smooth; dorsum brownish-olive, scales black-edged; longitudinal stripes on anterior half of body; black postocular stripe to neck.

4. TRICOLOURED RINGNECK *Liopeltis tricolor*, p. 281.
TL 560mm Body slender; head indistinct from neck; dorsals smooth; dorsum yellowish-olive; dark postocular streak to beyond neck.

5. DUSKY WOLF SNAKE *Lycodon albofuscus*, p. 281.
TL 2.7m Body slender, subcylindrical; head wider than neck; dorsals keeled.
(5a) Adult Dorsum unpatterned dark brown or brownish-black.
(5b) Juvenile 30–40 narrow white or yellow bands.

6. INDIAN WOLF SNAKE *Lycodon aulicus*, p. 281.
TL 800mm Body slender, subcylindrical; head flattened; dorsals smooth; dorsum brown or greyish-brown, with 12–19 white cross-bars, expanding laterally to enclose triangular patches.

7. BUTLER'S WOLF SNAKE *Lycodon butleri*, p. 282.
TL 1,000mm Body slender, subcylindrical; head flattened; dorsals weakly keeled; dorsum dark bluish-grey or blackish-brown with 40–50 irregular cross-bars in juveniles, suffused with dark pigments in adults.

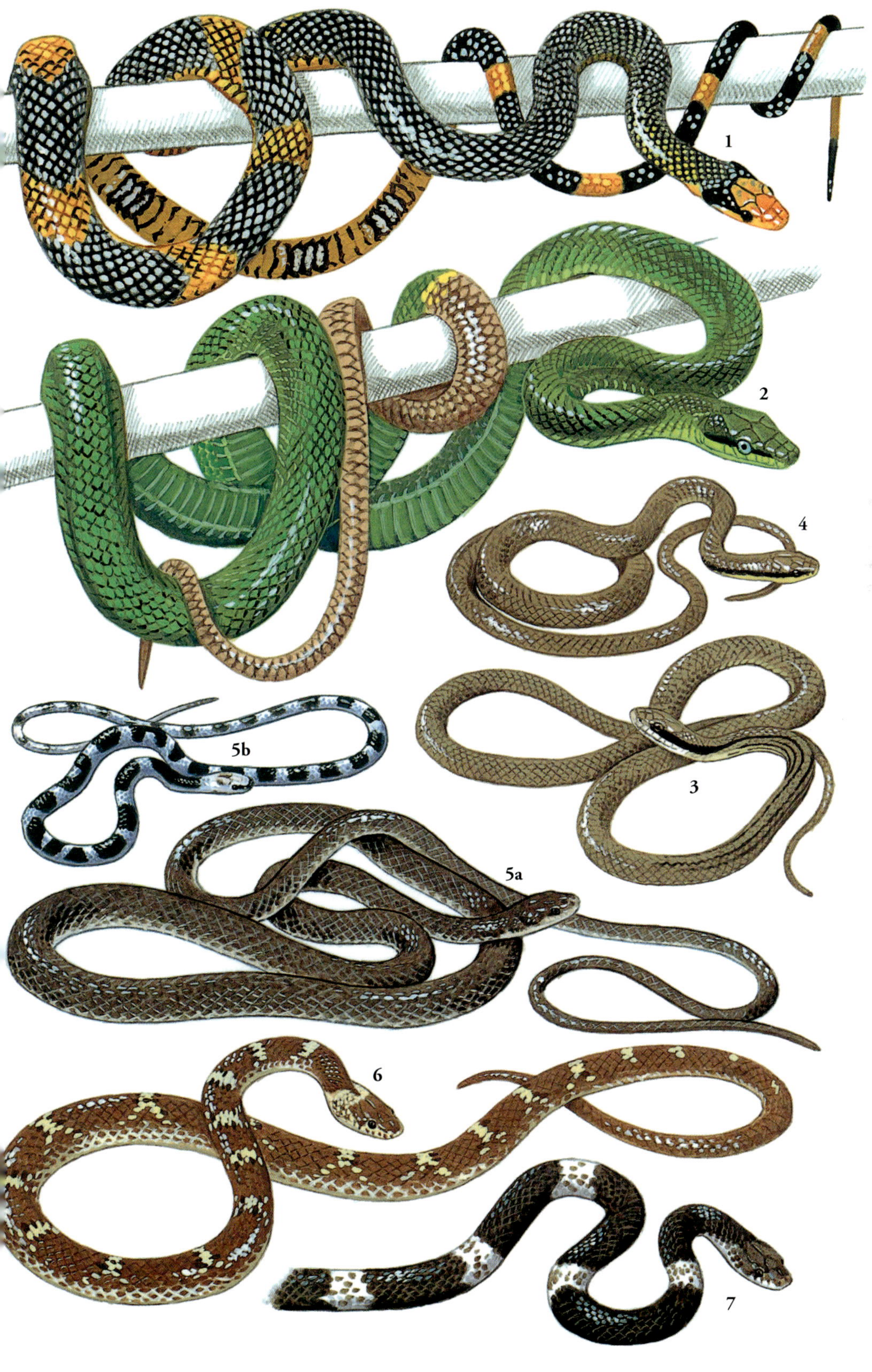

1
2
4
5b
3
5a
6
7

Plate 47. COLUBRIDAE

1. ISLAND WOLF SNAKE *Lycodon capucinus*, p. 282.
TL 760mm Body slender, subcylindrical; head flattened; dorsals smooth; dorsum brown, grey-brown or purple with narrow yellow or cream nuchal band; scales light-edged, forming indistinct cross-bars or reticulated pattern.

2. CARDAMOM MOUNTAINS WOLF SNAKE *Lycodon cardamomensis*, p. 282.
TL 316mm Body slender, subcylindrical; head flattened; dorsals weakly keeled; dorsum black with 12 white bands on body; forehead black except for pale sutures.

3. BROWN WOLF SNAKE *Lycodon effraenis*, p. 282.
TL 1,000mm Body slender, subcylindrical; head flattened; dorsals smooth or weakly keeled; dorsum reddish-brown or dark brown; 3 broad yellow or cream rings encircle body of juveniles; yellow or cream streaks on sides of head.

4. BANDED WOLF SNAKE *Lycodon fasciatus*, p. 282.
TL 895mm Body slender, subcylindrical; head flattened; dorsals weakly keeled; dorsum glossy black with 22–48 irregular cross-bars on body and tail, or a reticulated or spotted pattern.

5. LAOS WOLF SNAKE *Lycodon laoensis*, p. 283.
TL 500mm Body slender, subcylindrical; head flattened; dorsals smooth; dorsum shiny black with 13–29 white-edged yellow cross-bars on body; forehead and labials deep blue.

6. ANNAM WOLF SNAKE *Lycodon paucifasciatus*, p. 283.
TL 763mm Body slender; head depressed; dorsals smooth; dorsum black with 14 cream bands with irregular outline on body; cream bar at back of head.

7. RUHSTRAT'S WOLF SNAKE *Lycodon ruhstrati*, p. 283.
TL 940mm Body slender, subcylindrical; head flattened; dorsals smooth; dorsum black with 20–22 pale grey or white bands; head black with some pale grey or white areas.

8. BARRED WOLF SNAKE *Lycodon striatus*, p. 284.
TL 430mm Body slender, subcylindrical; head flattened; dorsals smooth; dorsum black to dark brown with transverse white or yellow marks; supralabials unpatterned white.

9. WHITE-BANDED WOLF SNAKE *Lycodon subcinctus*, p. 284.
TL 1,020mm Body slender; head flattened; dorsals weakly keeled; dorsum black or dark brown with 9–15 cream bands; pattern most distinct in juveniles.

10. ZAW'S WOLF SNAKE *Lycodon zawi*, p. 284.
TL 480mm Body slender; head flattened; dorsals smooth; dorsum brownish-black with narrow cream bands, less distinct posteriorly; labials pale brown.

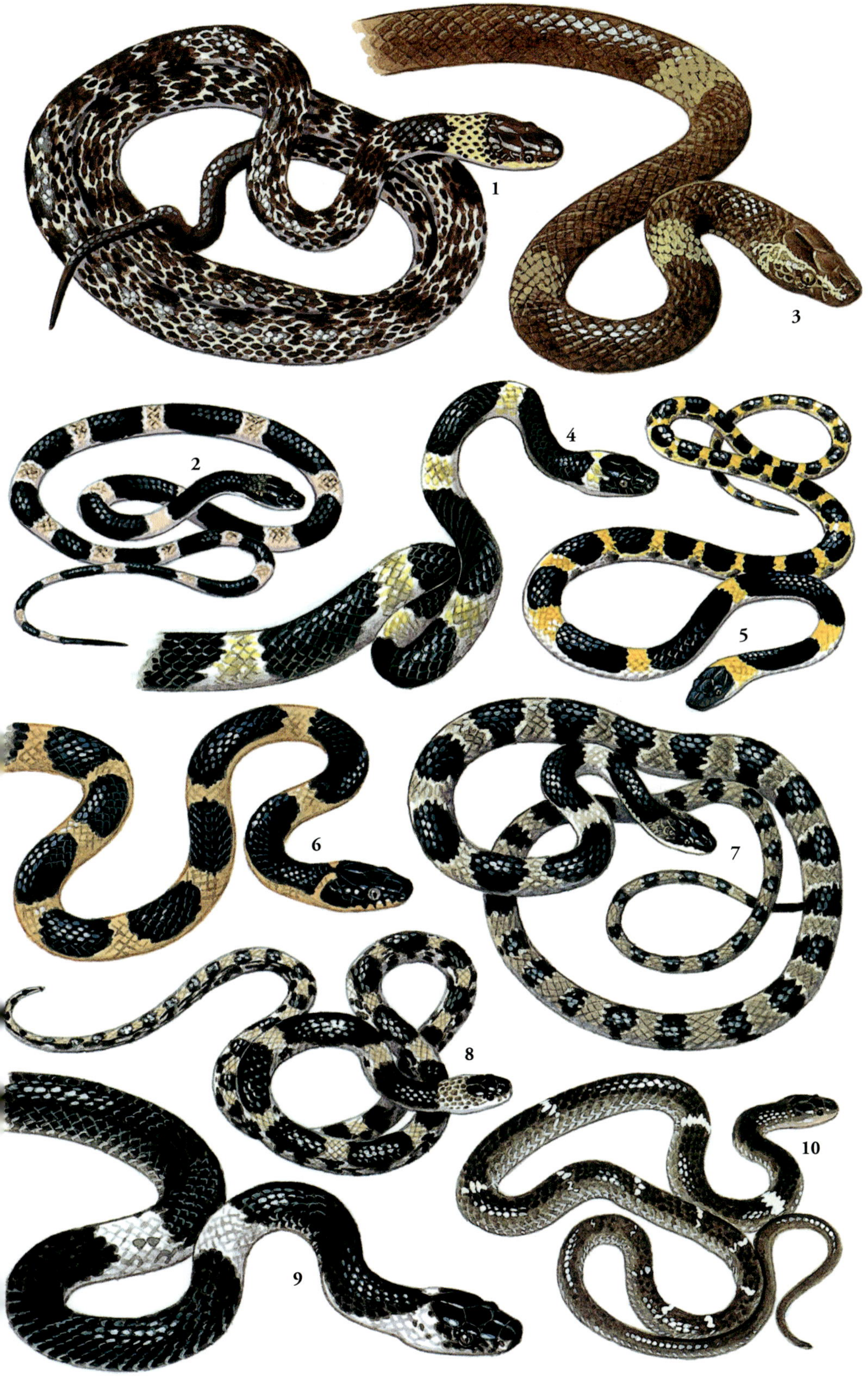
1
3
2
4
5
6
7
8
9
10

Plate 48. COLUBRIDAE

1. GENTING HIGHLANDS REED SNAKE *Macrocalamus gentingensis*, p. 284.
TL 378mm Body robust, subcylindrical; head small, wedge-shaped; loreal present; dorsals smooth; dorsum iridescent black with scattered yellow patches; head black with yellow postocular streak extending to neck; supralabials and infralabials yellow.

2. STRIPED REED SNAKE *Macrocalamus lateralis*, p. 285.
TL 298mm Body robust, subcylindrical; head small; no loreal; dorsals smooth; dorsum pale brown or yellowish-brown; dorsolateral row on anterior body with pale brown, dark-edged ocelli; paired dark brown ventrolateral stripes, separated by yellow or pale brown line.

3. SCHULZ'S REED SNAKE *Macrocalamus schulzi*, p. 285.
TL 399mm Body robust, subcylindrical; head small; loreal present; dorsals smooth; dorsum mid-brown, lacking ventrolateral stripes; outer dorsal scale rows pale yellow with brown mottling below.

4. TWEEDIE'S REED SNAKE *Macrocalamus tweediei*, p. 285.
TL 500mm Body robust, subcylindrical; head small; loreal present; dorsals smooth; dorsum uniformly black; head with yellow lateral marking extending ventrally; supralabials and infralabials yellow.

5. VOGEL'S REED SNAKE *Macrocalamus vogeli*, p. 285.
TL 192mm Body robust, subcylindrical; head small; loreal present; dorsals smooth; dorsum dark yellowish-brown; broad oblique yellow stripe behind head, followed by 2 narrow oblique lines.

6. DICE-LIKE TRINKET SNAKE *Maculophis bellus*, p. 285.
TL 927mm Body slender; head indistinct from neck; dorsals smooth or weakly keeled; dorsum pale brown with saddle-shaped yellow blotches or transverse or oblique cross-bars; forehead with Y-shaped black mark or lighter streak edged with black.

7. WHITE-BARRED KUKRI SNAKE *Oligodon albocinctus*, p. 286.
TL 1,015mm Body robust, subcylindrical; midbody scale rows 17 (rarely 19); dorsum brownish-red, sometimes with white, yellow or fawn black-edged cross-bars, numbering 19–27 on body; forehead with dark stripe from supralabials to orbit, V-shaped yellow or cream mark, edged with black, on forehead.

8. SPOTTED KUKRI SNAKE *Oligodon annulifer*, p. 286.
TL 450mm Body robust, subcylindrical; midbody scale rows 15; dorsum dark brown, lighter on flanks, with 20–26 orange-yellow blotches edged with greyish-brown.

9. BARRON'S KUKRI SNAKE *Oligodon barroni*, p. 286.
TL 401mm Body moderate, subcylindrical; midbody scale rows 17; dorsum light brown; series of transversely arranged 10–14 large brown or blackish-brown blotches on body; sometimes, 3 indistinct cross-bars between each spot.

10. JAVANESE MOUNTAIN KUKRI SNAKE *Oligodon bitorquatus*, p. 287.
TL 370mm Body robust, subcylindrical; midbody scale rows 17; dorsum dark brown, purple or greyish-brown with small red or yellow spots arranged to form bands; median series of larger spots sometimes present; forehead with dark stripes and yellow or grey bands.

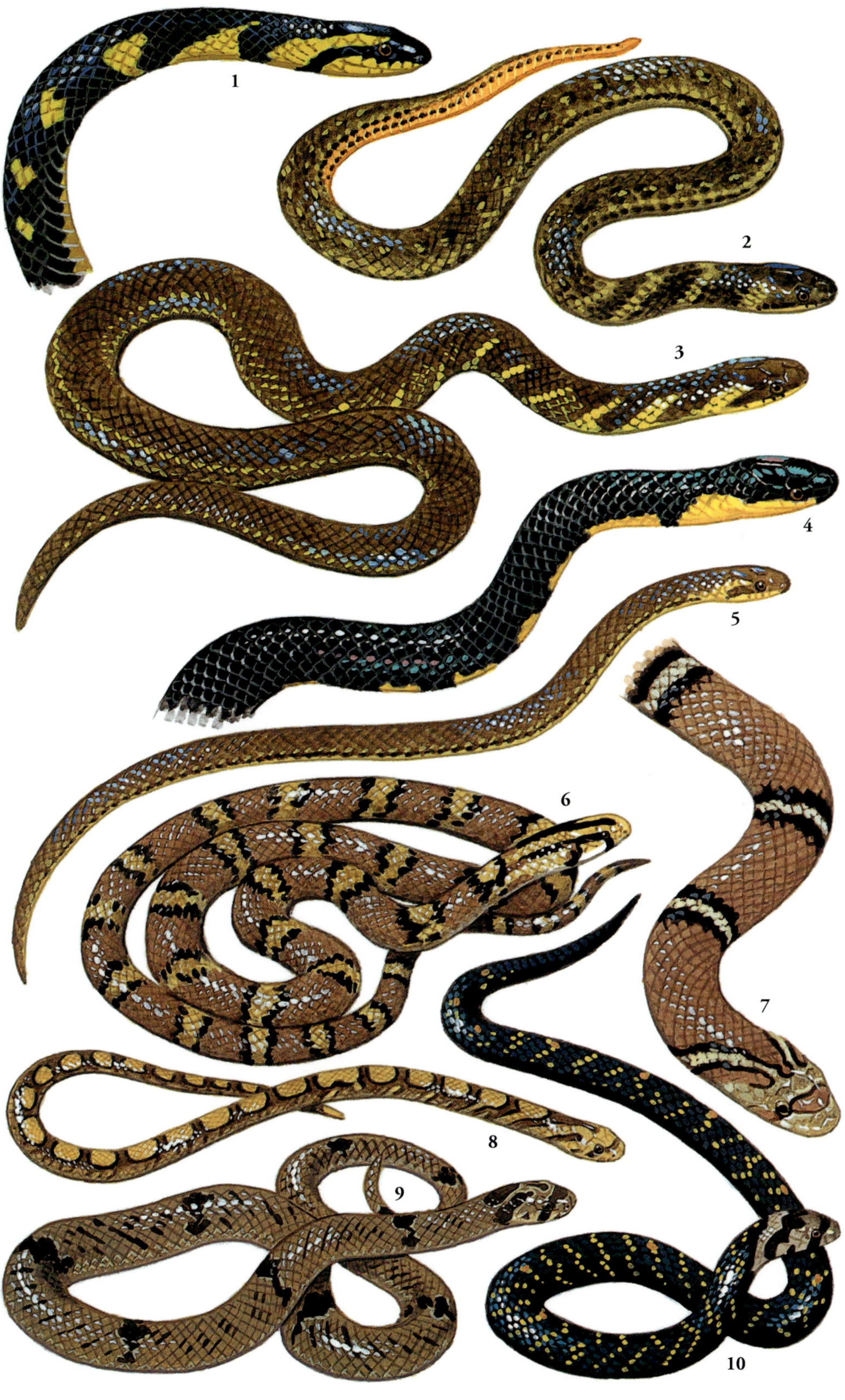
1
2
3
4
5
6
7
8
9
10

Plate 49. COLUBRIDAE

1. BOO LIAT'S KUKRI SNAKE *Oligodon booliati*, p. 287.
TL 510mm Body robust, subcylindrical; midbody scale rows 17; dorsum and flanks deep maroon-red; 19–22 indistinct transverse brown bars from nape along body, fading towards tail.

2. SPOT-TAILED KUKRI SNAKE *Oligodon dorsalis*, p. 288.
TL 628mm Body robust, subcylindrical; midbody scale rows 15.
(2a) Body Dorsum dark brown or purple with light vertebral stripe, sometimes dark-edged or containing small black spots.
(2b) Anterior body Second stripe along second and third dorsal scale rows.

3. CHINESE KUKRI SNAKE *Oligodon chinensis*, p. 287.
TL 496mm Body robust, subcylindrical; midbody scale rows 17; dorsum greyish- or reddish-brown with dorsal series of narrow, elongated dark brown or blackish spots, edged with black, numbering 9–18 on body; a red vertebral stripe occasionally present.

4. CHAIN-BANDED KUKRI SNAKE *Oligodon catenatus*, p. 287.
TL 640mm Body robust, subcylindrical; midbody scale rows 13.
(4a) Body Dorsum purplish-brown.
(4b) Head Brown with dark brown snout-tip; crescentic mark between labials, through orbit and across snout.

5. GREY KUKRI SNAKE *Oligodon cinereus*, p. 287.
TL 730mm Body robust, subcylindrical; midbody scale rows 17; dorsum reddish-brown or red, unpatterned, or with white or grey, black-edged cross-bars.

6. PEGU KUKRI SNAKE *Oligodon cruentatus*, p. 287.
TL 410mm Body robust, subcylindrical; midbody scale rows 17; dorsum olive-brown with 4 indistinct longitudinal dark stripes, median rows separated by 3 scale rows, lateral stripe extending to tail-tip; incomplete dark collar.

7. CANTOR'S KUKRI SNAKE *Oligodon cyclurus*, p. 288.
TL 940mm Body robust, subcylindrical; midbody scale rows 19; dorsum yellowish-brown to dark brown, with dark reticulations and black-edged cross-bars or transverse oval dark spots; head with V-shaped dark markings; second stripe along second and third dorsal scale rows.

8. JEWELLED KUKRI SNAKE *Oligodon everetti*, p. 289.
TL 420mm Body moderate, subcylindrical; midbody scale rows 15; dorsum pinkish-red or greyish-brown with 3 blackish-brown stripes; broad vertebral stripe encloses white and red or orange bars; dark stripe on lower flanks encloses white spots.

9. SMALL-BANDED KUKRI SNAKE *Oligodon fasciolatus*, p. 289.
TL 882mm Body robust, subcylindrical; midbody scale rows 21 or 23; dorsum yellowish-olive with 13–18 transverse blotches separated by 3–4 dark cross-bars, or with a reticulated pattern; sometimes a pale dorsolateral stripe.

10. BEAUTIFUL KUKRI SNAKE *Oligodon formosanus*, p. 289.
TL 750mm Body robust, subcylindrical; midbody scale rows 19.
(10a) Body Dorsum greyish- or reddish-brown; dorsal pattern reticulated with irregular black cross-bars; broad tan vertebral stripe.
(10b) Forehead Dark brown chevron mark extends to neck; another incomplete mark anteriorly; supralabials and infralabials cream with brown mottling.

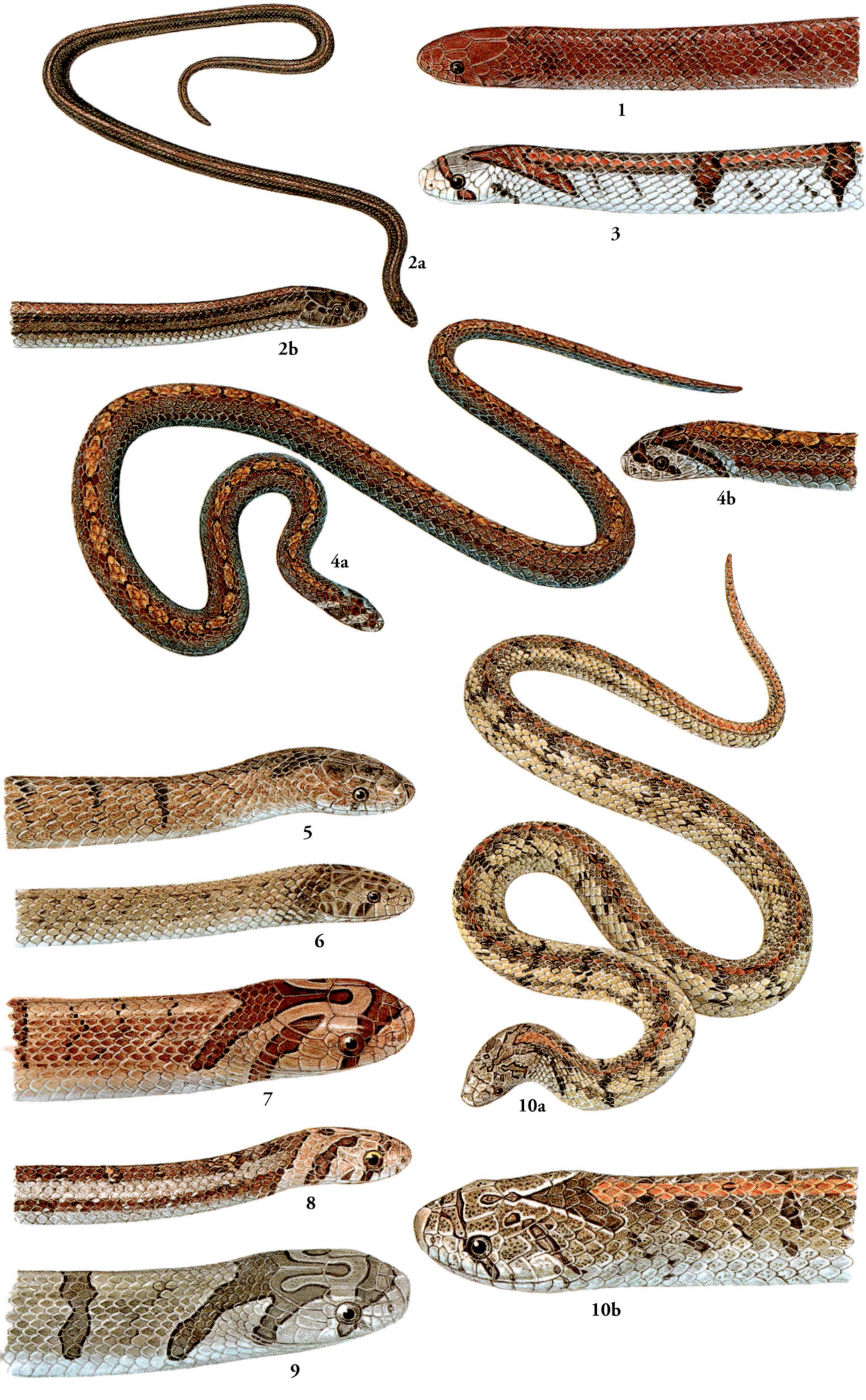
1
2a
3
2b
4b
4a
5
6
7
10a
8
10b
9

Plate 50. COLUBRIDAE

1. UNICOLOURED KUKRI SNAKE *Oligodon inornatus*, p. 290.
TL 580mm Body robust, subcylindrical; midbody scale rows 15; dorsum unpatterned brown or dull red with indistinct black cross-bars; forehead brown.

2. JOYNSON'S KUKRI SNAKE *Oligodon joynsoni*, p. 290.
TL 865mm Body robust, subcylindrical; midbody scale rows 17; dorsum purplish-brown with dark reticulations forming cross-bars, alternating with 50 transverse black spots.

3. LACROIX'S KUKRI SNAKE *Oligodon lacroixi*, p. 290.
TL 636mm Body robust, subcylindrical; midbody scale rows 15.
(3a) **Body** Dorsum dark grey or purplish-brown; sometimes with vertebral series of black-edged orange spots, numbering 11–12 on body; 4 longitudinal dark stripes.
(3b) **Head** Dark grey or purplish-brown with chevron behind eyes.

4. EIGHT-LINED KUKRI SNAKE *Oligodon octolineatus*, p. 292.
TL 700mm Body slender, subcylindrical; midbody scale rows 17.
(4a) **Body** Dorsum black or dark brown with a red vertebral stripe and 5–7 longitudinal light stripes.
(4b) **Head** Pale grey or brown with dark brown bar across eyes.

5. SPLENDID KUKRI SNAKE *Oligodon splendidus*, p. 293.
TL 830mm Body robust, subcylindrical; midbody scale rows 21.
(5a) **Body** Dorsum pale brown, each scale with dark centre; 14–17 median blotches on body; flanks with smaller dark spots.
(5b) **Head** Pale brown with dark chevron on nuchal region.

6. FOUR-LINED KUKRI SNAKE *Oligodon quadrilineatus*, p. 293.
TL 402mm Body robust, subcylindrical; midbody scale rows 19; dorsum greyish-brown; narrow greyish-yellow vertebral stripe; brown paravertebral stripe, sometimes intersected with dark brown subrectangular blotches.

7. STRIPED KUKRI SNAKE *Oligodon taeniatus*, p. 294.
TL 447mm Body moderate, subcylindrical; midbody scale rows 19.
(7a) **Body** Dorsum brownish-grey or greyish-tan; dorsals edged with dark brown posteriorly; 2 longitudinal dark paravertebral stripes edge yellow vertebral stripe, and 2 narrow dorsolateral stripes.
(7b) **Head** Brownish-grey or greyish-tan with transverse dark marking across snout, and on frontal and temporals, and an arrow-shaped nuchal one.

8. EYED KUKRI SNAKE *Oligodon ocellatus*, p. 291.
TL 852mm Body robust, subcylindrical; midbody scale rows 19; dorsum orange, brown to yellow-ochre, with 11–14 rhomboid orange or brown vertebral blotches.

9. HALF-KEELED KUKRI SNAKE *Oligodon subcarinatus*, p. 294.
TL 390mm Body slender, subcylindrical; midbody scale rows 17; dorsum greyish-brown with 20–30 dark-edged pale brown, reddish-yellow or cream cross-bars.

10. PURPLE KUKRI SNAKE *Oligodon purpurascens*, p. 293.
TL 950mm Body robust, subcylindrical; midbody scale rows 19 or 21; dorsum brownish-purple with wavy dark bands, or transverse dark-edged yellow or cream bands or blotches, sometimes separated by dark cross-bars.

11. MANDALAY KUKRI SNAKE *Oligodon theobaldi*, p. 294.
TL 437mm Body slender, subcylindrical; midbody scale rows 17; dorsum light brown with narrow, closely set transverse or angular dark cross-bars; 4 longitudinal dark stripes along dorsum.

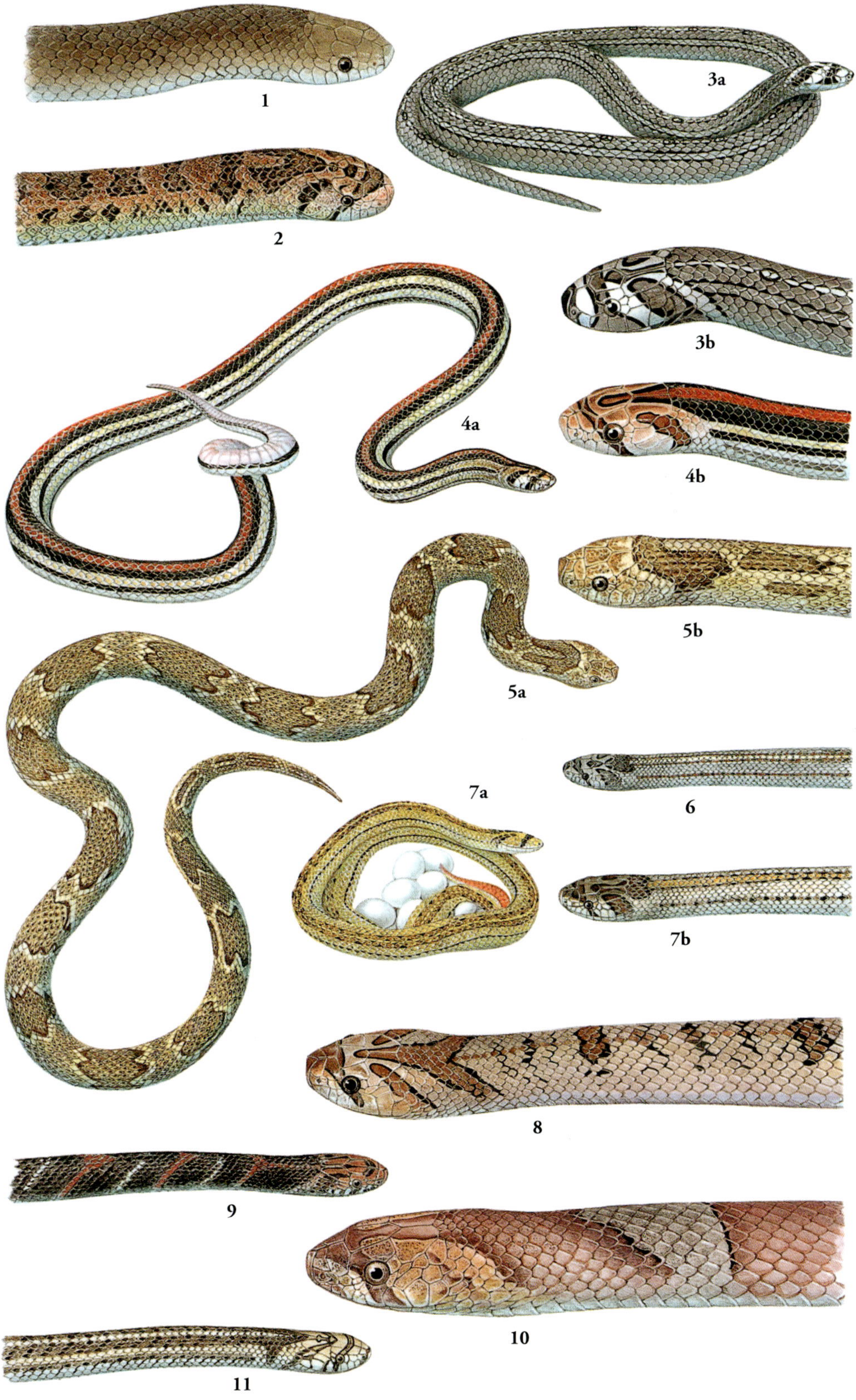
1
2
3a
3b
4a
4b
5a
5b
6
7a
7b
8
9
10
11

Plate 51. COLUBRIDAE

1. MOUNTAIN REED SNAKE *Oreocalamus hanitschi*, p. 295.
TL 570mm Body robust, subcylindrical.
(**1a**) **Body** Dorsum brownish-tan with dark brown scales forming zigzag pattern; on lower flanks, dark scales join to form continuous line.
(**1b**) **Head** Short, indistinct from neck.

2. RED BAMBOO TRINKET SNAKE *Oreocryptophis porphyraceus*, p. 295.
TL 1,250mm Body slender; head elongated, slightly distinct from neck; snout rounded; dorsum deep reddish-brown or reddish-orange; dark flank stripe along posterior half of body.
(**2a**) *O. p. coxi* Ventrals 213; subcaudals 62; tongue rosy-pink.
(**2b**) *O. p. laticinctus* Ventrals 191–201; subcaudals 62–67; tongue reddish-brown.
(**2c**) *O. p. vaillanti* Ventrals 187–209; subcaudals 55–75; tongue reddish-brown.

3. CAVE RACER *Orthriophis taeniurus*, p. 295.
TL 2m Body slender, elongated; head long, distinct from neck; dorsals weakly keeled; dorsum greyish-brown, blue-grey or greyish-black; cream or tan vertebral stripe; forehead olive or blue-grey.
(**3a**) *O. t. yunnanensis* Supralabials IV–V or V–VI contact orbit; midbody scale rows 23; ventrals 236–260; subcaudals 89–120; dorsum yellowish-brown.
(**3b**) *O. t. taeniurus* Supralabials 8–9; Supralabials V–VI, IV–V or IV–VI contact orbit; midbody scale rows 23–25; ventrals 225–255; subcaudals 84–112; dorsum yellowish-brown with some orange areas.
(**3c**) *O. t. grabowskii* Supralabials V–VI contact orbit; midbody scale rows 25 or 27; ventrals 275–285; subcaudals 102–114; dorsum lead-grey to greyish-brown
(**3d**) *O. t. ridleyi* Supralabials V–VI contact orbit; midbody scale rows 25; ventrals 285–305; subcaudals 105–122; dorsum beige to ochre.
(**3e**) **Head** Subspecies *O. t. grabowski*.

4. ANNAM KEELBACK *Parahelicops annamensis*, p. 296.
TL 558mm Body slender; head distinct from neck; median rows feebly keeled anteriorly, becoming strongly keeled posteriorly; dorsum iridescent purplish-brown; flanks brown; orange speckles on forehead; orange dorsolateral stripe and indistinct longitudinal rows of dark brown spots.

5. EASTERN TRINKET SNAKE *Orthriophis cantoris*, p. 295.
TL 2m Body slender; head elongated, slightly distinct from neck; dorsum greyish-brown to yellowish-brown; dark or light scale borders produce reticulated pattern; on top dark brown or reddish-brown blotches form transverse bands towards posterior of body.

6. MOELLENDORFF'S TRINKET SNAKE *Orthriophis moellendorffi*, p. 295.
TL 2.5m Body robust; head elongated, distinct from neck; dorsals weakly keeled; flank scales smooth; dorsum greenish-grey or greyish-pink with saddle-like rusty-brown or pale grey blotches edged with black; flanks with smaller rounded blotches.

7. COMMON BLOTCH-NECKED SNAKE *Plagiopholis nuchalis*, p. 297.
TL 450mm Body robust, rounded; head indistinct from neck; dorsals smooth or keeled; dorsum blackish-brown, mid-brown or reddish-brown; scales edged with black; dorsolateral series of rounded black spots, connected by pale cross-bars or series of oblique light brown cross-bars or elongated spots.

8. FUJIAN BLOTCH-NECKED SNAKE *Plagiopholis styani*, p. 297.
TL 396mm Body robust, rounded; head short, distinct from neck; dorsals smooth; dorsum mid-brown, dotted with black; black nuchal blotch or cross-bars sometimes present.

9. INDO-CHINESE SAND SNAKE *Psammophis indochinensis*, p. 297.
TL 1,075mm Body slender; head oval, distinct from neck; dorsals smooth; dorsum olive-green or buff with 4 dark brown stripes.

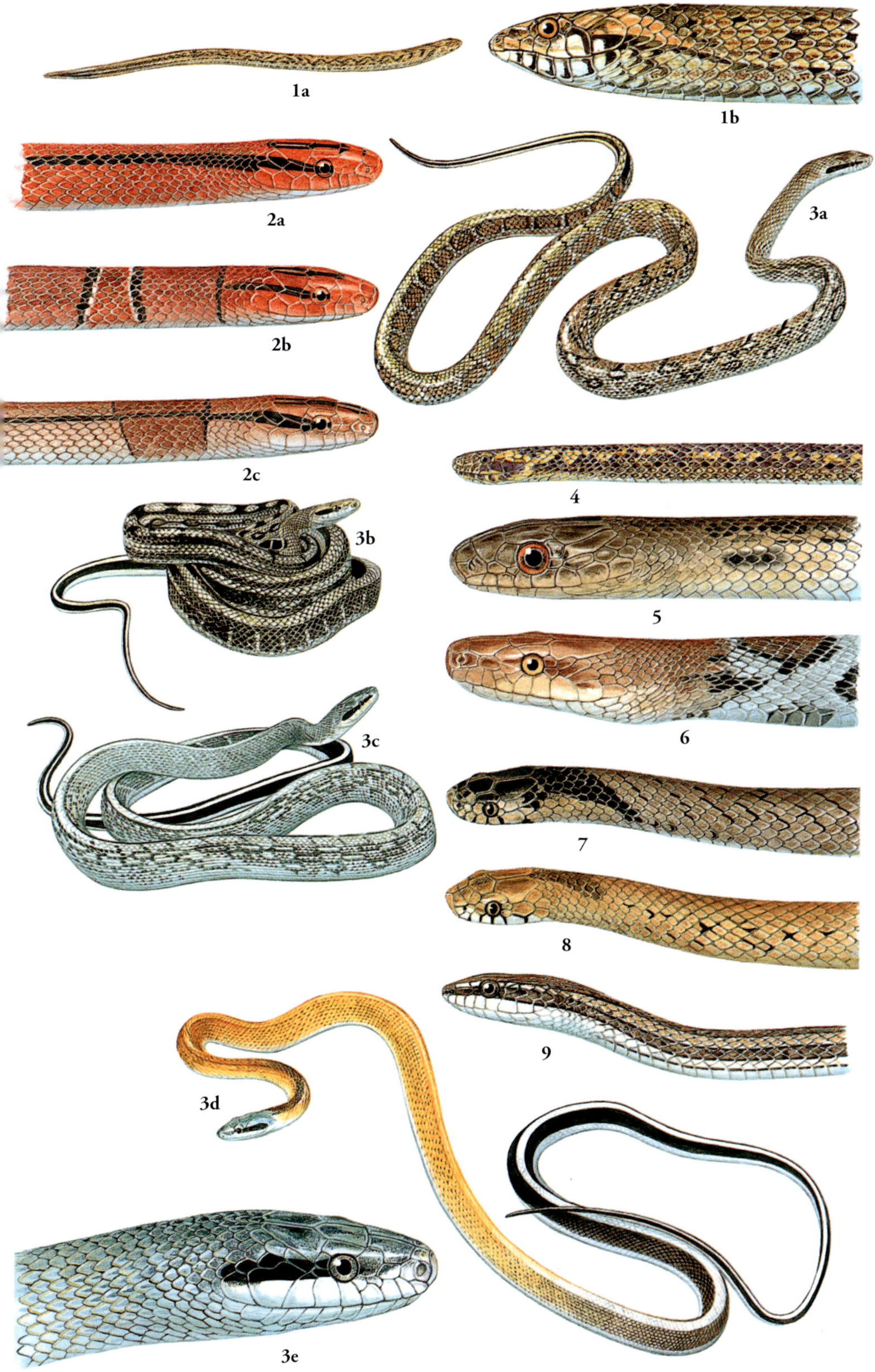
1a
1b
2a
3a
2b
2c
4
3b
5
6
3c
7
8
9
3d
3e

Plate 52. COLUBRIDAE

1. WHITE-COLLARED REED SNAKE *Pseudorabdion albonuchalis*, p. 297.
TL 270mm Body slender; head indistinct from neck; nostril between 2 nasals; preocular absent.
(**1a**) **Body** Dorsum iridescent black or brown.
(**1b**) **Head** Broad yellow or red collar over half of parietals and band of 4 scales behind parietals.

2. DWARF REED SNAKE *Pseudorabdion longiceps*, p. 298.
TL 230mm Body slender; head indistinct from neck; nostril in single nasal; single preocular; dorsum iridescent black or brown; narrow white or yellow collar and yellow spot above angle of mouth.

3. MOCQUARD'S REED SNAKE *Pseudorabdion collaris*, p. 298.
TL 250mm Body slender; head indistinct from neck; nostril between 2 nasals; no preocular; dorsum shiny black; red or yellow collar sometimes present.

4. MALAYAN KEELED RAT SNAKE *Ptyas carinata*, p. 298.
TL 4m Body slender; head distinct from neck; dorsals smooth, except 2–4 median rows that are keeled; dorsum olive-brown to black anteriorly, sometimes with indistinct yellow cross-bars; posterior dorsum yellow with distinct black chequered pattern, terminating in black tail with yellow spots.

5. BLACK-STRIPED RAT SNAKE *Ptyas dhumnades*, p. 298.
TL 2.56m Body robust; head distinct from neck; 2–6 middorsal rows keeled, other dorsals smooth; dorsum with longitudinal black stripes; anterior half of body grey, posterior half slaty-black.
(**5a**) **Adult head** Snout elongated, pattern on nape showing less contrast.
(**5b**) **Juvenile head** Snout short, pattern on nape showing higher contrast.

6. WHITE-BELLIED RAT SNAKE *Ptyas fusca*, p. 299.
TL 3m Body slender; head distinct from neck; dorsals smooth; dorsum mid-brown, pinkish-brown or brownish-grey, nearly black anteriorly, lightening posteriorly; red vertebral stripe sometimes present.

7. INDIAN RAT SNAKE *Ptyas mucosa*, p. 299.
TL 3.7m Body slender; head elongated and distinct from neck; dorsals smooth or weakly keeled; dorsum yellowish-, reddish- or olivaceous-brown to black; posterior of body with dark bands or reticulated pattern.
(**7a**) **Forehead** Widened supraoculars 'hang' over eyes.
(**7b**) **Male combat** Two males entwined in combat.

8. JAVAN RAT SNAKE *Ptyas korros*, p. 299.
TL 2.68m Body robust; head elongated, distinct from neck; dorsals smooth anteriorly, weakly keeled posteriorly; anterior dorsum grey, reddish-brown or olive-brown, posteriorly darkening to nearly black, with white bands.

9. GREEN RAT SNAKE *Ptyas nigromarginata*, p. 299.
TL 1,915mm Body slender; head elongated; dorsals smooth, 4–6 median scale rows distinctly keeled.
(**9a**) **Body** Dorsum green or olive-green; scales black-edged; 4 longitudinal black stripes in juveniles, which are confined in adults to posterior third of body.
(**9b**) **Forehead** Reddish-brown.

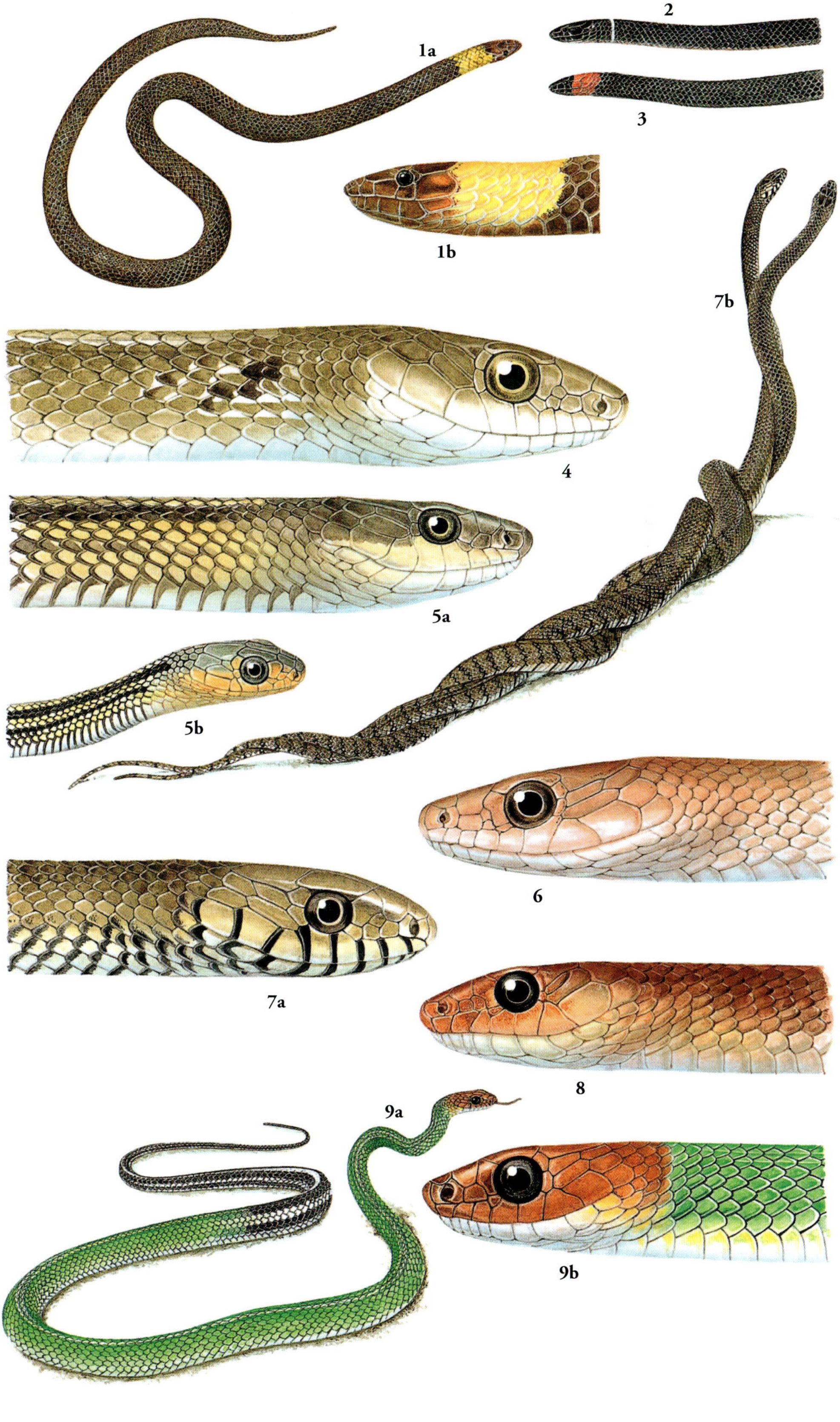
1a
2
3
1b
7b
4
5a
5b
6
7a
8
9a
9b

Plate 53. COLUBRIDAE

1. KHASI HILLS TRINKET SNAKE *Rhadinophis frenatus*, p. 300.
TL 1,500mm Body slender; snout-tip pointed and slightly arched forwards; dorsals weakly keeled, except outer 2–3 rows.
(1a) Body Dorsum green or turquoise.
(1b) Head Labials green or yellowish-green; black postocular stripe.

2. GREEN TRINKET SNAKE *Rhadinophis prasinus*, p. 300.
TL 1,200mm Body slender; snout long; dorsals weakly keeled, except outer 2–3 rows; dorsum green or turquoise; labials green or yellowish-green; faint dark postocular stripe.

3. VIETNAMESE HORNED SNAKE *Rhynchophis boulengeri*, p. 300.
TL 1,380mm Body slender; fleshy, scale-covered appendage on snout-tip; dorsals smooth.
(3a) Body Dorsum pale green or greyish-cream; interstitial skin dark.
(3b) Head Forehead and rostral appendage dark green or greyish-cream; supralabials yellow or pale green; narrow stripe between nostril and orbit; narrow dark postocular stripe; sometimes, dark streaks on forehead and nape.

4. BORNEAN BLACK SNAKE *Stegonotus borneensis*, p. 301.
TL 1,370mm Body robust, cylindrical; distinct vertebral ridge; dorsals smooth; dorsum grey-black, dark brown or black, without markings; supralabials grey with pink tinge.

5. TWIN-STREAKED BLACK-HEADED SNAKE *Sibynophis bistrigatus*, p. 300.
TL 300mm Body slender, cylindrical; dorsals smooth; dorsum pale reddish-brown or tan; series of black spots on vertebral region; black lateral stripe; forehead and nape black; paired yellow spots on nuchal region.

6. CHINESE BLACK-HEADED SNAKE *Sibynophis chinensis*, p. 300.
TL 694mm Body slender, cylindrical; dorsals smooth; dorsum pale tan, greyish-brown or dark brown; head black; yellow or cream nuchal collar.

7. COLLARED BLACK-HEADED SNAKE *Sibynophis collaris*, p. 301.
TL 760mm Body slender, cylindrical; dorsals smooth; dorsum brown or greyish-brown; black vertebral stripe comprising black spots; dotted light dorsolateral line sometimes present.

8. STRIPED BLACK-HEADED SNAKE *Sibynophis geminatus*, p. 301.
TL 1,000mm Body slender, cylindrical; dorsals smooth; dorsum orange-yellow, darker caudally; bright orange nuchal collar present or absent; dark vertebral stripe, sometimes edged with red or reddish-brown stripe.

9. WHITE-LIPPED BLACK-HEADED SNAKE *Sibynophis melanocephalus*, p. 301.
TL 600mm Body slender, cylindrical; dorsals smooth; dorsum reddish-brown, grey or brown with short black cross-bars over lighter band; lower flanks sometimes with small yellow spots.

10. TRIANGLED BLACK-HEADED SNAKE *Sibynophis triangularis*, p. 301.
TL 700mm Body slender, cylindrical; dorsals smooth; dorsum greyish-brown anteriorly, darker posteriorly, with pale brown spots forming dorsolateral stripe.

11. FRUHSTORFER'S MOUNTAIN SNAKE *Tetralepis fruhstorferi*, p. 302.
TL 500mm Body robust, cylindrical; dorsals smooth; dorsum dark reddish-brown; black vertebral stripe.

12. ORNATE BROWN SNAKE *Xenelaphis ellipsifer*, p. 302.
TL 2.51m Body robust; dorsals smooth; subcaudals 124–134; dorsum orange-red with 18–20 large elliptical or squarish brown blotches on body.

13. MALAYAN BROWN SNAKE *Xenelaphis hexagonotus*, p. 302.
TL 2m Body robust; dorsals smooth; subcaudals 140–179; dorsum brown, dark green or greenish-olive with vertical narrow black bars, their apices reaching lower flanks; dark forehead contrasts strongly with pale sides and throat.

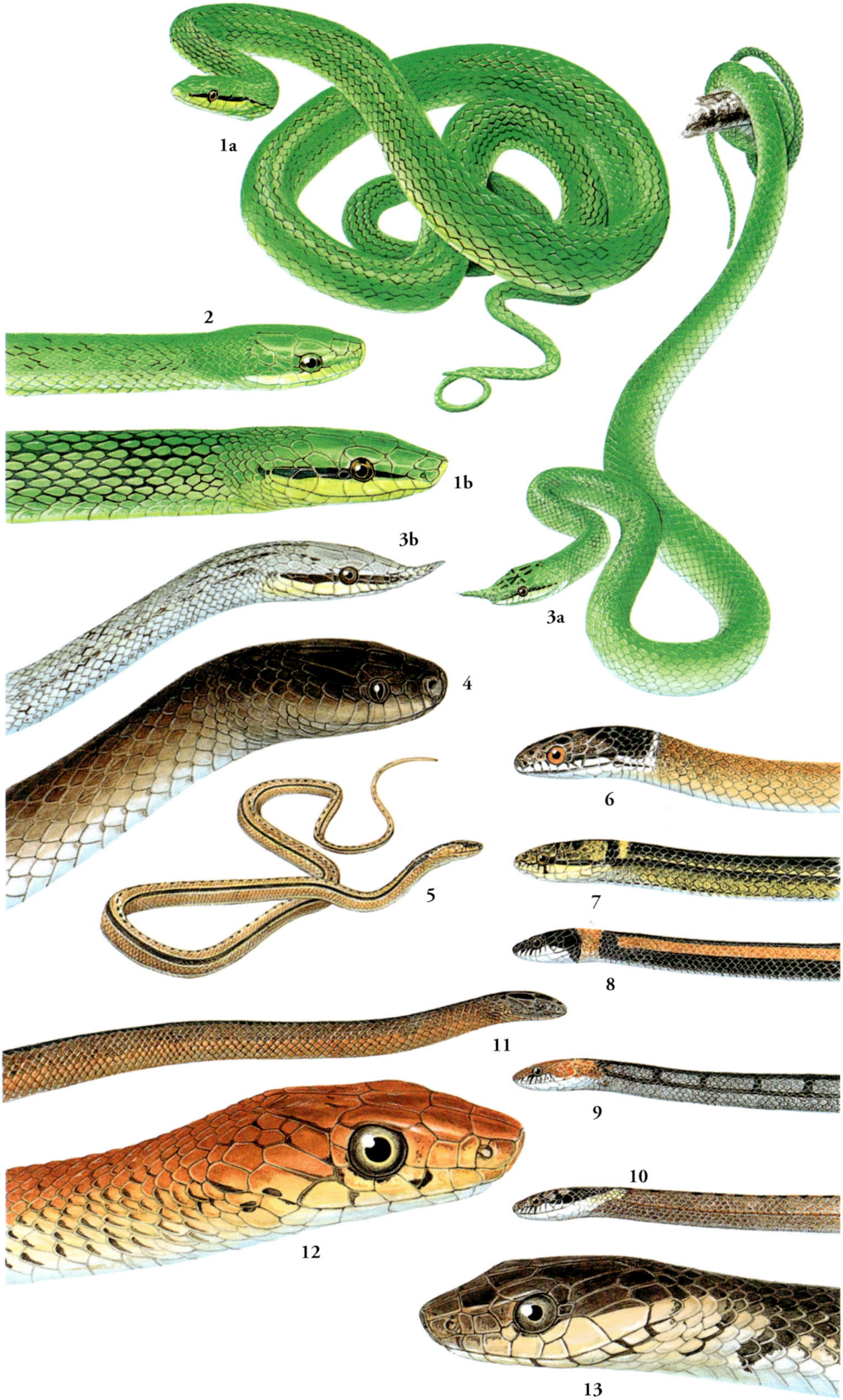
1a
2
1b
3b
3a
4
6
7
5
8
11
9
10
12
13

Plate 54. VIPERIDAE

1. MANGROVE PIT VIPER *Cryptelytrops purpureomaculatus*, p. 304.
TL 1,040mm Body slender; head distinct from neck; dorsals keeled.
(**1a**) **Body** Dorsum olive, grey or brownish-purple with darker mottling.
(**1b**) **Head** Forehead olive with brown mottling.

2. KANBURI PIT VIPER *Cryptelytrops kanburiensis*, p. 304.
TL 667mm Body slender; head distinct from neck; dorsals keeled, except for lowest row.
(**2a**) **Body** Dorsum olive or greyish-green; white or blue spots on first dorsal scale row.
(**2b**) **Head** With brown blotches; white or blue spots on labials.

3. BEAUTIFUL PIT VIPER *Cryptelytrops venustus*, p. 304.
TL 580mm Body slender; head distinct from neck; dorsals keeled.
(**3a**) **Body** Dorsum olive or bluish-green in males, grass-green in females; brown dorsal bands; pale spot on first dorsal scale row.
(**3b**) **Head** With brown blotches; labials green with brown patch; dark postorbital stripe.

4. WHITE-LIPPED PIT VIPER *Cryptelytrops albolabris*, p. 303.
TL 1,040mm Body robust; head relatively long; dorsals smooth or weakly keeled; dorsum bright green or lime-green; sides of head below eye, including labials, yellow, white or pale green; tail reddish-brown.

5. SPOT-TAILED PIT VIPER *Cryptelytrops erythrurus*, p. 303.
TL 1,050mm Body slender; head relatively long; dorsals strongly keeled.
(**5a**) **Body** Dorsum dark green.
(**5b**) **Head** Dark green; sides of head below eye, including labials, yellow.

6. LESSER SUNDA WHITE-LIPPED PIT VIPER *Cryptelytrops insularis*, p. 303.
TL 930mm Body slender; head distinct from neck; dorsals keeled; dorsum bright green, olive or blue with transverse dark bands; supralabials yellow to greenish-white.

7. LARGE-EYED PIT VIPER *Cryptelytrops macrops*, p. 304.
TL 710mm Body slender; head distinct from neck; dorsals keeled, except first row; dorsum pale green, sometimes with pale blue lateral stripe; labials pale bluish-green; forehead pale green.

8. MALAYAN PIT VIPER *Calloselasma rhodostoma*, p. 303.
TL 1,450mm Body robust; snout acute and slightly upturned.
(**8a**) **Body** Dorsum reddish-brown or purplish-brown; flanks paler and speckled with dark brown; 19–31 subtriangular dark brown marks.
(**8b**) **Head** Dark postocular patch, scalloped ventrally, extends beyond angle of jaws.

9. SHARP-NOSED VIPER *Deinagkistrodon acutus*, p. 304.
TL 1,490mm Body robust; head distinct from neck; dorsals keeled; scale tips bituberculate.
(**9a**) **Body** Dorsum grey or brown with 15–23 pairs of subtriangular dark marks on each side that meet or alternate.
(**9b**) **Head** Snout upturned with rostral raised upwards; narrow dark brown or black postocular stripe.

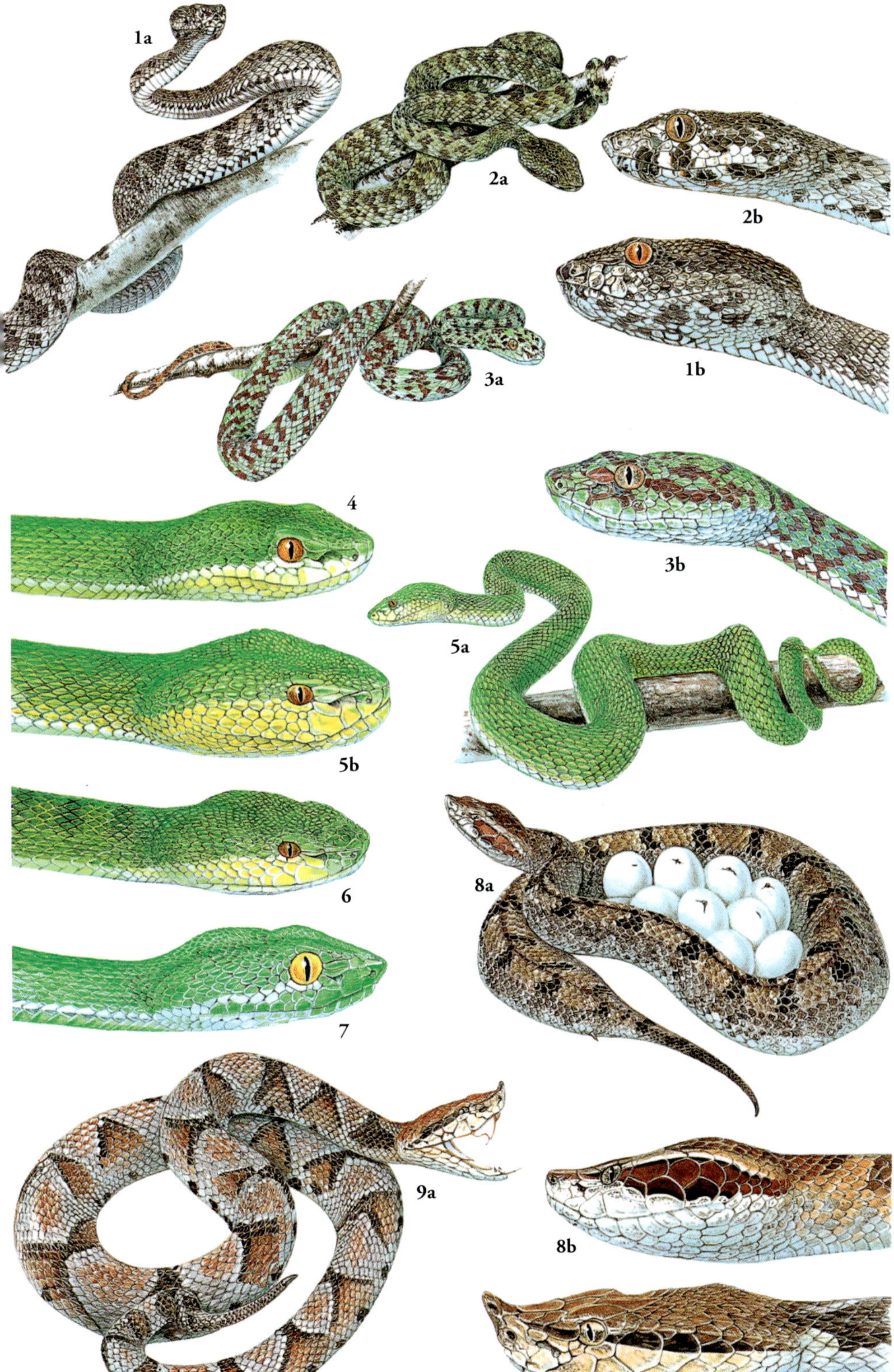

1a
2a
2b
1b
3a
4
3b
5a
5b
6
8a
7
9a
8b
9b

Plate 55. VIPERIDAE

1. KINABALU BROWN PIT VIPER *Garthius chaseni*, p. 305.
TL 690mm Body robust; head triangular; dorsals strongly keeled anteriorly, weakly keeled posteriorly.
(1a) Body Dorsum dark or reddish-brown with irregular dark brown blotches in paired rows anteriorly, joining to form bands towards posterior; smaller dark blotches on lower flanks.
(1b) Head Dark postocular stripe edged ventrally by line of cream or yellow scales, extends to neck.

2. MONTANE PIT VIPER *Ovophis monticola*, p. 305.
TL 1,250mm Body robust; head triangular; dorsals smooth or weakly keeled.
(2a) Body Dorsum brownish-grey or yellowish-pink with dark brown blotches; flanks with smaller dark brown spots.
(2b) Head Forehead dark; yellow, brown or pale tan postocular stripe reaches to neck.

3. MALAYAN BROWN PIT VIPER *Ovophis convictus*, p. 305.
TL 710mm Body robust; head triangular; dorsals keeled; dorsum mid-brown or brownish-yellow with subrectangular dark brown or black blotches; yellow, brown or pale tan postocular stripe reaches to neck.

4. TONKIN PIT VIPER *Ovophis tonkinensis*, p. 305.
TL 561mm Body robust; head triangular; dorsals smooth; dorsum grey, tan or pale brown with indistinct dark brown mottling and large black-edged dark blotches; on flanks, smaller brown blotches; yellow, brown or pale tan postocular stripe reaches to neck.

5. HAGEN'S GREEN PIT VIPER *Parias hageni*, p. 306.
TL 1,160mm Body robust; head distinct from neck; dorsals weakly keeled.
(5a) Body Dorsum bright or pale green; forehead and body scales not edged with black; pale line, bordered ventrally by dark line or series of dark spots, extends along dorsal rows 1–2.
(5b) Head Forehead scales not edged with black.
(5c) Juvenile Pale sides of head; sometimes, preocular and postocular stripes.

6. KINABALU GREEN PIT VIPER *Parias malcolmi*, p. 306.
TL 1,330mm Body robust; head narrow, distinct from neck; dorsals weakly keeled.
(6a) Body Dorsum green with indistinct black bands; juveniles with white flank stripe.
(6b) Head Forehead black with green-centred scales; edges of supralabials black.

7. SUMATRAN PIT VIPER *Parias sumatranus*, p. 306.
TL 1,355mm Body robust; head distinct from neck; dorsals weakly keeled.
(7a) Body Dorsum green with black cross-bars.
(7b) Head Forehead scales edged with black; supralabials light blue.
(7c) Juvenile Forehead scales not edged with black.

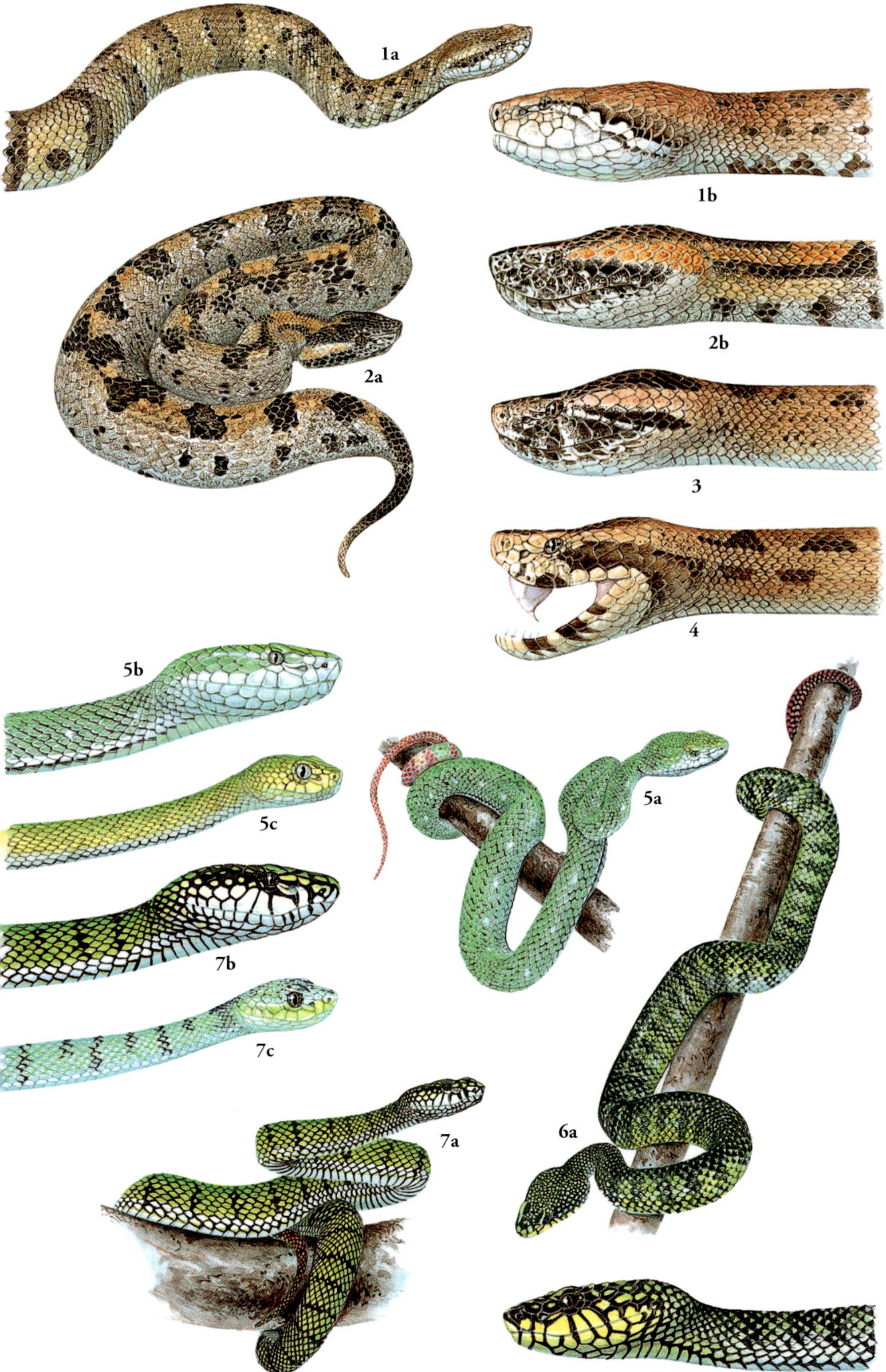
1a
1b
2a
2b
3
4
5b
5a
5c
7b
7c
7a
6a
6b

Plate 56. VIPERIDAE

1. SUMATRAN GREEN PIT VIPER *Popeia barati*, p. 306.
TL 729mm Body slender in males, robust in females; head distinct from neck; dorsum green, lacking cross-bars.
(1a) Body Ventrolateral stripe in males, absent in females (illustrated), stripes reddish-brown below, white above.
(1b) Head Green, lacking postocular streak.

2. PULAU TIOMAN PIT VIPER *Popeia buniana*, p. 306.
TL 783mm Body slender; head triangular.
(2a) Body Dorsum pale turquoise with spots sometimes arranged to form 81 transverse maroon bands on body and 39 brown bands on tail.
(2b) Head Maroon postorbital stripe; forehead darker.

3. THAI PENINSULA PIT VIPER *Popeia fucata*, p. 306.
TL 868mm Body robust; head triangular; dorsum green with rusty or reddish-brown cross-bars; sometimes, a white or white and reddish-brown postocular streak and white dots on vertebral region of adult males; bright ventrolateral stripe, orange or red ventrally, white dorsally in males; thin white ventrolateral stripe in females.

4. CAMERON HIGHLANDS PIT VIPER *Popeia nebularis*, p. 307.
TL 1,002mm Body robust; head distinct from neck; dorsals keeled; dorsum bright green with hint of blue; supralabials bluish-green; ventrolateral stripe typically absent.

5. POPES'S PIT VIPER *Popeia popeiorum*, p. 307.
TL 1,050mm Body slender in males, robust in females; head distinct from neck; dorsals smooth or weakly keeled.
(5a) Body Dorsum bright green or bluish-green, lacking cross-bars; bicoloured ventrolateral stripe, red ventrally, white dorsally in males (illustrated); stripe white in females.
(5b) Head Postocular streak narrow and white ventrally, broad and red dorsally in males; postocular streak narrow and white or absent in females (illustrated).

6. SABAH GREEN PIT VIPER *Popeia sabahi*, p. 307.
TL 810mm Body slender in males, robust in females; head triangular; dorsals keeled; dorsum bright green, lacking cross-bars or postocular streak; ventrolateral stripe red or rusty-red in males, white or yellow in females; iris red or orange in males and females, orange or yellowish-green in juveniles.

7. FAN-SI-PAN HORNED PIT VIPER *Protobothrops cornutus*, p. 307.
TL 696mm Body moderate; head distinct from neck; horn-like enlarged supraocular scale; dorsals with single unserrated keel; dorsum greyish-brown with series of 48–51 brown blotches or cross-bars; lateral series of greyish-brown spots without dark borders.

8. KAULBACK'S PIT VIPER *Protobothrops kaulbacki*, p. 308.
TL 1,340mm Body slender; head elongated, massive; dorsum yellowish-grey or olive-green with large, rhombohedral dark blotches that may be fused; small spots on flanks; yellow lines on head.

9. JERDON'S PIT VIPER *Protobothrops jerdonii*, p. 308.
TL 1,090mm Body slender; head distinct from neck; dorsals strongly keeled.
(9a) *P. j. bourreti* TL 1,090mm; dorsum yellowish-green or olive with black bands; forehead black with large A-shaped yellow mark; ventrals 189–192; subcaudals 65–72.
(9b) *P. j. jerdoni* TL 990mm; dorsum greenish-yellow or olive, with reddish-brown blotches edged with black; forehead black with fine yellow lines; ventrals 164–193; subcaudals 44–78.

10. BROWN-SPOTTED PIT VIPER *Protobothrops mucrosquamatus*, p. 308.
TL 1,174mm Body slender; head distinct from neck; scales strongly keeled posteriorly.
(10a) Body Dorsum greyish-olive or brown with series of large brown spots with dark edges; series of dark brown spots on flanks.
(10b) Head Dark postocular streak present.

11. THREE-HORNED PIT VIPER *Triceratolepidophis sieversorum*, p. 308.
TL 1,255mm Body slender; head subtriangular; raised horn-like multiple supraoculars; dorsal scales with 3 keels; dorsum greyish-brown with darker blotches edged finely with yellow; flanks with smaller blotches; dark brown postocular stripe; venter yellowish-brown to beige, partially spotted with grey-brown.

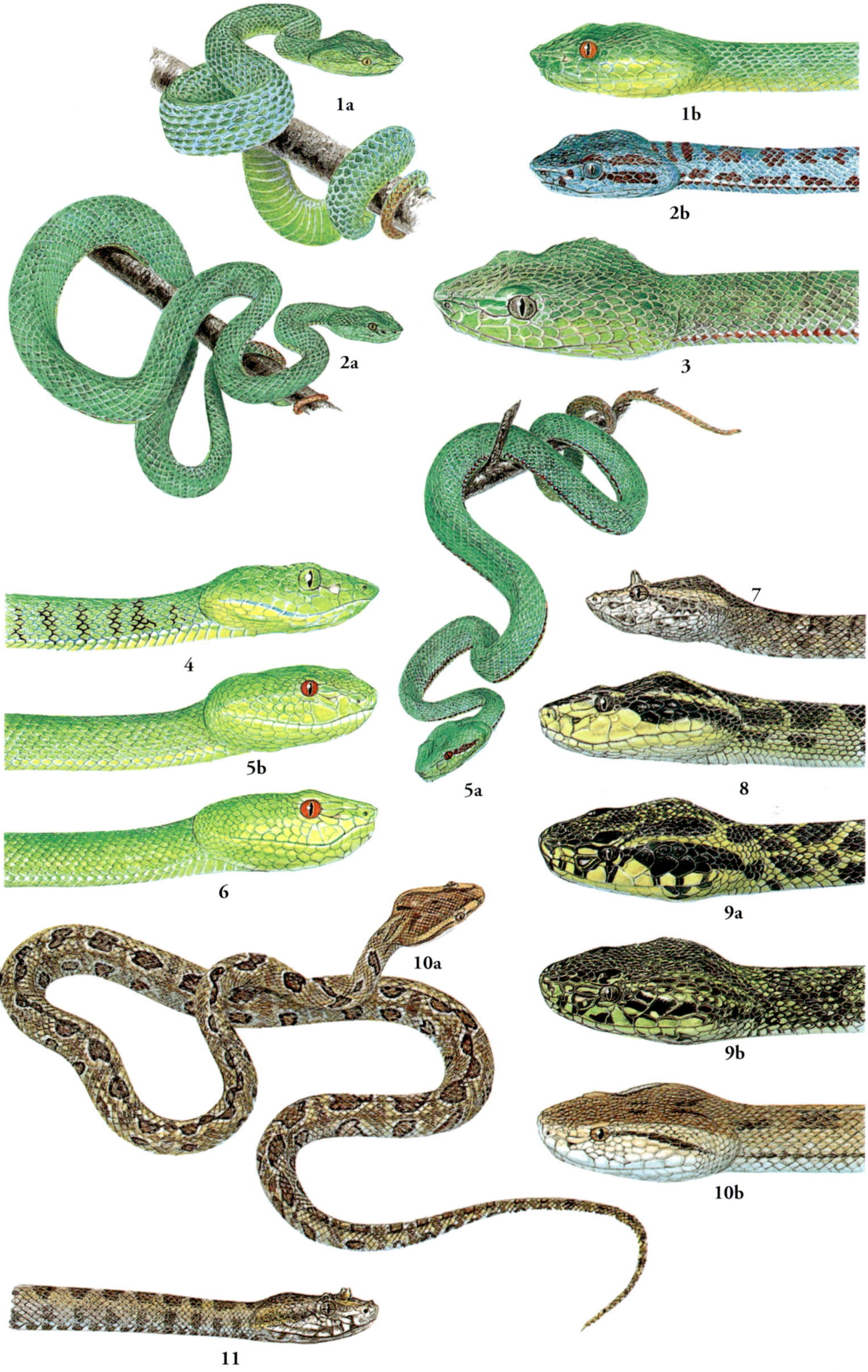
1a
1b
2b
2a
3
4
5b
5a
6
7
8
9a
9b
10a
10b
11

Plate 57. VIPERIDAE

1. SUMATRAN PALM PIT VIPER *Trimeresurus andalasensis*, p. 309.
TL 809mm Body robust; head triangular; dorsals smooth or weakly keeled.
(1a) Body Dorsum purplish-brown with 17–25 pale grey cross-bar.
(1b) Head Dark purplish-brown.

2. WIROT'S PALM PIT VIPER *Trimeresurus wiroti*, p. 310.
TL 889mm Body slender in males, stout in females; head distinct from neck; dorsals moderately keeled or smooth.
(2a) Body Dorsum dark greyish-brown in males (illustrated), dark brown in females, with 22–35 dark cross-bands; dark brown dorsolateral blotches.
(2b) Head Pale postocular bar.

3. BRONGERSMA'S PALM PIT VIPER *Trimeresurus brongersmai*, p. 309.
TL 410mm Body slender; head triangular; 4–5 narrow supraoculars, strongly erect and divergent; dorsum dark greyish-brown with 25–30 dark cross-bars, light centrally and darker on edges; smaller subrectangular or subtriangular dark greyish-brown blotches on flanks.

4. TRUONG SON PIT VIPER *Trimeresurus truongsonensis*, p. 309.
TL 642mm Body slender; head subtriangular; dorsals strongly keeled, first row with weak keel; dorsum greenish-blue in males (illustrated), light brown in females, with 66–70 reddish-brown cross-bars; pale ventrolateral stripe with reddish-brown ventral edge; reddish-brown postocular stripe.

5. BORNEAN PALM PIT VIPER *Trimeresurus borneensis*, p. 309.
TL 830mm Body robust; head triangular; dorsals smooth or weakly keeled.
(5a) Body Dorsum mottled light brown, medium brown with saddle-like dark brown pattern comprising 20–30 blotches or cross-bars, to bright yellow with darker mottling.
(5b) Head Oblique pale postocular stripe to neck.

6. JAVANESE PALM PIT VIPER *Trimeresurus puniceus*, p. 309.
TL 920mm Body slender in males, stout in females; head distinct from neck; dorsals smooth or weakly keeled.
(6a) Body Dorsum grey or yellowish-brown with 20–30 dark cross-bands, which may be complete or incomplete.
(6b) Head No postocular stripe in adults.

7. BORNEAN KEELED GREEN PIT VIPER *Tropidolaemus subannulatus*, p. 310.
TL 963mm Body slender in juveniles, relatively thick in adults; head distinct from neck; dorsals feebly keeled in males, distinctly keeled in females.
(7a) Body Dorsum green or greenish-blue with dark cross-bars in females; white spots, or white and red spots or stripes on green dorsum in males and juveniles (illustrated).
(7b) Adult female head Cream or yellow area below dark postocular stripe.
(7c) Adult male/juvenile head White and red stripes.

8. WAGLER'S KEELED GREEN PIT VIPER *Tropidolaemus wagleri*, p. 310.
TL 920mm Body slender in juveniles and males, robust in adult females; head triangular; dorsals feebly keeled in males, distinctly keeled in females.
(8a) Adult female body Dorsum black with yellow cross-bars (yellow with white cross-bars in juvenile females).
(8b) Head Black postocular stripe.
(8c) Adult male/juvenile body Green with white spots.

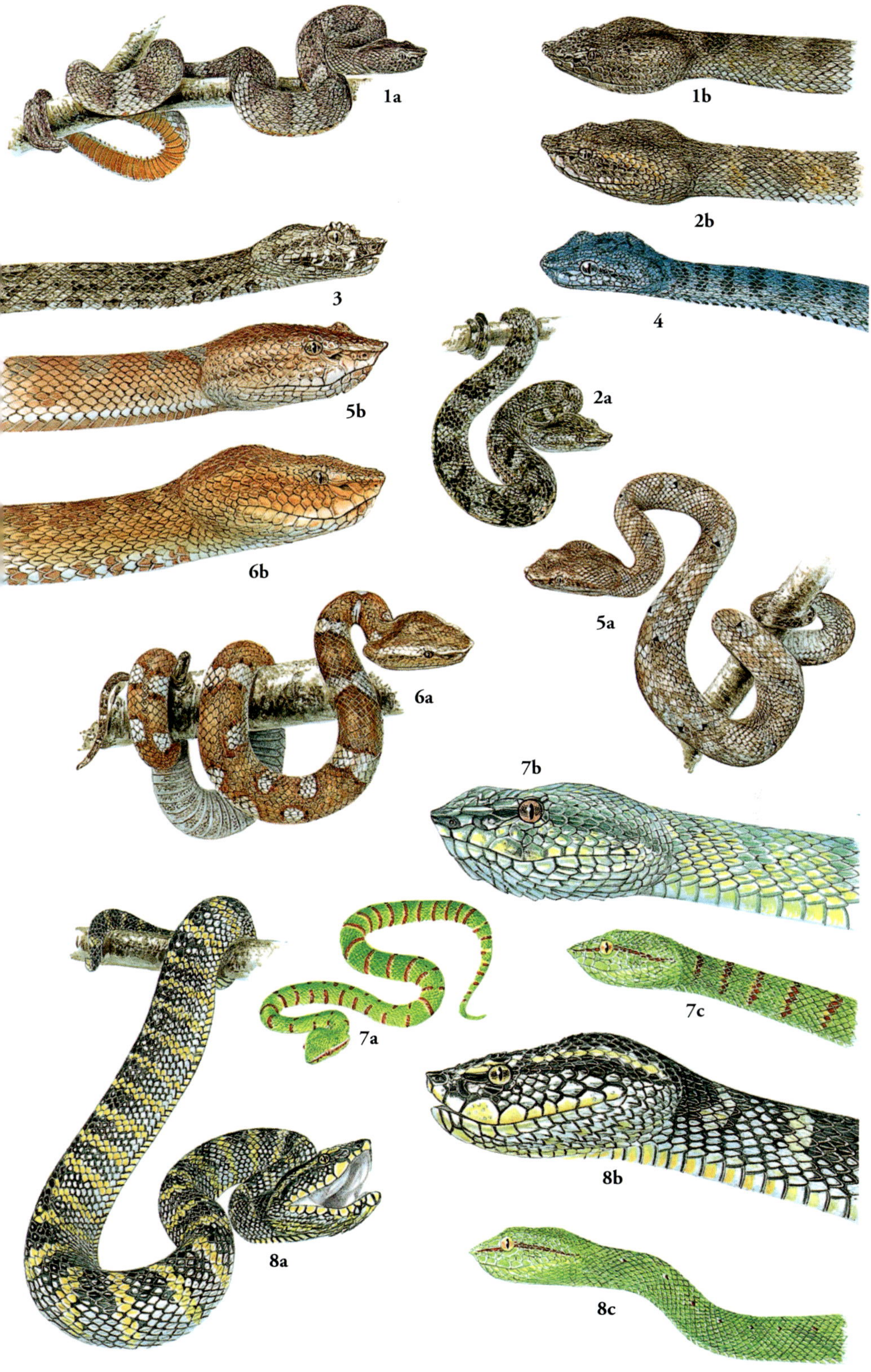
1a
1b
2b
3
4
5b
2a
6b
5a
6a
7b
7c
7a
8b
8a
8c

Plate 58. VIPERIDAE

1. GUMPRECHT'S PIT VIPER *Viridovipera gumprechti*, p. 310.
TL 1,280mm Body slender, elongated, triangular; head distinct from neck; dorsals moderately keeled.
(1a) Male Dorsum bright pale green; 3 small white vertebral spots at posterior of body; interstitial skin black; bicolored postocular streak, red dorsally, white ventrally; ventrolateral stripe white dorsally, red ventrally.
(1b) Female Dorsum dark green; 3 small white vertebral spots at posterior of body; interstitial skin black; postocular area pale; ventrolateral stripe white or blue.

2. MEDO PIT VIPER *Viridovipera medoensis*, p. 311.
TL 677mm Body slender, elongated, cylindrical; head distinct from neck; dorsals with obtuse keels on vertebral region and flanks.
(2a) Body Dorsum dark green, sometimes edged with turquoise-blue, with bicoloured white and red ventrolateral stripe.
(2b) Head Green without postocular stripe; labials yellow.

3. STEJNEGER'S PIT VIPER *Viridovipera stejnegeri*, p. 311.
TL 1,120mm Body slender, elongated, triangular; head distinct from neck; dorsals weakly keeled.
(3a) Male body Dorsum bright green; interstitial skin dark grey or greyish-brown; venter pale green; ventrolateral stripe orange, brown or red ventrally, white dorsally.
(3b) Male head and neck Red ventrolateral stripe.
(3c) Female head and neck No red ventrolateral stripe.

4. VOGEL'S PIT VIPER *Viridovipera vogeli*, p. 311.
TL 1,120mm Body slender, elongated; head distinct from neck; dorsals smooth or strongly keeled; dorsum dark or grass-green; ventrolateral line of adult males with lower red component, or mostly white; interstitial skin blue; males with white vertebral flecks and yellow or pale orange iris; females with yellowish-green ventrolateral stripe; venter yellowish-green.

5. YUNNAN PIT VIPER *Viridovipera yunnanensis*, p. 311.
TL 750mm Body robust; head distinct from neck; dorsum dark green; bicoloured ventrolateral stripe, ventrally orange or brown, and dorsally white, in males, and white only or absent in females, across outermost scale row and portion of second row; iris red in males; golden yellow in females (illustrated).

6. FEA'S VIPER *Azemiops feae*, p. 311.
TL 925mm Body robust, cylindrical; sensory pit absent; head distinct from neck; forehead covered with large symmetrical shields; nostril large; dorsals smooth.
(6a) Body Dorsum blackish-brown with 14–15 narrow white, red or pink cross-bars, sometimes interrupted middorsally, or alternating laterally.
(6b) Head Forehead yellowish-orange, sometimes with pair of dark brown to black stripes from prefrontals to end of neck.

7. RUSSELL'S VIPER *Daboia russelii*, p. 312.
TL 1,850mm Body robust; head distinct from neck; forehead with small scales; nostril enlarged; all except outermost row of midbody scale rows strongly keeled.
(7a) Adult body Body pale brown with 5–7 rows of large dark-margined blotches along dorsum, and smaller ones on flanks.
(7b) Juvenile body Slender.
(7c) Head Forehead with small scales; nostril enlarged, located in large nasal shield.

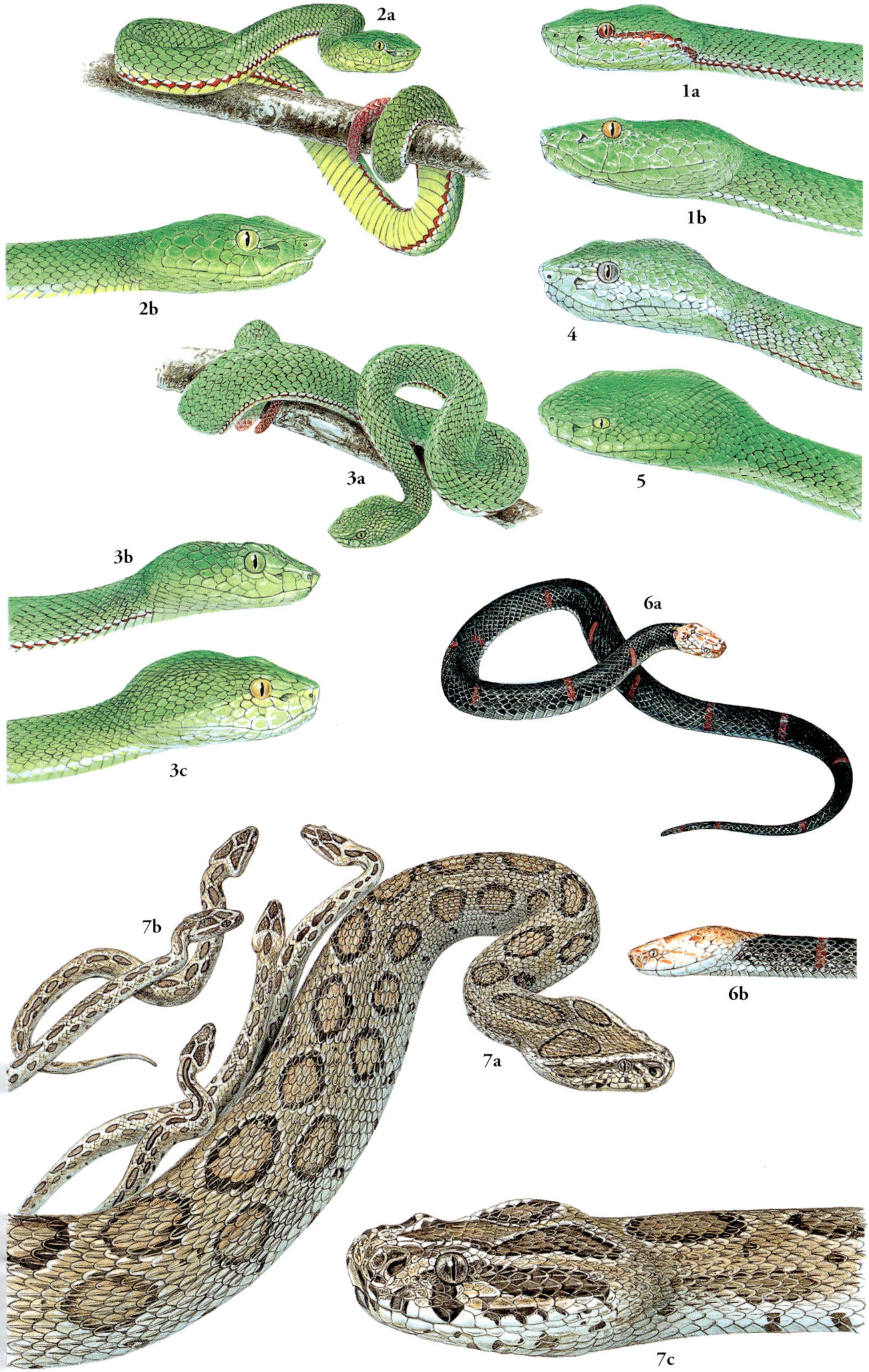
2a
1a
1b
2b
4
3a
5
3b
6a
3c
7b
6b
7a
7c

Plate 59. CYLINDROPHIIDAE & ELAPIDAE

1. ENGKARI PIPE SNAKE *Cylindrophis engkariensis*, p. 312.
TL 485mm Body robust; head long; dorsals smooth; dorsum black; short light postocular streak; irregular rows of paravertebral spots; tail black with black and white mottling; venter with black and white cross-bars, divided in midline.

2. LINED PIPE SNAKE *Cylindrophis lineatus*, p. 312.
TL 982mm Body robust; head long, blunt; dorsals smooth; dorsum yellowish-white; series of longitudinal red or yellow stripes from back of head to base of tail; forehead sometimes with scattered dark spots; venter with black cross-bars.

3. COMMON PIPE SNAKE *Cylindrophis ruffus*, p. 312.
TL 900mm Body robust; head short, blunt; dorsals smooth; dorsum iridescent dark brown, grey or black with pale yellow or cream collar and cream bands; venter with black cross-bars.

4. HIMALAYAN KRAIT *Bungarus bungaroides*, p. 313.
TL 1,400mm Body robust; dorsals smooth; midbody scale rows 15; dorsum black, blue-black or dark brown with transverse yellowish-cream lines or narrow bars; white stripe across snout, another across nape and a third from eye to end of jaws.

5. MALAYAN KRAIT *Bungarus candidus*, p. 313.
TL 1,600mm Body robust; dorsals smooth; midbody scale rows 15 (rarely 17); dorsum black or bluish-black with 20–35 white cross-bars on body, sometimes all black; sometimes indistinct light chevron on nuchal region.
(**5a**) **Banded morph A** From Thailand.
(**5b**) **Banded morph B** From Vietnam.
(**5c**) **Melanistic morph** From Java and Bali.

6. BANDED KRAIT *Bungarus fasciatus*, p. 313.
TL 2.25m Body robust; dorsals smooth; midbody scale rows 15; dorsum yellow or pale brown with black bands; forehead with V-shaped pale marking.

7. RED-HEADED KRAIT *Bungarus flaviceps*, p. 313.
TL 2.07m Body robust; dorsals smooth; midbody scale rows 13.
(**7a**) *B. f. flaviceps* Head reddish-orange; body without black rings.
(**7b**) *B. f. baluensis* Head red (rarely yellow); posterior half of body and tail with 5–8 pairs of black rings, separated by narrow white ring.

8. SPLENDID KRAIT *Bungarus magnimaculatus*, p. 314.
TL 1,300mm Body robust; dorsals smooth; midbody scale rows 15; dorsum black or bluish-black with 11–14 white cross-bars; white preocular spot.

9. MANY-BANDED KRAIT *Bungarus multicinctus*, p. 314.
TL 1,354mm Body robust; dorsals smooth; midbody scale rows 15; dorsum jet-black, dark brown or bluish-black with 27–44 light cross-bars on body.

10. RED RIVER KRAIT *Bungarus slowinskii*, p. 314.
TL ca 1,350mm Body robust; dorsals smooth, midbody scale rows 15; dorsum iridescent blue-black; incomplete V-shaped nuchal mark; incomplete narrow light stripe across snout; dorsum with 20–33 pale rings.

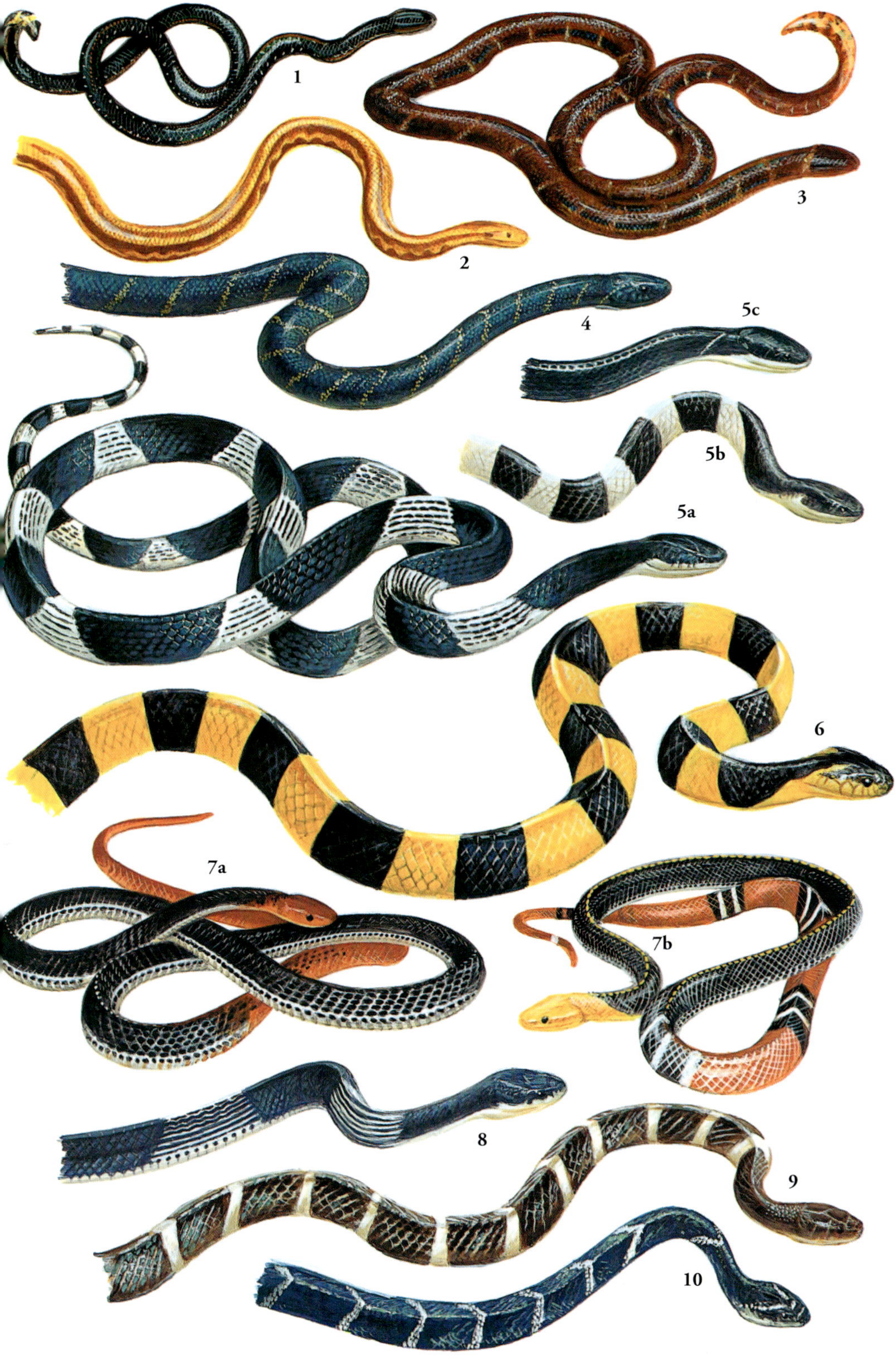
1
3
2
4
5c
5b
5a
6
7a
7b
8
9
10

Plate 60. ELAPIDAE

1. BLUE CORAL SNAKE *Calliophis bivirgatus*, p. 314.
TL 1,850mm Body slender; dorsals smooth; dorsum dark blue to blue-black, some populations with distinct stripes along flanks; head, tail and venter coral-red.
(1a) *C. b. bivirgatus* No pale flank stripe; sometimes a narrow white paravertebral stripe; head, tail and venter coral-red or orange.
(1b) *C. b. flaviceps* Pale blue flank stripe; no white paravertebral stripe.
(1c) *C. b. tetrataenia* Cream paravertebral and flank stripes, outer broadest.

2. SPOTTED CORAL SNAKE *Calliophis gracilis*, p. 315.
TL 740mm Body slender; dorsals smooth; dorsum pale brown, reddish-brown or greyish-brown with narrow black vertebral stripe; paired oval black marks, fused on nape and tail; black spots on flanks.

3. MALAYAN STRIPED CORAL SNAKE *Calliophis intestinalis*, p. 315.
TL 710mm Body slender; dorsals smooth; dorsum reddish-brown, dark brown or black; flanks with white, tan or red stripes.
(3a) *C. i. intestinalis* Dorsum black; narrow yellow or white lines on vertebral region and lower flanks; venter with broad, transverse black and white bands, black cross-bands in contact with black on sides; forehead black with Y-shaped cream mark.
(3b–d) *C. i. lineata* Dorsum greyish-brown; narrow white lines on vertebral region and on lower flanks, edged with black, extend to tail-tip; lower flanks with narrow white line edged with black; venter with broad, transverse black and white bands, black cross-bands not in contact with black on sides; forehead brown mottled with black; subcaudals orange with 2 narrow bars. Three colour morphs depicted:
(3b) Representative of population from Bukit Fraser, Peninsular Malaysia.
(3c) Representative of population from Sumatra, Indonesia.
(3d) Representative of population from Kernanam, Terengganu, east coast of Peninsular Malaysia.
(3e) *C. i. thepassi* Dorsum brown; broad black vertebral stripe extends to tail; dorsolateral stripes 2 scales wide; venter with broad, transverse black and white bands, the black cross-bands not in contact with black on sides; forehead rufous-brown.

4. SPECKLED CORAL SNAKE *Calliophis maculiceps*, p. 315.
TL 480mm Body slender; dorsals smooth; head and nape black with yellow occipital spot. Three colour morphs depicted:
(4a) Dorsum brownish-yellow with black vertebral stripe; no black spots in longitudinal series along each side; pale band across forehead.
(4b) Dorsum brownish-yellow without black vertebral stripe; black spots in longitudinal series along each side; no pale band across forehead.
(4c) Dorsum reddish-brown without black vertebral stripe; black spots in longitudinal series along each side; no pale band across forehead.

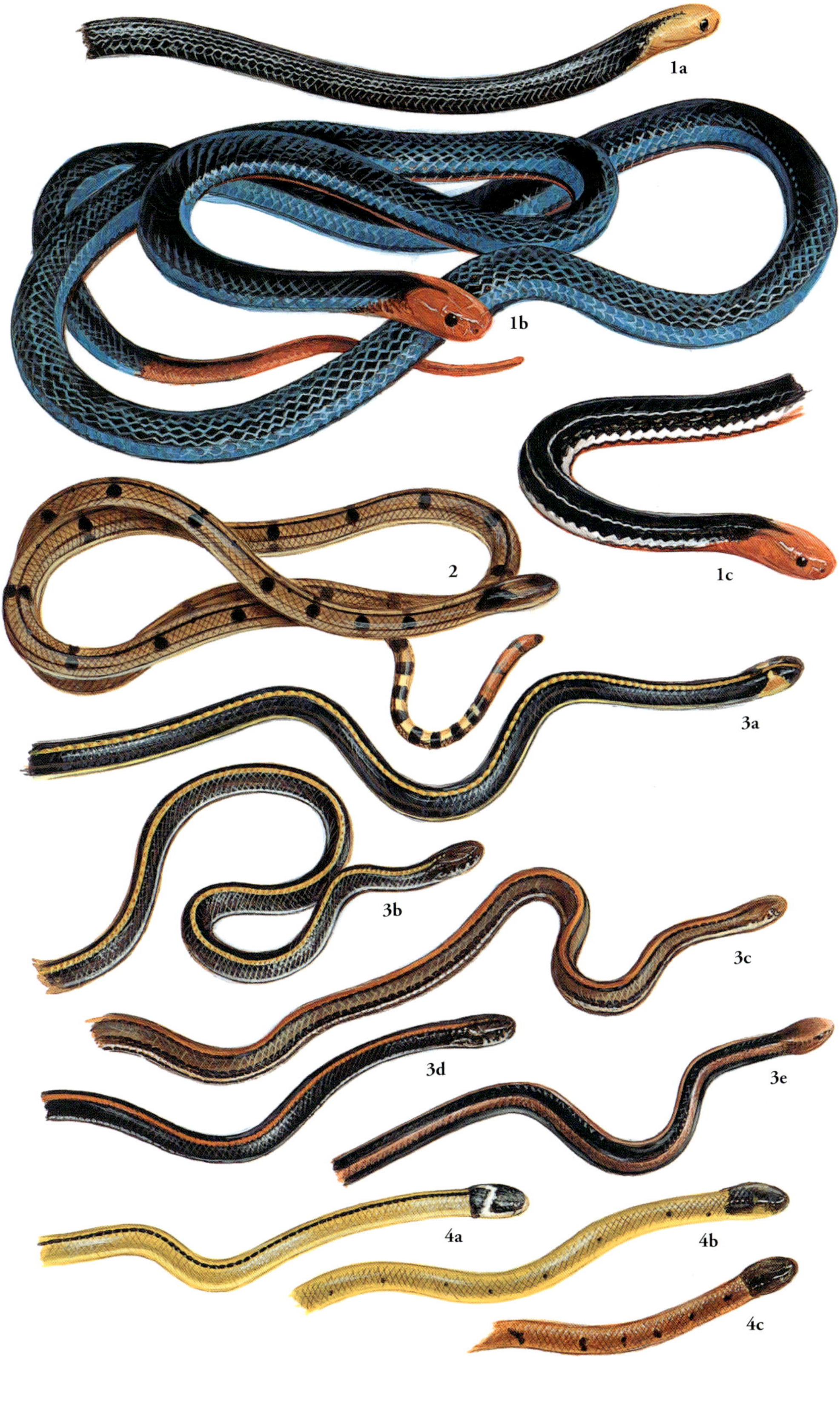
1a
1b
1c
2
3a
3b
3c
3d
3e
4a
4b
4c

Plate 61. ELAPIDAE

1. CHINESE COBRA *Naja atra*, p. 316.
TL 1,650mm Body moderate; head large; hood short.
(1a) Body Dorsum black, grey or brown, sometimes with narrow white cross-bars, split into double or quadruple bands.
(1b) Hood Mark consists of a mask, a monocle, a horseshoe or an O-shape.

2. MONOCLED COBRA *Naja kaouthia*, p. 316.
TL 2.3m Body robust; head large; hood rounded.
(2a) Body Dorsum brown, greyish-brown, blackish-brown or pale yellow; some with darker bands.
(2b) Hood Mark consists of a light circle or mask-shape with a dark centre; 1–2 dark spots sometimes present in pale oval portion.

3. MANDALAY COBRA *Naja mandalayensis*, p. 316.
TL 1,400mm Body robust; head large; hood oval-elongate.
(3a) Body Dorsum mid-brown to dark brown with pale interstitial skin; light cross-bars.
(3b) Hood Mark consists of an indistinct spectacle, especially in juveniles, or hood is unpatterned.

4. INDO-CHINESE SPITTING COBRA *Naja siamensis*, p. 316.
TL 1,600mm Body robust; head large; hood oval.
(4a) Body Dorsum contrasting black and white pattern, to black or grey with white speckling; dark dorsum may be interrupted by light cross-bars.
(4b) Hood Marking absent, or U-, V- or H-shaped.

5. EQUITORIAL SPITTING COBRA *Naja sputatrix*, p. 316.
TL 1,500mm Body robust; head large; hood elongated; dorsum blackish-grey, silvery or brown; hood pattern chevron-shaped, or occasionally mask-, horseshoe- or spectacle-shaped, or unpatterned.

6. SUMATRAN COBRA *Naja sumatrana*, p. 317.
TL 1,500mm Body robust; head large; hood rounded in adults, more elongated in juveniles; dorsum bluish-black.
(6a) Adult hood Unpatterned dark brown or bluish-black; or with narrow pale cross-bars.
(6b) Juvenile hood Dark brown or bluish-black, unbanded or with 12 narrow pale cross-bars.

7. KING COBRA *Ophiophagus hannah*, p. 317.
TL 5.85m Body robust in adults, slender in juveniles; head large; hood elongated; juveniles with 27–84 yellow bands.
(7a) Adult From a population showing bands.
(7b) Adult From a population nearly unpatterned.
(7c) Hatchling head Forehead unpatterned.
(7d) Juvenile head Forehead with distinct pale bands.

8. KELLOG'S CORAL SNAKE *Sinomicrurus kelloggi*, p. 317.
TL 800mm Body slender; head short and rounded; dorsum reddish-brown with 17 dark cross-bars on body; forehead black with pale crescentic mark on snout; chevron at back of head.

9. MacCLELLAND'S CORAL SNAKE *Sinomicrurus macclellandi*, p. 317.
TL 840mm Body slender, cylindrical; head short and rounded; dorsum reddish-brown with 23–40 yellow or pale brown-edged black stripes; head black with cream or yellow band behind eyes.

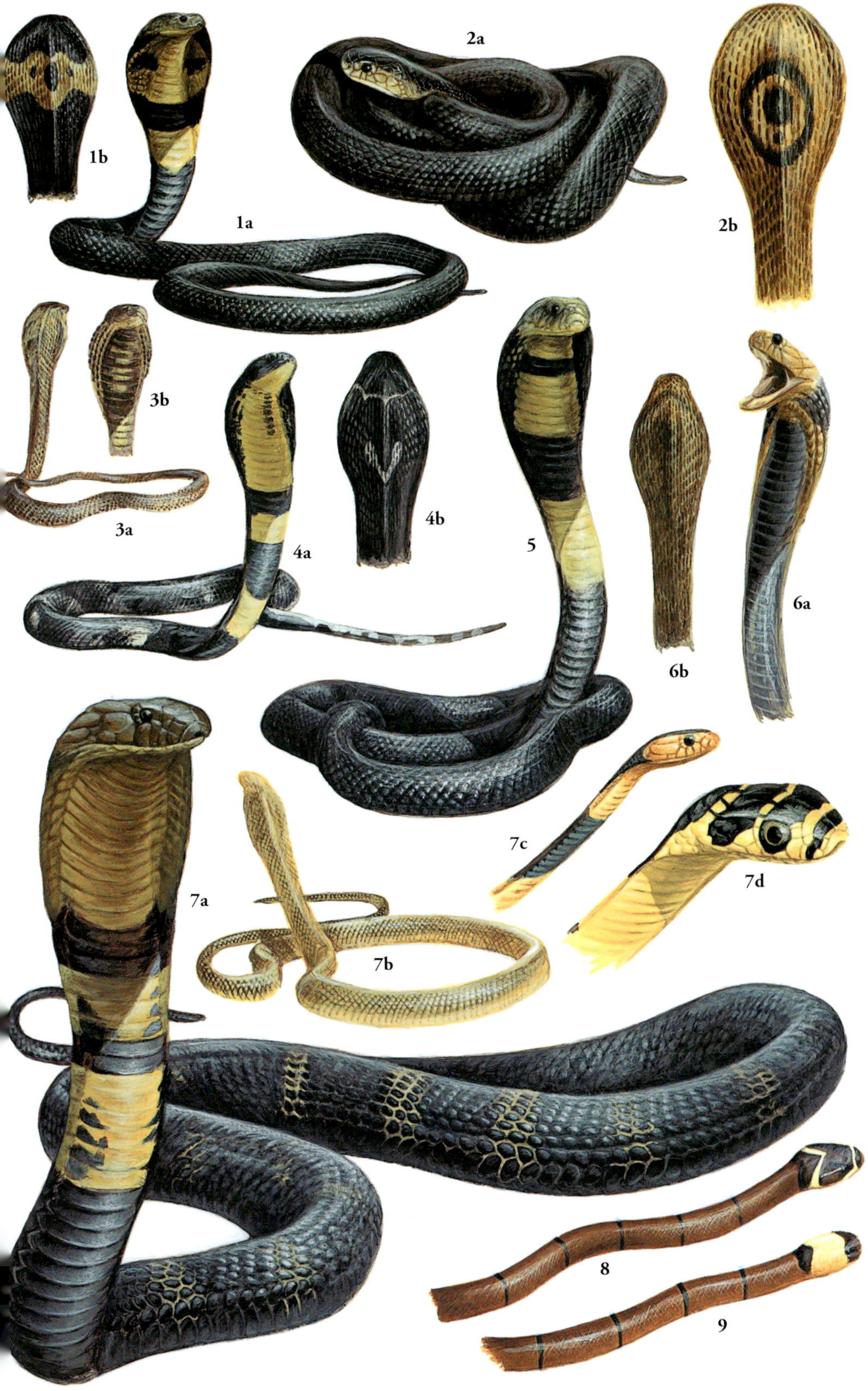
1a
1b
2a
2b
3a
3b
4a
4b
5
6a
6b
7a
7b
7c
7d
8
9

Plate 62. ELAPIDAE

1. HORNED SEA SNAKE *Acalyptophis peronii*, p. 318.
TL 1,250mm Body robust; head small, shields symmetrical, some with spines on posterior edges; dorsals keeled.
(1a) Body Dorsum light brown with dark bands encircling body, widest on vertebral region.
(1b) Head Forehead pale brown.

2. BEADED SEA SNAKE *Aipysurus eydouxii*, p. 318.
TL 1,500mm Body slender; head shields symmetrical, some with spines on posterior edges; dorsals smooth.
(2a) Body Dorsum brownish-olive with 44–55 tan or yellowish-cream bands, which may be broken up on vertebral region.
(2b) Head Forehead dark brown or olive.

3. STOKES'S SEA SNAKE *Astrotia stokesii*, p. 318.
TL 1,800mm Body short, robust; head enlarged; dorsals imbricate and keeled; dorsum yellowish-grey or pale brown with 24–37 black annuli on body; head black, dark olivaceous or yellow.

4. BEAKED SEA SNAKE *Enhydrina schistosa*, p. 318.
TL 1,580mm Body robust; head narrow; dorsals keeled.
(3a) Body Dorsum greyish-olive or silvery-grey with indistinct darker markings forming 40–60 transverse dark bands.
(3b) Head Rostral scale extends ventrally to form 'beak'; forehead dark.

5. AAGAARD'S SEA SNAKE *Hydrophis aagaardi*, p. 318.
TL 1,030mm Body moderately robust; head elongated; dorsals keeled, tuberculate; dorsum greyish-yellow or greenish-grey with 48–76 dark olive or black bands.

6. BLACK-HEADED SEA SNAKE *Hydrophis atriceps*, p. 319.
TL 1,200mm Body robust posteriorly; head very small; slender neck and forebody; dorsals with central keel; dorsum dark olive to black with yellow spots on sides, sometimes connected to form 50–75 cross-bars; posteriorly grey; yellow spot between nostril and eye or behind eye.

7. CAPTAIN BELCHER'S SEA SNAKE *Hydrophis belcheri*, p. 319.
TL 932mm Body robust to moderate; dorsals imbricate and hexagonal or rounded, keeled; dorsum olive-green with 48–64 dark bands, broad dorsally and narrow on flanks; head black flecked with olive.

8. RAJAH BROOK'S SEA SNAKE *Hydrophis brookii*, p. 319.
TL 1,035mm Body robust; head small; dorsals keeled; dorsum blackish-grey with 60–80 dark grey cross-bars, twice broader than pale interspaces; head greyish-black with curved yellow mark across snout.

9. BLUE-GREY SEA SNAKE *Hydrophis caerulescens*, p. 319.
TL 1,090mm Body moderately robust; head small; dorsals keeled.
(9a) Body Dorsum bluish-white or bluish-grey with 40–60 black bands, narrow on lower flanks.
(9b) Head Dark, upper jaw projecting.

10. ANNULATED SEA SNAKE *Hydrophis cyanocinctus*, p. 320.
TL 1,885mm Body elongated, moderately robust anteriorly, thickening posteriorly; dorsals strongly keeled.
(10a) Body Dorsum olive or yellow with numerous transverse bluish-black bands encircling body.
(10b) Head Small, yellowish-green.

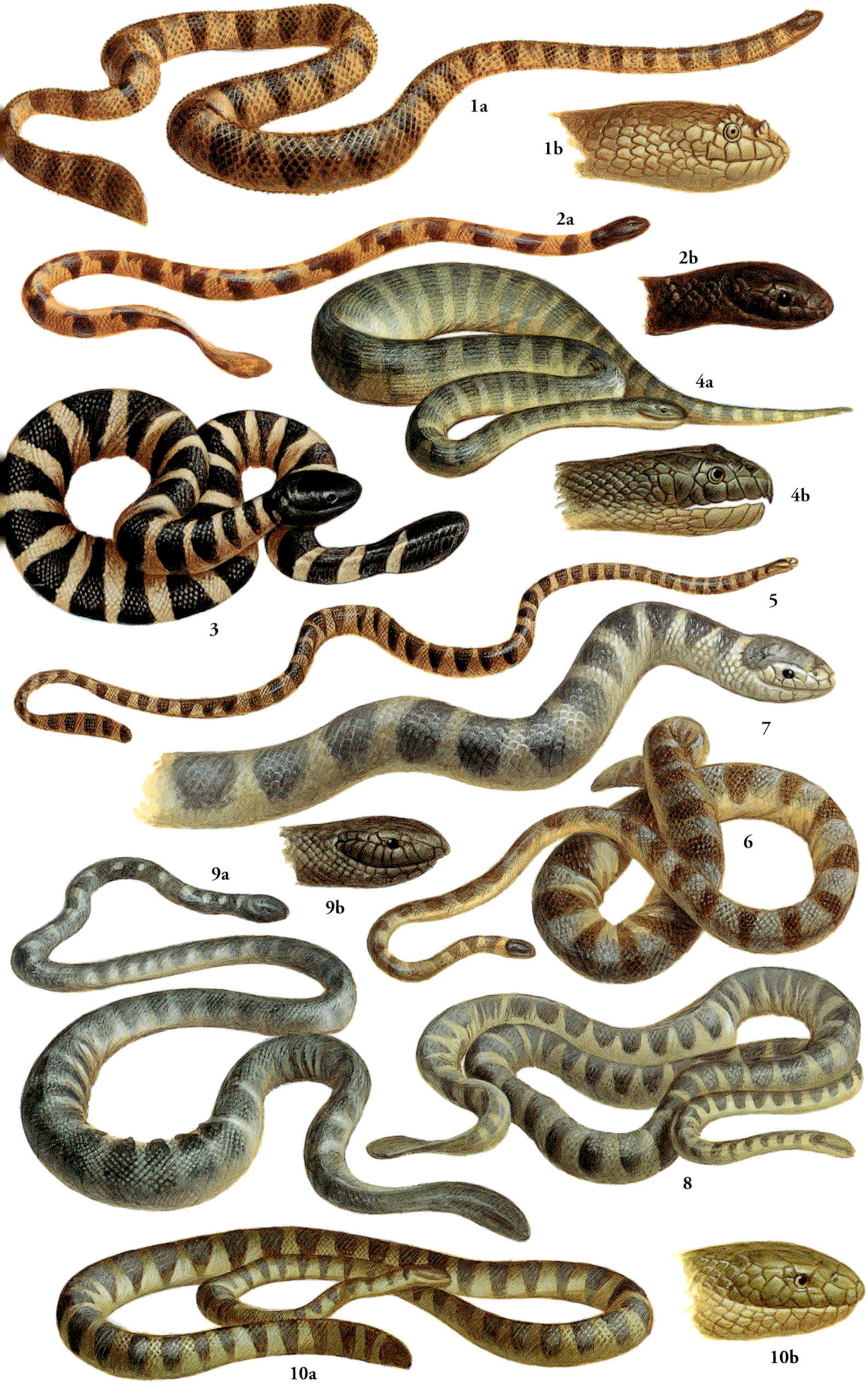

1a
1b
2a
2b
4a
4b
3
5
7
6
9a
9b
8
10a
10b

Plate 63. ELAPIDAE

1. BANDED SEA SNAKE *Hydrophis fasciatus*, p. 320.
TL 1,100mm Body slender anteriorly, thickening posteriorly; head small; dorsals juxtaposed or slightly imbricate and keeled; dorsum beige, dark olive to black with oval pale yellow spots on flanks, sometimes connected as cross-bars; posteriorly more grey in some individuals; rhomboidal dark spots along flanks may form annuli in juveniles.

2. NARROW-HEADED SEA SNAKE *Hydrophis gracilis*, p. 320.
TL 950mm Body slender anteriorly, thickening posteriorly; dorsals keeled.
(2a) Body Dorsum bluish-grey with 40–60 dark blue-black bands or lateral blotches.
(2b) Head Blue-black; upper jaw projecting.

3. KLOSS'S SEA SNAKE *Hydrophis klossi*, p. 320.
TL 1,190mm Body robust; head small; dorsals smooth or weakly keeled; dorsum blackish-grey to olivaceus-yellow with 50–75 dark cross-bars; bars broadest dorsally and broader than their interspaces; black vertebral stripe sometimes present; forehead black or olivaceous.

4. PERSIAN GULF SEA SNAKE *Hydrophis lapemoides*, p. 321.
TL 960mm Body robust; head moderate; dorsals weakly tuberculate or with short keel; dorsum grey with 29–52 rhombic black spots, wider on vertebral region and narrower on flanks; head dark dorsally with curved yellow or white mark from forehead to back.

5. LESSER DUSKY SEA SNAKE *Hydrophis melanosoma*, p. 321.
TL 1,390mm Body robust; head moderate; dorsals strongly keeled; dorsum blackish-grey with 50–70 wide white or yellow bands, subequal to their interspaces; head and body black with yellow subovoid marks; forehead with yellow speckles.

6. ORNATE SEA SNAKE *Hydrophis ornatus*, p. 321.
TL 1,150mm Body robust; head large; dorsals tuberculate or with short keel; dorsum greyish or olive to unpatterned cream, with 30–60 dark bars or rhomboidal spots separated by narrow interspaces on body.

7. BROAD-HEADED SEA SNAKE *Hydrophis pachycercos*, p. 322.
TL 1,110mm Body robust; head shields large and regular; dorsals keeled; dorsum light yellow with light brown bands; head black or dark grey; supralabials cream.

8. SIBAU RIVER SEA SNAKE *Hydrophis sibauensis*, p. 322.
TL 735mm Body slender; head narrower than body; dorsal scales with median keel; dorsum grey-brown, darkening posteriorly, with 49–58 yellow to light orange bands; forehead black with yellow spots and arrow-shaped marking.

9. SPIRAL SEA SNAKE *Hydrophis spiralis*, p. 322.
TL 2.75m Body moderately robust anteriorly, thickening posteriorly; head and neck slender; dorsals smooth or keeled; dorsum olive-yellow to olive-brown, with 35–50 dark bands narrower than interspaces; sometimes a black dorsal spot between bands; flanks yellowish-cream.

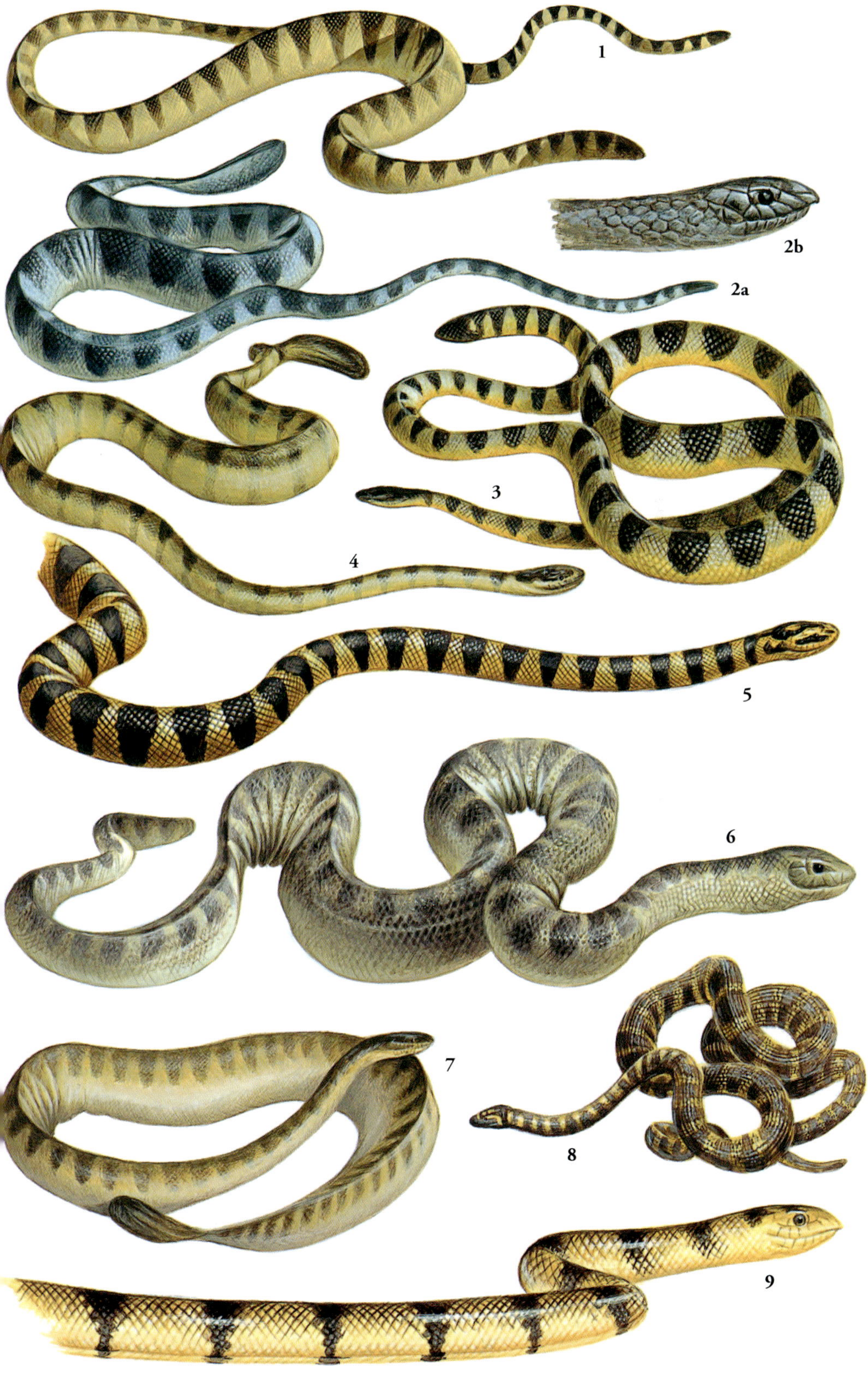
1
2b
2a
3
4
5
6
7
8
9

Plate 64. ELAPIDAE

1. NARROW-NECKED SEA SNAKE *Hydrophis stricticollis*, p. 322.
TL 1,050mm Body slender; head small; scales feebly imbricate or juxtaposed, and keeled; dorsum greyish-olive with 45–65 dark bands in juveniles, broadest dorsally and narrow on flanks; head black or olive; snout and sides of head with small yellow patches.

2. GARLANDED SEA SNAKE *Hydrophis torquatus*, p. 322.
TL 895mm Body moderately robust; head moderate; dorsals keeled; dorsum greyish- or greenish-grey with over 50 dark brown bands on body and tail; forehead dark greyish-tan or dark olive, sometimes with horseshoe-shaped yellow mark.

3. SADDLE-BACKED SEA SNAKE *Kerilia jerdoni*, p. 323.
TL 1,000mm Body robust; head short; dorsals keeled and imbricate; midbody scale rows 21–23 (rarely 19); ventrals 200–278.
(3a) Body Dorsum yellowish-olive with black dorsal spots or rhomboidal marks forming bands.
(3b) Head Forehead dark grey; sometimes, a black band across neck.

4. ANNANDALE'S SEA SNAKE *Kolpophis annandalei*, p. 323.
TL 520mm Body robust; head small, narrower than widest part of body; fragmented head scales; dorsals keeled; dorsum greyish-purple with 35–46 dark bands, with bands broader than pale interspaces; head dark greyish-purple.

5. SHORT SEA SNAKE *Lapemis curtus*, p. 323.
TL 972mm Body robust; head broad and short; scales squarish or hexagonal, lowermost rows especially in males with a short keel; dorsum in juveniles (illustrated) brownish-grey, yellow or olive with 35–55 olive to dark grey bands that taper to a point on flanks; adults unpatterned olive to dark grey.

6. YELLOW-LIPPED SEA KRAIT *Laticauda colubrina*, p. 323.
TL 1,710mm Body robust, especially in adult females; dorsals smooth; midbody scale rows 21–25.
(6a) Body Dorsum blue-grey with 24–64 black bands.
(6b) Head Snout yellow; supralabials yellow.

7. LARGE-SCALED SEA KRAIT *Laticauda laticaudata*, p. 323.
TL 1,100mm Body robust in adult females; dorsals smooth; midbody scale rows 19; dorsum bright blue with 20–70 black bands; snout tan; supralabials dark brown or black.

8. YELLOW-BELLIED SEA SNAKE *Pelamis platura*, p. 324.
TL 1,000mm Body slender; head elongated, bill-like and slightly flattened; top half of dorsum black or dark brown, bottom half bright yellow or cream; venter light brown or yellow with black spots or bars; tail with diamond-shaped cream or yellow pattern.

9. GREY SEA SNAKE *Thalassophis viperina*, p. 324.
TL 925mm Body robust; head short, depressed; forehead scales entire; dorsals keeled; dorsum unicoloured grey, or with lighter mottling or 25–35 dark cross-bars or spots; forehead grey or black.

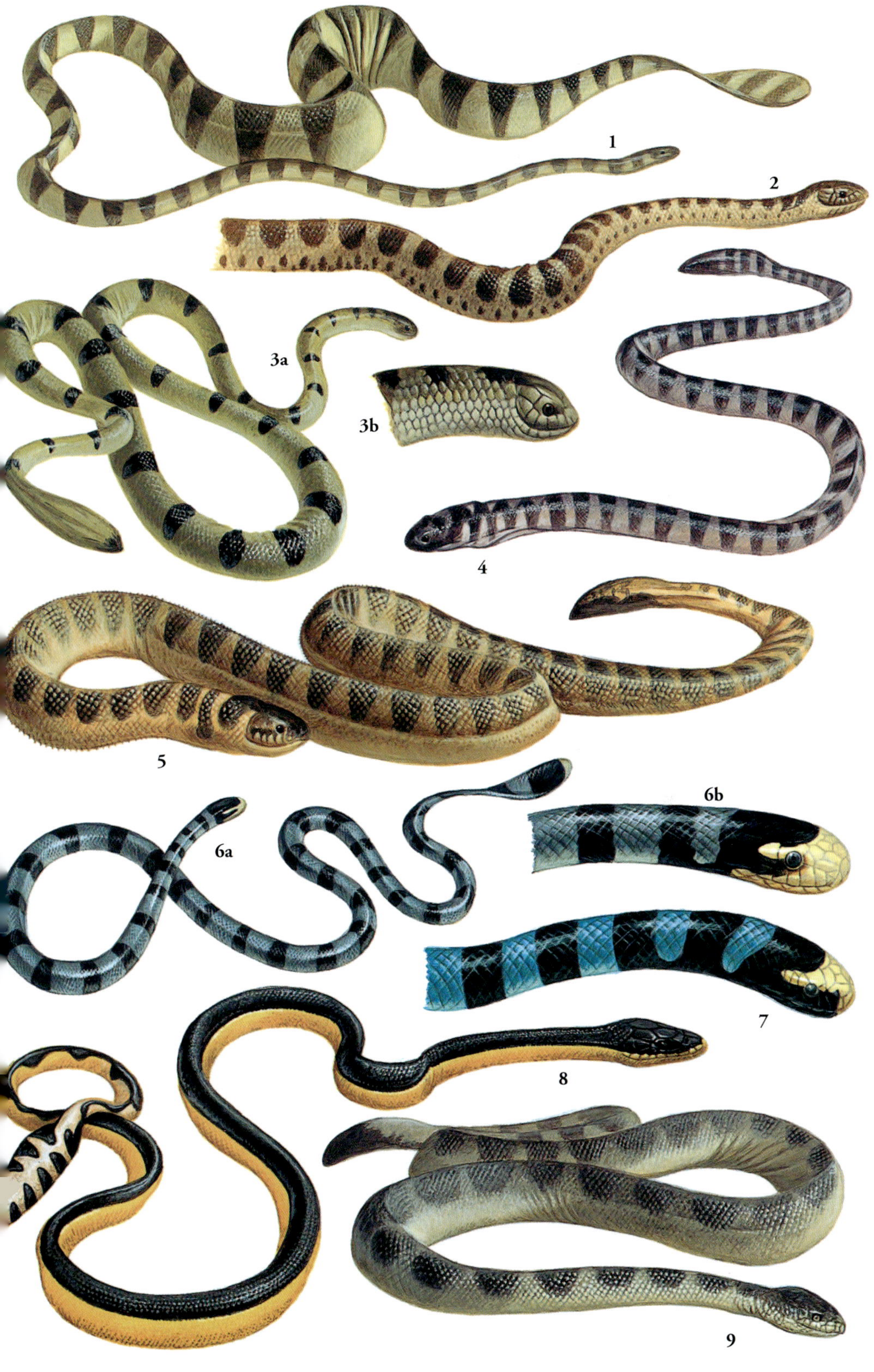
1
2
3a
3b
4
5
6b
6a
7
8
9

Plate 65. HOMALOPSIDAE

1. KEEL-BELLIED WATER SNAKE *Bitia hydroides*, p. 324.
TL 800mm Body slender anteriorly, robust on posterior two-thirds; head narrow in juveniles, moderate in adults; dorsals smooth; dorsum greyish-blue, pale brown or greyish-yellow with 40–42 dark grey or black cross-bars on body; head dark brown or dark grey with 1–2 small dark brown spots on each scute.

2. YELLOW-BANDED MANGROVE SNAKE *Cantoria violacea*, p. 325.
TL 1,200mm Body slender; head indistinct from neck; dorsals smooth; dorsum dark blackish-grey or black with transverse yellow bars, narrower than interspaces; head white-spotted or with 2 yellow cross-bars.

3. DOG-FACED WATER SNAKE *Cerberus rynchops*, p. 325.
TL 1,270mm Body moderately robust; head long and distinct from neck; dorsals strongly keeled; dorsum dark grey or greyish-green with dark blotches; dark postocular stripe on neck.

4. BENNETT'S WATER SNAKE *Enhydris bennetti*, p. 325.
TL 610mm Body moderate; head large; dorsals smooth.
(4a) Body Dorsum greyish-brown with irregular black or greyish-black cross-bars, sometimes forming zigzag pattern; white stripe adjacent to ventrals.
(4b) Head Greyish-brown; throat pale yellow.

5. BOCOURT'S WATER SNAKE *Enhydris bocourti*, p. 325.
TL 1,100mm Body robust; head large; dorsals smooth; dorsum reddish-brown or dark brown, with transverse black-edged, reddish-brown bars, narrowing on flanks to meet ventrals; head reddish-brown; venter yellow.

6. CHINESE WATER SNAKE *Enhydris chinensis*, p. 326.
TL 610mm Body moderate; head rounded; dorsals smooth; dorsum olive-brown or grey with irregular scattered black spots; sometimes, cream or yellow zigzag stripe on lower flanks.

7. MARQUIS DORIA'S WATER SNAKE *Enhydris doriae*, p. 326.
TL 700mm Body robust; head small; forehead scales fragmented; dorsals smooth; dorsum reddish-brown to greyish-brown; labials blotched with grey or cream.

8. RAINBOW WATER SNAKE *Enhydris enhydris*, p. 326.
TL 810mm Body robust; head small; dorsals smooth; dorsum dark brown, greyish-brown or olive-green with 2 pale brown paravertebral stripes from upper surface of head to tail.

9. GYI'S WATER SNAKE *Enhydris gyii*, p. 326.
TL 766mm Body robust; head small; dorsals smooth; dorsum dark brown, except for reddish-brown lowest scale rows; anterior supralabials greyish-black; dark stripe from nape to angle of jaws.

10. BLACK-SPOTTED WATER SNAKE *Enhydris innominata*, p. 327.
TL 175mm Body robust; head short; dorsals smooth; dorsum brownish-grey, sometimes with black spots in 3 longitudinal rows; flanks yellowish-white or tan with 38–39 vertical dark brown blotches; forehead with black spots.

11. JAGOR'S WATER SNAKE *Enhydris jagorii*, p. 327.
TL 560mm Body robust; head short; dorsals smooth; dorsum greyish-brown with black occelated spots arranged in linear series; supralabials yellowish-cream.

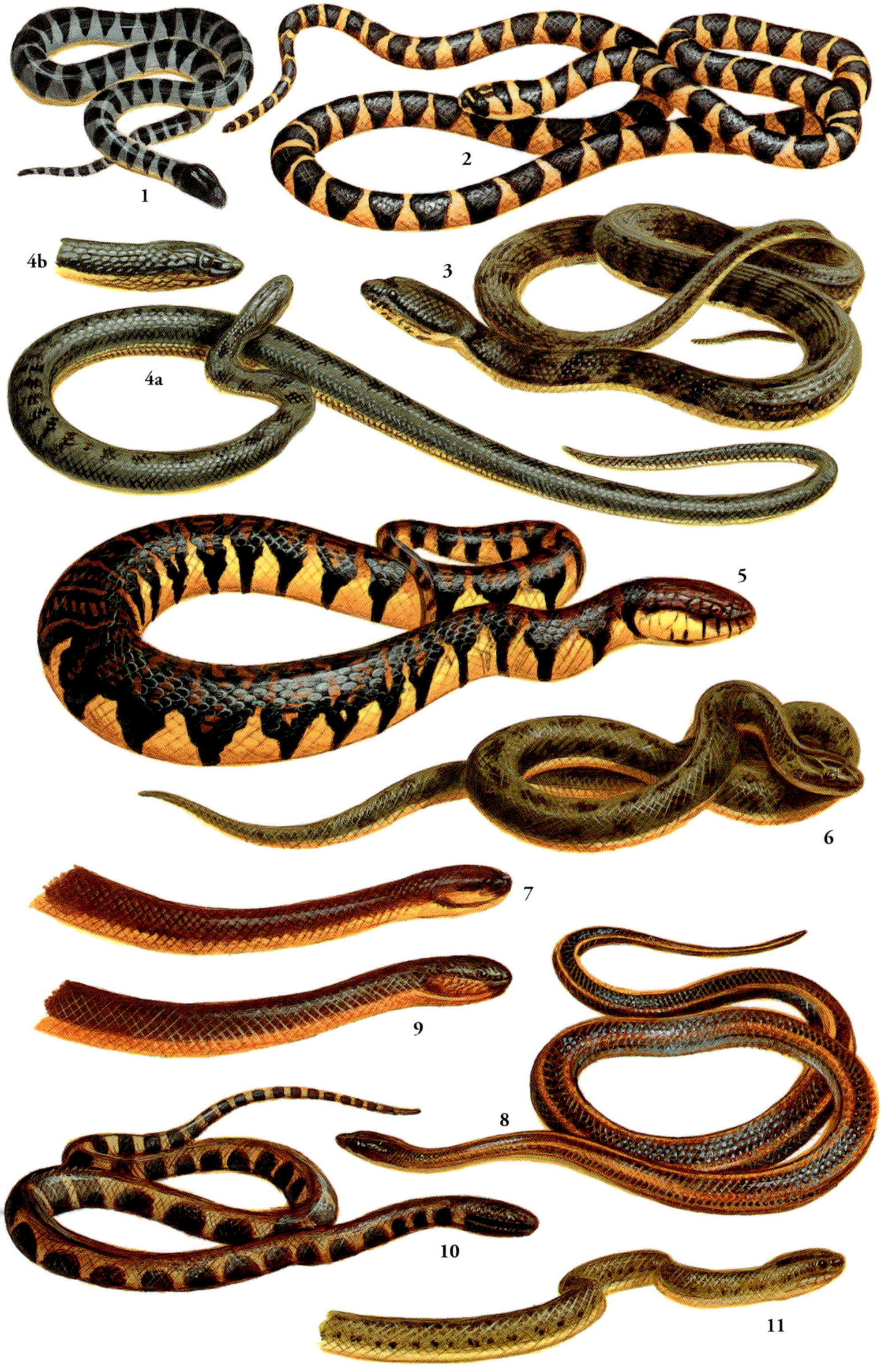
1
2
4b
3
4a
5
6
7
9
8
10
11

Plate 66. HOMALOPSIDAE

1. GREY WATER SNAKE *Enhydris plumbea*, p. 328.
TL 480mm Body robust; head short; dorsals smooth; dorsum grey or greyish-olive; scales dark brown or black edged; supralabials cream or yellow.

2. SPOTTED WATER SNAKE *Enhydris punctata*, p. 328.
TL 730mm Body robust; head short; dorsals smooth; dorsum greyish-black to dark brown; yellow cross-bar on occipital region; sometimes, transverse rows of small yellow spots, of which 6–7 anterior rows form transverse bands.

3. INDO-CHINESE WATER SNAKE *Enhydris subtaeniata*, p. 328.
TL 870mm Body moderate; head indistinct from neck; dorsals smooth; dorsum mid-brown with black ventolateral stripe covering scale rows 1–3; indistinct postocular stripe.

4. TENTACLED SNAKE *Erpeton tentaculatus*, p. 329.
TL 770mm Body slender; head small; dorsals strongly keeled.
(4a) Body Dorsum olive, grey or brown with 2 indistinct longitudinal dark paravertebral stripes; broad dark lateral stripe from snout, across orbit, to along flanks.
(4b) Head Long and scaly flexible rostral appendages.

5. CRAB-EATING MANGROVE SNAKE *Fordonia leucobalia*, p. 329.
TL 950mm Body robust; head short; dorsals smooth. Three colour morphs shown:
(5a) Dark morph Dark grey or brown.
(5b) Yellow morph Light yellow with dark spots.
(5c) Orange morph Orange with white spots.

6. GLOSSY MARSH SNAKE *Gerarda prevostiana*, p. 329.
TL 530mm Body slender; head small; dorsals smooth.
(6a) Body Dorsum grey, greyish-green or brown; sometimes, cream or yellow dorsolateral stripe.
(6b) Head Labials edged with dark grey or olive.

7. PUFF-FACED WATER SNAKE *Homalopsis buccata*, p. 329.
TL 1,400mm Body robust; head large and distinct from neck; dorsals keeled; dorsum greyish, dark brown or black with 19–51 narrow, black-edged yellow cross-bars.

8. CAMBODIAN PUFF-FACED WATER SNAKE *Homalopsis nigroventralis*, p. 330.
TL ca 1,400mm Body robust; head large and distinct from neck; dorsals keeled; interrupted cream ventrolateral stripe connects dorsal body bands; forehead black or grey; venter black with scattered cream or yellow spots in juveniles, yellowish-olive or olive-brown with cream spots in adults.
(8a) Adult Dorsum black or grey, sometimes with yellowish-cream cross-bars.
(8b) Juvenile Dorsum black or grey with yellow or pale orange cross-bars; yellowish-tan bars or patch across head; X-shaped white mark on chin.

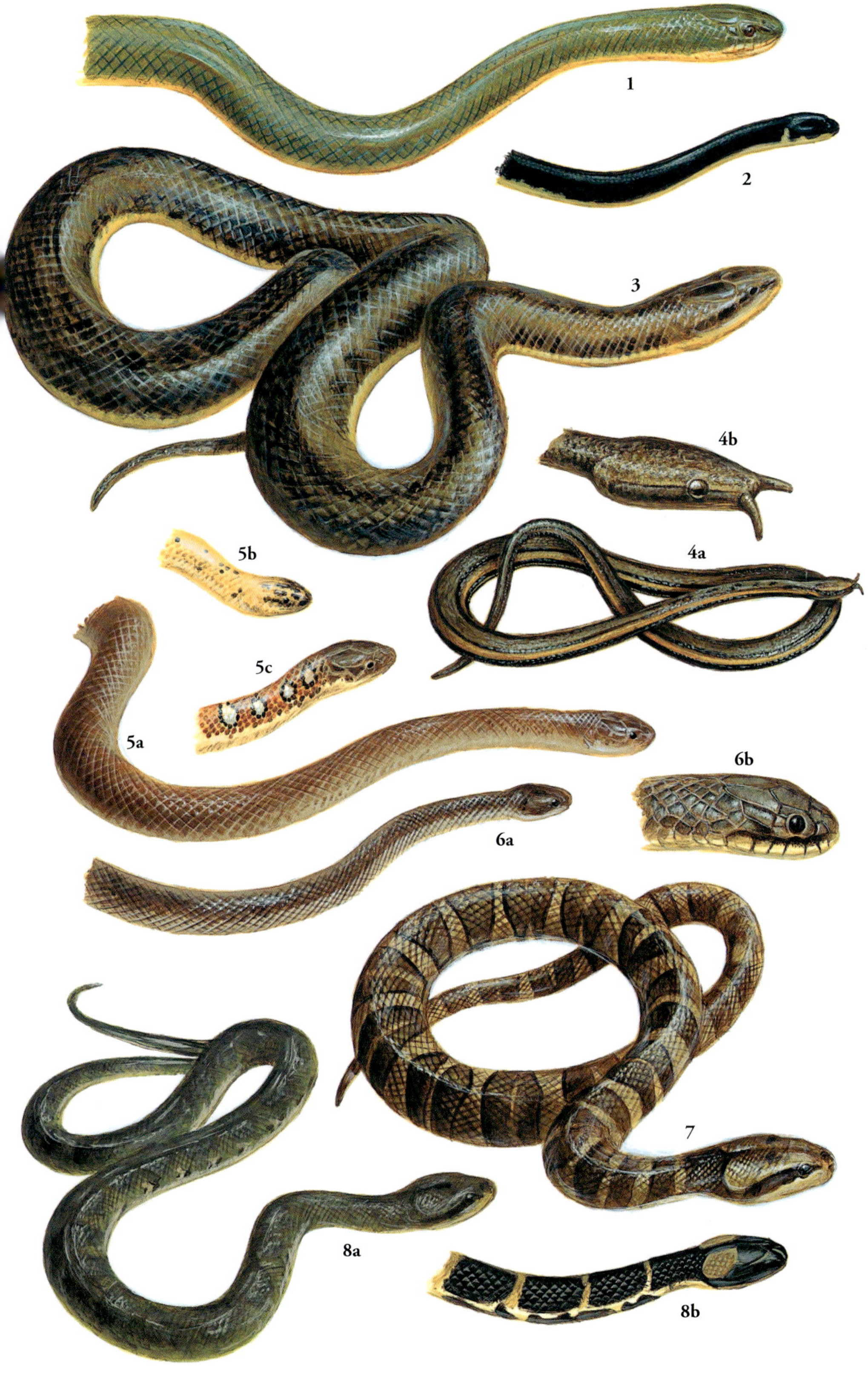
1
2
3
4b
4a
5b
5c
5a
6b
6a
7
8a
8b

Plate 67. NATRICIDAE

1. ANDREA'S KEELBACK *Amphiesma andreae*, p. 330.
TL 608mm Body slender; head distinct from neck; dorsals keeled; dorsum mid-brown; head and neck with pale dark-edged blotches, turning into pale black-edged bars on anterior body, indistinct at midbody.

2. MOUNTAIN KEELBACK *Amphiesma atemporale*, p. 330.
TL 500mm Body slender; head distinct from neck; dorsals keeled; dorsum reddish-brown to grey; scales edged with black; 2 pale dorsolateral stripes, sometimes broken up into spots; pale collar.

3. TWO-STRIPED KEELBACK *Amphiesma bitaeniatum*, p. 330.
TL 708mm Body moderate; head distinct from neck; dorsals keeled; dorsum dark brownish-grey or ochre-brown with beige-yellow dorsolateral stripe; forehead greyish-brown; dark brown postocular.

4. BOULENGER'S KEELBACK *Amphiesma boulengeri*, p. 331.
TL 877mm Body slender; head distinct from neck; dorsals weakly keeled, except smooth outermost rows; dorsum bluish-brown, greyish-black or brown with pair of white dorsal stripes on head that turn pinkish-brown on body; forehead brownish-grey with grey vermiculation; white postocular streak to nape.

5. KUATUN KEELBACK *Amphiesma craspedogaster*, p. 331.
TL 635mm Body slender; head distinct from neck; dorsals keeled; dorsum dark brown; rusty-red streak along flanks, with yellow spots; flanks with indistinct black spots; labials cream or yellow with black bars.

6. WHITE-FRONTED KEELBACK *Amphiesma flavifrons*, p. 331.
TL 750mm Body slender; head distinct from neck; dorsals keeled; dorsum greyish-brown or olive-grey with darker markings; series of white blotches on flanks to paravertebral region; white to yellowish-cream spot on snout; juveniles with paired white spots along midback.

7. GROUNDWATER'S KEELBACK *Amphiesma groundwateri*, p. 332.
TL 450mm Body slender; head distinct from neck; dorsals smooth anteriorly, weakly keeled posteriorly; dorsum black or dark brown with dorsolateral stripe comprising yellow spots; labials yellow with black edges; forehead dark; V-shaped yellow mark on neck.

8. GUNUNG INAS KEELBACK *Amphiesma inas*, p. 332.
TL 615mm Body slender; head distinct from neck; dorsals keeled; dorsum dark olive-brown with indistinct dorsolateral spots; flanks with yellow spots; forehead brown variegated with black.

9. KHASI HILLS KEELBACK *Amphiesma khasiense*, p. 332.
TL 600mm Body slender; head distinct from neck; dorsals keeled; dorsum reddish-brown with greyish-black vertebral stripe and dark reddish-brown lower flanks; forehead greyish-brown with pale brown marks.

10. WHITE-LIPPED KEELBACK *Amphiesma leucomystax*, p. 332.
TL 772mm Body slender in males, robust in females; head distinct from neck; first 2 dorsal scale rows smooth, rest keeled; dorsum brownish-grey; loose network on dorsum and flanks; indistinct pale beige dorsolateral stripe; head dark brown with irregular pale vermiculation and scattered beige dots; distinct white subocular stripe extends to neck.

11. GÜNTHER'S KEELBACK *Amphiesma modestum*, p. 333.
TL 600mm Body slender; head distinct from neck; dorsals keeled; dorsum mid-brown with indistinct small black and yellow spots on flanks that may form stripes; labials cream with dark margins or entirely dark; sometimes, series of yellow spots on flanks that may be fused.

12. MOUNT EMEI KEELBACK *Amphiesma optatum*, p. 333.
TL ca 650mm Body slender; head distinct from neck; dorsals weakly keeled; dorsum reddish-brown or bluish-black with 18–30 white cross-bars; flanks with vertical yellow bars; white preocular, postocular and subocular stripes.

13. STRIPED KEELBACK *Amphiesma parallelum*, p. 333.
TL 635mm Body robust; head distinct from neck; dorsals keeled except for outer row; dorsum reddish-brown, olive-brown or greyish-brown with dorsolateral stripes or series of spots; forehead brown; pale vertebral streak and black streak from eye to angle of mouth

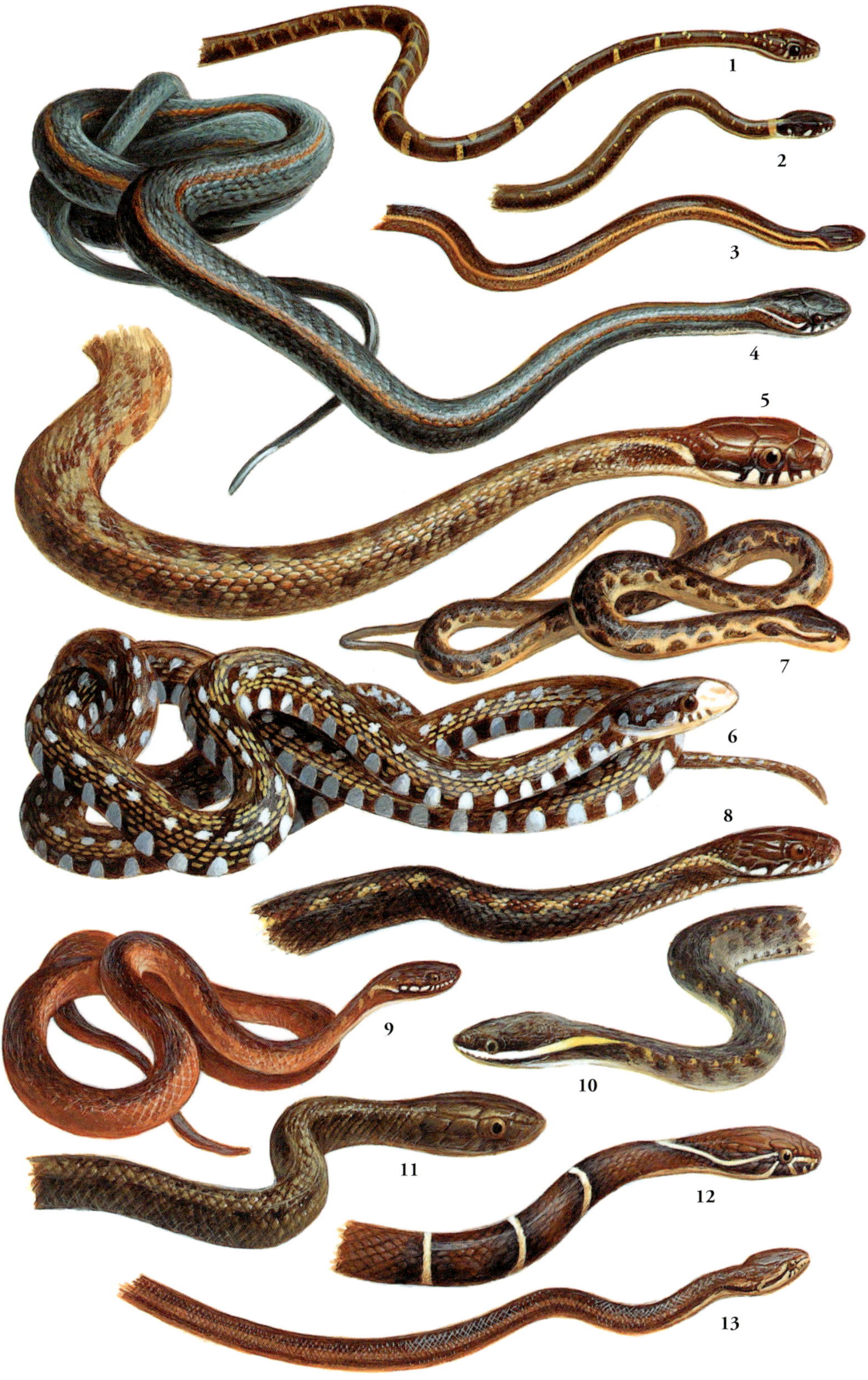
1
2
3
4
5
7
6
8
9
10
11
12
13

Plate 68. NATRICIDAE

1. PETERS'S KEELBACK *Amphiesma petersii*, p. 333.
TL 600mm Body slender; head distinct from neck; dorsals keeled; dorsum yellowish-pink or dark brown with rounded dark grey blotches; forehead dark olive with black vermiculation; labials yellow with black sutures.

2. RED MOUNTAIN KEELBACK *Amphiesma sanguineum*, p. 334.
TL 600mm Body slender; head distinct from neck; dorsals keeled; dorsum crimson with vertebral band comprising 4–5 rows of olive- and diamond-shaped black marks; flanks with 2 alternating rows of black spots; forehead dark olive; black-edged white postocular stripe; labials pale pink or cream with black sutures.

3. SARAWAK KEELBACK *Amphiesma saravacense*, p. 334.
TL 780mm Body slender; head distinct from neck; dorsals keeled; dorsum olive to reddish-brown, back with black squarish markings; supralabials yellow or cream.

4. SAUTER'S KEELBACK *Amphiesma sauteri*, p. 334.
TL 401mm Body slender; head distinct from neck; dorsals keeled; dorsum dark brown with small black and pink spots; indistinct reddish-brown streak on flanks; small pale spots on anterior flanks.

5. SIEBOLD'S KEELBACK *Amphiesma sieboldii*, p. 334.
TL 943mm Body slender; head distinct from neck; dorsals keeled; dorsum unpatterned olive-green or brown; sometimes, dorsolateral series of white spots; forehead brown; pale occipital spots and postparietal streak; supralabials bordered by dark stripe, forming nuchal crescent.

6. BUFF-STRIPED KEELBACK *Amphiesma stolatum*, p. 334.
TL 800mm Body robust; head distinct from neck; dorsals keeled; dorsum reddish-brown, olive-grey to greenish-grey; pale yellow or orange-yellow dorsolateral stripes on fifth to seventh scale rows.

7. VENNING'S KEELBACK *Amphiesma venningi*, p. 335.
TL 780mm Body slender; head distinct from neck; dorsals weakly keeled, outer scale rows smooth; dorsum olive-brown indistinctly chequered with black; anteriorly with dorsolateral ochre spots; head with lighter vermicular marks.

8. STRANGE-TAILED KEELBACK *Amphiesma xenura*, p. 335.
TL 660mm Body slender; head distinct from neck; dorsals keeled; dorsum olive-brown to nearly black with paired series of reddish-orange, pale brown, yellow or white spots on flanks; adjacent spots may be connected by faint black cross-lines; labials white with dark lines on sutures.

9. YELLOW-SPOTTED WATER SNAKE *Hydrablabes periops*, p. 336.
TL 530mm Body slender; head distinct from neck; dorsals smooth; dorsum olive-brown with a pale yellow lateral stripe and a dark grey-black lateroventral one, or mostly unpatterned.

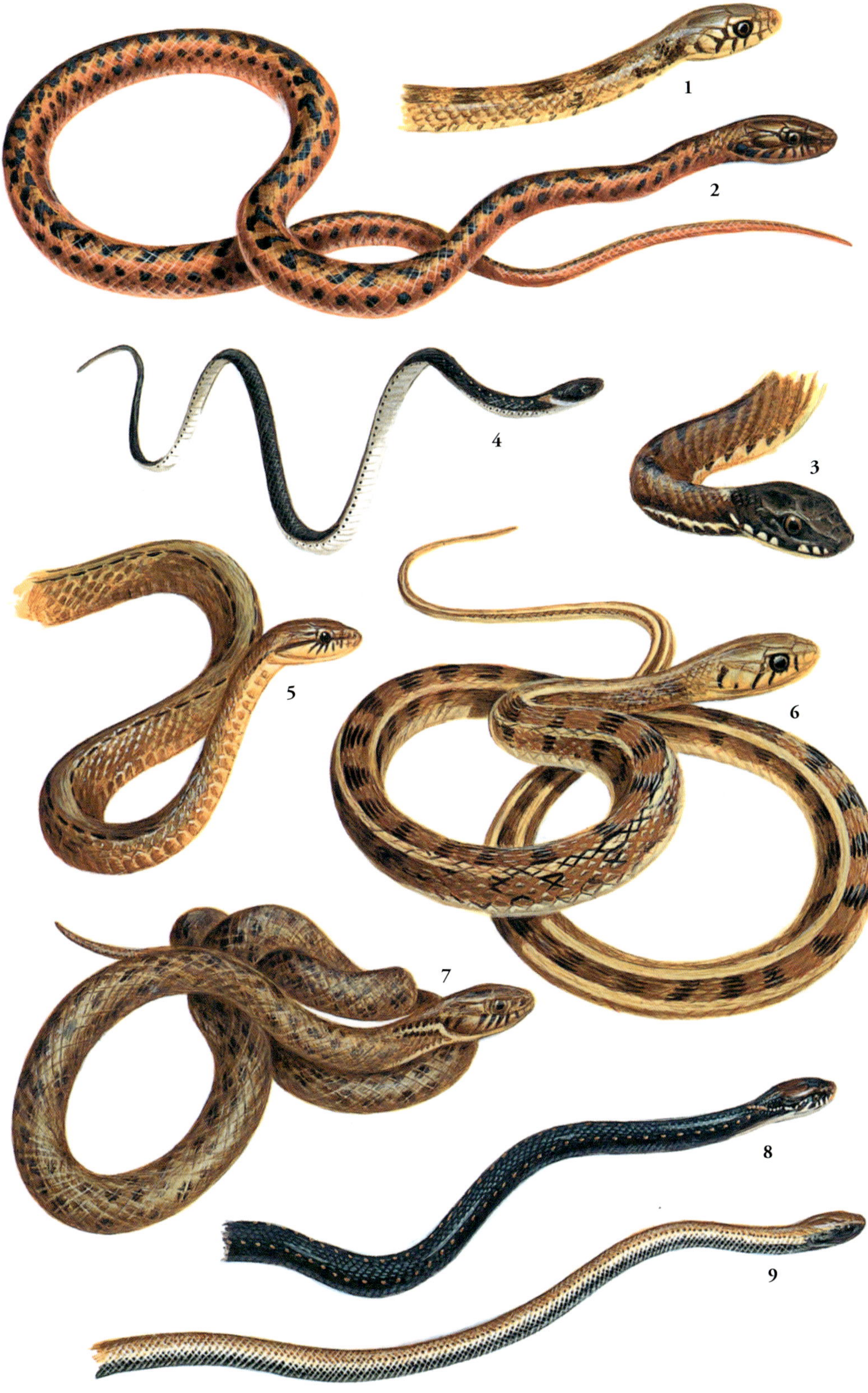
1
2
3
4
5
6
7
8
9

Plate 69. NATRICIDAE

1. ORANGE-LIPPED KEELBACK *Macropisthodon flaviceps*, p. 336.
TL 850mm Body robust; head distinct from neck; dorsals keeled; dorsum greyish-black with faint light cross-bars, narrow at vertebral region and wide on flanks; forehead light brown, yellowish-brown or olive; rusty-orange supralabials; black-edged orange nuchal loop, especially in juveniles.

2. OLIVE KEELBACK *Macropisthodon plumbicolor*, p. 336.
TL 485mm Body robust; head distinct from neck; dorsals strongly keeled.
(2a) Adult Dorsum grass-green with a yellow tinge on lower flanks.
(2b) Juvenile Large V-shaped mark on neck, followed by similar smaller one, with intervening areas of yellow or orange; black postocular stripe to angle of jaws; blue temporal region; black spots or cross-bars on dorsum.

3. BLUE-NECKED KEELBACK *Macropisthodon rhodomelas*, p. 336.
TL 750mm Body robust; head distinct from neck; dorsals keeled; dorsum reddish-brown; black vertebral stripe enters nape as inverted chevron; posteriorly, light blue.

4. ANDERSON'S STREAM SNAKE *Opisthotropis andersonii*, p. 337.
TL 500mm Body slender; head small, depressed, indistinct from neck; dorsals weakly keeled; dorsum greyish-black, green or olive-brown with indistinct fine black lines crossing scales; lowest row of dorsals bright yellow.

5. HAINANESE STREAM SNAKE *Opisthotropis balteata*, p. 337.
TL 1,050mm Body slender; head small, depressed, indistinct from neck; dorsals smooth; dorsum yellowish-orange with numerous paired black bands extending along flanks and meeting on venter; head mottled with black, with vertical yellow markings in front of and behind orbit, and one at angle of jaws.

6. DAO VAN TIEN'S STREAM SNAKE *Opisthotropis daovantieni*, p. 337.
TL 578mm Body slender; head small, depressed, indistinct from neck; dorsals smooth; dorsum pale brown or olive-grey; faint flank stripe located in dark dorsal coloration, not sharply separating dark dorsum from pale venter.

7. MAN-SON MOUNTAIN STREAM SNAKE *Opisthotropis lateralis*, p. 337.
TL 500mm Body moderate; head small, depressed, indistinct from neck; upper dorsals typically keeled, lower dorsals smooth; dorsum dark grey or greyish-brown; supralabials and infralabials yellow; faint dark lateral stripe on lower flanks sharply separates dark dorsum from pale venter.

8. SPOTTED MOUNTAIN STREAM SNAKE *Opisthotropis maculosus*, p. 338.
TL 520mm Body slender; head small, depressed, indistinct from neck; dorsals smooth; dorsum glossy black, scales with yellow spots; yellow spots larger on flank scales; forehead glossy black with scattered yellow flecks near sutures; labials yellow with black sutures.

9. CORRUGATED WATER SNAKE *Opisthotropis typica*, p. 338.
TL 502mm Body slender; head small, depressed, indistinct from neck; dorsals strongly keeled; dorsum unpatterned brownish- or blackish-grey.

10. PAINTED MOCK VIPER *Psammodynastes pictus*, p. 339.
TL 550mm Body slender; head flattened, distinct from neck; dorsals smooth, lacking apical pits; anal entire; dorsum brown or tan, sometimes black; transverse dark-edged light bands; dark streak along eye; venter cream with brown speckles.

11. MOCK VIPER *Psammodynastes pulverulentus*, p. 339.
TL 770mm Body slender; head flattened, distinct from neck; dorsals smooth, lacking apical pits; anal divided; dorsum reddish-brown to yellowish-grey, to nearly black, with small dark spots or streaks; sometimes, longitudinal stripe along middorsal region and 3 longitudinal stripes along flanks.

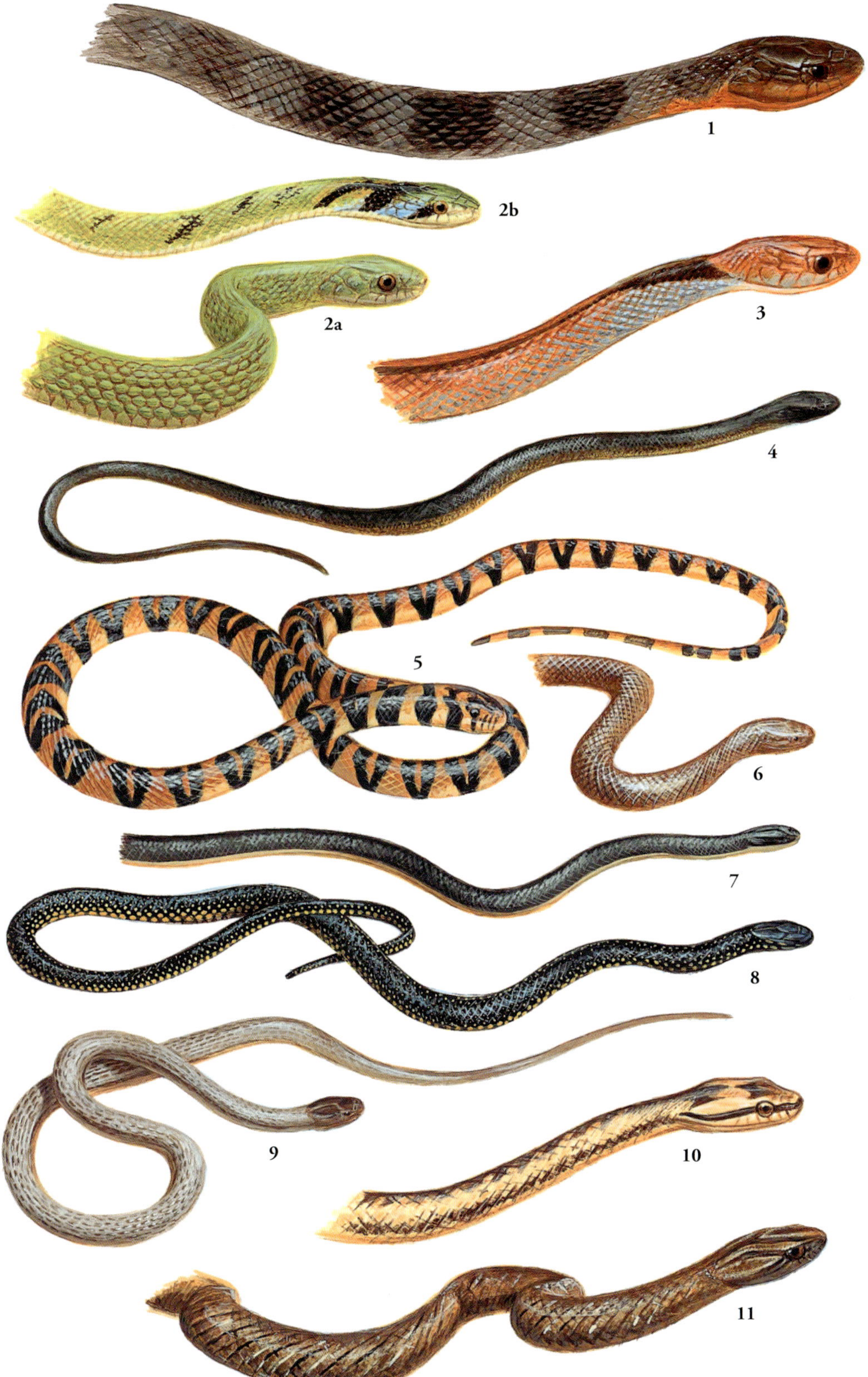
1
2b
2a
3
4
5
6
7
8
9
10
11

Plate 70. NATRICIDAE

1. SAHUL KEELBACK *Rhabdophis chrysargoides*, p. 339.
TL 870mm Body moderate; head distinct from neck; dorsals keeled.
(1a) Adult Dorsum dark olive, black or blackish-brown.
(1b) Juvenile Pale orange-yellow vertebral stripe, scalloped edged anteriorly; white spot on sides of black head and occiput; white band across frontal, supraoculars and preoculars; supralabials yellow or pinkish-white, sutures black.

2. SPECKLE-BELLIED KEELBACK *Rhabdophis chrysargos*, p. 340.
TL 980mm Body moderate; head distinct from neck; dorsals keeled; dorsum olive-grey or olive-brown; supralabials yellow or cream with darker smudges; cream, reddish-brown or orange chevron on neck; rest of back with yellow and brown oblong marks, within darker bands.

3. RED-BELLIED KEELBACK *Rhabdophis conspicillatus*, p. 340.
TL 550mm Body moderate; head distinct from neck; dorsals keeled, except outer rows; dorsum brown to reddish-brown; sides of head with cream postocular stripe that curves downwards; supralabials cream; nape and neck with 2 narrow cream collars.

4. HIMALAYAN KEELBACK *Rhabdophis himalayanus*, p. 340.
TL 1,250mm Body moderate; head distinct from neck; dorsals keeled; dorsum olive, olive-brown or dark brown; 2 dorsolateral rows of widely separated, orange-yellow spots; anterior body chequered with vermilion spots; neck with bright yellow collar, edged with black at back; labials cream or yellow edged with black; black subocular stripe.

5. LEONARD'S KEELBACK *Rhabdophis leonardi*, p. 340.
TL 1,060mm Body moderate; head distinct from neck; dorsals keeled, except 2–3 lower rows; dorsum olive-brown; head olive-brown turning greyish-cream near labials; narrow black subocular stripe; sometimes, narrow reddish-orange cross-bar on nuchal region; indistinct dark vertebral stripe, darkening posteriorly.

6. GUNUNG MURUD KEELBACK *Rhabdophis murudensis*, p. 340.
TL 873mm Body moderate; head distinct from neck; dorsals keeled, except outer rows; dorsum brownish-grey with indistinct dark cross-bars; row of light spots on edges of cross-bars; supralabials yellow, bright red or brown.

7. BLACK-BANDED KEELBACK *Rhabdophis nigrocinctus*, p. 341.
TL 950mm Body moderate; head distinct from neck; dorsals keeled.
(7a) Adult Dorsum olive-green turning brown posteriorly, with indistinct, narrow black cross-bars; 2 oblique black stripes on flanks; forehead copper-brown, lighter on sides; oblique black subocular and postocular stripes, another on nape.
(7b) Juvenile Forehead paler than that of adult; yellowish-cream nuchal bar.

8. COLLARED KEELBACK *Rhabdophis nuchalis*, p. 341.
TL 620mm Body moderate; head distinct from neck; dorsals keeled, except outer row; dorsum light brown chequered with pale reddish-brown spots; forehead brown speckled with red; alternate rows of body scales red and brown; interstitial skin bluish-black; juvenile with reddish-yellow collar, posteriorly with reddish tinge.

9. RED-NECKED KEELBACK *Rhabdophis subminiatus*, p. 341.
TL 1,300mm Body moderate; head distinct from neck; dorsals keeled, except outer rows.
(9a) Adult Dorsum olive-brown or green, unpatterned or with black and yellow reticulation; forehead yellowish-olive; nape with yellow and red band; indistinct oblique dark subocular bar.
(9b) Juvenile Dorsum with oval black spots; forehead grey; black nuchal patch; oblique dark subocular bar.

10. JAPANESE KEELBACK *Rhabdophis tigrinus*, p. 341.
TL 1,013mm Body robust; head distinct from neck; dorsals keeled; dorsum greenish-olive anteriorly, orange-red elsewhere, with series of large rectangular black spots; curved black spot on sides of neck.

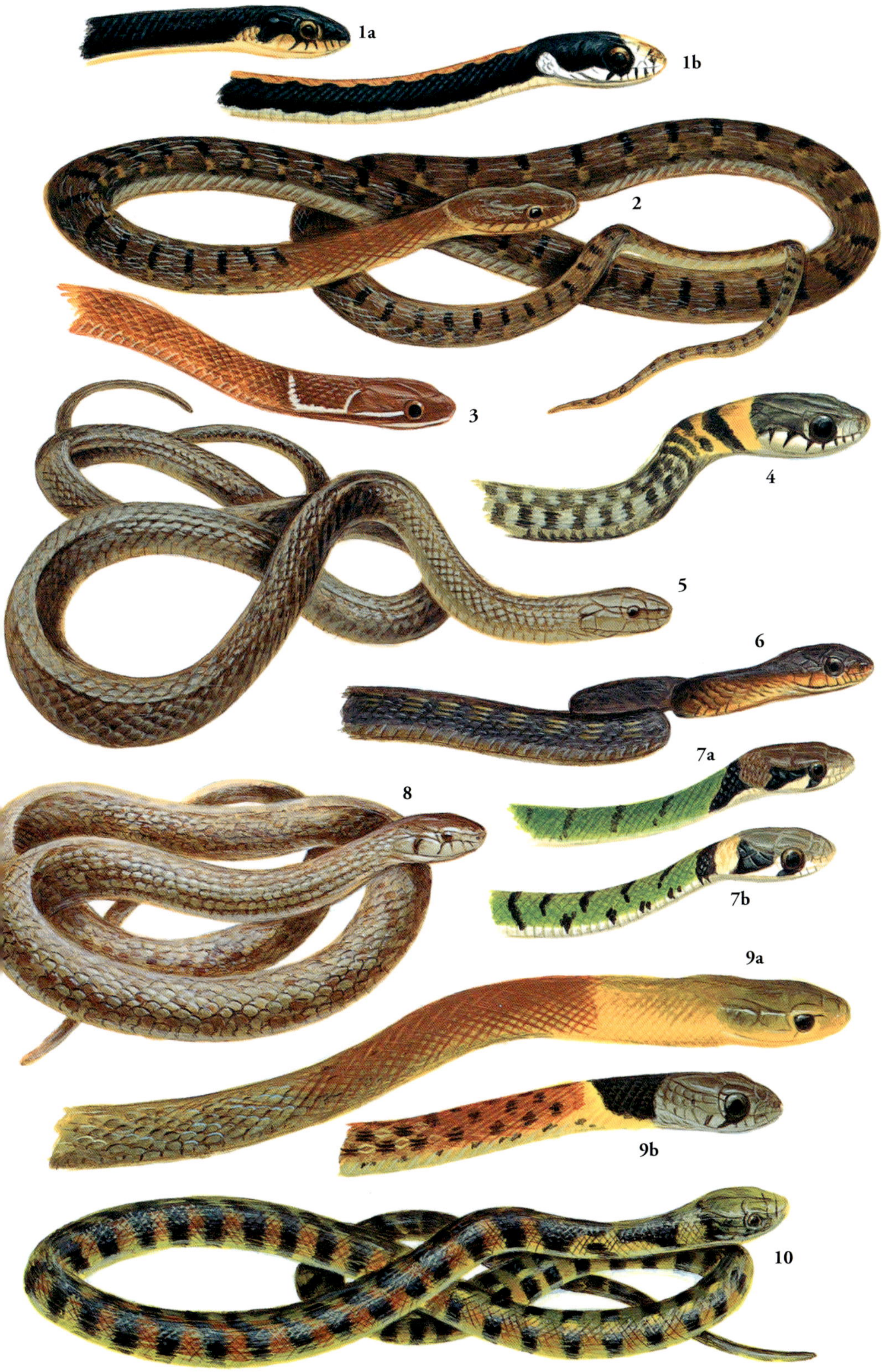
1a
1b
2
3
4
5
6
7a
7b
8
9a
9b
10

Plate 71. NATRICIDAE

1. CHINESE SPOTTED KEELBACK WATER SNAKE *Sinonatrix aequifasciata*, p. 342.
TL 1,420mm Single supralabial enters eye; ventrals 142–153; subcaudals 67–76; males with tubercles on chin shields and on Infralabial I; dorsum greyish-olive with 16–21 bands encircling body and 7–12 on tail.

2. RINGED KEELBACK WATER SNAKE *Sinonatrix annularis*, p. 342.
TL 941mm Ventrals 145–163; subcaudals 58–69; body with dark cross-bars, numbering 34–46 on body and 16–27 on tail; venter red between dark cross-bars.
(2a) Adult Dorsal pattern obscure.
(2b) Juvenile Dorsal pattern distinct.

3. CHINESE KEELBACK WATER SNAKE *Sinonatrix percarinata*, p. 342.
TL 1,100mm Head large, distinct from neck; ventrals 133–157; subcaudals 68–85; dorsum and forehead olive-grey with light-edged black bars on flanks; rest of head yellowish-cream.
(3a) Adult Dorsal pattern obscure.
(3b) Juvenile Dorsal pattern distinct.

4. YUNNAN KEELBACK WATER SNAKE *Sinonatrix yunnanensis*, p. 342.
TL 498mm Ventrals 156–165; subcaudals 61–83; dorsum brown or brownish-black with transverse black lines that form X-pattern on flanks.

5. YELLOW-SPOTTED KEELBACK WATER SNAKE *Xenochrophis flavipunctatus*, p. 342.
TL 974mm Keels become more distinct posteriorly and lack apical pits; ventrals 122–143; subcaudals 70–91; dorsum olivaceous or greenish-grey with black spots, turning into reticulated pattern posteriorly.
(5a) Yellow phase.
(5b) Red phase.

6. MALAYAN SPOTTED KEELBACK WATER SNAKE *Xenochrophis maculatus*, p. 343.
TL 1,000mm Two anterior temporals; ventrals 140–156; subcaudals 95–117; dorsum brownish-orange with 4 longitudinal series of small dark squarish marks; sometimes, paired row of yellow spots.

7. JAVAN KEELBACK WATER SNAKE *Xenochrophis melanzostus*, p. 343.
TL 975mm Tail relatively short in females; ventrals 128–142; subcaudals 66–83; 2 colour morphs, one with elongated blotches on dorsum, the other with broad longitudinal dark stripes on dorsum.
(7a) Adult Dorsum brown.
(7b) Juvenile Dorsum orange.

8. CHEQUERED KEELBACK WATER SNAKE *Xenochrophis piscator*, p. 343.
TL 1,020mm Dorsal scales strongly keeled; ventrals 132–151; subcaudals 68–99; dorsum olive-brown with black spots arranged in 5–6 rows; black stripe from eye to upper lip and from postoculars to edge of mouth.

9. ST JOHN'S KEELBACK WATER SNAKE *Xenochrophis sanctijohannis*, p. 343.
TL 710mm Supralabial IV touches eye; dorsal scales feebly keeled or nearly smooth; ventrals 139–154; subcaudals 84–94; dorsum pale olive, uniform or with indistinct dark spots, sometimes with double series of cream spots along body.

10. RED-SIDED KEELBACK WATER SNAKE *Xenochrophis trianguligerus*, p. 343.
TL 1,350mm Head large, distinct from neck; ventrals 134–145; subcaudals 86–97; dorsum blackish-brown with orange-red triangles on nape and front of body; triangle-shaped dark marks on forebody.

11. STRIPED KEELBACK WATER SNAKE *Xenochrophis vittatus*, p. 344.
TL 700mm Supralabials IV–VI contact eye; ventrals 140–151; subcaudals 53–84; dorsum mid-brown with 3 black stripes; head with dense mottling or dark areas.

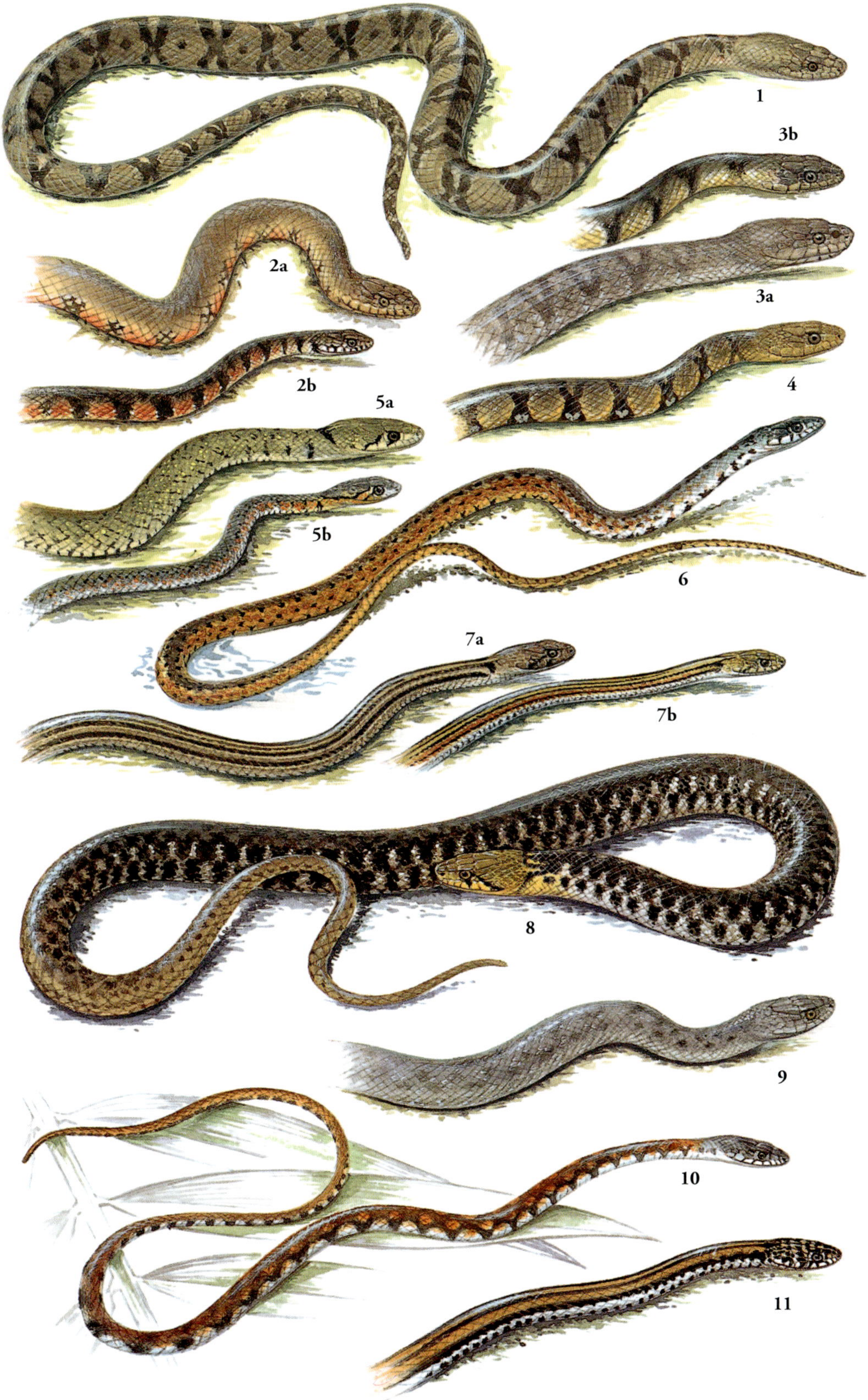
1
3b
2a
3a
2b
4
5a
5b
6
7a
7b
8
9
10
11

Plate 72. PAREATIDAE & PSEUDOXENODONTIDAE

1. BLUNT-HEADED SLUG-EATING SNAKE *Aplopeltura boa*, p. 344.
TL 850mm Body slender, laterally compressed; head short, rounded, distinct from neck; dorsals smooth; dorsum brown to greyish-brown with dark-edged saddle; flanks sometimes with large cream spots; forehead dark brown; cream subocular patch with dark subtriangular area.

2. SMOOTH SLUG-EATING SNAKE *Asthenodipsas laevis*, p. 344.
TL 600mm Body slender, laterally compressed; head short, rounded, distinct from neck; dorsals smooth; dorsum brown with dark vertical bars, extending to venter; forehead darker than dorsum, lacking lines.

3. MALAYAN SLUG-EATING SNAKE *Asthenodipsas malaccanus*, p. 344.
TL 450mm Body robust, laterally compressed; head short, rounded, distinct from neck; dorsals with weak keels; dorsum pale brown, mid-brown to nearly black with brownish-grey cross-bars, sometimes edged with white; forehead darker than dorsum; dark brown or black band on neck and forebody.

4. MOUNTAIN SLUG-EATING SNAKE *Asthenodipsas vertebralis*, p. 345.
TL 771mm Body slender, laterally compressed; head short, rounded, distinct from neck; dorsals smooth; dorsum greyish-brown with dark brown spots and sometimes indistinct dark cross-bars; interrupted yellow vertebral stripe occasionally present.

5. KEELED SLUG-EATING SNAKE *Pareas carinatus*, p. 345.
TL 600mm Body slender, laterally compressed; head short, rounded, distinct from neck; dorsals weakly keeled on 2 median rows; dorsum olive-brown, yellow or reddish-brown with indistinct transverse black bars on anterior body; dark postocular streak.

6. HAMPTON'S SLUG-EATING SNAKE *Pareas hamptoni*, p. 345.
TL 705mm Body slender, laterally compressed; head short, rounded, distinct from neck; dorsals smooth; dorsum light brown with vertical black cross-bars; forehead with dense black spots.

7. SPOTTED SLUG-EATING SNAKE *Pareas macularius*, p. 345.
TL 450mm Body slender, laterally compressed; head short, rounded, distinct from neck; median dorsals keeled; dorsum grey with dark spots or black cross-bars bordered with white; labials pale; pale nuchal collar may be present.

8. WHITE-SPOTTED SLUG-EATING SNAKE *Pareas margaritophorus*, p. 345.
TL 450mm Body slender, not laterally compressed; head short, rounded, distinct from neck; dorsals smooth; dorsum grey with black cross-bars bordered with white; labials cream with black mottling; white or yellow nuchal collar may be present.

9. MONTANE SLUG-EATING SNAKE *Pareas monticola*, p. 346.
TL 610mm Body slender, laterally compressed; head short, rounded, distinct from neck; dorsals weakly keeled; dorsum mid-brown with blackish-brown bars on flanks; black postocular stripe to nape; black postocular streak; forehead brown with dense black spots.

10. BAMBOO FALSE COBRA *Pseudoxenodon bambusicola*, p. 346.
TL 530mm Body robust; head distinct from neck; dorsals keeled; dorsum light brownish-grey or yellow-brown with 15–24 distinct black cross-bars; black or dark brown chevron on forehead.

11. KARL SCHMIDT'S FALSE COBRA *Pseudoxenodon karlschmidtii*, p. 347.
TL 1,730mm Body robust; head distinct from neck; dorsals keeled; dorsum greyish-black to dark brown; dark chevron in juveniles; vertebral region with 24 grey or yellow spots; anterior flanks with rows of spots composed of black-bordered scales.

12. LARGE-EYED FALSE COBRA *Pseudoxenodon macrops*, p. 347.
TL 1,400mm Body robust; head distinct from neck; dorsals keeled, except lower rows; dorsum brownish-grey, red or olivaceous; series of yellow, reddish-brown or orange cross-bars or spots; nape with dark chevron; throat pale yellow.

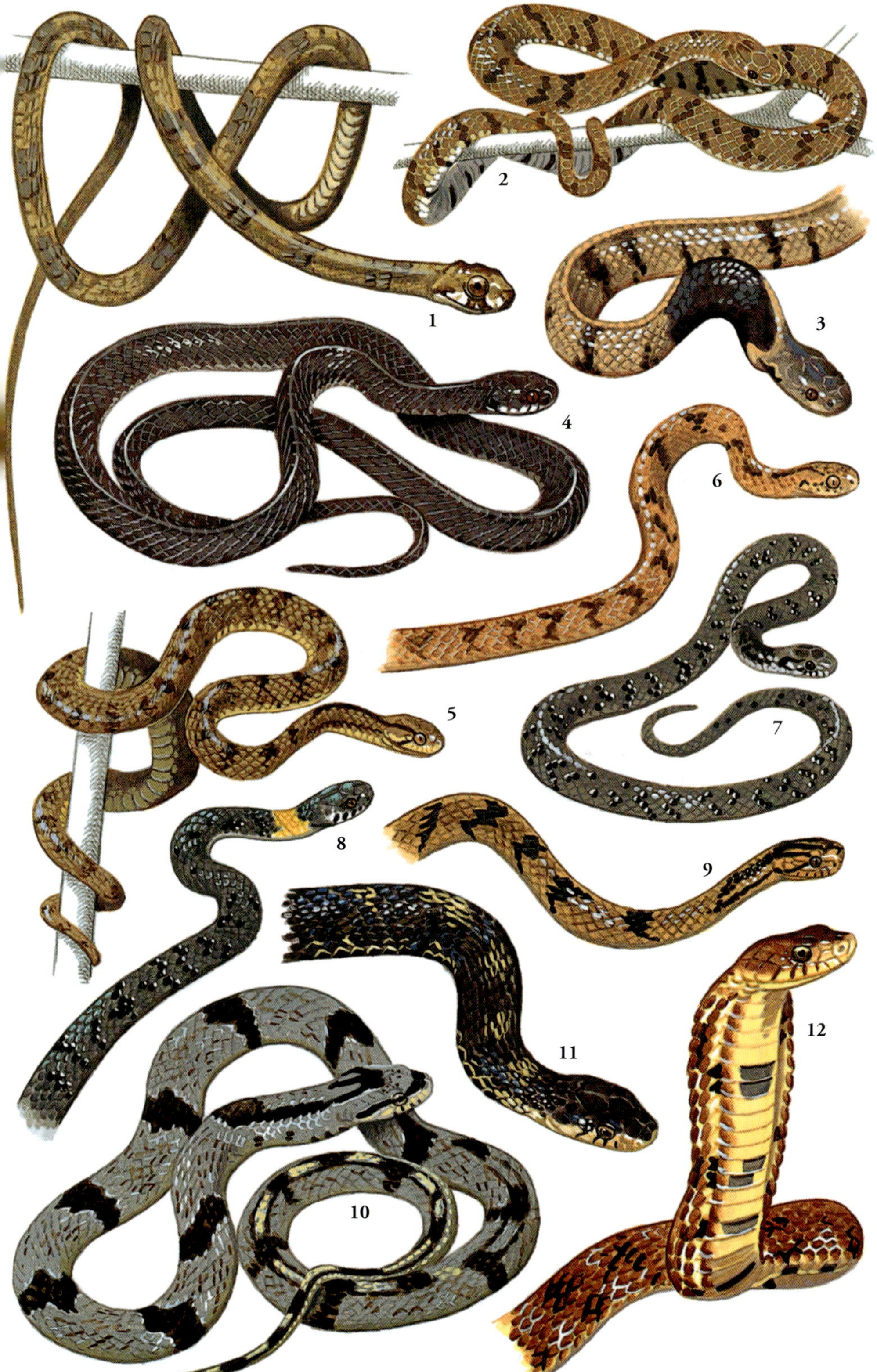
1
2
3
4
5
6
7
8
9
10
11
12

Plate 73. TYPHLOPIDAE

1. WHITE-HEADED BLIND SNAKE *Ramphotyphlops albiceps*, p. 348.
TL 302mm Body slender; head indistinct from neck, spatulate; midbody scale rows 20; dorsum dark brown; forehead red or buff; chin, gular region and tail cream.

2. BRAHMINY BLIND SNAKE *Ramphotyphlops braminus*, p. 348.
TL 180mm Body moderate; head indistinct from neck; midbody scale rows 20; dorsum black or brown; snout and tip of tail paler.

3. LINED BLIND SNAKE *Ramphotyphlops lineatus*, p. 348.
TL 480mm Body moderate; head indistinct from neck; midbody scale rows 22–24; dorsum cream, pinkish-brown or yellow with 10 longitudinal brown stripes from head to tail-tip.

4. BLACK BLIND SNAKE *Typhlops ater*, p. 348.
TL 166mm Body slender; head indistinct from neck; midbody scale rows 18; dorsum black or dark brown; chin and anal region cream.

5. DIARD'S BLIND SNAKE *Typhlops diardii*, p. 349.
TL 430mm Body moderate; head indistinct from neck; midbody scale rows 24–26 (rarely 28).
(5a) Dorsum Dark brown, each scale with indistinct transverse light streak; flanks light brown; gradual transition between dark dorsum and pale venter.
(5b) Caudal spine Sharp.

6. JERDON'S BLIND SNAKE *Typhlops jerdoni*, p. 349.
TL 280mm Body moderate; head indistinct from neck; midbody scale rows 20–22; dorsum dark brown to nearly black; snout and anal region cream.

7. MÜLLER'S BLIND SNAKE *Typhlops muelleri*, p. 350.
TL 540mm Body moderate; head indistinct from neck; midbody scale rows 24–26 (rarely 28–30); dorsum dark brown, purple or black; clear line of demarcation between dark dorsum and yellow, cream or gold venter.

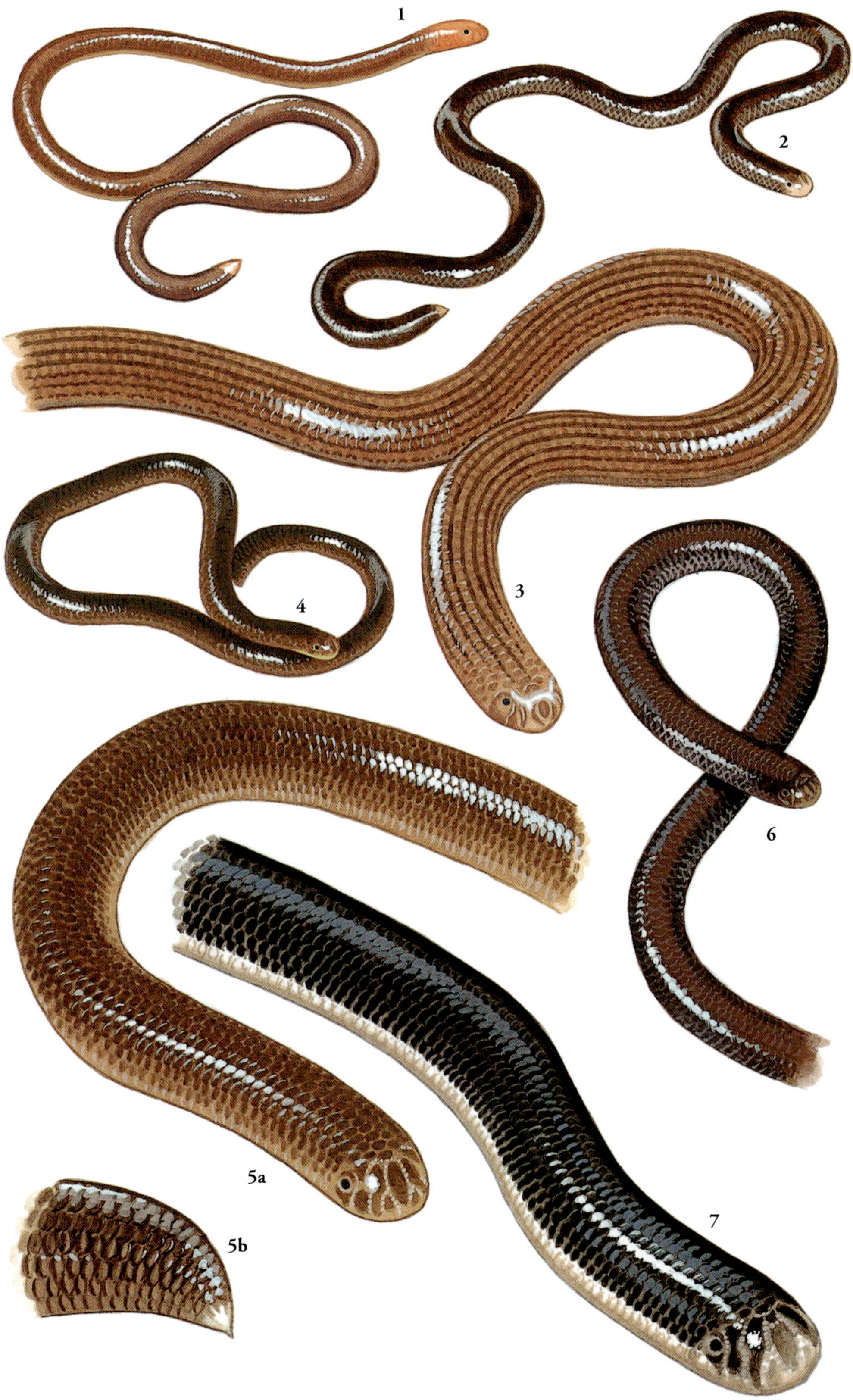
1
2
3
4
5a
5b
6
7

Plate 74. XENODERMATIDAE, XENOPELTIDAE & XENOPHIDIIDAE

1. BLACK BURROWING SNAKE *Achalinus ater*, p. 351.
TL 401mm Body slender, cylindrical; head indistinct from neck; dorsals keeled; midbody scale rows 23; subcaudals 56–63; dorsum iridescent black or dark brown; sides of ventrals paler.

2. RUFOUS BURROWING SNAKE *Achalinus rufescens*, p. 351.
TL 450mm Body slender, cylindrical; head indistinct from neck; dorsals strongly keeled; midbody scale rows 23 or 25; subcaudals 54–82; dorsum reddish-brown or greyish-brown; head and body with iridescent sheen.

3. GREY BURROWING SNAKE *Achalinus spinalis*, p. 351.
TL 412mm Body slender, cylindrical; head indistinct from neck; dorsals strongly keeled; midbody scale rows 21 or 23; subcaudals 39–62; dorsum iridescent walnut-brown, rufous medially; dark vertebral line extends to tail-tip; nuchals with black central spot.

4. KLOSS'S ROUGH WATER SNAKE *Fimbrios klossi*, p. 351.
TL 395mm Body slender, cylindrical; head long; dorsals keeled.
(4a) Body Dorsum uniform dark grey, olivaceous or purple brown, lacking pale blotches and stripes.
(4b) Head Pale grey; loreal large; preocular absent.

5. STOLICZKA'S STREAM SNAKE *Stoliczkia borneensis*, p. 352.
TL 750mm Body slender, laterally compressed; sharp ridge on vertebral region; dorsals strongly keeled.
(5a) Body Dorsum dark bluish-black or dark greyish-brown with short transverse dark bands, as broad as or broader than interspaces.
(5b) Head Large; nostrils large and flaring; eyes small and beady; forehead with 2 rows of small scales in front of eyes, separating prefrontals from frontal.

6. ROUGH-BACKED LITTER SNAKE *Xenodermus javanicus*, p. 352.
TL 650mm Body slender, compressed; head large, distinct from neck; 3 rows of large keeled scales on dorsum; exposed skin between scales.
(6a) Body Dorsum unpatterned grey; ridges on scales cream.
(6b) Head Snout pale grey; nasal scales enlarged; nostrils flaring and pointed forwards.

7. HAINAN SUNBEAM SNAKE *Xenopeltis hainanensis*, p. 352.
TL 628mm Body robust, cylindrical; head slightly distinct from neck; single postocular; dorsum iridescent bluish-brown; 2 series of white spots in longitudinal series.

8. SUNBEAM SNAKE *Xenopeltis unicolor*, p. 353.
TL 1,140mm Body robust, cylindrical; head slightly distinct from neck; postoculars 2; supralabials 8; dorsals smooth.
(8a) Body Dorsum iridescent brown, each scale light-edged.
(8b) Head Snout rounded and depressed; large interparietal in middle of 4 parietals.
(8c) Juvenile Yellow, white or cream collar.

9. SCHÄFER'S SPINY-JAWED SNAKE *Xenophidion schaeferi*, p. 353.
TL 263mm Body slender, moderately compressed; head slightly distinct from neck; dorsals keeled; dorsum dark brown, iridescent, with undulating greyish-whitish stripe on paravertebral region, from neck to tail-tip.

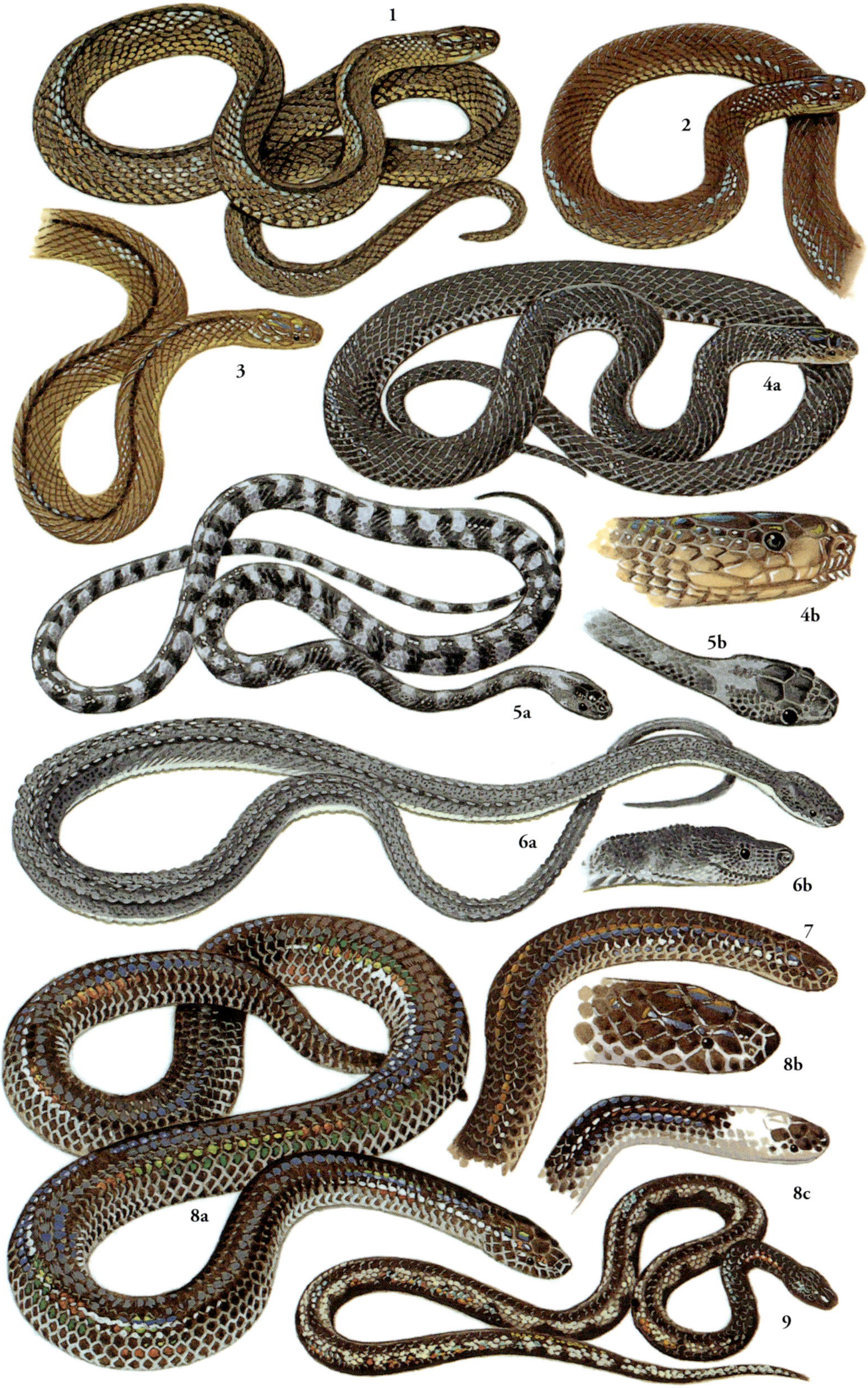
1
2
3
4a
4b
5b
5a
6a
6b
7
8b
8c
8a
9

Order CROCODYLIA

Family CROCODYLIDAE CROCODILES

Crocodiles are characterized by heavy armour, an elongated snout, heavy jaws, a robust body and osteoderms in the dorsal scales. They are aquatic, and most are inhabitants of freshwaters, although a few inhabit saltwater and brackish water habitats. All have webbed feet, nostrils capable of closing via valves and eyes equipped with a transparent membrane. Other characteristics of the crocodile family include a muscular partition separating the pectoral and abdominal cavities, alveoli in the lungs and a four-chambered heart. Crocodilians are carnivorous, and many also scavenge. They are oviparous, and their diet includes large vertebrates and, in juveniles, arthropods and fish. Some large-growing individuals may pose a danger to humans and livestock. The sex of all crocodiles is determined not by genetics, but by the incubation temperature of their eggs.

MARSH CROCODILE
Crocodylus palustris   PLATE 1
Measurement TL 5m **Identification** Forehead concave; snout relatively broad and heavy; ridges in front of eyes absent; dorsal scales in 16–17 rows on trunk; postoccipital scutes present; 13–14 pairs of teeth on upper jaw and 14–15 pairs on lower jaw. **Coloration** Juveniles light tan or brown with dark cross-bands on body and tail; adults grey to brown, usually without dark bands. **Habitat and Behaviour** Inhabits freshwaters including rivers, lakes, dams and reservoirs, typically away from tidal influence. Juvenile diet includes insects and small vertebrates such as fish and frogs; adults consume larger animals such as deer and goats, water birds, fish, snakes, lizards and turtles. A hole-nester, laying 10–50 eggs at a time; 2 clutches are produced at intervals of 30–57 days, with eggs measuring 63.8–83.8 x 39.3–50.5mm. Incubation period 60–97 days. Sex is determined by incubation temperature. Hatchlings 25–40mm. **Distribution** Myanmar. Also eastern Iran, Pakistan, India, Nepal, Bangladesh and Sri Lanka. **Status** Vulnerable. This species is common in extralimital areas, where it is threatened by hunting for its skin and by the alteration of habitats.

SALTWATER CROCODILE
Crocodylus porosus   PLATE 1
Measurement TL 6.2m **Identification** Large head; heavy snout; pair of ridges running from orbit to centre of snout; scales on back more oval than those of most other crocodiles; anterior nuchals typically absent. **Coloration** Juveniles more brightly coloured than adults, being black banded, spotted or blotched on a pale yellow or grey background; adults turn greyish-olive, showing less contrast. **Habitat and Behaviour** Inhabits rivers and coasts, especially mangrove forests, and occasionally found in standing bodies of water, at up to 1,130km inland. Juveniles feed on crabs, shrimps, insects, fish, lizards and snakes; adults can take turtles, birds and mammals. Large individuals have occasionally been known to attack humans. The female constructs a mound nest, in which 37–80 eggs are deposited. The nest is guarded by the adult until the eggs hatch. Hatchlings 30mm.

Distribution Myanmar, Thailand, Cambodia, Vietnam, Peninsular Malaysia, Singapore, Borneo, Sumatra, Java and Bali. Also east coast of India, Seychelles, Sri Lanka, Bangladesh, New Guinea, the Philippines, Australia and the South Pacific. **Status** Lower Risk/Least Concern.

BORNEAN SWAMP CROCODILE
Crocodylus raninus   NOT ILLUSTRATED
Measurement TL Unknown **Identification** Large head; broad snout; lachrymal ridges weakly developed; palatine-pterygoid suture transverse; 4 well-developed postoccipital scales; longitudinal interorbital ridge absent (distinct in Siamese Crocodile, *C. siamensis*); transverse ventral scale rows 25 (vs 29–35 in Saltwater Crocodile, *C. porosus*, and 29–33 in Siamese Crocodile); transverse throat scale rows 38–39 (vs 49–53 in Siamese Crocodile). **Coloration** Unknown, but probably similar to that of Siamese Crocodile. **Habitat and Behaviour** Apparently restricted to swamps, such as peat swamps. Diet and reproductive habits unstudied. **Distribution** Borneo. **Status** Not Evaluated.

SIAMESE CROCODILE
Crocodylus siamensis   PLATE 1
Measurement TL 3.5m **Identification** Large head; broad snout; pair of ridges running from orbit to centre of snout; similar to Saltwater Crocodile (*C. porosus*), but differing in broader snout, lack of longitudinal ridges anterior to eyes, raised crest behind eyes and webbing on feet not reaching tips of digits. **Coloration** Juveniles more brightly coloured than adults; dorsum pale yellow or grey with black spots or blotches; adults greyish-olive with indistinct bands. **Habitat and Behaviour** Inhabits freshwaters, especially those associated with forested habitats, and also marshes. Fish reported as food, although the broad snout suggests generalization for aquatic animals such as birds and small mammals. A mound nester. Clutches comprise 12–50 eggs, measuring 75–80 x 50mm. **Distribution** Thailand, Cambodia, Laos (probably extinct) and Vietnam. **Status** Critically Endangered.

Family GAVIALIIDAE GHARIALS

Gharials possess heavy armour and a narrow to very narrow snout. They inhabit freshwaters including large rivers with sandy banks and blackwater lakes surrounded by peat swamp forests. Characteristics shared with crocodiles include webbed feet, nostrils capable of closing via valves, eyes with a transparent membrane, a muscular partition separating the pectoral and abdominal cavities, alveoli present in the lungs and a four-chambered heart. The adult diet of gharials, while primarily consisting of fish, includes birds, turtles and mammals up to the size of monkeys, while juveniles may ingest insects, crustaceans and frogs. The family is currently represented by two living species. The fossil record shows a diverse fauna from North and South America, Africa and Asia, indicating that gharials were once near cosmopolitan in distribution.

GANGES GHARIAL
Gavialis gangeticus PLATE 1
Measurement TL 7m **Identification** Body slender, elongated; snout slender, parallel-sided, tip with distinctive knob in adult males; ca 100 sharp interlocking teeth; neck armour continuous with back armour. **Coloration** Dorsum olive to tan with dark blotches or bands on body and tail; venter cream; juveniles dark brown. **Habitat and Behaviour** Inhabits large rivers with sandbanks. A specialized fish-eater, especially of catfish, although other, smaller river animals may occasionally be ingested, and scavenging is also known. Nests on sandbanks or alluvial deposits. Clutches comprise 7–97 eggs, measuring 59–64mm. Incubation period 60–92 days. Hatchlings 350–393mm. **Distribution** Ayeyarwady River in Myanmar. Also Pakistan, India, Nepal and Bangladesh. Extinct in Bhutan. **Status** Critically Endangered.

MALAYAN FALSE GHARIAL
Tomistoma schlegelii PLATE 1
Measurement TL 5.5m **Identification** Body elongated; snout narrow, tapering gradually from skull (and ending with a rounded tip), differing from Ganges Gharial (*G. gangeticus*), in which a sharp demarcation exists between skull and snout; nuchal and dorsal scutes form continuous shield of 22 transverse series. **Coloration** Dorsum brown with black spots and bands, especially on flanks; tail with broad black bands; venter cream; juveniles bright yellow with dark bands; iris yellowish-brown. **Habitat and Behaviour** Inhabits freshwaters such as rivers, swamps and lakes overgrown with vegetation. Diet consists of fish, although much larger prey, including monkeys, has been recorded. Makes a mound nest on peat, or a mixture of peat and humus, within or close to roots of large trees ca 60cm high. Clutches comprise 20–60 eggs, measuring 70 x 100mm. Incubation period 75–90 days. **Distribution** Southern Thailand, Peninsular Malaysia, Sumatra, Borneo and Java. Possibly also Sulawesi, Indonesia. Subfossil remains (ca 17th century) from China. **Status** Endangered.

Order CHELONII

Family GEOEMYDIDAE ASIAN HARDSHELLED TURTLES

Members of this family have an oval to oblong, depressed or domed carapace bearing 11 pairs of sutured peripherals, and a nuchal lacking costiform processes. The large plastron is occasionally hinged, there is no mesoplastron and the plastral buttresses are typically firmly articulated with the costals of the carapace. These turtles are capable of withdrawing their neck vertically into their shell, and their pelvic girdle articulates flexibly with the plastron. Most species are aquatic, inhabiting standing and flowing water bodies inland, although a few are found in brackish waters or along sea coasts; terrestrial species are in a minority. A majority are omnivorous, and herbivorous adults tend to be carnivorous as juveniles. A few species have specialized diets (which include snails), while many others may scavenge. All species are oviparous. The family is widely distributed in Asia, as well as in southern Europe, and central and northern South America.

SOUTHERN RIVER TERRAPIN
Batagur affinis PLATE 2
Measurement SCL 600mm **Identification** Carapace domed, heavily buttressed; long plastron; head small with upturned snout; plastron lacks hinge; forehead covered with small scales; 4 claws on each forelimb. **Coloration** Carapace olive-grey or brown; head similar in colour but lighter on sides; plastron unpatterned yellow; head dark, in mature males turning black; iris white. **Habitat and Behaviour** Inhabits mouths of rivers under tidal influence. Diet consists of fruits of *Sonneratia*; leaves, stems and fruits also consumed, plus molluscs, crustaceans and fish. Migrates upriver to nest on beaches away from tidal influence. Clutches comprise 19–37 eggs, measuring 68 x 40mm. Incubation period 61–66 days. **Distribution** Southern Peninsular Malaysia and Sumatra. Identity of populations of the genus *Batagur*

from northern Peninsular Malaysia, Singapore (now extinct), Cambodia and southern Vietnam (now extinct) unknown. **Status** Critically Endangered.

NORTHERN RIVER TERRAPIN
Batagur baska NOT ILLUSTRATED
Measurement SCL 600mm **Identification** Carapace domed, heavily buttressed; long plastron; plastron lacks hinge; head small with upturned snout; forehead covered with small scales; 4 claws on each forelimb. **Coloration** Carapace olive-grey or brown; head similar in colour but lighter on sides; plastron unpatterned yellow; head and neck black in mature males; tip of snout blue; base of neck crimson; forelimbs carmine; iris greenish-yellow. **Habitat and Behaviour** Inhabits mangrove forests at mouths of rivers. Diet consists of fruits of *Sonneratia*. Nesting takes place on sandy sea beaches. Diet and reproductive habits unstudied. **Distribution** Myanmar (lower reaches of Ayeyarwady, Sittang and Salween Rivers). Also coasts of eastern India and Bangladesh. Historically also southern Pakistan. **Status** Critically Endangered.

PAINTED TERRAPIN
Batagur borneoensis PLATE 2
Measurement SCL 600mm **Identification** Carapace oval, flattened; marginals smooth; in juveniles shell is flattened and vertebral keel is distinct; plastron lacks hinge; 5 claws on each forelimb. **Coloration** Carapace light brown or olive with 3 longitudinal black stripes; head of females olive, non-breeding males have a grey head; adult males in breeding season have white forehead and red stripe between eyes; plastron unpatterned cream. **Habitat and Behaviour** Inhabits coastal areas and estuaries. Herbivorous, consuming aquatic macrophytes; shrimps, crabs and fish occasionally ingested. A nesting migration occurs, with adults travelling as far as 3km downriver to nest on sea beaches. Clutches comprise 10–16 elongated eggs, measuring 68–76 x 36–44mm; 2 clutches produced annually. Incubation period 61–82 days. **Distribution** Southern Thailand, Peninsular Malaysia, Sumatra and Borneo; records from Myanmar require verification. **Status** Critically Endangered.

BURMESE PAINTED TERRAPIN
Batagur trivittata PLATE 2
Measurement SCL 580mm **Identification** Carapace low with distinct vertebral keel; plastron narrow, truncated anteriorly, notched posteriorly; plastron lacks hinge; head with pointed, slightly upturned snout; feet with broad webbing; 5 claws on each forelimb. **Coloration** Male carapace olive-green with 3 longitudinal black stripes, which may be fused posteriorly; plastron yellow; forehead crimson with lozenge-shaped black area; neck yellow. Female carapace dark brown; forehead and neck greenish-olive with black crown patch; plastron yellow. **Habitat and Behaviour** Inhabits large rivers including estuaries, choosing banks with vegetation. Herbivorous. Clutches comprise 10–30 eggs, measuring 65–75 x 40–32mm; 3 clutches produced annually. Incubation period ca 70 days. **Distribution** Endemic to Myanmar (Ayeyarwady, Sittang and Salween River Drainages). **Status** Endangered.

MALAYAN BOX TURTLE
Cuora amboinensis PLATE 2
Measurement SCL 216mm **Identification** Carapace high-domed and smooth; vertebral keel in adults; juveniles with 2 additional keels laterally; plastron with transverse moveable hinge behind pectoral and abdominal scutes; posterior anal notch absent on plastron. **Coloration** Carapace olive, brown or nearly black; plastron yellow or cream with single black blotch on each scute; face with longitudinal yellow stripes. **Subspecies** Four subspecies are recognized, of which three occur in the region. *C. a. kamaroma*, shell relatively high-domed, narrow, without well-defined margin; posterior plastral lobe not flared; carapace without black middorsal stripe. *C. a. lineata*, shell relatively high-domed, broad; lateral margins reverted; posterior plastral lobe flared; carapace with black middorsal stripe. *C. a. cuoro*, shell relatively low-domed with or without well-defined margin; posterior plastral lobe not flared; carapace without black middorsal stripe. **Habitat and Behaviour** Inhabits both standing and slow-flowing water bodies, including rivers, lakes, marshes and mangrove swamps, as well as agricultural areas; juveniles are more aquatic than the amphibious adults. Omnivorous, preferring water plants and fungi; worms and aquatic insects also consumed. Clutches comprise 1–6 elongated eggs, measuring 40–55 x 25–34mm; 2 clutches produced annually. Incubation period 1.5–3 months. **Distribution** Myanmar (*C. a. lineata*), Thailand, Laos, Cambodia, Vietnam, Peninsular Malaysia, Singapore, Borneo (*C. a. kamaroma*), Sumatra, Java and Bali (*C. a. cuoro*). Also northeastern India, Bangladesh (*C. a.* probably *lineata*), Maluku and Sulawesi (eastern Indonesia), and the Philippines (*C. a. amboinensis*). **Status** Vulnerable.

VIETNAMESE BOX TURTLE
Cuora cyclornata NOT ILLUSTRATED
Measurement SCL 297mm **Identification** Carapace rounded, depressed, oval; juveniles more domed; posterior marginals serrated; keels absent; plastron with transverse moveable hinge behind pectoral and abdominal scutes; plastron with median notch. **Coloration** Carapace brown, lacking radiating pattern or blotches on marginals; black vertebral stripe fails to reach nuchal; forehead olive or orange-brown; iris blue-green; plastron yellow with central black pattern. **Subspecies** Two subspecies are recognized. *C. c. cyclornata*, carapace marginals not flared; carapace reddish-brown; posterior marginals serrated. *C. c. meieri*, carapace marginals weakly flared in males; carapace chestnut-brown; posterior marginals not serrated. **Habitat and Behaviour** Inhabits lowland forests, including limestone-dominated evergreen and semitropical monsoon forests, at elevations of 200–800m asl. Diet and reproductive habits

unstudied. **Distribution** South-central Vietnam, eastern Laos (*C. c. cyclornata*) and northern Vietnam (*C. c. meieri*). Also southern China (*C. c. cyclornata* and *C. c. meieri*). **Status** Not Evaluated.

INDO-CHINESE BOX TURTLE
Cuora galbinifrons PLATE 2

Measurement SCL 190mm **Identification** Carapace high-domed and smooth; vertebral keel present in juveniles; plastron with transverse moveable hinge behind pectoral and abdominal scutes; posterior anal notch absent on plastron. **Coloration** Carapace with narrow yellow or cream vertebral stripe; head yellow, pale green or grey, juveniles with dark speckles; narrow dark stripe on face; plastron dark brown or black with yellow pigments on seams; undersurface of marginals sometimes with yellow spot. **Habitat and Behaviour** Inhabits forests in the highlands. Diurnal and terrestrial. Diet consists of vegetation and animal matter. Reproductive habits unstudied. **Distribution** Laos and northern Vietnam. Also south-eastern China. **Status** Critically Endangered.

KEELED BOX TURTLE
Cuora mouhotii PLATE 2

Measurement SCL 180mm **Identification** Carapace elongated, distinctly flat-topped; marginals serrated posteriorly and sometimes anteriorly, with 3 prominent keels on carapace; weak transverse hinge across plastron in adult females; digits half webbed; tail long in juveniles. **Coloration** Carapace dark or light brown; plastron yellow or light brown with dark brown blotches on each scute. **Habitat and Behaviour** Inhabits evergreen hill forests, and associated with moist leaf litter. Omnivorous. Clutches comprise 1–5 brittle-shelled eggs, measuring 40–56 x 24.8–27mm. Incubation period 90–101 days. **Distribution** Myanmar, northern Thailand, Laos, Cambodia and Vietnam. Also north-eastern India and southern China. **Status** Endangered.

THREE-STRIPED BOX TURTLE
Cuora trifasciata PLATE 2

Measurement SCL 230mm **Identification** Carapace elevated, arched and elongated, posteriorly flared; posterior marginals not serrated; plastron with transverse moveable hinge between pectoral and abdominal scutes; posterior anal notch distinct on plastron. **Coloration** Carapace light brown with 3 longitudinal black stripes; vertebral stripe reaches nuchal; scutes with thin radiating black pattern; black blotches on marginals; head bright yellow with broad black postorbital stripe that encloses elongated brown or olive triangle behind eye and narrow brown or olive bar extending from tympanum; plastron yellow; scutes black, pale centrally in gular region, yellow bordered, producing triangular mark; tail pinkish-orange with 2 black lateral stripes. **Habitat and Behaviour** Inhabits clear mountain streams and ponds in subtropical and evergreen monsoon forests, at elevations of 100–850m asl. Diurnal and nocturnal, and semi-terrestrial, with significant aquatic activities. Diet consists of fish, frogs, insects and crustaceans; plants and fallen fruits are also consumed. Clutches comprise 2–8 eggs, measuring 45–54 x 25–32mm. The eggs are deposited in a nest 10–15cm deep, in a flask-shaped chamber that has been excavated in soil or under leaves. Hatchlings 40–45mm. **Distribution** Laos and Vietnam. Also southern China (Hong Kong, Guangdong, Fujian and Guangxi Provinces). **Status** Critically Endangered.

BLACK-BRIDGED LEAF TURTLE
Cyclemys atripons PLATE 3

Measurement SCL 236mm **Identification** Carapace ovoid, elongated, depressed, bearing 3 keels; enlarged scales on forehead; plastron with hinge in adults; femoral midseam shorter or equal to anal midseam; anal notch narrow to wide. **Coloration** Carapace chestnut-brown, unpatterned or with fine radiating black lines; head and neck striped; forehead spotted or speckled with black; plastron yellow with radiating dark pattern; ocellated pattern along submarginal seams in juveniles. **Habitat and Behaviour** Inhabits evergreen hill forests at elevations of up to ca 700m asl. Nocturnal, associated with hill streams with rocky and sandy substratum. Diet and reproductive habits unstudied. **Distribution** South-eastern Thailand, and south-western and eastern Cambodia. **Status** Not Evaluated.

COMMON LEAF TURTLE
Cyclemys dentata PLATE 3

Measurement SCL 210mm **Identification** Carapace oval, depressed, bearing 3 keels; enlarged scales on forehead; plastron with hinge in adults around 23–25cm in shell length; femoral midseam shorter than anal midseam; anal notch narrow, acute-angled. **Coloration** Carapace dark brown, unpatterned or with fine black lines; forehead not speckled; temples and neck with no dark stripes; plastron yellow with radiating dark pattern; plastron of juveniles mottled with black. **Habitat and Behaviour** Inhabits the plains and low hills, although more common in the plains. Lives in small rivers, streams and ponds. Omnivorous, feeding on figs and invertebrates. Clutches of 2–4 elongated hardshelled eggs laid in nests dug underground. Incubation period 75 days. **Distribution** Southern Peninsular Malaysia, Sumatra, Borneo and Java. Also Palawan and the Sulu Archipelago; introduced into Leyte (the Philippines). **Status** Lower Risk/Near Threatened.

SUNDA LEAF TURTLE
Cyclemys enigmatica NOT ILLUSTRATED

Measurement SCL 235mm **Identification** Carapace oval, depressed, bearing 3 keels; enlarged scales on forehead; plastron with hinge in adults; femoral midseam greater or equal to anal midseam; anal notch wide, obtuse-angled. **Coloration** Carapace dark brown, typically with reddish tinge; forehead copper to brown, lighter than temporal region; neck dark, lacking stripes; plastron dark brown to black, sometimes with radiating dark lines. **Habitat and**

Behaviour Inhabits the plains and low hills. Lives in streams and ponds. Omnivorous, feeding on fallen fruits and invertebrates. Reproductive habits unstudied. **Distribution** Southern Peninsular Malaysia, Sumatra, Borneo and Java. **Status** Not Evaluated.

GRAY LEAF TURTLE
Cyclemys fusca PLATE 3
Measurement SCL 242mm **Identification** Carapace oval, depressed, bearing 3 keels; enlarged scales on forehead; plastron with hinge in adults; femoral midseam greater or equal to anal midseam; anal notch wide, obtuse-angled. **Coloration** Carapace dark brown, sometimes with radiating dark lines; forehead greenish-yellow, lighter than temporal region; neck dark, lacking stripes; plastron dark brown to black, sometimes with radiating dark lines. **Habitat and Behaviour** Inhabits streams and ponds. Diet and reproductive habits unstudied. **Distribution** Northern and central Myanmar. Possibly also Bangladesh and north-eastern India. **Status** Not Evaluated.

OLDHAM'S LEAF TURTLE
Cyclemys oldhamii PLATE 3
Measurement SCL 254mm **Identification** Carapace rectangular, depressed, bearing 3 keels; enlarged scales on forehead; plastron with hinge in adults; femoral midseam greater or equal to anal midseam; anal notch wide, obtuse-angled. **Coloration** Carapace dark brown; forehead speckled; neck with or without stripes; plastron dark brown to black, sometimes with radiating dark lines; juvenile plastron brown or yellow with large central plastral figure and ocellated pattern along submarginal seams. **Habitat and Behaviour** Inhabits the plains and low hills. Lives in streams and ponds. Diet and reproductive habits unstudied. **Distribution** Myanmar, Thailand, Laos, northern Cambodia and Vietnam. Possibly also southern China. **Status** Not Evaluated.

VIETNAMESE LEAF TURTLE
Cyclemys pulchristriata PLATE 3
Measurement SCL 227mm **Identification** Carapace oval, elongated, depressed, bearing 3 keels; enlarged scales on forehead; plastron with hinge in adults; femoral midseam shorter or equal to anal midseam; anal notch narrow to wide. **Coloration** Carapace chestnut-brown or light brown with wide, radiating dark lines or thick black speckling; forehead with dark speckling; temples and neck with distinct stripes; plastron yellow with short, thick radiating dark lines; juvenile plastron with large dark specks and ocellated pattern along submarginal seams. **Habitat and Behaviour** Inhabits streams and ponds. Diet and reproductive habits unstudied. **Distribution** Central and southern Vietnam, and eastern Cambodia. **Status** Not Evaluated.

BLACK-BREASTED LEAF TURTLE
Geoemyda spengleri PLATE 3
Measurement SCL 125mm **Identification** Carapace depressed, tricarinate; anterior marginals serrated; plastron lacks hinge; eyes large; upper jaw hooked; outer face of forelimbs with enlarged scales; digits half webbed; tail long with soft spines basally. **Coloration** Carapace dark reddish-orange, orange-yellow, olive or light brown, with black lines or wedges over keels; venter dark brown, edges yellow; head brown with yellow postocular stripe to neck; snout, tympanic region and throat yellow-spotted. **Habitat and Behaviour** Inhabits temperate forests in the mid-hills. Crepuscular and terrestrial, occasionally entering streams. Diet consists of worms, as well as small insects and other arthropods. Clutches comprise 1 brittle-shelled egg, measuring 42–45 x 18mm; 3 clutches produced annually. Incubation period 66–73 days. Hatchlings 30mm. **Distribution** Vietnam. Also south-eastern China. **Status** Endangered.

YELLOW-HEADED TEMPLE TURTLE
Heosemys annandalii PLATE 4
Measurement SCL 506mm **Identification** Carapace elongated, raised, flat-topped in adults; plastral notch distinct; plastron lacks hinge; head small; 2 large cusps on upper jaw; tail very short; hatchlings' carapace elevated with distinct vertebral keel; shell outline rounded; posterior marginals serrated. **Coloration** Carapace black with orange lines near marginals and on vertebral keel in juveniles; plastron, bridge and lower marginals pale orange with grey vermiculations, turning pale grey and eventually black in adults; forehead and jaws pale yellow. **Habitat and Behaviour** Inhabits freshwater ponds, rivers and irrigation canals. Adults diurnal; juveniles active in late afternoon and after dusk. Diet consists of macrophytes, as well as overhanging vegetation and fruits. Clutches comprise 4–6 (occasionally 8) brittle, hardshelled eggs, measuring 47–60 x 34–40mm, which are laid in nests dug in soil 18–25cm deep. Incubation period 134 days. Hatchlings 53cm. **Distribution** Thailand (Chao Phya and Mae Klong Drainages, and other areas in southern peninsula), Laos, Cambodia, Vietnam and northern Peninsular Malaysia. Records from southern Myanmar require verification. **Status** Endangered.

ARAKAN FOREST TURTLE
Heosemys depressa PLATE 4
Measurement SCL 242mm **Identification** Carapace depressed; vertebral region flattened with an obtuse keel; posterior marginals serrated; plastron lacks hinge, truncated anteriorly, tapering posteriorly; anal notch distinct; bridge well developed; head with blunt snout; fingers half webbed; toes with rudiments of webs. **Coloration** Carapace light brown, sometimes with dark brown mottling; plastron yellowish-brown with black blotches or radiating lines on each scute; head grey; neck and limbs yellowish-brown; enlarged scales on limbs brown margined black. **Habitat and Behaviour** Inhabits evergreen forests containing extensive tracts of bamboo (*Melocanna bambusoides*). Diet and reproductive habits unstudied. **Distribution** Endemic to Myanmar (Rakhine Hills). **Status** Critically Endangered.

GIANT ASIAN POND TURTLE
Heosemys grandis PLATE 4
Measurement SCL 480mm **Identification** Carapace tricarinate, elevated in adults, depressed in juveniles; lateral keel does not extend across fourth costal; plastron lacks hinge; forehead with enlarged scales; front faces of forelimbs with transversely enlarged scales. **Coloration** Carapace dark brown, lighter around marginals; keels on carapace yellowish-orange or brown; forehead with orange or red vermiculation; plastral scutes with radiating pattern emerging from darker blotch in each scute in juveniles. **Habitat and Behaviour** Inhabits standing bodies of water. Diet includes macrophytes, molluscs, insects, worms and fish. Reproductive habits unstudied. **Distribution** Myanmar, Thailand, Laos, Cambodia, Vietnam and Peninsular Malaysia. **Status** Vulnerable.

SPINY HILL TURTLE
Heosemys spinosa PLATE 4
Measurement SCL 220mm **Identification** Carapace oval, arched in adults; strong vertebral keel; marginals of adults smooth; carapace of juveniles with greatly expanded marginals bearing distinct spines; plastron with weak hinge in females; fingers and toes partially webbed; hind limbs club-shaped. Adult males possess relatively longer and thicker tails than adult females. **Coloration** Carapace brown, usually with a pale vertebral keel; small, yellowish-orange or red spot behind eye; plastron brown, each scute with radiating lines. **Habitat and Behaviour** Inhabits tropical forests, from the lowlands to middle elevations. Diet includes both plants and animals. Clutches comprise 3 elongated, hardshelled eggs. Hatchlings 63mm. **Distribution** Southern Myanmar, Thailand, Peninsular Malaysia, Singapore, Sumatra, Pulau Bangka, Pulau Natuna, the Batu Archipelago and Borneo. Also the Sulu Archipelago and Tawi-Tawi (the Philippines). **Status** Endangered.

MALAYAN SNAIL-EATING TURTLE
Malayemys macrocephala PLATE 4
Measurement SCL 300mm **Identification** Carapace tricarinate; scutes thin; plastron with strong buttresses; head large, especially in old females; posterior plastron notched in males, more rounded in females; plastron lacks hinge. **Coloration** Carapace dark brown with black areoli and yellow rim; 4 or fewer nasal stripes; infraorbital stripe wide at loreal seam and does not extend or extends slightly superior to loreal seam; plastron yellow with large black blotch in each scute. **Habitat and Behaviour** Inhabits the lowlands, especially those associated with large river basins, and also coastal areas and wet agricultural fields, at elevations of up to 300m asl. Diurnal (with some nocturnal activity) and sedentary. Diet consists primarily of freshwater snails; fish, crabs, shrimps and leeches are also consumed. Clutches comprise 3–6 brittle-shelled eggs, measuring 32–44 x 20–24.5mm. Incubation period 99–225 days. Hatchlings 32.9–35.3mm. **Distribution** South-eastern Thailand (Chao Phraya and Mae Klong Basins) and northern Peninsular Malaysia (Kedah and Perlis States). **Status** Not Evaluated.

MEKONG SNAIL-EATING TURTLE
Malayemys subtrijuga PLATE 4
Measurement SCL 210mm **Identification** Carapace tricarinate; scutes thin; plastron with strong buttresses; head large, especially in old females; plastron lacks hinge. **Coloration** Carapace dark brown with black areoli and yellow rim; 6 or more nasal stripes; infraorbital stripe narrow at loreal seam, extending completely superior to loreal seam, and joins supraorbital stripe; plastron yellow with large black blotch in each scute. **Habitat and Behaviour** Inhabits the lowlands including ponds, agricultural fields and other wetlands. Diet includes freshwater snails. Reproductive habits unstudied. **Distribution** Eastern Thailand, Cambodia, Laos and southern Vietnam (Mekong Basin); historically recorded from Java. **Status** Vulnerable.

VIETNAMESE POND TURTLE
Mauremys annamensis PLATE 4
Measurement SCL 170mm **Identification** Carapace oval, low, tricarinate in juveniles; lateral keels indistinct in adults; posterior marginals weakly serrated; plastron lacks hinge; plastral buttresses strongly developed. **Coloration** Carapace dark grey-brown; plastron yellow or yellowish-orange, each scute with large black blotch; head dark brown with narrow pale yellow stripes across nostrils and eyes, extending up to neck, where it broadens; other parallel lines ventrally on head. **Habitat and Behaviour** Inhabits slow-moving streams and marshes in the lowlands. Diet and reproductive habits unstudied. **Distribution** Endemic to Vietnam (Buon Loy, Ban Loi, Gia Lai Province, Phuc Son, Da Nang Province and Fai-Fo, Quang Nam Province). **Status** Critically Endangered.

ASIAN YELLOW POND TURTLE
Mauremys mutica PLATE 4
Measurement SCL 194mm **Identification** Carapace oval, depressed, tricarinate; vertebral keel low; nuchal, small axillary and inguinals present; plastron lacks hinge; median notch of upper jaw shallow. **Coloration** Carapace yellow-brown or brown; marginals dark; plastron yellow, scutes with square or rectangular dark blotches; broad yellow postocular stripe to sides of neck; dark pigments on chin and neck reduced or absent. **Habitat and Behaviour** Inhabits ponds, marshes and swamps in the lowlands. Carnivorous, preferring fish. Eggs brittle-shelled, measuring 38 x 21mm. Incubation period ca 94 days. Hatchlings 32.4mm. **Distribution** Vietnam (central Annam). Also southern China (*M. m. mutica*), the Ryukyu Archipelago and central Japan (*M. m. kami*). **Status** Endangered.

CHINESE STRIPE-NECKED TURTLE
Mauremys sinensis PLATE 4
Measurement SCL 240mm **Identification** Carapace oval, moderately flattened, but not depressed; juvenile

carapace tricarinate, keels low and discontinuous, keels disappear in adults; Vertebrals IV and V wider than long; plastron lacks hinge. **Coloration** Carapace reddish-brown to greyish-brown; plastron yellow, each scute with large blackish-brown spot; head, neck, limbs and tail with numerous dark-bordered yellow stripes. **Habitat and Behaviour** Inhabits marshes, swamps, ponds and canals in the lowlands. Diet consists of fish, frogs and macrophytes. Clutches comprise 3 eggs, measuring 40 x 25mm. Hatchlings 35mm. **Distribution** Northern Vietnam. Also eastern China. **Status** Endangered.

INDIAN BLACK TURTLE
Melanochelys trijuga PLATE 5
Measurement SCL 280mm (subspecies *M. t. edeniana*) **Identification** Carapace elongated, fairly high in adults, depressed in juveniles, tricarinate; plastron lacks hinge; head moderate with short snout; upper jaw notched; toes fully webbed. **Coloration** Carapace brown or blackish-grey; plastron dark red with pale yellow border (lost in old individuals); head colour variable, and forms the basis of subspecific differentiation. **Subspecies** Five subspecies are recognized, of which one (*M. t. edeniana*) occurs in the region. *M. t. edeniana* has a brown head, sometimes with olive-brown or yellow reticulations. **Habitat and Behaviour** Inhabits standing bodies of water with aquatic vegetation, and also rivers and streams. Omnivorous, consuming freshwater prawns, grass, Common Water Hyacinths (*Eichhornia crassipes*) and fruits. Also scavenges a long distance from water. Reproductive habits of this subspecies unstudied. **Distribution** Myanmar and north-western Thailand (Mae Hong Song and Tak Provinces). Also eastern Pakistan (*M. t. trijuga*), India (*M. t. trijuga, M. t. coronata, M. t. indopeninsularis* and *M. t. thermalis*), Sri Lanka (*M. t. parkeri* and *M. t. thermalis*), the Maldives (probably introduced), Nepal (*M. t. indopeninsularis*) and Bangladesh (*M. t. indopeninsularis*). **Status** Lower Risk/Near Threatened.

BURMESE EYED TURTLE
Morenia ocellata PLATE 5
Measurement SCL 155mm **Identification** Carapace domed; smooth-shelled; low vertebral keel, especially in juveniles, becoming indistinct in adults; posterior marginals unserrated; plastron lacks hinge; head small with short pointed snout; digits entirely webbed; tail short. **Coloration** Dorsum greenish-brown or olive; vertebrals and costals with large yellow ocelli with dark brown centres; plastron unpatterned yellow; head olive or brown with 3 thin yellow stripes on each side, above and behind eyes and over jaws. **Habitat and Behaviour** Inhabits slow-moving and standing water bodies. Diet and reproductive habits unstudied. **Distribution** Endemic to Myanmar (Ayeyarwady Drainage). **Status** Vulnerable.

MALAYAN FLAT-SHELLED TURTLE
Notochelys platynota PLATE 5
Measurement SCL 400mm **Identification** Carapace flat with low, interrupted vertebral keel; 6–7 vertebrals; weak transverse hinge on plastron; toes fully webbed. **Coloration** Carapace olive, yellowish-brown or brick-red; hatchlings have bright green carapace; head brown; juveniles with 2 longitudinal yellow stripes; plastron yellowish-orange, each scute with black blotch. **Habitat and Behaviour** Inhabits shallow, macrophyte-dominated water bodies, including swamps, marshes and forest streams. Diet consists of aquatic macrophytes. Clutches comprise 3 hardshelled eggs, measuring 56 x 27–28mm. **Distribution** Southern Thailand, Vietnam, Peninsular Malaysia, Singapore, Borneo, Sumatra and Java. **Status** Vulnerable.

MALAYAN GIANT TURTLE
Orlitia borneensis PLATE 5
Measurement SCL 800mm **Identification** Carapace humped in juveniles, turning smooth with growth; adults with relatively narrow shell; plastron lacks hinge; head large; band-like scales on outer faces of forelimbs; fingers and toes extensively webbed. **Coloration** Carapace unpatterned black, brown or grey; plastron yellowish-orange or brown, sometimes with dark flecks; head dark with pale line from mouth to back of head. **Habitat and Behaviour** Inhabits large rivers and lakes, especially blackwater habitats. Diet in the wild unknown, but captives are omnivorous. Eggs are elongated with brittle hard shells, and measure 80 x 40mm. Hatchlings 60mm, with rough shell texture and markedly serrated marginals at carapace posterior. **Distribution** Peninsular Malaysia, Sumatra and Borneo. **Status** Endangered.

BEALE'S FOUR-EYED TURTLE
Sacalia bealei PLATE 5
Measurement SCL 143mm **Identification** Carapace elongated, unicarinate, slightly depressed; posterior marginals not serrated; vertebral keel low; plastral buttresses weak; forehead smooth; plastron lacks hinge. **Coloration** Carapace yellowish-brown to chocolate-brown, anterior margin with extensive black or dark brown speckling; plastron yellow to light olive, sometimes with dark vermiculations; forehead finely spotted or vermiculated with black; 2 distinct ocelli on posterior part of forehead; anterior pair of ocelli less distinct than posterior pair; front face of forearm yellow, lateral aspects brown; plastron pale with black blotches or streaks. **Habitat and Behaviour** Inhabits mountain streams in temperate forests at elevations of 100–400m asl. Diet in the wild unstudied. Clutches comprise 2–6 eggs. **Distribution** Vietnam. Also eastern China. **Status** Endangered.

FOUR-EYED TURTLE
Sacalia quadriocellata PLATE 5
Measurement SCL 140mm **Identification** Carapace elongated, unicarinate, slightly depressed; posterior marginals not serrated; vertebral keel low; plastral buttresses weak; forehead smooth; plastron lacks hinge. **Coloration** Carapace dark brown with radiating dark lines, anterior region lacks dark speckling or speckling much reduced, compared with that occurring in Beale's

Four-eyed Turtle (*S. bealei*); plastron yellow with numerous black dots, some elongated; forehead unpatterned olive or chocolate-brown; anterior pair of ocelli as distinct as posterior pair; no contrasting yellow and brown colour on forearm; plastron pale with black blotches or streaks. **Habitat and Behaviour** Inhabits mountain streams that are situated within temperate forests. Diet in the wild unstudied. Clutches comprise 2 eggs. **Distribution** Laos and Vietnam. Also southern China. **Status** Endangered.

BLACK MARSH TURTLE
Siebenrockiella crassicollis PLATE 5
Measurement SCL 200mm **Identification** Carapace oval with serrated posterior marginals; vertebral region of adults flattened; juveniles with 3 keels on carapace; adults with 1 keel; plastron lacks hinge. **Coloration** Carapace dark grey or nearly black; plastron pale grey with large dark areas in each scute; juveniles with light head spots that are retained by adult females, but fade with growth in males. **Habitat and Behaviour** Inhabits standing or sluggish water bodies, including marshes, swamps, ponds, streams and lakes. Diet includes macrophytes and aquatic animals; also known to scavenge. Clutches comprise 1–2 eggs, measuring 45 x 19mm; 3–4 clutches produced annually. Hatchlings 52mm. **Distribution** Southern Myanmar (Tenasserim, Tanintharyi Division), Thailand, Laos, Cambodia, Vietnam, Peninsular Malaysia, Singapore, Borneo, Sumatra and Java. **Status** Vulnerable.

Family CHELONIIDAE SEA TURTLES

Members of this family have a large head that is non-retractable into the shells, a smooth carapace bearing scutes, a large plastron, a secondary palate, elongated and flipper-like forelimbs, and short and rounded hind limbs. The family includes all the world's marine turtles, except the sole living member of the Dermochelyidae (see page 174). Cosmopolitan in the world's oceans and exclusively aquatic, only the adult females generally come ashore, although in some parts of the world adult males may ascend beaches to bask. The diet of sea turtles includes a variety of marine organisms, ranging from jellyfish to corals. All species are oviparous, and their eggs are laid in deep holes excavated in sandy beaches. Most populations of sea turtles are threatened by anthropogenic factors, including the harvesting of adult turtles for flesh, incidental capture in fishing gear, sand mining and modifications to nesting habitats.

LOGGERHEAD SEA TURTLE
Caretta caretta PLATE 6
Measurement SCL 1,200mm **Identification** Carapace elongated with tapering end; 5 pairs of costals, costal I contacting nuchal; head massive; 3–4 infralabial scutes that lack pores; 13 marginal scutes. Can be differentiated from Olive Ridley Turtle (*Lepidochelys olivacea*) in showing 5 (vs 6) costals and bridge with 3 (vs 4) inframarginals. **Coloration** Carapace reddish-brown; plastron yellowish-brown or yellowish-orange. Can be differentiated from *L. olivacea* in showing reddish-brown carapace (vs olive-green or greyish-olive). **Habitat and Behaviour** Associated with warm subtropical seas, bays, lagoons and estuaries. Diet consists of molluscs and crustaceans. Clutches comprise 23–178 eggs, measuring 34.7–55.2mm. Incubation period 49–80 days. **Distribution** Cosmopolitan. **Status** Endangered.

GREEN TURTLE
Chelonia mydas PLATE 6
Measurement SCL 1,400mm **Identification** Carapace heart-shaped; pair of prefrontal scales on forehead; scutes of carapace not overlapping; upper jaw without hook; forelimbs with single claw. **Coloration** Carapace olive or brown with radiating dark pattern; plastron pale yellow. English name comes from colour of turtle's fat, which was once in demand for making turtle soup. **Habitat and Behaviour** Associated with tropical regions, and common around oceanic islands and along coasts with wide sandy beaches. Juveniles are carnivorous, while adults consume only seagrass and seaweeds. Eggs are softshelled and spherical, and each nest contains 98–172 eggs, measuring 41.4–42.1mm. Up to 11 nests may be laid within a nesting season. Incubation period ca 60 days. **Distribution** Cosmopolitan. **Status** Endangered.

HAWKSBILL SEA TURTLE
Eretmochelys imbricata PLATE 6
Measurement SCL 1,000mm **Identification** Carapace heart-shaped; scutes of carapace with 4 pairs of imbricate costal scutes; 2 pairs of prefrontal scales; upper jaw relatively narrow, elongated; upper jaw forwards projecting to form bird-like beak. **Coloration** Carapace olive-brown; juveniles with darker blotches; plastron pale in adults, dark in juveniles. **Habitat and Behaviour** Inhabits reefs, bays, estuaries and lagoons. Diet includes sponges, algae, corals and shellfish. Clutches comprise 96–177 eggs, measuring 30–35mm. Incubation period 57–65 days. **Distribution** Cosmopolitan. **Status** Critically Endangered.

OLIVE RIDLEY SEA TURTLE
Lepidochelys olivacea PLATE 6
Measurement SCL 800mm **Identification** Carapace broad, heart-shaped, posterior marginals serrated, with juxtaposed costal scutes; 5–9 pairs of costals; bridge with 4 inframarginals, each with pore; adult shell smooth; hatchling shell tricarinate; upper jaw hooked. **Coloration** Carapace olive-green or greyish-olive; plastron greenish-yellow; juveniles grey-black dorsally; venter cream. **Habitat and Behaviour** Associated with sea beaches in the vicinity of

mangroves. Breeding tends to take place en masse, involving hundreds and even thousands of females, in a phenomenon referred to as 'arribadas'. Clutches comprise 50–160 eggs, measuring 34–43mm. Incubation period 45–60 days. Hatchlings 37.9–49.9mm. **Distribution** Pacific and Indian Oceans. **Status** Vulnerable.

Family DERMOCHELYIDAE LEATHERBACK SEA TURTLES

Species with a broad, ridged shell lacking epidermal scutes. In leatherbacks the dermal bones of the shell are replaced by a mosaic of small platelets, the limbs are paddle-shaped, lacking claws, the forelimbs are enlarged and the hind limbs are broadly connected to the tail via a web. The sole living species inhabits the world's oceans and feeds on jellyfish, as well as crustaceans, molluscs, fish and seaweeds. It frequently wanders into the cold waters of the Arctic, presumably in search of jellyfish.

LEATHERBACK SEA TURTLE
Dermochelys coriacea PLATE 6
Measurement SCL 2.5m **Identification** Carapace elongated, tapered towards end; 7 ridges on carapace and 5 on plastron; shell covered with skin in adults; distinct scales present on hatchling shells; limbs paddle-like and clawless. **Coloration** Dorsum black or blackish-blue with paler flecks; plastron pale grey or pinkish-grey. **Habitat and Behaviour** This inhabitant of the open oceans nests on oceanic islands as well as on wide mainland beaches. Diet consists primarily of jellyfish; in northern temperate waters it is capable of diving up to 1,200m below the surface in search of food. It nests more deeply than other sea turtles. Clutches comprise 90–130 eggs, measuring 50–54mm. **Distribution** Cosmopolitan. **Status** Critically Endangered.

Family EMYDIDAE NEW WORLD HARDSHELLED TURTLES

Species in this family are characterized by an oval to oblong domed carapace with 11 pairs of sutured peripherals around the margin, and a nuchal lacking costiform processes. The plastron is large, sometimes hinged and lacks a mesoplastron, and the plastral buttresses typically articulate with the costals of the carapace. The neck withdraws vertically, and the pelvic girdle articulates flexibly with the plastron. These turtles are primarily found in freshwater habitats, including ponds, marshes and rivers, although a few species inhabit coastal or brackish water habitats. Their diet tends to be omnivorous, and only exceptionally carnivorous or herbivorous. The natural distribution of the New World hardshelled turtles includes North and South America, and Europe to the Ural Mountains and adjacent regions; in South-East Asia there is a single introduced species.

RED-EARED SLIDER
Trachemys scripta PLATE 7
Measurement SCL 280mm **Identification** Carapace rounded with nearly smooth outline; plastron lacks hinge; females much larger than males, and males possess elongated claws on fingers. **Coloration** Carapace green with yellow lines, turning darker with growth; plastron bright yellow with large black mark on each scute; orange or red patch on temples; neck with yellow stripes. **Habitat and Behaviour** Omnivorous, feeding on leaves, fruits, fish, frogs and carrion. Clutches comprise 2–25 eggs, measuring 30–42 x 19–29mm. Incubation period 65–75 days. Hatchlings 30–33mm. **Distribution** A native species of the Mississippi River Drainage, USA. In South-East Asia, naturalized populations are known from Thailand, Malaysia, Singapore and Indonesia. **Status** Lower Risk/Near Threatened.

Family PLATYSTERNIDAE ASIAN BIG-HEADED TURTLES

Species with a massive head that cannot be retracted into the flattened shell, which is subrectangular. The plastron is reduced and connected to the carapace by ligamentous tissue. The tail is long, especially in juveniles, in which it may equal or exceed the carapace length. These turtles inhabit mountain streams and are nocturnal, prowling after dark for small invertebrates. The family contains a single living species from Indo-China.

BIG-HEADED TURTLE
Platysternon megacephalum PLATE 7
Measurement SCL 184mm **Identification** Carapace oval, flattened; vertebral keel present; head large and covered with undivided scales; jaws hooked; throat with flattened rounded tubercles; tail covered with squarish scales. **Coloration** Carapace yellowish-brown, grey to olive, each scute with radiating dark pattern; plastron and bridge yellow, sometimes with darker pattern. **Subspecies** Three subspecies are recognized, of which one (*P. m. peguense*), showing dark plastral seams and a black-bordered pale postorbital stripe, occurs in the region. **Habitat and Behaviour** Inhabitant of cool hill streams that

clambers onto rocks and is even capable of climbing trees. Carnivorous, feeding on insects, snails and worms. Clutches comprise 2–3 eggs, measuring 33–37mm. Hatchlings 38–40mm. **Distribution** Southern Myanmar, Thailand, Laos and western Vietnam (*P. m. peguense*). Also southern and eastern China (*P. m. megacephalum*), and northern Vietnam (*P. m. shiui*). **Status** Endangered.

Family TESTUDINIDAE LAND TORTOISES

Species possessing a high shell, head and limbs with heavy scales, columnar forelimbs, club-shaped hind limbs, and a head and limbs that are completely retractable inside the shell. While the family reaches its greatest diversity in the world's arid regions, the South-East Asian species (with one exception) are linked to relatively moist habitats. A bulk of their diet consists of vegetation, although carrion is sometimes ingested, probably as a source of calcium and protein. The eggs are buried in the ground or laid in leaf litter mounds. Tortoises are cosmopolitan in distribution, except in Australia.

BURMESE STAR TORTOISE
Geochelone platynota PLATE 7
Measurement SCL 260mm **Identification** Carapace high, convex; nuchal scute absent; posterior marginals weakly serrated. **Coloration** Shell black or dark brown, each vertebral and costal scute with yellow centre and radiating lines; plastron yellow, with plastron scutes with black or dark brown patches; head and limbs yellow or tan. **Habitat and Behaviour** Restricted to dry scrub and semi-desert scrub regions. Ecologically similar to the more familiar Indian Star Tortoise (*G. elegans*), feeding on herbs and fallen fruits, and producing rounded eggs that measure 40 x 55mm and are laid in shallow nests excavated in earth. **Distribution** Endemic to central Myanmar. **Status** Critically Endangered.

ELONGATED TORTOISE
Indotestudo elongata PLATE 7
Measurement SCL 330mm **Identification** Carapace domed, highest point in Vertebral III, flattened dorsally with arching sides; posterior marginals slightly flared and serrated (especially in juveniles), shell broadest posteriorly; plastron elongated with deep notch posteriorly; limbs heavily scaled, club-like, bearing 5 claws each; nuchal scute long and narrow. **Coloration** Carapace and plastron yellowish-brown or olive with scattered black blotches, though plastron sometimes unpatterned. **Habitat and Behaviour** Inhabits deciduous and evergreen forests. Primarily herbivorous, with diet including leaves, flowers, fruits, fungi, dead animal matter and slugs. Eggs are hardshelled, spherical or elongated. Clutches comprise 1–7 brittle eggs, measuring 50 x 37mm. Incubation period 96–165 days. Hatchlings 49mm. **Distribution** Myanmar, Thailand, Cambodia, Laos, Vietnam and northern Peninsular Malaysia (Perlis State). Also eastern and north-eastern India, Bangladesh and Nepal. **Status** Endangered.

ASIAN GIANT TORTOISE
Manouria emys PLATE 7
Measurements SCL 500mm (*M. e. emys*); 580mm (*M. e. phayrei*) **Identification** Carapace low; vertebral region depressed; distinct growth rings on carapace scutes; posterior marginals weakly serrated; upper jaw hooked; outer surfaces of forelimbs bear large scales; pair of tuberculate scales on thighs; pectoral scutes separated or fused with each other. **Coloration** Shell light or dark brown; plastron lighter; head and limbs brown. **Subspecies** Two subspecies are recognized. *M. e. emys*, shell medium brown; pectoral scutes small and separated. *M. e. phayrei*, shell blackish-brown; pectoral scutes large and fused with each other. **Habitat and Behaviour** Inhabits primary forests in the mid-hills. Diet consists of vegetation, although insects and frogs are also eaten. Constructs a mound nest by sweeping leaf litter in which 23–56 hardshelled spherical eggs of diameter 51–54mm are deposited. Thereafter it guards the nest. Incubation period 60–75 days. Hatchlings 60–66mm. **Distribution** Southern Thailand, Vietnam, Cambodia, Laos, Peninsular Malaysia, Sumatra and Borneo (*M. e. emys*), Myanmar and northern Thailand (*M. e. phayrei*). Also north-eastern India and Bangladesh. **Status** Endangered.

IMPRESSED TORTOISE
Manouria impressa PLATE 7
Measurement SCL 302mm **Identification** Carapace low; vertebral and costal regions slightly concave; carapace scutes mostly smooth; posterior marginals strongly serrated; upper jaw hooked; outer surfaces of forelimbs bear large scales; single tuberculate scale on each thigh. **Coloration** Scutes on carapace translucent brown to yellowish-orange, with brown, black or orange streaks; plastron yellowish-brown with radiating dark lines; head yellow; limbs brown. Juveniles yellowish-brown with fine black speckles. **Habitat and Behaviour** Inhabits mid-elevation evergreen and bamboo forests at elevations of 900–1,200m asl. Diet consists of vegetation, especially mushrooms. Clutches comprise 17–20 eggs, measuring 44 x 40mm, which are laid in shallow nests. Hatchlings 39g. **Distribution** Myanmar (Kachin State and Karen Hills), Thailand, Laos, Cambodia (Cardamom Mountains), northern Vietnam and northern Peninsular Malaysia. Also southern China (record from market). **Status** Vulnerable.

Family TRIONYCHIDAE SOFTSHELL TURTLES

Species in this family possess a skin-clad, flattened shell (with two regional exceptions) with reduced bony elements, three claws on each limb and nostrils located on a fleshy proboscis. One genus has femoral flaps, which protect the hind limbs of the turtle after it has retracted into the shell. The family includes some of the world's largest freshwater turtle species – they are primarily inhabitants of freshwaters, with only one invading brackish waters or sea coasts. Their diet consists mostly of small invertebrates and fish, and they are also known to scavenge.

ASIAN SOFTSHELL TURTLE
Amyda cartilaginea PLATE 8
Measurement SCL 750mm **Identification** Carapace rounded or oval; differentiated from Malayan Softshell Turtle (*Dogania subplana*) by the rounded (vs straight) sides to its carapace and its relatively narrower head; distinct tubercles at anterior carapace margin. **Coloration** Dorsum greenish-grey or olive, sometimes with yellow-bordered black spots or radiating streaks, which tend to disappear with growth; plastron cream in males, grey in females; head with light blotches on dark grey or olive background in northern populations, nearly unpatterned in populations from the Sundas. **Habitat and Behaviour** Inhabits side streams of large muddy rivers, peat swamps and marshes. Nocturnal and carnivorous, feeding on fish, frogs, shrimps and water insects. Clutches comprise 4–8 eggs, measuring 21–33mm, which are laid in nests excavated in riverbanks. Incubation period 130–140 days. Hatchlings 37.2–49.5mm. **Distribution** Myanmar, Thailand, Peninsular Malaysia, Singapore, Laos, Cambodia, Vietnam, Borneo, Sumatra, Java and Bali. Also north-eastern India, and Lombok and Sulawesi (Indonesia), the latter two localities probably based on introductions. **Status** Vulnerable.

NARROW-HEADED SOFTSHELL TURTLE
Chitra chitra PLATE 8
Measurement SCL 1,220mm **Identification** Carapace rim smoothly joins cartilaginous part to skin of neck; head small with tiny proboscis; eyes situated close to snout-tip; neural bones narrow; carapace does not end abruptly; lower arm with 2–4 scales. **Coloration** Dorsum brown or yellowish-brown with bright yellow, greenish-yellow or tan dark-edged stripes, including inverted chevron-like marking on anterior of carapace and neck; neck with 5 lines; plastron cream or pale pink. **Subspecies** Two subspecies are recognized. *C. c. chitra*, pale coloration; midline and lateral vertebral carapacial stripes present; bell-shaped mark on carapace anterior indistinct; no X-shaped figure between eyes; ocelli in eye region; indistinct dark speckling and ocelli on chin; broad costal markings. *C. c. javanensis*, dark coloration, especially in juveniles; midline and lateral vertebral carapacial stripes typically missing; bell-shaped mark on carapace anterior distinct; X-shaped figure between eyes; ocelli in eye region absent; bolder black speckling and ocelli on chin; narrower, more elongated costal markings. **Habitat and Behaviour** Inhabits medium to large fast-flowing rivers with sandy bottoms. *C. c. javanensis* associated with tidal creeks. Piscivore. Clutches comprise 60–117 eggs, measuring 34–40mm. Incubation period 62–66 days. Hatchlings 40–42mm. **Distribution** Thailand, including Khwae, Mae Klong and Mae Ping River systems and isolated localities in Peninsular Malaysia (*C. c. chitra*), and central and eastern Java, including Pasuruan and Solo Rivers (*C. c. javanensis*), with a report from Danau Sentarum, western-central Borneo. **Status** Critically Endangered.

BURMESE NARROW-HEADED SOFTSHELL TURTLE
Chitra vandijki PLATE 8
Measurement SCL 412mm **Identification** Carapace rim smoothly joins cartilaginous part to skin of neck; head small with tiny proboscis; eyes situated close to snout-tip; bony shell broad with broad neural bones; thick-edged bony carapace ends abruptly. **Coloration** Carapace chocolate-brown or olive-green; indistinct dark-bordered head stripes with speckling in between; V-shaped neck marking; no dark vertebral stripe; 3 black-bordered stripes on neck; transverse light, dark-bordered bar connecting eyes, and 1–2 pairs of entire or nearly entire light, dark-bordered ocelli posterior to transverse bar between or behind eyes; plastron white or pinkish. **Habitat and Behaviour** Inhabits rivers with sandy bottoms. Diet consists of fish and presumably shellfish. Reproductive habits unstudied. **Distribution** Endemic to Myanmar (Ayeyarwaddy River system). **Status** Not evaluated.

MALAYAN SOFTSHELL TURTLE
Dogania subplana PLATE 8
Measurement SCL 350mm **Identification** Carapace flat, oval, with distinctly straight sides; head large, bearing down-turned snout; adults with carapace hinge. **Coloration** Carapace dark olive or brown with dark median stripe and 2–3 pairs of black-centred, eye-like spots; pattern most distinct in juveniles, fades with growth; plastron cream or grey; juveniles with reddish-brown blotch behind eyes. **Habitat and Behaviour** Encountered in highlands and associated with clear, rocky and shallow streams with sandy bottoms. Large head adaptive for cracking shells of aquatic snails; fish, prawns and crabs also eaten. **Distribution** Southern Myanmar (Tennasserim, Taninthayi Division and the Mergui Archipelago), Thailand, Peninsular Malaysia, Singapore, Java, Sumatra, Pulau Singkep, Pulau Natuna, Borneo and Java. Also Palawan (Philippines). **Status** Lower Risk/Least Concern.

INDIAN FLAPSHELL TURTLE
Lissemys punctata PLATE 8
Measurement SCL 370mm **Identification** Carapace oval, domed; plastron with 7 callosities, its hinged

anterior lobe closing completely; pair of flaps covers hind limbs when retracted; entoplastral callosity small in adults. **Coloration** Carapace olive-green with dark yellow blotches; head with yellow spots; plastron cream or pale yellow. **Subspecies** Two subspecies are recognized, of which one (*L. p. andersoni*) occurs in the region. **Habitat and Behaviour** Inhabits rivers, ponds, oxbow lakes, streams, rice fields and canals. Active by both day and night, while feeding at dusk. Diet consists of frogs, tadpoles, fish, crustaceans, snails, earthworms, insects, carrion and water plants. Clutches comprise 5–14 brittle, hardshelled eggs, measuring 24–30mm; more than a single clutch may be laid in a season. Incubation period 9 months. Hatchlings 42mm. **Distribution** Myanmar (*L. p. andersoni*). Also Pakistan (*L. p. andersoni*), India (*L. p. punctata* and *andersoni*), Nepal (*L. p. andersoni*), Sri Lanka (*L. p. punctata*) and Bangladesh (*L. p. andersoni*). **Status** Lower Risk/Least Concern.

BURMESE FLAPSHELL TURTLE
Lissemys scutata PLATE 8
Measurement SCL 370mm **Identification** Carapace oval, domed; plastron with 7 callosities, its hinged anterior lobe closing completely; pair of flaps covers hind limbs when retracted; entoplastral callosity enlarged in adults. **Coloration** Carapace brownish-olive, sometimes with fine black spots or reticulations; head with dark postocular stripe edged with a paler one; plastron unpatterned yellow. **Habitat and Behaviour** Inhabits both flowing water bodies (such as rivers) and standing waters (especially ponds caused by flooding of rivers) with dense macrophytic growth. Presumably carnivorous and nocturnal. Reproductive habits unstudied. **Distribution** Endemic to Myanmar (Ayeyarwady and Salween Rivers). **Status** Data Deficient.

BURMESE SOFTSHELL TURTLE
Nilssonia formosus PLATE 9
Measurement SCL 650mm **Identification** Carapace rounded, smooth in adults and with longitudinal rows of tubercles in juveniles; series of enlarged blunt tubercles above neck; preneural bone absent. **Coloration** Carapace olive-grey to brown with dark reticulations; juveniles with 4 dark-centred, light-bordered occelli, which disappear in adults; forehead with dark vermiculations on light background, less distinct in large individuals; paired elongated yellow spots at back of head; other spots on temples, corners of jaws and chin; plastron unpatterned cream. **Habitat and Behaviour** Inhabits large rivers, including their upper reaches. Diet and reproductive habits unstudied. **Distribution** Endemic to Myanmar (Ayeyarwadi, Sittang and lower Salween Rivers). **Status** Endangered.

WATTLE-NECKED SOFTSHELL TURTLE
Palea steindachneri PLATE 9
Measurement SCL 426mm **Identification** Carapace oval; juveniles with longitudinal rows of raised tubercles, becoming smooth in adults; anterior rim of carapace with tubercle-bearing ridges; preneural bone absent. **Coloration** Carapace unpatterned olive, brown or grey; head and limbs olive or brown; black stripes from orbits of eyes; pale yellow postocular stripe to neck; corners of jaws with yellow spot; plastron cream or yellow, typically unpatterned. **Habitat and Behaviour** Inhabits marshes and small rivers at elevations of up to 1,500m asl. Diet in captivity dominated by invertebrates and small vertebrates; aquatic macrophytes also consumed. Clutches comprise 3–28 spherical eggs, measuring 22mm. Hatchlings 54–58mm. **Distribution** Vietnam. Also southern and eastern China. Introduced into Hawaii (USA). **Status** Endangered.

ASIAN GIANT SOFTSHELL TURTLE
Pelochelys cantorii PLATE 9
Measurement SCL 1,500mm **Identification** Carapace flattened, relatively more elongated in juveniles and oval in adults; head short; eyes close to tip of snout; juveniles with numerous tubercles on carapace and low vertebral keel; proboscis very short and rounded. **Coloration** Carapace olive or brown, sometimes spotted or streaked with lighter or darker shades, with lighter outer edge; plastron unpatterned white or with pink blotches. **Habitat and Behaviour** Mainly associated with coasts and large rivers. Diet consists of fish, shrimps, crabs and molluscs, and sometimes aquatic plants. Clutches comprise 24–70 eggs, measuring 30mm. Hatchlings 42mm. **Distribution** Myanmar, Thailand, Laos, Cambodia, Vietnam, Peninsular Malaysia, Singapore, Sumatra and Borneo. Also India, Bangladesh, southern China, the Philippines and possibly Sulawesi. **Status** Endangered.

CHINESE SOFTSHELL TURTLE
Pelodiscus sinensis PLATE 9
Measurement SCL 250mm **Identification** Carapace oval, slightly longer than wide, smooth in adults and with longitudinal rows of low tubercles in juveniles; preneural bones absent. **Coloration** Dorsum unpatterned olive to greyish-green; juveniles with light-bordered spots; fine radiating black lines around eyes; plastron unpatterned cream, grey or yellow in adults, pinkish-red with black blotches in hatchlings. **Habitat and Behaviour** Inhabits ponds, marshes and rivers in the lowlands and mid-hills up to an elevation of at least 600m asl. Rarely basks. Diet consists of insects, crustaceans and fish; aquatic macrophytes also consumed. Clutches comprise 9–28 eggs, measuring 20–24mm. Incubation period 40–80 days. Hatchlings 27mm. **Distribution** Vietnam. Also southern China, Korea, Japan and Russia. Introduced into Thailand, Peninsular Malaysia, Singapore, Sumatra and other extralimital areas (e.g. the Philippines, Timor, Hawaii and the Bonin Islands). **Status** Vulnerable.

INDO-CHINESE GIANT SOFTSHELL TURTLE
Rafetus swinhoei PLATE 9
Measurement SCL 600mm **Identification** Carapace oblong; single neural separates first pair of costals; eighth pair of costals reduced and fails to meet at midline; 2 plastral callosities. **Coloration** Dorsum olive-

green, with numerous yellow spots encircled by yellow dots, sometimes forming stripes, especially in juveniles; head, neck and chin olive with yellow spots; plastron unpatterned pale olive-grey. **Habitat and Behaviour** Inhabits large water bodies such as lakes and reservoirs. Diet includes molluscs, insects, crustaceans and marsh plant seeds. Clutches comprise 130 eggs, measuring 20mm in diameter. **Distribution** Northern Vietnam. Also eastern China. **Status** Critically Endangered.

Order SQUAMATA

Family AGAMIDAE AGAMID LIZARDS

In the agamid lizards the large overlapping dorsals lack osteoderms, the pectoral girdle has a T-shaped or cruciform interclavicle, the clavicles are curved and rod-shaped, and the tail and limbs are well developed. Cranial characters include acrodont dentition, pterygoid lacking teeth and a developed cranial crest. There is no fracture plane in the caudal vertebrae. Agamid lizards are related to iguanas and chamaeleons, and are included in the superfamily Iguania. They are terrestrial, inhabiting subtropical and tropical regions, with many species entering montane limits. All species in the family are oviparous, producing eggs with leathery shells. The diet of most small species includes arthropods. Larger species are partially herbivorous, their diet including flowers, leaves, petals and seeds, although they are insectivorous when young. Members of the Agamidae are found in Asia, Europe, Africa, New Guinea and Australia.

GREATER SPINY LIZARD
Acanthosaura armata PLATE 10
Measurement SVL 140mm **Identification** Body robust, compressed; superciliary and occipital spines reach level of nuchal crest; nuchal crest separated from or joined to dorsal crest; supralabials 11–13; infralabials 13–15; small gular pouch in both sexes; tail longer than head-body length. **Coloration** Dorsum grey, brown to nearly black with darker marbling on back and flanks; large diamond-shaped mark on axilla and triangular patch on sides of head that encloses eye; forehead greenish-yellow, sometimes with cross-bars; flanks with pale green or grey patches; tail dark-banded; venter in shades of green, brown or red, sometimes spotted with black. **Habitat and Behaviour** Inhabits lowland forests, from mangrove swamps at sea level to dipterocarp forests at elevations of 850m asl. Diurnal and arboreal. Diet consists of large insects. Clutches comprise 12–16 eggs, measuring 11 x 19.5mm. Incubation period 191–193 days. Hatchlings 30mm. **Distribution** Myanmar, Thailand, Peninsular Malaysia, Singapore, Sumatra and Pulau Anamba. Also eastern China. **Status** Not Evaluated.

INDO-CHINESE SPINY LIZARD
Acanthosaura capra PLATE 10
Measurement SVL 137.9mm **Identification** Body robust, compressed; occipital spine between tympanum and nuchal crest; supralabials 10–12; infralabials 10–12; tympanum one-third orbit diameter; dorsals granular, slightly keeled, intermixed with larger ones; distinct antehumeral fold; small gular pouch; nuchal crest comprises broad lanciform scales; dorsal crest not joined to nuchal crest, comprising shorter scales that reduce in height caudally, becoming a ridge on tail; tail longer than head-body length; lamellae under Toe IV 22–24. **Coloration** Dorsum green or olive with black spots, and yellow spots encircled with black; head greenish-yellow in males, with broad green postocular band to axilla; nuchal crest yellow; green stripe on nape; throat black; in males, large yellow gular pouch with green streaks; in females, head green, gular pouch bluish-green; venter greenish-cream; iris reddish-brown. **Habitat and Behaviour** Inhabits lowland and mid-hill evergreen forests at elevations of up to 500m asl. Diurnal and arboreal, associated with the canopy. Diet consists of large insects and small lizards. Clutches comprise 19–20 eggs. Incubation period 169–190 days. **Distribution** Eastern Cambodia (Mondolikiri Province) and southern Vietnam (Dong Nai, Khanh Hoa and Lam Dong Provinces). Records from Laos require verification. **Status** Not Evaluated.

CROWNED SPINY LIZARD
Acanthosaura coronata PLATE 10
Measurement SVL 137.5mm **Identification** Body robust, compressed; occipital spine between tympanum and nuchal crest; tympanum subequal to orbit; nuchal and dorsal crests continuous, low, comprising triangular scales, extending to base of tail; gular pouch absent; antehumeral fold present; tail subequal to head-body length. **Coloration** Dorsum light green with grey-brown mottling in males, brownish-red in females; irregular dark brown cross-bars on dorsum and limbs; light green band, edged with black, across superciliaries; short, oblique white and light green band, edged with brown, from below orbit to supralabials; radiating brown streaks from orbit; tail with brown and pinkish-orange bands; orange-red bar on pectoral region; venter greyish-white with dark flecks. **Habitat and Behaviour** Inhabits evergreen forests at elevations of 480–700m asl. Diurnal and both terrestrial and arboreal, associated with the ground, and ascending trees when threatened; sleeps on saplings at night. Diet and reproductive habits unstudied. **Distribution** Eastern

Cambodia (O'Rang and Pichrada Districts) and southern Vietnam (Lam Dong and Dong Nai Provinces). **Status** Not Evaluated.

MASKED SPINY LIZARD
Acanthosaura crucigera PLATE 10
Measurement SVL 140mm **Identification** Body robust, compressed; supralabials 9–14; infralabials 9–13; nuchal crest tall, separated from a lower dorsal crest; spines of nuchal and dorsal crests broad basally; gular pouch present; tail longer than head-body length. **Coloration** Dorsum greenish-yellow, typically with dark indistinct network of enclosing pale yellow spots; dark facial mark; sides of head dark brown to black; rhomboidal nape patch, dark brown or black, present or absent; tail dark-banded. **Habitat and Behaviour** Inhabits lowland evergreen, deciduous and montane forests, as well as forest clearings, at elevations of 200–1,800m asl. Diurnal and arboreal, associated with trees, although it conceals itself on the ground. Diet unstudied. Clutches comprise 10–18 eggs, measuring 12 x 20mm. **Distribution** Southern Myanmar (Dawei, Tenasserim, Taninthayi Division), Thailand, Cambodia, Vietnam and northern Peninsular Malaysia (Pulau Langkawi, Kedah State). **Status** Not Evaluated.

SCALE-BELLIED SPINY LIZARD
Acanthosaura lepidogaster PLATE 10
Measurement SVL 111mm **Identification** Body robust, slightly compressed; head subtriangular; supralabials 11–12; infralabials 9–10; tympanum exposed; low dorsal crest; gular pouch absent; antehumeral fold present; dorsal scales heterogenous; nuchal crest comprises 6 conical scales; dorsal crest low, composed of subtriangular scales, decreasing caudally; tail longer than head-body length; lamellae under Toe IV 22. **Coloration** Dorsum bright green changeable to dark brown; throat and neck dark; forehead black; indistinct diamond-shaped black mark on nape. **Habitat and Behaviour** Inhabits lowland and submontane forests at elevations of 700–1,400m asl. Diurnal and arboreal. Diet consists of insects and other arthropods. Reproductive habits unstudied. **Distribution** Myanmar (Bago Division, Kayah and Rakhine States), northern Thailand (Khao Yai National Park and Phu Khieo National Park), Laos, Cambodia, and northern and central Vietnam. Also eastern China. **Status** Not Evaluated.

NATALIA'S SPINY LIZARD
Acanthosaura nataliae PLATE 10
Measurement SVL 158mm **Identification** Body robust, compressed; single postorbital spine; occipital spine above tympanum absent; large keeled scales intermixed with smaller scales on dorsals and flanks; supralabials 11; infralabials 11–12; nuchal and dorsal crests distinct, comprising lanceolate scales pointing posteriorly, separated from each other; row of rounded, enlarged scales below dorsal scales; gular pouch large in both males and females; 2–3 rows of papillar scales on midline of venter; 5–6 enlarged preanal scales; lamellae under Toe IV 20–27. **Coloration** Dorsum of males yellowish-brown changeable to shades of red, brown and yellow; head, nuchal and dorsal crests, and limbs red; gular pouch red with 3–5 oblique dark brown bands; black mask on head covering orbit and tympanic region; dark brown postocular stripe; tail yellowish-brown with 9–10 wide dark bands; venter light grey; iris red, blue ring around pupil; dorsum of females emerald-green, other colours similar to those of males; juveniles similar to adult males, with black spots. **Habitat and Behaviour** Inhabits wet evergreen forests in the mid-hills and submontane limits at elevations of 350–1,400m asl. Diurnal and arboreal; adults associated with trunks of tall trees, juveniles with bushes and areas close to substratum. Diet includes flowers and fruits. Clutches comprise 16 eggs. **Distribution** Southern Laos (Saravan and Xekong Provinces) and central Vietnam (Gia Lai, Kon Tum, Quang Nam, Da Nang, Thua Thien-Hue and Quang Tri Provinces). **Status** Not Evaluated.

LONG-SNOUTED SHRUB LIZARD
Aphaniotis acutirostris PLATE 10
Measurement SVL 72mm **Identification** Body slender, compressed; snout acute, longer than eye diameter; projecting convex scale above rostral; limbs long; dorsal scales small with larger scattered scales; row of tuberculate scales on paravertebral region; gular pouch weak; males with weak nuchal crest. **Coloration** Dorsum brown with darker variegation; radiating dark lines from orbit of eye; throat sometimes with dark spots; rest of venter pale brown; males with distinct yellow gular pouch; inner lining of mouth blue. **Habitat and Behaviour** Inhabits lowland rainforests. Diurnal and arboreal, associated with shrubs. Diet presumably consists of small insects and other invertebrates. Clutches comprise 2 eggs, measuring 7 x 12mm. Incubation period 48–63 days. **Distribution** Borneo (western Kalimantan Province), Sumatra, Pulau Simeuleu, Pulau Nias, the Mentawai Archipelago and Pulau Berhala. **Status** Not Evaluated.

BROWN SHRUB LIZARD
Aphaniotis fusca PLATE 10
Measurement SVL 67mm **Identification** Body slender, compressed; snout rounded, not longer than diameter of eye; limbs long; Toe V longer than Toe I; dorsal scales small with larger scattered scales; nuchal crest reduced, composed of short triangular scales; gular pouch weak; tail long, slender. **Coloration** Dorsum dark brown or brownish-olive; venter pale olive-brown; 2 dark interorbital bars; inner lining of mouth dark blue; gular pouch in adult males black with oval yellow spots. **Habitat and Behaviour** Inhabits primary and lightly disturbed lowland forests and the mid-hills. Diurnal and arboreal, associated with stems of saplings. Diet consists of caterpillars, beetles, millipedes, cockroaches and termites. Clutches comprise 1–2 eggs, measuring 7 x 18mm. Hatchlings 23–43mm. **Distribution** Southern Thailand, Peninsular Malaysia, Singapore, Sumatra,

Pulau Simalur, Pulau Nias, Pulau Berhala, Borneo, Singkep and the Natuna Islands. **Status** Not Evaluated.

ORNATE SHRUB LIZARD
Aphaniotis ornata   PLATE 10
Measurement SVL 57mm **Identification** Body slender, compressed; snout as long as eye diameter; snout-tip with fleshy conical appendage that points backwards, covered with keeled scales; tympanum hidden; males with low nuchal crest comprising erect scales; dorsal scales small with large scattered scales; gular pouch and fold weak; precloacal and femoral pores absent. **Coloration** Dorsal surface medium brown to brownish-red; small yellow spot on each side of eyelid; rostral appendage brown; lining of mouth pale blue; gular pouch and venter yellowish-cream. **Habitat and Behaviour** Inhabits lowland rainforests and the mid-hills at elevations of 150–900m asl. Diurnal and arboreal, associated with tree trunks and saplings. Diet consists of insects. Clutches comprise 2 eggs, measuring 7 x 15mm. **Distribution** Endemic to eastern Borneo. **Status** Not Evaluated.

CRESTED GREEN LIZARD
Bronchocela cristatella   PLATE 11
Measurement SVL 130mm **Identification** Body slender to moderate, distinctly compressed; head long; nuchal crest small, erect, flattened and continuous with dorsal crest, which is a serrated ridge not extending to tail; gular pouch small in males, sometimes absent in females; antehumeral skin absent; scales unequal, smooth; supralabials 8–10; infralabials 7–12; tympanum large, greater than half orbit diameter; ventral scales 1–5 times as large as dorsals, in 10–12 rows; dorsals unequal, smooth; midbody scale rows 53–100; lamellae under Toe IV 34–35. **Coloration** Dorsum bright green, sometimes with white or light blue spots that may form bars, or with white bars, changeable to brown, with indistinct dark bands on body and tail; pale grey stripe across eye; tympanum greyish-brown; venter yellowish-green; pupil with narrow yellow ring, iris dark brown, eyelids greenish-yellow; anterior half of tail green with indistinct grey bands; posterior half brownish-grey. **Habitat and Behaviour** Inhabits the lowlands, especially forest edges, and frequently seen in parks and gardens. Diurnal and arboreal. Capable of making short glides between trees. Diet includes mayflies, beetles, flies and ants, in addition to skinks. Clutches comprise 1–4 spindle-shaped eggs with pointed ends, measuring 30–35.8 x 8.1–11mm. **Distribution** Myanmar, Thailand, Peninsular Malaysia, Singapore, Sumatra, the Natuna Archipelago, Pulau Berhala, Pulau Nias, Borneo and Java. Also the Lesser Sundas, Pulau Buru, Halmahera, Patani, Pulau Ambon, Ternate and Maluku (Indonesia), Luzon, Mindoro, Mindanao, the Sulu Archipelago, Cebu, Samar, Negros, Panay, Palawan, Bohol and Leyte (the Philippines), the Nicobar Archipelago (India) and New Guinea. **Status** Not Evaluated.

HAYEK'S FOREST LIZARD
Bronchocela hayeki   PLATE 11
Measurement SVL 120mm **Identification** Body moderate, compressed; nuchal crest comprises crescentic scales joined to a low dorsal crest of slightly enlarged scales; supralabials 9–10; infralabials 9–10; tympanum subequal to orbit; scales on top of body directed posteriorly and upwards, those on lower body posteriorly and downwards; gulars enlarged; scales on gular pouch slightly smaller than ventrals; ventrals enlarged, in 8 rows; midbody scale rows 70–72. **Coloration** Dorsum pale green with scattered white patches; forehead and sides of head pale yellow, cream, tan or green; narrow brownish-grey areas on supralabials and infralabials, extending to tympanum; area around orbit brownish-black; gular pouch yellow changeable to green; venter greenish-yellow. **Habitat and Behaviour** Inhabits the low hills and submontane forests in areas dominated by conifers, at elevations of 300–1,400m asl. Diurnal and arboreal, associated with trees; juveniles found in high grass and bushes. Diet unstudied. Clutches comprise 2 spindle-shaped eggs. **Distribution** Endemic to northern Sumatra (Berastagi, Takengon, Bukit Lawang, Lake Toba and Aceh). **Status** Not Evaluated.

MANED FOREST LIZARD
Bronchocela jubata   PLATE 11
Measurement SVL 140mm **Identification** Body robust, compressed; nuchal crest large with falciform scales, directed posteriorly; dorsal crest lower, extending to tail; supralabials 9–10; infralabials 8–9; tympanum large, over half orbit diameter; gular pouch large; midbody scale rows 43–59; dorsals keeled; row of large scales along chin, parallel to labials; ventral scales large, strongly keeled. **Coloration** Dorsum green changeable to brown or black, with yellow or red spots or vertical bars; venter pale. **Habitat and Behaviour** Inhabits lowland forests and relatively open areas. Diet consists of insects. Clutches comprise 2 eggs, measuring 44–53 x 9.5–12mm. Incubation period 84 days. Hatchlings 162mm. **Distribution** South-central Thailand (Prachin Buri, Prachinburi Province and nr Korat, Nakhon Ratchasima Province), Cambodia, Pulau Nias, Borneo, Java, Bali and Pulau Singkep. Also the Sulawesi, Karakelang and Salibabu Archipelagos (Indonesia), and Mindanao (the Philippines). **Status** Not Evaluated.

ORLOV'S FOREST LIZARD
Bronchocela orlovi   NOT ILLUSTRATED
Measurement SVL 109mm **Identification** Body robust, compressed; supralabials 7–9; infralabials 9–10; 4 scales between nasal and anterior border of orbit along canthus rostralis; no enlarged supratympanic scales; 3–4 compressed, erect scales behind superciliary edge form short crest; 3 enlarged scales reach upper level of tympanum; gular pouch small; nuchal crest large, spines sickle-shaped and directed posteriorly; dorsal crest low, scales directed posteriorly; large, weakly keeled dorsolateral and lateral scales; ventrals enlarged and strongly keeled; midbody scale rows 43. **Coloration** Dorsum blue in

preservative, probably green in life; large brown patch from axilla to midbody or to level of elbow; tympanum and labials dark brown; venter grey. **Habitat and Behaviour** Inhabits primary forests in the mid-hills at an elevation of 750m asl, on the Tay Nguyen Plateau. Diurnal and arboreal. Diet and reproductive habits unstudied. **Distribution** Endemic to central Vietnam (Buon Luoi, Gia Lai Province). **Status** Not Evaluated.

CHANTHABUN FOREST LIZARD
Bronchocela smaragdina   PLATE 11
Measurement SVL 113mm **Identification** Body slender, compressed; tail long; scales between orbit and tympanum subequal; tympanum over half orbit diameter; nuchal crest low, composed of small erect scales; dorsal crest absent; gular pouch absent; gular region with slightly enlarged and keeled scales; antehumeral fold absent; ventrals keeled; Toe V shorter than Finger IV; midbody scale rows 45–53. **Coloration** Dorsum emerald-green; tympanum brown; white or yellow stripe along lower flanks that extends to tail-base separates colour of dorsum from that of the pale green venter; tail lacks bands. **Habitat and Behaviour** Inhabits hill evergreen forests at an elevation of 600m asl. Diurnal and arboreal. Diet and reproductive habits unstudied. **Distribution** Cambodia (Mondolkiri Province) and southern Vietnam (Lam Dong Province). **Status** Not Evaluated.

VIETNAMESE FOREST LIZARD
Bronchocela vietnamensis   PLATE 11
Measurement SVL 122mm **Identification** Body slender, compressed; tympanum subequal to or less than orbit diameter; limbs and tail slender; supralabials 8; infralabials 9; nuchal crest comprises 6–12 small erect scales; dorsal crest absent; gular pouch small in males; gular region covered with little enlarged keeled scales; ventrals twice dorsals, keeled; Toe V shorter than Finger IV; midbody scale rows 47–54. **Coloration** Dorsum green; yellow band on belly to inguinal region and posterior side of femur and lateral side of tail; forehead green, radiating dark lines from posterior corner of orbit; tail dark-banded; venter pale blue. **Habitat and Behaviour** Inhabits submontane forests at elevations of 750–900m asl. Diurnal and arboreal, associated with tall trees. Diet and reproductive habits unstudied. **Distribution** Endemic to Vietnam (Dong Nai, Gia Lai and Phu Yen Provinces). **Status** Not Evaluated.

COLLARED FOREST LIZARD
Calotes chincollium   PLATE 11
Measurement SVL 142.9mm **Identification** Body robust, slightly compressed dorsoventrally; head broad; scales small, homogeneous, feebly keeled; upper dorsolateral scales point backwards and upwards; lower flank scales point backwards; midbody scale rows 59–74; dorsal and dorsolateral scales nearly equal in size to ventrals; enlarged temporal spine; supralabials 9–12; infralabials 8–11; nuchal crest composed of erect compressed scales; dorsal crest follows without gap; dorsal crest spinous; crescentic patch of granular scales and distinct oblique dermal fold in axilla; Toe IV with 23–28 subdigital lamellae; tail-base swollen in males. **Coloration** Dorsum grey with 4 irregular, saddle-like brown markings, more distinct in females and lighter and faded posteriorly in males; head grey, snout darker; males with dark brown patch over eye and tympanic region; females with brown postocular patch on tympanic region; tail light olive-brown with dark and light bars towards tip, gular pink; inner scales of oblique axillary fold black; venter grey. **Habitat and Behaviour** Known from areas under shifting cultivation and secondary forests at elevations of 737–1,940m asl. Diurnal and primarily terrestrial, but can climb the bases of trees. Diet includes newly sprouting crops of corn. Reproductive habits unstudied. **Distribution** Endemic to Myanmar (Chin State and Ponnyadaung Range). **Status** Not Evaluated.

FOREST CRESTED LIZARD
Calotes emma   PLATE 11
Measurement SVL 115mm **Identification** Body robust, compressed; head rather short; dorsal scales point backwards and upwards; cheek swollen in adult males; supralabials 9–12; infralabials 9–12; long spine above eye and 2 spines above tympanum; fold in front of shoulder; nuchal and dorsal crests continuous; large rectangular supraoculars; crescentic patch of granular scales in axilla; midbody scale rows 45–65. **Coloration** Dorsum olive-brown with dark brown dorsal bars or transverse spots; radiating dark lines from eye; light dorsolateral stripe; venter greyish-cream; breeding males have red gular pouch. **Subspecies** Two subspecies are recognized. *C. e. emma*, no enlarged elongated scales on supraoculars. *C. e. alticristatus*, enlarged elongated scales on supraoculars; 2 enlarged chin shields behind postmentals; 5 postrostral scales; chin grey, gular with black interstitial skin; tail-base not swollen. **Habitat and Behaviour** Inhabits open and evergreen forests in the lowlands and mid-hills, at elevations of 100–1,000m asl. Diurnal and arboreal, ascending trees to a minimum of 10m above substratum. Diet consists of insects. Clutches comprise 4–12 eggs, measuring 11 x 17mm. Incubation period 62–100 days. Hatchlings 32mm. **Distribution** Southern Thailand, northern Peninsular Malaysia (*C. e. emma*), Myanmar, northern Thailand (Chang Mai Province), Laos, Cambodia, Vietnam and northern Peninsular Malaysia (*C. e. alticristatus*). Also Bangladesh, Assam and Meghalaya (India) (*C. e. emma*), and southern China (*C. e. alticristatus*). **Status** Not Evaluated.

HTUNWIN'S FOREST LIZARD
Calotes htunwini   PLATE 12
Measurement SVL 91.4mm **Identification** Body robust, compressed; head distinct from neck; scales on sides of trunk point obliquely upwards; keeled scales on sides of neck and axilla horizontal; pair or cluster of spines in supratympanic region; anterior one is dorsolaterally directly above anterior half of

tympanum, separated by 6 scales; posterior one is level with posterior edge of tympanum, separated by 3 scales; scales comprising dorsal crest small. **Coloration** Dorsum beige to light tan; forehead speckled with dark brown; pair of dark brown, cream-centred nuchal spots with cream centres; dark brown lines radiate from orbit; indistinct middorsal stripe and broad light dorsolateral stripe; 8 dark brown dorsal blotches edged laterally by pale dorsolateral stripe, each blotch separated by narrow light bar; laterally neck dark above and lighter below, flanks darker; venter cream with beige tint; throat faded orange. **Habitat and Behaviour** Inhabits lowland forests and scrubby woodland. Diurnal. Diet and reproductive habits unstudied. **Distribution** Endemic to Myanmar (Sagaing and Magway Divisions). **Status** Not Evaluated.

AYEYARWADY FOREST LIZARD
Calotes irawadi PLATE 12
Measurement SVL 106.8mm **Identification** Body robust, compressed; head distinct from neck; scales on sides of trunk point obliquely upwards; tympanum subequal to orbit, naked; pair or clusters of spines above tympanum; adult females SVL to 77.9mm; adult males SVL to 106.8mm. **Coloration** Dorsum bronze-brown; forehead speckled with dark brown; pair of dark brown nuchal spots, 1 on each posterolateral edge of interparietal scale; 2 brown lines radiate posteroventrally from orbit; dusky-brown cheek patch; tympanum and shoulder spotted light green; indistinct brown dorsal blotches across middorsal crest, 6 on trunk laterally; neck light above and dark below; tail dark-banded; venter light dusky-beige with faded palmate striping on throat. **Habitat and Behaviour** Inhabits open forests, fence rows, gardens and degraded forests in the dry zone, from the lowlands to ca 1,000m asl. Diurnal, associated with shrubs growing up to 10m above substratum. Diet presumably consists of arthropods. Reproductive habits unstudied. **Distribution** Myanmar (Central Dry Zone, in Sagaing and Mandalay Divisions). **Status** Not Evaluated.

JERDON'S FOREST LIZARD
Calotes jerdoni PLATE 12
Measurement SVL 100mm **Identification** Body robust, compressed; head rather large; supralabials 9–11; infralabials 9–11; dorsal crest present, nuchal crest weak; crescentic patch of granular scales and distinct fold in axilla; dorsal scales larger than ventrals; 2 parallel rows of compressed scales above tympanum; midbody scale rows 45–57. **Coloration** Dorsum bright green, sometimes with pair of black-edged brown bands, and yellow, orange or brown blotches; tail dark banded or spotted. **Habitat and Behaviour** Diurnal and arboreal. Diet consists of insects. Clutches comprise 7–12 eggs. Hatchlings 70mm. **Distribution** Myanmar (Chin State). Also Khasi Hills (north-eastern India), Bangladesh and southern China. **Status** Not Evaluated.

KINGDON-WARD'S FOREST LIZARD
Calotes kingdonwardi NOT ILLUSTRATED
Measurement SVL 40mm (juvenile males; adults unknown) **Identification** Body robust, compressed; scales point backwards and downwards, except for upper 2–3 rows, which point slightly upwards or straight backwards; supralabials 7; infralabials 7; cephalic spines absent; dorsal crest reduced; tail weakly compressed, males having a swollen tail-base; midbody scale rows 45. **Coloration** Dorsum greyish-green with dark brown marks, including longitudinal ones on neck and 4 angular and larger ones on vertebral region; dark postocular streak; tail dark-banded; venter greyish-cream with small dark spots. **Subspecies** Two subspecies are recognized, of which one (*C. k. kingdonwardi*) occurs in the region. **Habitat and Behaviour** Inhabits montane forests at elevations of ca 2,134m asl. Diet and reproductive habits unstudied. **Distribution** Northern Myanmar (Adung Valley, Kachin State) (*C. k. kingdonwardi*). Also southern China (*C. k. bapoensis*). **Status** Not Evaluated.

MARIA'S FOREST LIZARD
Calotes maria NOT ILLUSTRATED
Measurement SVL 120mm **Identification** Body robust, compressed; head rather large, especially in adult males; scales on body point backwards and upwards; 2 parallel rows of compressed scales above tympanum; gular pouch absent; nuchal and dorsal crests moderately developed and shorter posteriorly; midbody scale rows 53–63. **Coloration** Dorsum green with red streaks on flanks and red spots on limbs and tail; head of breeding males red; venter greenish-white. **Habitat and Behaviour** Inhabits the low hills. Diurnal and arboreal. Diet consists of arthropods. Reproductive habits unstudied. **Distribution** Northern Myanmar. Also north-eastern India (Meghalaya State). **Status** Not Evaluated.

BLUE FOREST LIZARD
Calotes mystaceus PLATE 12
Measurement SVL 140mm **Identification** Body robust, compressed; head relatively large; cheeks swollen; dorsal and nuchal crests continuous; crest scales short in nape area; 2–3 spines behind eye; supralabials 9–11; infralabials 9–11; crescentic patch of granular scales and distinct fold in axilla; tail-base not swollen in males; midbody scale rows 47–57. **Coloration** Dorsum greyish-brown, turning bright blue to turquoise with 3–5 large dark spots on flanks during breeding season; pale stripe from supralabials to axilla; large orange-red dorsolateral spots irregularly edged with cream-white, numbering 1–3; venter greyish-cream. **Habitat and Behaviour** Inhabits evergreen forests at mid-altitudes, as well as parks and gardens in the lowlands and submontane forests, at elevations of 180–1,500m asl. Diurnal and arboreal, active on high tree trunks. Diet consists of insects. Clutches comprise 7 eggs, measuring 10–11 x 15–18mm. Incubation period 60–70 days. Hatchlings 26mm. **Distribution** Myanmar, Thailand, Laos,

Cambodia, Vietnam and Peninsular Malaysia. Also north-eastern India (Nagaland, Arunachal Pradesh, Manipur and Mizoram States) and southern China (Yunnan Province). **Status** Not Evaluated.

GARDEN LIZARD
Calotes versicolor PLATE 12
Measurement SVL 95mm (South-East Asian populations) **Identification** Body robust, compressed; head rather large, especially in adult males; scales on body point backwards and upwards; no antehumeral fold or pit in front of shoulder; 2 separated spines above tympanum; midbody scale rows 40–50. **Coloration** Variable and also changeable, head becoming orange or bright red, with a black patch on throat appearing in displaying males, fading to dull grey at other times; females too may become yellow, changing to dull greyish-olive after mating. **Habitat and Behaviour** Inhabits open forests and disturbed habitats such as parks and gardens, ascending to 3,000m asl on the mainland. Diet consists of insects and other invertebrates, although some vegetarian matter, such as unripe seeds, is also consumed. Sexually mature in 9–12 months. Clutches comprise 6–9 eggs in southern India, and up to 23 in northern India, with eggs measuring 10–11 x 4–5mm. Incubation period 42–67 days. **Distribution** Myanmar, Thailand, Laos, Cambodia, Vietnam and northern Peninsular Malaysia. Introduced into Singapore, Borneo and Java. Also Afghanistan, Pakistan and north-western India (*C. v. farooqi*), rest of India, Iran, Sri Lanka, Bangladesh, Bhutan, Nepal and the Maldives (*C. v. versicolor* or unassigned subspecies). Introduced into Florida, USA. **Status** Not Evaluated.

KINABALU FOREST LIZARD
Complicitus nigrigularis PLATE 12
Measurement SVL 75.5mm **Identification** Body robust, compressed; snout tapering, rounded; tail relatively long; distinct gular pouch in males; slight oblique fold in front of shoulder; midbody scale rows 60; middorsal crest slightly large, strongly keeled; anterior 9 scales of crest elongated and compressed. **Coloration** Dorsum dark brownish-grey with 2 broad, transverse white bands on trunk; white subocular spots; neck and shoulders mottled with white; gular pouch black with a pair of white spots near distal end; limbs with several indistinct white bands; tail with 11 broad, transverse pale grey bands; venter white with dark mottling. **Habitat and Behaviour** Inhabits submontane forests at an elevation of 1,500m asl. Arboreal and diurnal. Diet and reproductive habits unstudied. **Distribution** Endemic to Borneo (Gunung Kinabalu, Sabah State). **Status** Not Evaluated.

BOULENGER'S TREE LIZARD
Dendragama boulengeri PLATE 12
Measurement SVL 73mm **Identification** Body robust, compressed; on occiput, oblique bony ridge that is covered with large scales, and extends from above tympanum to orbit; short spine behind super- ciliary edge; short dorsal crest of 13–14 scales; nuchal crest with 6–7 erect scales; gular pouch small; body with heterogeneous scales; hind limbs long. **Coloration** Dorsum bluish-green or olive-grey with transverse dark bands; tail with dark bands; oval dark patch on sides of neck edged by yellow stripe; venter pink with brown spots. **Habitat and Behaviour** Inhabits montane forests at elevations of 1,200–2,800m asl. Diurnal and arboreal. Diet unstudied. Clutches comprise 4 eggs, measuring 8 x 14mm. **Distribution** Endemic to Sumatra (Binjai, nr Medan, Pis-Pis, Batakland and Gunung Singalang). **Status** Not Evaluated.

BLANFORD'S FLYING LIZARD
Draco blanfordii PLATE 13
Measurement SVL 130mm **Identification** Body slender; nostrils oriented dorsally; tympanum large; supralabials 7–9; no thorn-like scale on supraciliary edge; gular pouch elongated, slightly swollen, covered with large scales; low caudal crest in males; dorsals 183–224; subcaudals larger than adjacent scales; patagial ribs typically 5, exceptionally 6. **Coloration** Dorsum bluish-grey or brownish-grey; patagia of males with longitudinal white lines, of females with black cross-bars, which may be restricted to black reticulation at outer margin; head and shoulders with small dark spots; venter yellow; gular pouch yellowish-cream; patagium with 5 transverse dark bands with pale spots. **Habitat and Behaviour** Inhabits evergreen forests. Diurnal, presumably feeding on ants. Reproductive habits unstudied. **Distribution** Myanmar (Tanintharyi, Bago and Sagaing Divisions, and Kayah, Shan and Kachin States), Thailand, northern Peninsular Malaysia (Kedah, Perlis, Perak and Pahang States) and Indonesia (Pulau Batu). Also north-eastern India and Bangladesh. **Status** Not Evaluated.

HORNED FLYING LIZARD
Draco cornutus PLATE 13
Measurement SVL 88.9mm **Identification** Body slender; nostrils oriented laterally; distinct thorn-like scale over eye; tympanum scaleless; gular pouch subtriangular, covered with small scales; nostrils oriented laterally; dorsal crest absent; reduced nuchal crest consisting of triangular scales; dorsals 140–177; patagial ribs 6. **Coloration** Dorsum bright green to greenish-brown in males, tan or light brown in females; patagium reddish-orange with dark spots or bands; dark interorbital spot. **Habitat and Behaviour** Inhabits the plains and mid-hills, and common in mangrove forests. Diet consists of black ants. Clutches comprise 3–4 eggs. **Distribution** Sumatra, Borneo, Java and the Bangunan Archipelago. Also the Sulu Archipelago (the Philippines). **Status** Not Evaluated.

CRESTED FLYING LIZARD
Draco cristatellus PLATE 13
Measurement SVL 90mm **Identification** Body slender; nostrils oriented laterally or slightly obliquely; supralabials 9–10; tympanum large; gular pouch subtriangular, tapering gradually, covered with small

scales; males with low caudal crest; patagial ribs 5–6. **Coloration** Dorsum brownish-grey with transverse dark bars or spots; patagium dark brown with lighter lines; venter blue with dark spots; gular pouch bright yellow. **Habitat and Behaviour** Inhabits primary forests in the lowlands and mid-hills. Diurnal and arboreal, associated with large trees. Diet and reproductive habits unstudied. **Distribution** Peninsular Malaysia and Borneo. **Status** Not Evaluated.

FRINGED FLYING LIZARD
Draco fimbriatus PLATE 13
Measurement SVL 132mm **Identification** Body robust; nostrils oriented laterally; supralabials 11–12; spinous projection over eye; tympanum large, scaleless; males with low nuchal sail; gular pouch subtriangular with small scales; dorsals 155–226; caudal crest comprising small triangular scales; fringe-like scales on posteriors of thighs and tail-base; patagial ribs 5. **Coloration** Dorsum and patagium greyish-brown with grey and pale green markings. **Habitat and Behaviour** Inhabits the lowlands and mid-elevation forests, on large trees up to 20m. Diet presumably consists of arthropods. Clutches comprise 2–4 eggs, measuring 16–17 x 10–11mm. Incubation period 40–68 days. **Distribution** Southern Thailand, Peninsular Malaysia, Singapore, Sumatra, the Mentawai Archipelago, Borneo and Java. Also Mindanao (the Philippines). **Status** Not Evaluated.

BEAUTIFUL FLYING LIZARD
Draco formosus PLATE 13
Measurement SVL 101mm **Identification** Body moderate; nostrils oriented dorsally; supralabials 8–12; gular pouch widened distally in males; tympanum covered with small scales; thorn-like scale on supraciliary edge absent; gular pouch, lateral pouches and caudal crest in males; dorsals 146–204; patagial ribs 5. **Coloration** Dorsum bluish- or brownish-grey; gular region maroon or crimson, especially in males; patagia olive-grey dorsally with indistinct or obscure greyish-black or brown cross-bars or spots. **Habitat and Behaviour** Inhabits open forests in the lowlands and mid-hills at elevations of up to 900m asl. Diet consists of ants. Clutches comprise 3–4 eggs. **Distribution** Southern Thailand, Peninsular Malaysia and Pulau Siberut. **Status** Not Evaluated.

RED-BEARDED FLYING LIZARD
Draco haematopogon PLATE 13
Measurement SVL 94mm **Identification** Body slender; nostrils oriented dorsally; supralabials 11–12; tympanum large, skin-covered; row of keeled scales on snout; gular pouch elongated, subtriangular, tapered, covered with small scales; no thorn-like scale above eye; caudal crest absent; dorsals 148–184; tail crest absent; patagial ribs 5. **Coloration** Dorsum dark olive or brownish-grey with indistinct lighter and darker spots; patagium black with yellow spots. **Habitat and Behaviour** Inhabits the mid-hills and submontane forests. Diurnal and arboreal, associated with tall trees. Diet probably consists of ants. Clutches comprise 2–3 eggs. **Distribution** Peninsular Malaysia, Sumatra, Borneo and Java. **Status** Not Evaluated.

INDO-CHINESE FLYING LIZARD
Draco indochinensis PLATE 13
Measurement SVL 107.8mm **Identification** Body moderate; nostrils dorsal, oriented upwards; supralabials 8–11; no thorn-like scale on superciliary edge; tympanum exposed; broad transverse band across posterior of gular pouch, across lateral pouches; tip of gular pouch with enlarged scales; dorsals 189–210; patagial ribs 5. **Coloration** Dorsum brownish-grey with small dark spots; interorbital spot present or absent; patagium light brown towards body grading to dark brown on edges, with 6 transverse light-edged bars; gular pouch yellowish-cream with frontal part of base bluish-grey, posterior of pouch black; venter yellow or pink, ventral surface of patagium yellowish-brown. **Habitat and Behaviour** Inhabits hilly evergreen forests at elevations of up to 500m asl. Diurnal and arboreal, sighted 2m above ground to canopy level. Diet and reproductive habits unstudied. **Distribution** South-eastern and eastern Cambodia (Kamche Mountains and O'Rang District), and southern Vietnam (Kon Tum, Khanh Hoa, Lam Dong, Dong Nai and Tay Ninh Provinces). **Status** Not Evaluated.

ORANGE-WINGED FLYING LIZARD
Draco maculatus PLATE 14
Measurement SVL 82mm **Identification** Body moderate; nostrils lateral, oriented outwards; supralabials 8–9; tympanum scaly; gular pouch elongated, tapering basally, with tip broad with small scales; dorsals 117–152; males with low caudal crest; subcaudals subequal to lateral tail scales; patagial ribs 5. **Coloration** Dorsum pale blue or olive-grey; male patagium yellow, sometimes with orange outer edge, and narrow white lines and large yellow blotches; ventral surface with large black spots. **Subspecies** Three subspecies are recognized. *D. m. haasei*, patagium unpatterned ventrally or with up to 5 spots; base of gular pouch yellowish-cream; no blue spot on each side of base of gular pouch. *D. m. whiteheadi*, patagium unpatterned ventrally; base of gular pouch brown; no blue spot on each side of base of gular pouch. *D. m. maculatus*, patagium yellow ventrally, unpatterned or with 1–4 black spots; blue spot on each side of base of gular pouch. **Habitat and Behaviour** Inhabits primary and secondary forests in the lowlands and mid-hills at elevations of 200–1,000m asl. Diurnal and arboreal, associated with large trees. Glides of up to 10m by this species have been recorded. Diet consists of ants. Reproductive habits unstudied. **Distribution** Myanmar, western and southern Thailand, Peninsular Malaysia, Singapore (*D. m. maculatus*), eastern Thailand, Laos, Cambodia, central and southern Vietnam (*D. m. haasei*), and northern Vietnam (*D. m. whiteheadi*). Also north-eastern India (Nagaland and Arunachal Pradesh), Bangladesh, southern China (*D.*

m. maculatus) and eastern China (*D. m. whiteheadi*). **Status** Not Evaluated.

LARGE FLYING LIZARD
Draco maximus PLATE 14
Measurement SVL 140mm **Identification** Body robust; nostrils dorsal; supralabials 14–15; no spinous projections above eye; males with nuchal sail; gular pouch covered with small scales; dorsals 226–286, mostly homogenous, with series of large scales at base; fringe-like scales on posterior margins of thighs and tail-base; patagial ribs 6. **Coloration** Dorsum green or greyish-olive with brownish-olive pattern of bands; patagium red or yellowish-orange with black bands. **Habitat and Behaviour** Inhabits river edges at locations ranging from the lowlands to 1,000m asl. Diurnal and arboreal, associated with tree trunks. Diet presumably consists of ants, as well as other insects. Clutches comprise 1–5 eggs, measuring 11 x 18mm. **Distribution** Peninsular Malaysia, Sumatra, Borneo and the Natuna Islands. **Status** Not Evaluated.

BLACK-BEARDED FLYING LIZARD
Draco melanopogon PLATE 14
Measurement SVL 93mm **Identification** Body slender; nostrils dorsal; no spinous projections above eye; gular pouch elongated, scales slightly enlarged; lateral row of large scales present; fringe-like scales on posteriors of thighs and tail-base; dorsals 164–218; ventrals keeled, larger than dorsals; patagial ribs 5. **Coloration** Dorsum olive or green with brownish-grey bands or diamond-shaped spots; patagium black, sometimes with scattered yellow-orange spots. **Habitat and Behaviour** Inhabits lowland forests. Diet consists of ants, although small beetles, millipedes, isopods and termites are also consumed. Clutches comprise 2 eggs with pointed ends, measuring 6.5–7.6 x 14–15.7mm. **Distribution** Peninsular Thailand, Peninsular Malaysia, Singapore, Sumatra, Borneo and the Natuna Islands. **Status** Not Evaluated.

OBSCURE FLYING LIZARD
Draco obscurus PLATE 14
Measurement SVL 113mm **Identification** Body moderate; nostrils dorsal; supralabials 9–10; spinous projections on supraciliary absent; gular pouch oval-elongated, tapering distally, expanded at tip; tympanum entirely or partially covered with scales; dorsals 146–174, heterogenous, mostly smooth; fringe-like scales on posteriors of thighs and tail-base; patagial ribs 5. **Coloration** Dorsum grey-brown with scattered dark brown flecks; 5 indistinct butterfly-shaped blue marks across vertebral region; patagium yellowish-olive or yellowish-orange with 5–6 broad dark bands enclosing yellowish-olive or tan-grey areas; narrow dark bar across back of head joins orbits; nape with 2 oval black spots; gular pouch grey in males, olive-yellow in females. **Habitat and Behaviour** Inhabits lowland dipterocarp forests and the mid-hills at elevations of up to 1,000m asl. Diurnal and arboreal, associated with tree trunks. Diet consists of ants. Clutches comprise 1–5 eggs. **Distribution** Southern Thailand, Peninsular Malaysia, Sumatra, Borneo and the Natuna Islands. **Status** Not Evaluated.

FIVE-BANDED FLYING LIZARD
Draco quinquefasciatus PLATE 14
Measurement SVL 110mm **Identification** Body slender; nostrils dorsal; supralabials 12–14; spinous projections on supraciliary absent; dorsals heterogenous, mostly smooth; fringe-like scales on posterior margins of thighs and tail-base; dorsals 178–213; ventrals subequal to dorsals, keeled; males with low nuchal sail; patagial ribs 6. **Coloration** Dorsum bright green in males, brownish-olive or olive-brown in females, with dark speckling; patagium olive-brown, yellow or orange-red above with 5 dark brown or black cross-bars. **Habitat and Behaviour** Inhabits forests from the lowlands to the mid-hills, including peat swamps. Diurnal and arboreal. Diet includes ants. Clutches comprise 1–4 eggs, measuring 16.8–17.2 x 9.6–10.2mm. Hatchlings 30.3–30.4mm. **Distribution** Southern Thailand, Peninsular Malaysia, Sumatra, Pulau Sinkep, Pulau Belitung and Borneo. **Status** Not Evaluated.

COMMON FLYING LIZARD
Draco sumatranus PLATE 14
Measurement SVL 85mm **Identification** Body slender; nostrils lateral; supralabials 9–10; gular pouch subtriangular, covered with small scales; nuchal crest present, comprising 6–20 compressed triangular or keeled scales in males, with up to 14 scales in females; dorsals 103–166; tail crest absent; patagial ribs typically 6 (rarely 5 or 7). **Coloration** Dorsum light brown or yellowish-brown with dark brown blotches; gular pouch bright yellow with black dots at base; males have blue forehead when displaying. **Habitat and Behaviour** Inhabits lowland (to 300m) open forests, plantations and gardens. Diet includes ants, small beetles and termites. Clutches comprise 1–5 rounded eggs. **Distribution** Thailand, Peninsular Malaysia, Singapore, Sumatra, the Mentawai, Natuna and Riau Archipelagos, and Borneo. Also Palawan (the Philippines). **Status** Not Evaluated.

NARROW-LINED FLYING LIZARD
Draco taeniopterus PLATE 14
Measurement SVL 83mm **Identification** Body slender; nostrils dorsal; supralabials 7–8; tympanum large; subcaudals larger than adjacent tail scales; gular pouch tapered near base, then expands, tip broad, covered with enlarged scales; dorsals 138–173, subequal, smooth or faintly keeled; patagial ribs 5. **Coloration** Dorsum pale grey or greenish-brown; patagium pale grey or greenish-yellow with 5 irregular but distinct black bands; venter pale greenish-yellow; patagium yellowish-grey; throat and gular pouch deep crimson, with gular pouch dull yellow in males, females having paler gular pouch. **Habitat and Behaviour** Inhabits open evergreen forests. Diurnal and arboreal, associated with large trees 3–10m above substratum. Diet consists of ants and other small insects. Reproductive habits unstudied. **Distribution** Myanmar (Tenasserim, Taninthayi Division and the

Mergui Archipelago; Shan and Mon States), Thailand, Cambodia and northern Peninsular Malaysia (Perlis State). **Status** Not Evaluated.

JAVANESE FLYING LIZARD
Draco volans   PLATE 14
Measurement SVL 96mm **Identification** Body slender; nostrils lateral; supralabials 6–12; gular pouch subtriangular, covered with slightly enlarged scales; nuchal crest present, comprising 7–17 compressed triangular scales in males, 2–14 in females; dorsals 90–137, unequal, smooth or keeled; tail crest absent; patagial ribs typically 6 (rarely 5 or 7). **Coloration** Dorsum yellowish-tan or dark brown with indistinct dark and pale markings; interorbital spot present; distinct black edge of pale yellow-tan or yellow-brown patagium in males; gular region with fine spots; gular pouch yellow or orange in males, basally with small dark spots, bluish-grey in females; venter bluish-grey. **Habitat and Behaviour** Inhabits lowland and submontane forests, as well as avenues in towns and cities, at sea level to 1,500m asl. Diurnal and arboreal, associated with trunks of large trees. Diet consists of ants and possibly other small arthropods. Clutches comprise 2–6 eggs, measuring 12.5–13 x 6.5–7mm. Incubation period 26–29 days. **Distribution** Java and Bali. **Status** Not Evaluated.

ABBOTT'S ANGLE-HEADED LIZARD
Gonocephalus abbotti   PLATE 15
Measurement SVL 143mm **Identification** Body robust; dorsal crest lower than nuchal crest; nuchal crest composed of low, overlapping and slightly crescentic scales; superciliary border raised; no spine-like scales on crest in males. **Coloration** Dorsum greenish-olive changeable to reddish-brown under stress; several radiating dark lines around orbit of eye; iris bright red; venter cream; tail with 8 dark bands. **Habitat and Behaviour** Inhabits lowland forests. Diurnal and arboreal, associated with large trees. Diet and reproductive habits unstudied. **Distribution** Southern Thailand and Peninsular Malaysia. **Status** Data Deficient.

BLUE-NECKED ANGLE-HEADED LIZARD
Gonocephalus bellii   PLATE 15
Measurement SVL 150mm **Identification** Body robust; dorsal crest comprises tall scales that decrease in size progressively. **Coloration** Head and dorsum greyish-brown in males, reddish-brown in females; sides of body with net-like, dark grey-black pattern, enclosing pale area with cream centre; elongated dark patch on sides of neck; gular pouch of adult males pale green at base, tip indigo blue; gular region of females with pink flecks; tail grey-brown with yellow-cream bands. **Habitat and Behaviour** Inhabits lowland forests. Diurnal and arboreal, associated with large trees. Diet presumably consists of arthropods. Clutches comprise 3–5 eggs, measuring 11–11.5 x 18–20.5mm, which are produced at intervals of 2–3 months. Incubation period 8 months. **Distribution** Thailand, Peninsular Malaysia and Singapore. **Status** Not Evaluated.

BEYSCHLAG'S ANGLE-HEADED LIZARD
Gonocephalus beyschlagi   PLATE 15
Measurement SVL 126mm **Identification** Body robust; snout longer than orbit; nuchal and dorsal crests continuous, slightly notched at nape; those on dorsum long, sail-like and interconnected to form lance-like scales; regular longitudinal series of enlarged tubercles on flanks. **Coloration** Dorsum greenish-olive with longitudinal yellow streak on nape, above tympanum; another oblique streak present near axilla; iris brown; flanks with dark network that includes large yellow spots; venter yellowish-cream; limbs and tail with narrow pale bars. **Habitat and Behaviour** Inhabits lowland dipterocarp forests. Diurnal and arboreal. Diet and reproductive habits unstudied. **Distribution** Endemic to northern Sumatra. **Status** Not Evaluated.

BORNEAN ANGLE-HEADED LIZARD
Gonocephalus bornensis   PLATE 15
Measurement SVL 136mm **Identification** Body robust; nuchal and dorsal crests continuous, highly developed in males, comprising elongated lanceolate scales running up to level of tail and decreasing in height posteriorly; females with high nuchal crest but low dorsal crest; dorsal crest fused in juvenile males; dorsals intermixed with larger scales; no large scales below orbit of eye; ventrals smooth, more convex on pectoral region. **Coloration** Dorsum bright green with 5 dark bands; sides of head and flanks spotted with green; flanks with oval light spots; nuchal and body crests brown and yellow; gular pouch pale with broken dark stripes. **Habitat and Behaviour** Inhabits primary forests at elevations of up to 1,100m asl. Diurnal and arboreal. Diet consists of ants and spiders. Clutches comprise 4 eggs, measuring 4.2–4.3 x 9–9.9mm. Hatchlings 35–37mm. **Distribution** Endemic to Borneo. **Status** Not Evaluated.

TIOMAN ANGLE-HEADED LIZARD
Gonocephalus chamaeleontinus   PLATE 15
Measurement SVL 170mm **Identification** Body robust; head massive with distinctive curved eyebrows; dorsal crest comprises triangular to lanceolate scales; large tuberculate scales and small granular ones on flanks. **Coloration** Dorsum green with variable yellow pattern on dorsum and head, including spots, cross-bars and reticulated pattern; iris brownish-red, narrow golden ring around pupil; gular pouch with blue stripes. **Habitat and Behaviour** Inhabits lowland forests from sea level to ca 500m asl. Diurnal and arboreal, associated with trees. Diet consists of arthropods. Clutches comprise 3–6 eggs, measuring 12–13 x 21–28mm, with 4 clutches produced annually. Incubation period 81–119 days. **Distribution** West coast of Peninsular Malaysia (Pulau Tioman, Pahang State), Sumatra, the Mentawai and Natuna Archipelagos, and Java. **Status** Not Evaluated.

MARQUIS DORIA'S ANGLE-HEADED LIZARD
Gonocephalus doriae   PLATE 15
Measurement SVL 163mm **Identification** Body robust; dorsal crest as high as nuchal crest; nuchal crest

composed of low, overlapping and slightly crescentic scales; superciliary border raised; no spine-like scales on crest in males. **Coloration** Dorsum green with indistinct wavy grey pattern, or with large areas of orange changeable to reddish-brown with dark and light flecks; iris pink; gular pouch yellow, orange or grey with 6 bluish-grey stripes; lower flanks with 8 transverse bars; dorsum of females and juveniles green with dark blotches; venter bright yellow or cream; limbs and tail with pale green or grey and orange bands. **Habitat and Behaviour** Inhabits lowland or hill forests. Diurnal and arboreal, associated with low tree trunks and shrubs. Juveniles sleep on leaves of saplings close to streams and rivers. When threatened, this species opens its mouths without attempting to bite. Diet consists of arthropods. Breeding habits unknown, apart from the fact that it is oviparous. **Distribution** Endemic to Borneo. **Status** Not Evaluated.

GIANT ANGLE-HEADED LIZARD
Gonocephalus grandis PLATE 15
Measurement SVL 160mm **Identification** Body robust; males with high nuchal and dorsal crests, which are separated; females lack dorsal crest but have nuchal sail; dorsals subequal, not with larger scales; ventrals smooth. **Coloration** Dorsum greenish-brown, olive to grey-black, darker in stressed individuals; flanks blue with yellow spots in males; females brownish-green to nearly black. **Habitat and Behaviour** Inhabits primary forests at elevations of up to 1,400m asl. Diurnal and arboreal. Abundant near streams and small rivers, into which the lizards jump when threatened. Associated with tree trunks, although females and juveniles may also be found on the rocky banks of streams. Sleeps at night on the leaves of saplings or the tips of twigs overhanging water bodies. Diet consists of caterpillars, beetles, grasshoppers, ants, flies, cockroaches and spiders. Clutches comprise 1–6 eggs, measuring 10–11 x 21–26mm, which are laid several times a year. Incubation period 75–90 days. Hatchlings 32–35mm. **Distribution** Southern Thailand, south-eastern Laos, Vietnam (Gia Lai Province), Peninsular Malaysia, Sumatra, the Mentawai Archipelago and Borneo. **Status** Not Evaluated.

HERVEY'S ANGLE-HEADED LIZARD
Gonocephalus herveyi NOT ILLUSTRATED
Measurement SVL 145mm **Identification** Body robust; nuchal and dorsal crests continuous; scales at base of crest smooth; ventrals smooth or weakly keeled; enlarged scales on flanks; tail compressed. **Coloration** Dorsum greenish-olive; occipital region and nape sometimes with black vermiculation; venter cream; tail with dark bands. **Habitat and Behaviour** Known from forested offshore islands. Diurnal and arboreal. Diet and reproductive habits unstudied. **Distribution** Endemic to the Natuna Archipelago. **Status** Not Evaluated.

KLOSS'S ANGLE-HEADED LIZARD
Gonocephalus klossi PLATE 15
Measurement SVL 165mm **Identification** Body robust; lance-like scales form nuchal crest; nuchal and dorsal crests separated by gap; tail laterally compressed. **Coloration** Dorsum olive-brown with pale brown flecks and dark stripe on shoulder; iris powder-blue; head with greyish-black mask; labials pale brown; gular pouch cream; limbs bear indistinct grey bands; limbs and tail olive-brown with faint traces of cross-bars; venter lighter. **Habitat and Behaviour** Inhabits lowland forests. Diurnal and arboreal. Diet and reproductive habits unstudied. **Distribution** Endemic to Jambi Province and western Sumatra. **Status** Not Evaluated.

KUHL'S ANGLE-HEADED LIZARD
Gonocephalus kuhlii PLATE 15
Measurement SVL 100mm **Identification** Body robust; head with distinctive curved eyebrows; nuchal crest low; small dorsal scales mixed with large tubercular ones on flanks; ventral scales convex. **Coloration** Dorsum greenish-olive with cream or yellow flecks on head, especially on posterior corner of eye; iris brown; cream band on shoulder that is variable in size and form; vertical red or yellow bands sometimes present on dorsum; limbs greenish-olive with weak indications of cross-bars; tail with distinct dark brown bands. **Habitat and Behaviour** Inhabits forests in the plains and mid-hills at elevations of up to 1,600m; more common in the uplands. Diurnal and arboreal, associated with large trees. Diet presumably consists of arthropods. Clutches comprise 4 eggs, measuring 12–14 x 21–22mm. **Distribution** Sumatra and Java. **Status** Not Evaluated.

SIKULIKAP ANGLE-HEADED LIZARD
Gonocephalus lacunosus PLATE 15
Measurement SVL 145mm **Identification** Body robust; dorsal crest of males oriented towards posterior, and separate from nuchal crest; tail laterally compressed. **Coloration** Dorsum olive-green or olive-grey; enlarged flank scales greenish-olive or yellow; iris dark blue; head and gular pouch with turquoise markings, especially around the eye and on postocular region; pale green scales on lips and adjacent regions; tail with dark olive-grey and yellow-olive bands; females sometimes with vertical brownish-red bars. **Habitat and Behaviour** Inhabits montane forests at elevations of more than 1,000m. Diurnal and arboreal, although may shelter in holes. Diet presumably consists of arthropods. Clutches comprise 7–8 eggs. **Distribution** Known only from Sikulikap Waterfall area, nr Berastagi, northern Sumatra. **Status** Not Evaluated.

BLUE-EYED ANGLE-HEADED LIZARD
Gonocephalus liogaster PLATE 15
Measurement SVL 140mm **Identification** Body robust; nuchal and dorsal crests continuous, comprising lanceolate scales; in females nuchal crest long, dorsal crest low, ridge-like; supraciliary border

rounded; dorsals intermixed with larger scales; row of large scales below orbit of eye. **Coloration** Males brown or green with reticulated dark pattern on flanks; females with yellow cross-bars; eye of males bright blue, the skin surrounding the orbit reddish-orange; in females, eye brown. **Habitat and Behaviour** Inhabits lowland forests, including peat swamp forests. Diurnal and arboreal, associated with tree trunks, generally in the vicinity of streams and small rivers. Diet consists of arthropods. Clutches comprise 1–4 eggs, measuring 11 x 23mm. Incubation period 97 days. **Distribution** Peninsular Malaysia, Sumatra and Borneo. **Status** Not Evaluated.

LARGE-SCALED ANGLE-HEADED LIZARD
Gonocephalus megalepis PLATE 15
Measurement SVL 140mm **Identification** Body robust; nuchal and dorsal crests separated, comprising lanceolate scales that are directed posteriorly; dorsal scales sometimes large and form a longitudinal series at base of dorsal crest; ventral scales smooth; tail laterally compressed. **Coloration** Dorsum greenish-olive or pale green with indistinct transverse brown bands; iris indigo blue; some rounded scales on body bright yellow; pinkish-cream stripe along body, below dorsal crest; stripe of similar colour below eye; gular pouch pale green with cream stripes; venter yellow-olive; limbs and tail pale green with yellowish-green bands. **Habitat and Behaviour** Inhabits primary forests from the lowlands up to 1,200m asl. Diurnal and arboreal. Diet and reproductive habits unstudied. **Distribution** Known only from western Sumatra, and possibly also northern Sumatra. **Status** Not Evaluated.

MJÖBERG'S ANGLE-HEADED LIZARD
Gonocephalus mjobergi PLATE 16
Measurement SVL 88mm **Identification** Body robust; supraciliary border not raised; tympanum equal to eye diameter; nostril within single nasal; forehead scales feebly keeled; supralabials 8–9; infralabials 8; single large and flat scale below tympanum; gular pouch small, its edge feebly serrated and covered with small scales; nuchal crest present; dorsal crest a small ridge **Coloration** Dorsum pale green, changeable to brownish-grey, with narrow reticulated grey pattern that on lower flanks encloses yellow spots; venter pale green. **Habitat and Behaviour** Inhabits montane forests at elevations of 2,134–2,250m asl. Diurnal and arboreal, associated with trunks of large trees. Diet presumably consists of insects. Breeding habits unstudied. **Distribution** Endemic to north-western Borneo (Gunung Murud, Sarawak State). **Status** Not Evaluated.

ROBINSON'S ANGLE-HEADED LIZARD
Gonocephalus robinsoni PLATE 16
Measurement SVL 152mm **Identification** Body slender, compressed; nuchal and dorsal crests continuous, decreasing in height caudally; gular pouch large; flanks with scattered enlarged scales; supralabials 7–8; infralabials 7–8; tympanum small; midbody scale rows 72–80. **Coloration** Dorsum green with oblique dark cross-bars and yellow spots forming bars; labials white with dark bars on sutures; black postocular streak extends to tympanum; antehumeral fold black; gular pouch pink in males, yellow in females; tail dark-banded. **Habitat and Behaviour** Inhabits sub-montane forests. Diurnal and arboreal. Diet consists of arthropods. Breeding habits unknown, apart from the fact that it is oviparous. **Distribution** Endemic to Peninsular Malaysia (Gunung Tahan and Cameron Highlands, Pahang State). **Status** Not Evaluated.

BECCARI'S HORNED MOUNTAIN LIZARD
Harpesaurus beccarii PLATE 16
Measurement SVL 86mm **Identification** Body slender, compressed, with longitudinal series of smooth rhomboidal scales; double rostral appendage comprising longer anterior one and shorter posterior one, with an 86mm individual sporting a 10mm appendage; supralabials 7–8; infralabials 7; dorsal crest comprises 13 broad lanceolate scales, the first 3 recurved and separated from nuchal crest, which is composed of 4 low scales and 4 erect scales; tail long, compressed, upper margin bearing serrated ridge. **Coloration** Dorsum bright green; 2 oblique white stripes, one from sides of head to axilla, and one dorsal to supralabials before level of eye to beginning of forelimb; 2 large dark spots on flanks; iris light blue. **Habitat and Behaviour** Inhabits montane forests at an elevation of 1,500m asl. Diurnal and arboreal. Diet consists of insects and other arthropods. Reproductive habits unstudied. **Distribution** Endemic to Sumatra (Sijunjung and Lubuksulasih, Sumatera Barat). **Status** Not Evaluated.

BORNEAN HORNED LIZARD
Harpesaurus borneensis PLATE 16
Measurement SVL 59mm **Identification** Body slender, compressed; cylindrical rostral appendage in males surrounded by 4 large petal-like scales; nuchal crest developed, dorsal and caudal crests in males; gular pouch present; gulars acute with keeled tips; tail long and prehensile. **Coloration** Dorsum reddish-brown with black spots in 7 oblique series and yellow-brown bands; radiating dark brown lines from orbit of eye; rostral appendage greyish-cream; tympanum black; tail ringed with black and reddish-brown bands; venter white. **Habitat and Behaviour** Inhabits lowland rainforests. Diurnal and arboreal, associated with tree trunks. Diet probably consists of insects. Clutches comprise 2 eggs, measuring 10mm. **Distribution** Endemic to Borneo (Sarawak State and Kalimantan Province). **Status** Not Evaluated.

WERNER'S HORNED LIZARD
Harpesaurus ensicauda NOT ILLUSTRATED
Measurement SVL 60mm **Identification** Body slender, compressed; rostral appendage shorter than head, curved posteriorly, surrounded basally by 2 enlarged scales; forehead scales enlarged, especially those on occipital region; gulars smooth or weakly keeled; gular pouch absent; ventrals smooth; dorsal crest absent; tail compressed, with a denticulated crest.

Coloration Dorsum mid-brown; supralabials, and gular and pectoral regions cream. **Habitat and Behaviour** Inhabits lowland rainforests. Diet and reproductive habits unstudied. **Distribution** Endemic to Pulau Nias. **Status** Not Evaluated.

MODIGLIANI'S HORNED LIZARD

Harpesaurus modiglianii PLATE 16

Measurement SVL 83mm **Identification** Body slender; rostral horn on snout-tip up to 6mm long, gular pouch present; high nuchal crest or spinous scales; less spinose scales from nape to mid-tail; large scales on sides of body. **Coloration** Male dorsum azure-blue; lower sides olive mixed with cream; dark brown patch on fore flanks, behind forelimb; head blue, superlabials cream; gular pouch brown with light patches; neck brownish with light patches on tympanum up to base of forearm; light spots on upper arms; tail olive-blue with weak yellow-cream bands; rostral horn grey. **Habitat and Behaviour** Inhabits submontane forests at an elevation of 1,200m asl. Diet and reproductive habits unstudied. **Distribution** Endemic to northern Sumatra (Si Rambé, Riau Province). **Status** Not Evaluated.

THREE-BANDED HORNED LIZARD

Harpesaurus tricinctus PLATE 16

Measurement SVL 64mm **Identification** Body slender; rostral horn 21mm longer than head, curved upwards and surrounded at base by a few large scales; head with small scales that are slightly tubercular; large and prominent triangular scale on snout; gular scales tubercular; body compressed; dorsal crest a low serrated ridge; dorsal scales smooth and equal; tail compressed, longer than head-body length, with dorsal crest; limb scales keeled. **Coloration** Dorsum brown with 3 broad transverse yellow bands on body, the anterior band narrowest on shoulder, the others descending on flanks to venter. **Habitat and Behaviour** Nothing known of its natural history. **Distribution** Endemic to Java. **Status** Not Evaluated.

KINABALU CRESTED DRAGON

Hypsicalotes kinabaluensis PLATE 16

Measurement SVL 145mm **Identification** Body robust; limbs and tail long; sides of face with large scale below tympanum; nuchal and dorsal crests separate, continuing to tail; midbody scale rows 51–54; large lanceolate scales forming median row on gular pouch in males; gular pouch distinct in males; dorsal scales heterogenous. **Coloration** Dorsum green with chocolate-brown and black spots that form bands; gular pouch pale red with black and white stripes in anterior margin; venter brown with green spots; dorsum changeable to brown when threatened. **Habitat and Behaviour** Inhabits submontane and montane forests at elevations of 900–1,600m asl on Gunung Kinabalu. Diurnal and arboreal, associated with trunks of trees, including tree ferns. Slow-moving. Diet and reproductive habits unstudied. **Distribution** Endemic to northern Borneo (Gunung Kinabalu, Sabah State). **Status** Not Evaluated.

CHAPA MOUNTAIN LIZARD

Japalura chapaensis PLATE 16

Measurement SVL 59.6mm **Identification** Body robust, slightly compressed; head large, snout short; supralabials 7–8; infralabials 7; 35 scales on dorsal crest; nuchal crest present; midbody scale rows 45–47; lamellae under Toe IV 28–30. **Coloration** Live colours unknown. Preserved specimen pale greyish-tan with 5 transverse dark brown bars; pale dorsolateral stripe; forehead dark brown with pale interorbital bar; venter unpatterned. **Habitat and Behaviour** Inhabits primary forests at up to submontane limits. Probably diurnal and terrestrial, similar to congeners. Diet and reproductive habits unstudied. **Distribution** Endemic to northern Vietnam (Cao Bang Province, Sa Pá, Lào Cai Province and Hai Guong Province). **Status** Not Evaluated.

BANDED MOUNTAIN LIZARD

Japalura fasciata NOT ILLUSTRATED

Measurement SVL 75mm **Identification** Body robust, compressed; head large; tympanum concealed; nuchal crest present; supralabials 7–9; infralabials 7–9; enlarged dorsals not in regular longitudinal series; transverse gular fold present; hind limbs relatively long; midbody scale rows 35–38; lamellae under Toe IV 23–28. **Coloration** Dorsum brown with white cross-bar; large, hourglass-shaped pale blue mark enclosed by 2 broad bands; head green with dark variegation. **Habitat and Behaviour** Inhabits primary forests. Probably diurnal and terrestrial, similar to congeners. Diet and reproductive habits unstudied. **Distribution** Northern Vietnam (Lang Son Province). Also southern China. **Status** Not Evaluated.

HAMPTON'S MOUNTAIN LIZARD

Japalura hamptoni NOT ILLUSTRATED

Measurement SVL 77.2mm **Identification** Body robust, compressed; head large; tympanum concealed; enlarged dorsals not in regular longitudinal series; small gular pouch and weak gular fold; hind limbs relatively long. **Coloration** Dorsum brown or olive-grey with broad white stripe on anterior flanks; forehead green marbled with black; gular region with black streaks; venter greyish-white. **Habitat and Behaviour** Inhabits forested hills. Probably diurnal and terrestrial, similar to congeners. Diet and reproductive habits unstudied. **Distribution** Endemic to north-central Myanmar (Mogok, Mandalay Division). **Status** Not Evaluated.

KAULBACK'S MOUNTAIN LIZARD

Japalura kaulbacki NOT ILLUSTRATED

Measurement SVL 70mm **Identification** Body robust, compressed; snout long; cheeks swollen; forehead scales keeled; occipital scales with spinose tubercles; tympanum exposed; dorsal scales subequal; ventrals keeled, smaller than dorsals; antehumeral fold present; gular pouch poorly developed, consisting of triangular patch with small scales; nuchal crest low, composed of 9 lanceolate scales; dorsal crest reduced to ridge; limbs swollen. **Coloration** Dorsum brown in males, with indistinct transverse green bars; gular region bright yellow; gular patch pale blue anteriorly, red posteriorly; venter bluish-green; females green dorsally, with darker mottlings and cross-bars; labials and gular region yellow. **Habitat and Behaviour** Inhabits wet subtropical primary forests at elevations of ca 1,067m asl. Probably diurnal and terrestrial, similar to congeners. Diet and reproductive habits unstudied. **Distribution** Endemic to northern Myanmar (Nam Tamai Valley). **Status** Not Evaluated.

CRESTLESS MOUNTAIN LIZARD
Japalura planidorsa PLATE 16
Measurement SVL 53mm **Identification** Body robust; tympanum concealed; occipital region with numerous spinous tubercles; temporal spines conical, oriented at right angle to skin surface; body subquadrangular, with the back flattened, bordered on each side by ridge of enlarged keeled scales, and crossed at intervals by 6–7 V-shaped series of similar scales; flanks of body with numerous enlarged and scattered, strongly keeled scales; gular pouch absent; short antehumeral fold; limbs rather slender; tail compressed, covered with keeled scales; on dorsal surface of tail, large scales intermixed with smaller ones; midbody scale rows 52–61; lamellae under Toe IV 17–20. **Coloration** Dorsum yellowish-brown, lighter below; distinct series of dark streaks across back; superlabials light yellow, stripe extending to neck; orange or red gular region; light dorsolateral stripe may or may not be present. **Habitat and Behaviour** Inhabits primary forests. Diurnal and terrestrial. Locomotion on forest floor consists of hopping. Diet and reproductive habits unstudied. **Distribution** Myanmar (Chin State). Also north-eastern India (Assam and Meghalaya States). **Status** Not Evaluated.

ARROW-HEADED MOUNTAIN LIZARD
Japalura sagittifera NOT ILLUSTRATED
Measurement SVL 63.9mm **Identification** Body robust, not depressed; head broad, short; supralabials 7–8; infralabials 7–8; nuchal crest absent; dorsal crest a serrated ridge; 3–4 enlarged scales behind orbit; tympanum concealed; gular fold absent; antehumeral fold present; midbody scale rows 53–58; lamellae under Toe IV 25–27. **Coloration** Dorsum bright green in males; scarlet spot on gular pouch; longitudinal dark brown stripe on neck; dark brown postocular stripe to angle of jaws, edged anteriorly by white; female dorsum dark brown, enlarged scales blue; venter cream with or without brown gular and abdominal speckling in both sexes. **Habitat and Behaviour** Inhabits primary submontane forests at elevations of ca 1,166–1,230m asl. Probably diurnal and terrestrial, similar to congeners. Diet and reproductive habits unstudied. **Distribution** Endemic to northern Myanmar (Pangnamdim and Dadung, Kachin State). Also north-eastern India. **Status** Not Evaluated.

SWINHOE'S MOUNTAIN LIZARD
Japalura swinhonis PLATE 16
Measurement SVL 82.7mm **Identification** Body robust, compressed; dorsals on upper surface oriented upwards, those on flanks oriented posteriorly; supralabials 8–10; maxillary teeth 12 pairs; mandibular teeth 13 pairs; nuchal crest present; dorsal crest low, extending to part of tail; antehumeral fold present; ventrals strongly keeled; hind limbs short. **Coloration** Dorsum brownish-grey with light dorsolateral stripe; broad and angular dark cross-bars in intervening areas; forehead and flanks reticulated with brown; tail dark-banded; throat unpatterned cream; venter pale grey. **Habitat and Behaviour** Inhabits primary forests. Probably diurnal and terrestrial, similar to congeners. Diet and reproductive habits unstudied. **Distribution** Northern Vietnam (Sa Pá, Lào Cai Province). Also eastern China (Taiwan). Records from region may be based on Chapa Mountain Lizard (*J. chapaensis*). **Status** Not Evaluated.

YUNNAN MOUNTAIN LIZARD
Japalura yunnanensis PLATE 16
Measurement SVL 75mm **Identification** Body robust, compressed; tympanum concealed; tubercle on each side of back of head between orbit and commencement of nuchal crest; back covered with small keeled scales, intermixed with larger and more strongly keeled scales, arranged in vertical series; gular pouch small; indistinct antehumeral fold; low nuchal crest comprising separated triangular spines; dorsal crest a serrated ridge; limbs moderately strong; fourth toe distinctly longer than third; tail compressed, covered with keeled subequal scales. **Coloration** Dorsum light olive-green with dark green markings and pale cross-bars; pale dorsolateral stripe; forehead with dark bars; from eye to corner of mouth, dark streak that is bordered by a lighter streak; tail olive marked with light and dark bands. **Subspecies** Two subspecies are recognized, of which one (*J. y. popei*) occurs in the region. **Habitat and Behaviour** Inhabits primary forests. Probably diurnal and terrestrial, similar to its congeners. Diet and reproductive habits unstudied. **Distribution** Cambodia (*J. y. popei*). Also southern China (*J. y. yunnanensis*). **Status** Not Evaluated.

LUDEKING'S CRESTED LIZARD
Lophocalotes ludekingi PLATE 17
Measurement SVL 92mm **Identification** Body robust; smooth subdigital lamellae on fingers and toes; relatively large tympanum and large head, expanded below tympanum; males with larger head and more prominent dorsal crest, and longer occipital and nuchal spines than females. **Coloration** Males bright green or dark brown with large white patch behind tympanum, as well as indistinct transverse olive bands on body; females lack light vertebral band. **Habitat and Behaviour** Inhabits montane forests at elevations of ca 1,000–2,800m asl. Diurnal and arboreal, associated with bushes; may ascend to 1m above ground. Diet and reproductive habits unstudied.

Distribution Endemic to Sumatra (Barisan Range). **Status** Not Evaluated.

PHU WUA LIZARD
Mantheyus phuwuanensis PLATE 17
Measurement SVL 90mm **Identification** Body slender, depressed; snout elongated; dorsals granular or small and subtriangular; gular pouch surrounded by a U-shaped fold encompassed by a second U-shaped fold; transverse gular fold formed by posterior portion of outer fold being connected on either side to oblique folds in front of axilla, and continuing dorsally and posteriorly from axilla; femoral glands present; gular pouch in males; limbs and digits elongated; ventrals smooth; midbody scale rows 108–130. **Coloration** Dorsum dark brown or olive-green with green spots; flanks brown or bluish with black speckles and pale orange stripes; gular pouch basally orange-yellow, turning red apically; venter bluish-violet and yellow in males, unpatterned pale yellow in females. **Habitat and Behaviour** Inhabits rocky biotope within semi-evergreen lowland forests at elevations of 200–300m asl. Capable of dorsolateral flattening of its body, possibly to wedge itself within narrow rock crevices. Diurnal and nocturnal, and saxicolous, associated with boulders along streams. Diet consists of arthropods including insects and spiders. Clutches comprise 4 eggs, measuring 10.7 x 18mm, which are stuck on rock crevices. Incubation period 78 days. **Distribution** West-central Laos (Phou Khao National Biodiversity Conservation Area, Bolikhamxay Province) and north-eastern Thailand (Phu Wua Wildlife Sanctuary, Nong Khai Province). **Status** Not Evaluated.

BORNEAN SHRUB LIZARD
Phoxophrys borneensis PLATE 17
Measurement SVL 66mm **Identification** Body robust, short; spine above eye absent; nasal in contact with supralabials; continuous row of infraorbitals; gular scales sharply keeled; incomplete transverse gular fold; scales on sides of tail sharply keeled; 4 rows of keeled subcaudals near tail-base. **Coloration** Dorsum brown to greyish-brown; yellowish-tan bands; 2 dark interorbital bars, anterior narrower; superlabials cream; venter cream. **Habitat and Behaviour** Inhabits montane forests at elevations of 1,300–1,800m asl. Diet presumably consists of insects. Clutches comprise 2 eggs. **Distribution** Endemic to northern Borneo. **Status** Not Evaluated.

LARGE-HEADED SHRUB LIZARD
Phoxophrys cephalus PLATE 17
Measurement SVL 84mm **Identification** Body robust, short; spine above eye absent; nuchal crest comprises 7–8 thick conical scales; vertebral scale row comprises large, widely separated scales posteriorly; nasal in contact with supralabials; 2 continuous rows of infraorbitals; ventrals keeled. **Coloration** Dorsum pale green with wavy dark green or greyish-green bands, changeable to dark brown; large scales on nape yellow; 2 enlarged cream spines on sides of lower jaw; venter cream with dark green interscale markings joining to form stripes on greenish-cream throat. **Habitat and Behaviour** Inhabits submontane and montane forests at elevations of 1,300–2,100m asl. When threatened, it reveals the blue inner lining of its mouth, and becomes immobile. Diet unstudied. Clutches comprise 4 eggs, measuring 15.2–16.2 x 8.8–9.5mm. **Distribution** Endemic to northern Borneo. **Status** Not Evaluated.

BLACK-LIPPED SHRUB LIZARD
Phoxophrys nigrilabris PLATE 17
Measurement SVL 58mm **Identification** Body robust, short; spine above eye absent; nuchal crest comprises 6–12 compressed scales; vertebral scale rows with continuous series of large raised scales posteriorly; nasal separated from supralabials; gular scales distinctly keeled; 4 rows of large scales near tail-base. **Coloration** Dorsum of males brown, green or olive with or without transverse blue bands; females and juveniles brown to olive; venter yellowish-cream with dark brown vermiculations; inner lining of mouth pale blue. **Habitat and Behaviour** Inhabits lowland dipterocarp forests. Diurnal and arboreal, associated with low tree trunks and shrubs. When threatened, it opens its mouth to reveal the blue-black lining of its mouth, and sometimes attempts to bite. Diet includes insects and other arthropods. Reproductive habits unstudied. **Distribution** Borneo and Pulau Sirhassen in the Natuna Archipelago. **Status** Not Evaluated.

SPINY-HEADED SHRUB LIZARD
Phoxophrys spiniceps PLATE 17
Measurement SVL 60.3mm **Identification** Body robust, short; enlarged spine above eye; nasal in contact with second supralabial; infraorbitals in single continuous row; gular scales keeled; vertebral scale row with large, widely separated scales posteriorly; 2 rows of enlarged keeled subcaudals. **Coloration** Dorsum greenish-grey with brown bars and pale stripes, changeable to brown, with thin, pale transverse lines meeting medially; venter spotted with grey; gular pouch bright yellow. **Habitat and Behaviour** Inhabits montane forests at elevations of 1,200–1,800m asl. When threatened, it opens its mouth to reveal the deep blue mouth lining. Diet presumably consists of arthropods. Clutches comprise 2 eggs, measuring 12 x 7mm. **Distribution** Endemic to north-western Borneo (Gunung Murud and Gunung Mulu, Sarawak State). **Status** Not Evaluated.

SUMATRAN SHRUB LIZARD
Phoxophrys tuberculatus NOT ILLUSTRATED
Measurement SVL 43mm **Identification** Body robust, short; spine above eye absent; supraciliaries raised into crest equal to half orbit diameter; gulars keeled; 2 continuous rows of infraorbitals; scales on sides of tail keeled. **Coloration** Dorsum green with darker variegations; venter paler. As in conspecifics, dorsal colours presumably turn brown under stress. **Habitat and Behaviour** Inhabits montane forests.

Diet and reproductive habits unstudied. **Distribution** Endemic to western Sumatra (Gunung Singgalang, Sumatera Barat Province). **Status** Not Evaluated.

INDO-CHINESE WATER DRAGON
Physignathus cocincinus PLATE 17
Measurement SVL 250mm **Identification** Body moderate in juveniles, robust and compressed in adults; head enlarged in males; snout long; cheeks swollen; dorsal scales small and uniform, or larger scales intermixed; nuchal and dorsal crests continuous, separated from caudal crest; gular pouch absent; distinct gular fold; gulars 9–12; tympanum distinct, partially covered with scales; tail compressed; femoral pores in males 8 pairs. **Coloration** Dorsum bright green changeable to brownish-green; forehead and upper surfaces of limbs dark green; dark postocular stripe in adults; in juveniles, flanks with 3–5 narrow, oblique bluish-green stripes; in adults, stripes obscure or absent; gular region reddish-yellow; venter white with bright blue patches; iris brown with inner golden ring. **Habitat and Behaviour** Inhabits lowland forests and the mid-hills in deciduous and evergreen forests, at elevations of 190–700m asl. Diurnal and arboreal, associated with sandy soil. Runs bipedally with its forelimbs folded to its sides. Diet consists of insects and other arthropods. Clutches comprise 5–16 eggs, measuring 14–16 x 25–28mm. Incubation period 60–100 days. **Distribution** Myanmar, south-eastern Thailand, Laos, Cambodia and Vietnam. Also southern and eastern China (Yunnan and Guangdong Provinces). **Status** Not Evaluated.

SHORT-FOOTED LONG-HEADED LIZARD
Pseudocalotes brevipes PLATE 17
Measurement SVL 77.5mm **Identification** Body slender; head large; tail more than twice head-body length; forehead scales larger than body scales; supralabials 8–10; infralabials 7–9; tympanum exposed, ca half orbit diameter; nuchal crest comprising 6–7 erect compressed scales; enlarged spinose scale above tympanum; low denticulate dorsal crest; heterogenous dorsal scales; midbody scale rows 65–80; lamellae under Toe IV 22. **Coloration** Dorsum brown; head brown with dark flecks; radiating lines from orbit; cheeks sometimes bluish-grey; chin and anal regions yellowish-grey; gulars blackish-grey; venter with yellowish-grey band anteriorly and bluish-grey band posterior to it. **Habitat and Behaviour** Inhabits submontane forests at elevations of 914–1,219m asl. Diurnal and arboreal. Diet and reproductive habits unstudied. **Distribution** Northern Vietnam (Lang Son, Vinh Phuc, Ha Tay, Cao Bang and Hai Duong Provinces). Also southern China. **Status** Not Evaluated.

DRING'S LONG-HEADED LIZARD
Pseudocalotes dringi NOT ILLUSTRATED
Measurement SVL 70.3mm **Identification** Body slender; head length less than twice width; supralabials 8; infralabials 7–8; dorsal and lateral scales rectangular, feebly keeled, larger than ventrals; midbody scale rows 48–52; lamellae under Toe IV 26. **Coloration** Dorsum fawn with dark brown cross-bars; tail cross-barred greyish-fawn and dark brown; rhombic dark mark on nape; pale blue midline on throat and venter, rest of venter cream; gular pouch with oval blue-black ocellus enclosing pinkish-purple patch; radiating dark lines from gular pouch to infralabials. **Habitat and Behaviour** Inhabits montane forests at elevations of 1,981–2,194m asl. Diurnal and arboreal, associated with bushes, and also rocks and probably trees. Diet and reproductive habits unstudied. **Distribution** Endemic to Peninsular Malaysia (Gunung Tahan, Pahang State and Gunung Lawit, Terengganu State). **Status** Not Evaluated.

YELLOW-THROATED LONG-HEADED LIZARD
Pseudocalotes flavigula PLATE 17
Measurement SVL 72mm **Identification** Body robust; head elongated; supralabials 6; infralabials 8; gular pouch developed; no axillary fold; nuchal crest composed of 6 small spines; dorsal crest a low ridge; large lateral scales; tail not prehensile; midbody scale rows 38–40; lamellae under Toe IV 31–32. **Coloration** Dorsum light olive-green changeable to brown, with saddle-like brown patches; forehead olive-green speckled with brown; labials and throat white, pale coloration continuing towards axilla; tympanum dark brown; large white shoulder spot; flanks spotted with white; gular pouch bright yellow in males; gular region with mustard-yellow patch in females; tail dark-banded; thoracic region with light iridescent blue patch; venter ivory-white. **Habitat and Behaviour** Inhabits montane forests at elevations of 1,524–2,012m asl. Diurnal and arboreal, associated with tall trees. Diet unstudied. Clutches comprise 3 eggs, measuring 15 x 8mm. **Distribution** Peninsular Malaysia (Cameron Highlands, Pahang State). **Status** Not Evaluated.

FLOWER'S LONG-HEADED LIZARD
Pseudocalotes floweri PLATE 17
Measurement SVL 98mm **Identification** Body slender, compressed; head long; scales along canthus between nasal and supraciliaries 6; eye diameter greater than tympanum diameter; snout length twice orbit diameter; 6 scales between nasal and supraorbital; supralabials 8–12; infralabials 8–10; nuchal crest with 8 erect spines; scales under Finger IV 16–19; midbody scale rows 50–61. **Coloration** Dorsum brown; enlarged dorsal scales blue; spines light brown; radiating dark brown lines around eye extending to supraorbital ridge; limbs yellow with irregular dark brown bands; dark brown gular spot; tail with 10 dark brown spots. **Habitat and Behaviour** Inhabits submontane forests, including secondary forests, at elevations of 1,200–2,134m asl. Diurnal and arboreal. Diet and reproductive habits unstudied. **Distribution** South-eastern Thailand (Chanthaburi and Khao Sebab), Cambodia (Cardamom Mountains) and Vietnam (Kon Tum Province). **Status** Not Evaluated.

KAKHIEN HILLS LONG-HEADED LIZARD
Pseudocalotes kakhienensis PLATE 18
Measurement SVL 125mm **Identification** Body robust, compressed; head large and broad; tympanum at least half orbit diameter; gular pouch absent; supralabials 7–9; infralabials 7–9; nuchal crest present, composed of 7–9 separated spines; irregularly arranged small keeled scales intermixed with groups or pits of larger keeled scales on dorsal surface; midbody scale rows 50–66. **Coloration** Dorsum pale greenish-olive with light and dark brown variegations; supralabials and infralabials with dark bars on sutures; dark postocular stripe to tympanum; venter green speckled and streaked with dark brown. **Habitat and Behaviour** Inhabits submontane forests at elevations of 1,200–1,400m asl. Diurnal and arboreal. Diet unstudied. Clutches comprise 5 eggs. **Distribution** Myanmar (Karen Hills). Also southern China (Yunnan Province). **Status** Not Evaluated.

KHAO NAN LONG-HEADED LIZARD
Pseudocalotes khaonanensis NOT ILLUSTRATED
Measurement SVL 104.5mm **Identification** Body robust; supralabials 8; infralabials 9; 2 tubercles posterior to and in line with supraorbital ridge, dorsal and ventral scales rhomboidal; gulars distinct in males; 9 nuchal crest spines connected to indistinct dorsal crest, with weakly keeled scales; subdigital lamellae of Toe III bicarinate, anterior and posterior borders of lamellae equally developed; midbody scale rows 72–75. **Coloration** Dorsum rust-brown with irregularly arranged darker and lighter scales and 3 indistinct dark bands; supralabials cream to white; radiating dark rust-brown lines from orbit; gular pouch red with cream scales; dark greyish-black postocular stripe edged dorsally and ventrally with light rust-brown scales; tail dark-banded. **Habitat and Behaviour** Inhabits montane scrub cloud forests at elevations of more than 1,100m asl. Diurnal and arboreal, associated with stunted trees with dense epiphytic growth up to 4m above ground. Diet and reproductive habits unstudied. **Distribution** Endemic to peninsular Thailand (Khao Nan National Park). **Status** Not Evaluated.

BUKIT LARUT LONG-HEADED LIZARD
Pseudocalotes larutensis PLATE 18
Measurement SVL 77.3mm **Identification** Body slender; head large; snout long, somewhat pointed; limbs slender, elongated; supralabials 10; infralabials 8–9; dorsal scales keeled; nuchal crest comprises 5 erect compressed scales; antehumeral fold absent; midbody scale rows 51–53. **Coloration** Dorsum pale tan-yellow with greenish cast and 3 broad dark bands; gular pouch with yellow central spot, flanked laterally by plum-coloured patches; tail banded with reddish-brown and greenish-tan; venter cream; iris golden-brown. **Habitat and Behaviour** Inhabits submontane forests at elevations of ca 1,000m asl. Diurnal and arboreal, associated with trees at heights of at least 7–15m above substrate. Diet and reproductive habits unstudied. **Distribution** Endemic to Peninsular Malaysia (Bukit Larut, Perak State). **Status** Not Evaluated.

SMALL-SCALED LONG-HEADED LIZARD
Pseudocalotes microlepis NOT ILLUSTRATED
Measurement SVL 85mm **Identification** Body slender; head large; snout long, somewhat pointed; body compressed; tail relatively short; dorsal scales feebly keeled, pointing backwards; gular pouch small; nuchal crest in males low; midbody scale rows 65–72. **Coloration** Dorsum light brown with black flecks; colours become intensely dark when stressed; radiating dark lines from eyes. **Habitat and Behaviour** Inhabits submontane forests at elevations of up to 1,200m asl. Diurnal and arboreal. Diet and reproductive habits unstudied. **Distribution** Southern Myanmar (Mount Muleyit), northern and north-eastern Thailand (Chiang Mai, Chiang Rai, Chaiyaphum and Loei Provinces), Laos (Phong Saly Province) and Vietnam (Thua Thien-Hue, Da Nang, Quang, Nam, Lam Dong, Kon Tum, Gia Lai, Phu Yen and Dac Lac Provinces). Also southern China (Hainan Province) and north-eastern India (Manipur State). **Status** Not Evaluated.

POILANE'S LONG-HEADED LIZARD
Pseudocalotes poilani NOT ILLUSTRATED
Measurement SVL 89.4mm **Identification** Body robust, compressed, with small subequal dorsals, pointing posteriorly and downwards, those near crest pointing posteriorly and upwards; head long, dorsally flattened; snout elongated; forehead scales granular; supralabials 9; infralabials 9; nuchal crest well developed, composed of 9 scales, median ones longest, separated from low dorsal crest; gular pouch and postorbital spine absent; antehumeral fold present; gular scales granular; tympanum small, one-third orbit diameter; Fingers III and IV subequal; Toes III and IV subequal; tail long, compressed; femoral and preanal pores absent; midbody scale rows 54–60. **Coloration** Dorsum brownish-red with darker transverse bands; narrow lines on neck; venter pink with small, scattered black spots; orange mark with narrow, heart-shaped black edges on gulars; tail with 8–10 dark bands. **Habitat and Behaviour** Inhabits primary forests in submontane areas. Diurnal and arboreal, perches on branches over small streams. Diet and reproductive habits unstudied. **Distribution** Endemic to Laos (between Muang Pakxon and Pakxé and Boloven Highlands). **Status** Not Evaluated.

BORNEAN LONG-HEADED LIZARD
Pseudocalotes sarawacensis PLATE 18
Measurement SVL 82mm **Identification** Body robust; head large; snout obtuse; tail long, distinctly swollen; 2 projecting, compressed spine-like scales on temporal region, the larger one equal to tympanum in diameter; nuchal crest comprises 7 tall flexible scales; prefrontal scales slightly large, forming an inverted V. **Coloration** Dorsum dark brown with ill-defined green areas; head slightly lighter, lacking radiating dark lines from eye; nuchal crest and throat with orange-red wash; venter cream; tail with faint dark

blotches or bands, its base green. **Habitat and Behaviour** Inhabits lowland rainforests at an elevation of 240m asl. Diurnal and arboreal, found on vine growing 2m from substratum. Diet and reproductive habits unstudied. **Distribution** Endemic to northeastern Borneo (Lanjak-Entimau Wildlife Sanctuary, Sarawak State). **Status** Not Evaluated.

JAVANESE LONG-HEADED LIZARD
Pseudocalotes tympanistriga PLATE 18
Measurement SVL 80mm **Identification** Body slender; compressed; flexible scales in temporal region; denticulate nuchal crest in both sexes; forehead scales strongly keeled; gular pouch small; tympanum subequal to orbit diameter; midbody scale rows 43–64; supralabials 10–13; infralabials 9–11. **Coloration** Dorsum grass-green with or without olive-brown cross-bars; venter unpatterned cream. **Habitat and Behaviour** Inhabits the low hills and submontane forests at elevations of 1,300–1,500m. Diurnal and arboreal. Diet consists of arthropods. Clutches comprise 2 eggs, measuring 16.5–18 x 7.5–8mm. Incubation period 70–71 days. **Distribution** Sumatra and Java. **Status** Not Evaluated.

KON TUM LONG-HEADED LIZARD
Pseudocophotis kontumensis PLATE 18
Measurement SVL 87.8mm **Identification** Body robust, laterally compressed; head large; enlarged dorsal and lateral scales; postorbital spine absent; nuchal crest comprises 6 erect scales; low dorsal crest; smaller keeled ventral scales; tympanum concealed; gular fold present; gular pouch absent; tail short and prehensile; midbody scale rows 38. **Coloration** Dorsum brown with some green areas on head; supralabials and infralabials grey; light green stripe below eye extends to neck, turning yellow-brown; venter greyish-green. By day, dorsum with 2 transverse white bands between limbs and on neck; 6 indistinct bands on tail; night coloration darker. **Habitat and Behaviour** Inhabits montane forests at elevations of 1,200–1,250m asl. Diurnal and arboreal, associated with bushes and small trees near forest streams. Diet consists of insects. Clutches comprise 6–8 eggs. **Distribution** Endemic to Vietnam (Kon Tum Province). **Status** Not Evaluated.

SUMATRAN LONG-HEADED LIZARD
Pseudocophotis sumatrana NOT ILLUSTRATED
Measurement SVL 88mm **Identification** Body robust, compressed; scales large, in irregular rows; tympanum not visible externally; narrow soft rostral horn in males; nuchal crest comprises 7 spines separated from low dorsal crest; short prehensile tail; midbody scale rows 32–34. **Coloration** Dorsum green; venter pale. **Habitat and Behaviour** Inhabits submontane and montane forests. Possibly diurnal and arboreal. Diet and reproductive habits unstudied. **Distribution** Sumatra and Java (Gunung Pengalengan and Bogor). **Status** Not Evaluated.

MOUNT VICTORIA FAN-THROATED LIZARD
Ptyctolaemus collicristatus PLATE 18
Measurement SVL 91.3mm **Identification** Body slender, compressed; head long and slender; dorsal crest absent; femoral glands absent; distinct gular pouch in males, but pouch absent in females, which show faint gular folds; males with higher nuchal crest than females, comprising 15–16 large, flattened and triangular scales in adult males; femoral pores present; tail short. **Coloration** Dorsum pale greyish-brown, mottled or forming irregular dark brown saddles; flanks greyish-brown anteriorly, becoming yellowish-brown posteriorly, with brown reticulations from above axilla, continuing posteriorly; forehead pale greyish-brown; 2 parallel brown stripes radiating from orbit; dark anterior stripe reaches posterior margin of jaw; gular pouch bright yellow medially, greenish-yellow and dark brown laterally; forelimbs mottled with light and dark brown; venter cream. **Habitat and Behaviour** Inhabits the mid-hills to dry deciduous hardwood and pine forests and subtropical evergreen forests, at elevations of 790–1,940m asl. Diurnal, associated with secondary forests near human habitation. Arboreal, active on bushes and trees 1–4m above ground or on ground. Diet and reproductive habits unstudied. **Distribution** Endemic to northern Myanmar (Mount Victoria, Chin State). **Status** Not Evaluated.

GREEN FAN-THROATED LIZARD
Ptyctolaemus gularis PLATE 18
Measurement SVL 80mm **Identification** Body slender, compressed; head long and slender; dorsal scales keeled; dorsal crest absent; 3 longitudinal folds on each side of throat that curve to meet on back; gular pouch extendible when excited; femoral pores absent; low nuchal crest comprising 17–30 small triangular scales; tail long; midbody scale rows 64–65. **Coloration** Dorsum olive-brown, fold on back deep blue; 5 broad transverse bands on body; green dorsolateral band sometimes present on sides of front of body; flanks with network of dark brown, enclosing rounded areas of green. **Habitat and Behaviour** Inhabits submontane forests, and also found in urban habitats, at elevations of 172–1,220m asl. Diurnal and arboreal. Diet consists of insects, spiders and soil arthropods. Clutches comprise 14–15 eggs, measuring 7–12.5mm. **Distribution** Myanmar (Kachin State, Sagaing Division and Chin State). Also northeastern India, Bangladesh and Tibet (China). **Status** Not Evaluated.

SUMATRAN HORNED LIZARD
Thaumatorhynchus brooksi PLATE 18
Measurement SVL 60mm **Identification** Body slender, compressed; tympanum distinct; long cylindrical rostral appendage; dorsal and nuchal crests present, not continuous; gular pouch small; gular fold absent; femoral and precloacal pores absent; tail more than twice head-body length; supralabials 12–13; infralabials 11–12. **Coloration** Dorsum dark green, flanks with cream stripes; dark patches on forehead,

especially on supraocular regions; lobes of nuchal crest blue; broad blue bands on fore- and hind limbs; venter cream marbled with brown. **Habitat and Behaviour** Nothing known of its natural history. **Distribution** Endemic to Sumatra (Lebongtandai, nr Bengkulu, Bengkulu Province). **Status** Not Evaluated.

Family ANGUIDAE GLASS LIZARDS

Lizards in this family lack limbs. They possess a heavily armoured body with non-overlapping scales underlain by osteoderms, and have a longitudinal ventrolateral fold that separates the dorsal armour from the ventral armour. Related to lineages in central America, they are terrestrial and/or subfossorial, and occur in both open areas and forests. The tail of glass lizards is capable of autotomy, and their diet consists of arthropods and perhaps also small vertebrates. Glass lizards are known from America and Eurasia, as well as from South-East Asia.

BORNEAN GLASS SNAKE
Ophisaurus buettikoferi PLATE 19
Measurement TL 375mm **Identification** Body elongated; frontonasal much narrower than width of frontal; auricular opening as large as nostril; dorsals in 16–18 longitudinal rows; number of transverse rows of dorsal scales along side 98–105; central 12 rows of dorsals keeled to form unbroken straight lines. **Coloration** Dorsum brown with dark lateral band edged with light above, continuing to tail; small blue interparietal spot; front portion of back with irregular transverse series of blue spots, edged anteriorly with black; 5 oblique dark lines across labials; labials and venter pale yellow; pupil black, rounded, iris golden-yellow. **Habitat and Behaviour** Inhabits lowland forests up to submontane limits at elevations of 300–1,600m asl. Diurnal, and terrestrial and probably semi-fossorial. Diet consists of insects, and is likely to also include scavenging. Reproductive habits unstudied. **Distribution** Endemic to Borneo. **Status** Not Evaluated.

INDIAN GLASS SNAKE
Ophisaurus gracilis PLATE 19
Measurement TL 589.5mm **Identification** Body elongated; frontonasal narrower than frontal width; auricular opening as large as nostril; transverse rows of dorsal scales along side 88–94; dorsal scales across back 14 or 16; 3 scales in a line between nostril and prefrontals. **Coloration** Juveniles light brown; adults reddish-brown or yellow dorsally, with transverse rows of black-edged blue spots on forepart of dorsum; lower parts orange; pupil black, rounded, iris golden-yellow. **Habitat and Behaviour** Inhabits submontane and montane forests. Diurnal and terrestrial. When threatened, it shams death. Diet consists of arthropods. Oviparous; clutches comprise 4–7 eggs, measuring 18 x 12mm. Hatchlings 176.9–186.7mm. **Distribution** Myanmar, northern and north-eastern Thailand, Laos and Vietnam. Also northern, eastern and north-eastern India, Bangladesh and southern China. **Status** Not Evaluated.

HART'S GLASS SNAKE
Ophisaurus harti PLATE 19
Measurement TL 596mm **Identification** Body elongated; frontonasal narrower than frontal width; auricular opening minute, smaller than nostril; 2 scales in a line between rostral and prefrontals; transverse rows of dorsal scales along side 94–100; dorsal scales across back 16 (rarely 14 or 18). **Coloration** Dorsum brown or pale olive with transverse blue bars, or unpatterned; head dotted with black or unicoloured; venter cream; pupil black, rounded, iris golden-yellow. **Habitat and Behaviour** Inhabits submontane forests at an elevation of 900m asl. Diurnal and terrestrial, associated with grasslands. Diet and reproductive habits unstudied. **Distribution** Northern Vietnam (Sa Pá, Lao Cai Province, Chieng Di, Son La Province, Tam Dao, Vinh Phú Province and Ngan-Son, Cao Bang Province). Also southern China (Fujian and Sichuan Provinces, and Taiwan). **Status** Not Evaluated.

LUDOVIC'S GLASS SNAKE
Ophisaurus ludovici NOT ILLUSTRATED
Measurement TL 140mm **Identification** Body elongated; auricular opening minute, smaller than nostril; internasals 3; paired prefrontals; interparietal triangular; supraoculars 5; dorsal scales across back 16; ventrals smooth. **Coloration** Dorsum greyish-yellow with small dark spots; venter brown. **Habitat and Behaviour** Nothing known of its natural history. **Distribution** Endemic to northern Vietnam (Bao Lac, upper Tonkin) Also southern China. **Status** Not Evaluated.

SOKOLOV'S GLASS SNAKE
Ophisaurus sokolovi PLATE 19
Measurement TL 176mm **Identification** Body elongated; frontonasal equal to width of frontal; auricular opening three times as large as nostril; number of transverse rows of dorsal scales along side 88–92. **Coloration** Dorsum brownish-beige; olive cross-bars across body; longitudinal row of indistinct dark spots on forepart of dorsum; from flanks, approximately from middle of body, an indistinct dark stripe crosses over to tail, where it is dark brown; dark grey smudges on lower lips and on undersurface of body; pupil black, rounded, iris golden-yellow. **Habitat and Behaviour** Inhabits submontane forests. Diet and reproductive habits unstudied. **Distribution** Endemic to southern Vietnam (Kon Tum Province). **Status** Not Evaluated.

WEGNER'S GLASS SNAKE
Ophisaurus wegneri   NOT ILLUSTRATED
Measurement TL 450+mm **Identification** Body elongated; frontonasal narrower than width of frontal; auricular opening a little larger than nostril; supraoculars 5; supralabials 12; number of transverse rows of dorsal scales along side 98. **Coloration** Dorsum monochrome, lacking dark spots; pupil black, rounded, iris golden-yellow. **Habitat and Behaviour** Inhabits montane forests at an elevation of 1,600m asl. Diet and reproductive habits unstudied. **Distribution** Endemic to western Sumatra (Gunung Sago). **Status** Not Evaluated.

Family DIBAMIDAE WORM LIZARDS

Worm lizards lack limbs – apart from small, flap-like hind limbs in males – as well as external ear openings. Other characteristics that they share include a vestigial eye covered with a scale, a single fused parietal bone, pleurodont dentition, no pterygoid teeth, large, plate-like scales on the snout and mandible, and smooth and cycloid dorsals. They lack osteoderms, and possess more than 26 presacral vertebrae. They are capable of employing tail autotomy, and are oviparous, producing eggs with a calcareous shell. A relationship with gekkotans (including geckos and eyelid geckos) has been reported, both lineages possessing an egg-tooth. Worm lizards are poorly known, inhabit tropical forests, and are subfossorial or live in rotting tree trunks. They have a disjunct distribution, with a lineage in Mexico, and one in South-East Asia and western New Guinea.

ALFRED'S WORM LIZARD
Dibamus alfredi   NOT ILLUSTRATED
Measurement SVL 98mm **Identification** Body robust; postoculars 2; precloacal pores 4; midbody scale rows 20. **Coloration** Dorsum purplish-brown; parts of head and chin cream or ivory-white; venter similar to dorsum or paler. **Habitat and Behaviour** Inhabits lowland forests at the base of Bukit Besar. Fossorial. Presumably a predator of small invertebrates such as earthworms and insect larvae. Reproductive habits unstudied. **Distribution** Endemic to southern Thailand (Na Pradoo, Pattani Province). **Status** Not Evaluated.

BOO LIAT'S WORM LIZARD
Dibamus booliati   PLATE 19
Measurement SVL 102.7mm **Identification** Body robust; postocular single; rostral suture absent; interparietal posteriorly bordered by 4 nuchal scales; Supralabial I bordering ocular ventrally; scales bordering posterior edge of infralabial 3; ventrals 180–209; subcaudals 24–39. **Coloration** Dorsum brownish-red, scales dark-edged; snout-tip paler; nuchal band cream; venter pale brown; snout-tip and throat pinkish-white. **Habitat and Behaviour** Inhabits limestone regions within secondary forests at an elevation of 121m asl. Fossorial, burrowing in limestone rubble and under debris. When threatened it feigns death, curving its body into a circle, belly up, head slightly raised and the scales of its body perpendicular to the body. Diet and reproductive habits unstudied. **Distribution** Endemic to Peninsular Malaysia (Batu Gua Madu, Kelantan State). **Status** Not Evaluated.

BOURRET'S WORM LIZARD
Dibamus bourreti   NOT ILLUSTRATED
Measurement SVL 154mm **Identification** Body slender; tail long (more than 40 per cent of head-body length); complete rostral and nasal sutures; single postocular; precloacal pores 4 or none; midbody scale rows 20–24. **Coloration** Dorsum brown; venter pale brown; tail-tip white. **Habitat and Behaviour** Inhabits the mid-hills at elevations of 450–900m asl. Subfossorial, associated with leaf litter and also hiding under fallen trees and rocks. Presumably a predator of small invertebrates such as earthworms and insect larvae. Reproductive habits unstudied. **Distribution** Endemic to northern Vietnam (Tam Dao, Vinh Phúc Province). **Status** Not Evaluated.

DE HARVENG'S WORM LIZARD
Dibamus deharvengi   NOT ILLUSTRATED
Measurement SVL 92mm **Identification** Body slender; snout elongated, flattened; median rostral suture incomplete; nasal suture complete; labial suture incomplete, single postocular; scales bordering posterior edge of Infralabial I 2; postoculars 2; body scales smooth, subcycloid or subhexagonal, subequal; midbody scale rows 16; ventrals 179; subcaudals 57; presacral vertebrae 120; postsacral vertebrae 36; preanal pores 2 pairs. **Coloration** Colour in life unknown; in preservative, both dorsum and venter of body and tail pale, each scale with fine black dots. **Habitat and Behaviour** Inhabits lowland forests. Subfossorial. Presumably a predator of small invertebrates such as earthworms and insect larvae. Reproductive habits unstudied. **Distribution** Endemic to southern Vietnam (Binh Châu, Ba Ria-Vung Tau Province). **Status** Not Evaluated.

DE ZWAAN'S WORM LIZARD
Dibamus dezwaani   PLATE 19
Measurement SVL 123.1mm **Identification** Body slender; snout elongated; postoculars 2; frontonasal entire; incomplete rostral sutures; labial and nasal sutures complete; precloacal pores absent; tail relatively short; midbody scale rows 22; ventrals 178; subcaudals 37. **Coloration** Dorsum purplish-brown, scales edged with dark brown; nuchal and body bands absent; venter slightly paler; tip of snout, supralabials, throat and anal region cream. **Habitat and Behaviour**

Inhabits lowland rainforests at an elevation of 360m asl. Diet and reproductive habits unstudied. **Distribution** Endemic to Pulau Nias (Indonesia). **Status** Not Evaluated.

GREER'S WORM LIZARD
Dibamus greeri PLATE 19

Measurement SVL 82mm **Identification** Body slender; incomplete medial-rostral nasal suture; single postocular; supralabials 2; frontal enlarged; posteromedial edge of infralabials bordered by long narrow scale; precloacal pores absent; midbody scale rows 20; subcaudals 54. **Coloration** Dorsum dark brown, sometimes with 3 bright blue rings, each up to 9 scales wide, 2 on body and 1 on tail; venter purplish-brown. **Habitat and Behaviour** Inhabits primary forests in the mid-hills at elevations of 650–850m asl. Subfossorial, hiding in forest leaf litter, but capable of ascending trees under moss-covered branches and roots. Presumably a predator of small invertebrates such as earthworms and insect larvae. Reproductive habits unstudied. **Distribution** Endemic to southern Vietnam (A Luoi, Thua Thien-Hue Province, and K Bang, Kon Cha Rang, Buon Luoi and Tram Lap, Gia Lai Province). **Status** Not Evaluated.

INGER'S WORM LIZARD
Dibamus ingeri PLATE 19

Measurement SVL 96mm **Identification** Body robust; snout bluntly rounded, distinctly conical; nasal and labial sutures complete, extending from ocular to nostril; frontonasal divided; interparietal single, narrower than frontonasal and frontal, posteriorly bordered by 3 slightly smaller nuchal scales; postoculars 2; Supralabial I bordering ocular; infralabial lanceolate; midbody scale rows 20; ventrals 163; subcaudals 36; hind limbs reduced; large median scale on preanal region; precloacal pores absent. **Coloration** Dorsum unpatterned brown, except on posterior half of body, which has pale linear blotches; venter slightly paler; snout-tip, sides of head, including supralabials, throat, hind limbs and preanal region yellowish-cream; wide cream nuchal band; back of head with a few linear brown blotches. **Habitat and Behaviour** Inhabits submontane forests at an elevation of 1,180m asl. One was found inside a rotting log in a disturbed forest. Presumably a predator of small invertebrates such as earthworms and insect larvae. Reproductive habits unstudied. **Distribution** Endemic to Borneo (Mendolong, Sabah State). **Status** Not Evaluated.

CONDAO WORM LIZARD
Dibamus kondaoensis NOT ILLUSTRATED

Measurement SVL 112.4mm **Identification** Body slender; median rostral suture incomplete; nasal suture complete; labial suture complete; postoculars 2; scales bordering posterior edge of Infralabial I 3; postoculars 2; body scales smooth, subcycloid or subhexagonal, subequal; midbody scale rows 23; presacral vertebrae 140. **Coloration** Colour in life unknown; in preservative both dorsum and venter of body and tail and hind limbs of males unpatterned cream. **Habitat and Behaviour** Inhabits tropical forests on a lowland offshore island at an elevation of 500m asl. Presumably a predator of small invertebrates such as earthworms and insect larvae. Reproductive habits unstudied. **Distribution** Endemic to Vietnam (Con Dao). **Status** Not Evaluated.

WHITE-TAILED WORM LIZARD
Dibamus leucurus PLATE 19

Measurement SVL 136mm **Identification** Body slender; single postocular; nasal suture complete; rostral sutures incomplete; midbody scale rows 20–23; subcaudals 41–52; frontal greater than frontonasal; 3 small scales bound infralabials; single supralabial; large mental scale; tail-tip rounded. **Coloration** Dorsum medium brown, each scale with a darker edge; tips of snout and tail pale brown; no pale nuchal or body bands; venter pale brown. **Habitat and Behaviour** Inhabits lowland hilly dipterocarp forests, as well as coconut plantations, at elevations of 15–667m asl. Hides under coconut husks and rotting logs, and in the duff and humus of the forest floor, 4–10cm under the surface of the soil. One was found 90cm above ground, under tree bark. Presumably a predator of small invertebrates such as earthworms and insect larvae. Reproductive habits unstudied. **Distribution** Sumatra, Pulau Weh and Borneo. Also the Philippines. **Status** Not Evaluated.

MONTANE WORM LIZARD
Dibamus montanus NOT ILLUSTRATED

Measurement SVL 130mm **Identification** Body slender; single postocular; rostral sutures complete; single median suture above snout-tip; preanal scales enlarged; midbody scale rows 22; subcaudals 43–49. **Coloration** Dorsum chocolate-brown; dark pigment located in anterior part of dorsals, forming reticulated pattern; venter paler. **Habitat and Behaviour** Inhabits submontane forests. Subfossorial. Presumably a predator of small invertebrates such as earthworms and insect larvae. Reproductive habits unstudied. **Distribution** Endemic to central Vietnam (Daban, Langbian Plateau, Lam Dong Province). **Status** Not Evaluated.

SMITH'S WORM LIZARD
Dibamus smithi NOT ILLUSTRATED

Measurement SVL 108mm **Identification** Body slender; single postocular; rostral sutures converge on top of snout to form median suture in juveniles, reduced to grooves in adults; precloacal pores 2 in males, 1 in females; midbody scale rows 10–19; subcaudals 59–61. **Coloration** Dorsum brown; dark pigment located in anterior part of dorsals, forming reticulated pattern; venter paler. **Habitat and Behaviour** Inhabits submontane forests at elevations of 1,212–1,515m asl. Presumably a predator of small invertebrates such as earthworms and insect larvae. Reproductive habits unstudied. **Distribution** Endemic to central Vietnam (Daban, Langbian Plateau, Lam Dong Province). **Status** Not Evaluated.

SOMSAK'S WORM LIZARD
Dibamus somsaki NOT ILLUSTRATED
Measurement SVL 106.6mm (in juveniles; adults unknown) **Identification** Body slender; rostral suture complete, joining to form single median suture above snout-tip; nasal and labial sutures complete; scales bordering posterior edge of Infralabial I 2–3; single postocular; midbody scale rows 18–19; subcaudals 57–59. **Coloration** Dorsum and venter, including head and tail, unpatterned purplish-brown; snout-tip and lower jaw slightly paler. **Habitat and Behaviour** Known from a secondary forest in the lowlands. Fossorial, has been dug up from ground covered with duff and humus under a rotten log. Presumably a predator of small invertebrates such as earthworms and insect larvae. Reproductive habits unstudied. **Distribution** Endemic to south-eastern Thailand (Khao Soi Dao National Park, Chanthaburi Province). **Status** Not Evaluated.

TIOMAN WORM LIZARD
Dibamus tiomanensis PLATE 19
Measurement SVL 92.5mm **Identification** Body slender; snout blunt; rostral sutures incomplete; nasal and labial sutures complete; scales bordering posterior edge of Infralabial I 4; single postocular; midbody scale rows 25; subcaudals 45–48. **Coloration** Dorsum dark brown dorsally and ventrally; snout and jaws light brown; juveniles with cream snout and jaws. **Habitat and Behaviour** Inhabits lowland forests. When threatened, raises its scales in a possible mimicry of earthworms. Fossorial. Presumably a predator of small invertebrates such as earthworms and insect larvae. Reproductive habits unstudied.

Distribution Endemic to Peninsular Malaysia (Pulau Tioman and Pulau Tulai, Pahang State). **Status** Not Evaluated.

VORIS'S WORM LIZARD
Dibamus vorisi PLATE 19
Measurement SVL 89.2mm **Identification** Body slender; snout bluntly rounded; rostral suture incomplete; nasal suture complete, extending from ocular to nostril; labial suture absent; posterior border of rostral nearly straight; frontal single; frontonasal entire; interparietal single, narrower than frontonasal and frontal, posteriorly bordered by 3 slightly smaller nuchal scales; postoculars 2; supralabial 1; infralabial lanceolate, separated by smaller trapezoid mental; scales bordering posterior edge of infralabial, 3 bilaterally; head slightly distinct from neck and body; tail short; midbody scale rows 20; ventrals 147; subcaudals 11–33; large median scale on preanal region; precloacal pores absent; postanal scales not reduced. **Coloration** Dorsum unpatterned brown; venter slightly paler; snout-tip, sides of head including supralabials, throat, hind limbs and preanal region yellowish-cream; nuchal band absent; pale brown body band on anterior half of body both dorsally and ventrally; tail-tip cream. **Habitat and Behaviour** Inhabits hilly dipterocarp forests at elevations of under 300m asl. One was found 5cm below the soil surface within the confines of tree buttresses, another under dead leaves. Presumably a predator of small invertebrates such as earthworms and insect larvae. Reproductive habits unstudied. **Distribution** Endemic to Borneo (Lembah Danum, Sabah State). **Status** Not Evaluated.

Family EUBLEPHARIDAE EYELID GECKOS

Species with paired premaxillaries, the supratemporal bone present or absent, procoelous vertebrae, a single parietal bone, fleshy eyelids with no spectacles, and soft skin with numerous small juxtaposed scales. Their forelimbs and hind limbs are well developed, with narrow digits lacking subdigital setae. They are capable of employing tail autotomy, and are oviparous, producing elongated eggs with leathery or parchment-like shells. Although they are considered to be related to the Gekkonidae, relations between the two families are poorly resolved. Eyelid geckos include arboreal, terrestrial and saxicolour forms, and all are insectivorous. They are found in Asia, North America and Africa.

CAT GECKO
Aeluroscalabotes felinus PLATE 20
Measurement SVL 122mm **Identification** Body slender; eyelids fleshy; dorsal surface covered with granular scales; tail capable of being curled laterally; axillary pockets absent; large retractable claw on each digit between a dorsal and 2 lateral scales; transverse enlarged lamellae restricted to bases of digits. **Coloration** Dorsum tan-brown, more brightly coloured in juveniles, sometimes with orange or white spots and dark or pale vertebral stripe on body and tail; tail-tip sometimes white. **Habitat and Behaviour** Inhabits lowland rainforests and peat swamp forests at elevations of up to ca 1,000m asl. Nocturnal and arboreal, associated with low vegetation and dead logs. Diet consists of arthropods such as crickets and cockroaches. Clutches comprise 1–2 elongated eggs with parchment shells, measuring 9.5–13 x 17.0–21.2mm. Incubation period 35–64 days. Hatchlings 78–81mm. **Distribution** Thailand, Peninsular Malaysia, Singapore, Sumatra and Borneo. **Status** Not Evaluated.

VIETNAM LEOPARD GECKO
Goniurosaurus araneus PLATE 20
Measurement SVL 124mm **Identification** Body and limbs slender; eyelids fleshy; dorsal surface covered with granular scales; deep axillary pockets; claws within 4 scales; 18–22 precloacal pores. **Coloration** Dorsum dull yellow-grey, bands nearly immaculate and lack mottling, comprising wide dorsal bands between limb insertions and a nuchal loop posteriorly

protracted; venter unpatterned cream. **Habitat and Behaviour** Inhabits dry rocky areas, especially near caves. Diet and reproductive habits unstudied. **Distribution** Endemic to northern Vietnam (Cao Bang Province). **Status** Not Evaluated.

CAT BA LEOPARD GECKO
Goniurosaurus catbaensis　　　NOT ILLUSTRATED
Measurement SVL 111.5mm **Identification** Body and limbs slender; dorsals granular; internasal scales absent; enlarged row of supraorbital tubercles; outer surface of upper eyelid composed of granular scales subequal to those on forehead; long row of 6–9 enlarged tubercles; 8–11 scales surrounding dorsal tubercles; deep axillary pockets; claws within 4 scales; 16–21 precloacal pores. **Coloration** Dorsum grey-brown to pale brown with dark brown blotches; iris orange-brown; narrow, tapered yellow nuchal loop edged with dark brown; 3 thin yellow bands on body and 1 on tail-base, edged by narrow dark brown bands; venter unpatterned cream; indistinct brown lateral spots on gular region. **Habitat and Behaviour** Inhabits primary forests at elevations of 10–30m asl. Nocturnal, active on both forest floor and limestone cliffs at up to 2–3m from substrate. Diet and reproductive habits unstudied. **Distribution** Endemic to Vietnam (Cat Ba Island). **Status** Not Evaluated.

HUU LIAN LEOPARD GECKO
Goniurosaurus huuliensis　　　NOT ILLUSTRATED
Measurement SVL 117.34mm **Identification** Body slender, elongated; nasals 6–8, surrounding nostril; eyelids fleshy; eyelid fringe scales 41–44; dorsal surface covered by large granular tubercles and small granular scales; supralabials 10–11; infralabials 9–11; postmentals 2–3, bordered by 6–8 small gular scales; dorsum covered by large granular tubercles and small granular scales; 12–13 granular scales surround large granular tubercles; paravertebral tubercles between limb insertions 34–36; deep axillary pockets; long thin digits with wide subdigital lamellae; lamellae under Toe IV 15–16; claws within 4 scales; 25–28 precloacal pores. **Coloration** Dorsum grey-brown with dark brown blotches on lower flanks, near contact of dorsal surface and light grey venter and on head; nuchal loop pink or orange; 3 unpatterned pink or orange bands between limb insertions, 1 at tail-base, each band edged with irregular grey-black lines; tail grey-brown; venter light grey with indistinct spots on margins of venter of body and in gular region; iris red-brown. **Habitat and Behaviour** Inhabits karst-dominated forests at elevations of 300–370m asl. Nocturnal and saxicolous, hiding under leaf litter of limestone rocks. Diet and reproductive habits unstudied. **Distribution** Endemic to north-eastern Vietnam (Huu Lian Nature Reserve, Lang Son Province). **Status** Not Evaluated.

LICHTENFELDER'S LEOPARD GECKO
Goniurosaurus lichtenfelderi　　　PLATE 20
Measurement SVL 83mm **Identification** Body robust; eyelids fleshy; dorsal surface covered with granular scales; dorsals granular; 43–56 eyelid fringe scales; 24 paravertebral tubercles; axillary pockets absent; precloacal pores 28–30. **Coloration** Dorsum dark or light brown; 2 dorsal body bands, black-edged with wide border, between insertions of limbs; nuchal loop rounded posteriorly; venter pale grey. **Habitat and Behaviour** Inhabits lowland forests at elevations of 100–600m asl. Nocturnal and saxicolous, associated with rocky streams dominated by granite. Diet and reproductive habits unstudied. **Distribution** Islands in the northern-most Gulf of Tonkin, and north-eastern Vietnam (Lang Son, Quang Ninh, Bac Giang, Ha Bac and Hai Hung Provinces). **Status** Not Evaluated.

CHINESE LEOPARD GECKO
Goniurosaurus luii　　　PLATE 20
Measurement SVL 119mm **Identification** Body and limbs slender; dorsals granular; internasal scales present; supranasals do not meet in midline behind rostral suture; small granular scales on upper eyelid one-half size of those on forehead; rows of enlarged tubercles on dorsum absent; deep axillary pockets; precloacal pores 21–29; nasal scales surrounding nostril 6–8; supralabials 9–12; infralabials 8–11. **Coloration** Dorsum grey-brown to cream with dark and light mottling; nuchal loop wide and less tapered backwards (V-shaped); dark head pattern comprises dark brown blotches; iris bright orange. **Habitat and Behaviour** Inhabits secondary forests in the low hills at elevations of ca 770m asl. Nocturnal and saxicolous, associated with karst landscape. Diet and reproductive habits unstudied. **Distribution** Northern Vietnam (Cao Bang Province). Also south-eastern China (Guangxi Province and probably Hainan Province). **Status** Not Evaluated.

Family GEKKONIDAE GECKOS

Species with well-developed limbs, a single parietal bone, amphicoelous vertebrae, soft skin with numerous small, usually juxtaposed scales, a large eye covered with a spectacle, no temporal arches and tail showing autotomy. Geckos typically possess subdigital adhesive setae that provide their feet with traction, correlating with their arboreal lifestyle. They are found in a variety of habitats, and several are human commensals. A majority are nocturnal, a few diurnal or crepuscular. Their diet consists of arthropods, with some of the larger species being capable of consuming small vertebrates. They are oviparous, and typically produce two eggs at a time that have brittle, highly calcareous hard shells. A cosmopolitan group, geckos are represented by more than 1,200 species worldwide, in temperate, subtropical and tropical regions.

PENANG DAY GECKO
Cnemaspis affinis PLATE 20
Measurement SVL 50.8mm **Identification** Body slender; supralabials 8–11; infralabials 8–10; large lateral postmentals separated by 1–2 postmentals; forearm, subtibials, ventrals, subcaudals and dorsal tubercles keeled; paravertebral tubercles 20–24; small tubercles on flanks not linearly arranged; ventrolateral caudal tubercles present anteriorly; femoral pores absent; discontinuous row of 4–6 preanal scales bearing pores; 1–3 postcloacal tubercles; enlarged metatarsal scales absent; lamellae under Toe IV 25–32. **Coloration** Dorsum dark grey to brown; flanks with light markings; dark shoulder patch encloses white or yellow ocellus; prominent wide, yellow or white postscapular band; distinct dark caudal bands; venter dull beige; subcaudal region pigmented. **Habitat and Behaviour** Inhabits the mid-hills of primary forests. Diurnal, associated with rocks. Diet consists of small arthropods. Reproductive habits unstudied. **Distribution** Endemic to an island off the west coast of Peninsular Malaysia (Pulau Pinang, Penang State). **Status** Not Evaluated.

GUNUNG LAWIT DAY GECKO
Cnemaspis argus NOT ILLUSTRATED
Measurement SVL 65.3mm **Identification** Body robust; supralabials 8–10; infralabials 10–12; fingers IV and V subequal; ventrals keeled; midventrals ca 60; precloacal pores 10; median subcaudals not enlarged; lamellae under Toe IV 23–24. **Coloration** Dorsum dark brown and fawn with pale grey paravertebral blotches; oblique radiating yellow lines from orbit; flanks with vertical yellow bands; tail dark-banded; venter pinkish-grey; iris bright copper. **Habitat and Behaviour** Inhabits submontane forests dominated by palms at an elevation of 790m asl. Possibly diurnal and saxicolous. Diet and reproductive habits unstudied. **Distribution** Endemic to Peninsular Malaysia (Gunung Lawit, Terengganu State). **Status** Not Evaluated.

HON DAT HILL DAY GECKO
Cnemaspis aurantiacopes NOT ILLUSTRATED
Measurement SVL 56.5mm **Identification** Body slender; tubercles on flanks arranged in linear series; supralabials 9–10; infralabials 8–10; ventrals smooth; precloacal and femoral pores absent; dorsal, lateral and ventrolateral longitudinal rows of caudal tubercles; smooth subcaudals with an enlarged median row; cloacal tubercles 1–2; enlarged submetatarsal scales beneath first metatarsal; lamellae under Toe IV 28–30. **Coloration** Dorsum yellowish-grey with reddish-brown pattern; rhomboidal marks from occiput to base of tail, enclosing series of 8 yellowish-grey blotches; forehead reddish-brown; dorsal surface of limbs saffron with light mottling; tail reddish-brown, lacking bands; 3 reddish-brown postorbital stripes, uppermost extending to axilla and contacting blotch on nape; venter unpatterned dull orange; labials reddish-brown. **Habitat and Behaviour** Known from a cave entrance in a secondary semi-deciduous forest at an elevation of 30m asl. Nocturnal and saxicolous. Diet and reproductive habits unstudied. **Distribution** Endemic to southern Vietnam (Hon Dat Hill, Kien Giang Province). **Status** Not Evaluated.

BAUER'S DAY GECKO
Cnemaspis baueri PLATE 20
Measurement SVL 34.9mm **Identification** Body robust, elongated; head large, depressed; supralabials 14–16; infralabials 10–12; supranasals in contact; auricular opening narrow, slit-like; dorsal surface with enlarged tubercles in linear series; precloacal and femoral pores absent in males; Fingers V greater than IV; median subcaudals enlarged; lamellae under Toe IV 26–27. **Coloration** Dorsum dark brownish-olive with greyish-black spots on vertebral region; dorsal tubercles lack white tips; forehead with dark spots and broken lines; 2 radiating dark lines from orbit to back of head and across auricular opening to axilla; limbs and tail not dark-banded; venter unpatterned brown; iris orange. **Habitat and Behaviour** Inhabits lowland dipterocarp forests. Diurnal and saxicolous, associated with boulders. Diet unstudied. Lays eggs communally. **Distribution** Endemic to an island off the east coast of Peninsular Malaysia (Pulau Tulai, Johor State). **Status** Not Evaluated.

GUA BAYU DAY GECKO
Cnemaspis bayuensis NOT ILLUSTRATED
Measurement SVL 46.1mm **Identification** Body slender; supralabials 9–10; infralabials 8–9; dorsal tubercles keeled; paravertebral tubercles 25–30; tubercles on flanks, relatively small and not arranged in linear series; median subcaudals not enlarged or keeled; rest of subcaudals keeled; 2 postcloacal tubercles; femoral pores absent; row of 5–9 preanal scales bearing pores; subtibial scales not shield-like; lamellae under Toe IV 27–30. **Coloration** Dorsum brown with irregular dark spots; light markings on forehead; rostrum yellowish with dark streaks; thin dark postorbital stripes extend onto nape, not contacting medially; white mark on nape followed by alternating white paravertebral blotches to tail-base; transversely elongated white markings on flanks; gulars with pale brown reticulation; venter dull unpatterned beige, darker laterally. **Habitat and Behaviour** Inhabits a cave within a karst area at an elevation of 120m asl. Nocturnal and saxicolous. Diet unstudied. Clutches comprise 2 eggs. **Distribution** Endemic to Peninsular Malaysia (Gua Bayu Cave, Kelantan State). **Status** Not Evaluated.

TWO-EYED DAY GECKO
Cnemaspis biocellata NOT ILLUSTRATED
Measurement SVL 40.1mm **Identification** Body slender, elongated; anterior forearm with weakly keeled scales; ventrals smooth; femoral pores absent; 8–12 precloacal pores; subcaudals smooth with enlarged median row; lamellae under Toe IV 29–37. **Coloration** Dorsum dull yellow; 5 butterfly-shaped yellow vertebral blotches from axillary to inguinal regions; flanks and limbs with small yellow blotches; 2

white occipital ocelli; black occipital band bordering large pale spots that form nuchal band and small black shoulder patch enclosing a single white to yellow ocellus; tail lacks dark bands; venter unpatterned beige. **Habitat and Behaviour** Restricted to karst formations in lowland dipterocarp forests and disturbed areas. Diurnal, active in low-light areas of crevices. Diet and reproductive habits unstudied. **Distribution** Southern Thailand and northern Peninsular Malaysia (Perlis State). **Status** Not Evaluated.

BOULENGER'S DAY GECKO
Cnemaspis boulengerii   NOT ILLUSTRATED
Measurement SVL 66mm **Identification** Body robust; snout broad; rostral contacts nostril; supralabials 8–10; infralabials 7–8; dorsal surface with keeled tubercles in 8–10 longitudinal rows; enlarged series of subtibial scales; 6–8 enlarged series of femoral scales in both sexes; males lack preanofemoral pores; enlarged series of subcaudals. **Coloration** Dorsum grey with large black spots on nape and axilla, and sometimes indistinct spots on midback; tubercles on body cream; venter unpatterned greyish-white. **Habitat and Behaviour** Inhabits a lowland offshore island. Diet and reproductive habits unstudied. **Distribution** Endemic to Vietnam (Gia Lai, K Bang: Son Lang and Ba Ria-Vung Tau, Con Dao Provinces). **Status** Not Evaluated.

HON TRE ISLAND DAY GECKO
Cnemaspis caudanivea   NOT ILLUSTRATED
Measurement SVL 44mm **Identification** Body slender; dorsal surface with tubercles not arranged in linear series on flanks; supralabials 8–9; infralabials 8; ventrals smooth; femoral pores absent; 2 widely spaced precloacal pores; paravertebral and lateral longitudinal rows of caudal tubercles; smooth subcaudals without enlarged median row; cloacal tubercles 1–2; large, shield-like subtibial scales; lamellae under Toe IV 24–29. **Coloration** Dorsum grey with irregularly shaped reddish-brown mark on forehead and snout; squarish blotches on neck and body separated by dull white blotches; irregularly shaped bands on limbs; reddish-brown postorbital stripes edged ventrally by white that extends to sides of neck; venter unpatterned dull beige; gular region with dark smudges and light spots; tail-tip white. **Habitat and Behaviour** Known from a single granitic island within secondary semi-deciduous forest, at an elevation of 100m asl. Diurnal as well as nocturnal, and saxicolous. Diet and reproductive habits unknown, except for the fact that it nests communally, with sites such as overhangs of boulders carrying 400–500 eggs and egg scars. **Distribution** Endemic to southern Vietnam (Hon Tre Island, Kien Giang Province). **Status** Not Evaluated.

CHANTHABURI DAY GECKO
Cnemaspis chanthaburiensis   PLATE 20
Measurement SVL 41mm **Identification** Body slender; snout short; dorsal surface with subequal granular scales, lacking tubercles; smooth ventral and subcaudal scales; 7–9 precloacal pores in males; hyperphalangy of Digits II and V of manus and Digit I of pes; midventrals 37; lamellae under Toe IV 17–20. **Coloration** Dorsum chocolate-brown or greyish-olive; white paravertebral markings from nape to tail; markings on nape and between shoulders oval and coalesced at midline or separated by a pale stripe; lateral row of indistinct light blotches between axilla and groin on flanks; venter mottled light brown. **Habitat and Behaviour** Inhabits dry evergreen forests at an elevation of 850m asl. Diurnal and arboreal. Diet and reproductive habits unstudied. **Distribution** South-western Thailand (Chanthaburi and Chon Buri Provinces) and Cambodia. **Status** Not Evaluated.

DE ZWAAN'S DAY GECKO
Cnemaspis dezwaani   NOT ILLUSTRATED
Measurement SVL 31.4mm **Identification** Body slender; snout large; supralabials 8; infralabials 7; 2 semicircular supranasals, separated by a scale; 3 postnasals bound nasal; 4 scale rows separate orbit from supralabials; posteriorly, each postmental bounded by 3 smooth, rounded and juxtaposed scales; scattered spinose paravertebral rows of tubercles on dorsal surface; pectoral and abdominal scales not elongated, imbricate, with 1 keel; median subcaudals enlarged, unicarinate; males with 4–6 pairs of precloacal pores and 3 femoral pores; lamellae under Toe IV 18–19. **Coloration** Dorsum chocolate-brown with dark brown marbling; lips dark-barred; pale brown vertebral stripe; dorsal spines located within pale brown areas; limbs and tail dark-banded; dark canthal stripe; 3 radiating dark brown lines from eyes; venter unpatterned cream-brown. **Habitat and Behaviour** Inhabits a low-lying island at an elevation of 360m asl. Diet and reproductive habits unstudied. **Distribution** Endemic to Pulau Nias (Indonesia). **Status** Not Evaluated.

DRING'S DAY GECKO
Cnemaspis dringi   NOT ILLUSTRATED
Measurement SVL 45.5mm **Identification** Body slender; 5 postnasals; postmentals reduced; pectoral and abdominal scales distinctly elongated, imbricate and smooth; midventrals 39–40; median subcaudals tricarinate; males with 3 pairs of precloacal pores, lacking femoral pores; no precloacal groove; no postcloacal spur. **Coloration** Dorsum pale brown with dark brown lines along paravertebral region from nape to beyond caudal constriction; irregular markings along vertebral midline; flanks with distinct white patches; venter dark. **Habitat and Behaviour** Inhabits lowland forests. Diet and reproductive habits unstudied. **Distribution** Endemic to northern Borneo (Kapit, Sarawak State) **Status** Not Evaluated.

YELLOW-BELLIED DAY GECKO
Cnemaspis flavigaster   NOT ILLUSTRATED
Measurement SVL 50.1mm **Identification** Body slender, elongated; scales on anterior forearm keeled; precloacal pores 7–8; femoral pores absent; ventrals smooth; median subcaudal scales not enlarged or keeled; lamellae under Toe IV 29–34. **Coloration**

Dorsum brown or dark grey with a series of black spots; paired paravertebral spots on nape; flanks with white markings; venter orange. **Habitat and Behaviour** Inhabits lowland forests and limestone caves. Diurnal and saxicolous, seen on undersides of boulders. Diet and reproductive habits unstudied. **Distribution** Endemic to Peninsular Malaysia (Selangor State). **Status** Not Evaluated.

TITI WANGSA DAY GECKO
Cnemaspis flavolineatus PLATE 20
Measurement SVL 41.2mm **Identification** Body slender, flattened; supralabials 8–10; infralabials 7–10; lateral postmentals separated by 1–2 smaller postmentals; forearm scales, subtibials, ventrals, subcaudals and dorsal tubercles keeled; paravertebral tubercles 22–24; flank tubercles small, not linearly arranged; median subcaudals not enlarged; femoral pores absent; continuous row of 5 precloacal, pore-bearing scales; postcloacal tubercles 2–3; subtibials not shield-like; lamellae under Toe IV 23–26. **Coloration** Dorsum yellowish-brown or olive-brown; some individuals with wide yellow vertebral stripe; dark spots on neck or white markings on flanks absent; dark caudal bands absent; venter dull beige; subcaudal region pigmented. **Habitat and Behaviour** Inhabits the mid-hills within primary forests. Diurnal, associated with rocks. Diet and reproductive habits unstudied. **Distribution** Endemic to Peninsular Malaysia (Titi Wangsa Range at Cameron Highlands, Bukit Fraser and Gunung Benom, Pahang State). **Status** Not Evaluated.

JACOBSON'S DAY GECKO
Cnemaspis jacobsoni NOT ILLUSTRATED
Measurement SVL 30.5mm **Identification** Body slender; snout elongated; supralabials 7–8; infralabials 7–8; 2 semicircular supranasals that are separated by 2 scales, a large anterior one that contacts rostral and a small posterior one; 4 postnasals bound nasal; posteriorly, each postmental is bounded by 4 smooth, rounded and juxtaposed scales; no paravertebral rows of tubercles on dorsal surface; pectoral and abdominal scales distinctly elongated and imbricate, bearing a single keel; spinous processes on lateral surface of body; males lack precloacal and femoral pores, as well as a precloacal depression; lamellae under Toe IV 14. **Coloration** Dorsum light brown with brownish-yellow vertebral stripe, interrupted by 6 dark brown cross-bars forming chevron-like markings; radiating dark lines from eyes; venter unpatterned greyish-brown. **Habitat and Behaviour** Inhabits a low-lying island. Diet and reproductive habits unstudied. **Distribution** Endemic to Pulau Simeulue (Indonesia). **Status** Not Evaluated.

KARST-DWELLING DAY GECKO
Cnemaspis karsticola NOT ILLUSTRATED
Measurement SVL 48.1mm **Identification** Body slender; supralabials 7–8; infralabials 8–9; dorsal surface with keeled tubercles; paravertebral tubercles 17–19; tubercles on flanks large, conical, not arranged in linear series; median subcaudals not enlarged; subcaudals keeled; femoral pores absent; continuous row of 7–8 precloacal, pore-bearing scales; subtibial scales not shield-like; no enlarged submetatarsal scales; lamellae under Toe IV 27–30. **Coloration** Dorsum yellowish-brown with irregular white markings; rostrum yellow, lacking dark streaks; no dark postorbital stripe; 3 radiating dark lines on occiput; 5 pairs of dark paravertebral markings extend from nape to tail-base; venter pale. **Habitat and Behaviour** Inhabits karst areas of Gunung Reng at an elevation of 113m asl. Nocturnal and saxicolous. Diet and reproductive habits unstudied. **Distribution** Endemic to Peninsular Malaysia (Gunung Reng, Kelantan State). **Status** Not Evaluated.

KENDALL'S DAY GECKO
Cnemaspis kendallii PLATE 20
Measurement SVL 80mm **Identification** Body slender; canthal ridge well developed; postnasals 6; postmentals reduced; pectoral and abdominal scales distinctly elongated, imbricate and unicarinate; tail with median row of pointed, semi-erect scales below; midventrals 40; dorsal surface with larger and smaller scattered scales; median subcaudals tricarinate; males lack precloacal or femoral pores, or precloacal groove. **Coloration** Dorsum pale brown with oblong dark brown spots forming 7 interrupted bands; venter cream. **Habitat and Behaviour** Inhabits lowland forests, peat swamps and the mid-hills at between sea level and 1,310m asl. When alarmed, its tail curls over. Diet includes ants, earthworms, beetles and millipedes. Clutches comprise 2 eggs. **Distribution** Southern Peninsular Malaysia, Singapore, the Riau and Natuna Archipelagos, and north-western Borneo. **Status** Not Evaluated.

KUMPOL'S DAY GECKO
Cnemaspis kumpoli PLATE 20
Measurement SVL 52mm **Identification** Body robust; limbs long; snout elongated; supralabials 11; infralabials 10–11; auricular opening vertical; males with 8 precloacal pores; median subcaudals enlarged; midventrals ca 40; lamellae under Toe IV 32. **Coloration** Dorsum olive-grey with 4 transverse blackish-grey bars; dark postocular stripe meets at back of head, edged by olive-grey line, creating a horseshoe-shaped pattern on forehead; tail dark-banded; venter greyish-cream, each ventral with darker pigments. **Habitat and Behaviour** Inhabits lowland forests. Diurnal and saxicolous, associated with granite boulders. Diet and reproductive habits unstudied. **Distribution** Southern Thailand (Khao Chong, Trang Province) and northern Peninsular Malaysia (Wang Kelian, Perlis State). **Status** Not Evaluated.

LIM'S DAY GECKO
Cnemaspis limi PLATE 20
Measurement SVL 88.2mm **Identification** Body robust, elongated; head large, depressed; supralabials 14–16; infralabials 10–12; supranasals in contact;

auricular opening large; dorsal surface with enlarged smooth tubercles; precloacal and femoral pores absent in males; Fingers IV and V subequal; median subcaudals enlarged; lamellae under Toe IV 26–27. **Coloration** Dorsum dark green with oval-elongated black paravertebral spots enclosed within pale bands, extending to beyond sacral region; green flecks on dorsum and in areas between dark stripes on sides of neck; forehead green with dark spots; 2 thick, radiating dark postocular stripes, one extending to back of head, the other across auricular opening to axilla; cream tubercles on dorsum; limbs and tail dark-banded; venter unpatterned brown; iris orange. **Habitat and Behaviour** Inhabits lowland dipterocarp forests to hill- and ridge-top forests. Nocturnal and saxicolous, associated with boulders. Diet consists of larvae of geometrid lepidopterans, grasshoppers and beetles, and also alates of ants. Reproductive habits unstudied. **Distribution** Endemic to islands off the east coast of Peninsular Malaysia (Pulau Tioman and Pulau Tulai, Pahang State). **Status** Not Evaluated.

McGUIRE'S DAY GECKO
Cnemaspis mcguirei NOT ILLUSTRATED
Measurement SVL 65.2mm **Identification** Body slender; supralabials 7–9; infralabials 7–8; dorsal surface with keeled tubercles; 26–32 paravertebral tubercles; tubercles on flanks relatively small, not arranged in linear series; median subcaudals not enlarged; subcaudals keeled; 2–3 postcloacal tubercles; femoral pores absent; discontinuous or continuous row of 5–10 preanal scales bearing pores; subtibials not shieldlike; lamellae under Toe IV 30–35. **Coloration** Dorsum grey to brown with irregular dark blotches; forehead with light markings; dark postorbital stripe extends to nape and contacts dark, anteriorly projecting median stripe; large black shoulder patches enclose 2 yellow ocelli; shoulder patch edged posteriorly by wide yellow and white postscapular band; irregularly shaped white paravertebral markings on dorsum extending to tail-base; transversely elongated, distinct yellow markings on flanks; venter unpatterned dull beige, darkening laterally. **Habitat and Behaviour** Inhabits hill dipterocarp and lower montane forests at elevations of 800–1,351m asl. Diurnal and saxicolous, and also associated with vegetation. Diet and reproductive habits unstudied. **Distribution** Endemic to Peninsular Malaysia (Bintang Range at Gunung Inas, Gunung Bubu and Bukit Larut, Perak State). **Status** Not Evaluated.

MODIGLIANI'S DAY GECKO
Cnemaspis modiglianii NOT ILLUSTRATED
Measurement SVL 33.7mm **Identification** Body slender; supralabials 8–9; infralabials 6–8; supranasals separated by 1 scale; 5 postnasals bound nasal; 2 scale rows separate orbit from supralabials; postmentals bounded by 3 smooth, rounded and juxtaposed scales; paravertebral rows of tubercles absent; spinous processes on flanks; males with paired precloacal pores, no precloacal depression and 4 pairs of femoral pores; lamellae under Toe IV 16–18. **Coloration** Dorsum light brown with paired dark triangular marks on occipital region; lighter intervening region; 4 large, pale inverted chevrons on vertebral region; radiating dark lines from eyes; venter unpatterned pinkish-brown. **Habitat and Behaviour** Inhabits a low-lying island. Diet and reproductive habits unstudied. **Distribution** Endemic to Pulau Enggano (Indonesia). **Status** Not Evaluated.

GADING DAY GECKO
Cnemaspis nigridia PLATE 21
Measurement SVL 69.8mm **Identification** Body robust; canthal ridge developed; pectoral and abdominal scales imbricate and tricarinate; midventrals 68; median subcaudals smooth; dorsal surface with small granular scales and scattered large keeled tubercles; males lack precloacal or femoral pores, or precloacal groove; tail without median row of pointed scales below. **Coloration** Dorsum brownish-olive with black blotches; 2 pairs of elongated dark brown spots on nape and axilla; single elongated spot on vertebral region; some scales on middorsum green; venter unpatterned grey. **Habitat and Behaviour** Inhabits granite and limestone hills at elevations of 500–1,100m asl. Diurnal and saxicolus. Diet consists of spiders and presumably other arthropods. Clutches comprise 2 eggs, measuring 9.8–11.2 x 8.4–10.2mm, which are laid communally on rocky substrates such as rock crevices. **Distribution** North-western Borneo and Pulau Natuna (Indonesia). **Status** Not Evaluated.

NUI CAM HILL DAY GECKO
Cnemaspis nuicamensis NOT ILLUSTRATED
Measurement SVL 47.5mm **Identification** Body slender; tubercles not arranged in linear series on flanks; supralabials 8; infralabials 6–7; femoral pores absent; precloacal pores 4–6; subcaudals with slightly enlarged median row; 2 cloacal tubercles; enlarged submetatarsal scales beneath first metatarsal; lamellae under Toe IV 30–32. **Coloration** Dorsum dull yellow with large reddish-brown blotches and black paravertebral blotches; 4 indistinct reddish-brown caudal bands; reddish-brown reticulum that encloses white blotches and scattered dark spots on forehead and snout; 3 narrow reddish-brown postorbital stripes edged ventrally with white; venter unpatterned grey. **Habitat and Behaviour** Inhabits secondary semi-deciduous forests near streams at an elevation of 100m asl. Diurnal and saxicolous, associated with rocky biotope, concealing itself in rock cracks. Diet and reproductive habits unstudied, although it is known that its eggs are laid communally beneath the overhangs of large boulders. **Distribution** Endemic to Vietnam (Niu Cam Hill, An Giang Province). **Status** Not Evaluated.

PEMANGGIL DAY GECKO
Cnemaspis pemanggilensis PLATE 21
Measurement SVL 76mm **Identification** Body slender; supralabials 13–15; supranasals in contact; no ridge of tubercles bordering anterior margin of

auricular opening or extending from auricular opening to nape; infralabials 10–14; dorsal surface with tubercles showing multiple ridges; precloacal and femoral pores absent; median subcaudal scales enlarged and keeled; lamellae under Toe IV 27–31. **Coloration** Dorsum unpatterned black to grey with darker spots and stripes; dark bifurcating stripe on snout; 1 dark preorbital stripe; 3 dark postorbital stripes; dorsal stripe extends to nape forming tripartite band, middle stripe to nape forming second band, ventral-most stripe to forelimb base; rows of dark spots from nape to tail-base; light spots on flanks; venter beige with darker pattern. **Habitat and Behaviour** Inhabits granite boulders on a severely degraded island, at elevations of 0–250m asl. Diurnal and nocturnal. Diet and reproductive habits unstudied. **Distribution** Endemic to an island off the coast of Peninsular Malaysia (Pulau Pemanggil, Johor State). **Status** Not Evaluated.

PULAU PERHANTIAN DAY GECKO
Cnemaspis perhentianensis NOT ILLUSTRATED
Measurement SVL 47.2mm **Identification** Body slender; supralabials 8–10; infralabials 7–8; ventrals weakly keeled; femoral pores absent; 8 precloacal pores in males; flank tubercles not arranged in linear series; keeled subcaudals lack enlarged median row; lamellae under Toe IV 28–31. **Coloration** Dorsum grey to brown with irregular white markings on forehead; postorbital stripes absent; squarish white marking on neck; 6 irregular white paravertebral marks on dorsum extend from axilla to tail-base, alternating with transversely elongated white markings on flanks; 7 pairs of dark paravertebral blotches from nape to anterior tail; venter unpatterned beige, immaculate; gular region smudged with dark stippling. **Habitat and Behaviour** Inhabits an offshore island with hilly topography dominated by dipterocarp forests, at an elevation of 40m asl. Associated with granitic biotope. Diurnal and saxicolous. Diet and reproductive habits unstudied. **Distribution** Endemic to an island off the east coast of Peninsular Malaysia (Pulau Perhentian Besar, Terengganu State). **Status** Not Evaluated.

PHUKET DAY GECKO
Cnemaspis phuketensis PLATE 21
Measurement SVL 29.1mm **Identification** Body slender, elongated; supralabials 6–7; infralabials 6–7; 2 semicircular supranasals separated by a single scale; 3 postnasals border nasal; gulars and pectorals unicarinate; spines present on flanks; precloacal and femoral pores absent; median subcaudals not enlarged; midventrals 26–32; lamellae under Toe IV 16–17. **Coloration** Dorsum olive with sinuous dark greyish-brown markings on nape and body; labials with dark bars; spines on flanks pale; dark greyish-brown canthal stripe extends to axilla; throat and subcaudals with brown mottling; venter unpatterned cream; iris buff yellow. **Habitat and Behaviour** Inhabits lowland forests, and linked with waterfalls. Nocturnal and both arboreal and terrestrial, associated with tree trunks, saplings and substratum of forest floor near stream banks. Diet and reproductive habits unstudied. **Distribution** Endemic to southern Thailand (Phuket Island). **Status** Not Evaluated.

THAI DAY GECKO
Cnemaspis siamensis PLATE 21
Measurement SVL 42mm **Identification** Body slender; snout broad; supralabials 8–11; infralabials 8–10; dorsal surface with tubercles in 12–20 longitudinal rows; 1 tubercle on each side of vent; males with 2–8 precloacal pores in angular series; pores occasionally absent; median subcaudals enlarged; midventrals ca 30. **Coloration** Dorsum olive-grey with pale mottling; dark brown mottling may form cross-bars; postocular stripes meet at pale spot on nape; 2 pale spots at back of head, separated by brown spot; tail dark-banded; venter yellowish-white. **Habitat and Behaviour** Inhabits forests in the mid-hills. Diurnal and primarily saxicolous, associated with rocks and also tree buttresses. Diet consists of spiders and possibly other arthropods. Clutches comprise 2 eggs. **Distribution** Southern Myanmar (Tenasserim, Taninthayi Division) and southern Thailand (Chumphon, Nakhon Si Thammarat and Chanthaburi Provinces). **Status** Not Evaluated.

TUC DUP HILL DAY GECKO
Cnemaspis tucdupensis NOT ILLUSTRATED
Measurement SVL 51mm **Identification** Body slender; supralabials 10; infralabials 8–9; ventrals smooth; precloacal and femoral pores absent; flank tubercles arranged in linear series; smooth subcaudals with enlarged median row; enlarged submetatarsal scales beneath first toe; lamellae under Toe IV 27–30. **Coloration** Dorsum grey with large reddish-brown and yellow spots; limbs with reddish-brown and pale yellow alternating bands; 7 reddish-brown bands on tail; forehead and snout with dark reticulum within which are white blotches; 3 narrow reddish-brown postorbital stripes edged ventrally with white, dorsal-most extending to axilla; narrow dark preorbital stripe edged ventrally with white; venter unpatterned yellowish-orange. **Habitat and Behaviour** Inhabits secondary semi-deciduous forests at an elevation of 100m asl. Active by day (in shade) and night, and saxicolous, associated with landscape characterized by granitic outcroppings and caves. Diet and reproductive habits unstudied. **Distribution** Endemic to Vietnam (Tuc Dup Hill, An Giang Province). **Status** Not Evaluated.

WHITTENS'S DAY GECKO
Cnemaspis whittenorum NOT ILLUSTRATED
Measurement SVL 31.5mm **Identification** Body slender; head large; supralabials 6–7; infralabials 7; supranasals separated by a scale; 2 postnasals bound nasal; postmental bounded by 4 smooth, rounded juxtaposed scales; paravertebral rows of tubercles absent; pectoral and abdominal scales elongated and imbricate, keeled; spines on flanks; median subcaudals enlarged, unicarinate; lamellae under Toe IV 18–19. **Coloration** Dorsum grey-brown; yellowish-brown vertebral stripe

from forehead to tail-base; radiating brownish-grey lines from orbit; dark canthal stripe; venter yellowish-grey. **Habitat and Behaviour** Inhabits undulating lowland rainforests with peat swamps. Diet and reproductive habits unstudied. **Distribution** Endemic to Pulau Siberut. **Status** Not Evaluated.

MON STATE BENT-TOED GECKO
Cyrtodactylus aequalis NOT ILLUSTRATED
Measurement SVL 90mm **Identification** Body slender; dorsal surface with 24 longitudinal rows of large keeled tubercles; limbs robust, digits long; paired enlarged postmental scales in broad contact behind mental; midventrals 24; precloacal groove absent; 9 small precloacal pores in females; 3–4 femoral pores separated from precloacal pores by diastema; subcaudals expanded; lamellae under Toe IV 14, 8 broad lamellae. **Coloration** Dorsum mid-brown with a series of pairs of dark brown markings bordered by narrow white lines; postocular meets at occipital as band; white line ventral to dark postocular-temporal stripe; series of 5 blotches across nape, pair blotches on axilla; 5 paired irregular blotches between limb insertions; forehead mid-brown with diffuse set of white vermiform marks; venter beige with scattered brown marks. **Habitat and Behaviour** Nothing known of its natural history. **Distribution** Endemic to southern Myanmar (Kyaik-Hti-Yo Wildlife Sanctuary, Mon State). **Status** Not Evaluated.

ANGLED BENT-TOED GECKO
Cyrtodactylus angularis PLATE 21
Measurement SVL 95mm **Identification** Body slender; head broad; auricular opening large; dorsal surface with small scales intermixed with larger, conical tubercles; supralabials 9–11; infralabials 8–10; precloacal pores 5–6; subdigital lamellae not enlarged; subcaudals enlarged; preanofemoral scales 25; 6 precloacals in angular series and femorals slightly enlarged and continuous with precloacals in males; midventrals 26–33. **Coloration** Dorsum grey-brown to mid-brown, 2 series of large angular spots connected mesially; forehead with indistinct angular spots; dark postocular stripe edged with white extends to beyond forehead; tail dark-banded. **Habitat and Behaviour** Inhabits the mid-hills. Diet and reproductive habits unstudied. **Distribution** Endemic to eastern Thailand (Dong Paya Fai Mountains). **Status** Not Evaluated.

ANNANDALE'S BENT-TOED GECKO
Cyrtodactylus annandalei NOT ILLUSTRATED
Measurement SVL 55mm **Identification** Body slender; dorsal surface with smooth scales intermixed with 16–18 rows of keeled tubercles; limbs and digits short; midventrals 43; precloacal groove absent; 8–12 precloacal pores in single series in both sexes; femoral pores 10–11 in males, separated from precloacal series by diastema; subcaudals with alternating rows forming wide transverse plates; lamellae under Toe IV 10. **Coloration** Dorsum pale brown with chocolate-brown bands edged with narrow cream or white, 1–2 scales wide; occipital band extends to orbit and under eye to loreal region, continuing to rostral; forehead unpatterned; nuchal collar extends to posterior border of auricular opening; 4 bands on body; on flanks, pale brown area between body bands with large chocolate-brown spots; venter cream tinged with light brown speckling. **Habitat and Behaviour** Inhabits deciduous hardwood forests. Nocturnal and arboreal. Diet and reproductive habits unstudied. **Distribution** Endemic to Myanmar (Sagaing and Magwe Divisions). **Status** Not Evaluated.

PULAU AUR BENT-TOED GECKO
Cyrtodactylus aurensis NOT ILLUSTRATED
Measurement SVL 95mm **Identification** Body robust; dorsal surface with low and conical unkeeled tubercles; transversely enlarged median subcaudal scales; proximal subdigital lamellae transversely expanded; abrupt transition between large posterior and small ventral femoral scales; enlarged femoral scales or pores absent; precloacal groove, enlarged preanal scales and precloacal pores present; midventrals 45–51. **Coloration** Dorsum brown with light brown mottling; 4 narrow, irregular pale bands between limb insertions; elongated white blotches on flanks between bands; 1 short, irregular white band on nape; elongated white blotches on sides of neck; narrow white nuchal loop; cream preorbital stripe; thin, broken white line from anterior margin of auricular opening to angle of jaw; body bands edged with narrow dark brown; venter beige with faint black stippling on each scale. **Habitat and Behaviour** Known from edges of plantations in rocky biotope at an elevation of 100m asl. Nocturnal and saxicolous. Diet and reproductive habits unstudied. **Distribution** Endemic to Peninsular Malaysia (Pulau Aur, Johor State). **Status** Not Evaluated.

AYEYARWADY BENT-TOED GECKO
Cyrtodactylus ayeyarwadyensis PLATE 21
Measurement SVL 78mm **Identification** Body slender; dorsal surface with 22–24 rows of keeled oblong dorsal tubercles; limbs long; pair of enlarged postmental scales in broad contact behind mental; midventrals 32–37; precloacal groove absent; 10–28 precloacal pores in single series or with scattered gaps of 1 poreless scale in males; subcaudals without enlarged midventral plates; lamellae under Toe IV include 6 widened and 10 narrow lamellae. **Coloration** Dorsum mid-brown with 10 transverse rows of rectangular dark brown patches from occiput to sacrum, comprising rows of paired paravertebral marks and pair of ill-defined lateral marks; forehead mid-brown with diffuse dark brown semicircle between posterodorsal corners of orbits; postocular brown bordered by cream, reaching neck; venter white with pink on flanks; iris olive to greenish-gold. **Habitat and Behaviour** Inhabits lowland forests. Nocturnal and arboreal. Diet unstudied. Clutches comprise 2 eggs. **Distribution** Endemic to Myanmar (Rakhine State). **Status** Not Evaluated.

BADEN BENT-TOED GECKO
Cyrtodactylus badenensis PLATE 21
Measurement SVL 74.1mm **Identification** Body slender; head depressed; dorsal surface with small scales and larger conical tubercles in ca 6 rows; femorals enlarged; femoral pores absent; infralabials 8–10; supralabials 10–13; internasals 2; midventrals 25–28; transversely enlarged subcaudals; lamellae under Toe IV 18–22. **Coloration** Dorsum dark chocolate-brown with 4 white cross-bars; cone-shaped marks, arranged in series of 3–5, between axilla and inguinal region; forehead yellow to brown; dark line from snout-tip, across eye, to neck; tail with white cross-bars; tail-tip white. **Habitat and Behaviour** Restricted to landscapes with strong slopes, inhabiting caves at elevations of ca 986m asl. Nocturnal and saxicolous. Diet and reproductive habits unstudied. **Distribution** Endemic to southern Vietnam (Mount Ba Den, Tay Ninh Province). **Status** Not Evaluated.

KINABALU BENT-TOED GECKO
Cyrtodactylus baluensis PLATE 21
Measurement SVL 86mm **Identification** Body slender; dorsal surface with small scales intermixed with larger tubercles in 21–24 rows; supralabials 10–12; infralabials 9–10; midventrals 40–45; precloacal groove absent; precloacal pores 9–10; femoral pores 6–9; sharp boundary of scale size between ventral scales; posterior scales granular; lamellae under Toe IV 21–23. **Coloration** Dorsum brown to yellowish-brown with irregular dark spots that may form dark cross-bars; head with dark brown lateral band from snout-tip to nape; limbs and tail dark-banded; venter cream. **Habitat and Behaviour** Inhabits dipterocarp rainforests to montane oak forests at elevations of 150–2,500m asl. Nocturnal and arboreal, foraging on leaves of saplings, tree trunks, buttresses and forest clearings. Diet consists of insects and other arthropods. Clutches comprise 2 eggs, measuring 12 x 15mm, which are deposited at the bases of trees. Hatchlings 31–32mm. **Distribution** Endemic to Borneo (Sabah, Sarawak State and Brunei Darussalam). **Status** Not Evaluated.

PULAU BESAR BENT-TOED GECKO
Cyrtodactylus batucolus NOT ILLUSTRATED
Measurement SVL 75.2mm **Identification** Body robust; dorsal surface with conical keeled tubercles on occiput, forelimbs, hind limbs and beyond base of tail; no transversely enlarged subcaudal scales; abrupt transition between postfemoral and ventral femoral scales; enlarged femoral and preanal scales with continuous series of 43–46 pore-bearing scales in males; precloacal groove absent; precloacal depression present; midventrals 38–42; lamellae under Toe IV 17–19. **Coloration** Dorsum dark brown; forehead with white blotches; pale line across occiput; 9 paired, square dark brown paravertebral blotches from nape to tail-base; blotches edged posteriorly with cream blotches; smaller dark brown, countershaded markings on flanks; venter lightly stippled with grey; iris green. **Habitat and Behaviour** Inhabits lowland forests at an elevation of 39m asl. Nocturnal and saxicolous, associated with boulders bearing cracks or exfoliations in which it hides. Also found on large trees and in buildings. Diet and reproductive habits unstudied. **Distribution** Endemic to offshore islands south of Peninsular Malaysia (Panti Putera and Pulau Besar). **Status** Not Evaluated.

SHORT-TOED BENT-TOED GECKO
Cyrtodactylus brevidactylus PLATE 21
Measurement SVL 88mm **Identification** Body robust; digits short; dorsal surface with 27 rows of enlarged keeled tubercles; supralabials 8–10; subcaudals not enlarged; precloacal pores 8; femoral pores absent; midventrals 45; lamellae under Toe IV 11–13. **Coloration** Dorsum pale greyish-brown to white with 3–4 large dark blotches edged with narrow dark brown; forehead chocolate-brown with white markings, including a Y-shaped mark; venter cream, scales with light brown areas. **Habitat and Behaviour** Inhabits dry subtropical forests. Nocturnal and arboreal. Diet and reproductive habits unstudied. **Distribution** Endemic to central Myanmar (Popa Mountains, Mandalay Division). **Status** Not Evaluated.

SHORT-FINGERED BENT-TOED GECKO
Cyrtodactylus brevipalmatus PLATE 21
Measurement SVL 72mm **Identification** Body robust; dorsal surface with tubercles in 14–18 rows separated by 1–5 granules; supralabials 12–14; Supralabial I contacts nostril; infralabials 10–11; flanks with tubercles; 9 enlarged precloacal pores; 6–7 enlarged femoral pores in males; precloacal groove absent; toes basally webbed; subcaudals enlarged; midventrals 35–44. **Coloration** Dorsum brown with small dark spots, not arranged in linear series; juveniles with W-shaped dorsal marks; tail dark-banded; venter cream. **Habitat and Behaviour** Inhabits the mid-hills at an elevation of 750m asl. Nocturnal and arboreal, associated with trees. Diet and reproductive habits unstudied. **Distribution** Endemic to Thailand (Khao Luang, Nakon Si Thammarat Province, Tak, Tak Province, Kaeng Krachan National Park, Phetchaburi Province and Huai Kha Khaeng, Uthai Thani Province). **Status** Not Evaluated.

BUCHARD'S BENT-TOED GECKO
Cyrtodactylus buchardi NOT ILLUSTRATED
Measurement SVL 33.5mm (juvenile male; adults unknown) **Identification** Body slender; head short; tail short, slender; dorsal surface with conical or weakly keeled tubercles, arranged in 25 rows; supranasals present; supralabials 13–14; infralabials 11–16; auricular opening vertical; precloacal groove absent; 3 pairs of enlarged precloacal scales; no enlarged femoral scales; median subcaudals not enlarged; midventrals 38; lamellae under Toe IV 12. **Coloration** Dorsum dark tan; flanks with 15 irregular dark chocolate-brown blotches, edged by small white scales in 1–2 rows, in 6 series; V-shaped dark chocolate-brown nuchal mark; forehead dark tan, paler on snout; sides of neck with white spots; tail

dark-banded; venter ivory-cream with dense dark dots or star-like marks. **Habitat and Behaviour** Inhabits lowland monsoon evergreen forests, the unique holotype having been found at 90–300m asl. Probably nocturnal, daytime retreat includes stones. Diet and reproductive habits unstudied. **Distribution** Endemic to Laos (north-west of Kiatngong Xepian National Biodiversity and Conservation Area, Champasak Province). **Status** Not Evaluated.

CAO VAN SUNG'S BENT-TOED GECKO
Cyrtodactylus caovansungi PLATE 21
Measurement SVL 94mm **Identification** Body slender; dorsal surface with rounded-triangular dorsal tubercles, not arranged in longitudinal rows; 9 precloacal pores, arranged in an angular series in males; 8 pairs of enlarged femoral scales; femoral pores located close to knee; tail weakly segmented. **Coloration** Dorsum brown with 4 transverse bands with irregular edges; in between are light bands of similar width; small irregular dark spots in middle part of light bands; dark nuchal loop from posterior edge of eye; forehead with irregular light spots, producing marbled or reticulated pattern; venter unpatterned cream. **Habitat and Behaviour** Inhabits primary forests in the mid-hills at elevations of ca 400m asl. Nocturnal and arboreal, associated with trees. The only individuals known were found ca 1.5m above water in a rocky stream in an evergreen forest. Diet and reproductive habits unstudied. **Distribution** Endemic to southern Vietnam (Vinh Hai, Ninh Thuan Province). **Status** Not Evaluated.

NIAH BENT-TOED GECKO
Cyrtodactylus cavernicolus PLATE 22
Measurement SVL 80.8mm **Identification** Body slender; precloacal grooves with 2 pairs of pores; dorsal surface with granular scales interspersed with 20–22 rows of trihedral or conical tubercles; precloacal pores and femoral scales absent; lamellae under Toe IV 22–26. **Coloration** Dorsum brown with dark-edged brown cross-bars, changeable to brown-black; dark postocular stripe to nape; venter unpatterned cream; tail dark-banded. **Habitat and Behaviour** Inhabits limestone-dominated lowland forests. Within caves, found on walls and on wooden man-made structures. Also found under fallen tree trunks on forest floor. Diet consists of flattened cave cockroaches and moths. Reproductive habits unstudied. **Distribution** Endemic to north-western Borneo (Niah Caves, Sarawak State). **Status** Not Evaluated.

CHANHOME'S BENT-TOED GECKO
Cyrtodactylus chanhomeae PLATE 22
Measurement SVL 79mm **Identification** Body slender, limbs and digits long, slender; dorsal surface with 16–18 rows of keeled tubercles; midventrals 36–38; precloacal groove absent; continuous series of 32–34 pore-bearing preanofemoral scales in males; median subcaudal scales form broad transverse plates; Toe IV with 7–9 broad basal lamellae and 14 narrow distal lamellae. **Coloration** Dorsum pale brown with 3 mid-brown body bands edged with thin cream band; brown nuchal loop with darker margins; forehead pale yellow with scattered darker markings; pale line bordering nape band; venter pale brown; iris greenish-brown. **Habitat and Behaviour** Restricted to Phraya Chat-tan and Thep Nimit Caves. Active by day and at night within limestone caves, 0–3m above ground. Diet and reproductive habits unstudied. **Distribution** Endemic to central Thailand (Saraburi Province). **Status** Not Evaluated.

CHANQUANG BENT-TOED GECKO
Cyrtodactylus chauquangensis PLATE 22
Measurement SVL 99.3mm **Identification** Body slender; dorsal surface with homogeneous rounded scales, not arranged in linear series; midventrals 36–38; supralabials 9–10; infralabials 9–11; 6–7 precloacal pores in both sexes; femoral pores and femoral scales absent; lamellae under Toe IV 19–23. **Coloration** Dorsum chocolate-brown with 5 blackish-brown bands; dark postocular stripe meets at nuchal region; black spot on occiput; 2 spots between neck and axilla; 3 bands on back; venter yellow; gular region buff-coloured; tail banded; tail-tip white. **Habitat and Behaviour** Associated with karst landscape at an elevation of 90m asl. Nocturnal and saxicolous, associated with crevices in limestone rocks and on cave walls. Diet and reproductive habits unstudied. **Distribution** Endemic to north-central Vietnam (Chau Quang Village, Nghe An Province). **Status** Not Evaluated.

SHAN STATE BENT-TOED GECKO
Cyrtodactylus chrysopylos PLATE 22
Measurement SVL 79mm **Identification** Body slender, elongated; dorsal surface with 16 longitudinal rows of keeled tubercles; limbs and digits long; paired enlarged postmental scales in broad contact behind mental; midventrals 37; precloacal groove and femoral pores absent; 10 precloacal pores in single series; single scale with pore posterior to precloacal series and separated by single enlarged scale without pores; lamellae under Toe IV 13, plus 6 broad lamellae. **Coloration** Dorsum mottled purplish-brown with chocolate-brown and pale orange bands; dark occipital band extends to orbit and under eye to nostril; broken white line extends to posterior supraciliaries, bordered posteriorly by thick white line across auricular opening and enters supralabials; 6 pairs of light and dark alternating bands across axilla and 2 around tail-base; forehead with 4 dark marks, especially at posterior margin of orbit; venter beige to light brown. **Habitat and Behaviour** Inhabits lowland forests at an elevation of 319m asl. Diet and reproductive habits unstudied. **Distribution** Endemic to Myanmar (Panlaung-Pyadalin Cave Wildlife Sanctuary, Shan State). **Status** Not Evaluated.

CON DAO BENT-TOED GECKO
Cyrtodactylus condorensis NOT ILLUSTRATED
Measurement SVL 80mm **Identification** Body robust; dorsal surface with granular scales intermixed

with larger tuberculate scales; rostral enters nostril; supranasals small, separated by a small scale; supralabials 10–11; infralabials 8–9; series of enlarged femoral scales; 4–7 precloacal pores; precloacal groove absent; midventrals 35–40. **Coloration** Dorsum greyish-brown with large irregular dark blotches; dark postocular stripe forms nuchal loop; tail dark-banded; venter unpatterned grey. **Habitat and Behaviour** Inhabits a lowland offshore island. Diet and reproductive habits unstudied. **Distribution** Endemic to Vietnam (Con Dao, Ba Ria-Vung Tau Province). **Status** Not Evaluated.

DAWEI BENT-TOED GECKO
Cyrtodactylus consobrinoides　　　NOT ILLUSTRATED
Measurement SVL 50mm **Identification** Body slender; dorsal surface with granular scales intermixed with scattered tuberculate scales, not regularly arranged; supralabials 10; infralabials 9; auricular opening small; digits short and stout; precloacal pores 4–5 in angular series; femoral pores absent in males; midventrals 24–30. **Coloration** Dorsum light brown; several dark brown spots on forehead; dark postocular stripe meets at back of head, forming nuchal loop; 7 narrow, short dark bars on body, reaching only some distance down flanks; tail dark-barred; venter pale brown. **Habitat and Behaviour** Nothing known of its natural history. **Distribution** Endemic to southern Myanmar (Dawei, Tenasserim, Taninthayi Division). **Status** Not Evaluated.

PETERS'S BENT-TOED GECKO
Cyrtodactylus consobrinus　　　PLATE 22
Measurement SVL 125mm **Identification** Body robust; dorsal surface with scattered tuberculate scales in 18–20 irregular rows; supralabials 10–16; infralabials 9–13; midventrals 58–70; femoral pores up to 6; males and females with 9–14 preanal scales forming a narrow angular series, pores present in males; precloacal groove absent. **Coloration** Dorsum dark chocolate-brown with 4–8 transverse white or yellow bands; dark-edged intervening areas; gular region dark grey; pectoral and abdominal region cream or smoke-grey. Juvenile dorsum blackish-grey to jet-black, with lemon-yellow bands, narrower than dark intervening areas; forehead with reticulated dark pattern; isolated yellow spots on sides of head at level below tympanum. **Habitat and Behaviour** Inhabits lowland dipterocarp forests at elevations of up to 1,100m asl. Associated with large trees and limestone caves. Diet consists of insects. Clutches comprise 2 eggs, measuring 14–17.6mm. Hatchlings 31mm. **Distribution** Peninsular Malaysia, Sumatra, Pulau Sinkep and Borneo. **Status** Not Evaluated.

DARK-COLLARED BENT-TOED GECKO
Cyrtodactylus cryptus　　　PLATE 22
Measurement SVL 90.8mm **Identification** Body slender, rounded; dorsal surface with granular scales and rounded conical tubercles in 20 rows; ventrolateral skin folds along body; enlarged lateral tubercles absent; tubercles present on forehead, limbs and tail; longitudinal rows of ventral scales at midbody 47–50; segmented tail with whorls, not depressed, tail-base not enlarged; 3 pairs of cloacal spurs; 9–11 precloacal pores in angular series in males; 16–27 enlarged preanal scales in both sexes; enlarged femoral scales and femoral pores absent; precloacal groove absent; subcaudals small, not transversally enlarged; lamellae under Toe IV 20–23. **Coloration** Dorsum brownish-olive; labials brown with white blotches; forehead with rounded brown blotches; nuchal band triangular, violet-brown, white margined, with wide stripe extending from outermost neck band to posterior margin of orbit; dorsum with 3–4 transverse violet-brown, buff-margined bands; anterior dorsal band located behind axilla; posterior dorsal band located in sacral region; irregular dark violet-brown blotches, narrow buff margined between dorsal bands; flanks greyish-brown with buff blotches; venter greyish-brown; tail with white and dark brown rings. **Habitat and Behaviour** Restricted to karst areas within primary forests at elevations of ca 520m asl. Nocturnal, active on rocks, tree trunks, branches and leaves up to 2m above substratum. Diet unstudied. Clutches comprise 2 eggs, measuring 12 x 14.3mm, which are deposited in holes on rocks. **Distribution** Endemic to central Vietnam (Phong Nha-Ke Bang National Park, Quang Binh Province). **Status** Not Evaluated.

EISENMAN'S BENT-TOED GECKO
Cyrtodactylus eisenmani　　　NOT ILLUSTRATED
Measurement SVL 89.2mm **Identification** Body slender; tail long; longitudinal rows of tubercles at midbody between ventrolateral folds 14; precloacal groove and pores absent; ventral scales between ventrolateral folds 44–45; enlarged femoral scales beneath thigh 4–5; subcaudal scales enlarged to form broad transverse plates; subdigital lamellae under Toe IV 17–18. **Coloration** Dorsum chocolate-brown with 5 narrow white bands; nuchal loop bordered anteriorly by darker chocolate, and laterally by narrow pale band; tail chocolate-brown; venter light pinkish-white. **Habitat and Behaviour** Inhabits caves within areas dominated by mango and cashew plantations, punctuated by granatic outcrops, at elevations of 150–200m asl. Nocturnal and saxicolous. Diet and reproductive habits unstudied. **Distribution** Endemic to south-western Vietnam (Hon Son Island, Kien Giang Province). **Status** Not Evaluated.

GUNUNG LAWIT BENT-TOED GECKO
Cyrtodactylus elok　　　PLATE 22
Measurement SVL 67.5mm **Identification** Body robust; dorsal surface with tubercles in 6–10 rows, separated by 4–9 granules; supralabials 8–11; infralabials 9–10; proximal subdigital lamellae greatly expanded; toes basally webbed; femoral pores absent in males; precloacal groove absent; flanks lacking tubercles; midventrals 44; lamellae under Toe IV 9–10. **Coloration** Dorsum dark brown; broad yellowish-orange to silver-brown middorsal band, crossed by 5–7 dark brown bars; white postocular streak to posterior supralabials; venter cream with dark speckling; iris white with vein-like dark pattern.

Habitat and Behaviour Inhabits lowland forests at elevations of 43–215m asl. Nocturnal and arboreal, associated with low vegetation. This species is capable of coiling its tail laterally. Diet and reproductive habits unstudied. **Distribution** Endemic to Peninsular Malaysia (Cameron Highlands, Bukit Fraser and Lakum Forest Reserve, Pahang State, and Seremban, Negeri Sembilan State). **Status** Not Evaluated.

FEAE'S BENT-TOED GECKO
Cyrtodactylus feae   NOT ILLUSTRATED
Measurement SVL 47mm **Identification** Body robust; head large; auricular opening small, oblique; supralabials 7–8; infralabials 8–9; limbs elongated; preanofemoral pores 32; midventrals 35. **Coloration** Dorsum dark brown with 4 black cross-bars edged with white tubercles; forehead brown with large black spots within a reticulated cream pattern; postocular black edged with white, crescentic, meeting forming nuchal loop; tail dark-banded; venter unpatterned light brown. **Habitat and Behaviour** Inhabits submontane forests at elevations of 975–1,036m asl. Nocturnal and arboreal. Diet and reproductive habits unstudied. **Distribution** Endemic to southern Myanmar (Puepoli, Karin Biapo, Kayah Hills). **Status** Not Evaluated.

TAMARIND BENT-TOED GECKO
Cyrtodactylus fumosus   PLATE 22
Measurement SVL 75mm **Identification** Body robust; head large; snout elongated; dorsal surface with granular scales mixed with scattered, rounded, smooth or weakly keeled tubercles; supralabials 9–10; infralabials 8–9; preocloacal groove in males; continuous series of 42–52 preanofemoral pores; midventrals 35–40. **Coloration** Dorsum greyish-brown or pinkish-brown with blackish-brown spots that may form irregular transverse bars; dark postocular streak to axilla; tail dark-banded; venter cream or brown, unpatterned or with black spots. **Habitat and Behaviour** Inhabits mid-hill and submontane forests. Nocturnal and arboreal, associated with trees. Diet unstudied. Clutches comprise 2 eggs. Hatchlings 28mm. **Distribution** Java. Also Sulawesi and Halmahera (Indonesia). If the latter population (from the type locality) proves non-conspecific with the Javan one, the Sunda species will require a new name. **Status** Not Evaluated.

GANS'S BENT-TOED GECKO
Cyrtodactylus gansi   PLATE 22
Measurement SVL 63mm **Identification** Body slender; dorsum with rounded conical tubercles in 20–25 rows; limbs and digits short; pair of enlarged postmental scales in broad contact behind mental; midventrals 36–40; ventrolateral folds absent; shallow preocloacal groove in males, 16–29 large precloacal pores in angled series; median subcaudals not enlarged; Toe IV with widened as well as 11 narrow subdigital lamellae. **Coloration** Dorsum light to mid-brown with transverse dark markings, irregular from nape to shoulder, forming 7 continuous cross-bars from axilla to inguinal region; bands terminate on flanks; irregular small dark spots at ventrolateral margins of cross-bands; forehead mid-brown with irregular dark brown spots; brown postocular streak to above auricular opening; dark spots on neck form broken nape band; tail dark banded; venter beige with scattered dark pigment, especially under thighs and around cloaca; orbit rim and lateral tubercles yellow. **Habitat and Behaviour** Inhabits the mid-hills and submontane forests at elevations of 750–1,300m asl. Nocturnal and arboreal. Diet and reproductive habits unstudied. **Distribution** Endemic to Myanmar (Chin State). **Status** Not Evaluated.

GRISMER'S BENT-TOED GECKO
Cyrtodactylus grismeri   NOT ILLUSTRATED
Measurement SVL 95mm **Identification** Body robust; dorsal surface with tubercles in 18–22 rows; tail long; paired enlarged postmental scales in broad contact; precloacal groove and pores absent; 19–22 slightly enlarged poreless scales anterior to vent; interorbital scales 16–19; midventrals 33–38; subcaudal scales enlarged to form broad transverse plates; lamellae under Toe IV 16–19; no enlarged scales on heel. **Coloration** Dorsum dark brown with 5 narrow white bands; 2 longitudinal yellow stripes on snout; forehead with dark mottling, with narrow white band across posterior corners of orbits; dark postorbital stripes from corner of orbit to auricular opening; yellowish-brown dorsal tubercles; tail brown with small scattered yellow spots; venter pinkish-white. **Habitat and Behaviour** Inhabits caves within lowland forests at elevations of ca 100m asl. Nocturnal and saxicolous. Diet and reproductive habits unstudied. **Distribution** Endemic to south-western Vietnam (Tuc Dup Rocky Hill, An Giang Province). **Status** Not Evaluated.

HON TRE BENT-TOED GECKO
Cyrtodactylus hontreensis   NOT ILLUSTRATED
Measurement SVL 88.9mm **Identification** Body slender; dorsal surface with low smooth tubercles in 14 irregular rows; supralabials 8–10; infralabials 10–11; precloacal groove absent; 7–8 precloacal pores in males; 2–5 enlarged femoral scales; midventrals 40–42; shallow middorsal groove; subcaudals enlarged; lamellae under Toe IV 17–19. **Coloration** Dorsum and head brown; forehead with 2 yellow preorbital stripes; broad dark brown nuchal loop edged with white; 3 brown dorsal bands on body edged with white, with indistinct dark spots in interspaces; tail brown, tinged with dark brown blotches laterally; venter pinkish-white; iris ring yellow. **Habitat and Behaviour** Inhabits caves within lowland forests of offshore island at elevations of ca 15–100m asl. Nocturnal and saxicolous, found in landscapes dominated by granite. Diet and reproductive habits unstudied. **Distribution** Endemic to south-western Vietnam (Kien Giang Biosphere Reserve, Kien Giang Province). **Status** Not Evaluated.

CHUA CHAN MOUNTAIN BENT-TOED GECKO
Cyrtodactylus huynhi NOT ILLUSTRATED
Measurement SVL 79.8mm **Identification** Body slender; dorsal surface with weakly keeled tubercles in 16–18 irregular rows; limbs slender; precloacal groove absent; 2 pairs of enlarged postmental scales; 43–46 ventral scales across venter; 7–9 precloacal pores arranged in angular series; 3–5 enlarged femoral scales. **Coloration** Dorsum light brown with 5–6 broad dark brown bands bordered by yellow or cream tubercles; dark brown nuchal loop; tail with dark brown bands. **Habitat and Behaviour** Inhabits lowland caves within secondary deciduous forests at elevations of 70–300m asl. Diet and reproductive habits unstudied. **Distribution** Endemic to Vietnam (Chua Chan Mountain, Dong Nai Province). **Status** Not Evaluated.

INGER'S BENT-TOED GECKO
Cyrtodactylus ingeri PLATE 22
Measurement SVL 80.2mm **Identification** Body slender; dorsal surface with large tuberculate scales in 17 irregular rows; rounded imbricate ventrals; supralabials 10–12; infralabials 8–10; midventrals 40–43; males with 7–9 precloacal pores forming narrow angular series; femoral pores and precloacal groove absent; lamellae under Toe IV 23–28. **Coloration** Dorsum grey or yellowish-brown; 5–6 irregular, diamond-shaped, paired dark brown paravertebral blotches; short, interrupted dark brown flank stripe; nape with Y- or V-shaped dark mark; pale brown postocular stripe to insertion of forearm; tail with dark bands; gular region yellow-cream; pectoral and abdominal regions cream. **Habitat and Behaviour** Inhabits riparian forests at elevations of 500–800m asl. Nocturnal and arboreal, associated with saplings. Diet presumably consists of arthropods. Clutches comprise 2 eggs, measuring 12 x 9mm. **Distribution** Endemic to northern Borneo (Brunei Darussalam and Sabah State). **Status** Not Evaluated.

NAM NAO BENT-TOED GECKO
Cyrtodactylus interdigitalis PLATE 22
Measurement SVL 80mm **Identification** Body slender; dorsal surface with smooth or weakly keeled tubercles in 18–22 rows; preanal and femoral scales enlarged; precloacal pores 14; femoral pores 8–9; midventrals 37–42; lamellae under Toe IV 9–11, flat and widened, followed by 3–5 rows of pointed ones and 4–6 terminal small ones. **Coloration** Dorsum reddish-brown with dark bands and sometimes, 2 longitudinal dark stripes extending from before axilla to tail-base, with pale edges; dark nuchal band joining posterior corners of orbit, edged with pale margin; tail dark-banded; forehead nearly unpatterned brown; venter pale brown. **Habitat and Behaviour** Inhabits evergreen forests at an elevation of 600m asl. Nocturnal, and both saxicolous and arboreal, associated with fig-tree root systems near limestone cave regions. Diet and reproductive habits unstudied. **Distribution** Endemic to central Thailand (Nam Nao National Park, Petchabun Province). **Status** Not Evaluated.

CARDAMOM MOUNTAINS BENT-TOED GECKO
Cyrtodactylus intermedius PLATE 22
Measurement SVL 85mm **Identification** Body slender; dorsal surface with granules and conical tubercles; rostral contacts nostril; supralabials 10–11; infralabials 10–11; auricular opening small; 6–12 enlarged femoral scales; 8–10 precloacal pores; precloacal groove absent; midventrals 40–50; subcaudals transversely enlarged. **Coloration** Dorsum greyish-brown or dark brown with 4–5 wide, unserrated dark brown bands with yellow edges, the first being a nuchal loop; venter unpatterned cream. **Habitat and Behaviour** Inhabits grasslands and deciduous and evergreen forests in the plains up to submontane limits at elevations of 220–1,000m asl. Nocturnal, and saxicolous and arboreal, associated with trees and rocks, especially near water bodies, ascending trees up to at least 2m above substratum. Diet and reproductive habits unstudied. **Distribution** South-eastern Thailand (Khao Sebab, Chantabun Province), Cambodia (Cardamom Mountains) and Vietnam (Ma Da, Dong Nai Province and Hon Chong, Kien Giang Province). **Status** Not Evaluated.

LARGE-SPOTTED BENT-TOED GECKO
Cyrtodactylus irregularis NOT ILLUSTRATED
Measurement SVL 79mm **Identification** Body robust, slightly depressed; dorsal surface granular with scattered larger tubercles; limbs short; tail thick basally, shorter than body; supralabials 11; infralabials 8–9; precloacal pores 5–7; enlarged femoral scales 7–8; midventrals 41–46. **Coloration** Dorsum greyish-brown with 5–7 broad dark brown cross-bars (sometimes rounded patches) with uneven margin, edged with white; pale spots irregularly arranged on midback; tail dark-banded; broad, dark U-shaped postocular stripe, edged with white, runs from posterior edge of eye above auricular opening to neck; forehead reticulated with irregular small, white-edged dark spots; venter cream. **Habitat and Behaviour** Inhabits primary forests at an elevation of 3,500m asl. Diet and reproductive habits unstudied. **Distribution** Endemic to southern Vietnam (Lang Bian Mountain, Lam Dong Province). **Status** Not Evaluated.

PULAU JARAK BENT-TOED GECKO
Cyrtodactylus jarakensis NOT ILLUSTRATED
Measurement SVL 67mm **Identification** Body slender; dorsal surface with low, rounded and weakly keeled tubercles on occiput, forelimbs, hind limbs and beyond tail-base; subcaudal scales small, subgranular; smooth transition between postfemoral and ventral femoral scales; no enlarged femoral or preanal scales or pore-bearing scales; no precloacal groove or depression; midventrals 61; lamellae under Toe IV 24. **Coloration** Dorsum straw-yellow; head and body with dark brown blotches edged with yellow; U-shaped dark blotch contacts posterior margin of eye; large dark brown parietal blotch flanked laterally by larger squarish occipital blotches; large dark medial nape blotch followed by 8 pairs of dark paravertebral

blotches; 3 dark blotches on anterior of tail; black and white bands on remainder of tail; venter unpatterned beige. **Habitat and Behaviour** Inhabits a densely forested island at an elevation of 20m asl. Nocturnal and saxicolous. Diet and reproductive habits unstudied. **Distribution** Endemic to Peninsular Malaysia (Pulau Jarak). **Status** Not Evaluated.

JARUJIN'S BENT-TOED GECKO
Cyrtodactylus jarujini PLATE 22
Measurement SVL 90mm **Identification** Body slender; dorsal surface granular, intermixed with larger tubercles in 18–20 rows; limbs and tail slender; preanofemoral pores 26 pairs; supralabials 12–16; infralabials 10–12; midventrals 32–38; median sub-caudals enlarged; lamellae under digits not enlarged; lamellae under Toe IV 9. **Coloration** Dorsum mid-brown with dark brown blotches, mostly fused medially to form irregular bands; forehead pale brown with dark brown spots; tail dark-banded; venter unpatterned cream. **Habitat and Behaviour** Inhabits forests in the mid-hills at an elevation of 380m asl. Nocturnal and arboreal, associated with large trees. Diet and reproductive habits unstudied. **Distribution** Endemic to north-eastern Thailand (Phu Wua Wildlife Sanctuary, Nong Khai Province). **Status** Not Evaluated.

KHASI HILLS BENT-TOED GECKO
Cyrtodactylus khasiensis PLATE 23
Measurement SVL 85mm **Identification** Body slender; snout relatively long; dorsal surface of body and limbs with small granular scales, intermixed with larger keeled tubercles; lateral fold of enlarged scales; males with 12–14 precloacal pores; femoral pores absent. **Coloration** Dorsum dark greyish-brown with dark spots; dark curved mark extends across nape to join eyes; forehead black-spotted; venter cream. **Habitat and Behaviour** Inhabits primary forests in the plains and mid-hills. Nocturnal, associated with rocks and low vegetation. Diet includes arthropods. Reproductive habits unstudied. **Distribution** Northern Myanmar. Also north-eastern India, Bangladesh and Bhutan. **Status** Not Evaluated.

SUMATRAN BENT-TOED GECKO
Cyrtodactylus lateralis NOT ILLUSTRATED
Measurement SVL 81mm **Identification** Body robust, depressed; dorsal surface with granular scales intermixed with small, pointed keeled tubercles; head large; auricular opening oblique, one-third orbit diameter; supralabials 11; infralabials 10; precloacal pores 13 in angular series in males; femoral pores absent; midventrals 62–64; enlarged lamellae on Toe IV 9. **Coloration** Dorsum greyish-brown with a series of large dark brown spots, edged with cream; dark temporal streak; supralabials and infralabials with dark bars; tail dark-banded; venter greyish-cream. **Habitat and Behaviour** Nothing known of its natural history, except for the fact that its eggs are spindle-like, and that clutches of 2 are produced. **Distribution** Endemic to Sumatra (Delitua or Medan, Sumatera Utara Province). **Status** Not Evaluated.

GUNUNG RAYA BENT-TOED GECKO
Cyrtodactylus macrotuberculatus NOT ILLUSTRATED
Measurement SVL 120mm **Identification** Body robust; large keeled tubercles on forehead, body, limbs and tail; ventrals 19–22; transversely enlarged median subcaudal scales; abrupt transition between posterior and ventral femoral scales; enlarged femoral and preanal scales with continuous series of 35–37 pore-bearing scales; precloacal groove present, precloacal depression absent; lamellae under Toe IV 21–23. **Coloration** Dorsum tan; wide dark brown nuchal band edged with thin white lines across backs of eyes; 4 such dorsal bands on body; dark brown band posterior to hind limbs; tail with dark bands; venter anteriorly smudged with brown; abdomen unpatterned beige except for darker sides. **Habitat and Behaviour** Inhabits lowland forests and the mid-hills at elevations of 61–700m asl. Nocturnal, associated with rocky karst and granite landscape. Diet and reproductive habits unstudied. **Distribution** Endemic to a single island off the west coast of northern Peninsular Malaysia (Pulau Langkawi). **Status** Not Evaluated.

MALAYAN BENT-TOED GECKO
Cyrtodactylus malayanus PLATE 23
Measurement SVL 117mm **Identification** Body robust; dorsal surface with keeled tubercles in 18–20 rows; head large; precloacal groove absent; precloacal pores indistinct in males; subcaudals large. **Coloration** Dorsum chestnut-brown; forehead with cream network of reticulations; narrow light cross-bars on dorsum of body; series of isolated dark spots along vertebral region; venter grey. **Habitat and Behaviour** Inhabits lowland dipterocarp forests. Nocturnal and arboreal, associated with large trees and may also occur on the walls of limestone cliffs, sometimes descending to 1.5m above ground. Diet and reproductive habits unstudied. **Distribution** Endemic to Borneo. **Status** Not Evaluated.

CLOUDED BENT-TOED GECKO
Cyrtodactylus marmoratus PLATE 23
Measurement SVL 74.4mm **Identification** Body robust, elongated; dorsal surface with granular scales intermixed with rounded, weakly keeled tubercles; head large; auricular opening large, one-third orbit diameter; supralabials 12; infralabials 10; precloacal pores 16; precloacal groove present; femoral pores 3–10; midventrals 40–45; lamellae under Toe IV 15. **Coloration** Dorsum light brown with dark brown spots that sometimes form cross-bars; forehead with irregular dark spots; a broad dark postocular stripe extends to sides of neck; tail dark-banded; venter yellowish-cream with dark brown blotches. **Habitat and Behaviour** Inhabits forested lowlands. Nocturnal and arboreal. Diet unstudied. Clutches comprise 2 eggs, measuring 13–14 x 11–12mm. Incubation period 93 days. **Distribution** Java and Bali. **Status** Not Evaluated.

MATSUI'S BENT-TOED GECKO

Cyrtodactylus matsuii PLATE 23

Measurement SVL 105mm **Identification** Body robust; dorsal surface with large tuberculate scales arranged in 18 irregular rows; supralabials 10–12; infralabials 10–11; midventrals 48–51; males with 7–8 precloacal pores forming angular series; precloacal groove absent; lamellae under Toe IV 22; subdigital lamellae widened. **Coloration** Dorsum yellowish-brown or pale brown with irregular dark crossbars; forehead with small dark spots; dark band on interorbital region joining behind eyes; upper surfaces of limbs and tail dark-banded; venter pale grey or brown. **Habitat and Behaviour** Inhabits forests at elevations of 900–1,600m asl. Diet consists of insects and other arthropods. Reproductive habits unstudied. **Distribution** Endemic to Borneo (Gunung Kinabalu and Crocker Range, Sabah State). **Status** Not Evaluated.

BLACK-EYED BENT-TOED GECKO

Cyrtodactylus nigriocularis NOT ILLUSTRATED

Measurement SVL 107.5mm **Identification** Body slender, elongated; head depressed; supralabials 13–14; infralabials 13–15; limbs moderately long, digits long; tail longer than body; large undivided subcaudals; femoral scales absent; 2 indistinct precloacal pores in males; midventrals 42–48; lamellae under Toe IV 17–21, narrow. **Coloration** Dorsum dark brown changeable to greyish-brown; 4 grey bands across body; forehead grey-brown; venter yellow to white, brighter in precloacal region; iris dark brown. **Habitat and Behaviour** Inhabits caves within forests. Nocturnal and saxicolous, inhabiting deeper parts of caves and climbing cave walls. Diet and reproductive habits unstudied. **Distribution** Endemic to southern Vietnam (Mount Ba Den, Tay Ninh Province). **Status** Not Evaluated.

OLDHAM'S BENT-TOED GECKO

Cyrtodactylus oldhami PLATE 23

Measurement SVL 65mm **Identification** Body robust; dorsal surface granular with scattered larger tuberculate scales; supralabials 10–13; infralabials 9–11; auricular opening large; precloacal pores 4 pairs in angular series, and enlarged femoral scales in males; subcaudals enlarged; midventrals 30–38. **Coloration** Dorsum brown with elongated dark-edged cream spots, which are arranged in 4 longitudinal series; dark postocular stripe, edged with white, meets at back of forehead; forehead unpatterned brown; tail dark-banded; venter unpatterned cream. **Habitat and Behaviour** Inhabits moist evergreen forests on mountain-tops at an elevation of 900m asl. Nocturnal and saxicolous, associated with lateritic soil. Diet and reproductive habits unstudied. **Distribution** Southern Myanmar (Mintao and Dawei, Tenasserim, Taninthayi Division) and south-western Thailand (Petchaburi Province and Phuket Island). **Status** Not Evaluated.

GUNUNG PANTI BENT-TOED GECKO

Cyrtodactylus pantiensis NOT ILLUSTRATED

Measurement SVL 77.2mm **Identification** Body slender; conical keeled tubercles on forehead, limbs and beyond tail-base; ventral scales 40–45; no transversely enlarged median subcaudal scales; proximal subdigital lamellae transversely expanded; abrupt transition between postfemoral and ventral femoral scales; no enlarged femoral and preanal scales; no femoral pores; 8–9 precloacal pores; no precloacal groove or depression; lamellae under Toe IV 22–23. **Coloration** Dorsum yellow; forehead mottled with dark brown; paired, semilunar-shaped dark blotches on upper nape edged with yellow; no wide dark ventrolateral stripes on flanks confluent with wide dark postorbital stripe; no white reticulum on head; dark blotches on body; venter spotted with grey. **Habitat and Behaviour** Inhabits vegetation at edges of streams associated with swamp forests, at an elevation of 20m asl. Nocturnal and a riparian species, associated with the roots of streamside vegetation. Diet and reproductive habits unstudied. **Distribution** Endemic to southern Peninsular Malaysia (Gunung Panti Forest Reserve, Johor State). **Status** Not Evaluated.

BUTTERFLY BENT-TOED GECKO

Cyrtodactylus papilionoides PLATE 23

Measurement SVL 93mm **Identification** Body slender; dorsal surface with 12–14 rows of tubercles; supralabials 10–11; infralabials 8–10; precloacal pores 4–6 in continuous series of 29–33 preanofemoral pores; midventrals 30–34; lamellae under Toe IV 10. **Coloration** Dorsum beige-brown to mid-brown with 4 butterfly-shaped dark patches edged with dull yellow and cream; dark postocular stripe reaches tympanum; dorsal tubercles pale on greenish-olive background; forehead brownish-grey with pale-edged dark spots, labials yellow with dark blotches; venter greyish-cream; gular region with brown network; iris golden. **Habitat and Behaviour** Inhabits forested mid-hills supporting agriculture at elevations of ca 400m asl. Diet in captivity includes arthropods such as cockroaches and woodlice; diet in the wild unstudied. Clutches comprise 2 (rarely 1) eggs, measuring 14.9 x 14mm. Incubation period 130 days. Hatchlings 31–38mm. **Distribution** Endemic to central Thailand (Thanon Khao Yai, Nakhon Ratchasima Province). **Status** Not Evaluated.

PARADOXICAL BENT-TOED GECKO

Cyrtodactylus paradoxus NOT ILLUSTRATED

Measurement SVL 84mm **Identification** Body robust; precloacal groove and femoral pores absent in males and females; precloacal pores present (up to 4) or absent in males; enlarged femoral scales distinct; midventrals 26–36; median series of transversely widened subcaudal scales; lamellae under Toe IV 17–23. **Coloration** Dorsum yellowish-brown with 7 irregular dark brown bands, with light edges that are interrupted in vertebral region; venter tan; tail dark-banded. **Habitat and Behaviour** Known from an offshore island and adjacent mainland within secondary dipterocarp forests at sea level. Nocturnal and saxicolous, associated with rocky biotope. Diet unstudied. Clutches comprise 2 eggs. Incubation period ca 55 days. Hatchlings 45–47mm. **Distribution** Endemic to southern Vietnam (Hon Thom Isle, Phu Quoc Island and Kien Giang Province). **Status** Not Evaluated.

PEGU BENT-TOED GECKO
Cyrtodactylus peguensis PLATE 23
Measurement SVL 85mm (*C. p. peguensis*); SVL 58.5mm (*C. p. zebraicus*) **Identification** Body robust; dorsal surface with keeled tubercles; head large; dorsal tubercles keeled; supralabials 9; infralabials 7–8; auricular opening large; precloacal groove absent; femoral pores present or absent; midventrals 43–45. **Coloration** Dorsum pale grey or mid-grey with either spots or cross-bars; venter unpatterned cream. **Subspecies** Two subspecies are recognized. *C. p. peguensis*, dorsum with irregular blackish-brown marks edged with cream, including a crescentic postocular stripe that forms a nuchal loop; 2 series of large irregular spots on dorsum and smaller spots on flanks; femoral pores absent. *C. p. zebraicus*, dorsum grey with 8 transverse stripes that are wider than the interspaces; femoral pores 8. **Habitat and Behaviour** Inhabits lowland forests. Nocturnal and arboreal. Diet and reproductive habits unstudied. **Distribution** Southern Myanmar (*C. p. peguensis*), Thailand (Ronpibon, Nakhon Si Thammarat Province) and northern Peninsular Malaysia (*C. p. zebraicus*). **Status** Not Evaluated.

PHONG NHA KE BANG BENT-TOED GECKO
Cyrtodactylus phongnhakebangensis PLATE 23
Measurement SVL 96.3mm **Identification** Body slender; dorsal surface with tubercles in 11–20 rows; preanofemoral pores 32–42; postcloacal tubercles 4–5; subcaudals enlarged; midventrals 32–42; lamellae under Toe IV 18–26. **Coloration** Dorsum grey with 4 wide, serrated brownish-grey bands with light yellow seams; U- or V-shaped nuchal band, bordered by bright yellow line that extends across head, over orbit; supralabials and infralabials grey with darker and lighter areas; tail dark-banded; venter unpatterned grey. **Habitat and Behaviour** Inhabits lowland forests dominated by limestone at elevations of 50–100m asl. Nocturnal and saxicolous, associated with karst landscape at heights of 1.5–4m above substratum. Diet consists of arthropods such as isopods, spiders, crickets and bugs. Clutches comprise 2 eggs, measuring 7.3mm. **Distribution** Endemic to central Vietnam (Phong Nha-Ke Bang National Park, Quang Binh Province). **Status** Not Evaluated.

FALSE LINED BENT-TOED GECKO
Cyrtodactylus pseudoquadrivirgatus PLATE 23
Measurement SVL 84mm **Identification** Body slender; interorbitals 37–55; dorsal surface with tubercles in 16–24 rows; interorbitals 37–55; longitudinal rows of ventral scales at midbody 41–57; precloacal pores in angular series 5–9; precloacal groove absent; enlarged femoral scales; transversely enlarged subcaudals; lamellae under Toe IV 16–25. **Coloration** Dorsum light brown with dark brown mottling, stripes or bands; neck band medially interrupted; limbs striped or mottled; tail dark-banded. **Habitat and Behaviour** Inhabits secondary evergreen forests at elevations of 400–1,100m asl. Nocturnal and arboreal, associated with granitic rocks. Diet consists of spiders and possibly also other arthropods. Reproductive habits unstudied, although it is known that this species produces a single large egg. **Distribution** Endemic to central Vietnam (Quang Tri, Thua Thien-Hue, Da Nang and Kon Tum Provinces). **Status** Not Evaluated.

GROOVED BENT-TOED GECKO
Cyrtodactylus pubisulcus PLATE 23
Measurement SVL 77mm **Identification** Body slender; dorsal surface with tubercles in irregular longitudinal rows; ventrolateral dermal fold present; supralabials 10–11; midventrals 43–55; precloacal groove present; precloacal pores 3–5 pairs; subcaudals not transversely arranged. **Coloration** Dorsum grey or brown with dark cross-bars or blotches, sometimes arranged in longitudinal series; venter cream. **Habitat and Behaviour** Inhabits lowland rainforests and peat swamps. Nocturnal and arboreal, associated with tree trunks and leaves of shrubs. Diet consists of insects such as cockroaches. Clutches comprise 2 eggs, measuring 10–13mm. **Distribution** Endemic to Borneo (Sarawak State and Brunei Darussalam). **Status** Not Evaluated.

BEAUTIFUL BENT-TOED GECKO
Cyrtodactylus pulchellus PLATE 23
Measurement SVL 115mm **Identification** Body robust; dorsal surface with small scales that are intermixed with larger rounded scales; head large and with small rounded tubercles that give it a granular appearance; no large tubercles in gular and throat regions; ventrolateral folds composed of small smooth scales; small tubercles and granular scales on limbs; smooth flat imbricate scales on abdomen; males with longitudinal groove in pubic region, within which are 4 precloacal pores; 15–20 femoral pores; midventrals 33–55; lamellae under Toe IV 19–20. **Coloration** Dorsum yellowish-brown with 4 dark brown cross-bars that are edged with yellow or cream, the first connecting the eyes; tail with dark encircling bands; venter cream or bluish-grey. **Habitat and Behaviour** Inhabits lowland forests. Nocturnal and saxicolous, associated with rocky habitats including the rocky edges of streams and walls. Diet consists of crickets and other large arthropods. Clutches comprise 2 eggs. Hatchlings 60–70mm. **Distribution** Myanmar, Thailand, Peninsular Malaysia and Singapore. **Status** Not Evaluated.

FOUR-STRIPED BENT-TOED GECKO
Cyrtodactylus quadrivirgatus PLATE 23
Measurement SVL 71mm **Identification** Body slender; dorsal surface with tubercles in regular longitudinal rows; supralabials 10; infralabials 10; midventrals 40; males with 4 precloacal pores; precloacal groove absent; median series of widened subcaudals absent. **Coloration** Dorsum grey, fawn or dark brown with 4 longitudinal black lines separated by lighter areas; no continuous black band joining eyes; venter cream; iris grey, ochre or dark brown. **Habitat and Behaviour** Inhabits primary and secondary forests in the lowlands. Nocturnal and arboreal, associated with tree trunks and saplings. Diet consists of arthropods. Clutches comprise 2 eggs. Hatchlings 25–26mm. **Distribution** Southern Thailand, Peninsular Malaysia, Singapore, northern Sumatra, the Mentawai Archipelago and north-western Borneo. Suspected to represent a species-complex. **Status** Not Evaluated.

RUSSELL'S BENT-TOED GECKO
Cyrtodactylus russelli PLATE 23
Measurement SVL 116mm **Identification** Body slender, elongated; dorsal surface with conical or keeled tubercles in 22 longitudinal rows; ventrolateral folds well developed; limbs stout; digits long; paired enlarged postmental scales in broad contact behind mental; midventrals 35–41; precloacal groove absent; 15 precloacal pores in single series; 16–19 pairs of femoral pores, separated from precloacal pores by diastema; precloacal and femoral pores absent in females; subcaudal scales with broad transverse plates; lamellae under Toe IV 13; 9 broad lamellae. **Coloration** Dorsum mid- to dark brown with indistinct, elongated dark paravertebral blotches and longitudinal bands on dorsolateral margins of trunk; flanks with lighter longitudinally oriented blotches; 3 grey spots on axilla; head darker than body; venter greyish-cream with dark pigment on limb margins and on throat and margins of jaws; tail dark-banded; iris bronze. **Habitat and Behaviour** Inhabits forested low hills at an elevation of 227m asl. Diet and reproductive habits unstudied. **Distribution** Endemic to Myanmar (Sagaing and Kachin States). **Status** Not Evaluated.

PENINSULAR MALAYSIAN BENT-TOED GECKO
Cyrtodactylus semenanjungensis PLATE 24
Measurement SVL 69mm **Identification** Body robust; dorsal surface with conical keeled tubercles; tubercles on forelimbs and beyond base of tail; midventrals 48–53; no transversely enlarged median subcaudal scales; proximal subdigital lamellae transversely expanded; abrupt transition between posterior and ventral femoral scales; enlarged femoral scales or pores absent; precloacal groove present; enlarged preanal scales. **Coloration** Dorsum grey with dark mottling; 7 irregular wide dark bands; ends of anterior-most body band join with dark lateral stripes on neck to connect with posterior portion of dark nuchal loop; dark nape spot enclosed between lateral stripes; large lateral spots between some bands; flanks with weak dark mottling and light tubercles; venter beige; gular scales with single black spot, pectoral and abdominal scales containing 2–3 spots, except on unpatterned midventral scales; flanks beige with faint black stippling on each scale. **Habitat and Behaviour** Inhabits lowland rainforests near freshwater swamps. Nocturnal and arboreal, associated with leaves of trees 1–1.5m above ground. Diet and reproductive habits unstudied. **Distribution** Endemic to Peninsular Malaysia (Gunung Panti and nr Jemaluang, Johor State). **Status** Not Evaluated.

SERIBUAT BENT-TOED GECKO
Cyrtodactylus seribuatensis PLATE 24
Measurement SVL 75mm **Identification** Body slender, elongated; dorsal surface with tubercles in 27–35 rows; precloacal groove absent; preanofemoral pores 40–44; midventrals 28–39; lamellae under Toe IV 19–22. **Coloration** Dorsum light grey with narrow dark bands; head and labials with fine dark spots; postocular tubercles dark; faint postorbital stripe; body with indistinct irregular bands; venter unpatterned white. **Habitat and Behaviour** Inhabits low-lying offshore islands off the west coast of Peninsular Malaysia. Nocturnal, active in intertidal regions, hiding by day under beach debris. Diet and reproductive habits unstudied. **Distribution** Endemic to islands off the east coast of Peninsular Malaysia (Pulau Seribuat, Pulau Sembilang, Pulau Sibu, Pulau Sibu Tengah, Pulau Lima Besar, Pulau Mentingi and Pulau Nangka Kecil). **Status** Not Evaluated.

SLOWINSKI'S BENT-TOED GECKO
Cyrtodactylus slowinskii PLATE 24
Measurement SVL 108mm **Identification** Body slender; dorsal surface with flattened, weakly keeled tubercles in 10–22 rows; precloacal groove absent; femoral pores 11; subcaudals with enlarged scales; midventrals 20–32; digits elongated; lamellae under Toe IV comprise 6 widened ones and 13 narrow ones. **Coloration** Dorsum mid-brown with 6–8 paired chocolate-brown patches, with dark brown border and edged with narrow yellowish-cream edges; forehead with pale lines forming reticulated pattern; dark postocular streak up to neck; tail dark-barred; venter cream with scattered pale patches. **Habitat and Behaviour** Inhabits evergreen forests. Nocturnal and arboreal. Diet and reproductive habits unstudied. **Distribution** Endemic to north-central Myanmar (Sagaing Division and southern Chin State). **Status** Not Evaluated.

STRESEMANN'S BENT-TOED GECKO
Cyrtodactylus stresemanni NOT ILLUSTRATED
Measurement SVL 95.5mm **Identification** Body slender; dorsal surface with conical tubercles in 13 rows; head distinct from neck; supralabials 13; infralabials 10; midventrals 63; deep precloacal groove; femoral scales and femoral pores not enlarged; tubercles on dorsal and ventral sides of tail. **Coloration** Dorsum olive-brown; labials with dark

and light areas; broad brown-olive preocular stripe; broad brown-olive postocular stripe; vertebral stripe grey-brown; neck with triangular, dark-bordered blackish-olive spot; 3 pairs of elongated brown-olive marks on dorsum; tail dark-banded; throat yellow-olive; venter olive. **Habitat and Behaviour** Inhabits the mid-hills at elevations of 800–900m asl. Thought to be nocturnal and arboreal. Diet and reproductive habits unstudied. **Distribution** Endemic to Peninsular Malaysia (Batang Padang Tal, Perak State). **Status** Not Evaluated.

SUMONTHA'S BENT-TOED GECKO
Cyrtodactylus sumonthai PLATE 24
Measurement SVL 70.66mm **Identification** Body slender, elongated; dorsal surface with tubercles in 12 rows; digits and tail long and slender; enlarged patch of preanal scales bearing 2 small precloacal pores; midventrals 33–36; lamellae under Toe IV 12. **Coloration** Dorsum yellowish-cream; 4 pale brown bands edged with dark brown, distinct dorsally and fading on flanks; forehead pale brown with darker diffuse markings; dark brown postocular stripe meets at back of head to form nuchal collar; tail dark-banded; venter cream tinged with light brown; iris golden-brown. **Habitat and Behaviour** Inhabits limestone caves in the Khao Wong range. Nocturnal and saxicolous, climbing high cave walls. Diet and reproductive habits unstudied. **Distribution** Endemic to south-eastern Thailand (Rayong Province). **Status** Not Evaluated.

SWORDER'S BENT-TOED GECKO
Cyrtodactylus sworderi PLATE 24
Measurement SVL 80mm **Identification** Body robust; large keeled tubercles on forehead, body and tail; midventral longitudinal scale rows 42–49; median subcaudal scales not transversely enlarged; smooth transition between posterior and ventral femoral scales; no enlarged femoral scales or femoral pores; precloacal groove absent; triangular series of enlarged preanal scales with 8–9 precloacal pores in males; lamellae under Toe IV 18. **Coloration** Dorsum dark brown; yellow mottling on forehead; supralabials white or mottled with dark brown; yellow spots in postorbital region extend across nape and meet at midline; spots continue as vertebral stripe or as separate spots to base of tail; indistinct paravertebral yellow spots extend from eye to tail-base; tubercles and some granular scales on flanks yellow; tail black with 10–11 bands; venter unpatterned beige. **Habitat and Behaviour** Inhabits lowland rainforests. Nocturnal and arboreal, associated with tree trunks and branches. Diet unstudied. Clutches comprise 2 eggs. **Distribution** Endemic to Peninsular Malaysia (Johor State). **Status** Not Evaluated.

TA KOU BENT-TOED GECKO
Cyrtodactylus takouensis NOT ILLUSTRATED
Measurement SVL 81.1mm **Identification** Body slender; dorsal surface with smooth tubercles in 9–10 rows; limbs slender; 2 pairs of enlarged postmental scales; midventrals 39–40; no precloacal groove; 3–5 enlarged femoral scales; median row of subcaudals transversely enlarged. **Coloration** Dorsum with 5 yellow bands alternating with dark brown bands; dark brown canthal stripe continuous with nuchal loop, bordered by narrow yellow edge, fading to mottled brownish crown; tail dark brown with narrow pale bands; venter cream. **Habitat and Behaviour** Inhabits the Hang To Cave within a deciduous forest at an elevation of 450m asl. Diet and reproductive habits unstudied. **Distribution** Endemic to Vietnam (Ta Kou Nature Reserve, Binh Thuan Province). **Status** Not Evaluated.

THIRAKHUPT'S BENT-TOED GECKO
Cyrtodactylus thirakhupti PLATE 24
Measurement SVL 80mm **Identification** Body slender; dorsal surface with keeled tubercles in 14 rows; limbs and digits long, slender; tail long; pair of enlarged postmental scales in broad contact; ca 37 ventral scales across belly; weakly developed ventrolateral folds; no precloacal groove; no precloacal or femoral pores in either sex; median subcaudal scales enlarged; lamellae under Toe IV 12, narrow. **Coloration** Dorsum brown with transverse grey bands edged with dark brown; first band across nape forms collar joining posterior upper border of orbits; second band on neck, 4 on body and 1 across base of tail; smaller ocelli or partial bands anterior and posterior to middle 2 body bands; forehead brown with irregular grey markings; grey postocular stripe, edged above with dark brown, passes above tympanum, to band on neck. Supralabials and infralabials brown with diffuse grey punctuations; tail with dark and light brown annuli; venter cream. **Habitat and Behaviour** Restricted to a single cave within a forested coastal limestone hill at an elevation of 30m asl. Nocturnal and saxicolous. Diet and reproductive habits unstudied. **Distribution** Endemic to southern Thailand (Khao Sonk, Surat Thani Province). **Status** Not Evaluated.

STRIPED BENT-TOED GECKO
Cyrtodactylus tigroides PLATE 24
Measurement SVL 83.2mm **Identification** Body slender; dorsal surface with keeled tubercles in 13 rows; limbs and digits long, slender; tail very long; midventrals 34; precloacal groove absent; 8–9 precloacal pores separated by diastema of 7–9 poreless scales; 5–7 pairs of femoral pores in both sexes; 7–8 broad basal lamellae; median subcaudal scales enlarged to form broad transverse plates; lamellae under Toe IV 12–15. **Coloration** Dorsum mid-brown; yellowish-cream body bands with slightly paler markings, each edged with dark brown border; pale band across nape, 4 across trunk between limb insertions; pattern less distinct on flanks; alternating light and dark pattern of dorsum continues to tail; distinct brown collar with darker margins; forehead pale brown with scattered darker markings; venter cream. **Habitat and Behaviour** Inhabits a limestone hill covered by bamboo forest. Nocturnal, associated with exposed

limestone 1–1.5m above substratum. Diet and reproductive habits unstudied. **Distribution** Endemic to western Thailand (Sai-Yok, Kanchanaburi Province). **Status** Not Evaluated.

PULAU TIOMAN BENT-TOED GECKO
Cyrtodactylus tiomanensis PLATE 24
Measurement SVL 86mm **Identification** Body robust; dorsal surface with keeled tubercles; head large; snout elongated; supralabials 8–11; infralabials 9–11; preanofemoral pores 19; midventrals 36–40; lamellae under Toe IV 20–22. **Coloration** Dorsum chocolate-brown; dark brown temporal patch; olive-yellow nuchal loop joins posterior edge of orbit, extending to nostrils; body with 4 olive-yellow bands edged with narrow greyish-brown bands; tail with yellowish-cream bands; venter unpatterned cream, scales with minute black spots. **Habitat and Behaviour** Inhabits lowland forests on the offshore island of Pulau Tioman at elevations of 50–150m asl. Nocturnal, and both saxicolous and arboreal, associated with boulders and trees. Diet consists of large insects. Reproductive habits unstudied. **Distribution** Endemic to an island off the east coast of Peninsular Malaysia (Pulau Tioman, Pahang State). **Status** Not Evaluated.

VARIEGATED BENT-TOED GECKO
Cyrtodactylus variegatus PLATE 24
Measurement SVL 71mm **Identification** Body slender; dorsal surface with granular scales intermixed with larger tuberculate scales; supralabials 10–11; infralabials 10; lateral fold with enlarged scales; preanofemoral pores 32; subcaudals enlarged; midventrals 22. **Coloration** Dorsum grey, spotted and marbled with black, edged with white; forehead dark-spotted; dark postocular stripe, edged with white, meets at back of head; tail dark-banded; venter cream, darkening posteriorly, the posterior third darkest. **Habitat and Behaviour** Inhabits forests in the low hills. Diet and reproductive habits unstudied. **Distribution** Endemic to Myanmar (Amherst District, probably Dawna Hills, Kayah State). **Status** Not Evaluated.

WAKES'S BENT-TOED GECKO
Cyrtodactylus wakeorum PLATE 24
Measurement SVL 64mm **Identification** Body slender; dorsal surface smooth textured with 24 longitudinal rows of oval to rounded keeled tubercles; limbs and digits short; paired enlarged postmental scales in broad contact behind mental; midventrals 31; precloacal groove absent; 12 precloacal pores in single series in females; femoral pores absent; subcaudal scales not expanded; lamellae under Toe IV 10, including 6 broad ones. **Coloration** Dorsum mid-brown; 5 chocolate-brown bands on body, posteriorly edged with narrow cream border; occipital band turns at right angle above and behind auricular opening and extends anteriorly to orbit and under eye to loreal region, portion on temporal region almost completely surrounded by narrow white border; nuchal markings paired; venter cream tinged with light brown speckling of scales; tail dark-banded; iris golden-brown. **Habitat and Behaviour** Inhabits forests at an elevation of 180m asl. Diet unstudied. Clutches comprise 2 eggs. **Distribution** Endemic to Myanmar (Rakhine State). **Status** Not Evaluated.

YOSHI'S BENT-TOED GECKO
Cyrtodactylus yoshii PLATE 24
Measurement SVL 96mm **Identification** Body robust; dorsal surface with small scales and tubercles in 17 rows; midventrals 50–58; males without femoral pores; precloacal pores 8–12, forming a narrow angular series; precloacal groove absent; subdigital lamellae not widened. **Coloration** Dorsum grey with 5 V-shaped dark cross-bars between nape and inguinal region; dark brown V extends from posterior corner of orbit to nape; lips white-spotted; dorsal surfaces of limbs with indistinct dark bands. **Habitat and Behaviour** Inhabits lowland forests. Found on tree trunks and walls of man-made structures within forests. Diet presumably consists of large insects. Reproductive habits unstudied. **Distribution** Endemic to northern Borneo (Sabah State). **Status** Not Evaluated.

ZIEGLER'S BENT-TOED GECKO
Cyrtodactylus ziegleri NOT ILLUSTRATED
Measurement SVL 93mm **Identification** Body slender; dorsal surface with enlarged, rounded and keeled tubercles in 20–24 rows; limbs moderately long; tail not thick; postmentals 2, first pair in broad contact, second pair ca half of first pair; rostral contacts nostril; male with 5–8 precloacal pores; 8–10 enlarged pairs of femoral scales; subcaudals lack transversely enlarged plates; midventrals 33–39. **Coloration** Dorsum light yellow to light brown with 4–6 irregular transverse dark brown bands, lacking contrasting pale edges; narrow U-shaped occipital dark band between auricular opening to orbit; forehead dark with irregular spots or other indistinct pattern; tail dark-banded; venter unpatterned grey. **Habitat and Behaviour** Inhabits submontane forests at an elevation of 900m asl. Nocturnal and arboreal. Diet and reproductive habits unstudied. **Distribution** Endemic to Vietnam (Chu Yang Sin National Park, Dac Lac Province). **Status** Not Evaluated.

ORANGE-TAILED GROUND GECKO
Dixonius hangseesom PLATE 25
Measurement SVL 42.12mm **Identification** Body slender, elongated; dorsals irregular, imbricating and scattered, in 12–14 rows; supralabials 8; infralabials 6; auricular opening oval, large; precloacal pores present; femoral scales and femoral pores absent in males; midventrals 22–26; lamellae under Toe IV 13; terminal scansors x 2 mid-digital toe width. **Coloration** Dorsum beige, grey to yellowish-tan with dark cross-bands and reticulations; dark postocular stripe extends to auricular opening and sides of neck; tail bright orange with indistinct dark bands, except in juveniles and old individuals, where it is darker and

shows less contrast to dorsum. **Habitat and Behaviour** Inhabits lowland forests of bamboo within limestone hills. Nocturnal and crepuscular, and saxicolous. Diet in captivity consists of arthropods. Diet in the wild and reproductive habits unstudied. **Distribution** Endemic to western Thailand (nr Ban Tha Sao and Sai Yoke National Park, Kanchanaburi Province). **Status** Not Evaluated.

DARK-SIDED GROUND GECKO
Dixonius melanostictus PLATE 25
Measurement SVL 50mm **Identification** Body moderate, cylindrical; enlarged trihedral or keeled tubercles, median rows of which are smaller than other rows, in 10–11 rows; supralabials 9; Supralabial VII subocular; infralabials 7; males with 9 precloacal pores; midventrals 22; lamellae under Toe IV 10. **Coloration** Dorsum yellowish-brown or lavender-grey; dorsal spots absent, or present in 4 longitudinal series; forehead lighter than dorsum; cream-coloured stripe from snout, across eye to axilla; black postocular stripe extends to axilla, where it widens and continues across flanks, narrowing on tail; tail unbanded; venter cream with fine black speckles. **Habitat and Behaviour** Inhabits lowland forests. Nocturnal, and terrestrial, concealing itself by day under logs and other fallen objects. Diet and reproductive habits unstudied. **Distribution** Thailand (Sara Buri Province) and Vietnam (Dong Nai Province). **Status** Not Evaluated.

SPOTTED GROUND GECKO
Dixonius siamensis PLATE 25
Measurement SVL 57mm **Identification** Body moderate, cylindrical; head wider than long; supranasals separated by 2 granular scales; supralabials 8; Supralabial VI suborbital; infralabials 6; flanks with tuberculate scales; tail cylindrical with indistinct segments; median subcaudal series transversely enlarged; lamellae under Toe IV 12–13; terminal scansors x 1.5 mid-digital toe width. **Coloration** Dorsum lavender-brown to grey, an indistinct series of 17 pink, buff or yellow dots extending from occipital region to tail-tip; supralabials and infralabials cream with dark spots; venter dull yellow; subcaudals grey with brown flecks. **Habitat and Behaviour** Inhabits primary and secondary forests at elevations of 0–700m asl. Nocturnal and terrestrial, hiding under stones and vegetation by day. Diet and reproductive habits unstudied. **Distribution** Myanmar, Thailand (Nakhon Ratchasima Province and Dong Phaya Fai Range, Sara Buri Province), Cambodia (Cardamom Mountains) and Vietnam (Ninh Thuan, Gia Lai, Lam Dong and Binh Phuoc Provinces). **Status** Not Evaluated.

VIETNAMESE GROUND GECKO
Dixonius vietnamensis PLATE 25
Measurement SVL 46.3mm **Identification** Body moderate, cylindrical; head wider than long; 2 supranasals in narrow contact; supralabials 7; Supralabials V–VI subocular; infralabials 6–7; median subcaudal series transversely enlarged; midbody scale rows 20; precloacal depression present; precloacal pores 6; lamellae under Toe IV 13. **Coloration** Dorsum olive-grey with brownish-olive blotches, a dark stripe from rostrum nearly meeting 2 broken bands at back of head; venter cream. **Habitat and Behaviour** Inhabits the lowlands to the mid-hills, from rocky coasts within coastal forest to deciduous and evergreen forests, at elevations of 2–700m asl. Nocturnal and terrestrial, associated with rocky biotope. Diet and reproductive habits unstudied. **Distribution** Eastern Cambodia (Pichrada District) and Vietnam (Nha Trang and Tay Ninh Provinces). **Status** Not Evaluated.

SLENDER-TAILED FOUR-CLAWED GECKO
Gehyra angusticaudata NOT ILLUSTRATED
Measurement SVL 57mm **Identification** Body slender; tail slender, accentuated; supralabials 10; infralabials 8–9; inner finger small, clawless; preanofemoral pores 15–18 in males; median subcaudals enlarged; midventrals ca 35. **Coloration** Dorsum dark brown; forehead paler; tail dark lavender with faint pale bands; venter yellowish-white, each scale with fine black dots; subcaudals dark, turning blackish apically. **Habitat and Behaviour** Inhabits lowland forests as well as bamboo-nipa structures. Diet and reproductive habits unstudied. **Distribution** Endemic to south-eastern Thailand (Siracha, Chon Buri Province). **Status** Not Evaluated.

BUTLER'S FOUR-CLAWED GECKO
Gehyra butleri NOT ILLUSTRATED
Measurement SVL 32mm **Identification** Body slender; tail depressed, with finely serrated edge; supralabials 8; infralabials 6–7; lamellae under Toe IV 7. **Coloration** Dorsum reddish-brown with cream dots and 3 longitudinal series of dark brown spots; dark brown streak on sides of head; venter pale. **Habitat and Behaviour** Inhabits the lowlands. Nocturnal and possibly arboreal. Diet and reproductive habits unstudied. **Distribution** Endemic to Peninsular Malaysia (Kuala Lumpur). **Status** Not Evaluated.

FEHLMANN'S FOUR-CLAWED GECKO
Gehyra fehlmanni PLATE 25
Measurement SVL 51mm **Identification** Body slender; tail narrow, tapering; supralabials 8; Supralabials VII–VIII midorbital; basal subcaudals slightly enlarged, and subsequently subcaudals greatly enlarged; 22 preanofemoral pores in an arched series; tail indistinctly segmented; midventrals ca 42. **Coloration** Dorsum pale brown with scattered pale flecks and dark bands; forehead similar; venter yellowish-cream, each scale with small black pigments; subcaudals with dense brown pigmentation. **Habitat and Behaviour** Inhabits the lowlands. Nocturnal, and possibly both arboreal and terrestrial. Diet and reproductive habits unstudied. **Distribution** Thailand (Kanchanaburi Province) and Vietnam (Ho Chi Minh City). **Status** Not Evaluated.

KANCHANABURI FOUR-CLAWED GECKO
Gehyra lacerata PLATE 25
Measurement SVL 55mm **Identification** Body robust; dorsal scales small, dorsum lacking tubercles; tail shorter than body, not distinctly segmented; median subcaudals not enlarged; supralabials 12; Supralabials X or X–XI midorbital; infralabials 10–11; single series of 17–20 precloacal pores; midventrals ca 48. **Coloration** Dorsum grey with scattered, large dark grey spots and double row of rounded pale spots; forehead with ca 20 smaller dark grey spots; venter unpatterned cream. **Habitat and Behaviour** Inhabits limestone regions. Nocturnal and saxicolous, and may conceal itself by day under rocks and logs. Diet and reproductive habits unstudied. **Distribution** Endemic to Thailand (Chonburi, Kanchanaburi, Khon Kaen, Nakhon Ratchasima, Phetchaburi and Sakaeo Provinces); records from Vietnam require verification. **Status** Not Evaluated.

COMMON FOUR-CLAWED GECKO
Gehyra mutilata PLATE 25
Measurement SVL 64mm **Identification** Body robust; head relatively large; skin delicate; tail flattened, widening at base, with sharp, somewhat denticulate edges; large flat scales on tail and venter; claw on inner digit absent; males with 25–41 preanofemoral pores. **Coloration** Dorsum pale, nearly translucent grey to pinkish-grey, typically with pale vertebral area; indistinct white band along face; venter pale pink. **Habitat and Behaviour** Associated with both human habitations and primary forests. Call a series of 6–8 monosyllablic 'tok', which increase in intensity. Nocturnal. Diet consists of insects. Clutches comprise 2–3 eggs, measuring 7.1–8.5 x 10.5–11.2mm, which are fused together. Hatchlings 17–23mm. **Distribution** Myanmar (Ayeyarwady Division), Thailand, Peninsular Malaysia, Singapore, Laos, Cambodia, Vietnam, Sumatra, the Natuna Archipelago, Borneo, Pulau Panjang, Java and Bali. Also India, Sri Lanka, eastern China, the Lesser Sundas, the Philippines and New Guinea; introduced into Mauritius, the Seychelles, Madagascar, Mexico, Cuba and Hawaii (USA). **Status** Not Evaluated.

BADEN'S GECKO
Gekko badenii PLATE 25
Measurement SVL 76.5mm **Identification** Body moderate; head broad, slightly depressed; dorsals granular, in 12–17 rows; interorbitals 30–37; internasals 1–3; supralabials 12–13; infralabials 13–14; males with 14–18 precloacal pores; femoral pores absent; tail slender; midventrals 29–35; lamellae under Toe IV 18–19. **Coloration** Dorsum brown; forehead with light spots; 4–8 narrow transverse light bands with dark edges, sometimes fragmented, especially on vertebral region; limbs with transverse bands of light spots; venter light. **Habitat and Behaviour** Inhabits forested mountain-tops at an elevation of 986m, as well as lowlands that are dominated by rice fields. Diet and reproductive habits unstudied. **Distribution** Endemic to southern Vietnam (Ba Den Mountains, Tay Ninh Province). **Status** Not Evaluated.

CHINESE GECKO
Gekko chinensis NOT ILLUSTRATED
Measurement SVL 70mm **Identification** Body robust; head broad; nostrils contact rostral; single internasal exceeds supranasal in size; dorsum with tubercles; tubercles absent on limbs; supralabials 12–15; infralabials 10–12; males with 18–24 precloacal pores; single cloacal spur; midbody scale rows 118–140; lamellae under Toe IV 11–14. **Coloration** Dorsum brown with 4 indented cloudy cross-bars with light flecks in intervening areas; venter pale. **Habitat and Behaviour** Inhabits human habitations as well as lightly forested areas in the lowlands. Nocturnal and scansorial. Diet consists of small insects. Clutches comprise 2 eggs, which are attached to walls. Hatchlings 50mm. **Distribution** Northern Vietnam. Also southern China. **Status** Not Evaluated.

TOKAY GECKO
Gekko gecko PLATE 25
Measurement SVL 185mm **Identification** Body robust; head large; body thick-set, with granular scales interspersed with subconical tubercles arranged in 12 longitudinal series; supralabials 12–14; infralabials 10–12; supralabials and infralabials with weak keels; internasals 2–3; males with angular series of 10–24 precloacal pores; femoral pores present; lamellae under Toe IV 20–23. **Coloration** Dorsum slaty-grey or bluish-grey with red or orange spots; dorsum of juveniles with cream spots, sometimes coalescing into 7–8 pale bands or spots; bands sometimes evident in adults; tail dark-banded; venter cream, unpatterned or variegated with grey or spotted with pink; iris yellow. **Subspecies** Two subspecies are recognized, of which one (*G. g. gecko*) occurs in the region. **Habitat and Behaviour** Associated with human-modified environments, and common in houses in suburbia. Loud territorial calls supposed to be the source of the name 'gecko', syllabilized as 'tok-ay', uttered 4–9 times in slow succession, and heard by day and night. Diet includes moths, grasshoppers, beetles, spiders, other geckos, small mice and snakes. Clutches comprise 1–2 eggs, measuring 19–21 x 15–17mm, which are deposited in tree-holes. Incubation period 64 days. Hatchlings 39.8–42.3mm. **Distribution** Myanmar, Thailand, Cambodia, Laos, Vietnam, Peninsular Malaysia, Singapore, Sumatra, the Natuna Archipelago, Borneo, Java and Bali. Also eastern and northeastern India, Nepal, Bangladesh (*G. g. azhari*), southern China, the Lesser Sundas and Sulawesi (Indonesia), and the Philippines (ranges of *G. g. gecko* and other populations); introduced into Florida and Hawaii (USA), Martinique (West Indies) and Madagascar. **Status** Not Evaluated.

GROSSMANN'S GECKO
Gekko grossmanni PLATE 25
Measurement SVL 90mm **Identification** Body moderate; dorsals small, granular; supranasals 2 (rarely

3); internasals 0–1; supralabials 11–13; infralabials 9–12; tail long and slender; subcaudals enlarged; males with 12–14 precloacal pores; femoral pores absent; midventrals 27–31; lamellae under Toe IV 18–20. **Coloration** Dorsum yellowish-grey to greyish-black with several rows of more or less transversally arranged cream spots and flecks, as well as dark blue and grey marbling; tail dark and pale banded. **Habitat and Behaviour** Habitat in the wild unstudied. Nocturnal and presumably arboreal. In captivity, diet consists of insects and other arthropods. Diet in the wild unstudied. Clutches comprise 2 eggs, measuring 13 x 10.5mm. Incubation period 84–85 days. Hatchlings 29–30mm. **Distribution** Endemic to southern Vietnam (Nyachang, Khánh Hòa Province). **Status** Not Evaluated.

JAPANESE GIANT GECKO
Gekko japonicus　　　　　　　　　　　PLATE 25
Measurement SVL 74mm **Identification** Body robust, depressed; dorsals imbricate; internasals 1–2; supralabials 11; infralabials 9; auricular opening oval, less than half orbit diameter; toes basally webbed; males with 6–9 precloacal pores; spur-like tubercles on either side of vent; lamellae under Toe IV 15. **Coloration** Dorsum grey with 6 indented dark crossbands that are medially interrupted by light flecks; venter cream with minute black dots. **Habitat and Behaviour** Inhabits human dwellings in the lowlands and mid-hills. Nocturnal and scansorial, found on walls and within roof tiles. Diet comprises insects. Reproductive habits unstudied. **Distribution** Recorded from northern Vietnam; report in need of verification, and may be based on the Palmated Gecko (*G. palmatus*). Species known from eastern China, Japan, Korea and Tablas Island (the Philippines). **Status** Not Evaluated.

WARTY HOUSE GECKO
Gekko monarchus　　　　　　　　　　　PLATE 25
Measurement SVL 102mm **Identification** Body robust and tuberculate; dorsals with large tuberculate scales arranged in 16–17 longitudinal rows; scales on throat granular; midventrals 30–38; digits widened, basally webbed; preanofemoral pores 23–42; median row of subcaudals widened. **Coloration** Dorsum greyish-brown with dark brown blotches arranged in 7–9 pairs; venter cream. **Habitat and Behaviour** Associated with both human habitations and forest edges at elevations of up to 1,500m asl. Calls comprise 51 individual low 'tock-tock'. Diet includes insects and other arthropods. Clutches comprise 2 eggs, measuring 9.3–13.6 x 9.5–11.4mm, which are attached to rock crevices. Communal nesting is known in this species, with more than 50 eggs laid together. Incubation period 120 days. Hatchlings 25–30mm. **Distribution** Thailand, Peninsular Malaysia, Singapore, Sumatra, the Mentawai Archipelago, Pulau Simeulue, Pulau Berhala, Borneo and Java. Also Maluku (Indonesia) and the Philippines. **Status** Not Evaluated.

NUTAPHAND'S GECKO
Gekko nutaphandi　　　　　　　　　NOT ILLUSTRATED
Measurement SVL 116mm **Identification** Body robust; dorsum with 14 rows of large conical tubercles; rostral ca x 3 wider than deep, lacking rostral groove and excluded from contact with nostril rim; precloacal pores in a series of pale scales in females, in continuous series of 17–22; femoral pores absent; lamellae under Toe IV 15, enlarged. **Coloration** Dorsum greyish-brown to chestnut-brown; series of 8 transverse rows of bright white spots on back of head, body and tail-base; anterior margin of rows coincident with indistinct brown band; scattered white tubercles from auricular opening to posterior edge of orbit and across temporal region; 3 white spots form triangle at occiput; venter greyish-white, scales with diffuse minute dark speckles; iris brick-red. **Habitat and Behaviour** Known from a single limestone hill within a bamboo forest. Associated with walls of shallow caves and slender vegetation, particularly bamboo. Diet and reproductive habits unstudied. **Distribution** Endemic to central-western Thailand (Sai Yok Noi Waterfall, Kanchanaburi Province). **Status** Not Evaluated.

PALMATED GECKO
Gekko palmatus　　　　　　　　　　　PLATE 25
Measurement SVL 79mm **Identification** Body robust; head broad; dorsal surface with tubercles in ca 12 rows; nostrils contact rostral; internasals 0–2, smaller than supranasal; supralabials 11–14; infralabials 9–12; tubercles absent on limbs; toes half-webbed; males with 24–27 precloacal pores; cloacal spur single; midbody scale rows 132–149; lamellae under Toe IV 11–15. **Coloration** Dorsum greyish-tan; paired rounded or oval-elongated spots on occipital and nuchal region, followed by 4–6 middorsal spots; light middorsal stripe; tail dark-banded; venter cream-yellow, with dark dots especially on subcaudals. **Habitat and Behaviour** Inhabits the offshore island of Cu Lao Phon Vong, as well as the Man Son Mountains, at elevations of 914–1,219m asl. Nocturnal and arboreal. Diet and reproductive habits unstudied. **Distribution** Endemic to northern Vietnam (Bac Ky or Dông Kinh, and Cu Lao Phon Vong). **Status** Not Evaluated.

SANDSTONE GECKO
Gekko petricolus　　　　　　　　　　　PLATE 26
Measurement SVL 101mm **Identification** Body moderate; dorsal tubercles not greatly enlarged; tail slender, depressed; nostrils border rostral; scales on snout larger than those on dorsum of body; supralabials 12; infralabials 10–11; precloacal pores 9–10; femoral pores absent; dorsal tubercles not greatly enlarged; midventrals 27–32. **Coloration** Dorsum yellow; forehead lavender-grey; numerous rounded cream spots on head and body; venter unpatterned yellow or cream. **Habitat and Behaviour** Inhabits sandstone hills. Nocturnal and saxicolous, associated with rocks. Diet unstudied. Clutches comprise 2 eggs, measuring 8 x 9–11mm, which are

stuck onto rocky substrate. Incubation period 46–144 days. Hatchlings 30–33mm. **Distribution** Thailand and Laos. Suspected to occur in Cambodia. **Status** Not Evaluated.

SEVEN-SPOTTED BENT-TOED GECKO
Gekko scientiadventura PLATE 26
Measurement SVL 73mm **Identification** Body slender, tubercles absent; head broader than neck; head and body slightly depressed; weak lateral folds; fingers and toes slightly webbed basally; precloacal pores 5–8; nostrils touch rostral; internasals absent; posterior ciliaries spiny; lamellae under Toe IV 14–17. **Coloration** Dorsum yellow to brown, typically with 7 large light spots, sometimes expanded to lateral narrow wavy bands; tail with 7–10 light cross-bands; gulars and parts of venter marbled. **Habitat and Behaviour** Inhabits karst-dominated lowland forests at elevations of 50–150m asl. Nocturnal and arboreal. Call is a growl, heard at dawn and at night, but also on summer afternoons. Diet unstudied. Nests communally, and clutches comprise 2 eggs. **Distribution** Endemic to central Vietnam (Phong Nha-Ke Bang National Park, Quang Binh Province). **Status** Not Evaluated.

SIAMESE GECKO
Gekko siamensis PLATE 26
Measurement SVL 141mm **Identification** Body robust; nostrils separated from rostral; supralabials 17–21; auricular opening large; dorsal tubercles at midbody 16–19; precloacal pores 11–12 in males; femoral pores absent; lamellae under Toe IV 19–20. **Coloration** Dorsum purplish-grey, paler on snout; occipital region white-spotted; dark nuchal loop between auricular openings; V-shaped dark mark, also emerging from auricular opening and extending beyond axilla, to level of insertion of forelimbs; straight dark band between insertion of hind limbs; white spots form transverse rows posterior to dark body bands; tail with 7 white bands, each edged with narrow dark band anteriorly; venter unpatterned grey; iris green. **Habitat and Behaviour** Inhabits the lowlands in rocky biotope at elevations of ca 400m asl. Nocturnal and arboreal. Diet consists of large insects. Clutches comprise 2 eggs, measuring 21 x 21.5–22.5mm. Incubation period 93–182 days. Hatchlings 44–53mm. **Distribution** Endemic to north-central Thailand (Phetchabun and Nakhon Ratchasima Provinces). **Status** Not Evaluated.

SMITH'S GIANT GECKO
Gekko smithii PLATE 26
Measurement SVL 191mm **Identification** Body robust; head large; body thick-set with scattered tubercles on dorsum; males with 11–16 precloacal pores in short angular series. **Coloration** Dorsum greyish-brown with a transverse series of white spots; tail dark-banded; venter cream with grey patches; eye green. **Habitat and Behaviour** Inhabits forested habitats in the plains and low hills, including offshore islands. Call is reminiscent of the bark of a dog. Diet consists of arthropods. Clutches comprise 2 eggs, measuring 19 x 20mm, which are glued to tree trunks. **Distribution** Thailand, Peninsular Malaysia, Singapore, Sumatra, Pulau Nias, the Natuna Archipelago, Borneo and Java. Also the Nicobar Islands (India). **Status** Not Evaluated.

ULIKOVSKI'S GECKO
Gekko ulikovskii PLATE 26
Measurement SVL 140mm **Identification** Body robust; dorsals granular with evenly scattered tubercles not arranged in linear series; rostral contacts nostril; interorbitals 40–46; internasal 1; supralabials 14; males with 10–15 precloacal pores; femoral pores absent; lamellae under Toe IV 16–20. **Coloration** Dorsum yellow or yellowish-green with up to 8 narrow light bands. **Habitat and Behaviour** Inhabits rocky biotope within dry forests. Call described as a cackle. Nocturnal and saxicolous. Diet in captivity includes insects and mice; diet in the wild unstudied. Clutches comprise 2 eggs, measuring 18 x 15mm, with 3–4 clutches produced per season. Incubation period 64–88 days. Hatchlings 70mm. **Distribution** Endemic to central Vietnam (Kon Tum Province). **Status** Not Evaluated.

BOWRING'S HOUSE GECKO
Hemidactylus bowringii PLATE 26
Measurement SVL 51mm **Identification** Body robust, flattened; head relatively large; snout pointed; supralabials 7–12; infralabials 6–10; 18–27 preanofemoral pores in males; tail flattened, segmented, tail-base not swollen, lacking denticulated edges; lamellae under Toe IV 7–11. **Coloration** Dorsum light brown, mid-brown or tan with dark brown smudges and occasionally white spots; dark brown postocular streak; sometimes 4 longitudinal streaks along dorsum; tail with dark chevrons; venter unpatterned cream to light yellow; pelvic region and underside of tail light orange-beige. **Habitat and Behaviour** Inhabits human-modified habitats including houses, and also lowland forests. Nocturnal, and terrestrial in forested habitats. Daytime retreat is in leaf litter and under logs and bark. Insectivorous. Clutches comprise 2–3 eggs, which are stuck to walls. Incubation period ca 30 days. **Distribution** Myanmar, Laos and Vietnam (Ho Chi Minh City and northern Vietnam). Also eastern and north-eastern India, Bangladesh, Nepal, Japan (the Ryukyu Archipelago), and southern and eastern China (including Yunnan, Sichuan, Taiwan and Hainan Provinces). **Status** Not Evaluated.

BROOKE'S HOUSE GECKO
Hemidactylus brookii PLATE 26
Measurement SVL 65mm **Identification** Body robust, slightly flattened; head broad, oval; head scales small; dorsal surface with small granular scales and tubercles in 16–20 rows; tail tapering, plump with spine-like tubercles on dorsum; males with 7–12 preanofemoral pores; enlarged lamellae under Toe IV 7–8, subsequent 5–7 divided. **Coloration** Dorsum dark brown to light grey with dark spots, which are

usually arranged in groups; 2 dark lines along nostrils and eyes; venter cream; undersurface of tail orange-beige. **Habitat and Behaviour** Inhabits areas created through human activities, such as parks, gardens and houses, in addition to open forests. Nocturnal and scansorial. Call is a loud 'chuck-chuck-chuck'. Diet consists of small arthropods. Clutches comprise 2 eggs, measuring 7 x 9mm, and more than 1 clutch may be laid. Incubation period ca 43 days. **Distribution** Myanmar (north-central, south to southern Thaninthary) and Borneo (north-western Sarawak State). Also India, Pakistan, West Africa, southern and eastern China (Hong Kong and Macau, and Zhejiang Province), the West Indies and the Philippines. Some populations introduced through human agencies. **Status** Not Evaluated.

FRILLY FOREST GECKO
Hemidactylus craspedotus PLATE 26
Measurement SVL 62mm **Identification** Body slender; body and tail depressed; dorsum with scattered tubercles; skin frills on sides of body, tail and sides of throat, and along lateral edges of limbs; digits nearly entirely webbed. **Coloration** Dorsal surface greyish-brown with 2 rows of rectangular dark spots; dark streak along head; tail dorsum dark banded, its venter reddish-orange basally, greyish-yellow distally; venter bright yellow speckled with dark brown. **Habitat and Behaviour** Inhabits lowland forests, including swamp forests. Nocturnal and crepuscular, active on tree trunks. Known to both parachute and glide for distances of up to 3m between trees. Diet consists of arthropods. Reproductive habits unstudied. **Distribution** Southern Thailand (Koh Samui), Peninsular Malaysia, Singapore and Borneo. Suspected to occur on Java. **Status** Not Evaluated.

ASIAN HOUSE GECKO
Hemidactylus frenatus PLATE 26
Measurement SVL 67mm **Identification** Body robust, slightly flattened; head large; supralabials 9–12; infralabials 7–10; tail segmented, tapering; dorsal scales smooth; no webbing in fingers and toes; sides of tail with enlarged tubercles; no flaps of skin along flanks and at backs of hind limbs; males with 23–36 preanofemoral pores; lamellae under Toe IV 8–11 undivided, subsequent 5–8 divided. **Coloration** Dorsum greyish-brown or dusky brown, sometimes with darker markings; light brown streak, with lighter edge on top, across sides of head, sometimes continuing onto flanks; venter unpatterned cream or light beige. **Habitat and Behaviour** Associated with human habitations, and less often with forested areas. Call is a series of 4–5 loud stacatto notes. Diet consists of arthropods. Clutches comprise 2 eggs, measuring 8.3 x 9.8mm. **Distribution** Myanmar, Thailand, Laos, Cambodia, Vietnam, Peninsular Malaysia, Singapore, Sumatra, Borneo, Java and Bali. Also India, Bangladesh, Sri Lanka and southern China; introduced into Central and South America, Madagascar, eastern and southern Africa, Mauritius, New Guinea, Polynesia and Australia. **Status** Not Evaluated.

GARNOT'S HOUSE GECKO
Hemidactylus garnotii PLATE 26
Measurement SVL 65mm **Identification** Body robust; head large; supralabials 14–19; infralabials 10–18; dorsal scales small; tail segmented, depressed, with denticulate lateral edges; 14–19 enlarged femoral scales; preanofemoral pores absent; lamellae under Toe IV 6–8, distal-most undivided. **Coloration** Dorsum brownish-grey to yellowish-tan with 5 longitudinal rows of cream spots from nape to inguinal region, or longitudinal brown stripes; dark-brown postocular stripe to above auricular opening may extend to flanks; venter unpatterned cream to pale yellow. **Habitat and Behaviour** Associated with human habitations in the lowlands, and may be found on trees, as well as on walls of buildings, frequently entering houses. Nocturnal and scansorial. Diet consists of small arthropods. Parthenogenetic, producing 2 eggs, measuring 9 x 10mm. Hatchlings 27–28mm. **Distribution** Myanmar, Thailand, Laos, Cambodia, Vietnam, Peninsular Malaysia, Sumatra, Nias, Borneo and Java. Also eastern India, southern China, Ambon, Flores, Sumbawa, the Philippines, New Caledonia, the Solomon Islands, Fiji and Tahiti; naturalized in Florida and Hawaii (USA). **Status** Not Evaluated.

BURMESE SPOTTED GECKO
Hemidactylus karenorum NOT ILLUSTRATED
Measurement SVL 56mm **Identification** Body robust, flattened; head large and broad; dorsum with small granular scales and numerous small tubercles not arranged in longitudinal series; fingers and toes with slight or no webbing; dorsal and lateral scales small, subequal, juxtaposed tuberclulate, interspersed with smooth conical tubercles; supralabials 10–12, infralabials 7–11; 26–38 pairs of preanofemoral pores in males; tail flattened; lamellae under Toe IV 8–10. **Coloration** Dorsum medium brown to greyish-khaki; forehead unpatterned or mottled light and medium brown; irregular rectangular dark spots from nape to tail-base; row of 6 spots on midline of back and tail, and dorsolateral row and lateral row on trunk, may be fused into stripe. Dark phase individuals with dark brown postorbital stripe to flanks; dorsum nearly uniform beige or with faded markings on light background; venter unpatterned cream to light yellow. **Habitat and Behaviour** Inhabits lowland forests. Nocturnal, and terrestrial and semi-arboreal, associated with leaf litter along streams and tree bases. Diet and reproductive habits unstudied. **Distribution** Possibly endemic to south-central Myanmar (Sittaung Valley). Suspected to occur in eastern Bangladesh; records from north-eastern India (Assam Province) require verification. **Status** Not Evaluated.

FRILLY HOUSE GECKO
Hemidactylus platyurus PLATE 26
Measurement SVL 69mm **Identification** Body slender in juveniles, robust in adults, flattened; dorsum smooth with tiny granules; tail strongly depressed, lateral edges serrated with pointed scales; snout rather long; supralabials 9–11; infralabials 7–8;

fingers and toes half-webbed; ventrolateral dermal fringe on trunk, posterior edges of thighs and crus, and anterior edges of upper arms; males with 34–36 femoral pores; lamellae under Toe IV 6–9, undivided, subsequent 5–7 divided. **Coloration** Dorsum light grey to mid-brown, sometimes with darker variegation or elongated dark brown spots; typically dark grey postocular streak to axilla; venter unpatterned cream or light yellow. **Habitat and Behaviour** Associated with human habitations, but also encountered in lightly forested areas. Nocturnal and scansorial. Diet includes arthropods such as spiders, and ants. Eggs measure 10–10.6 x 8.5–8.9mm. Nesting sometimes communal, with eggs laid in retreat sites. Hatchlings 20.5–25mm. **Distribution** Myanmar, Thailand, Laos, Cambodia, Vietnam, Peninsular Malaysia, Singapore, Sumatra, Borneo, Java and Bali. Also Nepal, Bangladesh, eastern India, the Andaman and Nicobar Islands, Sri Lanka, Komodo, Flores, Sulawesi, the Philippines and China; introduced into New Guinea. **Status** Not Evaluated.

STEJNEGER'S HOUSE GECKO
Hemidactylus stejnegeri   PLATE 26
Measurement SVL 59.6mm **Identification** Body robust and depressed; dorsal surface without tubercles; snout elongated; supralabials 13–17; infralabials 10–12; precloacal and femoral pores indistinct within slightly enlarged preanal and femoral scales; midventrals 117; lamellae under Toe IV 12–14. **Coloration** Dorsum greyish-tan with indistinct cream spots on body and limbs; indistinct light bands on tail; venter unpatterned cream. **Habitat and Behaviour** Inhabits lowland forests, including human habitations. Nocturnal, scansoral and arboreal. Diet consists of small insects. Clutches comprise 2 (rarely 1) eggs, measuring 9.9–10.2 x 8.7–9mm. Incubation period 56 days. Hatchlings 18.5–20.6mm. **Distribution** Vietnam (Quang Ninh Province). Also Taiwan and Luzon in the northern Philippines. A triploid species complex whose identities and relations remain unresolved. **Status** Not Evaluated.

VANDEERMEERMOHR'S HOUSE GECKO
Hemidactylus vandermeermohri   NOT ILLUSTRATED
Measurement SVL 53mm **Identification** Body slender; snout elongated; rostral with median suture, in contact with nostril; supralabials 11; infralabials 8–9; dorsals with granules intermixed with enlarged tubercles; preanofemoral pores 28; midventrals 31–34; lamellae under Toe IV 10. **Coloration** Dorsum light grey with black spots; dark stripe along head, across eye and auricular opening, to axilla, beyond which it is indistinct; venter cream. **Habitat and Behaviour** Inhabits a low-lying offshore island. Nothing known of its natural history. **Distribution** Endemic to Pulau Berhala (Indonesia). **Status** Not Evaluated.

VIETNAMESE HOUSE GECKO
Hemidactylus vietnamensis   NOT ILLUSTRATED
Measurement SVL 58mm **Identification** Body robust; head large; supralabials 9–12; infralabials 8–11; interprimary and intersecondary postmental scales 6–16; lateral postmental scales 1–6; tail depressed, with sharp denticulated lateral edges; femoral pores 8 pairs; lamellae under Toe IV 5–6. **Coloration** Dorsum grey with cream spots of various sizes and shapes; venter orange-yellow; subcaudals pale red. **Habitat and Behaviour** Inhabits the lowlands in forest edges, and may be associated with man-made structures. Nocturnal and saxicolous, occurring on tree trunks and the walls of buildings. Diet and reproductive habits unstudied, although it is known that it is an all-female parthenogenetic species. **Distribution** Myanmar (Karen Mountain) and northern Vietnam (Bac-Khan, Tonkin and Kuk-Fiong Reserve, Hanam Ninh Province). Possibly also southern and eastern China, north-eastern India, Laos and Cambodia. **Status** Not Evaluated.

CHAPA WORM GECKO
Hemiphyllodactylus chapaensis   NOT ILLUSTRATED
Measurement SVL 33mm **Identification** Body slender, elongated; head distinct from neck; Supralabial VII suborbital; postmentals reduced; ventrals slightly larger than dorsals; tail shorter than head-body length; inner digits reduced, clawless, unwebbed; digits with widened lamellae; 6 enlarged lamellae under Toe IV, which follow another 6 that are narrow. **Coloration** Dorsum brownish-green; sinuous lines forming double-chained pattern on anterior and narrow indistinct bands on posterior; dark lateral stripe from orbit to axilla; venter greyish-cream. **Habitat and Behaviour** Inhabits submontane forests as well as human dwellings at elevations of ca 1,500m asl. Nocturnal and arboreal, sometimes found inside houses. Diet and reproductive habits unstudied. **Distribution** Endemic to northern Vietnam (Sa Pá, Lào Cai Province and Moc Chau City, Son La Province). **Status** Not Evaluated.

HARTERT'S WORM GECKO
Hemiphyllodactylus harterti   PLATE 27
Measurement SVL 45mm **Identification** Body slender, elongated; head barely distinct from neck; snout short; supralabials 9–10; infralabials 9; auricular opening small, oval; dorsals granular, in 30–36 rows, dorsum lacking tubercles; ventrolateral fold absent; ventrals slightly larger than dorsals; angular rows of 26–42 preanofemorals; 2–3 postcloacal tubercles; median subcaudals not enlarged; lamellae under Toe IV 17–18. **Coloration** Dorsum greyish-brown with irregular brown pattern; dark lateral stripe from head; tail pale grey; venter greyish-brown. **Habitat and Behaviour** Inhabits submontane and montane forests at elevations of 800–1,600m asl. Nocturnal, terrestrial and arboreal, inhabiting houses, and also encountered under stones and fallen logs. Diet and reproductive habits unstudied. **Distribution** Endemic to Peninsular Malaysia (Gunung Inas, Gunung Kledang and Bukit Larut, Perak State, and Cameron Highlands, Pahang State). **Status** Not Evaluated.

COMMON WORM GECKO
Hemiphyllodactylus typus PLATE 27
Measurement SVL 60mm **Identification** Body slender, elongated; head barely distinct from neck; granular dorsal scales; supralabials 10–14; infralabials 10–11; ventral scales smooth, rounded and imbricate; digits free; terminal phalange short, clawed; tail prehensile; males with 10–12 precloacal pores and 8–10 femoral pores; lamellae under Toe IV 3–6. **Coloration** Dorsum dark brown or yellowish-brown with dark brown blotches or specks that may form longitudinal stripes; dark brown stripe from nostril to shoulder; venter cream with dark brown speckles. **Subspecies** Two subspecies are recognized, of which one (*H. t. typus*) occurs in the region. **Habitat and Behaviour** Inhabits forested areas and urban centres, from the coast to nearly 1,000m. Sometimes seen on the walls of houses. Diet consists of small insects. Unisexual and parthenogenetic. Eggs measure 8 x 6mm, and are laid in pairs in the hollows of dead branches, inside rotting logs and in the axils of ferns. Incubation period 40 days. Hatchlings 14.3–17.7mm. **Distribution** Myanmar, Thailand, Vietnam, Peninsular Malaysia, Singapore, Sumatra, Borneo, Java and Bali (*H. t. typus*). Also the Andaman and Nicobar Islands (India), Sri Lanka, southern and eastern China, the Ryukyu Archipelago of Japan, the Philippines, New Guinea, the Solomon Islands, New Caledonia, Polynesia, Hawaii (USA), the Mascarene Islands (*H. t. typus*) and Komodo (*H. t. pallidus*). **Status** Not Evaluated.

YUNNAN WORM GECKO
Hemiphyllodactylus yunnanensis NOT ILLUSTRATED
Measurement SVL 41mm **Identification** Body slender; dorsals granular and subequal; supralabials 10–11; infralabials 9–10; chin shields distinct; auricular opening large; digits free; digit tips expanded; Digit I clawless; ventrals imbricate; preanal pores 12–22 in angular series in males. **Coloration** Dorsum mid-brown mottled with grey; ladder-like dark brown pattern on back; tail dark-banded; venter white with cloudy pattern. **Subspecies** Four subspecies are recognized, of which one (*H. y. yunnanensis*) occurs in the region. **Habitat and Behaviour** Inhabits primary forests in the mid-hills. Diet unstudied. Clutch size unknown, eggs measure 7 x 5mm. **Distribution** Northern Myanmar, northern Laos and Thailand (Phu Kading, Loei Province) (*H. y. yunnanensis*). Also southern and eastern China, including Yunnan Province (*H. y. yunnanensis, H. y. jinpingensis* and *H. y. longlingensis*) and Guizhou Province (*H. y. dushanensis* and *H. y. jinpingensis*). **Status** Not Evaluated.

COMMON MOURNING GECKO
Lepidodactylus lugubris PLATE 27
Measurement SVL 49mm **Identification** Body slender, elongated; head longer than broad; dorsals granular; tubercles on dorsum absent; supralabials 11–13; infralabials 9–12; tail wide and flattened; preanofemoral scales 25–31; lamellae under Toe IV 12–14. **Coloration** Dorsum cinnamon-brown, pale brown or greyish-brown with brownish-red tail; dark stripe along face; W-shaped cross-bars on body and tail; small black blotch on each side of pelvic region; venter cream. **Habitat and Behaviour** Associated with coastal and mangrove forests. Nocturnal, found on leaves and branches. Predator of small arthropods; nectar and plant juice are also lapped up. Some populations unisexual, parthenogenetic. Clutches comprise 2 eggs, measuring 7.0–8.7 x 9.5–10mm, which are laid on leaves, sometimes communally. Incubation period 60–117 days. Hatchlings 17.2–17.8mm. **Distribution** Vietnam, Peninsular Malaysia, Singapore, Sumatra, Pulau Berhala, Java and Borneo. Also the Andaman and Nicobar Islands (India), Sri Lanka, the Maldives, southern China, New Guinea and the south-west Pacific; introduced into Panama, Ecuador, the Galapagos Islands and Central America. **Status** Not Evaluated.

KINABALU MOURNING GECKO
Lepidodactylus ranauensis PLATE 27
Measurement SVL 47.7mm **Identification** Body slender, elongated, depressed; dorsals small, lacking large tubercles; supralabials 9; Supralabials VII–VIII subocular; infralabials 9–10; midbody scale rows 108; digits long, relatively narrow; tail wide; males with a continuous series of 35–37 preanofemoral pores; lamellae under Toe IV 15. **Coloration** Dorsum greyish-brown; forehead greyish-brown or with a slightly reddish-brown area; pair of triangular dark markings on dorsolateral part of tail-base; venter greyish-tan. **Habitat and Behaviour** Associated with human habitations at middle elevations and in the foothills of the Kinabalu Massif. Nocturnal, active in dark areas of walls. Diet presumably consists of insects. Clutches comprise 2 eggs, measuring 4.1 x 3.7 and 4.5 x 3.6mm. **Distribution** Endemic to Borneo (Gunung Kinabalu, Sabah State). **Status** Not Evaluated.

BROOKS'S CAMOUFLAGE GECKO
Luperosaurus brooksii NOT ILLUSTRATED
Measurement SVL 58mm **Identification** Body slender, elongated; dorsals flat, granular; auricular opening rounded, small; Supralabial I enters nostril; Supralabial XI subocular; subrictal tubercles absent; genials small; single intersupranasal contacts rostral; chin shields absent; limbs with dermal folds; digits strongly dilated, half webbed; tail flattened, with large, triangular and projecting scales laterally; uninterrupted series of 40 preanofemoral pores. **Coloration** Dorsum pale greyish-brown with brown dots forming 5 cross-bars; venter unpatterned white. **Habitat and Behaviour** Nothing known of its natural history. **Distribution** Endemic to Sumatra (Lebongtandai, nr Bengkulu, Bengkulu Province). **Status** Not Evaluated.

BROWN'S CAMOUFLAGE GECKO
Luperosaurus browni PLATE 27
Measurement SVL 66.4mm **Identification** Body slender, elongated; dorsal surface with rounded, flattened or slightly convex tubercles; skin folds

present on limbs; ventrals flat, subimbricate; preanofemoral pores 28–32; digits dilated and half-webbed; first digit unclawed; tail depressed with lateral spines; lamellae under Toe IV 16–19, entire. **Coloration** Dorsum light grey with minute black spots on head, body and limbs; dark stripe in supralabial region; 5 broken dark chevrons on middorsum; tail dark-banded; venter white with several dark spots on tail. **Habitat and Behaviour** Associated with lowland rainforests. Diet presumably consists of small insects. Clutches comprise 2 eggs, measuring 8.1–9.6 x 8.8–8.9mm. Hatchlings 28.3–29.3mm. **Distribution** Peninsular Malaysia, Singapore and Borneo. **Status** Not Evaluated.

CROCKER RANGE CAMOUFLAGE GECKO
Luperosaurus sorok PLATE 27
Measurement SVL 34.7mm **Identification** Body robust, elongated; head narrower than body; supralabials 14–15; Supralabial IV suborbital; infralabials 11; auricular opening oval-squarish; subrictal tubercles present; 2 intersupranasals contact rostral; rostral in contact with nostrils; dorsal body scales rounded, convex and granular; mid-ventrals 45; ventrolateral body tubercles spinose; femoral and precloacal pores absent in females; tail fringe with distinct serrations; lamellae under Toe IV 8–10. **Coloration** Dorsum pale grey with dark grey double chevrons fused middorsally, 1 on forehead, 4 on body; 6 dark bands on tail; dark areas edged with black; throat yellowish-cream with darker speckles; rest of venter yellowish-cream with darker variegation. **Habitat and Behaviour** Inhabits submontane forests. Probably a high-canopy species. Diet and reproductive habits unstudied. **Distribution** Endemic to Borneo (Crocker Range, Sabah State). **Status** Not Evaluated.

YASUMA'S CAMOUFLAGE GECKO
Luperosaurus yasumai PLATE 27
Measurement SVL 38.9mm **Identification** Body moderately robust, not elongated; head depressed; snout strongly tapered; supralabials 9–10; infralabials 10–11; dorsal tubercles conical or spinose; ventrals flat, subimbricate; lamellae entire; tail strongly depressed; lamellae under Toe IV 7. **Coloration** Dorsum brownish-tan; dorsal surfaces of head and tail yellowish-brown; numerous pale cloudy markings on head, body and tail; 2 distinct rounded ivory spots on middorsum; venter light grey with several dark dots in gular region; 8 indistinct broad dark bands on tail. **Habitat and Behaviour** Only known individual found in regenerating dipterocarp forest, where it was lying motionless on the ground. Diet and reproductive habits unstudied. **Distribution** Endemic to Borneo (Kalimantan Timur Province). **Status** Not Evaluated.

HORSFIELD'S PARACHUTE GECKO
Ptychozoon horsfieldii PLATE 27
Measurement SVL 80mm **Identification** Body robust; head large; dorsal tubercles absent; femoral pores 8–11; precloacal pores 10–11; femoral and precloacal pores in separated series; 21–22 denticulate tail lobes; tail lobe size reduction to tail-tip gradual. **Coloration** Dorsum grey or brown with black mottling; broad dark band from posterior corner of eye to beyond tympanum; 2 large oval blotches that reach insertion of forelimbs; butterfly-shaped dark mark on axilla; trunk with 3 other wavy bands; chin yellow with a few scattered brown spots. **Habitat and Behaviour** Inhabits lowland rainforests and frequents man-made structures. Diet consists of arthropods. Clutches comprise 2 eggs, measuring 13.7 x 11.9mm. Hatchlings 34mm. **Distribution** Myanmar, Thailand, Peninsular Malaysia, Singapore, Sumatra and Borneo. **Status** Not Evaluated.

KUHL'S PARACHUTE GECKO
Ptychozoon kuhli PLATE 27
Measurement SVL 107.8mm **Identification** Body robust; head large; dorsal surface with scattered granules, tubercles in 2–6 straight rows; supralabials 11–15; infralabials 10–12; tail depressed; tail lappets set at right angles to tail; tail terminates in broad flap; feet webbed; males with 18–28 precloacal pores, oriented in a curved line. **Coloration** Dorsum grey or reddish-brown with 4–5 transverse, wavy dark brown bands; dark brown line from eye to first dorsal band; venter unpatterned yellow. **Habitat and Behaviour** Inhabits lowland forests at least 35m above ground, and occasionally wooden structures in the vicinity of forests and houses. Nocturnal. Diet includes arthropods. Clutches comprise 2 spherical eggs, measuring 9.5–15mm, which are attached to tree trunks, branches or rocks. Incubation period 73–122 days. Hatchlings 34mm. **Distribution** Thailand, Peninsular Malaysia, Sumatra, the Mentawai and Natuna Archipelagos, Borneo and Java. **Status** Not Evaluated.

SMOOTH PARACHUTE GECKO
Ptychozoon lionotum PLATE 27
Measurement SVL 95mm **Identification** Body robust; head large; no scattered granules on dorsum of body; tail lappets narrow and directed backwards; tail terminates in broad flap; feet webbed; males with 16–25 precloacal pores, oriented in a curved line. **Coloration** Dorsum greyish-brown with transverse dark brown bands; dark brown line from eye to first dorsal band; venter unpatterned yellow. **Habitat and Behaviour** Associated with lowland forests and the mid-hills at elevations of up to 700m asl. Occasionally enters human dwellings. Diet and reproductive habits unstudied. **Distribution** Myanmar, Thailand, Cambodia, Vietnam and northern Peninsular Malaysia. Also north-eastern India. **Status** Not Evaluated.

KINABALU PARACHUTE GECKO
Ptychozoon rhacophorus PLATE 27
Measurement SVL 75mm **Identification** Body robust; dorsal surface spinose or with scattered thorny tubercles; skin flanks jagged and irregularly lobed; head lacks lateral skin fringes; short, sharply tapered tail lacks distinct lobes; terminal expansion of tail absent; reduced digital webbing; 17 precloacal pores. **Coloration** Dorsum brownish-green, unpatterned, or

with dark mottling and indistinct wavy bands; venter light brown speckled with dark brown. **Habitat and Behaviour** Inhabits submontane and montane forests at elevations of 600–1,600m asl. Observed on the walls of buildings and the trunks of large trees. Nocturnal. Known to eat moths. Reproductive habits unstudied. **Distribution** Borneo (Gunung Kinabalu, Sabah State). **Status** Not Evaluated.

THREE-BANDED PARACHUTE GECKO
Ptychozoon trinotaterra PLATE 27
Measurement SVL 71.3mm **Identification** Body robust, depressed; head wide; dorsum with or without midvertebral row of enlarged tubercles; supralabials 11; infralabials 9–11; 19–21 preanofemoral pores; 15–16 denticulate tail lobes; terminal tail flap present; lamellae under Toe IV 12–15. **Coloration** Dorsum light grey with 3 transverse dark bands; intervening areas with dark mottling; forehead dark brown; venter yellowish-cream with brown and grey spots. **Habitat and Behaviour** Inhabits dry evergreen and lowland dipterocarp forests at elevations of up to 900m asl. Nocturnal and arboreal, known from tree trunks. Diet and reproductive habits unstudied. **Distribution** Thailand (Sakaerat, Nakhon Ratchasima Province) and Vietnam (Yok Don National Park, Yok Don Province, Cat Tien, Dong Nai Province and Tram Lap, Gia Lai Province). **Status** Not Evaluated.

Family LACERTIDAE 'TYPICAL' LIZARDS

Species in this family are typified by an elongated body, paired nasals, postorbitals and squamosals, pleurodont dentition, a conical head on a distinct neck, a long, moderately thick tail, well-developed limbs and femoral pores. They have large forehead scales, granular scales dorsally on the neck and trunk, and large abutting scales on the ventral surface, with osteoderms absent on the body scales and present on the forehead scales. The Lacertidae can employ tail autotomy, and South-East Asian species are mostly arboreal and predators of small invertebrates. They are oviparous, producing eggs with leathery shells, and are found in Asia, Europe and Africa.

HAN'S GRASS LIZARD
Takydromus hani PLATE 28
Measurement SVL 79mm **Identification** Body slender, elongated; head relatively short, less than twice as long as wide; dorsal surface with large, smooth, plate-like scales; femoral pores 6–8. **Coloration** Dorsum green; flanks paler; thin black stripe across eye to tympanum; venter greenish-yellow; thin black stripe across eye to tympanum; pupil golden-yellow. **Habitat and Behaviour** Inhabits primary subtropical forests at elevations of 200–1,450m asl. Diurnal and arboreal, associated with undergrowth and streams. Diet and reproductive habits unstudied. **Distribution** Endemic to central Vietnam (Rao-An and Ngoc Linh Mountains). **Status** Not Evaluated.

KÜHNE'S GRASS LIZARD
Takydromus kuehnei PLATE 28
Measurement SVL 60mm **Identification** Body slender, elongated; head relatively short, as long as wide; dorsal surface with large, smooth, plate-like scales; femoral pores 4–5. **Coloration** Dorsum olive-brown; limbs and tail yellowish-olive; flanks dark olive-brown; pale line, edged with dark brown, across nostril and eye extends to axilla; venter greenish-cream; underneath of tail orange. **Subspecies** Two subspecies are recognized, of which one (*T. k. vietnamensis*) occurs in the region. It shows occipital and interparietal not in contact, gulars reaching third pair of chin shields and enlarged scales on lower flanks, especially in females. **Habitat and Behaviour** Inhabits primary lowland forests at elevations of 125–350m asl. Diurnal and arboreal, associated with vegetation 1–2m above substrate. Diet and reproductive habits unstudied. **Distribution** Northern Vietnam (*T. k. vietnamensis*). Also southern China and Taiwan (*T. k. kuehnei*). **Status** Not Evaluated.

LONG-TAILED GRASS LIZARD
Takydromus sexlineatus PLATE 28
Measurement SVL 61mm (*T. s. sexlineatus*); SVL 65mm (*T. s. ocellatus*) **Identification** Body slender, elongated; head at least twice as long as wide; dorsal surface with 4–6 large, smooth, plate-like scales; tail 3–5 times as long as body; sides of body with single row of large scales; transverse scale rows and plates 40; femoral pores 1–2. **Subspecies** Two subspecies are recognized. *T. s. sexlineatus*, 2 femoral pores on each side; forehead scales smooth. *T. s. ocellatus*, single femoral pore on each side; forehead rugose. **Coloration** Dorsal surface with coffee-brown vertebral stripe, yellow or cream stripe from orbit to flanks; paravertebral stripe extends beyond tail-base; dorsolateral stripe dark brown, unpatterned; flanks greenish-yellow in anterior first third; venter with pale brown sheen; males brighter than females; juveniles resemble adult females. **Habitat and Behaviour** Associated with open areas, especially grasslands and marshes in the lowlands and mid-hills, from sea level to 850m asl. Diurnal and arboreal, with some terrestrial activity. Diet consists of insects and millipedes. Clutches comprise 2–3 eggs, measuring 10–11 x 6–7mm. Incubation period 47–62 days. Hatchlings 23mm. **Distribution** Myanmar, Thailand (*T. s. sexlineatus* and *T. s. ocellatus*), Laos, Cambodia, Vietnam (*T. s. ocellatus*), Peninsular Malaysia, Sumatra, the Natuna Archipelago, Borneo, Java and Bali (*T. s. sexlineatus*). Also southern and eastern China (*T. s. ocellatus*), and north-eastern India

and Bangladesh (*T. s.* unknown subspecies). **Status** Not Evaluated.

WOLTER'S GRASS LIZARD
Takydromus wolteri PLATE 28
Measurement SVL 58.8mm **Identification** Body slender, elongated, weakly depressed; head relatively short, less than twice broad; dorsal surface with large, smooth, plate-like scales; nasals contact behind rostral; supraoculars 4; dorsal plates in 8 longitudinal series; ventrals 27–31; single femoral pore; lamellae under Toe IV 20–26. **Coloration** Dorsum olive with light dorsolateral streak and dark olive lateral band; dark-edged postocular streak to inguinal region; venter yellow or greenish-cream. **Habitat and Behaviour** Inhabits temperate forests. Diurnal and arboreal, active on rocky and grassy biotope including agricultural fields. Diet includes insects, earthworms and snails. Clutch size unknown. Incubation period 21–37 days. Hatchlings 19.03–24.63mm. Reproductive habits unstudied. **Distribution** Reported from northern Vietnam, but this requires verification. Also southern China and Korea. **Status** Not Evaluated.

Family LANTHANOTIDAE EARLESS MONITOR

Earless monitors have short limbs, reduced eyes, nostrils situated dorsally, fused nasals, supraoccipital in broad contact with parietal, trapezoidal frontals, and prefrontal and postfrontal with contact above orbit. They have heterogeneous scalation with few underlying osteoderms that are not fused to the skull, and lack hemibacula, an upper temporal bar, a parietal eye and ear openings. Osteological characters defining the family include pterygoid teeth, pleurodont dentition and nine cervical vertebrae. Together with the Varanidae, earless monitors are the closest relatives of snakes. They are subfossorial and piscivorous. The single extant species in the region is found in Borneo, where it is restricted to lowland sites within tropical forests, and is known to be oviparous.

BORNEAN EARLESS MONITOR
Lanthanotus borneensis PLATE 28
Measurement SVL 200mm **Identification** Body slender, elongated and cylindrical; limbs short; forehead covered with small granular scales; eye reduced, with moveable lids; lower eyelid with clear window; tongue forked; males with blunt rectangular jaws; females with relatively pointed jaws; 6 parallel rows of large tuberculate scales on dorsum. **Coloration** Dorsum unpatterned brownish-orange, or with dark vertebral stripe; venter yellow. **Habitat and Behaviour** Associated with lowland streams and marshes, and also agricultural land. Nocturnal, hiding by day in burrows up to 30cm deep, which are situated along riverbanks or under rocks and fallen logs. Diet consists of earthworms and crustaceans. Clutches comprise 2–5 oval eggs, measuring 30mm. **Distribution** Endemic to Borneo (Sarawak State and Kalimantan Province). **Status** Not Evaluated.

Family LEIOLEPIDAE BUTTERFLY LIZARDS

Butterfly lizards are characterized by a robust, depressed body lacking nuchal and dorsal crests and a gular pouch, and possessing a distinct gular fold, a rounded tail and femoral pores. The family includes sand-dwelling species found in coastal areas and lowland forests of South-East Asia. Butterfly lizards are diurnal and terrestrial, and live in burrows. The flanks of their body are frequently brightly coloured and exposed during displays. The single genus is oviparous, and has bisexual (*Leiolepis belliana, L. guttata, L. reevesii* and *L. peguensis*) as well as parthenogenetic representatives, including triploid (*L. triploida* and *L. guentherpetersi*) and diploid (*L. boehmei*) types.

BELL'S BUTTERFLY LIZARD
Leiolepis belliana PLATE 28
Measurement SVL 170mm **Identification** Body robust, depressed; dorsal crest absent; dorsal scales small; skin on sides loose, expandable; supralabials 7–11; infralabials 8–11; tympanum large; scales across undersurface of tibia 7–12; femoral pores 13–20 pairs; midventrals 10–18, as broad as 3–4 dorsals; tail long, rounded and a little depressed; lamellae under Toe IV 32–41. **Coloration** Dorsum greyish-olive or blackish-brown with black-edged pale yellow spots and 3 longitudinal stripes; flanks bluish-black with 7–9 broad vertical orange bars; forehead dark olive-brown; venter uniform cream; females less brightly coloured; dorsal surfaces of limbs patterned with yellowish-cream spots. **Habitat and Behaviour** Inhabits open forests and other lowlands such as areas with grassland and sandy soil, including scrub forests and coconut groves. Diurnal and terrestrial. Lives in excavated burrows, 30 cm deep and 70 cm long, in loose soils. Territorial: threat display involves extending its ribs to reveal the bright colours on its butterfly-like 'wings'. Diet includes vegetation, crabs, insect larvae, grasshoppers, butterflies and other insects. Clutches comprise 6 eggs, measuring ca 25 x 13mm. **Distribution** Southern Thailand, Laos, Cambodia, Vietnam, the west coast of Peninsular Malaysia, Sumatra and Pulau Bangka. Introduced into Florida (USA). **Status** Not Evaluated.

BÖHME'S BUTTERFLY LIZARD
Leiolepis boehmei PLATE 28
Measurement SVL 126mm **Identification** Body slender; dorsal crest absent; rostral ca x 3 as wide as high, bordered by 2 labials and 7 smooth postrostrals, followed by 10 feebly keeled scales; supralabials 9–11; infralabials 11–12; skin on sides loose, expandable; enlarged scales across lower part of tibia 12–14; femoral pores 15–19; dorsal scales between longitudinal dorsolateral stripes 42–46; tail long, rounded and a little depressed; midventrals 17–19; lamellae under Toe IV 29–35. **Coloration** Dorsum blackish-olive with 2 pale grey, uninterrupted lateral stripes, 6–7 body scales wide, which fade in neck region; a third (median) stripe absent; between 2 stripes, series of 6–7 oval grey spots aligned in indistinct transverse rows on dorsum; flanks with indistinct oblique light stripes; forehead olive, sides darker, with small yellow dot on lower eyelid; chin, throat and chest grey with vertical white bars. **Habitat and Behaviour** Terrestrial, living in burrows in open forested habitat. Diet and reproductive habits unstudied. **Distribution** Southern Thailand (Songkhla Province). **Status** Not Evaluated.

PETERS'S BUTTERFLY LIZARD
Leiolepis guentherpetersi NOT ILLUSTRATED
Measurement SVL 156mm **Identification** Body slender; rostral ca x 3 as wide as high, bordered by 2 supralabials and 6 smooth postrostrals; scales on snout and interorbital area distinctly keeled; supralabials 8–9; infralabials 9–10; skin on sides loose, expandable; scales between dorsolateral stripes 37–43; midventrals as broad as 2 dorsals, 17–20; series of 21–23 femoral pores separated medially by 12 pubic scales; tail long, rounded and a little depressed; lamellae under Toe IV 40–43, bicarinate or tricarinate; 4 scales with enlarged triangular spurs at base of Toe III. **Coloration** Dorsum blackish-olive with 2 indistinct cream dorsolateral stripes, each 7–8 scales wide, which extend from level of forelimbs to hind limbs and tail-base; dorsolateral stripes enclose 4–5 oval cream spots, their olive-grey edges forming a broad reticulated pattern; forehead olive, sides of head darker; small yellow spots on lower eyelid and under eye; venter, including chin, grey with indistinct spots. **Habitat and Behaviour** Terrestrial, inhabiting coastal dunes dominated by bushes and grasses. Diet consists of flies and butterflies, as well as vegetation. Clutches comprise 1–3 eggs. **Distribution** Endemic to central Vietnam (Binh Tri Thien and Danang Provinces). **Status** Not Evaluated.

SPOTTED BUTTERFLY LIZARD
Leiolepis guttata PLATE 28
Measurement SVL 184mm **Identification** Body robust, depressed; dorsal crest absent; head small; supralabials 8–11; infralabials 9–13; scales small; tympanum large; skin on sides loose, expandable; scales across undersurface of tibia 12–14; femoral pores 19–22 pairs; tail long, rounded and a little depressed; midventrals 18–27, as broad as 2 dorsals; lamellae under Toe IV 37–45. **Coloration** Dorsum pale greyish-olive with pink spots and 3 pale dorsolateral stripes; flanks bluish-black with 7 vertical broad white bars; forehead pale olive-brown; nape and dorsal surfaces of limbs brick-red in males; venter blue. **Habitat and Behaviour** Terrestrial, living in excavated burrows in loose sandy soil. Diet includes the crocus flower (*Kaempferia candida*), as well as various types of arthropod. Reproductive habits unstudied. **Distribution** Endemic to southern Vietnam (Thua Thien Hue, Da Nang, Binh Dinh, Khanh Hoa, Ninh Thuan and Binh Thuan Provinces). **Status** Not Evaluated.

EYED BUTTERFLY LIZARD
Leiolepis ocellata PLATE 28
Measurement SVL 156mm **Identification** Body robust, depressed; dorsal crest absent; scales small; tympanum large; skin on sides loose, expandable; tail long, rounded and a little depressed. **Coloration** Dorsum greyish-olive or blackish-grey with black-edged pale yellow spots and 3 longitudinal stripes, 1 median and 2 lateral, which extend to pelvic region; dorsum covered with oval light ocelli, the dark edges of which form vertical bars; flanks orange with vertical black bands; forehead dark olive-brown; venter uniform cream; females less brilliantly coloured than males. **Habitat and Behaviour** Nothing known of its natural history, except the fact that it may reach an elevation of 1,085m asl. **Distribution** South-eastern Myanmar (Kokariet, Tenasserim River, Taninthayi Division and Dawna Hills, Kayah State), and north-western and north-eastern Thailand (Un Pang, unlocated, Doi Suthep, Chiang Mai Province, Ban Mae Chai Nua, Chiang Rai Province and Doi Angkhang, Chiang Mai Province). **Status** Not Evaluated.

PEGU BUTTERFLY LIZARD
Leiolepis peguensis NOT ILLUSTRATED
Measurement SVL 136mm **Identification** Body robust, depressed; dorsal crest absent; head small; scales small; supralabials 10–12; infralabials 10; tympanum large; skin on sides loose, expandable; femoral pores reduced, typically indistinct, numbering over 18 pairs; tail long, rounded and a little depressed; lamellae under Toe IV 39. **Coloration** Dorsum pale greyish-olive with contrasting pattern of short pale and dark cross-stripes that do not contact each other on flanks at level behind forelimbs; sides of neck unpatterned; pale vertical stripe on dorsum from base of tail; venter blue. **Habitat and Behaviour** Nothing known of its natural history. **Distribution** Myanmar (nr Mawlamyine, Mon State, Sagaing, Magwe and Mandalay Divisions, and Chin and Shan States). **Status** Not Evaluated.

REEVES'S BUTTERFLY LIZARD
Leiolepis reevesii PLATE 28
Measurement SVL 151mm **Identification** Body robust, depressed; dorsal crest absent; scales small; supralabials 6–9; infralabials 6–10; tympanum large; skin on sides loose, expandable; tail long, rounded and

a little depressed; midventrals 13–18. **Coloration** Distinguished from Bell's Butterfly Lizard (*L. belliana*) by its diffuse, buff brown-grey dorsum, with spot markings on the dark brown dorsal surface that produce a fishnet pattern of broad reticulations and ocelli with light centres and grey edges; flanks orange with vertical black bars; dorsal surfaces of limbs and tail spotted with yellowish-cream spots. **Habitat and Behaviour** Inhabits sandy coastal regions with grass cover. Diurnal and terrestrial, living in burrows in arid habitats. Diet consists of invertebrates and vegetation. Clutches comprise 2–8 pliable-shelled eggs, which are laid in deep (to 80cm) holes. Hatchlings 33.6–44.2mm. **Distribution** Northern Vietnam (Thanh Hoa, Nghe An, Ha Tinh, Quang Tri and Thua Thien-Hue Provinces). Also southern and eastern China. **Status** Not Evaluated.

RED-BANDED BUTTERFLY LIZARD
Leiolepis rubritaeniata PLATE 28
Measurement SVL 134mm **Identification** Body robust, depressed; dorsal crest absent; scales small; supralabials 9; infralabials 8–9; tympanum large; skin on sides loose, expandable; femoral pores 15–16; scales across undersurface of tibia 7–12; 50–65 (typically 54–60) whorled scales around tail; tail long, rounded and a little depressed. **Coloration** Dorsum greyish-olive or blackish, with oval black-edged, bluish-grey spots within a polygonal black network; females with pale ocelli in network; irregular orange dorsolateral stripes evident above level of forelimbs and hind limbs; spots on dorsum lack longitudinal or linear arrangement, and show a dark reticulated pattern; flanks with 8 vertical black bars in males, unicolored in females; intervening areas orange; forehead dark olive-brown; venter uniform cream; dorsal surfaces of limbs and tail with cream spots. **Habitat and Behaviour** Nothing known of its natural history. **Distribution** South-western part of Laos, north-eastern Thailand and southern Vietnam (Gia Lai and Kien Giang Provinces). **Status** Not Evaluated.

TRIPLOID BUTTERFLY LIZARD
Leiolepis triploida PLATE 28
Measurement SVL 148mm **Identification** Body robust, depressed; dorsal crest absent; scales small; tympanum large; skin on sides loose, expandable; femoral pores 17–21; tail long, rounded and a little depressed; midventrals 17–22; lamellae under Toe IV 34–39. **Coloration** Dorsum olive-brown with 3 dull-bordered dorsal stripes that extend from axilla to inguinal regions, enclosing 2 rows of oval eye-like areas, created by yellow spot within grey area, within reticulated pattern; flanks darker than dorsum, with 5 yellowish-cream cross-bars; forehead olive-brown; dorsal surfaces of limbs and tail olive-grey, with pale areas inside reticulated pattern, but pattern here less distinct generally than that on dorsum of body. **Habitat and Behaviour** Inhabits mixed dipterocarp forests. Terrestrial and diurnal, living in burrows. Diet and reproductive habits unstudied. **Distribution** Southern Thailand and northern Peninsular Malaysia (Kedah State). **Status** Not Evaluated.

Family SCINCIDAE SKINKS

Members of the skink family have a smooth and slender body, and a forehead covered with enlarged scales, often with osteoderms. They have paired nasals and squamosals, and a single fused parietal. Osteological characters include pleurodont dentition, a secondary palate ranging from partial to complete and, sometimes, pterygoid teeth. Skinks can employ tail autotomy as a means of defence against predators. Several lineages have independently lost their limbs, a fact that is correlated with their fossorial habits. The tail is typically well developed, and more so in species showing arboreal habits. Most species are diurnal and insectivorous, with a few consuming small vertebrates. Slightly over half of extant species are ovoviviparous; those producing eggs lay small clutches in holes in the earth or among vegetation. This cosmopolitan family includes more than 1,200 species.

STRIPED BORNEAN TREE SKINK
Apterygodon vittatum PLATE 29
Measurement SVL 96mm **Identification** Body robust; pterygoid teeth absent; dorsals keeled; prefrontals separated from each other; tympanum present; midbody scale rows 30; 2 large heel scales in males; tail rounded in cross-section, tapering to a fine point. **Coloration** Forehead and anterior of dorsum of body black; rest of dorsum brownish-grey with dark and light spots; light cream or yellow stripe from snout-tip to back of head; pale stripe from above eye along body; venter pale green. **Habitat and Behaviour** Inhabits lowland rainforests, as well as open, lightly forested areas such as parks and gardens within cities. Diurnal and arboreal, occurring on trees up to ca 37m, and generally observed close to bases of trees. Diet consists of ants and other small insects. Oviparous; clutches comprise 2–4 thick-shelled eggs. **Distribution** Endemic to Borneo. **Status** Not Evaluated.

CHINESE SLENDER SKINK
Ateuchosaurus chinensis PLATE 29
Measurement SVL 83.8mm **Identification** Body elongated; snout short; lower eyelid scaly; supranasals absent; frontal long, constricted in middle; parietals absent; nostril in nasal; auricular opening distinct, tympanum sunk; midbody scale rows 30; limbs well developed, pentadactyle; tail long; lamellae under Toe IV 16–17. **Coloration** Dorsum greyish-brown or reddish-brown, scales with dark central spot; flanks pale, nearly uniform or spotted with black and white,

with black spot preceding white; lateral neck dark brown or black, white spotted; gular and pectorals black spotted; venter cream or yellowish-white; tail-tip with dark marbling posteriorly. **Habitat and Behaviour** Inhabits montane evergreen forests. Diurnal and nocturnal, and terrestrial, associated with moist leaf litter. Diet consists of termites, cockroaches and earthworms. Oviparous; clutches comprise 2–5 eggs, measuring 3.6–4.6mm. **Distribution** Northern Vietnam (Ha Giang, Lang Son and Bac Giang Provinces). Also central, southern and eastern China. **Status** Not Evaluated.

BORNEAN LIMBLESS SKINK
Brachymeles apus PLATE 29
Measurement SVL 131mm **Identification** Body elongated and limbless; lower eyelid with single large scale; auricular opening absent; postnasals absent; prefrontals separated from each other; frontal longer than wide; supralabials 6; infralabials 5; single large postmental; dorsals and ventrals subequal; midbody scale rows 22–24; ventrals 108–113; enlarged preanal scales 5–8; tail thick, rounded, blunt-tipped. **Coloration** Dorsum reddish-brown, darkening posteriorly; snout-tip lighter; tail-tip dark brown to reddish-black; ocular scales darker; venter unpatterned cream. **Habitat and Behaviour** Inhabits montane forests and agricultural areas at elevations of 1,300–1,520m asl. Diurnal and fossorial. Diet unstudied. Ovoviviparous; clutches comprise 4 neonates, measuring 42–43mm. **Distribution** Endemic to northern Borneo (Gunung Murud, Sarawak State, and Gunung Kinabalu and Crocker Range, Sabah State). **Status** Not Evaluated.

BALINESE SNAKE-EYED SKINK
Cryptoblepharus balinensis PLATE 29
Measurement SVL 43mm **Identification** Body slender; snout acute; orbit surrounded by granular scales; lower eyelid coalesced with rudimentary upper eyelid; supraciliaries 5–6; Supralabial IV contacts orbit; midbody scale rows 20–28; lamellae under Toe IV 20–25. **Coloration** Dorsum olive with 4 black areas, separated by 3 light olive lines and edged ventrally by olive on flanks; dark stripes do not enter tail, breaking up into spots caudally; tail olive. **Habitat and Behaviour** Nothing known of its natural history. **Distribution** Java, Madura and Bali. Also Sumbawa and Lombok (Indonesia). **Status** Not Evaluated.

BOUTON SNAKE-EYED SKINK
Cryptoblepharus boutonii NOT ILLUSTRATED
Measurement SVL 43mm **Identification** Body slender; snout acute; orbit surrounded by granular scales; lower eyelid coalesced with rudimentary upper eyelid; supraoculars 5; supraciliaries 5–6; Supralabials V–VI subocular; auricular opening small and rounded, half orbit diameter; supranasals absent; paired nuchals; tail longer than head-body length; midbody scale rows 26 (rarely 24 or 28). **Coloration** Dorsum brown or olive-brown with indistinct pale lateral stripes. **Subspecies** Eleven subspecies are recognized, of which one (*C. boutonii*) occurs in the region. **Habitat and Behaviour** Inhabits sand dunes at river mouths on continental islands, as well as beaches of offshore islands. Diurnal and terrestrial. Diet and reproductive habits unstudied. **Distribution** Southern Java and Bali. Range includes the Indo-Pacific, from the coast of Africa to the South Pacific; there are also isolated records from the west coast of South America (several subspecies are described within this range: *C. b. ahli*, *C. b. ater*, *C. b. boutonii*, *C. b. cognatus*, *C. b. degrijsi*, *C. b. mayottensis*, *C. b. nigropunctatus*, *C. b. quinquetaeniatus*, *C. b. schlegelianus* and *C. b. voeltzkowi*). **Status** Not Evaluated.

BEACH SNAKE-EYED SKINK
Cryptoblepharus cursor PLATE 29
Measurement SVL 43mm **Identification** Body slender; snout acute; orbit surrounded by granular scales; lower eyelid coalesced with rudimentary upper eyelid; supraoculars 5; supraciliaries 5–6; Supralabial V or VI subocular; auricular opening small and rounded, half orbit diameter; supranasals absent; paired nuchals; tail longer than head-body length; midbody scale rows 26. **Coloration** Dorsum light greenish-olive with 2 narrow dark lines that do not cover vertebral scale rows (or that contact outermost edges of such scales), the lines meeting at tail-base and subsequently disappearing; silvery-white lateral stripe, edged with white-spotted dark stripe, from above orbit to tail-base; flanks with dark spots; tail light olive-green, brownish-green apically; venter silvery-white. **Habitat and Behaviour** Inhabits sandy substratum on the lowlands of small islands. Diurnal and terrestrial. Diet and reproductive habits unstudied. **Distribution** Bali. Also Lombok (Indonesia). **Status** Not Evaluated.

BLUE-TAILED SNAKE-EYED SKINK
Cryptoblepharus renschi PLATE 29
Measurement SVL 50mm **Identification** Body slender, depressed; snout acute; lower eyelid coalesced with rudimentary upper eyelid; tail longer than head-body length; supraciliaries 4–5; lower eyelid transparent; Supralabial IV in subocular position; limbs reduced; midbody scale rows 21–24; lamellae under Toe IV 18–26. **Coloration** Dorsum black to brownish-black with 5 bluish-white or yellow bands; median stripe commences from snout-tip; paravertebral bands start from posterior orbit; venter bluish-white. **Habitat and Behaviour** Inhabits small islands within savannah habitats close to the sea, at elevations under 100m asl. Diurnal and arboreal, associated with trees and found under tree bark; terrestrial activity also known. Diet consists of spiders, termites and beetles. Oviparous; clutch size unknown. Hatchlings 17mm. **Distribution** Bali. Also Sumba, Padar, Komodo and Pulau Longo (Indonesia). **Status** Not Evaluated.

GREY TREE SKINK
Dasia grisea PLATE 29
Measurement SVL 130mm **Identification** Body slender; snout elongated; pterygoid teeth present; 3 strong keels on dorsal scales; prefrontals in broad contact; supranasals in broad contact; paravertebrals not widened; ventrals 57–69; midbody scale rows 26–30; lamellae under Toe IV 16–19. **Coloration** Dorsum light or dark brown with 8–14 narrow dark rings; dark-banded tail; banded pattern most distinct in juveniles, turning obscure with growth; venter bright green. **Habitat and Behaviour** Inhabits lowland dipterocarp forests. Found 2–5m up on tree trunks. Diet consists of ants, termites, beetles and snails, as well as fruits. Oviparous; clutches comprise 2–6 eggs, measuring 22.5–24 x 14.5–15.5mm. Hatchlings 46–52mm. **Distribution** Peninsular Malaysia, Singapore, Sumatra and Borneo. Also the Philippines. **Status** Not Evaluated.

OLIVE TREE SKINK
Dasia olivacea PLATE 29
Measurement SVL 115mm **Identification** Body robust; snout elongated; pterygoid teeth present; scales under tail not enlarged; auricular opening small; enlarged paired nuchals; midbody scale rows 28–30; vertebrals 41–46; 2 large heel scales in males; lamellae under Toe IV 17–22. **Coloration** Dorsum yellowish-olive or greenish-brown, sometimes black-spotted; venter unpatterned cream; juveniles golden-yellow with 13–16 transverse dark bands, each 3 scales wide, broader than the orange-yellow bands. **Habitat and Behaviour** Inhabits open forests at elevations of up to 1,200m. Diurnal and arboreal, associated with large trees, especially at the edges of clearings, sheltering under peeling bark. Diet consists of bees, beetles, ants, flies and other arthropods. Oviparous; clutches comprise 6–14 eggs, measuring 18–19.5 x 10–12mm. More than one clutch may be produced annually. Nests made in tree-holes or within arboreal ferns. Hatchlings 32–38mm. **Distribution** Myanmar, Thailand, Cambodia, Vietnam, Peninsular Malaysia, Singapore, Sumatra, Pulau Siberut, Pulau Berhala, Pulau Natuna and Borneo. Also the Nicobar Islands (India). **Status** Not Evaluated.

HALF-BANDED TREE SKINK
Dasia semicincta PLATE 29
Measurement SVL 130mm **Identification** Body robust; snout elongated; pterygoid teeth present; head large; postorbital bone absent; vertebrals not enlarged, numbering 59–63; 3 weak keels on dorsal scales; prefrontals typically separated, but occasionally in contact; supranasals in broad contact; 2 large heel scales in males; midbody scale rows 28–30; lamellae under Toe IV 17–22. **Coloration** Dorsum grey-black; adults with 5–8 broad dark rings between axilla and groin; dark bands broader than intervening light areas in juveniles, which have a glossy coal-black body with orange-yellow bars from snout to tail-tip; limbs, including digits, barred with yellow; transverse bands on head; dark caudal rings complete ventrally. **Habitat and Behaviour** Inhabits submontane forests at elevations of more than 1,000m asl. Diurnal and arboreal. Diet and reproductive habits unstudied. **Distribution** Borneo (Kelabit Highlands, Sarawak State). Also Mindanao (the Philippines). **Status** Not Evaluated.

MIRIAM'S LIMBLESS SKINK
Davewakeum miriamae NOT ILLUSTRATED
Measurement SVL 114mm **Identification** Body slender, limbless; head not distinct from neck; upper eyelid absent; nostril in single nasal; loreals 2; tympanum absent, supranasals absent; single frontonasal; prefrontals, frontoparietals and parietals paired; interparietal present; supralabials 6; Supralabial I largest; Supralabial IV contacts orbit; infralabials 5; midbody scale rows 20–22, dorsals smooth; ventrals 93–113; enlarged preanal scales 5; subcaudals 79–101. **Coloration** Dorsum tan, typically with dark brown spots on each scale that form series of lines; venter pale with indistinct dark lines. **Habitat and Behaviour** Inhabits dry evergreen forests, including those partially cleared. Nocturnal, active on soil surface. Daytime concealment sites include places under rocks and in tree buttresses, 1–15cm underground. Diet and reproductive habits unstudied. **Distribution** Endemic to central Thailand (region between Khorat Plateau and Central Valley, Nakhon Ratchasima Province). **Status** Not Evaluated.

MANGROVE SKINK
Emoia atrocostata PLATE 29
Measurement SVL 97.5mm **Identification** Body elongated, slender; eye large; nostrils not large; limbs and tail well developed; snout short, tapering; prefrontals narrowly in contact or separate; interparietal distinct, narrow; auricular lobules present; supralabials 6–8; infralabials 6–7; lamellae under Toe IV 30–42. **Coloration** Dorsum greyish-olive flecked with dark brownish-grey; dark lateral stripe to tail; venter bluish-grey to cream with dark pigmentation on throat. **Habitat and Behaviour** Associated with coastal regions, including sandy and rocky beaches, sheltering inside hollow tree trunks during high tide. Diurnal. Diet consists of small crabs, termites, fish and other lizards. Oviparous; clutches comprise 1–3 eggs, measuring 8 x 20mm, which are deposited inside piles of driftwood and tree-holes. **Distribution** Vietnam (Quan dao Hoàng Sa, Da Nang Province), Peninsular Malaysia, Singapore, Sumatra, the Mentawai and Natuna Archipelagos, Pulau Berhala, Borneo and Java. Also New Guinea, the Solomon Islands and northern Australia. **Status** Not Evaluated.

COMMON BLUE-TAILED SKINK
Emoia caeruleocauda PLATE 29
Measurement SVL 65mm **Identification** Body robust; snout short, tapering; frontonasal broader than long, in contact with rostral and with frontal; interparietal distinct; 1 pair of enlarged nuchals; tympanum smaller than eye; midbody scale rows 16,

smooth. **Coloration** Dorsum with dark vertebral stripe in males, yellow in females; tail venter blue. **Habitat and Behaviour** Inhabits sandy beaches of offshore islands. Diurnal and terrestrial. Diet unstudied. Oviparous; clutches comprise 2 eggs. Hatchlings 22.4mm. **Distribution** Borneo (islands east of Sabah State). Also Sulawesi (Indonesia), New Guinea, Fiji and the Solomon Islands. **Status** Not Evaluated.

BLUE-TAILED SKINK
Emoia cyanura PLATE 29

Measurement SVL 61mm **Identification** Body slender; snout short, tapering; prefrontals separated; interparietal fused with frontoparietals; paravertebral scale rows 52–64; midbody scale rows 25–36. **Coloration** Dorsum brownish-black or dark brown with distinct pale vertebral and dorsolateral stripes, ground colour fading posteriorly to blue or brown; pale midlateral stripe; venter bluish-cream to cream, or greyish-tan; tail green, blue or tan without black spots or bars. **Habitat and Behaviour** Inhabits coastal scrub, gardens, plantations and secondary growth. Diurnal and generally terrestrial, active on leaf litter, palm trash and plantation rubble, and can climb low vegetation. Diet includes insects, spiders, woodlice and earthworms. Oviparous; eggs measure 10–13 x 6–9mm. Incubation period 40–51 days. Hatchlings 22mm. **Distribution** Borneo (Pulau Derawan, off east coast). Also Sulawesi, Seram, the Sula Archipelago and Halmahera (Indonesia), Oceania, Hawaii (USA), Clipperton Island, Easter Island, Polynesia, Melanesia, and the Bismarck Archipelago; introduced to the west coast of Central and South America. **Status** Not Evaluated.

LAO BAO FOREST SKINK
Emoia laobaoensis NOT ILLUSTRATED

Measurement SVL 74mm **Identification** Body elongated, slender; snout short, tapering; eye small; nostrils large, located posteriorly on nasal; supranasals large; supplementary shield present above postnasal; limbs and tail well developed; supralabials 6–7; Supralabials V or VI contact orbit; dorsals smooth; midbody scale rows 38–42, smooth; lamellae under Toe IV 30–32. **Coloration** Dorsum mid-brown with scattered black spots; flanks with dark brown bands, 3–4 scale rows wide; gular region and subcaudals dark grey; venter mid-brown. **Habitat and Behaviour** Inhabits primary forests at an elevation of 250m asl, located ca 75km from the coast. Diet and reproductive habits unstudied. **Distribution** Endemic to Vietnam (Lao Bao). **Status** Not Evaluated.

INDO-PACIFIC MOLE SKINK
Eugongylus rufescens PLATE 29

Measurement SVL 143mm **Identification** Body robust, elongated; lower eyelid scaly; dorsals smooth in adults, weakly keeled in juveniles; limbs short; supranasals separate; frontal not broader than supraocular region; supraoculars 5; Supralabial V subocular; parietals in contact; auricular opening oval, small; auricular lobules present; paired enlarged nuchals; tail nearly as thick as body; preanal scales not enlarged; midbody scale rows 28–30; lamellae under Toe IV 16–19, smooth. **Coloration** Dorsum of adults glossy brown, unpatterned or with transverse dark bars; supralabials and infralabials with dark bars on sutures; venter yellowish-cream; juveniles dark brown with numerous narrow, transverse cream bars; V-shaped mark on gular region; tail yellowish-brown. **Habitat and Behaviour** Inhabits primary forests from the lowlands (ca 300m asl) to submontane limits. Nothing known of its natural history. **Distribution** Sumatra. Also islands of the Maluku and New Guinea groups. **Status** Not Evaluated.

CHAPA GROUND SKINK
Eutropis chapaensis NOT ILLUSTRATED

Measurement SVL 72mm **Identification** Body robust; lower eyelid with clear window; supralabials 7; Supralabials V–VI subocular; supranasals absent; auricular opening large, rounded, lacking lobules; nuchals absent; midbody scale rows 34–36; 2 enlarged preanal scales; dorsals larger than laterals; lamellae under Toe IV 18. **Coloration** Dorsum bronze; brown stripe along flanks, dark dorsally, the two colours separated by narrow line commencing from nostrils and continuing to tail; row of small white spots, edged with black, on lower flanks; labials cream with dark sutures; venter cream. **Habitat and Behaviour** Inhabits submontane forests at elevations of ca 1,500m asl. Nocturnal and arboreal. Diet and reproductive habits unstudied. **Distribution** Endemic to northern Vietnam (Sa Pá, Lào Cai Province). **Status** Not Evaluated.

DAREVSKI'S GROUND SKINK
Eutropis darevskii NOT ILLUSTRATED

Measurement SVL 50.5mm **Identification** Body robust; postnasal absent; additional scute between frontal and frontonasal; frontonasal broader than long; prefrontals separate; supranasals separate; dorsals and laterals with 5 sharp keels; midbody scale rows 30; lamellae under Toe IV 12. **Coloration** Dorsum olive-brown with irregular black spots; flanks with large white spots edged with black, forming indistinct light stripes from postocular region to axilla, and from auricular opening to forelimbs; limbs olive-brown with numerous small white spots; venter light with indistinct dark grey dots; throat and chin bright red. **Habitat and Behaviour** Only individual known was found among rock and brushwood piles in a cornfield situated on a mountain slope. Diet and reproductive habits unstudied. **Distribution** Endemic to northern Vietnam (Cao Pha, Son La Province). **Status** Not Evaluated.

STRIPED GROUND SKINK
Eutropis dissimilis PLATE 30

Measurement SVL 150mm **Identification** Body robust; head small; snout short; lower eyelid with clear window; auricular opening oval with 3–4 lobules; scales with 3 keels; ventral scales smooth; longitudinal scale rows at midbody 32–38; lamellae on Toe IV

12–17. **Coloration** Dorsum dark brown or pale olive with yellow stripes edged with black dots, which may be joined to form lines; sides of body with small white spots; white stripe along face, below eye; venter greenish-white. **Habitat and Behaviour** Inhabits damp grasslands. Diurnal and terrestrial, known to occupy holes of mole rats along riverbanks. Diet consists of insects, spiders and frogs. Oviparous; clutches comprise 3–8 eggs. **Distribution** Myanmar. Also Pakistan, northern India, Nepal, Bangladesh and Afghanistan. **Status** Not Evaluated.

DORIA'S GROUND SKINK
Eutropis doriae NOT ILLUSTRATED
Measurement SVL 98mm **Identification** Body robust; snout short, obtuse; lower eyelid with clear window; supranasals in contact; supraoculars 4; supraciliaries 7; parietals separated by interparietal; paired enlarged nuchals; Supralabials IV–V subocular; auricular opening large, slightly smaller than orbit, with 2–3 lobules; dorsals with 7–9 sharp keels; midbody scale rows 34. **Coloration** Dorsum pale olive with 4 longitudinal series of small dark spots; broad blackish-brown lateral stripe, starting from nostril and reaching flanks; venter cream. **Habitat and Behaviour** Inhabits the mid-hills. Nothing else known of its natural history. **Distribution** Endemic to south-eastern Myanmar (Minhla, Sagaing District). **Status** Not Evaluated.

PHILIPPINE GROUND SKINK
Eutropis indeprensa NOT ILLUSTRATED
Measurement SVL 67mm **Identification** Body robust; tail tapering; tympanum reduced; nostril in nasal; supranasals separated; frontonasal long and narrow, forming a suture with rostral and frontal; prefrontals in contact or separated; parietals typically divided by large interparietal; supraoculars 4; supralabials 6–7; infralabials 6; dorsal scales with 3 keels; midbody scale rows 27–34; vertebrals 40–48; lamellae under Toe IV 18–24. **Coloration** Dorsum brownish-tan to light brown, typically with 2–4 longitudinal rows of small dark brown blotches; flanks dark brown; narrow light stripe from labials to insertion of forelimbs; venter greyish-blue. **Habitat and Behaviour** Inhabits lowland forests and the mid-hills, especially in the vicinity of human settlements, at elevations of up to 1,200m asl. Diurnal and terrestrial. Diet unstudied. Oviparous. Hatchlings 22.5–24.7mm. **Distribution** Northern Borneo (Sabah State). Also islands in the central and southern Philippines. **Status** Not Evaluated.

LONG-TAILED GROUND SKINK
Eutropis longicaudata PLATE 30
Measurement SVL 140mm **Identification** Body slender; head elongated; lower eyelid scaly; frontoparietals in broad contact; supralabials 7; Supralabial IV subocular; paired enlarged nuchals; auricular opening small, with or without lobules anteriorly; middorsals slightly keeled; tail nearly three times as long as body; preanal scales enlarged; midbody scale rows 28. **Coloration** Dorsum mid-brown or reddish-brown, sometimes with 7 broken longitudinal black stripes; dark grey band on flanks from eye to beyond inguinal region; supralabials and infralabials yellow barred with black; venter greenish-yellow. **Habitat and Behaviour** Inhabits the lowlands around open forests, as well as close to human habitations, at elevations of up to ca 500m asl. Diurnal and terrestrial. Diet consists of large insects such as crickets and grasshoppers, and also earthworms. Oviparous; clutches comprise up to 16 hardshelled eggs, measuring 16mm, which are laid in holes on walls and on the ground. Incubation period ca 45 days. **Distribution** Thailand, Laos, Cambodia, Vietnam and Peninsular Malaysia. Also eastern China. **Status** Not Evaluated.

LARGE-EYED GROUND SKINK
Eutropis macrophthalma PLATE 30
Measurement SVL 108mm **Identification** Body moderately robust; head slightly distinct from neck; snout acute; eye large; lower eyelid scaly; auricular opening large with 2–3 pointed lobules on anterior edge; supranasals separated; postnasal present; supralabials 7; infralabials 6; prefrontals in broad contact; paired, enlarged keeled nuchals; dorsals with 3 keels; preanals slightly enlarged; subcaudals not enlarged; midbody scale rows 27; lamellae under Toe IV 19–22. **Coloration** Dorsum iridescent greenish-brown; nuchal region brown; 2 dorsal rows of 5 black-spotted scales; black lateral stripe, 3 scales wide, commencing as broad postocular stripe, narrowing at level of axilla, and continuing to about level of midbody; flanks grey to bluish-grey; venter white. **Habitat and Behaviour** Nothing known of its natural history. **Distribution** Endemic to Java. **Status** Not Evaluated.

LITTLE GROUND SKINK
Eutropis macularia PLATE 30
Measurement SVL 75mm **Identification** Body slender; head distinct; snout short; limbs well developed; lower eyelid scaly; auricular opening small with several indistinct lobules anteriorly; dorsal scales with 5–9 keels; midbody scale rows 28–34; lamellae under Toe IV 12–17. **Coloration** Dorsum bronze-brown, unicoloured or spotted; flanks darker, spotted with white, especially in juveniles and in males, brown or grey in females; venter unpatterned cream; breeding males have bright red lips and flanks. **Habitat and Behaviour** Inhabits deciduous and evergreen forests, and also habitats modified by humans, including plantations and secondary forests. Diet includes beetles and grasshoppers. Oviparous; clutches comprise 1–4 eggs, measuring 13–15 x 6.9–8.1mm, which are deposited under dead leaves or logs between June and September. More than a single clutch is produced. **Distribution** Myanmar, Thailand, Laos, Cambodia and Vietnam. Also Sri Lanka, Pakistan, India, Nepal, Bangladesh and Bhutan. **Status** Not Evaluated.

COMMON SUN SKINK
Eutropis multifasciata PLATE 30
Measurement SVL 137mm **Identification** Body robust; head distinct; snout short; lower eyelid scaly; paired enlarged nuchals; auricular opening small with small pointed lobules; dorsal scales with 3 (rarely 5) keels; midbody scale rows 29–35; vertebrals 42–48; lamellae under Toe IV 17–23. **Coloration** Dorsum bronze-brown, usually with yellow or red stripe along flanks; series of white spots or streaks along flanks; breeding males with bright orange or reddish-orange flanks; pale dorsolateral line; venter cream. **Subspecies** Two subspecies are recognized. *E. m. multifasciata*, Supralabial I and frontal not in contact; no reddish-brown streak on top of snout and reddish-orange flank stripes in males. *E. m. balinensis*, Supralabial I and frontal in contact; reddish-brown streak on top of snout and yellow flank stripes in males. **Habitat and Behaviour** Common in forest edges and around human settlements at elevations of up to 1,800m asl. Diurnal and terrestrial. Insectivorous, with diet also including centipedes. Ovoviviparous; clutches comprise 1–10 neonates, measuring 33.1–43mm. **Distribution** Myanmar, Thailand, Laos, Cambodia, Vietnam, Peninsular Malaysia, Singapore, Sumatra, the Anamba, Natuna, Riau and Mentawai Archipelagos, Borneo, Java (*E. m. multifasciata*) and Bali (*E. m. balinensis*). Also north-eastern India, the Nicobar Islands, southern China (Yunnan Province), Taiwan, the Lesser Sundas, Maluku, Sulawesi, Halmahera, the Sulu and Togian Archipelagos, and the Philippines (*E. m. multifasciata*); introduced into New Guinea, Australia and the southern USA. **Status** Not Evaluated.

NINE-KEELED GROUND SKINK
Eutropis novemcarinata NOT ILLUSTRATED
Measurement SVL 98mm **Identification** Body robust; head indistinct from neck; snout short; lower eyelid with clear window; supralabials 7; Supralabial V subocular; infralabials 6; paired enlarged nuchals; auricular opening oval or semicircular with 2–3 lobules; dorsal scales with 7–11 sharp keels; midbody scale rows 32–34; lamellae under Toe IV 17. **Coloration** Dorsum light brown with small dark brown spots in a longitudinal series; broad dark brown stripe along sides of head and body, edged with white on upper edge; venter pale green. **Habitat and Behaviour** Inhabits lowland forests, rarely reaching the mid-hills, and common in relatively open areas. Terrestrial and diurnal. Diet and reproductive habits unstudied. **Distribution** Myanmar (Sagaing Division and Chin State), southern Thailand (Nakhon Si Thammarat, Trang, Pattani, Narathiwat and Yala Provinces) and northern Peninsular Malaysia. Also north-eastern India. **Status** Not Evaluated.

FOUR-KEELED GROUND SKINK
Eutropis quadricarinata PLATE 30
Measurement SVL 50mm **Identification** Body robust; head indistinct from neck; snout short; lower eyelid scaly; postnasal present; auricular opening rounded, lacking lobules; paired enlarged nuchals; dorsal scales with 4 distinct keels; midbody scale rows 26–28. **Coloration** Dorsum olive-brown, unpatterned or with small black spots arranged longitudinally; flanks sometimes with dark lines; cream streak, edged with black, from below eye to auricular opening; venter and supralabials cream. **Habitat and Behaviour** Associated with open forests and clearings. Diet and reproductive habits unstudied. **Distribution** Myanmar (Kachin State, Sagaing Division and Mandalay Division). Also north-eastern India (Assam State). **Status** Not Evaluated.

FIVE-KEELED GROUND SKINK
Eutropis quinquecarinata NOT ILLUSTRATED
Measurement SVL 63mm **Identification** Body robust; head indistinct from neck; snout short; lower eyelid scaly; auricular lobules present; parietals in contact; postnasal present; supraoculars 4; dorsal scales with 2–3 keels; midbody scale rows 28. **Coloration** Dorsum dark brown, unpatterned or with 5–7 narrow cream stripes; supralabials and infralabials greenish-white with white spots; flanks brown with white spots anteriorly; gular scales with brown edges; venter greenish-white. **Habitat and Behaviour** Inhabits the lowlands and submontane forests at elevations of up to 1,200m asl. Diet and reproductive habits unstudied. **Distribution** Sumatra (Medan, Sumatera Utara, Sungeisalak, Riau Province and Gunung Raja, Sumatera Barat Province), Pulau Simeulue and eastern Java. **Status** Not Evaluated.

BLACK-BANDED GROUND SKINK
Eutropis rudis PLATE 30
Measurement SVL 120mm **Identification** Body robust; head indistinct from neck; snout short; forehead scales at posterior rugose; dorsal scales with 3 strong keels; middorsal scale rows 28–30. **Coloration** Dorsum olive-brown with light-edged dark brown line along sides of head and body; sides white-spotted; throat of adult males crimson, that of females unpatterned cream; venter greenish-white, throat sometimes dark-spotted. **Habitat and Behaviour** Associated with forested habitats. Diurnal and terrestrial, feeding on insects. Oviparous; clutches comprise 2–4 eggs. **Distribution** Borneo and Sumatra. Also Sulawesi (Indonesia), the Nicobar Islands (India) and the Sulu Archipelago (the Philippines). **Status** Not Evaluated.

RED-THROATED GROUND SKINK
Eutropis rugifera PLATE 30
Measurement SVL 65mm **Identification** Body robust; head indistinct from neck; snout short; lower eyelid scaly; paired enlarged nuchals; Supralabial V or VI subocular; auricular opening small with small pointed lobules; dorsal scales with 5 (rarely 7) distinct keels; midbody scale rows 24–28; lamellae under Toe IV 18–26. **Coloration** Dorsum dark brown or bronze-brown, lighter on flanks, with 5–7 longitudinal greenish-cream stripes that are sometimes broken up to form spots, giving it a grizzled

appearance; narrow yellowish-grey postocular stripe extends to axilla and to tail; venter cream or greenish-cream; throat dark spotted, turning bright red in breeding males. **Habitat and Behaviour** Associated with forests in the mid-hills. Diurnal and terrestrial, feeding on arthropods. Reproductive habits unstudied. **Distribution** Southern Thailand (Hala Bala Wildlife Sanctuary, Narathiwat Province and Bannang Sata, Yala Province), Peninsular Malaysia, Singapore, Sumatra, the Mentawai Archipelago, Borneo, Java and Bali. Also India (the Nicobar Islands). **Status** Not Evaluated.

EEL-LIKE LIMBLESS SKINK
Isopachys anguinoides NOT ILLUSTRATED
Measurement SVL 95mm **Identification** Body elongated, limbless; snout short; frontals and parietals broad; prefrontals small and widely separated; frontal and frontonasal subequal; nasals separated; paired nuchals; supralabials 5–6; Supralabial III contacts orbit; infralabials 5; tail thick; midbody scale rows 21–26; preanal scales enlarged. **Coloration** Dorsum pale greyish-brown with paired dark brown streaks on vertebral and dorsolateral regions; venter with brown lines between series of scales. **Habitat and Behaviour** Inhabits forested locations. Subfossorial; one was found under a log in dry sandy soil. Diet and reproductive habits unstudied. **Distribution** Endemic to Thailand (Nong Kae, Bangtaphan and Hua Hin, Prachuab Khiri Khan Province, Bang La Moung, Chon Buri Province and Koh Tao, Surat Thani Province). **Status** Not Evaluated.

NORTHERN LIMBLESS SKINK
Isopachys borealis NOT ILLUSTRATED
Measurement SVL 167mm **Identification** Body elongated, limbless; tail-tip rounded; parietal eye-spot absent; prefrontal absent; nasals separated; supraciliaries 4, Supraciliary I in contact with frontal and all projecting medially; infralabials 3–4; maxillary teeth 10; supraocular scales 1–2; pretemporal single; inner preanal scale overlaps outer; dentary teeth 13/12; midbody scale rows 20–22. **Coloration** Dorsum cream with 44 pairs of dark brown spots, which sometimes coalesce on paravertebral rows, forming a narrow band on body and tail; head dark brown with small pale brown patches; first pair of dorsal spots fused with dark colours of forehead, forming hourglass-shaped mark; snout yellowish-brown; venter dark brown. **Habitat and Behaviour** Known from tuber and fruit plantations at elevations of 80–300m asl. Subfossorial, associated with topsoil. Diet and reproductive habits unstudied. **Distribution** Myanmar (Mawlamyine, Mon State) and north-eastern, central and western Thailand (Dong Nai and Tap-Tan, Uthai-Thani Province, Mae-Sot, Tak Province, Taling Sung, Kamphaeng Phet Province, Khon-Kaen, Khon Kaen Province and Nam Len, Phetchabun Province). **Status** Not Evaluated.

COUNT GYLDENSTOLPE'S LIMBLESS SKINK
Isopachys gyldenstolpei PLATE 31
Measurement SVL 220mm **Identification** Body elongated, limbless; rostral large; nasals in contact; frontal and frontonasal fused; supralabials 4; Supralabial II contacts orbit; pectoral girdle absent; pelvic girdle reduced; midbody scale rows 24–28. **Coloration** Dorsum dark grey-brown; yellow stripes on paravertebral stripe 3–4 scales wide; narrow vertebral stripe extends to tail-base; dark dorsal stripe with serrated edges caudally. **Habitat and Behaviour** Inhabits the mid-hills such as mountain foothills. Subfossorial, associated with dry sandy soil. Diet probably consists of earthworms. Reproductive habits unstudied. **Distribution** Endemic to Thailand (nr Kanchanaburi, Kanchanaburi Province, Nong Kae, Prachuab Khiri Khan Province, and Hua Hin and Koh Lak Paa, Prachuap Khiri Khan Province). **Status** Not Evaluated.

ROULE'S LIMBLESS SKINK
Isopachys roulei NOT ILLUSTRATED
Measurement SVL 106mm **Identification** Body elongated, limbless; snout flattened anteriorly; nasals in broad contact; frontonasal larger than frontal; supralabials 4–5; Supralabial III contacts orbit; infralabials 4; midbody scale rows 18. **Coloration** Dorsum pinkish-fawn; forehead darker with T-shaped lighter frontal and frontonasal region; 2 broad dark stripes from parietals, narrowing on dorsum and separated from each other by 2 entire scales and 2 half scales; on tail, dorsal line less distinct and fragmented; venter grey. **Habitat and Behaviour** Inhabits the lowlands and has been found in coconut groves. Subfossorial, associated with dry sandy soil underneath fallen coconut trunks and stumps. Diet and reproductive habits unstudied. **Distribution** Endemic to south-eastern Thailand (Ang Hin, Bang Lamung and Bang Saen, Chon Buri Province). **Status** Not Evaluated.

WHITE-SPOTTED TREE SKINK
Lamprolepis leucosticta NOT ILLUSTRATED
Measurement SVL 74mm **Identification** Body slender; snout obtusely pointed; supraoculars 4; supraciliaries 7; Supralabial IV subocular; paired enlarged nuchal; midbody scale rows 27–28; lamellae under Toe IV 17–19. **Coloration** Dorsum olive-brown with black flecks; pale vertebral stripe; tympanum and flanks with flecks; pale areas within black spots, especially on neck and flanks; forehead scales and dorsals with black edges; venter olive; palms, soles and lamellae olive-brown. **Habitat and Behaviour** Inhabits submontane forests at an elevation of 1,400m asl. Diet and reproductive habits unstudied. **Distribution** Endemic to western Java (Cibodas). **Status** Not Evaluated.

NIEUWENHUIS'S TREE SKINK
Lamprolepis nieuwenhuisii NOT ILLUSTRATED
Measurement SVL 72mm **Identification** Body slender; snout obtusely pointed; lower eyelid scaly;

supranasals present, failing to contact each other; body scales smooth; midbody scale rows 24–26. **Coloration** Dorsum metallic brownish-green with numerous scattered dark spots; pale dorsolateral band; no dark linear pattern; forehead more brown than rest of dorsum; sutures of scales dark; tail-base with transverse dark bands; venter light blue or green. **Habitat and Behaviour** Inhabits lowland forests at elevations of up to 915m asl. Diurnal and arboreal, associated with trees at ca 2m above substrate. Diet and reproductive habits unstudied. **Distribution** Endemic to Borneo. **Status** Not Evaluated.

VYNER'S TREE SKINK

Lamprolepis vyneri PLATE 31

Measurement SVL 66mm **Identification** Body slender; snout obtusely pointed; lower eyelid scaly; supranasals present, failing to contact each other; midbody scale rows 21–22; dorsals smooth; tail equal to head-body length; lamellae under Toe IV 20. **Coloration** Dorsum with 4 longitudinal stripes that consist of a series of black dorsal scales with a central olive-grey area; flanks with black-edged brown scales; forehead olive-grey; scales edged with black; tail greyish-olive; venter pale green. **Habitat and Behaviour** Inhabits low hills within dipterocarp forests. Diurnal and arboreal, associated with trees. Diet and reproductive habits unstudied. **Distribution** Endemic to Borneo. **Status** Not Evaluated.

BUKIT LARUT LIMBLESS SKINK

Larutia larutensis PLATE 31

Measurement SVL 191mm **Identification** Body elongated, slender; supralabials 7; second pair of chin shields separated by 2 gular scales; first pair of chin shields contact only 1 infralabial; longitudinal scale rows at midbody 25–26; well-defined digits with 5 subdigital lamellae. **Coloration** Dorsum blue-black or brown; single light nuchal band; opaque rostral, first supralabial, mental and Infralabial I scales; pale marks on supraoculars, prefrontals, frontoparietals and interparietals absent; venter pale brown. **Habitat and Behaviour** Inhabits the mid-hills to submontane forests at elevations of 240–1,200m asl, concealing itself under logs and rocks. Diet and reproductive habits unstudied. **Distribution** Endemic to Peninsular Malaysia (Fraser's Hill, Pahang State and Bukit Fraser, Perak State). **Status** Not Evaluated.

ONE-FINGERED SKINK

Larutia miodactyla NOT ILLUSTRATED

Measurement SVL 151mm **Identification** Body elongated, slender; supralabials 5; Supralabial III or IV subocular; single toe; 2 claws on hands and 1 claw on feet; first pair of chin shields contacts only 1 infralabial; longitudinal scale rows at midbody 20–22. **Coloration** Dorsum chocolate-brown; flanks light brown with 3–4 longitudinal series of dark brown spots, corresponding with scale rows, and may appear as stripes; nuchal bands absent; rostral, nasals, first supralabial, mental and Infralabial I dark brown and shiny; pale marks on supraoculars, prefrontals, frontoparietals and interparietals absent; weak subcaudal mottling; venter light brown. **Habitat and Behaviour** Inhabits submontane and montane forests at elevations of more than 667m asl. Lives in burrows on forest floor. Diet unstudied. Oviparous; produces a single egg, measuring 18.3 x 6.2mm. **Distribution** Endemic to Peninsular Malaysia (Cameron Highlands, Bukit Ulu Kali and Genting Highlands, Pahang State). **Status** Not Evaluated.

GUNUNG PUEH LIMBLESS SKINK

Larutia puehensis NOT ILLUSTRATED

Measurement SVL 141mm **Identification** Body elongated, slender; head elongated, triangular; eyelid covered by small, plate-like scales; 2 kidney-shaped frontoparietals; right frontoparietal overlaps left, supraoculars 4; supralabials 5; infralabials 5; smooth, cycloid imbricate scales, midbody scale rows 23; ventrals large, rounded; limbs small, covered by small round imbricate scales; 2 claws on hands and feet; Digits III and IV clawed, lamellae absent beneath digits; subcaudal scales subequal to dorsal caudals. **Coloration** Dorsum orange-brown with scattered dark brown flecks covering supraoculars and sides of head; nuchal region unpatterned dark brown with radiating light linear pattern; nuchal bands absent; pale marks on supraoculars, prefrontals, frontoparietals and interparietals absent; rostral, nasals, first supralabial, mental and Infralabial I dark brown and shiny; linearly arranged dark spots forming stripes extend from nuchal region to tail-tip; venter orange-brown. **Habitat and Behaviour** Inhabits the mid-hills at an elevation of 300m asl on Gunung Berumput, a sandstone massif. Likely to be fossorial, inhabiting leaf litter and feeding on small invertebrates such as arthropods and worms. Reproductive habits unstudied. **Distribution** Endemic to north-western Borneo (Gunung Beremput, Sarawak State). **Status** Not Evaluated.

SERIBUAT LIMBLESS SKINK

Larutia seribuatensis PLATE 31

Measurement SVL 115mm **Identification** Body elongated, slender; head indistinct from neck, elongated; snout slightly pointed; supralabials 5–6; infralabials 5; limbs small, covered by small rounded imbricate scales; forelimbs and hind limbs with 2 clawed digits; well-defined digits with 4 subdigital lamellae; longitudinal scale rows at midbody 24–26. **Coloration** Dorsum chocolate-brown; paired light-coloured dorsolateral stripes from axilla to tail-tip; 2 light spots at base of frontoparietal; opaque rostral, first supralabial, mental and Infralabial I scales; light supraocular markings; light markings on rostrum; 3 yellow nuchal bands, first contacts posterior edge of eye; small pale marks on supraoculars, prefrontals, frontoparietals and interparietals; venter cream; flanks beige with brown mottling. **Habitat and Behaviour** Inhabits coastal forests, including secondary vegetation, at sea level. Fossorial, concealing itself in leaf litter and under rocks. Diet and reproductive habits unstudied. **Distribution** Endemic to west coast of Peninsular Malaysia (Pulau Tulai, Pahang State). **Status** Not Evaluated.

SUMATRAN LIMBLESS SKINK
Larutia sumatrensis PLATE 31
Measurement SVL 176mm **Identification** Body elongated, slender; head indistinct from neck; 2 claws on hands and feet; longitudinal scale rows at midbody 22–23. **Coloration** Dorsum brownish-grey; rostral, nasals, first supralabial, mental and Infralabial I dark brown and shiny; pale marks on supraoculars, prefrontals, frontoparietals and interparietals absent; nuchal bands absent; weak subcaudal mottling; venter pale brown. **Habitat and Behaviour** Inhabits submontane forests. Diet and reproductive habits unstudied. **Distribution** Endemic to Sumatra (Bukit Tinggi, Lolo and Agam). **Status** Not Evaluated.

THREE-LINED LIMBLESS SKINK
Larutia trifasciata PLATE 31
Measurement SVL 250mm **Identification** Body elongated, slender; second pair of chin shields separated by 2 gulars; well-defined digits with 4 subdigital lamellae; longitudinal scale rows at midbody 29–30. **Coloration** Dorsum brown; opaque rostral, first supralabial, mental and Infralabial I scales; 3 light nuchal bands, first complete with smooth borders, contacting posterior parietals, and second and third close behind, incomplete; small pale marks on supraoculars, prefrontals, frontoparietals and interparietals; weak subcaudal mottling; venter pale brown. **Habitat and Behaviour** Inhabits moss-covered verges of vegetable fields in and on banks along road cuts, at elevations of 1,200–1,600m asl. Found beneath logs and boulders. Diet and reproductive habits unstudied. **Distribution** Endemic to Peninsular Malaysia (Cameron Highlands, Pahang State). **Status** Not Evaluated.

OSELLA'S LIMBLESS SKINK
Leptoseps osellai NOT ILLUSTRATED
Measurement SVL 41mm **Identification** Body elongated, slender; head indistinct from neck; eye large; lower eyelid scaly; supranasals and prefrontals absent; loreal single; tympanum concealed; limbs small; forelimbs and hind limbs with 4 digits; dorsals smooth; midbody scale rows 18, smooth; tail terminates in sharply pointed spiny scale. **Coloration** Dorsum brown with longitudinal dark brown stripes, especially on middorsum, where median stripes are broader than outer stripes; pale marks on supraoculars, prefrontals, frontoparietals and interparietals absent; venter cream; precloacal region and undersurface of tail dark brown. **Habitat and Behaviour** Inhabits forested hills, including secondary forests that contain high grass. One was found under a dead tree. Diet and reproductive habits unstudied. **Distribution** Endemic to northern and north-eastern Thailand (Mae Kwuang, Chiang Mai Province and Nam Nao National Park, Petchabun Province). **Status** Not Evaluated.

POILANE'S LIMBLESS SKINK
Leptoseps poilani NOT ILLUSTRATED
Measurement SVL 43mm **Identification** Body slender, elongated; snout obtuse; auricular opening covered with scales; supranasals and prefrontals absent; frontoparietal double; large interparietal; preoculars 2; supraoculars 5; paired enlarged nuchals join smaller pair of nuchals; supralabials 7; Supralabials IV–VI subocular; midbody scale rows 18; dorsals subequal to ventrals; limbs small with distinct joints and reduced digits comprising 5 fingers and 4 toes; lamellae under Toe IV 3. **Coloration** Dorsum unpatterned brown with 2 brown stripes, commencing from back of head to tail-tip; flanks with 3–4 longitudinal black stripes; less distinct stripes on venter; stripes on subcaudals distinct, turning black. **Habitat and Behaviour** Inhabits forests in the mid-hills at an elevation of 800m asl. Probably subfossorial. Diet and reproductive habits unstudied. **Distribution** Endemic to Vietnam (Dong-Tam-Ve and Lang Tam, Quang Tri Province). **Status** Not Evaluated.

FOUR-FINGERED SKINK
Leptoseps tetradactylus NOT ILLUSTRATED
Measurement SVL 35mm **Identification** Body slender; supranasals absent; parietal eye absent; lower eyelid scaly; external auricular opening indicated by depression; loreal single; supralabials 6; Supralabials III–V subocular; infralabials 7; supraoculars 4; enlarged nuchals 6 in double transverse rows; limbs well developed with distinct joints; 4 well-developed fingers and 5 toes; tail over twice head-body length, tapering to a point; median subcaudals enlarged; enlarged paired preanal scales; midbody scale rows 20; lamellae under Toe IV 10. **Coloration** Dorsum mid-brown, iridescent; longitudinal dark stripes on centres of all scales on dorsum and flanks, starting from nuchals; dark postocular stripe extends to flanks; dorsum of tail with broad stripes; black spots on infralabials and mental region; venter unpatterned cream. **Habitat and Behaviour** Inhabits karst areas within primary evergreen monsoon forests at an elevation of 300m asl. Nocturnal and subfossorial. Diet and reproductive habits unstudied. **Distribution** Endemic to central Vietnam (Phong Nha-Ke Bang National Park, Quang Binh Province). **Status** Not Evaluated.

BORNEAN STRIPED SKINK
Lipinia inexpectata PLATE 31
Measurement SVL 40.6mm **Identification** Body slender, elongated; snout acute; lower eyelid with clear spectacle; auricular opening absent; supralabials 6; infralabials 6–7; subcaudals 68–74; midbody scale rows 20; lamellae under Toe IV 16–17. **Coloration** Dorsum tan-brown with a series of dark grey-brown stripes; paired paravertebral stripes from behind eye to tail-tip; paired dorsal stripes from temporals to inguinal region. **Habitat and Behaviour** Associated with the lowlands, including those on islands. Daytime microhabitat includes rotting logs. Diet unstudied. Oviparous; clutches comprise 2 eggs.

Distribution Endemic to Borneo and its offshore islands to the north. **Status** Not Evaluated.

PULAU MIANG STRIPED SKINK
Lipinia miangensis NOT ILLUSTRATED
Measurement SVL 39mm **Identification** Body slender, elongated; tail long; snout pointed; 2 vertebral series of large scales; parietals in contact with interparietal; supraoculars 4; midbody scale rows 24; supraoculars 4; nuchals 3 pairs; lamellae under toe IV 21, smooth. **Coloration** Dorsum golden-yellow with 2 longitudinal dark brown stripes from snout-tip, over eyes, to around mid-tail; dark postocular stripe extends to bases of forelimbs; limbs light brown with indistinct yellow dots; tail yellow dorsally; venter unpatterned greenish-white. **Habitat and Behaviour** Presumably an inhabitant of lowland rainforests. Diet and reproductive habits unstudied. **Distribution** Endemic to Pulau Miang (off eastern Borneo). **Status** Not Evaluated.

SARAWAK STRIPED SKINK
Lipinia nitens PLATE 31
Measurement SVL 33.6mm **Identification** Body slender, elongated; snout pointed; 2 enlarged paravertebral scale rows; dorsal and lateral scales smooth; auricular opening absent; limbs reduced, scarcely meeting when adpressed; midbody scale rows 20–22; lamellae under Toe IV 16. **Coloration** Dorsum metallic green; flanks spotted with black and green; pale yellow vertebral stripe with jagged edged black lines, one on each side from supraorbital to tail-base. **Habitat and Behaviour** Inhabits lowland forests. Diurnal and terrestrial. Diet includes ants. Reproductive habits unstudied. **Distribution** Endemic to western Borneo. **Status** Not Evaluated.

MENTAWAI STRIPED SKINK
Lipinia relicta NOT ILLUSTRATED
Measurement SVL 56mm **Identification** Body slender, elongated; lower eyelid with clear window; Supralabial IV or V subocular; interparietal separated from frontoparietals; tympanic region covered with scales and indicated by depression; 3 pairs of nuchals; tail thick, longer than head-body length; preanal scales enlarged; midbody scale rows 20–22; lamellae under Toe IV 10–17. **Coloration** Dorsum mid-brown; tail with slight reddish tint. **Habitat and Behaviour** Inhabits lowland forests on offshore islands and also on the continental island of Java. Probably diurnal and arboreal. Diet unstudied. Oviparous; clutches comprise 2 eggs. **Distribution** Pulau Simeulue, Pulau Nias, Pulau Enggano and western Java (Bantam). **Status** Not Evaluated.

MALAYAN STRIPED SKINK
Lipinia surda PLATE 31
Measurement SVL 50mm **Identification** Body slender, elongated; Supralabials IV–V subocular; 3 pairs of nuchals; auricular opening concealed, indicated by depression; midbody scale rows 20; lamellae under Toe IV 15. **Coloration** Dorsum dark brown, lacking vertebral or paravertebral stripes; venter pale grey. **Habitat and Behaviour** Inhabits lowland forests, including offshore islands. Diurnal and arboreal, associated with the roots of epiphytes and underneaths of palm fibres. Diet consists of small arthropods. Reproductive habits unstudied. **Distribution** Southern Thailand (Yala Province) and Peninsular Malaysia (Pulau Tioman, Pahang State, Sungai Buloh, Selangor State and Kuala Lumpur). **Status** Not Evaluated.

COMMON STRIPED SKINK
Lipinia vittigera PLATE 31
Measurement SVL 43mm (*L. v. vittigera*); 34mm (*L. v. kronfanum*) **Identification** Body slender, elongated; tail long, slender, tapering; snout elongated and acute; lower eyelid with transparent disc; auricular opening small; dorsals smooth; midbody scale rows 28–32, smooth; lamellae under Toe IV 25. **Coloration** Dorsum brownish-black with bright yellow vertebral stripe commencing from snout-tip; flanks with dark and pale spots; venter greenish-white. **Subspecies** Two subspecies are recognized. *L. v. vittigera*, midbody scale rows typically 28 (rarely 30 or 32); limbs long; dorsum with single light stripe. *L. v. kronfanum*, midbody scale rows 28–32 (rarely 28); limbs short; dorsum with 5 light stripes. **Habitat and Behaviour** Inhabits lowland forests, including relatively open areas. Diurnal, active on tree trunks and buttresses, sheltering under exfoliating tree bark. Diet consists of small insects. Oviparous; clutches comprise 2–4 eggs. **Distribution** Southern Myanmar, Thailand, Laos, Cambodia, Peninsular Malaysia, the Mentawai Archipelago, Sumatra, Borneo (*L. v. vittigera*) and Vietnam (Daban, Langbian Plateau, Lam Dong Province and upriver from Thua Thien-Hue Province) (*L. v. kronfanum*). **Status** Not Evaluated.

BACBO SKINK
Livorimica bacboensis NOT ILLUSTRATED
Measurement SVL 42mm **Identification** Body slender; lower eyelid scaly; prefrontals in broad contact; supranasals absent; supraoculars 4; supralabials 6; Supralabials IV–V contact orbit; infralabials 6; auricular lobules absent; paired enlarged nuchals; preanal scales enlarged; midbody scale rows 29–32; lamellae under Toe IV 13–15; pad-like lamellae on manus and pes. **Coloration** Dorsum yellowish-tan bearing golden sheen, with blackish-brown marks that form dots posteriorly; narrow blackish-brown dorsolateral stripe, 1–1.5 scales wide; venter unpatterned cream; iris dark blue. **Habitat and Behaviour** Known from the vicinity of just a single village; original habitat unknown. Presumably diurnal and terrestrial. Diet unstudied. Oviparous; clutches comprise 2–4 eggs, measuring 2.5–3.5mm. **Distribution** Endemic to northern Vietnam (Dong-Luong Village, Bac Can Province). **Status** Not Evaluated.

WHITE-SPOTTED SUPPLE SKINK
Lygosoma albopunctatum PLATE 32
Measurement SVL 65mm **Identification** Body elongated; head nearly indistinct from neck; tail rather thick, rounded, tapering to a narrow point; lower eyelid scaly; auricular opening rounded; midbody scale rows 26–28, smooth or feebly keeled; lamellae under Toe IV 12–15. **Coloration** Dorsum brown to reddish-brown, each scale with dark spot, forming a longitudinal series; sides of neck and flanks dark brown or black, spotted with white; venter unpatterned yellowish-white; tail of juveniles bright red; tail of adults brown. **Habitat and Behaviour** Inhabits the plains and low hills. Associated with rocky substrates. Diet consists of small insects. Reproductive habits unstudied. **Distribution** Myanmar and Vietnam (Dak Lak Province). Also northern India, Nepal and Bangladesh. **Status** Not Evaluated.

ANGEL'S SUPPLE SKINK
Lygosoma angeli NOT ILLUSTRATED
Measurement SVL 100mm **Identification** Body slender; lower eyelid scaly; thick tail subequal to head-body length; auricular opening punctiform; supranasals in contact; pair of slightly enlarged nuchals on each side; single frontoparietal; supralabials 7; Supralabials IV or IV–V subocular; dorsals smooth; midbody scale rows 30, smooth; paravertebral scales 110–115; lamellae under Toe IV 5. **Coloration** Dorsum unpatterned brown, each scale with black mark; venter cream with smaller black speckles. **Habitat and Behaviour** Inhabits lowland forests in riparian habitats. Terrestrial and subfossorial, concealing itself under bark of trees and in burrows in loose earth. Diet and reproductive habits unstudied. **Distribution** Laos and southern Vietnam (Trang Bom, nr Bien Hoa, Dong Nai Province). **Status** Not Evaluated.

BURMESE SUPPLE SKINK
Lygosoma anguineum NOT ILLUSTRATED
Measurement SVL 55mm **Identification** Body slender; limbs reduced; tail as thick as body; lower eyelid with clear window; supranasals entire; supralabials 7; Supralabial V subocular; paired enlarged nuchals; midbody scale rows 22; digits short, subcylindrical; lamellae under Toe IV 6–9. **Coloration** Dorsum brown with dark spots forming dorsolateral line; venter paler. **Habitat and Behaviour** Inhabits the lowlands and low hills. Nothing else known of its natural history. **Distribution** Southern Myanmar (Bago Division) and Thailand (Prachuap Khiri Khan and Chumphon Provinces). **Status** Not Evaluated.

BAMPFYLDE'S GIANT SKINK
Lygosoma bampfyldei PLATE 32
Measurement SVL 142.1mm **Identification** Body robust; snout obtuse; lower eyelid scaly; 3 large auricular lobules; limbs short; scales smooth; supranasals contact behind rostral; prefrontals small; supraoculars 4; supraciliaries 6; supralabials 6; infralabials 6; parietals form suture behind interparietal; no enlarged nuchals; preanal scales small, numbering 10; midbody scale rows 38–40, smooth; digits short; lamellae under Toe IV 14; deep groove from posterior corner of nostril to below eye; tail-tip acute. **Coloration** Dorsum pale grey-brown up to lower flanks; brownish-black band, 5–6 scales wide, over nuchals, extends ventrally to below tympanum; cream band anteriorly from back of eye; area around eye with dark patches; grey vertical bar across rostrum; flanks lack bands; upper surfaces of limbs dark brownish-grey; venter cream; tail dark brownish-grey dorsally, pale grey ventrally. **Habitat and Behaviour** Inhabits lowland forests up to the mid-hills at an elevation of 800m asl. Heavy body with short limbs suggests cryptic, fossorial lifestyle. Diet and reproductive habits unstudied. **Distribution** Peninsular Malaysia, north-western Sumatra (Indragiri, Riau Province) and Borneo (Sungei Rejang, Deramakot and Crocker Range, Sabah State). **Status** Not Evaluated.

BÖHME'S SUPPLE SKINK
Lygosoma boehmei PLATE 32
Measurement SVL 86mm **Identification** Body robust; scaly lower eyelid; dorsal scales with pseudokeels; paired frontoparietals; supralabials 7; infralabials 7; midbody scale rows 32; paravertebrals 66; smooth ventral scales in 81 transverse rows; median subcaudals 108, non-enlarged; lamellae under Toe IV 14, keeled. **Coloration** Dorsum reddish-brown or brownish-black; anterior supralabials and infralabials with sutures edged with greyish-black; chin and throat light orange; venter cream to light brown. **Habitat and Behaviour** Inhabits karst areas within primary evergreen monsoon forests at altitudes of 350–400m asl. Nocturnal and terrestrial. Diet probably includes earthworms. Reproductive habits unstudied. **Distribution** Endemic to central Vietnam (Phong Nha-Ke Bang National Park, Quang Binh Province). **Status** Not Evaluated.

BOWRING'S SUPPLE SKINK
Lygosoma bowringii PLATE 32
Measurement SVL 58mm **Identification** Body slender, elongated; head nearly indistinct from neck; lower eyelid scaly; auricular opening rounded; single frontoparietal; paired nuchals; paravertebral scales 52–58; midbody scale rows 24–28, smooth or weakly keeled; tail thick, rounded, tapering to a narrow point. **Coloration** Dorsum bronze-brown; flanks with dark band, within which are white and black spots forming longitudinal lines, in juveniles; venter unpatterned yellow; tail of juveniles bright red, that of adults grey or brown. **Habitat and Behaviour** Inhabits open areas in the plains, such as clearings in towns and cities, and also pine plantations and evergreen forests in the mid-hills. Diurnal, and terrestrial and subfossorial. Diet consists of small insects. Clutches comprise 2–4 eggs, measuring 7.3 x 12.4mm. **Distribution** Myanmar, Thailand, Laos, Cambodia, Vietnam, Peninsular Malaysia, Singapore, Sumatra,

Pulau Berhala, Pulau Weh, Borneo, Java and Bali. Also Bangladesh, eastern China, the Andaman and Nicobar Islands (India), Sulawesi (Indonesia) and the Sulu Archipelago (the Philippines); introduced to Christmas Island, Indian Ocean (Australia) and Mindanao (the Philippines). **Status** Not Evaluated.

KEELED SUPPLE SKINK
Lygosoma carinatum   NOT ILLUSTRATED
Measurement SVL 77mm **Identification** Body robust; supralabials 7; Supralabial V subocular; infralabials 6–7; midbody scale rows 38–40; paravertebral scales 81–85, scales bearing pseudo-keels; transverse rows of ventrals 92; median subcaudal scales 115; lamellae under Toe IV 16, keeled. **Coloration** Dorsum light brown or greyish-brown, lighter on posterior half, without distinctive pattern; flanks with dark spots; venter yellow, lighter than dorsum; sutures between head scales dark. **Habitat and Behaviour** Inhabits primary forests in the mid-hills of Tay Nguyen Plateau at elevations of 600–700m asl. Diurnal and terrestrial. Diet and reproductive habits unstudied. **Distribution** Endemic to Vietnam (Buon Luoy and Kannak, Gia Lai-Contum Province). **Status** Not Evaluated.

ANNAM SUPPLE SKINK
Lygosoma corpulentum   PLATE 32
Measurement SVL 170mm **Identification** Body robust; eyelids scaly; auricular opening lacks lobules; body scales smooth; supralabials 7; supraoculars 4; supranasals entire, in contact behind rostral; 2 frontoparietals behind frontal; loreals 2; midbody scale rows 34–38; lamellae under Toe IV 10–13. **Coloration** Dorsum light yellowish-brown or chocolate-brown, densely mottled on back and flanks with dark brown; supralabials and infralabials yellow edged with black; sides of neck and throat yellow; rest of venter yellowish-cream. **Habitat and Behaviour** Inhabits wet montane forests. Diet and reproductive habits unstudied. **Distribution** South-eastern Thailand (Khao Ang Ru Nai Wildlife Sanctuary, Chachaengsao Province and Khao Sao Doi Wildlife Sanctuary, Chantaburi Province), Laos (Champasak Province) and Vietnam (Da Lat, Lam Dong Province). **Status** Not Evaluated.

PYGMY SUPPLE SKINK
Lygosoma frontoparietale   NOT ILLUSTRATED
Measurement SVL 41mm **Identification** Body slender; anterior border of auricular opening with 2 lobules; single frontoparietal; loreals 2; supralabials 7; Supralabial V subocular; infralabials 6; nuchals 2; limbs reduced; midbody scale rows 20–30; enlarged median subcaudals 78; lamellae under Toe IV 78. **Coloration** Dorsum dark brown; paired dark brown dorsolateral stripe along flanks; venter cream; subcaudals grey. **Habitat and Behaviour** Inhabits forests in the low hills in the central part of the Khorat Plateau. Nothing else known of its natural history. **Distribution** Endemic to southern Thailand (Sara Buri Province). **Status** Not Evaluated.

HAROLD YOUNG'S SUPPLE SKINK
Lygosoma haroldyoungi   PLATE 32
Measurement SVL 141.3mm **Identification** Body robust, elongated; limbs reduced; head broad; tail long, rounded in cross-section; nostril in single nasal; frontonasal wider than long and laterally in contact with anterior loreal; nuchals absent; single occipital scale; supralabials 9; infralabials 9–10; midbody scale rows 40–42, smooth. **Coloration** Dorsum yellowish-brown with dark brown spots, small anteriorly, larger posteriorly; forehead dark brown; 27 dark brown bands on body, may form reticulated pattern; venter pale yellow with irregular dark patches and spots; 33 dark bands on tail. **Habitat and Behaviour** Inhabits submontane forests and the low hills. Subfossorial and probably crepuscular. Diet in captivity includes earthworms and insects. Reproductive habits unstudied. **Distribution** Northern Thailand (Chachoengsao, Chaiyaphum, Chanthaburi, Chiang Mai, Chiang Rai, Loei, Nakhon Ratchasima, Nong Khai, Phetchabun and Phitsanulok Provinces) and Laos. Suspected to occur in Myanmar and Cambodia. **Status** Not Evaluated.

HERBERT'S SUPPLE SKINK
Lygosoma herberti   PLATE 32
Measurement SVL 67mm **Identification** Body slender; snout obtusely pointed; lower eyelid scaly; supranasals present; nuchals absent; supraoculars 4; supralabials 7; Supralabial V subocular; infralabials 6; midbody scale rows 26–30, scales with 5 keels; lamellae under Toe IV 14. **Coloration** Dorsum bronze-brown; dark postocular stripe reaches axilla, where it is most distinct, and continues to flanks; flanks, sides of neck and tail with pale spots, each one scale wide; venter light brown. **Habitat and Behaviour** Inhabits the mountain foothills. Diet and reproductive habits unstudied. **Distribution** Southern Thailand (Ranong, Phang-nga, Phuket, Surat Thani and Khao Wang Hip, Nakon Si Thammarat Province) and northern Peninsular Malaysia (Sungai Menora, nr Kangsar, Perak State). **Status** Not Evaluated.

CENTRAL SUPPLE SKINK
Lygosoma isodactylum   NOT ILLUSTRATED
Measurement SVL 82.5mm **Identification** Body slender; lower eyelid scaly; supranasals fused anteriorly with nasals; single frontoparietal; tail length subequal to head-body length; middorsal scale rows 88–98; midbody scale rows 30–34. **Coloration** Dorsum dark yellow with dark brown mottling that forms dark blotches; scales on flanks with dark edge, forming oblique lines upwards and backwards; venter pale yellow or yellowish-cream with irregular brown speckles; labials, with the exception of the first, barred alternately with yellow and brown. **Habitat and Behaviour** Inhabits the lowlands, and known from sand quarries and gardens. Subfossorial, hiding in cracks in the earth. Diet and reproductive habits unstudied. **Distribution** Central Thailand (Sanam Cheng, N of Lopburi, Sam Kok and Chong Kae, from Ayutthia, Saraburi and Nakhon Sawan Provinces) and Cambodia. **Status** Not Evaluated.

KORAT SUPPLE SKINK
Lygosoma koratense PLATE 32
Measurement SVL 106mm **Identification** Body robust, elongated; lower eyelid scaly; auricular opening small but distinct; nasal-supranasal scales fused; limbs short; tail thick; frontoparietals 2; supraoculars 4; supralabials 8; Supralabials I–VI contact orbit; lamellae under Toe IV 10–14; midbody scale rows 30–34, smooth. **Coloration** Dorsum reddish-brown, dorsal scale-tip black; flanks greenish-yellow, lateral scale-tip black; head scales with central spots and dark edges; labials yellow with large black spots; venter yellowish-cream. **Habitat and Behaviour** Inhabits the lowlands at the foot of limestone mountains. Subfossorial, known from under fallen trees. Diet and reproductive habits unstudied. **Distribution** Eastern and central Thailand (Dong Phaya Fai, Nakhon Ratchasima Province, Lat Bua Kao, Korat Province and nr Muak Lek, Saraburi Province). **Status** Not Evaluated.

LINED SUPPLE SKINK
Lygosoma lineolatum PLATE 32
Measurement SVL 63mm **Identification** Body slender; limbs reduced; lower eyelid with clear window; tail as thick as body; supranasals entire; supralabials 7; Supralabial V subocular; paired enlarged nuchals; midbody scale rows 22; digits elongated; lamellae under Toe IV 8–10. **Coloration** Dorsum brown with dark dots forming dorsolateral line; sides of neck and anterior flanks sometimes with white spots. **Habitat and Behaviour** Inhabits the lowlands and low hills. Nothing else known of its natural history. **Distribution** Myanmar (Rakhine State and Mandalay Division). Also Bangladesh. **Status** Not Evaluated.

INDRAGIRI SUPPLE SKINK
Lygosoma opisthorhodum NOT ILLUSTRATED
Measurement SVL 93mm **Identification** Body slender; snout short; lower eyelid scaly; supraoculars 4; Supralabial IV or V subocular; auricular opening small, rounded; temporals absent; enlarged nuchals absent; preanal scales weakly enlarged; tail slightly longer than head-body length; midbody scale rows 30; lamellae under Toe IV 13. **Coloration** Dorsum blackish-brown anteriorly, paler posteriorly; light dorsolateral stripe from frontal, through upper eyelid, to flanks; broad dark lateral stripe, anteriorly white-spotted, from nostril, through orbit, to flanks; infralabials brown; tail light reddish-brown; venter yellowish-brown. **Habitat and Behaviour** Nothing known of its natural history. **Distribution** Endemic to north-western Sumatra (Indragiri, Riau Province). **Status** Not Evaluated.

SPOTTED SUPPLE SKINK
Lygosoma punctatum PLATE 33
Measurement SVL 85mm **Identification** Body slender, elongated; head nearly indistinct from neck; lower eyelid with undivided transparent disk; supralabials 7; Supralabial V subocular; auricular opening rounded; tail thick, rounded, tapering to a narrow point; midbody scale rows 24–28, smooth; lamellae under Toe IV 11–14. **Coloration** Dorsum bronze-brown with 4–6 rows of black spots, the lateral ones more distinct, forming 4–6 longitudinal stripes on juvenile dorsum; broad cream stripe along body; venter unpatterned cream; tail of juveniles bright red, that of adults brown or pink. **Habitat and Behaviour** Inhabits the hills and plains. Diurnal, and active on surface of leaf litter. Diet consists of small insects. Oviparous; clutches comprise 2–4 eggs. **Distribution** Myanmar and Vietnam. Also India, Sri Lanka, Bangladesh and Pakistan. **Status** Not Evaluated.

SHORT-LIMBED SUPPLE SKINK
Lygosoma quadrupes PLATE 33
Measurement SVL 96mm **Identification** Body extremely elongated; tail as thick as body; rostral in contact with frontonasal; single frontoparietal; nasal single; postnasals absent; supraoculars 4; supralabials 7; infralabials 6; auricular opening narrow, partially covered with scales; midbody scale rows 24–26, smooth; lamellae under Toe IV 5. **Coloration** Dorsum yellowish-brown with dark lines bordering edges of all dorsals, which enter tail; forehead and supralabials darker than rest of dorsum; venter and subcaudals pale pink. **Habitat and Behaviour** Inhabits lowland forests, as well as open areas such as agricultural land and forest edges, at elevations of up to 700m asl. Subfossorial, associated with leaf litter. The reduced limbs are folded back to its body during rapid locomotion. Diet and reproductive habits unstudied. **Distribution** Thailand, Laos, Cambodia, Vietnam, Peninsular Malaysia, Sumatra and Java. **Status** Not Evaluated.

FALSE STRIPED SKINK
Paralipinia rara PLATE 33
Measurement SVL 45mm **Identification** Body slender; supranasals absent; prefrontals in broad contact; postnasals absent; nuchals 3 pairs; lower eyelid with clear window; limbs short; 2 median vertebrals enlarged; midbody scale rows 24, smooth; double row of basal subdigital lamellae. **Coloration** Dorsum and flanks golden-brown with black spots arranged in 2 linear rows; dark lateral band across eye, widening on axilla and continuing on to tail; venter unpatterned cream. **Habitat and Behaviour** Inhabits primary forests in the mid-hills at an elevation of 750m asl. Diurnal and arboreal, suspected to be associated with the forest canopy. Diet and reproductive habits unstudied. **Distribution** Endemic to southern Vietnam (Buon Lai, Gia Lai Province). **Status** Not Evaluated.

CHINESE BLUE-TAILED SKINK
Plestiodon chinensis PLATE 33
Measurement SVL 134mm **Identification** Body robust; limbs and tail well developed; snout short, rounded; postnasals absent; supralabials 7 (rarely 9); infralabials 7; frontal in contact with 2 supraoculars; auricular opening with 3–4 small lobules; paired

nuchals (rarely 1 or 3); midbody scale rows 24 (rarely 22 or 26); lamellae under Toe IV 17. **Coloration** Dorsum in juveniles black with cream stripe from interparietal; dorsolateral stripe from last supraocular; tail blue; bright juvenile pattern lost with growth, head turning reddish-brown and dorsum olive or brownish-olive, with scattered red or orange blotches on flanks; throat cream, scales edged with grey; rest of venter unpatterned cream or yellow. **Subspecies** Five species are recognized, of which one (*P. c. chinensis*) occurs in the region. **Habitat and Behaviour** Inhabits the lowlands, including agricultural areas, at elevations of up to ca 150m asl. Diurnal and terrestrial, active in grassy habitats and hiding under stones. Diet includes arthropods, earthworms and snails. Oviparous; clutches comprise 8–13 eggs. Incubation period 27–45 days. Hatchlings 23–33mm. **Distribution** Vietnam (*P. c. chinensis*). Also southern China (*P. c. chinensis*), northern China (*P. c. pulcher*), the islands of Zhejiang Province (*P. c. daishanensis*) and Taiwan (*P. c. formosensis* and *P. c. leucostictus*). **Status** Not Evaluated.

ELEGANT BLUE-TAILED SKINK
Plestiodon elegans NOT ILLUSTRATED
Measurement SVL 74.3mm **Identification** Body robust; dorsals not larger than scales on flanks; limbs and tail well developed; supranasals in broad contact; supraoculars 4; supralabials 7; Supralabial V subocular; infralabials 7; single postmental; postnasals absent; paired enlarged nuchals; triangular upper secondary temporal; paired keeled postcloacal scales; midbody scale rows 25–28. **Coloration** Dorsum olive-brown with 5 longitudinal pale green stripes continuing up to tail; flanks dark brown; venter cream; abdomen and lower flanks with blue areas. **Habitat and Behaviour** Inhabits the open lowlands, especially rocky areas, including agricultural fields. Diurnal, with activity peaking during midday, and terrestrial. Diet includes arthropods such as spiders, beetles, cockroaches, grasshoppers, cicadas and aphids, in addition to earthworms and plant seeds. Oviparous; clutches comprise 3–8 eggs, measuring 16.7 x 11.9mm. Incubation period 20–44 days. Hatchlings 25.7–30mm. **Distribution** Northern Vietnam (Bac Kan, Vin Phuc and Thanh Hoa Provinces). Also southern and eastern China, and the Ryukyu Archipelago (Japan). **Status** Not Evaluated.

FOUR-LINED BLUE-TAILED SKINK
Plestiodon quadrilineatus PLATE 33
Measurement SVL 73mm **Identification** Body slender; 2 median dorsals larger than scales on flanks; limbs and tail well developed; snout short; 1 postnasal; nuchals 2–3; supralabials 7–8; supraoculars 4; midbody scale rows 20; lamellae under Toe IV 19–20. **Coloration** Dorsum dark grey-brown; greenish-white stripe along flanks, starting from snout-tip and running along second scale row to tail; light lateral line from labials crosses level of lower auricular opening and continues to inguinal region; head yellowish-brown; venter unpatterned cream or grey. **Habitat and Behaviour** Inhabits submontane forests at elevations of 900–1,200m asl. Diurnal and terrestrial, associated with grassy and rocky areas. Diet comprises cockroaches, beetles, grasshoppers and earthworms. Oviparous; clutch size unknown. Hatchlings ca 50mm. **Distribution** Northern and central Thailand (Muak Lek, Saraburi Province and Doi Nang Na Mountains, Chiang Mai Province), Cambodia and Vietnam (Cao Pha, Son La Province, Po-mu Mountain, Mau Son Mountains, Phong-Nha and other unverified localities). Also southern and eastern China. **Status** Not Evaluated.

TAMDAO BLUE-TAILED SKINK
Plestiodon tamdaoensis PLATE 33
Measurement SVL 129mm **Identification** Body slender; snout pointed in adults, short, more rounded in juveniles; supranasals in contact; 1 postnasal; loreals 3; Supralabial V or VI subocular; infralabials 7; frontoparietals in contact with prefrontals; enlarged nuchals 2–3; auricular opening large, vertical, with small lobules anteriorly; 2 postmentals, second larger; paravertebrals 38–44; midbody scale rows 22–24; scales in longitudinal series 38–40; 2 enlarged preanal scales; median subcaudals enlarged; lamellae under Toe IV 17–19. **Coloration** Dorsum bronze- or copper-brown; narrow pale dorsolateral stripes; broad dark stripe from nostril to along flanks, edged ventrally by narrow pale stripe; supralabials pale; pale postocular and median stripe in juveniles; tail bright blue; venter unpatterned grey or cream. **Habitat and Behaviour** Inhabits submontane forests. Diurnal and terrestrial, associated with grassy areas. Diet and reproductive habits unstudied. **Distribution** Northern Vietnam (Da Nang and Dong Nai Provinces). Also eastern China (Hong Kong). **Status** Not Evaluated.

MARQUIS DORIA'S GROUND SKINK
Scincella doriae NOT ILLUSTRATED
Measurement SVL 59mm **Identification** Body robust; prefrontals separate or with narrow contact; upper postocular wide; enlarged nuchals 3–5 pairs; lower eyelid with clear window; limbs robust; midbody scale rows 26–32; ventrals 70–79; lamellae under Toe IV 15–18. **Coloration** Dorsum mid-brown with small dark brown spots that do not form vertebral band; dark lateral stripe extends from nostril to flanks, where it is broken up by pale spots; margins of labials edged with brown; venter yellowish-white. **Habitat and Behaviour** Inhabits primary forests from the mid-hills to submontane limits. Diurnal and terrestrial, associated with moist leaf litter. Diet and reproductive habits unstudied. **Distribution** North-eastern Myanmar (Kakhien Hills and Bhamo, Kachin State), northern and north-eastern Thailand (Doi Inthanon and Doi Suthep, Chiang Mai Province, and Phu Luang Wildlife Sanctuary, Loei Province), and Vietnam. Also southern China (Yunnan Province). **Status** Not Evaluated.

KOHTAO GROUND SKINK
Scincella kohtaoensis   NOT ILLUSTRATED
Measurement SVL 44.1mm **Identification** Body robust; upper postocular wide; prefrontals in wide contact; auricular opening smaller than orbit diameter; midbody scale rows 29–31; ventrals 59–62; lamellae under Toe IV 14–17. **Coloration** Dorsum pale brown; small, irregularly arranged brown spots on dorsum, in vertebral region; dark brown stripe narrow on head, widening on flanks, composed of spots; venter unpatterned cream. **Habitat and Behaviour** Inhabits forests on island in Gulf of Thailand. Diurnal and terrestrial. Diet and reproductive habits unstudied. **Distribution** Endemic to Thailand (Koh Tao, Surat Thani Province). **Status** Not Evaluated.

BLACK-SPOTTED GROUND SKINK
Scincella melanosticta   PLATE 33
Measurement SVL 57.4mm **Identification** Body robust; head small; lower eyelid with clear window; limbs long; upper postocular wide; prefrontals in wide contact; enlarged nuchals absent; auricular lobules absent; dorsals and laterals subequal; midbody scale rows 34–38; ventrals 64–72; lamellae under Toe IV 14–20. **Coloration** Dorsum olive-, bronze- or golden-brown; large brown or black spots concentrated along vertebral region; dark brown or black stripe starts from nostril, narrow on head, widening on flanks, where it is broken up by cream spots; lower flanks white with small black specks; supralabials and infralabials cream with black sutures; breeding males with bright orange head, neck and throat; venter unpatterned yellow or cream. **Habitat and Behaviour** Inhabits evergreen forests in the mid-hills at elevations of 100–1,219m asl. Diurnal and terrestrial, active on rocks and logs. Diet and reproductive habits unstudied. **Distribution** Eastern Myanmar (Mount Mulayit and Tenasserim, Taninthayi Division), Thailand (Upper Mekong, Dong Phaya Fai Range, Huey Sapon, Nakon Sri Thamarat and Chanthabun), Cambodia (Cardamom Mountains) and Vietnam (Langbian Plateau, Lam Dong Province). **Status** Not Evaluated.

TAWNY GROUND SKINK
Scincella ochracea   NOT ILLUSTRATED
Measurement SVL 51mm **Identification** Body slender; head small; lower eyelid with clear window; tail longer than head-body length; limbs small; prefrontals in broad contact; frontal shorter than frontoparietals; upper postocular wide; 3–6 pairs of nuchals; auricular opening vertical oval; 2–4 small auricular lobules present; supralabials 7; Supralabials V–VI subocular; infralabials 6–7; dorsals larger than laterals; enlarged paired preanal scales; median subcaudals enlarged; midbody scale rows 30–32; lamellae under Toe IV 15–20. **Coloration** Dorsum yellow-bronze; black vertebral stripe composed of juxtaposed dark spots extends to tail; brown dorsolateral stripe along upper neck and flanks, sprinkled with dark scales, continuing to tail-tip; small dark spots on lower flanks; venter unpatterned cream. **Habitat and Behaviour** Inhabits primary forests in the mid-hills. Diet unstudied. Oviparous; clutches comprise 4 eggs, measuring 7.5 x 4mm. **Distribution** Laos and northern Vietnam. Also possibly southern China (Yunnan Province). **Status** Not Evaluated.

SPOT-LINED GROUND SKINK
Scincella punctatolineata   NOT ILLUSTRATED
Measurement SVL 38.1mm **Identification** Body slender, elongated; head small; upper postocular wide; enlarged nuchals 1–2 pairs; auricular lobules absent; forelimbs and hind limbs short; midbody scale rows 22–28; ventrals 58–69; lamellae under Toe IV 13–15. **Coloration** Dorsum pale brown with small dark brown spots forming streaks; dark lateral band, 2 scales wide, with larger dark brown spots; lower flanks with indistinct streaks; venter unpatterned cream. **Habitat and Behaviour** Nothing known of its natural history. **Distribution** Myanmar (Karen Hills) and western Thailand (questionably from Tasan, Chumphon Province and Bong Tee Valley, Kanchanaburi Province). **Status** Not Evaluated.

REEVES'S GROUND SKINK
Scincella reevesii   PLATE 33
Measurement SVL 57.4mm **Identification** Body slender, elongated; head small; lower eyelid with clear window; upper postocular wide; prefrontals in narrow contact; auricular lobules absent; no enlarged nuchals; ventrals 57–73; midbody scale rows 29–42; lamellae under Toe IV 13–21. **Coloration** Dorsum bronze-brown; black spots on dorsum concentrated in vertebral region; dark brown stripe narrow on head, widening on flanks, where it is broken up by cream spots; venter unpatterned cream to yellowish-white. **Habitat and Behaviour** Inhabits forested hills at elevations of up to 1,500m asl. Diurnal and terrestrial, associated with rocky biotope. Diet includes crickets, termites, beetle larvae and woodlice. Ovoviviparous; clutches comprise 2–5 neonates, measuring 25mm. **Distribution** North- and south-eastern Myanmar, Thailand and Vietnam. Suspected to occur in Laos and Cambodia. Also Nepal, Bangladesh, Korea, and southern and eastern China. **Status** Not Evaluated.

ROCK-DWELLING GROUND SKINK
Scincella rupicola   NOT ILLUSTRATED
Measurement SVL 44mm **Identification** Body slender; snout obtuse; lower eyelid with clear window; paired, slightly enlarged nuchals; supraoculars 4; supranasals absent; upper postocular wide; auricular opening large, lacking lobules; midbody scale rows 32–34; lamellae under Toe IV 17–18. **Coloration** Dorsum light brown; series of large, irregular black spots on vertebral region; spots paired on nape; dark brown or black postocular stripe broadens on flanks and extends to tail-base; sutures on supralabials and infralabials with faint dark spots; tail pink; venter cream. **Habitat and Behaviour** Inhabits forested habitats. Diurnal and terrestrial, may be associated with streams. Diet and reproductive habits unstudied. **Distribution** Eastern Thailand, Laos and southern Vietnam. **Status** Not Evaluated.

MOUNT VICTORIA GROUND SKINK
Scincella victoriana PLATE 33
Measurement SVL 76.7mm **Identification** Body robust; dorsal keels distinct; dorsal and lateral scales subequal; upper postocular narrow; supralabials 7; infralabials 6; limbs well developed; ventrals 53–56; lamellae under Toe IV 15–16. **Coloration** Dorsum mid-brown; sometimes golden and dark brown spots form longitudinal stripes on back; thick dark postocular stripe with several cream spots within extends along flanks and continues along tail, anteriorly reaching snout-tip; dark grey on lower flanks; venter cream or grey. **Habitat and Behaviour** Inhabits montane forests of pine and mixed hardwood at elevations of 1,954–2,800m asl. Diurnal and presumably insectivorous. Reproductive habits unstudied. **Distribution** Endemic to western Myanmar (Mount Victoria and Laiva Forest Reserve, Chin State). **Status** Not Evaluated.

OAK FOREST SKINK
Sphenomorphus aesculeticola PLATE 34
Measurement SVL 43mm **Identification** Body slender; head not wider than neck; limbs short; prefrontals separated; parietals contact supraoculars; supraoculars 4; frontonasal absent; supraciliaries 8–10; loreals 2; supralabials 6; infralabials 5; preanal scales not enlarged; midbody scale rows 28–30, smooth; lamellae under Toe IV 6–10. **Coloration** Dorsum brown; many scales dark spotted, forming dark lines or chequered pattern; dark lateral band; venter unpatterned cream. **Habitat and Behaviour** Inhabits submontane forests dominated by oak at elevations of 1,350–1,650m asl. Diurnal and semi-fossorial, hiding under rocks and logs. Diet presumably consists of small insects and their larvae. Oviparous; clutches comprise 2 eggs. Hatchlings 15mm. **Distribution** Endemic to northern Borneo (Gunung Kinabalu and Crocker Range, Sabah State). **Status** Not Evaluated.

ALFRED'S FOREST SKINK
Sphenomorphus alfredi NOT ILLUSTRATED
Measurement SVL 33mm **Identification** Body slender, elongated; snout short; lower eyelid scaly; limbs short; tail thick basally, tapering to a point; prefrontals large, in contact; supraoculars 4; Supralabials IV–VI subocular; auricular opening large, subequal to orbit; enlarged nuchals absent; preanal scales not enlarged; midbody scale rows 28–30; lamellae under Toe IV 7–12. **Coloration** Dorsum reddish-brown; dark brown spots on nape; dark lateral stripe along sides of head crosses flanks; gular region with brown spots; venter unpatterned cream. **Habitat and Behaviour** Inhabits primary forests in the low hills. Diet and reproductive habits unstudied. **Distribution** Endemic to northern Borneo (Keningau and Penambo Range). **Status** Not Evaluated.

SUMATRAN FOREST SKINK
Sphenomorphus anomalopus NOT ILLUSTRATED
Measurement SVL 70mm **Identification** Body slender; snout short; supralabials 7; Supralabials V–VI or VI–VII subocular; infralabials 6–7; prefrontals in contact; parietals contact supraoculars; supraoculars 5; loreals 2; no enlarged nuchals; lower eyelid scaly; auricular opening oval, small, lacking lobules; midbody scale rows 38–39, smooth; preanal scales enlarged; lamellae under Toe IV 16–17, strongly keeled and greatly elongated. **Coloration** Dorsum mid-brown with reddish-brown cross-bars; dark lateral stripe from rostral, across flanks, where it is broken into a series of spots by transverse pale bands; supralabials with dark brown sutures; infralabials cream; venter unpatterned white or with series of small black spots on each side. **Habitat and Behaviour** Inhabits lowland forests, including those on an offshore island. Diet and reproductive habits unstudied. **Distribution** West coast of northern Peninsular Malaysia (Pulau Pinang, Penang State) and Sumatra (Pulau Nias). **Status** Not Evaluated.

BUON LOI FOREST SKINK
Sphenomorphus buenloicus NOT ILLUSTRATED
Measurement SVL 56mm **Identification** Body slender; snout short, obtuse; supralabials 7; Supralabials IV–VI subocular; infralabials 7; paired nuchals; auricular opening large; rostral and frontonasal not separated; prefrontals in contact; supraoculars 4; parietals in contact behind interparietal; limbs well developed; midbody scale rows 32–34; 2 enlarged preanal scales. **Coloration** Dorsum greyish-brown with indistinct dark spots; indistinct dark temporal stripes from parietals along edges of back; flanks light; chin, gular and abdominal region reddish-orange; iris black; rest of venter light. **Habitat and Behaviour** Inhabits primary tropical forests. Diurnal and terrestrial, associated with clearings near roadsides and banks of water reservoirs, hiding in forest litter. Diet unstudied. Oviparous; clutches comprise 2–3 eggs. Hatchlings 54mm. **Distribution** Endemic to southern Vietnam (Buon Loi, Kon Tum Province). **Status** Not Evaluated.

BÜTTIKOFER'S FOREST SKINK
Sphenomorphus buettikoferi NOT ILLUSTRATED
Measurement SVL 35mm **Identification** Body slender; snout short, obtuse; tail thick; lower eyelid scaly; frontonasal broader than long; supraoculars 4; supraciliaries 9; large nuchals absent; auricular opening rounded, lacking lobules; Supralabials III–V in midorbital position; dorsal scales smooth; midbody scale rows 24, smooth; limbs well developed; lamellae under Toe IV 21–23. **Coloration** Dorsum reddish-brown with 4 longitudinal rows of darker spots, 2 in paravertebral region, 1 on each flank; venter grey. **Habitat and Behaviour** Inhabits primary rainforests in the mid-hills. Diet and reproductive habits unstudied. **Distribution** Endemic to Borneo (Bukit Liang Kubung, Kalimantan Province). **Status** Not Evaluated.

BUKIT FRASER FOREST SKINK
Sphenomorphus bukitensis NOT ILLUSTRATED
Measurement SVL 44mm **Identification** Body slender; supralabials 6; infralabials 5; midbody scale

rows 31–33, smooth; paravertebral scales 73–74; ventrals 61–74; supraoculars 4; parietals contact supraoculars; prefrontals in contact; loreal scales 2; 2 enlarged preanal scales; lamellae under Toe IV 12–13, keeled. **Coloration** Dorsum light brown to dull orange with dark mottling; orange dorsolateral stripe above forelimbs, turning indistinct on body and bordered below by thick dark brown stripe, breaking up into speckles posterior to axilla. **Habitat and Behaviour** Inhabits hill dipterocarp forests at elevations of 1,046–1,239m asl. Retreats include logs buried in leaf litter. Diet and reproductive habits unstudied. **Distribution** Endemic to Peninsular Malaysia (Bukit Fraser, Pahang State). **Status** Not Evaluated.

BUTLER'S FOREST SKINK
Sphenomorphus butleri   NOT ILLUSTRATED
Measurement SVL 44mm **Identification** Body slender; supraoculars 4; parietals contact supraoculars; prefrontals in contact; loreals 1–2; supralabials 6; infralabials 5–6; midbody scale rows 31–33; enlarged preanal scales; lamellae under Toe IV 11–12, smooth. **Coloration** Dorsum mid-brown with distinct dorsolateral stripe; distinct dark temporal stripe. **Habitat and Behaviour** Inhabits submontane forests. Diurnal and terrestrial. Diet and reproductive habits unstudied. **Distribution** Southern Thailand (Khao Rama, Nakhon Si Thammarat Province) and Peninsular Malaysia (Temengor Forest Reserve and Bukit Larut, Perak State, and Telon Valley, Pahang State). **Status** Not Evaluated.

CAMERON HIGHLANDS FOREST SKINK
Sphenomorphus cameronicus   NOT ILLUSTRATED
Measurement SVL 70mm **Identification** Body elongated; snout rounded; lower eyelid scaly; tympanum half eye diameter; parietals contact supraoculars; supraoculars 4; supralabials 7; infralabials 5; loreals 2; preanal scales enlarged; midbody scale rows 38, smooth; lamellae under Toe IV 20–22. **Coloration** Dorsum dark brown with black spots; dark flank stripe from neck to tail; lower flanks white-spotted; venter unpatterned white. **Habitat and Behaviour** Inhabits montane forests. Diurnal and probably terrestrial. Diet and reproductive habits unstudied. **Distribution** Endemic to Peninsular Malaysia (Cameron Highlands, Pahang State). **Status** Not Evaluated.

GUNUNG TAHAN FOREST SKINK
Sphenomorphus cophias   NOT ILLUSTRATED
Measurement SVL 46mm **Identification** Body slender, elongated; snout short; lower eyelid scaly; parietals contact supraoculars; frontoparietals paired; supraoculars 4; supralabials 7; Supralabials IV–V subocular; infralabials 7; loreals 2; enlarged nuchals absent; 2 enlarged preanal scales; tympanum concealed, its presence indicated by depression; limbs small; midbody scale rows 22–24, smooth; lamellae under Toe IV 7–12. **Coloration** Dorsum dark brown with dark brown mottling; dark dorsolateral stripe; venter unpatterned cream. **Habitat and Behaviour** Inhabits submontane forests at elevations of ca 1,000m asl. Diet and reproductive habits unstudied. **Distribution** Endemic to Peninsular Malaysia (Gunung Tahan, Pahang State). **Status** Not Evaluated.

THICK FOREST SKINK
Sphenomorphus crassus   NOT ILLUSTRATED
Measurement SVL 82mm **Identification** Body robust; lower eyelid scaly; tympanum deeply sunk; auricular lobules absent; limbs short, pentadactyle; supraoculars 4; prefrontals widely separated; frontoparietals distinct; supraciliaries 8; supralabials 7; infralabials 7; postmental larger than mental; 2 pairs of large chin shields, first pair in contact medially; median subcaudals enlarged; median preanal scales enlarged; midbody scale rows 32; lamellae under Toe IV 18–19. **Coloration** Dorsum and flanks medium brown with faint dark lines formed by dark centres of most dorsal scales; trunk spotted with yellow, especially on flanks and tail; dark lateral band absent; venter unpatterned cream. **Habitat and Behaviour** Inhabits forests in the mid-hills at an elevation of 670m asl. Fossorial, known from under dead leaves in selectively logged forest. Diet and reproductive habits unstudied. **Distribution** Endemic to Borneo (Mendolong, Sabah State). **Status** Not Evaluated.

EARLESS FOREST SKINK
Sphenomorphus cryptotis   PLATE 34
Measurement SVL 82mm **Identification** Body slender; snout short, obtuse; limbs moderate; scales smooth, dorsals larger than ventrals; supraoculars 6; supraciliaries 10–12; supralabials 7; Supralabials V–VI subocular; auricular opening concealed; enlarged nuchals 6; limbs moderate; paired enlarged preanals; midventrals 36–39; tail long, compressed; lamellae under Toe IV 16–21. **Coloration** Dorsum dark brown with golden sheen; indistinct dorsolateral stripe, edged dorsally and ventrally with 2 rows of cream ocellate or comma-shaped spots, extending to tail; flanks pale grey with indistinct cream and dark streaks; labials with dark bars; temporal region mottled with black; neck and throat with small black spots; venter unpatterned white; lower third of tail black; iris bright yellow. **Habitat and Behaviour** Inhabits lowland forests, reaching submontane limits, especially the vicinity of streams, at elevations of 250–1,000m asl. Diurnal and aquatic, known to both swim and dive; night-time retreat consists of small branches overhanging streams. Diet and reproductive habits unstudied. **Distribution** Endemic to northern Vietnam (Quang Ninh, Nghe An and Lao Cai Provinces). **Status** Not Evaluated.

BLUE-THROATED FOREST SKINK
Sphenomorphus cyanolaemus   PLATE 34
Measurement SVL 60mm **Identification** Body slender; limbs relatively long; auricular opening lacking lobules; prefrontals in broad contact with each other; anterior loreals 2, superimposed; nuchals absent; supraoculars 6; supraciliaries 12–15; supralabials 6–8; infralabials 6; mental as wide as

rostral; 2 large preanal scales; midbody scale rows 37–42, smooth; lamellae under Toe IV 16–19. **Coloration** Dorsum bronze- or olive-brown with 2 rows of yellow-brown spots; dark grey-brown dorsolateral stripe; head, throat and pectoral region of adult males deep indigo blue, in females, light blue; lower part of abdomen pale blue in males, yellowish-orange in females; lips blue with dark brown bars. **Habitat and Behaviour** Inhabits lowland forests. Diurnal, and terrestrial and semi-arboreal, capable of climbing trees to at least 5m. Nocturnal retreat includes leaves of saplings overhanging hill streams, 30–50cm above water. Diet presumably consists of insects. Oviparous; clutches comprise 2 eggs, measuring 6 x 11mm, which are deposited in ant heaps within the buttresses of trees. Hatchlings 25–26mm. **Distribution** Peninsular Malaysia, Sumatra and Borneo. **Status** Not Evaluated.

GRASSHOPPER-EATING FOREST SKINK
Sphenomorphus devorator   PLATE 34
Measurement SVL 58mm **Identification** Body slender; snout acute; limbs moderate; supraoculars 4 enlarged, followed by a small one; superciliaries 8; supralabials 7; Supralabials V–VI subocular; enlarged nuchals 6; auricular opening large, deeply sunk, subequal to orbit, lacking lobules; dorsals smooth, enlarged, paired rows in vertebral region; midbody scale rows 30; lamellae under Toe IV 17–19. **Coloration** Dorsum silver-grey; dark vertebral stripe (2 scales wide) disappears on tail; dark brown lateral stripes from nostril and from posterior corner of orbit to tail, edged ventrally by numerous small black spots; labials with numerous black spots; forehead with irregular black spots; venter unpatterned cream. **Habitat and Behaviour** Inhabits primary forests in the mid-hills at an elevation of 600m asl. Diurnal and arboreal, associated with trees. Diet includes arboreal grasshoppers (genus *Gigantettix*) found on trunks of large trees. Reproductive habits unstudied. **Distribution** Endemic to north-eastern Vietnam (Uong Bi, Quang Ninh Province). **Status** Not Evaluated.

FLORES FOREST SKINK
Sphenomorphus florensis   NOT ILLUSTRATED
Measurement SVL 71mm **Identification** Body moderate, elongated; snout short; lower eyelid scaly; auricular opening oval, with 6 (rarely 4, 5 or 7) lobules; supranasals absent; supraoculars 6–7; no enlarged nuchals; supralabials 7; Supralabials VI–VII subocular; scales smooth; paired enlarged preanal scales; midbody scale rows 44–50; subdigital lamellae under Toe IV 24–29, smooth or feebly keeled. **Coloration** Dorsum light brown with metallic sheen, sometimes with narrow, paler middorsal stripe; flanks with dark brown vertical bars or dark flank stripe from posterior corner of orbit, across body, up to level of mid-tail; gular region with white spots on black; venter unpatterned cream or greenish-yellow. **Subspecies** Three subspecies are recognized, of which one (*S. f. florensis*) occurs in the region. **Habitat and Behaviour** Inhabits a variety of mostly open-area habitats, from rocky islands to moist forests, at elevations of up to 1,200m asl. Diurnal (with some nocturnal activity), and both terrestrial and arboreal; able to climb trees up to ca 10m above substrate. Diet includes beetle larvae, termites, ants, grasshoppers and spiders. Oviparous; clutches comprise 5 eggs. **Distribution** Java (*S. f. florensis*). Also the Lesser Sundas, Timor (*S. f. florensis*), Wetar (*S. f. barbouri*) and Damma (*S. f. weberi*). **Status** Not Evaluated.

GRANDISON'S FOREST SKINK
Sphenomorphus grandisonae   NOT ILLUSTRATED
Measurement SVL 30mm **Identification** Body robust; scales slightly elevated but not keeled; prefrontals in contact; frontoparietals 2; loreals 2; supralabials 6; infralabials 5; nuchals absent; tympanum large, superficial; limbs short; midbody scale rows 34; subcaudals not enlarged; lamellae under Toe IV 12. **Coloration** Dorsum light brown with minute dark flecks; supralabials and infralabials dark-edged; light dorsolateral stripe on sides of neck and axilla, bordered ventrally by brown stripe; venter unpatterned cream. **Habitat and Behaviour** Nothing known of its natural history. **Distribution** Endemic to northern Thailand (Ban Tong Pheung). **Status** Not Evaluated.

HAAS'S FOREST SKINK
Sphenomorphus haasi   NOT ILLUSTRATED
Measurement SVL 57mm **Identification** Body slender; auricular opening lacking lobules; snout obtuse; midbody scale rows 41–42, smooth; tail thick basally, acute terminally; lamellae under Toe IV 16–18. **Coloration** Dorsum greyish-brown with pale olive blotches; sclera of eye pale blue; dark dorsolateral band absent; venter cream. **Habitat and Behaviour** Inhabits lowland forests. Diurnal and terrestrial. Diet presumably consists of small arthropods. Reproductive habits unstudied, although it is known to be oviparous. **Distribution** Endemic to Borneo (Sarawak State). **Status** Not Evaluated.

HALLIER'S FOREST SKINK
Sphenomorphus hallieri   NOT ILLUSTRATED
Measurement SVL 57mm **Identification** Body slender; large upper temporal wedged between last supraocular and parietal; parietals reduced; no enlarged nuchals or preanal scales; loreals 2; tympanum slightly sunk, lacking auricular lobules; supraoculars 5–6; superciliaries 9–10; infralabials 5; preoculars 3–4; midbody scale rows 36–39; lamellae under Toe IV 10–15. **Coloration** Dorsum brown, typically with red or rust-coloured postocular streak to behind axilla that is broken up into spots or bars on trunk; flanks light brown or olive with numerous small yellow or green spots; throat of males light blue with dense black spots, sides of head nearly black; females with pink or yellow throat; venter yellow. **Habitat and Behaviour** Inhabits the lowlands and mid-hills in dipterocarp forests. Diet and reproductive habits unstudied. **Distribution** Endemic to Borneo. **Status** Not Evaluated.

HELEN'S FOREST SKINK

Sphenomorphus helenae NOT ILLUSTRATED
Measurement SVL 28mm **Identification** Body slender; head distinct from neck; auricular opening distinct, subequal to orbit diameter; lower eyelid scaly; supranasals absent; supraoculars 4; nuchals enlarged; midbody scale rows 30; lamellae under Toe IV 17. **Coloration** Dorsum yellowish-brown with scattered dark brown spots and dark median stripe from back of head to tail; dark brown lateral stripe from snout-tip, across sides of head, widens at axilla, and fragments into transverse spots on flanks to tail; labials and sides of head grey-spotted; venter unpatterned white. **Habitat and Behaviour** Nothing known of its natural history. **Distribution** Endemic to north-eastern and south-central Thailand (Phu Luang, Loei Province, Pak Chong and Dong Phaya Fai, Nakhon Ratchasima Province, and Nonthaburi, Nonthaburi Province). **Status** Not Evaluated.

INDIAN FOREST SKINK

Sphenomorphus indicus PLATE 34
Measurement SVL 97mm **Identification** Body slender; head distinct from neck; snout short; limbs moderate; supralabials 7; infralabials 7; tympanum deeply sunk; loreals 2; dorsal scales smooth; preanal scales 4; midbody scale rows 34–38; lamellae under Toe IV 14–18, rounded or slightly keeled. **Coloration** Dorsum brown, unpatterned or with dark brown spots arranged to form longitudinal lines; dark greyish-black stripe on flanks extends from eye to tail; lips yellow with black bars; venter unpatterned yellow or cream. **Habitat and Behaviour** Inhabits lowland evergreen forests to submontane forests at elevations of up to 1,250m asl. Diurnal and terrestrial, associated with leaf litter and fallen trees, and frequently with rocky biotope. Diet consists of cockroaches, grasshoppers and earthworms. Ovoviviparous; clutches comprise 4–11 neonates, measuring ca 30mm. **Distribution** Myanmar, Thailand, Vietnam, Laos, Cambodia and Peninsular Malaysia. Also eastern India, Bhutan, Bangladesh, and southern and eastern China. **Status** Not Evaluated.

ISHAK'S FOREST SKINK

Sphenomorphus ishaki PLATE 34
Measurement SVL 41mm **Identification** Body slender; limbs small; prefrontals in narrow contact; 2 pretemporals; suboculars 11; supralabials 6; infralabials 5; postsupralabials 3; uppermost secondary temporal and parietal separate; loreals 2; midbody scale rows 30–32, smooth; lamellae under Toe IV 11. **Coloration** Dorsum dark brown or greyish-brown, scale centres lighter, producing spotted appearance; postocular stripe yellow, extending to axilla, and becoming indistinct thereafter, edged ventrally by thick dark brown stripe; forehead with light spots; labials dark banded; venter grey. **Habitat and Behaviour** Inhabits cloud-forest section of the hill dipterocarp forests of Gunung Kajang at elevations of 850–915m asl. Diurnal, although active in low light, and terrestrial and associated with leaf litter. Diet and reproductive habits unstudied. **Distribution** Endemic to Peninsular Malaysia (Pulau Tioman, Pahang State). **Status** Not Evaluated.

GUNUNG KINABALU FOREST SKINK

Sphenomorphus kinabaluensis PLATE 34
Measurement SVL 58mm **Identification** Body slender; limbs long; prefrontals in wide contact with each other; supraoculars 5; parietals not in contact with supraoculars; supralabials 7; infralabials 7; nuchals absent; midbody scale rows 32–38; tail thick. **Coloration** Dorsum light to dark brown; longitudinal rows of dark brown to yellow spots, and sometimes dark brown speckles; black dorsolateral stripe with small yellowish flecks; throat sometimes speckled with dark brown; venter cream, yellow or grey. **Habitat and Behaviour** Inhabits submontane and montane forests at elevations of 1,600–2,200m asl. Diet consists of insects. Oviparous; clutches comprise 1–2 eggs, measuring 9.8–13.8 x 7.9–9.4mm. Sometimes nests within ant nests, in tunnels in fallen trees of moss forests. Hatchlings 20.1–21.5mm. **Distribution** Endemic to Borneo (Gunung Kinabalu and Crocker Range, Sabah State). **Status** Not Evaluated.

PULAU LANGKAWI FOREST SKINK

Sphenomorphus langkawiensis NOT ILLUSTRATED
Measurement SVL 36mm **Identification** Body slender; supraoculars 4; parietals contact posterior-most supraocular; 1 medially projecting superciliary scale; loreals 2; enlarged preanal scales; midbody scale rows 34–37, smooth; paravertebrals 60–72; lamellae under Toe IV 12, keeled. **Coloration** Dorsum light brown to dull orange with faint mottling, lacking bands; dorsolateral stripe from nuchals to axilla, bordered ventrally by dark brown stripe that breaks up into speckled pattern posterior to axilla; head light-spotted; labial scales banded; venter unpatterned cream. **Habitat and Behaviour** Inhabits stunted, wind-blown scrub vegetation and lowland dipterocarp forests on islands off the west coast of Peninsular Malaysia, at elevations of up to at least 517m asl. Diet and reproductive habits unstudied. **Distribution** Endemic to northern Peninsular Malaysia (Pulau Langkawi and Pulau Singa Besar, Kedah State). **Status** Not Evaluated.

SPOTTED-LINED FOREST SKINK

Sphenomorphus lineopunctulatus NOT ILLUSTRATED
Measurement SVL 84mm **Identification** Body slender; head distinct from neck; snout short; supranasals absent; single frontonasal; prefrontals separate; nuchals absent; tympanum as large as orbit; tympanum deeply sunk; dorsal scales smooth; midbody scale rows 38; preanal scales enlarged; lamellae under Toe IV 22. **Coloration** Dorsum olive-brown; forehead darker than body, with small black spots; black spots on axilla; dark flank stripe, 2–3 scales wide; sides of neck grey; chin and throat with dark grey flecks; venter cream. **Habitat and Behaviour** Nothing known of its natural history. **Distribution** Endemic to Thailand (Ubon Province). **Status** Not Evaluated.

SPOTTED FOREST SKINK
Sphenomorphus maculatus PLATE 35
Measurement SVL 65mm **Identification** Body slender; head distinct from neck; snout short; tympanum on surface, not deeply sunk; supralabials 7; Supralabials V–VI subocular; no enlarged nuchals; dorsal scales smooth; auricular lobules absent; midbody scale rows 38–42; lamellae under Toe IV 16–22. **Coloration** Dorsum bronze-brown or brownish-pink, unpatterned or with dark green spots; 2 dark median series of spots; dark lateral band on flanks, spotted with white; venter cream, turning yellow during breeding season. **Habitat and Behaviour** Inhabits primary evergreen forests and more open areas, from sea coasts and the edges of mangrove swamps, to the mid-hills, at elevations of up to 800m asl. Diurnal and terrestrial, associated with leaf litter close to streams and rivers. Diet consists of spiders, crickets and moths. Oviparous; clutches comprise 4–5 eggs. **Distribution** Myanmar, Thailand, Laos, Cambodia and Vietnam. Also eastern India, the Andaman Islands and southern China. **Status** Not Evaluated.

WHITE-THROATED FOREST SKINK
Sphenomorphus maculicollus NOT ILLUSTRATED
Measurement SVL 49mm **Identification** Body slender; forehead scales smooth, iridescent; frontonasal twice as broad as long, not contacting frontal; prefrontals separated by azygous scale; supraciliaries 15–16; supraoculars 8; supralabials 5; infralabials 6; single preocular; nuchals absent; auricular lobules absent; tympanum slightly sunken; midbody scale rows 36; 2 enlarged preanal scales. **Coloration** Dorsum light brown speckled with minute dark brown spots; scales with dark spot on each side of neck; dorsal surfaces of limbs light brown with irregular white spots; chin, throat, venter and undersurface of tail-base unpatterned white; posterior two-thirds of tail with dark mottling. **Habitat and Behaviour** Inhabits the low hills and mid-hills up to submontane limits. Diet and reproductive habits unstudied. **Distribution** Endemic to north-western Borneo (Sungei Pesu, Sarawak State). **Status** Not Evaluated.

MALAYAN FOREST SKINK
Sphenomorphus malayanus NOT ILLUSTRATED
Measurement SVL 65mm **Identification** Body slender; snout obtuse; limbs short; lower eyelid scaly; supraoculars 4; prefrontals in contact; supranasals absent; supralabials 7; Supralabials V–VI suborbital; infralabials 7–8; loreals 2; no enlarged nuchals; auricular lobules absent; dorsal scales smooth; preanal scales enlarged; midbody scale rows 32–33; lamellae under Toe IV 12–13. **Coloration** Dorsum dark brown with yellow reticulations; postocular stripe dark brown edged with yellow spots, extends along flanks to tail; flanks with light brown spots; venter yellow. **Habitat and Behaviour** Inhabits montane forests at an elevation of 1,500m asl. Diet and reproductive habits unstudied. **Distribution** Peninsular Malaysia and western Sumatra (Gunung Tuju and Gunung Singalang). **Status** Not Evaluated.

DWARF FOREST SKINK
Sphenomorphus mimicus PLATE 35
Measurement SVL 36mm **Identification** Body slender; lower eyelid scaly; prefrontals in contact; supranasals absent; frontoparietals 2; nuchals 2; supralabials 7; Supralabials V–VI enlarged; infralabials 6; midbody scale rows 30; subcaudals enlarged. **Coloration** Dorsum brown with indistinct dark brown spots, especially around shoulders; supralabials and infralabials with dark brown spots; sides of neck and flanks with fine brown flecks; venter unpatterned cream. **Habitat and Behaviour** Nothing known of its natural history. **Distribution** Endemic to northern Thailand (Dong Phaya Fai Range, Nakorn Ratchasima Province). **Status** Not Evaluated.

MODIGLIANI'S FOREST SKINK
Sphenomorphus modiglianii NOT ILLUSTRATED
Measurement SVL 41mm **Identification** Body slender; snout short; lower eyelid scaly; parietals contact supraocular; supraoculars 4; Supralabials IV–V subocular; prefrontals in contact; no enlarged nuchals; auricular opening oval, smaller than orbit; dorsal scales smooth; midbody scale rows 32; lamellae under Toe IV 15. **Coloration** Dorsum mid-brown, flanks darker with pale spots; supralabials and infra-labials with dark sutures; venter unpatterned cream. **Habitat and Behaviour** Inhabits low-lying island off Sumatra, presumably in lowland forests. Diet and reproductive habits unstudied. **Distribution** Endemic to Pulau Sipura (Indonesia). **Status** Not Evaluated.

MANY-SCALED FOREST SKINK
Sphenomorphus multisquamatus PLATE 35
Measurement SVL 69mm **Identification** Body robust; snout short; body longer than tail; auricular lobules absent; prefrontals in broad contact; nuchals absent; supraoculars 6–7; supraciliaries 14–15; supralabials 6–7; infralabials 5; midbody scale rows 40–49; lamellae under Toe IV 18–23. **Coloration** Dorsum dark greyish-brown with 2–4 rows of squarish black spots, with or without dark dorsolateral bands; bright yellow ring around eye; black cervical spot absent; labials without black bars; males usually with blue throat and sides of neck; venter unpatterned cream. **Habitat and Behaviour** Inhabits lowland rainforests and peat swamps. Diurnal and terrestrial. Diet consists of small insects. Reproductive habits unstudied. **Distribution** Endemic to Borneo. **Status** Not Evaluated.

GUNUNG MURUD FOREST SKINK
Sphenomorphus murudensis PLATE 35
Measurement SVL 50.4mm **Identification** Body slender; snout rounded; lower eyelid scaly; auricular lobules absent; supranasals absent; nostril in single nasal; prefrontals in contact; supraoculars 6; parietals in contact with supraoculars; supraciliaries 8; supralabials 6; Supralabial IV midorbital; midbody

scale rows 30–32, scales smooth; ventrals larger than dorsals; preanal scales enlarged; lamellae under Toe IV 16. **Coloration** Dorsum dark brown, sometimes with black spots; dark band on flanks; hatchlings and juveniles with reddish-brown tails; adult tail dark brown. **Habitat and Behaviour** Inhabits submontane and montane forests at elevations of 1,500–2,400m asl. Diurnal and terrestrial. Diet unstudied. Oviparous; clutches comprise 2 eggs, measuring 17.3–18 x 10.5–10.7mm, which are deposited in deep crevices. **Distribution** Endemic to north-western Borneo (Gunung Murud, Sarawak State). **Status** Not Evaluated.

BOGOR FOREST SKINK
Sphenomorphus necopinatus   NOT ILLUSTRATED
Measurement SVL 44mm **Identification** Body slender; snout short; lower eyelid scaly; parietals contact supraoculars; suture between frontonasal and rostral over half rostral width; supraoculars 4; supralabials 6; Supralabials III–V suborbital; infralabials 4–5; loreals 2; no enlarged nuchals; auricular opening smaller than orbit; Toe IV longer than Toe III; preanal scales slightly enlarged; midbody scale rows 28–32, smooth; lamellae under Toe IV 11–15. **Coloration** Dorsum pale brown; each scale with a dark centre, producing 8 longitudinal dark stripes; white-spotted, greyish-black lateral stripe, edged dorsally by brown, from posterior of orbit, along flanks to tail; lower flanks cream with greyish spots creating longitudinal lines; labials with cream centres; forehead mottled with dark brown; venter cream; subcaudals with grey spots creating longitudinal stripes. **Subspecies** Two subspecies are recognized. *S. n. necopinatus*, SVL 38mm; dorsals 63–66; lamellae under Toe IV 11–13. *S. n. garutense*, SVL 44mm; dorsals 66–75; lamellae under Toe IV 13–15. **Habitat and Behaviour** Inhabits forests in the mid-hills. Diurnal and terrestrial. Diet unstudied. Oviparous; clutches comprise 1–4 eggs. **Distribution** Endemic to western Java: Bogor (*S. n. necopinatus*) and Garut (*S. n. garutense*). **Status** Not Evaluated.

BUKIT LARUT FOREST SKINK
Sphenomorphus praesignis   PLATE 35
Measurement SVL 110mm **Identification** Body robust; snout short; parietals contact supraoculars; prefrontals separate; supraoculars 4; supralabials 7; infralabials 7; loreals 2; auricular opening lacks lobules; preanal scales enlarged; midbody scale rows 26–28, smooth; lamellae under Toe IV 20–26. **Coloration** Dorsum brownish-grey or chestnut-brown with black mottling; 5–6 large black patches on sides of neck and flanks; chin and throat grey-green; venter unpatterned pale blue; undersurfaces of limbs cream. **Habitat and Behaviour** Inhabits submontane forests at elevations of ca 1,280m asl. Diet and reproductive habits unstudied. **Distribution** Southern Thailand (Khao Wang Hip, Nakhon Si Thammarat Province) and Peninsular Malaysia (Gunung Lawit, Terengganu State, and Tanah Rata, Cameron Highlands and Fraser's Hill, Pahang State). **Status** Not Evaluated.

BLACK-SPOTTED FOREST SKINK
Sphenomorphus puncticentralis   NOT ILLUSTRATED
Measurement SVL 45mm **Identification** Body slender; snout short; lower eyelid scaly; supraoculars 5; supralabials 7; infralabials 7; auricular opening small, lacking lobules; paired enlarged nuchals; limbs well developed; scales unkeeled, bearing striae; paravertebral scales slightly broader than adjacent scale rows; paired enlarged preanal scales; median subcaudals enlarged; midbody scale rows 29; lamellae under Toe IV 25. **Coloration** Dorsum light brown, iridescent, with 14 black spots, 3–5 scales wide, on paravertebral region; lateral stripe commencing from posterior corner of orbit and continuing along flanks; flanks and dorsal surfaces of limbs blackish-brown with small white and light brown spots; forehead darker than dorsum; supraoculars dark-edged; supralabials and infralabials with dark spots; venter unpatterned light yellow or cream. **Habitat and Behaviour** Sole individual known from open secondary forest at an elevation of 700m asl. Diurnal and terrestrial. Diet and reproductive habits unstudied. **Distribution** Endemic to central Java (Baturaden, on southern slope of Gunung Slamet). **Status** Not Evaluated.

RED-TAILED FOREST SKINK
Sphenomorphus rufocaudatus   NOT ILLUSTRATED
Measurement SVL 51mm **Identification** Body slender; snout short, obtuse; supralabials 7; Supralabials IV–VI subocular; infralabials 7; rostral and frontonasal separated; prefrontals typically separated; supraoculars 4; parietals in contact behind interparietal; paired nuchals; auricular opening large; limbs well developed; midbody scale rows 31–34; 2 enlarged preanal scales. This taxon has been allocated to the genus *Scincella* by some authorities. **Coloration** Dorsum light brown with small dark spots on vertebral region; narrow dark brown temporal stripes start from nostrils, and cross orbit, along sides of body to tail-base; flanks greyish-white; gular and venter light orange in males; tail-base and subcaudals light red; in females, bright colours replaced with shades of bluish-white or cream; iris light grey. **Habitat and Behaviour** Inhabits primary evergreen and dipterocarp forests at elevations of 400–542m asl. Diurnal and terrestrial, associated with both closed-canopy and open areas. Diet unstudied. Oviparous; clutches comprise 2–3 eggs. Hatchlings 46mm. **Distribution** Laos, south-western and eastern Cambodia (Cardamom Mountains and O'Rang District), and southern Vietnam (Buon Lai, Kon Tum Province). **Status** Not Evaluated.

SABAH FOREST SKINK
Sphenomorphus sabanus   PLATE 35
Measurement SVL 58mm **Identification** Body robust; lobules in auricular opening absent; prefrontals in broad contact or separated; parietals in contact with supraoculars; supraoculars usually 6; supraciliaries 14–17; supralabials 7; infralabials 5–7; midbody scale rows 38–42; lamellae under Toe IV

18–22. **Coloration** Dorsum pinkish-brown or olive-yellow with indistinct light spots; forehead orange-brown, scales with dark grey areas; dark dorsolateral band absent; sides of neck and flanks of males ringed with orange; lips barred with black; males usually with an orange flush on flanks; gular and pectoral regions unpatterned cream; venter region pale yellow. **Habitat and Behaviour** Inhabits the lowlands to submontane forests. Diurnal and low-shade active, and terrestrial but capable of climbing tree trunks, often associated with tree trunks and buttresses. Diet consists of beetles, spiders, ants, cockroaches, grasshoppers, moths and flies. Oviparous; clutches comprise 2–3 eggs, measuring 10–12mm. **Distribution** Endemic to Borneo. **Status** Not Evaluated.

YELLOW-LINED FOREST SKINK
Sphenomorphus sanctus NOT ILLUSTRATED
Measurement SVL 48mm **Identification** Body slender; snout elongated; prefrontals in contact; parietals contact supraoculars; supranasals absent; supraoculars 5–7; supralabials 7; Supralabials V–VI suborbital; infralabials 6; auricular lobules absent; loreals 2; nuchals absent; tail slender, apically pointed, longer than head-body length; preanal scales enlarged; midbody scale rows 32–34, keeled; lamellae under Toe IV 26–27. **Coloration** Dorsum greyish-brown; silver vertebral stripe extends from forehead to tail-tip, edged by 2 series of black spots; postocular stripe black with cream spots, extends along flanks up to inguinal region; venter pale green or cream. **Subspecies** Two subspecies are recognized. *S. s. sanctus*, 2 continuous dark vertebral stripes; pale vertebral stripe less bright than dorsolateral pair. *S. s. tenggeranus*, 4 continuous dark vertebral stripes; pale vertebral stripes as bright as dorsolateral pair. **Habitat and Behaviour** Inhabits submontane forests at elevations of up to 1,200m asl. Diurnal and terrestrial. Diet and reproductive habits unstudied. **Distribution** Sumatra (*S. s. sanctus*) and Java (*S. s. sanctus* and *S. s. tenggeranus*). Also Maluku. **Status** Not Evaluated.

SELANGOR FOREST SKINK
Sphenomorphus scotophilus PLATE 35
Measurement SVL 50mm **Identification** Body slender; snout obtusely pointed; eye large; lower eyelid scaly; nostril in a single nasal; prefrontal meeting in midline; supraoculars 5; nuchals absent; supralabials 7; Supralabials V–VI subocular; infralabials 6; auricular lobules absent; midbody scale rows 30, smooth; lamellae under Toe IV 23. **Coloration** Dorsum dark brown with brownish-black and yellow spots; dorsolateral series of rounded cream spots; supralabials and infralabials white with black spots; venter unpatterned cream. **Habitat and Behaviour** Inhabits lowland forests, including those on offshore islands, as well as limestone caves. Diurnal, and arboreal and scansorial, associated with tree trunks and rocks. Diet consists of small arthropods. Reproductive habits unstudied. **Distribution** Southern Thailand (Khao Chong, Trang Province) and Peninsular Malaysia (Pulau Aur, Pulau Pemanggil, Pulau Tioman, Pulau Tulai, Seberang Peri, Pulau Penang and Kepong). **Status** Not Evaluated.

SHELFORD'S FOREST SKINK
Sphenomorphus shelfordi NOT ILLUSTRATED
Measurement SVL 67mm **Identification** Body slender; snout short, obtuse; lower eyelid scaly; nostril in a single nasal; supranasals absent; rostral as long as broad, in contact with frontonasal; frontal narrow, as long as frontoparietals and interparietal; supraoculars 4; supraciliaries 7; midbody scale rows 30–34, smooth; enlarged pair of anals; lamellae under Toe IV 27–29. **Coloration** Dorsum olive-brown irregularly spotted with black; black stripe from snout to inguinal region, broken into spots on flanks; axilla bright red in males; venter grey. **Habitat and Behaviour** Inhabits submontane forests at an elevation of 1,219m asl. Diet and reproductive habits unstudied. **Distribution** Endemic to north-eastern Borneo (Gunung Penrissen, Sarawak State). **Status** Not Evaluated.

PULAU SIBU FOREST SKINK
Sphenomorphus sibuensis NOT ILLUSTRATED
Measurement SVL 17.7mm in juvenile (adult size unknown) **Identification** Body slender; supraoculars 4; parietals contact supraoculars; single loreal; supralabials 6; infralabials 5; nuchals absent; midbody scale rows 29, smooth; lamellae under Toe IV 9. **Coloration** Dorsum mid-brown; yellow dorsolateral stripe extends from posterior edge of eye to mid-tail; stripe interrupted on neck, where it is a linear row of spots; forehead with pale spots; labials banded; venter grey. **Habitat and Behaviour** Inhabits lowland coastal forests close to mangrove swamps. Diurnal and terrestrial. Both known individuals were discovered under rotten logs. Diet and reproductive habits unstudied. **Distribution** Endemic to Peninsular Malaysia (Pulau Sibu). **Status** Not Evaluated.

STARRED FOREST SKINK
Sphenomorphus stellatus PLATE 36
Measurement SVL 80mm **Identification** Body slender; snout pointed; head distinct from neck; scales smooth, iridescent; paravertebral scales large; upper posterior temporals large; frontonasal width less than twice length; prefrontals large, juxtaposed; frontal longer than parietals and interparietal; supraoculars 4; presuboculars 3; supraciliaries 8; supralabials 5–6; infralabials 6; auricular lobules absent; 2 enlarged preanal scales; midbody scale rows 22–24, smooth; lamellae under Toe IV 18–23. **Coloration** Dorsum yellowish-, greenish- or bronze-brown with small, star-like white areas; dorsolateral dark stripe present or absent; labials dark and, except for first labial, white-spotted; limbs bronze with white flecks; throat with 7–8 distinct longitudinal dark bands between scale rows; venter pale with traces of dark pigmentation; tail with white spots laterally. **Habitat and Behaviour** Inhabits forests in the mid-hills. Known from under dead bark of standing trees. Diet and reproductive habits unstudied. **Distribution** South-eastern Thailand (Phu Wiang, Khon Kaen Province and Khao

Sa Bab, Chantaburi Province), south-western Cambodia (Chum Noab), southern Vietnam (Quang Nam, Kon Tum and Lam Dong Provinces), Peninsular Malaysia (Bukit Larut, Perak State) and western Borneo (Sarawak State). **Status** Not Evaluated.

MONTANE FOREST SKINK
Sphenomorphus tanahtinggi NOT ILLUSTRATED
Measurement SVL 64mm **Identification** Body robust; limbs large; parietals contact supraoculars; supraoculars 5; supralabials 8–9; infralabials 7; 3 small superimposed anterior loreal scales; prefrontals narrowly in contact or separated; auricular lobules absent; no large nuchals; midbody scale rows 40–42; lamellae under Toe IV 16–17. **Coloration** Dorsum olive-brown, darker on tail, without markings or with widely scattered small dark spots on centres of a few scales; sides of head and flanks dark brown; light line along dorsal margin of dark lateral band; 10 scale rows between dark bands; ventral margin of lateral band gradually fades into venter colour; few small light spots in lower portion of band; venter pale unpatterned olive-grey. **Habitat and Behaviour** Known from selectively logged forests at elevations of 850–1,180m asl. Diet and reproductive habits unstudied. **Distribution** Endemic to northern Borneo (Gunung Lumaku and Gunung Trus Madi, Sabah State). **Status** Not Evaluated.

TEMMINCK'S FOREST SKINK
Sphenomorphus temminckii PLATE 36
Measurement SVL 56mm **Identification** Body slender; snout short; lower eyelid scaly; parietals contact supraoculars; suture between frontonasal and rostral half or less than half rostral width; supraoculars 4; supralabials 6; Supralabials III–V suborbital; infralabials 4–5; loreals 2; no enlarged nuchals; auricular opening smaller than orbit; Toes III and IV subequal; preanal scales not enlarged; midbody scale rows 30–37, smooth; lamellae under Toe IV 9–11. **Coloration** Dorsum greyish-brown or mid-brown with dark brown lateral stripe from posterior corner of eye to flanks; sides of head with large pale and dark areas; tail with dark spots; venter unpatterned yellow. **Habitat and Behaviour** Inhabits lowland to montane forests at elevations of ca 350–2,000m asl. Diet unstudied. Oviparous; clutches comprise 2 eggs, measuring 5–5.5 x 10–11mm. Incubation period 56 days. **Distribution** Sumatra (Sijunjung and Lubuksulasih, Sumatera Barat Province), Java (Bogor, Garot, Tasek Malaya and Gunung Gede) and Bali. **Status** Not Evaluated.

NARROW-NECKED FOREST SKINK
Sphenomorphus tenuiculus NOT ILLUSTRATED
Measurement SVL 46mm **Identification** Body slender; tail thick; limbs relatively long; parietals in contact with one another; supraoculars 4; supralabials 7; infralabials 6; preanal scales enlarged; midbody scale rows 26; lamellae under Toe IV 21–24. **Coloration** Dorsum brown, spots on dorsum not fused to form longitudinal lines; flanks dark with white spots; venter unpatterned yellow. **Habitat and Behaviour** Inhabits submontane forests at an elevation of 1,225m asl. Diet and reproductive habits unstudied. **Distribution** Endemic to northern Borneo (Gunung Kinabalu, Sabah State). **Status** Not Evaluated.

PALE FOREST SKINK
Sphenomorphus tersus PLATE 36
Measurement SVL 92mm **Identification** Body slender; snout obtusely pointed; lower eyelid scaly; nuchals absent; supraoculars 4; supralabials 7; Supralabials V–VI subocular; infralabials 7; auricular opening large, lacking lobules; preanal scales enlarged; midbody scale rows 34, smooth; lamellae under Toe IV 18–19. **Coloration** Dorsum dark brown, unpatterned or with indistinct dark brownish-black and black spots and variegations, spots arranged in longitudinal series; flanks lighter; supralabials and infralabials with black bars along sutures; venter cream. **Habitat and Behaviour** Inhabits lowland forests at elevations of ca 100–300m asl. Diurnal and terrestrial. Diet and reproductive habits unstudied. **Distribution** Southern Thailand (Khao Wang Hip, Nakhon Si Thammarat Province and Pak Chan Estuary at Tasan, Isthmus of Kra) and Peninsular Malaysia (Kepong, Selangor State). **Status** Not Evaluated.

THREE-TOED FOREST SKINK
Sphenomorphus tridigitus NOT ILLUSTRATED
Measurement SVL 35mm **Identification** Body slender, elongated; snout elongated; rostral large; frontonasal with 2 widely separated prefrontals; nasal fused to Infralabial I; loreal not fused to preocular; several pairs of enlarged nuchals; supralabials 6; Supralabials II–V subocular; single preocular; postoculars 2; supraoculars 5; tympanum and supranasals absent; 2 large parietals in contact behind interparietal; reduced limbs bearing 3 fingers and 5 toes; tail as thick as body; ventrals subequal to dorsals; 2 enlarged preanal scales; median subcaudals enlarged; midbody scale rows 20; lamellae under Toe IV 7. **Coloration** Dorsum brownish-red; flanks darker, separated from dorsum by narrow stripe; forehead vermiculate; venter cream with irregular dark spots. **Habitat and Behaviour** Inhabits primary submontane and open forests at elevations of 940–1,470m asl. Daytime spent in concealment inside fallen logs on forest floor. Diet and reproductive habits unstudied. **Distribution** Laos (Champasak Province) and Vietnam (Bach Ma, Quang Nam Province). **Status** Not Evaluated.

THREE-BANDED FOREST SKINK
Sphenomorphus tritaeniatus NOT ILLUSTRATED
Measurement SVL 47mm **Identification** Body slender; snout obtuse; forehead scales smooth; rostral in broad contact with frontonasal, which is wider than long; prefrontals separated; frontal as long as frontoparietals and joined; parietals in narrow contact behind interparietals; 3 scales edge parietal; supraoculars 4; supralabials 7; Supralabial V subocular;

supraoculars separated from orbit by series of scales; auricular opening lacks lobules, larger than half orbit diameter; midbody scale rows 38, smooth; 2 enlarged preanal scales; tail gradually tapering, with widened subcaudals; Fingers III and IV subequal; lamellae under Toe IV 15. **Coloration** Dorsum brownish-grey with 3 discontinuous brown stripes and scattered pale spots; median stripe commences on nape and becomes thick and distinct on midbody, disappearing beyond tail-base; dorsolateral stripes commence anterior to orbit, cross eye and terminate a third of the way down tail; venter pale, unpatterned. **Habitat and Behaviour** Inhabits montane forests. Diet and reproductive habits unstudied. **Distribution** Endemic to northern Vietnam (Tam Dao, Vinh Phú Province and Cuc Phuong, Ninh Phuong Province). **Status** Not Evaluated.

VAN HEURN'S FOREST SKINK

Sphenomorphus vanheurni NOT ILLUSTRATED

Measurement SVL 64mm **Identification** Body slender; snout short, obtuse; supranasals absent; lower eyelid scaly; auricular opening large, lacking lobules; loreals 2; supraoculars 4; supralabials 6; Supralabial IV suborbital; nuchals absent; midbody scale rows 31; digits short and thick; lamellae under Toe IV 12–15. **Coloration** Dorsum light brown with 6 longitudinal lines composed of dark brown spots; forehead brown mottled with dark brown; supralabials greyish-brown with pale spots; flanks variegated with greyish-brown and cream spots; gular region dark grey with faint cream mottling; venter mostly cream with isolated small grey dots. **Subspecies** Two subspecies are recognized. *S. v. vanheurni*, dorsals 81 between parietal up to line connecting backs of thighs; lamellae under Toe IV 14–15. *S. v. balicus*, dorsals 64 between parietal up to line connecting backs of thighs; lamellae under Toe IV 12. **Habitat and Behaviour** Inhabits the lowlands to montane limits at elevations of 55–1,900m asl. Diurnal and terrestrial. The holotype was collected from the shores of the crater lake of Taman Hidup. Diet and reproductive habits unstudied. **Distribution** Eastern Java (Taman Hidup, above Bremi, Gunung Hiyang) (*S. v. vanheurni*) and Bali (Gitgit) (*S. v. balicus*). **Status** Not Evaluated.

BAVI WATER SKINK

Tropidophorus baviensis PLATE 36

Measurement SVL 91mm **Identification** Body robust, dorsoventrally depressed; forehead scales smooth; supranasals absent; paired prefrontals; supraoculars 4; supraciliaries 7; frontoparietals in contact; loreals 2, rugose; supralabials 6; Supralabial IV subocular; infralabials 6; dorsals strongly keeled and slightly smaller than ventrals, arranged in 8 longitudinal rows, with lateral 2 rows strongly keeled; flank scales in 6–7 longitudinal rows; ventral scales in 8 longitudinal rows; 2 enlarged preanal scales; midbody scale rows 28–30; lamellae under Toe IV 18–22. **Coloration** Dorsum dark brown with cream spots and blotches forming broken bands; flanks dark brown with lighter markings; forehead dark brown; venter unpatterned cream. **Habitat and Behaviour** Inhabits forests in the mid-hills at an elevation of 400m asl. Typically found near hill streams, more rarely in burrows excavated on embankments. Diet and reproductive habits unstudied. Ovoviviparity and parental care of neonates suspected. **Distribution** Endemic to Vietnam (Bavi Mountains, Ha Tay Province and Cuc Phuong National Park, Ninh Binh Province). **Status** Not Evaluated.

BECCARI'S WATER SKINK

Tropidophorus beccarii PLATE 36

Measurement SVL 98mm **Identification** Body robust, rounded in adults, slender in juveniles; scales smooth at least in adults; forehead scales smooth; prefrontals in broad contact or separated; tympanum smaller than orbit of eye; midbody scale rows 28–30; lamellae under Toe IV 19–21. **Coloration** Dorsum dark brown or reddish-brown with dark brown blotches and cross-bars; sides of head and flanks with light spots; broad dark stripe on flanks, with vertical light bars or wedge-shaped marks. **Habitat and Behaviour** Inhabits rocky streams within dipterocarp forests at elevations of up to 1,000m asl. Diet comprises water insects. Ovoviviparous; clutches comprise 4 neonates, measuring 30mm. **Distribution** Endemic to Borneo. **Status** Not Evaluated.

BERDMORE'S WATER SKINK

Tropidophorus berdmorei PLATE 36

Measurement SVL 97mm **Identification** Body robust, rounded; forehead scales smooth; dorsals and laterals smooth or obtusely keeled; laterals directed backwards; frontonasal entire; supraciliaries 8; loreals 2; scales from parietals to above vent 64; scales from chin shields to vent 53; midbody scale rows 32–40. **Coloration** Dorsum brown with transverse dark and pale orange marks; subcaudals bright orange. **Habitat and Behaviour** Inhabits the forested mid-hills at elevations of ca 600–900m asl. Diurnal and aquatic. Diet and reproductive habits unstudied. **Distribution** Myanmar (Mergui, Tenasserim, Taninthayi Division, Bago, Bia-po, Kayin Hills and Bhamo, Kachin State), northern, north-eastern and western Thailand (S Utaradit, Doi Nga Chang and Me Wang) and Vietnam. Also southern China. **Status** Not Evaluated.

BROOKE'S WATER SKINK

Tropidophorus brookei PLATE 36

Measurement SVL 101mm **Identification** Body robust, rounded in adults, slender in juveniles; tympanum smaller than orbit; single postmental; midbody scale rows 32, keeled, forming 8 longitudinal ridges on dorsum and oblique ones on flanks. **Coloration** Dorsum dark brown with darker spots and blotches forming indistinct transverse bands; black spot on sides of neck; flanks with dark and white spots; venter cream. **Habitat and Behaviour** Inhabits lowland dipterocarp forests. Diurnal and aquatic, associated with small rocky hill streams. Both juveniles and adult females shelter by night on leaves and the stems of saplings overhanging hill streams, and in clefts in rock faces of waterfalls and buttresses. Adult males sleep inside rocky clefts within and along streams. Diet presumably consists of aquatic arthropods. Ovoviviparous; clutches comprise 1–5 neonates. **Distribution** Endemic to Borneo. **Status** Not Evaluated.

INDO-CHINESE WATER SKINK
Tropidophorus cocincinensis PLATE 36
Measurement SVL 80mm **Identification** Body robust, rounded; Supralabial V largest and suborbital in position; series of small scales between loreals and supralabials; scales keeled in juveniles, smooth in adults; caudals strongly keeled; laterals directed obliquely; midbody scale rows 30–32; preanal scales 2. **Coloration** Dorsum light brown or orangish-brown with indistinct darker markings; flanks brownish-black with small white spots or orange areas; venter cream; gular region with grey mottling. **Habitat and Behaviour** Inhabits lowland forests at elevations of 109–160m asl. Diurnal and aquatic, associated with rocky streams. Diet unstudied. Ovoviviparous; clutches comprise 7–9 neonates, measuring 26–30mm. **Distribution** North-eastern Thailand (Khao Phanom Dongrak Range), Laos (Xe Kong Province) and Vietnam (Kon Tum, Quang Binh, Thua-Thien-Hue, Da Nang and Thua Thien-Hue Provinces). **Status** Not Evaluated.

HAINAN WATER SKINK
Tropidophorus hainanus NOT ILLUSTRATED
Measurement SVL 52mm **Identification** Body robust, rounded; forehead scales strongly striated; frontonasal single; supraoculars 4; supralabials 6; Supralabial IV subocular; infralabials 5; dorsals and laterals with sharp keels; some scales in vertebral row bicarinate or, when unicarinate, smaller than adjacent scales; laterals directed obliquely; midbody scale rows 30–40; preanal scales 2. **Coloration** Dorsum reddish-brown with indistinct light-edged cross-bars, the anterior 2 V-shaped; flanks with large, dark-edged cream blotches; venter white with black speckles; throat with longitudinal white streaks. **Habitat and Behaviour** Inhabits submontane forests at elevations of 870–1,550m asl. Diurnal and semi-aquatic, associated with rocky streams. Diet includes termites. Reproductive habits unstudied. **Distribution** Vietnam. Also eastern China. **Status** Not Evaluated.

SPINY-TAILED WATER SKINK
Tropidophorus hangnam PLATE 36
Measurement SVL 78.2mm **Identification** Body robust, dorsoventrally flattened; forehead scales smooth; supranasals absent; prefrontals in contact; enlarged nuchals 1–2; single large loreal; lower eyelid scaly; preoculars 3; supralabials 5–6; Supralabial V subocular; infralabials 5–6; dorsals weakly keeled; flanks and dorsal surfaces, anterior and posterior of limbs keeled; tail with spinous keels laterally; paired enlarged preanal scales; median subcaudals not enlarged; midbody scale rows 28–31; lamellae under Toe IV 18–19. **Coloration** Dorsum blackish-brown with interrupted transverse yellow bars; forehead reddish-brown; venter yellowish-cream. **Habitat and Behaviour** Inhabits rocky banks of seasonal streams. Possibly diurnal, and scansorial and aquatic, resting sites including holes in rocks. Diet and reproductive habits unstudied. **Distribution** Endemic to north-eastern Thailand (Phu Khiew Wildlife Sanctuary, Chaiyaphum Province). **Status** Not Evaluated.

SUNGEI KAJAN WATER SKINK
Tropidophorus iniquus NOT ILLUSTRATED
Measurement SVL 96mm **Identification** Body slender, rounded; snout acute; forehead shields rugose; tympanum smaller than orbit; supraciliaries 6; supraoculars 5; Supralabial V largest, contacting orbit; supralabials 8; infralabials 5; large azygous postmental, followed by 2 pairs of large postmentals in contact; 3 swollen bands on nuchals, separated by grooves; dorsals bicarinate, grooved; midbody scale rows 34; single large preanal scale; subdigital lamellae smooth. **Coloration** Dorsum dark grey; venter unpatterned white; undersurfaces of tail and feet dark (preserved coloration; colours in life unknown). **Habitat and Behaviour** Inhabits the mid-hills in primary forests. Diet and reproductive habits unstudied. **Distribution** Endemic to Borneo (upper reaches of Sungei Kajan, Kalimantan Province). **Status** Not Evaluated.

LAOTIAN WATER SKINK
Tropidophorus laotus PLATE 36
Measurement SVL 75mm **Identification** Body robust, rounded; forehead scales smooth; frontonasal and usually also anterior loreal divided; supraoculars 4; supralabials 6; Supralabial IV subocular; infralabials 5; dorsal and lateral scales smooth in adults; laterals directed backwards; midbody scale rows 33–34; preanal scales 2; lamellae under Toe IV 18–20. **Coloration** Dorsum dark brown, sometimes with V-shaped, black-edged light bars; flanks with small white spots; venter cream; subcaudals with thick black spots. **Habitat and Behaviour** Inhabits submontane forests. Possibly nocturnal and crepuscular, and aquatic, associated with hill streams. Diet and reproductive habits unstudied. **Distribution** Thailand (Phu Luang and Phu Kradung, Loei Province, and Phu Wua, Nong Khai Province) and Laos (Muang Liep, nr Pak Lai). **Status** Not Evaluated.

BROAD-SCALED WATER SKINK
Tropidophorus latiscutatus NOT ILLUSTRATED
Measurement SVL 102mm **Identification** Body robust, dorsoventrally depressed; forehead scales smooth; supraciliaries 6–7; supraoculars 4; enlarged nuchals 3–4; frontonasal entire; supralabials 6; Supralabials IV subocular; infralabials 5 (rarely 4 or 6); paired enlarged preanals; median subcaudals enlarged; midbody scale rows 28–30; paravertebral scale rows smooth, enlarged, 58–63; scale rows at tenth subcaudal on tail 13; lamellae under Toe IV 18–22. **Coloration** Dorsum brown with irregularly shaped, transverse pale brown bands, numbering 2 on head, 9 on body and 10 on tail; labials with pale brown spots; gulars, venter and anterior subcaudals yellowish-ivory;

posterior subcaudals brownish-yellow. **Habitat and Behaviour** Inhabits dry open forests near marshes at elevations of ca 200m asl. Possibly nocturnal and aquatic, with saxicolous habits. Diet and reproductive habits unstudied. **Distribution** Endemic to north-eastern Thailand (Phu Wua Wildlife Sanctuary, Nonh Kai Province). **Status** Not Evaluated.

MATSUI'S WATER SKINK
Tropidophorus matsuii   NOT ILLUSTRATED
Measurement SVL 94mm **Identification** Body robust, dorsoventrally depressed; forehead scales smooth; frontonasal divided; supraciliaries 8; supraoculars 4; supralabials 6; Supralabial IV subocular; infralabials 6; paired enlarged preanals; median subcaudals enlarged; midbody scale rows 34; paravertebral scale rows smooth or weakly keeled, 65; scale rows at tenth subcaudal on tail 15; lamellae under Toe IV 22–23. **Coloration** Dorsum dark brown with transverse pale brown bands, numbering 3 on head, 9 on body and 23 on tail; flanks with several irregular brown spots; labials with small pale brown spots; gular and venter yellowish-ivory; subcaudals yellowish-ivory with indistinct dark flecks. **Habitat and Behaviour** Inhabits lowland humid evergreen forests at elevations of ca 350m asl. Associated with sandstone outcrops. Diet and reproductive habits unstudied. **Distribution** Endemic to north-eastern Thailand (Phu Pa Namtip, Roi Et Province). **Status** Not Evaluated.

SMALL-SCALED WATER SKINK
Tropidophorus microlepis   NOT ILLUSTRATED
Measurement SVL 83mm **Identification** Body robust, rounded; Supralabial V largest and subocular in position; series of small scales between loreals and supralabials; Infralabial I reduced; Infralabial II contacts postmental; forehead scales rugose; dorsals and laterals strongly keeled and spinous; midbody scale rows 28–32; preanal scales 3. **Coloration** Dorsum light brown with 6 dark brown blotches; indistinct anterolateral stripe anteriorly; indistinct black spots on anterior flanks; limbs and digits dark-banded; venter unpatterned cream. **Habitat and Behaviour** Associated with primary evergreen forests at elevations of 180–440m asl. Diurnal and semi-aquatic. Diet unstudied. Ovoviviparous; clutches comprise 7–9 neonates, measuring 56–60mm. **Distribution** South-eastern Thailand (Khao Sebab and Khao Soi Dao, Chanthaburi Province, and Pang Sida National Park, Sa Kaeo Province), Laos (Champasak Province), eastern Cambodia (O'Rang and Keo Seima Districts) and central Vietnam (Dran, Langbian Plateau, Lam Dong Province). **Status** Not Evaluated.

SMALL-FOOTED WATER SKINK
Tropidophorus micropus   NOT ILLUSTRATED
Measurement SVL 40mm **Identification** Body slender, rounded; scales on forehead striated; frontonasal as long as broad; superciliaries 7; scales on flanks small; ventrals smooth, larger than laterals; midbody scale rows 34; subdigital lamellae smooth. **Coloration** Dorsum dark brown with black spot on sides of neck; venter cream with irregular dark spots. **Habitat and Behaviour** Inhabits lowland forests. Retreat sites include fallen tree trunks and cracks in rocks. Defensive behaviour includes excretion of an unpleasant-smelling musk from its cloacal glands. Diet unstudied. Ovoviviparous; clutches comprise 3 neonates, measuring 26–27.4mm. **Distribution** Endemic to Borneo. **Status** Not Evaluated.

MOCQUARD'S WATER SKINK
Tropidophorus mocquardii   PLATE 37
Measurement SVL 95mm **Identification** Body slender; forehead scales smooth; tympanum smaller than orbit; midbody scale rows 34, smooth at least in adults; digits short with smooth lamellae. **Coloration** Dorsum brown with transverse dark bands; flanks with white spots; venter cream. **Habitat and Behaviour** Inhabits the mid-hills within primary forests. Diurnal and aquatic, concealing itself in rocky streams. Distress call a low squeak. Diet and reproductive habits unstudied. **Distribution** Endemic to Borneo (Gunung Kinabalu, Sabah State). **Status** Not Evaluated.

MURPHY'S WATER SKINK
Tropidophorus murphyi   PLATE 37
Measurement SVL 85.1mm **Identification** Body robust, dorsoventrally depressed; forehead scales smooth; supraciliaries 6–8; supraoculars 4; frontonasal single; 3 enlarged pairs of nuchals; supralabials 6; Supralabial IV midorbital; infralabials 5–6; paired enlarged preanals; median subcaudals enlarged; paravertebral scale rows smooth or weakly keeled, 55–67; midbody scale rows 30–32; scale rows at tenth subcaudal on tail 13; lamellae under Toe IV 24. **Coloration** Dorsum dark brown with transverse pale brown bands, numbering 3 on head, 7 on body and 17 on tail; labials with pale brown spots; gular and venter yellowish-ivory; subcaudals yellowish-ivory with indistinct dark flecks. **Habitat and Behaviour** Inhabits rocky banks of streams at elevations of 700–750m asl. Nocturnal and saxicolous. Diet unstudied. Ovoviviparous; clutches comprise 3–5 neonates, measuring 28mm. **Distribution** Endemic to northern Vietnam (Quang Thanh Village, Cao Bang Province). **Status** Not Evaluated.

NOGGE'S WATER SKINK
Tropidophorus noggei   PLATE 37
Measurement SVL 110.2mm **Identification** Body robust, distinctly depressed; head triangular; frontonasal and frontal undivided; supralabials 6; Supralabials IV and VI largest; postmental undivided; tympanum superficial, ovoid; dorsals enlarged, smooth, except outermost row; midbody scale rows 22–24; lamellae under Toe IV 18–20. **Coloration** Dorsum dark brown, scales edged with light brown; 3 pale bands on neck, 6–9 pale brown bands on dorsum; tail dark-banded; venter yellowish-beige to greyish with indistinct dark marbling. **Habitat and**

Behaviour Inhabits karst areas within primary forests at elevations of 300–400m asl. Shelters in rock crevices 0.3–1.5m above substratum. Diet includes worms, ants, termites, katydids and centipedes. Clutch size 2–7, and suspected of being ovoviviparous, like its congeners. **Distribution** Endemic to central Vietnam (Cha Noi, Phong Nha-Ke Bang National Park, Quang Binh Province). **Status** Not Evaluated.

PERPLEXING WATER SKINK
Tropidophorus perplexus PLATE 37
Measurement SVL 73mm **Identification** Body slender, rounded; dorsal and lateral scales large and strongly keeled; scales on forehead rugose; frontonasal divided, as broad as long; supraoculars 5; supraciliaries 5–6; supralabials 6; tympanum subequal to orbit; midbody scale rows 30; gulars keeled; ventrals large, cycloid, imbricate and smooth; lamellae under digits feebly keeled; tail long, slightly compressed, with keeled scales, upper rows spinose. **Coloration** Dorsum rich brown with paler narrow cross-bars; venter yellow. **Habitat and Behaviour** Inhabits primary forests in the mid-hills. Diurnal, associated with forest streams. Sites of concealment include rotten logs. Diet and reproductive habits unstudied. **Distribution** Endemic to Borneo (Sungei Tinjar and Sungei Mengiong, northern Sarawak State, and Mendolong and Marak Parak, Sabah State). **Status** Not Evaluated.

ROBINSON'S WATER SKINK
Tropidophorus robinsoni NOT ILLUSTRATED
Measurement SVL 75mm **Identification** Body robust, rounded; forehead scales rugose; frontonasal entire; supraoculars 4; loreals 2; supralabials 6; Supralabial IV subocular; dorsals and laterals sharply keeled; laterals directed backwards; midbody scale rows 30–40; lamellae under Toe IV 17–18; preanal scales 2. **Coloration** Dorsum dark brown or blackish-brown with black-edged, light brown cross-bars or spots; flanks and tail with small white spots; head dark; labials with white spots; gular region spotted with black; venter unpatterned yellowish-white. **Habitat and Behaviour** Inhabits lowland forests. Diurnal and aquatic. Diet unstudied. Ovoviviparous; clutches comprise 4–5 neonates. **Distribution** Myanmar (Tanintharyi Division) and peninsular Thailand (Tasan, Chumphon Province). **Status** Not Evaluated.

CHINESE WATER SKINK
Tropidophorus sinicus PLATE 37
Measurement SVL 65mm **Identification** Body robust, rounded; head narrow; forehead scales strongly striated; dorsals with sharp keels; divided postmentals and frontonasals; loreals 2; infralabials 5; laterals directed obliquely; midbody scale rows 28–30; preanal scales 2. **Coloration** Dorsum dark brown with 10 rusty-brown bars on body and tail; supralabials and infralabials cream with black bars; venter salmon-pink. **Habitat and Behaviour** Inhabits streams in montane forests. Presumably nocturnal and aquatic. Diet consists of termites, cockroaches, insect larvae and earthworms. Ovoviviparous; clutches comprise up to 6 neonates. **Distribution** Vietnam (Man Son Mountains, Bac Ky or Dong Kinh). Also eastern China. **Status** Not Evaluated.

THAI WATER SKINK
Tropidophorus thai NOT ILLUSTRATED
Measurement SVL 85mm **Identification** Body robust, rounded; forehead scales rugose; frontal and frontonasal divided; supraoculars 4; supralabials 6; Supralabial IV subocular; dorsals and laterals strongly keeled; laterals directed obliquely; midbody scale rows 38; 2 preanal scales; lamellae under Toe IV 18–19. **Coloration** Dorsum light brown with series of oblique, dark-edged, V-shaped yellow marks; flanks with small pale spots; gular region paler; labials black with cream spots; venter brownish-white. **Habitat and Behaviour** Inhabits primary forests at elevations of ca 600m asl. Diurnal and aquatic. Diet and reproductive habits unstudied. **Distribution** Endemic to north-western Thailand (Pa Meang, Me Wang, Chiang Mai Province). **Status** Not Evaluated.

ROUGH VIETNAMESE SKINK
Vietnascincus rugosus NOT ILLUSTRATED
Measurement SVL 82mm **Identification** Body robust; snout pointed; pterygoid teeth absent; eyelids well developed, lower eyelid with translucent disc; nuchals paired; nostril at posterior of undivided nasal, followed by small postnasal; supranasals present; forehead scales rugose; auricular opening small with 3 auricular lobules; tympanum depressed; dorsal scales with sharp keels; limbs developed; midbody scale rows 36; scales with 5–7 keels. **Coloration** Dorsum bluish-olive with 2 yellow dorsolateral stripes on anterior dorsum that gradually diffuse posteriorly; venter yellow or greenish-grey. **Habitat and Behaviour** Inhabits primary forests at an elevation of 700m asl. Diurnal and arboreal/semi-arboreal, associated with fallen tree trunks. Diet and reproductive habits unstudied. **Distribution** Endemic to central Vietnam (Buon Luoi, Gia Lai Province). **Status** Not Evaluated.

Family SHINISAURIDAE CROCODILE LIZARDS

In crocodile lizards, the temporal arches of the skull are strongly developed and large temporal openings are not roofed by the skull bones, which are rugose by fusion to the cranial osteoderms and have small head scales. Crocodile lizards are restricted to cold mountain streams in evergreen forests, and are predators of small aquatic invetebrates. The single extant species is known from northern Indo-China and eastern China.

CHINESE CROCODILE LIZARD
Shinisaurus crocodilurus   PLATE 37

Measurement SVL 300mm **Identification** Body robust; head short; males with enlarged neck scales and more developed cranial crest than that of females; tail longer than head-body length; supralabials 16–21; infralabials 10–16; enlarged collar scales 10–12; preanal scales 4–8; transverse ventral rows 31–42; lamellae under Toe IV 17–27. **Coloration** Dorsum grey or reddish-brown; red, orange and yellow blotches on head, throat and flanks of males; radiating dark lines from orbit; venter reddish-brown in males, cream or grey in females. Juveniles with distinct yellow spot on snout. **Habitat and Behaviour** Inhabits evergreen forests at elevations of 200–1,500m asl. Solitary, slow moving and arboreal, associated with dense vegetation on the banks of slow-moving mountain streams and ponds. Hibernates for 3–4 months annually. Sleeps on branches above shallow ponds and streams. Forages by day in water, and escapes from predators by swimming, being able to remain submerged for more than 30 minutes. Diet consists of insects, spiders, crustaceans, tadpoles, snails and probably also earthworms. Ovoviviparous; clutches comprise 2–12 neonates, measuring 10–15cm. **Distribution** North-eastern Vietnam (Yen Tu Nature Reserve, Quang Ninh Province). Also eastern China. **Status** Vulnerable.

Family VARANIDAE MONITOR LIZARDS

Monitor lizards are characterized by a small head, long neck and powerful long tail. They have a long and smooth bifid tongue, rounded pupils, small, juxtaposed cephalic scales, rounded or oval dorsals, quadrangular ventrals in transverse rows and preanal pores. They live in a variety of habitats, from wetlands and sea coasts to arid deserts, and feed on invertebrates and vertebrates (two extralimital species seasonally consume fruits). Monitor lizards are a sister group to the snakes, sharing with them a forked tongue and an ability to consume large prey. They are oviparous, and their eggs are softshelled. Extant species are found in the Old World tropics and subtropics, Asia, New Guinea, Australia and Africa.

BENGAL MONITOR LIZARD
Varanus bengalensis   PLATE 37

Measurement SVL 1,740mm **Identification** Body slender; snout somewhat elongated; nostril nearer eye than snout-tip; nostril an oblique slit; supraoculars not enlarged; nuchal scales rounded; crown scales larger than nuchal scales; midventral scales smooth; tail flattened. **Coloration** Dorsum of juveniles pale olive or dark grey, with yellow bands comprising spots in transverse series; snout unpatterned; adults unicoloured or with less distinct pattern; chin and throat with black spots; venter cream or yellow. **Habitat and Behaviour** Inhabits forests and forest edges, as well as plantations. Diurnal and largely terrestrial. Diet consists of insects, spiders, snails, crabs, frogs, small mammals, birds, lizards and snakes; also known to scavenge. Clutches comprise 12 eggs, measuring 29 x 15mm, which may be laid in termitaria. Hatchlings 94mm. **Distribution** Western Myanmar and north-western Vietnam. Also Afghanistan, Pakistan, India, Sri Lanka, Bangladesh, Nepal and southern China. **Status** Not Evaluated.

DUMÉRIL'S MONITOR LIZARD
Varanus dumerilii   PLATE 37

Measurement SVL 1,500m **Identification** Body robust; head small, flattened; snout short and broad; nostrils elongated; tympanum large, rounded; nuchal scales oval, flat, smooth or posteriorly feebly keeled; abdominal scales weakly keeled, in 37–41 rows; tail laterally compressed. **Coloration** Dorsum brownish-yellow or tan with dark temporal streak from eye to auricular opening; forehead orange or yellowish-orange, especially in juveniles, becoming tan or yellow with growth; vertical dark bars on lips; throat yellow with 6–8 orange stripes; venter yellow with transverse dark bars. **Habitat and Behaviour** Associated with lowland forests and the mid-hills, especially mangrove swamps. Diurnal and arboreal. Diet consists primarily of crabs, although ants, scorpions, beetle larvae, spiders, fish, other lizards, eggs and rodents may also be consumed. Clutches comprise 12–23 eggs. Incubation period 215–234 days. Hatchlings 81–83.5mm. **Distribution** Southern Myanmar, Thailand, Peninsular Malaysia, Singapore, Sumatra, Borneo, Pulau Bangka and Pulau Belitung. **Status** Not Evaluated.

SOUTH-EAST ASIAN MONITOR LIZARD
Varanus nebulosus   PLATE 37

Measurement SVL 1,200m **Identification** Body slender; snout somewhat elongated; nostril nearer eye than snout-tip; nostril an oblique slit; median supraoculars transversely enlarged; nuchal scales rounded, larger than crown scales; midventral scales smooth; tail flattened. **Coloration** Dorsum of juveniles dark grey with yellow bands; adults unicoloured dark grey or with reduced pattern; chin and thoat with transverse black bars; tail with transverse bands; venter cream or yellow. **Habitat and Behaviour** Inhabits the edges of open deciduous forests, and plantations. Diurnal, and terrestrial and arboreal. Diet consists of invertebrates and small vertebrates; also known to scavenge. **Distribution** Myanmar, Thailand, Laos, Cambodia, Vietnam, Peninsular Malaysia, Singapore, Sumatra and Java. **Status** Not Evaluated.

ROUGH-NECKED MONITOR LIZARD
Varanus rudicollis   PLATE 37
Measurement SVL 1,460m **Identification** Body slender; snout relatively long; nuchal scales strongly keeled; abdominal scales keeled, in 79–90 transverse rows; body and neck somewhat slender; limbs relatively thin. **Coloration** Dorsum blackish-grey in adults, with yellow tinge on neck and foreparts of body in juveniles; neck with 3 black stripes; flanks with yellow ocelli; melanistic individuals common; hatchlings with horizontal yellow and black bands on venter. **Habitat and Behaviour** Inhabits lowland forests. Diurnal and arboreal. Diet consists of ants, termites, stick insects, cockroaches, grasshoppers, spiders, scorpions and, less frequently, small mammals, frogs, fish and crabs. Clutches comprise 13–14 eggs, and up to 3 clutches are laid. Incubation period 180–184 days. Hatchlings 238–260mm. **Distribution** Southern Myanmar, Thailand, Peninsular Malaysia, Sumatra, Pulau Bangka and Borneo. **Status** Not Evaluated.

WATER MONITOR LIZARD
Varanus salvator   PLATE 37
Measurement SVL ca 800mm (*V. s. macromaculatus*); 505mm (*V. s. bivittatus*) **Identification** Body robust in adults, relatively slender in juveniles; snout depressed; nostrils rounded or oval, twice as far from orbit as from snout-tip; nuchal scales strongly keeled; crown scales large, flat, smooth, larger than nuchal scales; supraoculars well differentiated; midventral scales feebly keeled, numbering 148–153; tail strongly compressed with double-toothed crest above; caudal scales keeled dorsally and ventrally. **Coloration** Juveniles dark dorsally, yellow spotted or with ocelli in transverse series; snout black barred, especially on lips; venter yellow with narrow, vertical V-shaped black marks extending to sides of venter; dorsum darkens with growth – occasionally all vestiges of chain-like yellow pattern are lost. **Subspecies** Four subspecies are recognized, two of which occur in the region. *V. s. macromaculatus*, head long; dorsal spots not fused. *V. s. bivittatus*, head short; 1 dark band on sides of neck; first transverse row of dorsal spots in front of forelimbs a continuous light cross-bar. **Habitat and Behaviour** Inhabits mangrove swamps, riverbanks, canals, dipterocarp forests, towns and cities. Diurnal, and terrestrial and sometimes arboreal; also a strong swimmer. Diet includes large invertebrates and small vertebrates such as insects, fish, crabs, freshwater turtles, birds and their eggs, crocodiles, sea turtles, lizards and rats, in addition to carrion. Clutches comprise 5–30 eggs, measuring 32.3–42.9 x 64–82.6mm. Incubation period 180–327 days. Hatchlings 180–300mm. **Distribution** Myanmar, Thailand, Laos, Cambodia, Vietnam, Peninsular Malaysia, Singapore, Sumatra, Pulau Nias, Pulau Bangka, Pulau Belitung, Pulau Nias, Pulau Weh, the Mentawai Archipelago, Borneo (*V. s. macromaculatus*), Java and Bali (*V. s. bivittatus*). Also Sri Lanka (*V. s. salvator*), India, southern China (*V. s. macromaculatus*) and the Lesser Sundas (*V. s. bivittatus*). **Status** Not Evaluated.

Family ACROCHORDIDAE WART SNAKES

Species in this family have a heavy body with granular scales, loose, folded skin with bristle-tipped tubercles, valvular nostrils, dorsally directed eyes and a flap for closing the lingual opening of the mouth. They are aquatic snakes occupying freshwaters and sea coasts, and appear to be nocturnal, hiding under fallen logs and other debris underwater, and hunting for fish at night. They are found in South-East Asia, New Guinea and Australia. Some are killed for the belief that they destroy fish. Their flesh is consumed locally, and their skin has some commercial value.

WART SNAKE
Acrochordus granulatus   PLATE 38
Measurement TL 1,000mm **Identification** Body stout, compressed; head indistinct from neck, covered with small juxtaposed scales; supralabials 8–11; row of small scales separates supralabials from mouth; infralabials 12–18; eye tiny; vertically elliptical pupil; tail short, prehensile; fold of skin along middle of abdomen; rostral absent; chin shields absent; midbody scale rows 100, with those on vertebral region being largest. **Coloration** Dorsum olive, blue or blackish-grey with transverse cream bands, best marked in juveniles and sometimes disappearing in adults. **Habitat and Behaviour** Inhabits estuaries, mangrove swamps and sea coasts. Nocturnal and aquatic, associated with waters as deep as 20m, and can remain submerged for at least 139 minutes at a time. Diet consists of burrowing gobies, eels, crabs and other snakes. Ovoviviparous; clutches comprise 6–12 neonates, measuring 220–230mm. **Distribution** Myanmar, Thailand, Cambodia, Vietnam, Peninsular Malaysia, Sumatra, Borneo and Java. Also India, Sri Lanka, Bangladesh, Pakistan, China, the Philippines and the eastern Indonesian islands of Ambon, Flores, Jobi, Schouten, Sulawesi, Ternate, Timor and Pulau Weh, New Guinea and Australia. **Status** Not Evaluated.

ELEPHANT TRUNK SNAKE
Acrochordus javanicus   PLATE 38
Measurement TL 2m **Identification** Body extremely stout, slightly compressed; head indistinct from neck; supralabials 22–36; infralabials 31–34; forehead scales small and rough; eye small; dorsals keeled; midbody scale rows 120–150, with those around vertebrals largest; tail short and prehensile. **Coloration** Dorsum greyish-black with darker lines on head; 2 diffuse longitudinal stripes and elongated dark blotches on

flanks; venter cream. **Habitat and Behaviour** Inhabits ditches and canals in freshwater situations, including blackwater streams. Shelters under fallen logs underwater, or under the roots of trees growing in or along water, emerging at night to feed on fish such as eels and catfish. Ovoviviparous; clutches comprise 6–48 neonates, measuring 290–460mm. **Distribution** Thailand, Cambodia, Vietnam, Peninsular Malaysia, Singapore, Sumatra, Borneo and Java. **Status** Not Evaluated.

Family ANOMOCHILIDAE GIANT BLIND SNAKES

Members of this family are characterized by a subcylindrical body, and by the absence of a chin groove and teeth on pterygoid or palatine bones. They are found in forested habitats in South-East Asia, and are at least superficially similar to the Asian pipesnakes (Cylindrophiidae). The natural history of these snakes is among the least known among all tropical snakes, and populations appear localized. They are represented by three species, all of which are on Sundaland.

MALAYAN GIANT BLIND SNAKE
Anomochilus leonardi PLATE 38
Measurement TL 228mm **Identification** Body stout, rounded in cross-section; head small, indistinct from neck; forehead covered with large scales; an azygous parietofrontal; nostril in single nasal, in contact with Supralabial II; loreal and preocular absent; single postocular; eye small; mental groove absent; tail short and conical; dorsals smooth, slightly larger than ventrals at same level; midbody scale rows 17 or 19; ventral scales 214–252; subcaudals 6–7. **Coloration** Dorsum glossy black or purplish-brown with oval yellow spots; yellow bar covers most of frontal and part of preoculars; venter black; subcaudals red. **Habitat and Behaviour** Inhabits the plains and mid-hills at elevations of ca 250m asl. Nocturnal and subfossorial. Diet and reproductive habits unstudied. **Distribution** Peninsular Malaysia (Sungai Ngeram, nr Merapoh, Pahang State, and Ulu Gombak and Kepong, Selangor State) and Borneo (Sepilok, Sabah State). **Status** Data Deficient.

KINABALU GIANT BLIND SNAKE
Anomochilus monticola PLATE 38
Measurement TL 521.2mm **Identification** Body stout, rounded in cross-section; head small, indistinct from neck; forehead covered with large scales; an azygous parietofrontal; nostril in single nasal, in contact with Supralabial II; loreal and preocular absent; single postocular; eye small; mental groove absent; tail short and conical; dorsals smooth, slightly larger than ventrals at same level; midbody scale rows 19; ventrals 258–261; subcaudals 7–8. **Coloration** Dorsum blue-black, lacking light line along flanks and large pale blotches on either side of vertebral; transverse yellow bar across snout; series of isolated pale yellow scales on flanks; venter dark brown. **Habitat and Behaviour** Inhabits submontane forests at an elevation of 1,450m asl. Nocturnal and subfossorial. Diet includes arthropods. Reproductive habits unstudied. **Distribution** Endemic to northern Borneo (Gunung Kinabalu, Sabah State). **Status** Not Evaluated.

SUMATRAN GIANT BLIND SNAKE
Anomochilus weberi PLATE 38
Measurement TL 230mm **Identification** Body stout, rounded in cross-section; head small, indistinct from neck; forehead covered with large scales; paired parietofrontals; nostril in a single nasal, in contact with Supralabial II; loreal and preocular absent; single postocular; eye small; mental groove absent; tail short and conical; dorsals smooth, slightly larger than ventrals at same level; midbody scale rows 19; ventrals 242–248; subcaudals 6–8. **Coloration** Dorsum black with pale stripe along flanks and large pale blotches on either side of vertebral; venter black. **Habitat and Behaviour** Inhabits montane forests. Subfossorial. Diet unstudied. Oviparous; clutches comprise 4 eggs. **Distribution** Sumatra (Kayutanam, Padang Highlands and Tanangtalu, Ophir District) and Borneo (Kutai, Kalimantan Province). **Status** Not Evaluated.

Family PYTHONIDAE PYTHONS

Snakes in the python family have teeth on the premaxilla, a supraorbital bone on the dorsal margin of the orbit, rows of heat-sensing labial organs, and vestigial pelvic and hind-limb bones (externally visible as paired spurs on each side of the cloaca). The family includes some of the largest snakes known, reaching 10m in total body length. Pythons inhabit the lowlands to the mid-hills, and are not infrequently encountered within densely populated cities and towns. Their diet consists of small to very large vertebrates, and some of the large-growing species are potentially dangerous to humans. All known species are oviparous. They are found in the Old World tropics and subtropics, including Asia, Africa, New Guinea and Australia. Both the skin and the flesh of pythons are in demand, which has led to the extermination of some populations.

RETICULATED PYTHON
Broghammerus reticulatus PLATE 38
Measurement TL 9.83m **Identification** Body relatively elongated and slender, except in large individuals; head distinct from neck; rostral as broad as deep; supralabials 12–14; Supralabials I–IV with pits; Supralabials VII or VIII enter orbit; infralabials 23; 2–3 anterior and 5–6 posterior infralabials with pits; infralabial pits better defined than supralabial pits and set in a longitudinal groove; eye small; pupil vertical; midbody scale rows 69–79; ventrals 297–330; subcaudals 78–102, paired; anal entire; cloacal spur present in both sexes (more distinct in males). **Coloration** Dorsum yellow or brown with rhomboidal dark markings; black median line runs from snout to nape; oblique line from posterior of eye to corner of mouth; venter yellow with small brown spots. **Subspecies** Three subspecies have been described, of which one (*B. r. reticulatus*) occurs in the region. **Habitat and Behaviour** Inhabits forests and typically found at water's edge, where it ambushes deer and pigs. Also common in cities and towns in the region, where it inhabits sewers. Nocturnal and terrestrial, with some arboreal and aquatic activity. Diet consists of homoeotherms including mammals and birds, and lizards are also eaten; prey is killed by constriction. Clutches comprise 14–124 eggs, measuring 90–93 x 58–62mm. Incubation period 60–101 days. Hatchlings 600–750mm. **Distribution** Southern Myanmar (Tenasserim, Taninthayi Division), Thailand, Laos, Cambodia, Vietnam, Peninsular Malaysia, Singapore, Sumatra, Pulau Bangka, Pulau Belitung, Pulau Weh, Pulau Enggano, Pulau Nias, the Mentawai, Natuna and Riau Archipelagos, Borneo, Java, Bali (*B. r. reticulatus*), Pulau Selayar, south-west of Sulawesi (*B. r. saputrai*) and Pulau Tamahjampea and possibly also small islands off Sulawesi (*B. r. jampeanus*). Also the Nicobar Islands, islands off eastern Indonesia, including Ambon, the Anambas Archipelago, Babi, Batjan, Banda Besar, Bankak, Boano, Buru, Butung, Flores, Halmahera, Haruku, Lang, Lombok, Obira, Saparua, Seram, the Sula Archipelago, Sulawesi, Sumba, Sumbawa, Tanimbar, Ternate, Timor and Verlate and the Philippine Islands, including Basilan, Bohol, the Calamian Islands, Cebu, Leyte, Luzon, Mindanao, Mindoro, Negros, Palawan, Panay, Polillo, Samar, Tawi-Tawi and the Sulu Archipelago (*B. r. reticulatus*). **Status** Not Evaluated.

BORNEAN SHORT PYTHON
Python breitensteini PLATE 38
Measurement TL >2m **Identification** Body short, robust; head elongated, flattened, distinct from neck; vertebral region ridged; naso-preocular groove present, comprising series of granular scales; rostral broader than deep with a deep pit on each side; loreals large; anterior pair of parietals in broad contact at median suture; single supralabial contacts orbit; postoculars 1–4; supralabials 9–11, 2 of them pitted; Supralabial VI contacts orbit; infralabials 14–19; anterior and posterior infralabials with weak pits; supralabial pits better defined than infralabial pits; tail short; midbody scale rows 50–57; ventrals 154–165; subcaudals 27–33, paired; anal entire; cloacal spurs present in both sexes. **Coloration** Dorsum pale yellow or tan with subrectangular dark blotches subequal to body width, darkening posteriorly, or a dark dorsum, turning all black posteriorly; scattered pale spots on vertebral region, more numerous posteriorly, where they are elongated and form a vertebral stripe; flanks with smaller, dark-edged grey spots or wavy bands that rise halfway upwards and may coalesce with dorsal blotches on posterior flanks; forehead pale yellow or grey with a black stripe between internasals and occipital, coalescing with dark pattern on neck; sides of head darker than forehead, with dark flecks and a broad dark postocular pattern; pale postocular stripe to angle of jaws; chin and venter unpatterned cream or sometimes spotted with brown. **Habitat and Behaviour** Inhabits lowland rainforests and the mid-hills at elevations of 0–1,000m asl. Associated with edges of rivers, swamps and marshes. Nocturnal and terrestrial. Diet consists of small mammals and birds. Clutches comprise up to 12 eggs. **Distribution** Endemic to Borneo (Sabah and Sarawak States, Brunei and Kalimantan Province). **Status** Not Evaluated.

BRONGERSMA'S SHORT PYTHON
Python brongersmai PLATE 38
Measurement TL 2.6m **Identification** Body short, robust; head elongated, flat, distinct from neck; vertebral region ridged; naso-preocular groove present, comprising series of granular scales; anterior pair of parietals in broad contact in median suture; postoculars 1–3; supralabials 9–13; Supralabials VI–VII contact orbit; infralabials 17–22; supralabial pits better defined than infralabial pits; tail short; midventrals 53–61; ventrals 160–178; subcaudals 24–36, paired; anal entire; cloacal spurs present in both sexes. **Coloration** Dorsum red in populations from western Sumatra, reddish-brown, charcoal-grey, pale grey or brown elsewhere; faint, narrow dark stripe on middle of forehead between rostral and occipital; narrow pale line from posterior occipital to dark pattern on upper neck; supralabial region dark; narrow pale postocular stripe to angle of jaws; dorsal pattern comprises vertebral spots that sometimes coalesce to form elongated blotches or stripes; lower anterior flanks pale with longitudinal dark blotches; on posterior flanks, they become taller and may coalesce with the dark dorsum; dark blotches on flanks rounded, 2–6 scales wide, set within paler areas; venter anteriorly cream, posteriorly dark, with grey smudges and blotches. **Habitat and Behaviour** Inhabits lowland forests at elevations of 0–1,330m asl. Nocturnal and terrestrial, associated with streams. Diet consists of mammals and birds. Clutches comprise 10–15 eggs. Incubation period ca 15 days. **Distribution** Southern Thailand, Laos, Cambodia, Vietnam, Peninsular Malaysia, Singapore, Sumatra, Pulau Bangka and the Mentawai Archipelago. **Status** Not Evaluated.

SUMATRAN SHORT PYTHON
Python curtus PLATE 38
Measurement TL >2m **Identification** Body short, robust; head elongated, flat, distinct from neck; vertebral region ridged; naso-preocular groove present, comprising series of granular scales; anterior pair of parietals in broad median contact; postoculars 1–3; supralabials 9–12; Supralabial VI contacts orbit; infralabials 16–19; supralabial pits better defined than infralabial pits; tail short; midbody scale rows 55–61; ventrals 152–163, paired; subcaudals 28–33; anal entire; cloacal spurs present in both sexes. **Coloration** Dorsum brownish-grey with a series of irregular longitudinal, subrectangular dark blotches, sometimes as broad as body; flanks with longitudinal series of large blotches with black edges; most lateral blotches ascend halfway, and some coalesce with dorsal blotches; sides of snout with dark stripes; postocular pattern comprises triangular black blotches that widen to 4–6 scales at angle of jaws and coalesce with anterior labials; pale postocular stripe with dark smudges; chin and venter unpatterned cream or white. Large adults typically become melanistic. **Habitat and Behaviour** Inhabits lowland forests at up to submontane limits. Nocturnal and terrestrial, associated with streams and forest floor. Diet consists of small mammals and possibly also birds. Clutches comprise 10–12 eggs. Incubation period 60–65 days. **Distribution** Endemic to western and southern Sumatra. **Status** Not Evaluated.

INDIAN ROCK PYTHON
Python molurus PLATE 38
Measurement TL 7.6m **Identification** Body thick, cylindrical; head lance-shaped, distinct from neck; supralabials 11–13; supralabials separated from orbit by row of subocular scales; infralabials 20; sensory pits in rostral and first 2 supralabials and 14–18th infralabials; supralabial pits better defined than infralabial pits; mental groove present; spurs small; tail short, prehensile; midbody scale rows 60–75; ventrals 245–270; subcaudals 58–73, paired; anal divided; cloacal spurs present in both sexes. **Coloration** Dorsum dark brown or yellowish-grey with a series of 30–40 large, irregular and squarish dark chocolate-grey patches that are edged with black on dorsal surface and on flanks; dorsal and lateral spots dark, and dark grey subocular stripe present; venter grey with dark spots on outer scale rows. **Subspecies** Two subspecies are recognized, of which one (*P. m. bivittatus*) occurs in the region. **Habitat and Behaviour** Inhabits forests and sometimes towns and villages. Active by day and night, when it waits in ambush for warm-blooded prey. Diet consists of small and mid-sized mammals including monkeys, goats and calves. Clutches comprise 30–58 eggs, measuring 120 x 60mm, which are deposited in damp soil and guarded by mother for 50–90 days. Hatchlings 550mm. **Distribution** Myanmar, Thailand, Laos, Cambodia, Vietnam, Java and possibly also northern Peninsular Malaysia (*P. m. bivittatus*). Also eastern India, Bangladesh, Nepal, southern China, Sulawesi and Sumbawa; introduced into the south-eastern USA. The nominotypical subspecies (*P. m. molurus*) occurs in Pakistan, northern and southern India, and Sri Lanka. **Status** Lower Risk/Near Threatened.

Family COLUBRIDAE 'TYPICAL' SNAKES

Snakes in this family have large forehead scales, solid (not grooved) maxillary teeth, lateral nostrils, well-developed ventrals and no hypopophyses on the posterior dorsal vertebrae. The Colubridae includes a large number of genera, which are predators of invertebrates and small vertebrates. Many of the familiar snakes, such as those typically encountered in human-modified ecosystems, belong to this family. They may be oviparous or ovoviviparous, and are cosmopolitan in temperate, subtropical and tropical parts of the world.

SPECKLE-HEADED VINE SNAKE
Ahaetulla fasciolata PLATE 39
Measurement TL 1,690mm **Identification** Body slender; snout long, ending in curled rostral scale; single supraocular; loreals 2–3; single preocular; postoculars 2; supralabials 9; Supralabials IV–V or IV–VI contact orbit; 4 infralabials contact anterior chin shields; eye large, pupil horizontal; tail long with a prehensile tip; vertebrals enlarged; dorsals smooth; midbody scale rows 15; ventrals 211–240; subcaudals 178–197, paired; anal single. **Coloration** Dorsum light brown, grey or pinkish-tan with numerous narrow, oblique dark bands on anterior of body; forehead with elongated or curved dark markings; venter dark grey. **Habitat and Behaviour** Inhabits forested and semi-urban areas with an altitudinal distribution from sea level to 900m asl. Diurnal and arboreal, associated with thick undergrowth and other low vegetation. Diet includes lizards and frogs. Ovoviviparous. **Distribution** Southern Thailand (Nakhon Si Thammarat Province), Peninsular Malaysia, Singapore, Sumatra, the Natuna and Riau Archipelagos, and Borneo. **Status** Not Evaluated.

RIVER VINE SNAKE
Ahaetulla fronticincta PLATE 39
Measurement TL 980mm **Identification** Body slender; snout long; loreals 2; supralabials 7; Supralabial V contacts orbit; eye large; pupil horizontal; tail long with a prehensile tip; dorsals smooth; midbody scale rows 15; ventrals 168–196; subcaudals 139–148; anal divided. **Coloration** Dorsum either bright green or brownish-yellow; skin between scales black and white, forming oblique lines;

venter pale green or olive; white streak along lower flanks; forehead with or without black spots. **Habitat and Behaviour** Inhabits forested edges of creek and river mouths associated with mangroves. Diurnal and arboreal, associated with thick undergrowth and other low vegetation. Diet consists of surface-feeding fish, which are captured in ambush from overhanging vegetation. Ovoviviparous; clutches comprise 7 neonates. **Distribution** Endemic to coastal Myanmar. Reports from eastern and north-eastern India require verification. **Status** Not Evaluated.

MALAYAN VINE SNAKE
Ahaetulla mycterizans PLATE 39
Measurement TL 920mm **Identification** Body slender; snout elongated; groove along snout; loreals 3; single preocular; postoculars 2; single supraocular; supralabials 7–8 (rarely 9); Supralabials IV–V or IV–VI contact orbit; eye large; pupil horizontal; tail long with a prehensile tip; dorsals smooth; midbody scale rows 15; ventrals 186–195; subcaudals 132–163, paired; anal entire. **Coloration** Dorsum bright green, greyish-green or brown; in green morph, venter white with paired longitudinal green lines and sometimes green line along middle. **Habitat and Behaviour** Inhabits lowland forests. Diurnal and arboreal. Diet consists of lizards and birds. Reproductive habits unstudied. **Distribution** Southern Thailand, Peninsular Malaysia, Sumatra and Java. **Status** Not Evaluated.

INDIAN VINE SNAKE
Ahaetulla nasuta PLATE 39
Measurement TL 2m **Identification** Body elongated, slender; snout long with rostral appendage; groove in front of eyes; supraocular divided horizontally; loreal absent; single preocular; supralabials 8; Supralabials III–IV, IV–V or V–VI contact orbit; infralabials 9–10; eye large; pupil horizontal; tail long with a prehensile tip; dorsals smooth; midbody scale rows 15; ventrals 135–207; subcaudals 135–180, divided; anal divided. **Coloration** Dorsum bright green, or more rarely olive-brown, with a longitudinal yellowish line along outer margin of ventrals; venter pale green; iris yellow. **Habitat and Behaviour** Inhabits lightly forested habitats, including gardens, from the plains to the hills at elevations of up to 1,800m. Diurnal and arboreal, associated with thick undergrowth and other low vegetation, frequenting trees and bushes. Diet consists of tadpoles, lizards, birds and small mammals. Ovoviviparous; clutches comprise 3–23 neonates, measuring 200–440mm. **Distribution** Myanmar, northern and central Thailand, Laos, Cambodia and Vietnam. Also India, Bangladesh, Nepal and Sri Lanka. **Status** Not Evaluated.

ORIENTAL VINE SNAKE
Ahaetulla prasina PLATE 39
Measurement TL 1,970mm **Identification** Body slender; snout elongated; groove along snout; single supraocular; loreals 1–4; single preocular; postoculars 2; supralabials 9; Supralabials IV–VI contact orbit; 4 infralabials contact anterior chin shields; eye large; pupil horizontal; tail long with a prehensile tip; dorsals smooth; midbody scale rows 15; ventrals 194–235; subcaudals 141–207, paired; anal divided. **Coloration** Dorsum typically green, also brown, yellow, dark grey or golden-yellow, speckled with black; yellow stripe along lower flanks; venter light green or dark grey. **Subspecies** Two subspecies have been described, of which one (*A. p. prasina*) occurs in the region. **Habitat and Behaviour** Inhabits edges of forests from sea level to 2,100m asl. Diurnal and arboreal, associated with thick undergrowth and other low vegetation. Diet consists of lizards and birds. Ovoviviparous; clutches comprise 4–10 neonates, measuring 240–490mm. **Distribution** Myanmar, Thailand, Laos, Cambodia, Vietnam, Peninsular Malaysia, Singapore, Sumatra, the Mentawai, Riau and Natuna Archipelagos, Pulau Bangka, Pulau Belitung, Pulau Sibutu, Borneo, Java and Bali. Also Bhutan, north-eastern India, Bangladesh, China (*A. p. prasina*) and the Sulu Archipelago, the Philippines (*A. p. suluensis*). **Status** Not Evaluated.

IRIDESCENT SNAKE
Blythia reticulata PLATE 39
Measurement TL 514mm **Identification** Body slender, elongated; head scarcely distinct from neck; loreal and preocular absent; single postocular; supralabials 6 (rarely 5); Supralabials III–IV contact orbit; dorsals smooth, glossy, lacking apical pits; tail short with acute tip; midbody scale rows 13; ventrals 122–157; subcaudals 16–32, paired; anal divided. **Coloration** Dorsum olive to dark, highly iridescent; scales sometimes light-speckled or light-bordered; juveniles with yellowish-cream collar and a gap on the dark vertebral line; ventrals grey, each ventral with pale posterior edge. **Habitat and Behaviour** Inhabits wet evergreen forests at mid-altitudes to at least 1,040m asl. Subfossorial, associated with fallen logs and litter. Diet includes earthworms and possibly soil arthropods. Oviparous; clutches comprise up to 6 eggs. **Distribution** Myanmar. Also north-eastern India. **Status** Not Evaluated.

BENGKULU CAT SNAKE
Boiga bengkuluensis PLATE 39
Measurement TL 1,673mm **Identification** Body slender, elongated, laterally compressed; head large, distinct from neck; single preocular; postoculars 2; supralabials 8; Supralabials III–V contact orbit; infralabials 12–13; eye large; pupil vertical; vertebral scale rows enlarged; dorsals smooth; midbody scale rows 19; ventrals 261; subcaudals 146, paired; anal entire. **Coloration** Dorsum greenish-brown with 41 wide transverse cross-bars; interspaces with irregular dark green spots that extend ventrally; forehead dark green, sides of head with pale stripe from angle of jaws to back of head; chin cream; venter greenish-brown with dark speckling, a distinct dark stripe on each side of ventrals, and an indistinct stripe on midline of ventrals; subcaudals darker. **Habitat and Behaviour** Inhabits secondary forests at an elevation of 500m asl. Nocturnal and arboreal,

associated with dense vegetation. Diet and reproductive habits unstudied. **Distribution** Southern Thailand (Na Prado, Pattani Province) and Sumatra (Rejiang, Bengkulu Province). Possibly also Peninsular Malaysia. **Status** Not Evaluated.

BOURRET'S CAT SNAKE
Boiga bourreti   PLATE 39
Measurement TL 1,155mm **Identification** Body slender, elongated, laterally compressed; loreal present; preoculars 2–3; postoculars 2; supralabials 8; Supralabials III–V contact orbit; infralabials 11–12; eye large; pupil vertical; dorsals smooth; midbody scale rows 19; ventrals 236; subcaudals 106, divided; anal divided. **Coloration** Dorsum light greyish-brown; light-edged, blackish-brown postocular stripe extends to angle of jaws and towards the dark neck collar; dorsum pattern consists of a reddish-brown to black collar, 2 V-shaped reddish-brown to black bands that are sometimes fused, and a chequered pattern of light and dark blotches on top and sides of body; venter light with dark speckling; subcaudals with blackish-grey speckling; iris with brown streaks. **Habitat and Behaviour** Inhabits evergreen forests at an elevation of 550m asl. Nocturnal and terrestrial, associated with rocky biotope. Diet consists of birds' eggs. Oviparous; produces eggs measuring 14 x 5mm. Clutch size is unknown. **Distribution** Endemic to central Vietnam (Phong Nha-Ke Bang National Park, Quang Binh Province). **Status** Not Evaluated.

GREEN CAT SNAKE
Boiga cyanea   PLATE 39
Measurement TL 1,870mm **Identification** Body slender, elongated, laterally compressed; head large, distinct from neck; single preocular; postoculars 2 (rarely 3); supralabials 8–9; Supralabials III–V or IV–VI contact orbit; infralabials 10–12; eye large; pupil vertical; vertebral region with low ridge; vertebral scale row with enlarged scales; scales smooth with apical pits; midbody scale rows 21 or 23; ventrals 237–257; subcaudals 124–158, divided; anal entire. **Coloration** Dorsum emerald-green in adults, reddish-brown or olive with a yellowish-green forehead in juveniles; interstitial skin black; gular region sky-blue; venter greenish-white or greenish-yellow, unpatterned or spotted with dark green; iris brownish-grey. **Habitat and Behaviour** Inhabits forests as well as disturbed habitats, from the lowlands to montane limits, at elevations of 150–2,100m asl. Nocturnal and arboreal, found coiled on branches of trees overhanging water. Diet includes frogs, birds and their eggs, lizards, other snakes and small mammals. Oviparous; clutches comprise 4–10 eggs, measuring 40–48 x 15–21mm, which are produced several times a year. Incubation period 64–90 days. Hatchlings 350mm. **Distribution** Myanmar, Thailand, Laos, eastern, central and south-eastern Cambodia, Vietnam and northern Peninsular Malaysia (Pulau Langkawi, Kedah State). Also Bangladesh, Bhutan, Nepal, eastern and north-eastern India, the Nicobar Islands and southern China. **Status** Not Evaluated.

DOG-TOOTHED CAT SNAKE
Boiga cynodon   PLATE 39
Measurement TL 2.8m **Identification** Body slender, elongated, laterally compressed; head distinct from neck; snout short, rounded; loreal present; single preocular; postoculars 2; supralabials 8–10; Supralabials III–V, IV–V, IV–VI or V–VII contact orbit; infralabials 13–14; 4–5 infralabials contact anterior chin shields; anterior temporals 3; eye large; pupil vertical; vertebrals distinctly enlarged; dorsals smooth; midbody scale rows 23 or 25; ventrals 248–290; subcaudals 114–165, paired; anal entire. **Coloration** Dorsum brownish-tan or yellowish-brown, with dark brown or reddish-brown bands, darkening posteriorly; head sometimes more yellow than rest of body; dark postocular stripe; juveniles lighter. **Habitat and Behaviour** Inhabits lowland forests and edges. Nocturnal and arboreal, but with some terrestrial activity. Diet consists of small vertebrates, primarily lizards and birds and their eggs, as well as small mammals. Oviparous; clutches comprise 6–23 eggs, which are produced several times a year. **Distribution** Southern Thailand (Phang-Nga and Prachuap Khiri Khan Provinces), Peninsular Malaysia, Singapore, Sumatra, Pulau Nias, Pulau Belitung, Pulau Bangka, the Mentawai Archipelago, Borneo, Java and Bali. Also the Lesser Sundas and the Philippines. **Status** Not Evaluated.

MANGROVE CAT SNAKE
Boiga dendrophila   PLATE 40
Measurement TL 2.5m **Identification** Body large, robust, laterally compressed; head distinct from neck; snout short, rounded; loreal present; single preocular; postoculars 2; supralabials 8 (rarely 9); Supralabials III–V contact orbit; infralabials 11; eye large; pupil vertical; dorsals smooth; midbody scale rows 21 (rarely 23); ventrals 209–253; subcaudals 89–118, paired; anal entire. **Coloration** Dorsum black with 35–45 narrow yellow rings around body and 10 on tail; labials and gular region yellow; venter grey. **Subspecies** Eight subspecies are recognized, of which four occur in the region. *B. d. dendrophila*, gular scales uniformly yellow; ventrals with series of yellow spots medially. *B. d. annectens*, gular scales with black edges; narrow black sutures on supralabials; parietal without yellow spot. *B. d. melanota*, gular scales uniformly yellow; ventrals without series of yellow spots medially. *B. d. occidentalis*, gular scales with black edges; broad black sutures on supralabials; parietal with yellow spot. **Habitat and Behaviour** Inhabits lowland forests, including mangrove and peat swamps, and enters lowland mixed dipterocarp forests. Nocturnal, and both arboreal and terrestrial when foraging; arboreal when at rest; associated with trees. Foraging may also take place in dense undergrowth or along riverbanks. Diet consists of birds and their eggs and nestlings, amphibians, lizards, other snakes, mouse deer and tree shrews. Oviparous; clutches comprise 4–15 eggs, measuring 45.5–51 x 24.5–25mm. Incubation period 86 days. Hatchlings 340–360mm. **Distribution** Southern Thailand (Nakhon

Si Thammarat, Phattalung, Trang, Pattani and Narathiwat Provinces), Cambodia, Vietnam, Peninsular Malaysia, Singapore, eastern Sumatra, Pulau Belitung, the Batu Archipelago (*B. d. melanota*), western Sumatra, Pulau Nias (*B. d. occidentalis*), Borneo (*B. d. annectens*), Java and Bali (*B. d. dendrophila*). Also Luzon, Samar and Polillo, the Philippines (*B. d. divergens*), Mindanao, the Philippines (*B. d. latifasciata*), Palawan and Balabac, the Philippines (*B. d. multicincta*), and islands of western and central Indonesia, including Sulawesi (*B. d. gemmicincta*). **Status** Not Evaluated.

WHITE-SPOTTED CAT SNAKE
Boiga drapiezii PLATE 40
Measurement TL 2.1m **Identification** Body long and remarkably slender; head distinct from neck; loreals small or absent; single enlarged preocular; postoculars 2; supralabials 8; Supralabials III–V or IV–V contact orbit; infralabials 11; 5–6 infralabials contact chin shields; eye large; pupil vertical; vertebrals enlarged; dorsals smooth; midbody scale rows 19; ventrals 250–287; subcaudals 114–173, paired; anal entire. **Coloration** Dorsum varies from olive-grey to reddish-brown; vertebral region with paired pink spots anteriorly, sometimes fused to form bands; pink or cream spots on flanks; forehead with dark speckling. **Habitat and Behaviour** Inhabits lowland forests and the mid-hills at elevations of up to 1,000m asl. Nocturnal and arboreal, associated with trees and shrubs. Diet consists of birds, their eggs, frogs, lizards and large insects. Oviparous; clutch size unknown; eggs may be laid in termite-infested wood. **Distribution** Southern Thailand (Phangnga, Songkhla, Pattani and Narathiwat Provinces), Vietnam, Peninsular Malaysia, Singapore, Sumatra, Pulau Banka, the Mentawai Archipelago and Borneo. Also the central islands of Indonesia and the Philippines. **Status** Not Evaluated.

EASTERN CAT SNAKE
Boiga gokool PLATE 40
Measurement TL 1,200mm **Identification** Body slender, laterally compressed; head large, distinct from neck; single preocular; postoculars 1–2; supralabials 8; Supralabials III–V contact orbit; maxillary teeth 9–12 + 2; posterior genials in contact; dorsals smooth; eye large; pupil vertical; midbody scale rows 21 (rarely 17 or 19); ventrals 219–232; subcaudals 87–103, paired; anal entire. **Coloration** Dorsum yellowish-brown; series of vertical T- or Y-shaped marks on each side of dorsum; forehead with large, arrowhead-shaped brown mark, edged with black; black subocular stripe to angle of jaws; labials brown; venter cream with near continuous series of longitudinal stripes, composed of spots on each side of ventral. **Habitat and Behaviour** Inhabits the mid-hills and ascending montane forests. Nocturnal and arboreal, associated with bushes and shrubs. Diet consists of small rodents. Reproductive habits unstudied. **Distribution** Myanmar. Also eastern India and Bangladesh. **Status** Not Evaluated.

GUANGXI CAT SNAKE
Boiga guangxiensis PLATE 40
Measurement TL ca 2m **Identification** Body robust, laterally compressed; head large, distinct from neck; loreal present; preoculars 1–2; postoculars 2; 2–3 anterior temporals; 3 posterior temporals; supralabials 8; Supralabials III–V contact orbit; infralabials 11–12; eye large; pupil vertical; dorsals smooth; midbody scale rows 21; vertebrals enlarged; ventrals 263–270; subcaudals 119–147, divided; anal entire, with median fold. **Coloration** Dorsum brownish-tan or olive, becoming dark grey posteriorly, with irregular black cross-bars, 2–2.5 scales wide, which are more distinct anteriorly; bright red spots between cross-bars on anterior body; forehead greyish-black; supralabials pale yellow; venter greyish-white. **Habitat and Behaviour** Inhabits the forested mid-hills. Nocturnal and arboreal. Diet and reproductive habits unstudied. **Distribution** Northern and central Vietnam. Also eastern China (Guangxi Province). **Status** Not Evaluated.

JASPER CAT SNAKE
Boiga jaspidea PLATE 40
Measurement TL 1,500mm **Identification** Body slender, laterally compressed; head large, distinct from neck; loreal present; preoculars 1–2; postoculars 2; supralabials 8; Supralabials III–V contact orbit; infralabials 12; 4–5 infralabials contact anterior chin shields; eye large; pupil vertical; vertebrals enlarged; dorsals smooth; midbody scale rows 21; ventrals 243–267; subcaudals 128–166, paired; anal enlarged. **Coloration** Dorsum brown, reddish-brown or grey-brown with paired row of dark spots or bars on flanks; greyish-red vertebral stripe. **Habitat and Behaviour** Inhabits lowland forests and peat swamps up to submontane limits at an elevation of 1,524m asl. Nocturnal and arboreal, associated with trees and dense tangles of vegetation forming undergrowth. Diet includes small vertebrates such as geckos and other lizards; small mammals, birds and their eggs, and other snakes are also eaten. Oviparous; clutches comprise 6 eggs, measuring 38–39 x 18–19mm, which are deposited in the nests of tree-dwelling termites. Incubation period 101+ days. Hatchlings 390–400mm. **Distribution** Southern Thailand (Phuket, Phangnga and Pattani Provinces), Vietnam (Lam Dong Province), Peninsular Malaysia, Singapore, Sumatra, Pulau Nias, the Mentawai Archipelago, Pulau Bangka, Borneo and Java. **Status** Not Evaluated.

SQUARE-HEADED CAT SNAKE
Boiga kraepelini PLATE 40
Measurement TL 1,520mm **Identification** Body slender, laterally compressed; head large, distinct from neck; loreal present; preoculars 2–3; postoculars 1–3; supralabials 9–10; Supralabials III–V contact orbit; infralabials 12–13; eye large; pupil vertical; vertebral scale rows enlarged; dorsals smooth; midbody scale rows 21 or 23; ventrals 230–243; subcaudals 127–154; anal entire. **Coloration** Dorsum brownish-grey with 57 dark cross-bars that extend on flanks ca 4

scale rows from ventrals; lateral spots faint; forehead unpatterned brown; venter pale, median region dark grey and lateral line with irregular dark line. **Habitat and Behaviour** Inhabits the forested mid-hills at elevations of 150–400m asl. Nocturnal and arboreal. Diet consists of birds and their eggs. Oviparous; clutches comprise about 14 eggs, measuring 40 x 17mm. **Distribution** Vietnam (Lang Son, Cao Bang, Vinh Phuc, Nghe An, Ha Tinh and Thua Thien-Hue Provinces); possibly also Laos. Also eastern China (including Hainan Province and Taiwan). **Status** Not Evaluated.

MANY-SPOTTED CAT SNAKE
Boiga multomaculata PLATE 40
Measurement TL 1,870mm **Identification** Body slender, laterally compressed; head large, distinct from neck; loreal present; preoculars 1–2; postoculars 2–3; supralabials 8 (rarely 7 or 9); infralabials 11 (rarely 10 or 12); 2 posterior-most teeth enlarged; eye large; pupil vertical; dorsals smooth; midbody scale rows 17 or 19; ventrals 196–245; subcaudals 72–111, paired; anal entire. **Coloration** Dorsum grey-brown with black line from eye to jaws; 2 brown lines, edged with black, from snout to back of head; series of irregular brown blotches on dorsum; smaller brown marks on flanks; venter greyish-brown with small brown spots. **Habitat and Behaviour** Known from the lowlands and submontane forests at elevations of up to 1,500m asl. Nocturnal and arboreal, inhabiting short trees, bushes and bamboo groves. Diet consists of birds, and lizards including geckos. Oviparous; clutches comprise 4–8 eggs, measuring 26–32 x 11–12mm. Incubation period 60–67 days. Hatchlings 195–200mm. **Distribution** Myanmar, Thailand, Laos, Cambodia and Vietnam. Also north-eastern India, Bangladesh and eastern China. **Status** Not Evaluated.

BLACK-HEADED CAT SNAKE
Boiga nigriceps PLATE 40
Measurement TL 2m **Identification** Body robust, laterally compressed; head large, distinct from neck, loreal present; single preocular; postoculars 2; supralabials 8; Supralabials III–VI contact orbit; 4–5 infralabials contact anterior chin shields; eye large; pupil vertical; dorsals smooth; midbody scale rows 21; ventrals 240–293; subcaudals 134–164, paired; anal entire. **Coloration** Dorsum straw-brown, olive-brown or reddish-brown; forehead and tail often darker; labials cream or yellow; venter cream, darkening posteriorly. **Subspecies** Two subspecies are recognized. *B. n. nigriceps*, subcaudals 134–154. *B. n. brevicauda*, subcaudals 114–124. **Habitat and Behaviour** Inhabits lowland forests at elevations of up to 800m asl. Nocturnal and arboreal, associated with trees. A back-fanged snake with envenomings from bite reported. Diet includes birds and other snakes. Oviparous; clutches comprise 3 eggs, measuring 48 x 17mm. Incubation period 136 days. Hatchlings 390mm. **Distribution** Southern Thailand (Pattani and Surat Thani Provinces), Peninsular Malaysia, Sumatra, Pulau Nias, Pulau Simeulue, Borneo, Java (*B. n. nigriceps*) and the Mentawai Archipelago (*B. n. brevicauda*). **Status** Not Evaluated.

TAWNY CAT SNAKE
Boiga ochracea PLATE 40
Measurement TL 1,100mm **Identification** Body slender; head large, distinct from neck, preocular 1 (rarely 2); postoculars 2 (rarely 3); supralabials 8; Supralabials IV–VI contact orbit; eye large; pupil vertical; dorsals smooth with apical pits that are obliquely arranged; vertebral scale row greatly enlarged; midbody scale rows 19 or 21; ventrals 221–252; subcaudals 89–119, paired; anal entire. **Coloration** Dorsum reddish-brown, ochre or coral-red, unpatterned or with poorly defined transverse dark bars; dark streak from eye to angle of mouth; labials yellow or cream; venter yellow anteriorly, light grey posteriorly and on subcaudals. **Habitat and Behaviour** Inhabits forests in the mid-hills and submontane limits, as well as parks and gardens, at elevations of 350–1,400m asl. Nocturnal and crepuscular, and arboreal, associated with bushes and other undergrowth. Diet consists of birds and their eggs, small mammals and lizards. Oviparous. **Distribution** Myanmar and Thailand. Also Bangladesh, Bhutan, eastern and north-eastern India, and Nepal. **Status** Not Evaluated.

ASSAMESE CAT SNAKE
Boiga quincunciata NOT ILLUSTRATED
Measurement TL 1,550mm **Identification** Body slender, elongated; head distinct from neck; loreal present; single preocular, reaching dorsal surface of head; eye large; pupil vertical; dorsals smooth; vertebrals enlarged; midbody scale rows 19; ventrals 237–253; subcaudals 118–125. **Coloration** Dorsum yellow or greyish-brown finely speckled with dark brown; vertebral series of dark brown or black spots or blotches, each marking 5–8 scales wide, scales edged with white; flanks speckled or spotted with brown; nape with 3 longitudinal stripes; forehead brown, frontal and parietal scales black edged with white; black postocular stripe to angle of jaws; venter yellowish-white intensively speckled with brown. **Habitat and Behaviour** Inhabits wet evergreen forests. Nocturnal and arboreal, associated with bamboo internodes. Diet and reproductive habits unstudied. **Distribution** North-eastern Myanmar (Htingnan, north-east of Putao, Kachin State). Also north-eastern India (Assam State). **Status** Not Evaluated.

BANDED GREEN CAT SNAKE
Boiga saengsomi PLATE 40
Measurement TL 2.1m **Identification** Body slender, elongated; head distinct from neck; rostral small, triangular; nasals paired; internasals in contact; paired prefrontals contact single preocular; postoculars 2; supralabials 8; Supralabials III–V contact orbit; infralabials 12–13; eye large; pupil vertical; midbody scale rows 21; ventrals 231–245; subcaudals 116–127, paired; anal entire. **Coloration** Dorsum yellowish-

green with narrow yellow areas; interstitial skin black; forehead olive; supralabials yellow; tail scales black with a central yellow spot; gular region cream, rest of venter yellowish-cream. **Habitat and Behaviour** Inhabits lowland forests at an elevation of 170m asl. Probably crepuscular and nocturnal, and arboreal. Diet consists of small mammals and birds. Reproductive habits unstudied. **Distribution** Endemic to southern Thailand (Amphoe Sikao, Trang Province, and Nakhon Si Thammarat, Surat Thani and Ban Kanom, Krabi Province). **Status** Not Evaluated.

THAI CAT SNAKE
Boiga siamensis PLATE 40
Measurement TL 1,700mm **Identification** Body slender, elongated, laterally compressed; head distinct from neck; loreal present; single preocular; postoculars 2; supralabials 8; Supralabials III–V contact orbit; infralabials 12–13; anterior temporals 2; eye large; pupil vertical; dorsals smooth; midbody scales 23; vertebrals distinctly enlarged; ventrals 247–270; subcaudals 116–129, paired; anal entire. **Coloration** Dorsum light brown with 87–98 V-shaped dark brown bands, more distinct anteriorly; posteriorly, bands bar-shaped with little posterior extension; scales flecked with white, those comprising dark bands light at posterior tips; flanks with alternating dark and light spots; venter light brown; forehead mid-brown; dark streak from posterior margin of eye to beyond last supralabial; 2 black stripes on either side of vertebral row to first dark band and continuous with it; venter yellowish-brown or greyish-brown. **Habitat and Behaviour** Inhabits lowland evergreen forests up to submontane limits at an elevation of 1,780m asl. Nocturnal and arboreal, associated with trees as high as ca 6m above substrate. Diet consists of small rodents, and birds and their eggs. Reproductive habits unstudied. **Distribution** Myanmar (Toungoo and Myitkina), Thailand (Sakaerat, Nakhon Ratchasima Province, Dansai, Loei Province, Chiengmai University Campus, Chieng Mai Province, Bangtaphan and Me Ping River, Chumpon Province, and north of Raheng, Tak Province), Laos, Cambodia and Vietnam (Nam Cat Tien National Park, Lam Dong Province). Also eastern and north-eastern India (northern West Bengal, Assam and Arunchal Pradesh), and Bangladesh. **Status** Not Evaluated.

WALL'S CAT SNAKE
Boiga walli NOT ILLUSTRATED
Measurement TL 1,050mm **Identification** Body slender, elongated, laterally compressed; head distinct from neck; loreal present; single preocular; postoculars 2; supralabials 8; Supralabials IV–VI contact orbit; eye large; pupil vertical; vertebrals enlarged; dorsals smooth; midbody scale rows 19; ventrals 221–246; subcaudals 89–107, paired; anal entire. **Coloration** Dorsum yellowish-grey or pale grey with indistinct dark cross-bars; labials and gular region cream; venter pale grey. **Habitat and Behaviour** Inhabits lowland forests and the mid-hills. Nocturnal and arboreal, with some terrestrial activity. Diet and reproductive habits unstudied. **Distribution** Southern Myanmar (Tenasserim, Tanintharyi Division). Also the Andaman Islands (India). **Status** Not Evaluated.

PADANG REED SNAKE
Calamaria abstrusa NOT ILLUSTRATED
Measurement TL 200mm **Identification** Body slender, cylindrical; head short, indistinct from neck; loreal absent; single postocular; supralabials 5; Supralabials III–IV contact orbit; infralabials 5; 3 infralabials contact anterior chin shields; modified maxillary teeth 8–9; eye small; pupil rounded; tail short, tapering; dorsals smooth; midbody scale rows 13; ventrals 129–152; subcaudals 14–25, paired; anal entire. **Coloration** Dorsum brown, each scale lighter centrally; scattered scales with a central dark spot; scales on second row form near continuous stripe; dark spots on vertebral scale rows; yellow ring at back of head, 1–2 scales wide, separated from parietals; forehead dark brown with indistinct dark spots; venter yellow, ventrals with dark lateral margins; subcaudals yellow, sometimes with yellowish-brown streak. **Habitat and Behaviour** Nothing known of its natural history. **Distribution** Endemic to Pulau Nias and western Sumatra (Padang, Sumatera Barat Province). **Status** Not Evaluated.

WHITE-BELLIED REED SNAKE
Calamaria albiventer NOT ILLUSTRATED
Measurement TL 380mm **Identification** Body slender, cylindrical; head short, indistinct from neck; loreal absent; single preocular; single postocular; supralabials 5; Supralabials III–IV contact orbit; infralabials 5; mental contacts anterior chin shields; modified maxillary teeth 7; eye small; pupil rounded; tail short, tapering; dorsals smooth; midbody scale rows 13; ventrals 143–162; subcaudals 15–22, paired; anal entire. **Coloration** Dorsum reddish-brown with 4 narrow red stripes, flank with bluish-white stripe, all stripes edged with black; chin yellow, turning red on neck; venter and subcaudals red. **Habitat and Behaviour** Nothing known of its natural history. **Distribution** Peninsular Malaysia (Bukit Larut, Perak State and Pulau Pinang, Pinang State), Singapore and east-central Sumatra (Indragiri, Riau Province). **Status** Not Evaluated.

BENGKULU REED SNAKE
Calamaria alidae NOT ILLUSTRATED
Measurement TL 289mm **Identification** Body slender, cylindrical; head short, indistinct from neck; loreal absent; preocular absent; single postocular; supralabials 5; Supralabials III–IV contact orbit; infralabials 5; 3 infralabials contact anterior chin shields; modified maxillary teeth 9–10; eye small; pupil rounded; tail short, tapering; dorsals smooth; midbody scale rows 13; ventrals 185–231; subcaudals 17–27, paired; anal entire. **Coloration** Dorsum olive or dark brown, light dorsal scales with dark network; head brown; lower supralabials yellow; venter yellow; ventrals with dark brown spots laterally; subcaudals yellow with median and lateral dark brown stripes.

Habitat and Behaviour Nothing known of its natural history. **Distribution** Endemic to western Sumatra (Padang, Sumatera Barat Province, Kaba Wetan, Bengkulu Province and Lebongtandai, Bengkulu Province). **Status** Not Evaluated.

BATTERSBY'S REED SNAKE
Calamaria battersbyi NOT ILLUSTRATED
Measurement TL 92mm **Identification** Body slender, cylindrical; head short, indistinct from neck; loreal absent; single preocular; single postocular; supralabials 4; Supralabials II–III contact orbit; infralabials 5; 3 infralabials contact anterior chin shields; modified maxillary teeth 8; eye small; pupil rounded; tail short, abruptly tapered to a point; dorsals smooth; midbody scale rows 13; ventrals 171; subcaudals 16, paired; anal entire. **Coloration** Dorsum brown, scales yellow with brown margins; narrow dark stripe on adjacent edges of scale rows 1–2; wider dark stripe on adjacent edges of scale rows 2–3 and 4–5; narrow dark stripe on adjacent edges of scale row 6 and vertebral rows; narrow yellow ring behind head and one at tail base; head brown; lower half of supralabials yellow; venter with dark stippling anteriorly, unpatterned yellow posteriorly; subcaudals with dark median streak. **Habitat and Behaviour** Nothing known of its natural history. **Distribution** Endemic to south-eastern Borneo (Tanjong, Kalimantan Province). **Status** Not Evaluated.

BICOLOURED REED SNAKE
Calamaria bicolor PLATE 41
Measurement TL 450mm **Identification** Body slender, cylindrical; head short, indistinct from neck; loreal absent; single preocular; single postocular; supralabials 5; Supralabials III–IV contact orbit; infralabials 5; 3 infralabials contact anterior chin shields; nasal pointed forwards; eye small; pupil rounded; modified maxillary teeth 8–10; tail thick, tapering from base; dorsals smooth; midbody scale rows 13; ventrals 139–169; subcaudals 18–28, paired; anal entire. **Coloration** Dorsum blue-black or dark brown, unpatterned or with dark cross-bands; forehead dark brown, sometimes with 2 oblique dark bands crossing yellow labials; venter typically unpatterned yellow or spotted with black. **Habitat and Behaviour** Inhabits the mid-hills and reaches submontane limits. Terrestrial. Diet and reproductive habits unstudied. **Distribution** Borneo and Java. **Status** Not Evaluated.

BORNEAN REED SNAKE
Calamaria borneensis PLATE 41
Measurement TL 374mm **Identification** Body slender, cylindrical; head short, indistinct from neck; loreal absent; single preocular; single postocular; supralabials 4; Supralabials II–III contact orbit; infralabials 4; 2 infralabials contact anterior chin shields; modified maxillary teeth 8–9; eye small; pupil rounded; tail short; dorsals smooth; midbody scale rows 13; ventrals 126–192; subcaudals 13–26, paired; anal entire. **Coloration** Dorsum greyish-brown, each scale pale with dark reticulation; scattered dark spots or stripes on scales of middorsum; first scale row yellow, sometimes with dark cross-bars; head greyish-brown with indistinct dark spots; 1–3 yellow caudal rings; venter with dark stripe along ventral edges, or with chequered pattern of yellow and black. **Habitat and Behaviour** Known from the lowlands as well as submontane regions. Diet and reproductive habits unstudied. **Distribution** Endemic to Borneo. **Status** Not Evaluated.

BUCH'S REED SNAKE
Calamaria buchi NOT ILLUSTRATED
Measurement TL 466mm **Identification** Body slender, cylindrical; head short, indistinct from neck; loreal absent; single preocular; single postocular; supralabials 4; Supralabials II–III contact orbit; infralabials 5; 3 infralabials contact anterior chin shields; modified maxillary teeth 9; eye small; pupil rounded; rostral higher than wide; mental contacts anterior chin shields; 5 scales contact paraparietal; tail short; dorsals smooth; midbody scale rows 13; ventrals 221–236; subcaudals 13–14, paired; anal entire. **Coloration** Dorsum blackish, each dorsal scale with small light spots. **Habitat and Behaviour** Nothing known of its natural history. **Distribution** Endemic to central Vietnam (B'Lao, Lam Dong Province). **Status** Not Evaluated.

THICK REED SNAKE
Calamaria crassa NOT ILLUSTRATED
Measurement TL 440mm **Identification** Body slender, cylindrical; head short, indistinct from neck; loreal absent; single postocular; supralabials 5; Supralabials III–IV contact orbit; infralabials 5; 3 infralabials contact anterior chin shields; modified maxillary teeth 10; eye small; pupil rounded; tail short, tapering; dorsals smooth; midbody scale rows 13; ventrals 136–164; subcaudals 14–28, paired; anal entire. **Coloration** Dorsum brown with scattered dark spots forming longitudinal streaks; most dorsals with fine dark network; head brown with indistinct dark spots; chin yellow with dark spots on sutures; anterior venter yellow; dark areas of ventrals increase posteriorly; outer edges of subcaudal dark. **Habitat and Behaviour** Inhabits submontane forests at an elevation of 1,300m asl. Nothing else known of its natural history. **Distribution** Endemic to west-central Sumatra (Padang and Gunung Talakmau, Sumatera Barat Province). **Status** Not Evaluated.

DÖDERLEIN'S REED SNAKE
Calamaria doederleini NOT ILLUSTRATED
Measurement TL 290mm **Identification** Body slender, cylindrical; head very short, indistinct from neck; loreal absent; single preocular; single postocular; supralabials 5; Supralabials III–IV contact orbit; infralabials 5; 3 infralabials contact anterior chin shields; modified maxillary teeth 10; eye small; pupil rounded; tail short, tapering; dorsals smooth; midbody scale rows 13; ventrals 163; subcaudals 20, paired; anal entire. **Coloration** Dorsum light brown

with 23 narrow dark brown cross-bars on body and 2 on tail; each cross-bar one scale wide and separated from others by 5–9 scales; interspaces with dark network; supralabials and chin yellow; venter yellow with dark brown cross-bands, 1.5–2 ventrals wide, and separated by 2–5 ventrals. **Habitat and Behaviour** Nothing known of its natural history. **Distribution** Endemic to north-eastern Sumatra (Langkat, Aceh Province). **Status** Not Evaluated.

EISELT'S REED SNAKE
Calamaria eiselti   NOT ILLUSTRATED
Measurement TL 424mm **Identification** Body slender, cylindrical; head short, indistinct from neck; loreal absent; single preocular; single postocular; supralabials 5; Supralabials III–IV contact orbit; infralabials 5; 3 infralabials contact anterior chin shields; modified maxillary teeth 10; eye small; pupil rounded; tail short, tapering; dorsals smooth; midbody scale rows 13; ventrals 137–153; subcaudals 13–22, paired; anal entire. **Coloration** Dorsum dark brown, each scale finely speckled with yellow; a few scattered dark brown scales; forehead dark brown, unspotted; chin yellow with dark spots on sutures; supralabials yellow, except dorsal portion; yellow stripe, contiguous with labial stripe, extends along ventrolateral side of anterior body; anterior ventrals with black central area; posteriorly, venter black; subcaudals yellow with 1–3 black cross-bands. **Habitat and Behaviour** Nothing known of its natural history. **Distribution** Endemic to west-central Sumatra (Padang, Sumatera Barat Province). **Status** Not Evaluated.

EVERETT'S REED SNAKE
Calamaria everetti   PLATE 41
Measurement TL 330mm **Identification** Body slender; head short, indistinct from neck; loreal absent; single preocular; single postocular; supralabials 5; Supralabials III–IV contact orbit; infralabials 5; 3 infralabials contact anterior chin shields; mental fails to contact anterior chin shields; modified maxillary teeth 9; tail short; dorsals smooth; midbody scale rows 13; ventrals 136–157; subcaudals 16–25, paired; anal entire. **Coloration** Dorsum brown, scales with or without fine dark network; longitudinal stripe formed by dark areas on scales anteriorly; vertical pale brown or yellow bar at rear of head and collar of the same colour, 6–8 scales behind head, may be present; head brown with darker spots; venter yellow. **Habitat and Behaviour** Known from the forested lowlands at sea level to the mid-hills at elevations of up to 1,500m asl. Nothing else known of its natural history. **Distribution** Endemic to western and north-western Borneo. **Status** Not Evaluated.

FORCART'S REED SNAKE
Calamaria forcarti   NOT ILLUSTRATED
Measurement TL 290mm **Identification** Body slender; head short, indistinct from neck; loreal absent; single preocular; single postocular; supralabials 5; Supralabials III–IV contact orbit; infralabials 5; 3 infralabials contact anterior chin shields; eye small; pupil rounded; modified maxillary teeth 8; tail short, tapering; dorsals smooth; midbody scale rows 13; ventrals 176–200; subcaudals 16–30, paired; anal entire. **Coloration** Dorsum pale brown; dorsal scales with indistinct dark network and scattered dark spots, each spot 1–4 scales wide, arranged in transverse rows; yellow ring behind head, edged with blackish-brown bands, 1–2 scales wide; forehead brown with indistinct dark brown marks; venter yellow, unpatterned or with scattered brown spots; subcaudals yellow with dark brown lateral edges. **Habitat and Behaviour** Nothing known of its natural history. **Distribution** Endemic to Pulau Nias and northern Sumatra (Medan, Sumatera Utara Province). **Status** Not Evaluated.

GIMLETT'S REED SNAKE
Calamaria gimletti   NOT ILLUSTRATED
Measurement TL 269mm **Identification** Body slender, cylindrical; head short, indistinct from neck; rostrum visible from above, distinctly smaller than prefrontal suture; loreal absent; preocular absent; single postocular; supralabials 4; Supralabials II–III contact orbit; infralabials 5; 3 infralabials contact orbit; eye small; pupil rounded; modified maxillary teeth 9; tail short, not tapering, blunt-tipped; dorsals smooth; midbody scale rows 13; ventrals 161–249; subcaudals 10–20, paired; anal entire. **Coloration** Dorsum dark brown, dorsals with small white centres; cream ring around tail base; ventrals cream, the posterior ones with dark brown lateral edges; subcaudals brown with yellow median streak. **Habitat and Behaviour** Inhabits lowland to submontane forests. Diurnal and terrestrial. Diet and reproductive habits unstudied. **Distribution** Peninsular Malaysia (Gunung Pulai, Johor State, Bukit Fraser, Pahang State, Bukit Lagong, Selangor State, Tampin, Negri Sembilan State and Pulau Aor, Johor State, as well as Kelantan State) and the Riau Archipelago. **Status** Not Evaluated.

GRABOWSKY'S REED SNAKE
Calamaria grabowskyi   PLATE 41
Measurement TL 470mm **Identification** Body slender, cylindrical; head short, slightly distinct from neck; loreal absent; single preocular; single postocular; supralabials 5; Supralabials III–IV contact orbit; infralabials 5; 3 infralabials contact anterior chin shields; eye small; pupil rounded; modified maxillary teeth 8–10; tail long, tapering to a blunt tip; dorsals smooth; midbody scale rows 13; ventrals 150–190; subcaudals 20–29, paired; anal entire. **Coloration** Dorsum dark brown, each scale with dark network; scattered dark brown or yellow spots on back; labials yellow; lateral bands composed of elongated dark spots; venter unpatterned yellow or with varying amounts of dark pigmentation; subcaudals yellow with dark median band. **Habitat and Behaviour** Inhabits submontane forests at elevations of 1,000–1,400m asl. Nothing else known of its natural history. **Distribution** Endemic to Borneo. **Status** Not Evaluated.

SLENDER REED SNAKE
Calamaria gracillima NOT ILLUSTRATED
Measurement TL 290mm **Identification** Body slender, cylindrical; head short, indistinct from neck; loreal absent; preocular absent; single postocular; supraocular fused with postocular; supralabials 4; Supralabials II–III contact orbit; infralabials 5; 3 infralabials contact anterior chin shields; eye very small; pupil rounded; modified maxillary teeth 10; tail short, non-tapering, with a blunt tip; dorsals smooth; midbody scale rows 13; ventrals 290–304; subcaudals 12–15, paired; anal entire. **Coloration** Dorsum dark brown with a series of widely spaced, vertical yellow bars on anterior flanks and around tail base; venter dark brown, ventrals and subcaudals paler on posterior half. **Habitat and Behaviour** Nothing known of its natural history. **Distribution** Endemic to western Borneo (Gunung Matang and Tegora, Sarawak State). **Status** Not Evaluated.

LINED REED SNAKE
Calamaria griswoldi PLATE 41
Measurement TL 490mm **Identification** Body slender, cylindrical; head short, indistinct from neck; loreal absent; single preocular; single postocular; supralabials 5; Supralabials III–IV contact orbit; infralabials 5; 3 infralabials contact anterior chin shields; eye small; pupil rounded; modified maxillary teeth 9–10; tail short, tapering to a sharp point; dorsals smooth; midbody scale rows 13; ventrals 155–192; subcaudals 13–18, paired; anal entire. **Coloration** Dorsum dark brown with blackish-brown and yellow stripes; head dark brown; lower portions of supralabials yellow; oblique pale bar from parietals to gular region; ventrals unpatterned yellow; subcaudals yellow with indistinct zigzag mark medially. **Habitat and Behaviour** Inhabits submontane forests at elevations of 1,200–1,800m asl. Diurnal and terrestrial, associated with leaf litter. Defensive behaviour includes using its sharp tail-tip to poke adversary. Diet and reproductive habits unstudied. **Distribution** Endemic to northern Borneo (Gunung Kinabalu, Sabah State). **Status** Not Evaluated.

HILLENIUS'S REED SNAKE
Calamaria hilleniusi PLATE 41
Measurement TL 370mm **Identification** Body slender, cylindrical; head short, indistinct from neck; loreal absent; single preocular; single postocular; supralabials 5; Supralabials III–IV contact orbit; infralabials 5; 3 infralabials contact anterior chin shields; mental contacts anterior chin shields; modified maxillary teeth 10; tail short, tapering to tip gradually; dorsals smooth; midbody scale rows 13; ventrals 147–151; subcaudals 14–21, paired; anal entire. **Coloration** Dorsum brown, the dark pigmentation abruptly ending in centre of third scale row; forehead brown, dark pigments ending in oblique line from upper edge of second supralabial to lower third or fifth; supralabials yellow; throat yellow; venter unpatterned yellow. **Habitat and Behaviour** Known from lowland rainforests, reaching submontane limits. Diurnal and terrestrial, associated with forest litter. Diet and reproductive habits unstudied. **Distribution** Endemic to northern and eastern Borneo (Tuaran, Sabah State, Gunung Murud, Sarawak State and Samarinda, Kalimantan Timur Province). **Status** Not Evaluated.

INGER'S REED SNAKE
Calamaria ingeri PLATE 41
Measurement TL 177mm **Identification** Body slender, cylindrical; head short, indistinct from neck; loreal absent; single preocular; supralabials 5; Supralabials III–IV contact orbit; infralabials 5; 3 infralabials contact anterior chin shields; mental separated from anterior chin shields; 6 scales surrounding paraparietal; modified maxillary teeth 7; eye small; pupil rounded; tail short, thick and abruptly tapering; dorsals smooth, reduced to 4 rows on tail opposite last subcaudal anterior to terminal scute; midbody scale rows 13; ventrals 213–228; subcaudals 10–11, paired; anal entire. **Coloration** Dorsum dark brown with intermittent light scale spots; dark colour extends ventrally two-thirds down supralabial scales, half on first dorsal scale rows on body, and nearly down to level of first dorsal scale rows on tail; vertical light-coloured bar extends halfway up neck posterior to angle of jaw; 26 incomplete transverse light bands on body and tail; venter unpatterned cream. **Habitat and Behaviour** Inhabits lowland dipterocarp forest at an elevation of 98m asl on Pulau Tioman. Terrestrial. Diet and reproductive habits unstudied. **Distribution** Endemic to Peninsular Malaysia (Pulau Tioman, Pahang State). **Status** Not Evaluated.

JAVANESE REED SNAKE
Calamaria javanica NOT ILLUSTRATED
Measurement TL 192mm **Identification** Body slender, cylindrical; head short, indistinct from neck; loreal absent; preocular absent; single postocular; supralabials 4; Supralabials II–III contact orbit; infralabials 5; 3 infralabials contact anterior chin shields; modified maxillary teeth 8–9; eye small; pupil rounded; tail short, tapering gradually to a blunt point; dorsals smooth; midbody scale rows 13; ventrals 168–176; subcaudals 10–16, paired; anal entire. **Coloration** Dorsum dark brown, dorsal scales with small yellow specks; narrow vertical band, 6 scales wide, behind head may be present; head dark brown; supralabials and chin yellow; venter yellow; small brown spot at edges of ventrals on anterior of body; subcaudals yellow with dark brown edges and dark brown median stripe. **Habitat and Behaviour** Nothing known of its natural history. **Distribution** Java and Pulau Belitung. **Status** Not Evaluated.

WHITE-STRIPED REED SNAKE
Calamaria lateralis NOT ILLUSTRATED
Measurement TL 290mm **Identification** Body slender, cylindrical; head short, indistinct from neck; loreal absent; single preocular; single postocular; supralabials 5; Supralabials III–IV contact orbit; infralabials 5; 3 infralabials contact anterior chin

shields; modified maxillary teeth 8; eye small; pupil rounded; tail short, tapering gradually to a point; dorsals smooth; midbody scale rows 13; ventrals 146–151; subcaudals 26–23, paired; anal entire. **Coloration** Dorsum dark brownish-black with longitudinal white line from posterior to orbit to the length of body; venter dark brown. **Habitat and Behaviour** Nothing known of its natural history. **Distribution** Northern Borneo (Gunung Kinabalu, Sabah State) and Java. **Status** Not Evaluated.

COLLARED REED SNAKE
Calamaria leucogaster　　　　　　　　PLATE 42
Measurement TL 223mm **Identification** Body slender, cylindrical; head short, slightly distinct from neck; loreal absent; single preocular; single postocular; supralabials 5; Supralabials III–IV contact orbit; infralabials 5; 3 infralabials contact anterior chin shields; mental does not contact anterior chin shield; modified maxillary teeth 7–8; eye small; pupil rounded; tail tapers to a narrow point; dorsals smooth; midbody scale rows 13; ventrals 126–157; subcaudals 12–26, paired; anal entire. **Coloration** Dorsum highly variable, ranging from bright orange-red, through olive, to brown, typically with longitudinal dark stripes; black collar, 3–5 scales wide; half ring around tail base; venter cream. **Habitat and Behaviour** Inhabits lowland forests and the mid-hills. Terrestrial and fossorial, hiding under logs. Diet and reproductive habits unstudied. **Distribution** Sumatra and Borneo. **Status** Not Evaluated.

LINNAEUS'S REED SNAKE
Calamaria linnaei　　　　　　　　PLATE 42
Measurement TL 400mm **Identification** Body slender, cylindrical; head short, indistinct from neck; loreal absent; single preocular; single postocular; supralabials 4; Supralabials II–III contact orbit; infralabials 5; 3 infralabials contact anterior chin shields; modified maxillary teeth 10; eye small; pupil rounded; tail short, tapering abruptly near blunt tip; dorsals smooth; midbody scale rows 13; ventrals 130–166; subcaudals 7–22, paired; anal entire. **Coloration** Dorsum black to pale brown; dorsal scales light with dark network; sometimes isolated small scales on dorsum, spots uniting to form near continuous black stripes; narrow light chevron behind forehead; venter orange, yellow or cream with dark pigments. **Habitat and Behaviour** Inhabits the mid-hills up to submontane limits at elevations of ca 1,500m asl. Diet unstudied. Oviparous; clutches comprise 2–4 eggs, measuring 20–26 x 7–9mm. Incubation period 64–84 days. Hatchlings 92–120mm. **Distribution** Java and Pulau Bangka. **Status** Not Evaluated.

LOW'S REED SNAKE
Calamaria lovii　　　　　　　　NOT ILLUSTRATED
Measurement TL 320mm (*C. l. lovii*); 318mm (*C. l. ingermarxorum*); 315mm (*C. l. wermuthi*)
Identification Body slender, cylindrical; head short, indistinct from neck; loreal absent; single preocular; single postocular; supralabials 4; Supralabials II–III or III contact orbit; infralabials 4–5; 2 or 3 infralabials contact anterior chin shields; eye small; pupil rounded; modified maxillary teeth 8–10; tail short, thick, tapering abruptly; dorsals smooth; midbody scale rows 13; ventrals 205–256; subcaudals 11–26, paired; anal entire. **Coloration** Dorsum dark brown or greyish-blue; other patterns specific to subspecies. **Subspecies** Three subspecies are recognized. *C. l. lovii*, dorsum dark brown with yellow spots; sometimes narrow light stripes; complete or interrupted yellow ring around vent; head dark brown with indistinct light markings; venter cream anteriorly; posterior ventrals dark brown or yellow with irregular squarish black blotches; ventrals 190–254; subcaudals 11–26, paired. *C. l. ingermarxorum*, dorsum immaculate greyish-blue with light spots on each side of neck covering 4 scales; venter dark grey with light posterior edges; prefrontals longer than frontal; rostral portion visible from above more than half length of prefrontal suture; mental contacts anterior chin shields; dorsal scale reduction to 4 rows on tail opposite first to fifth subcaudal anterior to terminal scute; tail short, blunt and non-tapering; ventrals 205; subcaudals 23, paired. *C. l. wermuthi*, dorsum dark brown with longitudinal dark stripes; vertebral scale rows with yellow spots from back to head to base of tail; cream ring around vent; head dark brown with indistinct ligher markings; venter cream, posterior ventrals with dark brown lateral edges; subcaudals brown with yellow median streak; ventrals 256; subcaudals 11, paired. **Habitat and Behaviour** Inhabits the lowlands and mid-hills at an elevation of 750m asl. Nocturnal and terrestrial. Diet and reproductive habits unstudied. **Distribution** Southern Vietnam (Buon Lai, Gia Lai Province) (*C. l. ingermarxorum*), southern Thailand, Peninsular Malaysia, Borneo (*C. l. lovii*) and western Java (*C. l. wermuthi*). **Status** Not Evaluated.

VARIABLE REED SNAKE
Calamaria lumbricoidea　　　　　　　　PLATE 42
Measurement TL 640mm **Identification** Body moderately robust, cylindrical; head short, indistinct from neck; loreal absent; single preocular; postoculars 2; supralabials 5; Supralabials III–IV contact orbit; paraparietal surrounded by 4–5 scales; infralabials 5; 3 infralabials contact anterior chin shields; eye small; pupil rounded; modified maxillary teeth 9–11; tail short, thick and tapering abruptly to a narrow point; dorsals smooth; midbody scale rows 13; ventrals 137–210; subcaudals 14–26, paired; anal entire. **Coloration** Dorsum black with narrow cream or yellow rings; head red or pink in juveniles, turning black with growth; venter yellow with black ventral scales forming bands. **Habitat and Behaviour** Inhabits lowland forests and gardens up to submontane limits at an elevation of 1,676m asl. Terrestrial, living in leaf litter. Diet consists of earthworms and insect larvae. Reproductive habits unstudied. May possibly be a mimic of the venomous Red-headed Krait (*Bungarus flaviceps*) or Malayan Striped Coral Snake (*Calliophis bivirgatus*). **Distribution** Southern Thailand (Pattani and

Narathiwat Provinces), Peninsular Malaysia (Bukit Larut and Gunung Kledong, Perak State, Bukit Fraser and Cameron Highlands, Pahang State, and Kepong, Selangor State), Singapore, Sumatra, Pulau Nias, the Mentawai Archipelago, Borneo and Java. Also Mindanao, Basilan and Leyte (the Philippines). **Status** Not Evaluated.

LUMHOLZ'S REED SNAKE
Calamaria lumholzi   NOT ILLUSTRATED
Measurement TL 210mm **Identification** Body slender, cylindrical; head short, indistinct from neck; loreal absent; single preocular; single postocular; supralabials 5; Supralabials III–IV contact orbit; infralabials 5; 3 infralabials contact anterior chin shields; eye small; pupil rounded; modified maxillary teeth 8; tail short, tapering to a blunt tip; dorsals smooth; midbody scale rows 13; ventrals 167–171; subcaudals 13–15, paired; anal entire. **Coloration** Dorsum dark brown with longitudinal stripe on scale rows 2–3, bordered ventrally by dark line; head brown with black spots; venter white, sometimes with dark speckling along anterior margins. **Habitat and Behaviour** Nothing known of its natural history. **Distribution** Endemic to central Borneo (Sungei Merah and Tumbang Maruwai, Kalimantan Province). **Status** Not Evaluated.

STRIPE-NECKED REED SNAKE
Calamaria margaritophora   NOT ILLUSTRATED
Measurement TL 360mm **Identification** Body slender, cylindrical; head short, indistinct from neck; loreal absent; single preocular; single postocular; supralabials 5; Supralabials III–IV contact orbit; infralabials 5; 3 infralabials contact anterior chin shields; eye small; pupil rounded; modified maxillary teeth 8–9; tail short, tapering abruptly to a blunt tip; dorsals smooth; midbody scale rows 13; ventrals 147–163; subcaudals 8–17, paired; anal entire. **Coloration** Dorsum brown; dorsal scales above first row with fine light network; series of interrupted black lines formed by black spots in centre of scales, lines longest on neck; head dark brown with indistinct black marks; supralabials yellow with dark brown upper margins; chin and venter yellow with dark margin; subcaudals yellow. **Habitat and Behaviour** Inhabits submontane forests at an elevation of 945m asl. Nothing else known of its natural history. **Distribution** Southern Sumatra (Siulakderas, Jambi Province, Gunung Kerinchi, Sumatera Barat Province, Kaba Wetan, nr Kapahiang, Bengkulu Province, and Ampat Lawang, nr Tanjungraya, and Rimbopengadang, Sumatera Selatan Province); possibly also Java. **Status** Not Evaluated.

MECHEL'S REED SNAKE
Calamaria mecheli   NOT ILLUSTRATED
Measurement TL 250mm **Identification** Body slender, cylindrical; head short, indistinct from neck; loreal absent; preocular absent; single postocular; supralabials 5; Supralabials III–IV contact orbit; infralabials 5; 3 infralabials contact anterior chin shields; eye small; pupil rounded; modified maxillary teeth 8–9; tail short, tapering sharply to a point; dorsals smooth; midbody scale rows 13; ventrals 167–195; subcaudals 12–28, paired; anal entire. **Coloration** Dorsum brown; dorsal scales pale brown with fine dark network; narrow dark stripe along adjacent edges of ventrals and first scale row; some individuals with longitudinal dark stripes on vertebral region; head purplish-brown with obscure dark markings; venter cream; subcaudals cream, sometimes with narrow dark midventral stripe. **Habitat and Behaviour** Nothing known of its natural history. **Distribution** Endemic to Sumatra (Medan, Sumatera Utara Province, Air Maninjau, Sumatera Barat Province and Indragiri, Riau Province). **Status** Not Evaluated.

KAPUAS REED SNAKE
Calamaria melanota   NOT ILLUSTRATED
Measurement TL 260mm **Identification** Body slender, cylindrical; head short, indistinct from neck; loreal absent; single preocular; single postocular; supralabials 4; Supralabials II–III contact orbit; infralabials 5; 3 infralabials contact anterior chin shields; eye small; pupil rounded; modified maxillary teeth 8–9; tail short, tapering to a point; dorsals smooth; midbody scale rows 13; ventrals 121–154; subcaudals 16–26, paired; anal entire. **Coloration** Dorsum dark brown, each dorsal with light spot or pale reticulated tip; head dark brown; supralabials dark with yellow centres; chin yellow with dark areas; venter and subcaudals with dark anterior and pale posterior. **Habitat and Behaviour** Nothing known of its natural history. **Distribution** Endemic to southern and south-eastern Borneo (Gunung Mulu, Sarawak State, and Kuala Kapuas, Martapura, Sungei Pasir and Tanjong, Kalimantan Province). **Status** Not Evaluated.

YELLOW-SPOTTED REED SNAKE
Calamaria modesta   NOT ILLUSTRATED
Measurement TL 460mm **Identification** Body slender, cylindrical; head short, indistinct from neck; loreal absent; single preocular; single postocular; supralabials 5; Supralabials III–IV contact orbit; infralabials 5; 3 infralabials contact anterior chin shields; eye small; pupil rounded; modified maxillary teeth 6–9; tail short, tapering gradually to a point; dorsals smooth; midbody scale rows 13; ventrals 131–202; subcaudals 12–31, paired; anal entire. **Coloration** Dorsum dark brown or black; dorsal scales unpatterned or with light network and longitudinal dark streak; sometimes scattered yellow spots; forehead black with yellow spots or yellowish-brown with black spots; venter dark with pale areas medially and laterally. **Habitat and Behaviour** Inhabits submontane forests at elevations of 1,370–1,430m asl. Nothing else known of its natural history. **Distribution** Sumatra (Kayutanam, Gunung Singgalang, Air Mananjau and Padang, Sumatera Barat Province), Pulau Simeuleu, northern Borneo (Bundu Tuhan, Tenompok and Quoin Hill, Sabah State) and Java (numerous localities). **Status** Not Evaluated.

BROWN REED SNAKE
Calamaria pavimentata PLATE 42
Measurement TL 490mm **Identification** Body slender, cylindrical; head short, indistinct from neck; loreal absent; single preocular; single postocular; supralabials 4; Supralabials II–IV contact orbit; Supralabial I smaller than II; infralabials 4; 3 infralabials contact anterior chin shields; second pair of chin shields meets in midline; rostral as broad as high; rostral portion visible from above only one half to equal entire length of prefrontal suture; modified maxillary teeth 8–10; eye small; pupil rounded; tail moderately long, gradually tapering to a point; dorsals smooth; midbody scale rows 13; ventrals 133–168; subcaudals 13–20, paired; anal entire. **Coloration** Dorsum brown with narrow longitudinal dark stripes; solid black collar; ventrals with dark lateral tips. **Habitat and Behaviour** Inhabits hill forests. Nothing else known of its natural history. **Distribution** Myanmar (Chin and Kachin States), Thailand, Laos, Cambodia, Vietnam and Peninsular Malaysia. Also north-eastern India, southern and eastern China, and Japan (the Ryukyu Archipelago). **Status** Not Evaluated.

PRAKKE'S REED SNAKE
Calamaria prakkei NOT ILLUSTRATED
Measurement TL 260mm **Identification** Body slender, cylindrical; head short, indistinct from neck; loreal absent; single preocular; single postocular; supralabials 5; Supralabials III–IV contact orbit; infralabials 5; 3 infralabials contact anterior chin shields; modified maxillary teeth 6–7; eye small; pupil rounded; tail long and thick, tapering to a sharp point; dorsals smooth; midbody scale rows 13; ventrals 126–144; subcaudals 24–32, paired; anal entire. **Coloration** Dorsum brown; dorsal scales above first row with dark network; middorsal scales with dark centre, not forming longitudinal stripes; yellow ventrolateral stripe along first dorsal scale row and along adjacent halves of first and second scale rows; pale brown nuchal collar; forehead brown with darker spots; venter yellow with brown lateral edges; subcaudals yellow, typically with dark midventral stripe. **Habitat and Behaviour** Nothing known of its natural history. **Distribution** Singapore and north-eastern Borneo (Sandakan Bay, Sabah State). **Status** Not Evaluated.

REBENTISCH'S REED SNAKE
Calamaria rebentischi NOT ILLUSTRATED
Measurement TL 270mm **Identification** Body slender, cylindrical; head short, indistinct from neck; loreal absent; preocular absent; single postocular; supralabials 5; Supralabials III–IV contact orbit; infralabials 5; 3 infralabials contact anterior chin shields; eye small; pupil rounded; modified maxillary teeth 7; tail short, tapering; dorsals smooth; midbody scale rows 13; ventrals 140; subcaudals 29, paired; anal entire. **Coloration** Dorsum dark brown; dorsal scales yellowish-brown with dark network; forehead dark brown; supralabials yellow with brown spots; venter yellow with brown lateral edges; subcaudals yellow with a few small brown spots midventrally. **Habitat and Behaviour** Nothing known of its natural history. **Distribution** Endemic to western Borneo (Sinkawang, Kalimantan Province). **Status** Not Evaluated.

RED-HEADED REED SNAKE
Calamaria schlegeli PLATE 42
Measurement TL 450mm **Identification** Body slender, cylindrical; head short, indistinct from neck; nasal as large as eye; loreal absent; preocular present or absent; single postocular; supralabials 5; Supralabials III–IV contact orbit; infralabials 5; 3 infralabials contact anterior chin shields; eye small; pupil rounded; modified maxillary teeth 9–10; tail long and tapering; dorsals smooth; midbody scale rows 13; ventrals 130–180; subcaudals 19–44, paired; anal entire. **Coloration** Dorsum dark brown or black; venter unpatterned yellow. **Subspecies** Two subspecies are recognized. *C. s. schlegeli*, head red or orange. *C. s. cuvieri*, head dark brown. **Habitat and Behaviour** Inhabits lowland forests. Terrestrial, associated with leaf litter. Diet consists of frogs and slugs. Oviparous; clutch size unknown. **Distribution** Southern Thailand (Pattani Province), Peninsular Malaysia, Singapore, Sumatra, Borneo (*C. s. schlegeli*), Java and Bali (*C. s. cuvieri*). **Status** Not Evaluated.

SCHMIDT'S REED SNAKE
Calamaria schmidti PLATE 42
Measurement TL 280mm **Identification** Body slender, cylindrical; head short, indistinct from neck; loreal absent; preocular absent; single postocular; supralabials 4; Supralabials II–III contact orbit; infralabials 5; 3 infralabials contact anterior chin shields; mental does not contact chin shield; unmodified maxillary teeth 6–7; eye small; pupil rounded; tail short with a blunt point; dorsals smooth; midbody scale rows 13; ventrals 127–150; subcaudals 14–22, paired; anal entire. **Coloration** Dorsum unpatterned blackish-grey with green and blue iridescence, scales pale-margined; venter light grey or yellow, darkening posteriorly to purple. **Habitat and Behaviour** Inhabits montane forests at elevations of 1,370–1,570m asl. Nocturnal and terrestrial, active near streams, while hiding under fallen logs by day. Diet includes earthworms. Reproductive habits unstudied. **Distribution** Endemic to northern Borneo (Gunung Kinabalu, Sabah State). **Status** Not Evaluated.

NORTHERN REED SNAKE
Calamaria septentrionalis PLATE 42
Measurement TL 450mm **Identification** Body slender, cylindrical; head short, indistinct from neck; rostral portion visible from above less than one-third length of prefrontal suture; loreal absent; single preocular; single postocular; supralabials 4; Supralabials II–III contact orbit; infralabials 5; 3 infralabials contact anterior chin shields; eye small; pupil rounded; modified maxillary teeth 8–9; tail as thick as body, not tapering, with a broadly rounded

tip; dorsals smooth; midbody scale rows 13; ventrals 168–188; subcaudals 6–11, paired; anal entire. **Coloration** Dorsum dark brown or black, scales with numerous small pale dots, forming network; yellow nuchal loop; venter coral-red; subcaudals with narrow black line. **Habitat and Behaviour** Inhabits lowland forests, reaching submontane limits. Nocturnal and terrestrial. Diet consists of earthworms and possibly also ants and termites. Oviparous; clutch size unknown. **Distribution** Northern Vietnam (Ngan Son, Cao Bang Province, Gao Bang, Lao Kay, Thai Nguyen and Bac Ky or Dông Kinh region). Also eastern China. **Status** Not Evaluated.

YELLOW-BELLIED REED SNAKE
Calamaria suluensis　　　　　NOT ILLUSTRATED
Measurement TL 300mm **Identification** Body slender, cylindrical; head short, indistinct from neck; loreal absent; single preocular; single postocular; supralabials 5; Supralabials III–IV contact orbit; infralabials 5; 3 infralabials contact anterior chin shields; eye small; pupil rounded; modified maxillary teeth 7–8; tail long, tapering gradually to a blunt tip; dorsals smooth; midbody scale rows 13; ventrals 129–168; subcaudals 14–20, paired; anal entire. **Coloration** Dorsum brown, dorsal scales with fine dark network; scattered scales with dark central spots; head dark brown with scattered dark spots; venter yellow with dark lateral edges; subcaudals yellow with or without dark median line. **Habitat and Behaviour** Inhabits submontane forests at elevations of 915–1,430m asl. Nothing else known of its natural history. **Distribution** Western and northern Borneo (Sarikei, Sarawak State, Gunung Kinabalu, Sabah State, and Kutai, Puruk Tjau and Muara Teweh, Kalimantan Province). Also the Sulu Archipelago (the Philippines). **Status** Not Evaluated.

SUMATRAN REED SNAKE
Calamaria sumatrana　　　　　NOT ILLUSTRATED
Measurement TL 260mm **Identification** Body slender, cylindrical; head short, indistinct from neck; loreal absent; single preocular; single postocular; supralabials 5; Supralabials III–IV contact orbit; infralabials 5; 3 infralabials contact anterior chin shields; eye small; pupil rounded; modified maxillary teeth 8–9; tail thick, tapering; dorsals smooth; midbody scale rows 13; ventrals 126–175; subcaudals 10–20, paired; anal entire. **Coloration** Dorsum dark brown; dorsal scales with fine dark network or black central spot; yellow nuchal collar; forehead brown with indistinct black spots; venter cream or yellow with darker edges; subcaudals yellow with dark midventral stripe. **Habitat and Behaviour** Nothing known of its natural history. **Distribution** Endemic to Sumatra (Medan, Binjai, Sibolangit, Padang and Sukaranda, nr Pancurbatu, Sumatera Utara Province, and Tanjungampalu, Sumatera Tengah Province). **Status** Not Evaluated.

THANH'S REED SNAKE
Calamaria thanhi　　　　　NOT ILLUSTRATED
Measurement TL 455mm **Identification** Body slender, cylindrical; head short, indistinct from neck; preocular absent; supralabials 4; Supralabials II–III contact orbit; infralabials 5; mental not in contact with anterior chin shields; tail tapers gradually to a point; dorsals smooth; midbody scale rows 13; ventrals 198; subcaudals 21, paired; anal entire. **Coloration** Dorsum dark bluish-grey and iridescent with 4 yellow to beige zigzag-shaped bands; light dorsal markings on base and tip of tail; venter unpatterned yellow or beige. **Habitat and Behaviour** Inhabits a limestone cave within primary forest. Diet unstudied. Oviparous; clutches comprise at least 7 eggs, measuring 8–12.2mm x 2.3–4.5mm. **Distribution** Endemic to central Vietnam (Truong Son, in Annamite mountain range). **Status** Not Evaluated.

ULMER'S REED SNAKE
Calamaria ulmeri　　　　　NOT ILLUSTRATED
Measurement TL 280mm **Identification** Body slender, cylindrical; head short, indistinct from neck; loreal absent; single preocular; single postocular; supralabials 5; Supralabials III–IV contact orbit; infralabials 5; 3 infralabials contact anterior chin shields; unmodified maxillary teeth 10; eye small; pupil rounded; tail gradually tapering; dorsals smooth; midbody scale rows 13; ventrals 186; subcaudals 23+ (tail-tip of only individual known missing), paired; anal entire. **Coloration** Dorsum brown; dorsal scales with dark network and dark central spot; spots unite to form dark stripes along body and tail; head brown with small dark spots; venter yellow with dark brown lateral edges; subcaudals dark brown. **Habitat and Behaviour** Inhabits montane forests of Gunung Pusat Gayo at an elevation of 2,080m asl. Nothing else known of its natural history. **Distribution** Endemic to northern Sumatra (ca 40 km north-west of Blangkejeren, Aceh Province). **Status** Not Evaluated.

SHORT-TAILED REED SNAKE
Calamaria virgulata　　　　　PLATE 42
Measurement TL 370mm **Identification** Body slender, cylindrical; head short, indistinct from neck; loreal absent; single preocular; single postocular; supralabials 5; Supralabials III–IV contact orbit; infralabials 5; 3 infralabials contact anterior chin shields; eye small; pupil rounded; modified maxillary teeth 8–10; tail thick, tapering; dorsals smooth; midbody scale rows 13; ventrals 160–260; subcaudals 8–30, paired; anal entire. **Coloration** Dorsum dark brown; dorsal scales with light network, with or without longitudinal dark stripes; yellow nuchal collar; longitudinal dark stripes along body; head dark brown; supralabials typically yellow; venter cream or brownish-black, dark pigments on lateral edges of ventrals; subcaudals with dark edges and dark median stripe. **Habitat and Behaviour** Inhabits submontane forests. Nocturnal and terrestrial, associated with leaf litter. Diet unstudied. Oviparous; clutches comprise 3 eggs, measuring 26–30 x 8–8.5mm. **Distribution**

Sumatra (Gunung Sahilan, Bengkulu Province), northern and north-eastern Borneo (Gunung Kinabalu and Sapagaya Forest Reserve, Sabah State), and western Java (Gunung Salak). Also the Sulu Archipelago, Mindanao and Palawan (the Philippines), and Sulawesi (Indonesia). **Status** Not Evaluated.

YUNNAN REED SNAKE
Calamaria yunnanensis NOT ILLUSTRATED
Measurement TL 362mm **Identification** Body robust, cylindrical; head short, indistinct from neck; preocular absent; supralabials 4; Supralabials II–III contact orbit; infralabials 5; tail thick, non-tapering; dorsals smooth; midbody scale rows 13; ventrals 173–179; subcaudals 22, divided. **Coloration** Dorsum bluish-brown with 5 indistinct longitudinal brown stripes from eye to tail-tip; outermost stripe demarcates dorsal coloration from yellow venter; subcaudals unpatterned yellow. **Habitat and Behaviour** Inhabits primary forests. Terrestrial. Diet and reproductive habits unstudied. **Distribution** Laos (Phongsaly Province). Also southern China. **Status** Not Evaluated.

ORNATE FLYING SNAKE
Chrysopelea ornata PLATE 43
Measurement TL 1,400mm **Identification** Body slender; head depressed, distinct from neck; loreal present; single preocular; postoculars 2; supralabials 9; Supralabials IV–VI contact orbit; infralabials 10; eye large; pupil rounded; tail long, slender; ventrals with pronounced keels laterally; dorsals smooth or feebly keeled, with apical pits; lateral keels with sharp notch; midbody scale rows 17; ventrals 213–234; subcaudals 110–142, paired; anal divided. **Coloration** Dorsum greenish-yellow or pale green; head black dorsally with yellow and black cross-bars; scales with dark streak, creating an impression of longitudinal black stripes; venter pale green with a series of black lateral spots on each side. **Subspecies** Two subspecies are recognized, of which one (*C. o. ornatissima*) occurs in the region. **Habitat and Behaviour** Inhabits lowland forests associated with secondary vegetation and cultivated areas, and enters houses. Diurnal and arboreal, making long glides between trees, covering at least 50m. Diet consists of lizards, especially geckos and agamids, and also bats, rodents, birds and other snakes. Oviparous; clutches comprise 6–20 eggs, measuring 26–38 x 13–18mm. Incubation period 65–80 days. Hatchlings 115–260mm. **Distribution** Myanmar, Thailand, Laos, Cambodia, southern Vietnam and northern Peninsular Malaysia (Pulau Langkawi, Kedah State) (*C. o. ornatissima*). Also northern and eastern India, Bangladesh, Nepal (*C. o. ornatissima*), south-western India and Sri Lanka (*C. o. ornata*). **Status** Not Evaluated.

GARDEN FLYING SNAKE
Chrysopelea paradisi PLATE 43
Measurement TL 1,500mm **Identification** Body slender; head depressed, distinct from neck; loreal present; single preocular; postoculars 2; supralabials 9–10; Supralabials IV–VI or V–VI contact orbit; eye large; pupil rounded; tail long, slender; ventrals with pronounced keels laterally; dorsals smooth or weakly keeled with apical pits; midbody scale rows 17; ventrals 198–239; subcaudals 106–149, paired; anal divided. **Coloration** Dorsum black, centre of each scale with green spot; vertebral region of some individuals with row of 3–4 pink or red spots; forehead with yellow bands; venter green with black edges. **Subspecies** Three subspecies are recognized, of which one (*C. p. paradisi*) occurs in the region. **Habitat and Behaviour** Inhabits forested habitats in the lowlands and submontane areas at elevations of up to 1,524m asl. Arboreal, can make extended leaps between tree-tops, with body flattened into a ribbon-like shape. Diet consists of geckos. Oviparous; clutches comprise 5–8 eggs. **Distribution** Myanmar, Thailand, Peninsular Malaysia, Singapore, Sumatra, the Mentawai Archipelago, Borneo and Java (*C. p. paradisi*). Also Narcondum Island, the Andaman Archipelago (India) (*C. p.* unallocated subspecies), Sulawesi (Indonesia) (*C. p. celebensis*), Samar, Luzon, Mindanao and Masbate (the Philippines) (*C. p. variabilis*). **Status** Not Evaluated.

TWIN-BARRED FLYING SNAKE
Chrysopelea pelias PLATE 43
Measurement TL 740mm **Identification** Body slender; head depressed, distinct from neck; single preocular; postoculars 2; supralabials 9; Supralabials V–VI or IV–VI contact orbit; infralabials 10–11; tail long, slender; ventrals with pronounced keels laterally; dorsals smooth or weakly keeled with apical pits; midbody scale rows 17; ventrals 181–201; subcaudals 89–120, paired; anal divided. **Coloration** Dorsum red or orange with yellow or cream cross-bars, edged with black bars; forehead with 3 red cross-bars; venter pale. **Habitat and Behaviour** Inhabits open forests and enters human habitations around forest edges, from sea level to 600m asl. Diurnal and arboreal, known to glide and parachute. Diet consists of lizards. Reproductive habits unstudied. **Distribution** Myanmar, southern Thailand (Ton Nga Chang, Songkhla Province), Peninsular Malaysia, Singapore, Sumatra, Pulau Bangka, the Mentawai, Natuna and Riau Archipelagos, Borneo and Java. **Status** Not Evaluated.

ENGGANO RAT SNAKE
Coelognathus enganensis PLATE 43
Measurement TL 1,200mm **Identification** Body slender; snout relatively long; single preocular; postoculars 2; supralabials 8–9; Supralabials V–VI or IV–V contact orbit; infralabials 10–11; temporals horizontal; eye large; pupil rounded; tail long, slender; dorsals keeled; midbody scale rows 23 or 25; ventrals 236–243; subcaudals 101–109, paired; anal entire. **Coloration** Dorsum pale brown; forehead and nape dark brown; nape sometimes with indistinct black blotches and stripes; faint dark postocular stripe; dorsum either patternless or with rows of dark blotches on anterior; posterior body and tail

patternless; venter unpatterned yellow. **Habitat and Behaviour** Nothing known of its natural history. **Distribution** Endemic to Pulau Enggano. **Status** Not Evaluated.

PHILIPPINE TRINKET SNAKE
Coelognathus erythrurus PLATE 43
Measurement TL 1,670mm **Identification** Body slender; head elongated, distinct from neck; single preocular; supralabials 8–10; Supralabials IV–VI contact orbit; infralabials 10–11; eye large; pupil rounded; tail long, slender; dorsals keeled; midbody scale rows 19 or 21; ventrals 219–238; subcaudals 102–114 (for subspecies *C. e. philippina*). **Coloration** Dorsum brown to olive; narrow dark postocular stripe to angle of jaws; posterior third of body darker; tail reddish-brown; venter unpatterned cream or white; iris golden-brown. **Subspecies** Four subspecies are recognized, of which one (*C. e. philippina*) occurs in the region. **Habitat and Behaviour** Inhabits the lowlands to the mid-hills in open forests and forest edges, and also cultivated areas, at elevations of 0–850m asl. Diurnal and terrestrial. Diet consists of small rodents, birds and lizards. Oviparous; clutches comprise 6–10 eggs. Incubation period 90–120 days. Hatchlings ca 300mm. **Distribution** Borneo (Gunung Kinabalu and Kiulu, Sabah State) (*C. e. philippina*). Also Palawan and the Sulu Archipelago, the south-western Philippines (*C. e. philippina*), Sulawesi, Indonesia (*C. e. celebensis*), Mindanao, Samar and Leyte, the western and south-western Philippines (*C. e. erythrura*), Negros and other islands of the central Philippines (*C. e. psephenoura*), and Mindoro and Luzon (*C. e. manillensis*). **Status** Not Evaluated.

YELLOW-STRIPED TRINKET SNAKE
Coelognathus flavolineatus PLATE 43
Measurement TL 1,800mm **Identification** Body slender; head scarcely distinct from neck; snout long; loreal present; single preocular; postoculars 2; supralabials 8–9; Supralabials IV–VI contact orbit; infralabials 10–11; eye large; pupil rounded; tail long, slender; dorsals keeled; midbody scales 19; ventrals 193–242; subcaudals 80–116, paired; anal entire. **Coloration** Dorsum brownish-grey or brownish-olive; dark postocular stripe to above back of mouth, and another one along nape; several short dark stripes or elongated blotches on dorsum and flanks; venter pale yellow anteriorly, some ventrals with dark grey edges; tail darker posteriorly; subcaudals dark grey or black. **Habitat and Behaviour** Inhabits the forested lowlands, and disturbed habitats such as parks and gardens, from sea level to 900m asl. Terrestrial and arboreal. Diet includes rodents, birds, frogs and lizards. Oviparous; clutches comprise 5–12 eggs, measuring 51–62 x 23–25.5mm. Incubation period 75–109 days. Hatchlings 250–300mm. **Distribution** Southern Myanmar (Elphinstone, Mergui Archipelago and south of Tenasserim, Tanintharyi Division), southern Thailand (Phuket, Satun, Surat Thani and Narathiwat Provinces), Cambodia, Vietnam, Peninsular Malaysia, Singapore, Sumatra, Pulau Nias, the Mentawai Archipelago, Pulau Weh, the Riau Archipelago, Pulau Bangka, Pulau Belitung, Borneo and Java. Also the Andaman Islands (India) and Sulawesi (Indonesia). **Status** Not Evaluated.

COPPER-HEAD TRINKET SNAKE
Coelognathus radiatus PLATE 43
Measurement TL 2.3m **Identification** Body slender; head slightly distinct from neck; snout relatively long, narrow; loreal present; single preocular; postoculars 2; supralabials 8–9; Supralabials III–V contact orbit; infralabials 9–11; eye large; pupil rounded; tail long, slender; dorsals smooth anteriorly and on flanks, weakly keeled posteriorly; midbody scale rows 19; ventrals 207–250; subcaudals 80–108, paired; anal entire. **Coloration** Dorsum greyish-brown or yellowish-brown with 4 (2 broad, 2 narrow) black stripes along anterior of body; cream stripe runs along upper 2 stripes; lower stripes narrower and typically broken up; head coppery-brown with 3 radiating black lines from eyes; venter grey to yellowish-grey; iris golden-yellow; tongue dark brown to violet. **Habitat and Behaviour** Inhabits open grasslands and light forests in the plains up to submontane limits, ascending elevations of 1,400m asl. When threatened it rears its head and part of its body, throwing its neck and body into a coil. Diet consists of frogs, birds and rats. Oviparous; clutches comprise 5–23 eggs, measuring 40–53 x 20–26mm, and up to 4 clutches are produced annually. Incubation period 70–95 days. Hatchlings 250–300mm. **Distribution** Myanmar, Thailand, Cambodia, Laos, Vietnam, Peninsular Malaysia, Singapore, Sumatra, Pulau Bangka, Borneo and Java. Also India, Bangladesh, Nepal and southern and eastern China. **Status** Not Evaluated.

INDONESIAN TRINKET SNAKE
Coelognathus subradiatus PLATE 43
Measurement TL 2m **Identification** Body slender; snout relatively long; single preocular; postoculars 2; supralabials 8–9; Supralabials V–VI or IV–V contact orbit; infralabials 10–11; temporals oblique; eye large; pupil rounded; dorsals keeled; tail short, slender; midbody scale rows 23 or 25; ventrals 226–278; subcaudals 90–121; anal entire. **Coloration** Dorsum yellowish-brown to olive with either 4 narrow longitudinal stripes or saddle-like black blotches, or pattern showing both stripes and blotches; forehead yellow to dark brown with dark postocular stripe; supralabials unpatterned yellowish-white; venter yellow to cream with dusky edges and dark spot on edges of ventrals. **Habitat and Behaviour** Inhabits small islands. Diet consists of lizards, birds and small mammals. Reproductive habits unstudied. **Distribution** Suspected to occur on Bali. Also Komodo, Wetar, Rintja, Sumba, Flores, Sumbawa, Timor and Samao. **Status** Not Evaluated.

MOUNTAIN DWARF SNAKE
Collorhabdium williamsoni NOT ILLUSTRATED
Measurement SVL 300mm **Identification** Body slender, cylindrical; snout obtuse; head indistinct from

273

neck; loreal and anterior temporal shield absent; prefrontal does not contact orbit; single preocular; single postocular; supralabials 5; Supralabials III–IV contact orbit; infralabials 3–4; maxillary teeth 9; tail short, tip acute; dorsals smooth; midbody scale rows 15; ventrals 144–164; subcaudals 22–32; anal entire. **Coloration** Dorsum light brown with 5 dark stripes; nuchal pattern absent; 2 square-shaped yellow marks behind parietals; venter yellow. **Habitat and Behaviour** Inhabits submontane and montane forests at elevations of 1,036–1,829m asl. Subfossorial, hiding under fallen logs. Diet unstudied. Oviparous; clutch size unknown. Eggs elongated, measuring 23mm. **Distribution** Endemic to Peninsular Malaysia (Cameron Highlands, Pahang State and Bukit Larut, Perak State). **Status** Not Evaluated.

ANNAM BURROWING SNAKE
Cryptophidion annamense NOT ILLUSTRATED
Measurement TL unknown **Identification** Body slender; head small, indistinct from neck; snout depressed with pointed rostral; loreal absent; supralabials 8; Supralabials IV–V subocular; infralabials probably 8; large preorbital scute; eye small; nasals concave, separated from rostral; dorsals smooth, iridescent; tail short; ventrals 172; subcaudals 27; anal entire. **Coloration** Dorsum in preservative dark plumbeous or black, most shields with narrow silver posterior margins; venter unpatterned bluish-grey; live coloration unknown. **Habitat and Behaviour** Nothing known of its natural history, although morphological characters suggest a subfossorial lifestyle. **Distribution** Endemic to central Vietnam (west of Da Nang, Quang Nam-Da Nang Province). **Status** Not Evaluated.

DORIA'S GREEN SNAKE
Cyclophiops doriae NOT ILLUSTRATED
Measurement TL 910mm **Identification** Body moderately robust, cylindrical; head distinct from neck; snout shorter than that of Hampton's Green Snake (*C. hamptoni*); supralabials 7; Supralabials IV–V contact orbit; infralabials 6; eye large; pupil rounded; internasals truncate anteriorly; tail long, slender; midbody scale rows 15; ventrals 168–187; subcaudals 74–80, divided; anal entire. **Coloration** Dorsum green; supralabials and venter cream. **Habitat and Behaviour** Inhabits forested hills. Diurnal and arboreal. Diet and reproductive habits unstudied. **Distribution** Northern Myanmar. Also north-eastern India and southern China (Yunnan Province). **Status** Not Evaluated.

HAMPTON'S GREEN SNAKE
Cyclophiops hamptoni NOT ILLUSTRATED
Measurement TL 1,070mm **Identification** Body moderately slender, elongated, cylindrical; head distinct from neck; snout pointed; internasals truncate anteriorly; supralabials 7; Supralabials IV–V contact orbit; infralabials 6; eye large; pupil rounded; tail long, slender; dorsals smooth; midbody scale rows 15; ventrals 194; subcaudals 76, divided; anal entire. **Coloration** Dorsum green; supralabials and venter cream. **Habitat and Behaviour** Inhabits forested hills. Nothing else known of its natural history. **Distribution** Endemic to north-central Myanmar (Mogok, Mandalay Division). **Status** Not Evaluated.

GREATER GREEN SNAKE
Cyclophiops major PLATE 43
Measurement TL 1,200mm **Identification** Body moderately robust, cylindrical; head distinct from neck; snout elongated; supralabials 7; Supralabials IV–V contact orbit; infralabials 6; eye large; pupil rounded; internasals truncate anteriorly; tail long, slender; midbody scale rows 15, smooth, those posteriorly are keeled; ventrals 154–178; subcaudals 70–92, divided; anal divided. **Coloration** Dorsum bright green; scattered black spots in juveniles; venter greenish-yellow or cream. **Habitat and Behaviour** Inhabits the mid-hills and submontane forests. Diurnal and sometimes active at night, and terrestrial. Diet includes earthworms, caterpillars and insects. Oviparous; clutches comprise 4–13 eggs. **Distribution** Vietnam. Also eastern China. **Status** Not Evaluated.

MANY-BANDED GREEN SNAKE
Cyclophiops multicinctus PLATE 43
Measurement TL 1,070mm **Identification** Body moderately robust, cylindrical; head distinct from neck; snout more acute than that of Greater Green Snake (*C. major*); supralabials 7; Supralabials IV–V contact orbit; infralabials 6; eye large; pupil rounded; internasals narrow anteriorly; tail long, slender; midbody scale rows 15; ventrals 164–177; subcaudals 72–103, divided; anal divided. **Coloration** Dorsum green anteriorly, turning grey posteriorly; numerous cream cross-bars, edged with black, on posterior side of body and tail; cross-bars sometimes fused on vertebral, or may alternate; venter cream with green or grey dots, or entirely grey posteriorly. **Habitat and Behaviour** Inhabits forested hills. Nothing else known of its natural history. **Distribution** Northern Laos and Vietnam. Also eastern China. **Status** Not Evaluated.

STRIPE-TAILED BRONZEBACK TREE SNAKE
Dendrelaphis caudolineatus PLATE 44
Measurement TL 1,520mm **Identification** Body slender; head distinct from neck; snout bluntly rounded; loreal present; single preocular; postoculars 2; supralabials 9; Supralabials V and VI contact orbit; infralabials 11; eye large; dorsals smooth; tail long; midbody scale rows 13; ventrals 174–189; subcaudals 105–116, paired; ventrals and subcaudals with sharp keel on outer edges; anal divided. **Coloration** Dorsum reddish-brown or olive-brown; pale green stripe on lower flanks, edged dorsally by narrow black stripe, and ventrally by broad black stripe; stripes most conspicuous caudally; venter pale green. **Subspecies** Three subspecies have been described, of which one (*D. c. caudolineatus*) occurs in the region. **Habitat and Behaviour** Associated with lowland and submontane forests, as well as parks and gardens, from sea level to 1,524m asl. Diurnal, seen on shrubs. Diet includes

frogs and lizards. Oviparous; clutches comprise 5–8 eggs, measuring 12 x 48mm. Incubation period 54 days. Hatchlings 340mm. **Distribution** Southern Myanmar, southern Thailand, Laos, Peninsular Malaysia, Singapore, Sumatra, Pulau Babi, Pulau Bangka, the Batu, Natuna, Mentawai and Riau Archipelagos, Pulau Belitung, Pulau Nias and Borneo (*D. c. caudolineatus*). Also the Palawan group (*D. c. caudolineatus*), Luzon and the Camiguin Archipelago (*D. c. luzonensis*), and the Sulu Archipelago (the Philippines) (*D. c. flavescens*). **Status** Not Evaluated.

BLUE BRONZEBACK TREE SNAKE
Dendrelaphis cyanochloris PLATE 44
Measurement TL 1,430mm **Identification** Body slender; head distinct from neck; loreal present; single preocular; postoculars 2; supralabials 9 (rarely 8 or 10); Supralabials V and VI contact orbit; infralabials 9; Infralabial V largest; 4 infralabials contact anterior chin shields; eye small; pupil rounded; vertebrals enlarged; dorsals smooth; midbody scale rows 15; ventrals 189–206; subcaudals 137–156, paired; anal divided. **Coloration** Dorsum olive, scales edged with black; broad black temporal stripe to beyond neck at level of Ventral IV, breaking up into spots; all dorsals except first row with black anterior and lower edges; no pale ventrolateral stripe; venter yellow. **Habitat and Behaviour** Inhabits lowland forests and agricultural areas. Threat display includes expanding neck to reveal blue interstitial skin, resulting effect being blue with brown-barred neck and body. Diurnal and arboreal. Diet consists of lizards, birds and possibly frogs. Oviparous; clutches comprise 3–5 eggs. **Distribution** Myanmar, peninsular Thailand, Laos and Peninsular Malaysia (Pulau Pinang, Pinang State and Pulau Tioman, Pahang State). Also Bangladesh and eastern and north-eastern India. **Status** Not Evaluated.

BEAUTIFUL BRONZEBACK TREE SNAKE
Dendrelaphis formosus PLATE 44
Measurement TL 1,470mm **Identification** Body slender; head widened behind snout, distinct from neck; loreal fused with prefrontal; single preocular; postoculars 3; supralabials 9; Supralabials IV–VI contact orbit; infralabials 10; Sublabial I contacts over 2 infralabials; 5 infralabials contact anterior chin shields; eye large; pupil rounded; tail slender; vertebrals enlarged; midbody scale rows 15; ventrals 176–194; subcaudals 142–162, paired; anal divided. **Coloration** Dorsum blue; black postocular stripe from rostral covers temporal region and extends onto neck; 3 dark lateral stripes on posterior third of body; dorsals dark-edged; light ventrolateral stripe absent; subcaudals with a black point medially. **Habitat and Behaviour** Inhabits lowland forests and open areas. Diurnal and arboreal. Diet consists of frogs and lizards. Oviparous; clutches comprise 6–8 eggs, measuring 31–42.5 x 11.5–13mm. Incubation period 117 days. **Distribution** Peninsular Thailand, Peninsular Malaysia, Sumatra, the Mentawai Archipelago, Pulau Bangka, Pulau Belitung, Borneo and Java. **Status** Not Evaluated.

GORE'S BRONZEBACK TREE SNAKE
Dendrelaphis gorei NOT ILLUSTRATED
Measurement TL 900mm **Identification** Body slender; head distinct from neck; snout rounded; loreal present; single preocular; postoculars 2; supralabials 8–9; Supralabials IV–V contact orbit; infralabials 9; eye large; pupil rounded; dorsals smooth; vertebral scale rows enlarged; eye small; midbody scale rows 13; ventrals 187–199; subcaudals 139–154, paired; anal divided. **Coloration** Dorsum bronze-brown; yellow stripe along scale rows 1–2; dark postocular stripe extends to sides of neck; supralabials and chin yellow; venter pale green. **Habitat and Behaviour** Inhabits lowland forests at elevations of ca 120m asl. Diurnal and arboreal. Diet and reproductive habits unstudied. **Distribution** Southern Myanmar (Tongyi, nr Yangon, Yangon Division). Also eastern and north-eastern India (northern West Bengal, Assam, Nagaland and Arunachal Pradesh), and southern China. **Status** Not Evaluated.

HAAS'S BRONZEBACK TREE SNAKE
Dendrelaphis haasi NOT ILLUSTRATED
Measurement TL 945mm **Identification** Body slender; head distinct from neck; single loreal; single anterior temporal; single preocular; postoculars 1–2; supralabials 8–9; Supralabials V–VI contact orbit; infralabials 10–11 (rarely 9); 5 infralabials contact anterior chin shield; parietal extends to side of head; eye small; pupil rounded; tail long; vertebrals strongly enlarged, larger than dorsals of first row; dorsals smooth; midbody scale rows 15; ventrals 161–173; subcaudals 126–153, divided; anal divided. **Coloration** Dorsum olive-brown; narrow postocular stripe covers ventral part of lower temporals and ends at rear of jaw or extends onto neck; oblique black bars on sides of neck; light ventrolateral stripe; no black lines border ventrolateral stripe; venter light yellow or green. **Habitat and Behaviour** Inhabits lowland forests. One was observed on the bank of a small river by day. Diet and reproductive habits unstudied. **Distribution** Peninsular Malaysia, Sumatra, Pulau Nias, Pulau Belitung, the Mentawai Archipelago, Borneo and Java. **Status** Not Evaluated.

KOPSTEIN'S BRONZEBACK TREE SNAKE
Dendrelaphis kopsteini PLATE 44
Measurement TL 1,425mm **Identification** Body slender; head distinct from neck; 2 supralabials touch orbit; single loreal; eye moderate; pupil rounded; vertebral scales larger than lowest dorsal row; midbody scale rows 15; ventrals 167–181; subcaudals 140–154, paired; anal divided. **Coloration** Dorsum bronze-brown; dorsum weakly banded; black postocular stripe across lower half of temporal region to end of jaw; vertebral scales with broad black posterior margin; interstitial skin on anterior of body brick-red; venter grey. **Habitat and Behaviour** Inhabits lowland dipterocarp forests at elevations of up to 700m asl. Diurnal and arboreal, associated with shrubs and low tree branches. Diet includes geckos and possibly other

small vertebrates. Oviparous; clutches comprise 8 eggs. Incubation period 92 days. Hatchlings 270–300mm. **Distribution** Southern Thailand (Phang-Nga Province and nr Thung Song, Nakhon Si Tammarat Province), Peninsular Malaysia (base of Cameron Highlands, Pahang State and Sungai Endau, Johor State), Singapore, Sumatra (Padang and Medan), the Mentawai Archipelago and Borneo. **Status** Not Evaluated.

NGANSON BRONZEBACK TREE SNAKE
Dendrelaphis ngansonensis PLATE 44
Measurement TL 1,500mm **Identification** Body robust; head distinct from neck; loreal present; single preocular; postoculars 2; supralabials 9; Supralabials IV–VI contact orbit; eye moderate; pupil rounded; tail slender; dorsals smooth, enlarged; midbody scale rows 15; ventrals 165–199; anal divided. **Coloration** Dorsum and forehead bronze-brown; broad black postocular stripes extend to neck, becoming more bronze-brown; neck with black-grey lateral scales showing blue edges; yellow lateral stripe absent; ventrals dark green anteriorly, turning bronze-green; ventrals yellow-green medially. **Habitat and Behaviour** Inhabits primary semi-evergreen forests and forest clearings at elevations of 170–700m asl. Diurnal and arboreal, associated with trees. Diet unstudied. Oviparous; clutches comprise 4 eggs, measuring 38 x 13mm. **Distribution** Laos (Xieng-Khouang Province) and northern Vietnam (Ngan Son, Cao Bang Province, Tam Dao, Vinh Phú Province and Ha Tinh Province). Also China (Yunnan and Hainan Provinces). **Status** Not Evaluated.

PAINTED BRONZEBACK TREE SNAKE
Dendrelaphis pictus PLATE 44
Measurement TL 1,250mm **Identification** Body slender; head distinct from neck; loreal present; single preocular; postoculars 2; supralabials 9 (rarely 8); Supralabials V–VI contact orbit; infralabials 9–15; temporals 5–7; eye large; pupil rounded; vertebrals smaller than or equal to dorsals on first row; dorsals smooth; midbody scales 15; ventrals 167–200; subcaudals 109–169, paired; anal divided. **Coloration** Dorsum bronze-brown or brownish-olive; yellow or cream ventrolateral stripe edged with black along flanks; forehead brown with black postocular stripe that covers over half temporal region and extends to neck; blue or greenish-blue patch on neck displayed when excited; iris golden. **Habitat and Behaviour** Inhabits lowland and submontane forests, and also encountered in parks, gardens, plantations and human habitation, from sea level to 1,524m asl. Diurnal and arboreal. Diet consists of frogs and lizards, especially geckos. Oviparous; clutches comprise 3–8 eggs, measuring 22–38.5 x 8.5–11mm, with several clutches being produced annually. Incubation period 85–126 days. Hatchlings 202–303mm. **Distribution** Myanmar, Thailand, Laos, Cambodia, Vietnam, Peninsular Malaysia, Singapore, Sumatra, Pulau Belitung, the Mentawai Archipelago, Borneo, Java and Bali. Also north-eastern India, Bangladesh, southern and eastern China, and the Philippines. **Status** Not Evaluated.

STRIATED BRONZEBACK TREE SNAKE
Dendrelaphis striatus PLATE 44
Measurement TL 1,020mm **Identification** Body slender; head distinct from neck; snout short, rounded; loreal present; single preocular; postoculars 2; supralabials 9 (rarely 8 or 10); Supralabials V–VI or IV–VI contact orbit; infralabials 12; eye moderate; vertebral scales slightly enlarged; dorsals smooth; midbody scale rows 15; ventrals 152–163; subcaudals 103–142, paired; anal divided. **Coloration** Dorsum bronze-brown; labials and throat yellow; narrow dark stripe between nostrils to orbit; dark postocular stripe covers temporal region and extends onto neck; neck yellow when inflated; oblique black bars laterally on body; interstitial skin blue. **Habitat and Behaviour** Inhabits lowland forests. Diurnal and arboreal. Diet and reproductive habits unstudied. **Distribution** Southern Thailand, Peninsular Malaysia, Sumatra and Borneo. **Status** Not Evaluated.

MOUNTAIN BRONZEBACK TREE SNAKE
Dendrelaphis subocularis PLATE 44
Measurement TL 880mm **Identification** Body slender; head distinct from neck; snout rounded; loreal present; single preocular; postoculars 2; supralabials 7–8; Supralabial V contacts orbit; infralabials 10–11; eye small; pupil rounded; dorsals smooth; vertebral scales weakly enlarged; midbody scale rows 15; ventrals 153–175; subcaudals 74–105, paired; anal divided. **Coloration** Dorsum bronze-brown, scales with black edges; lower flanks beyond scale row 2 bright cream, olive or greenish-white; head and neck olive-green; dark postocular stripe extends to sides of neck, breaking up thereafter into bars; scales on neck sometimes yellow. **Habitat and Behaviour** Inhabits mid-hill forests. Diurnal and arboreal. Diet and reproductive habits unstudied. **Distribution** Southern Myanmar (Bhamo), Thailand, Laos, Cambodia and Vietnam. Also southern China. **Status** Not Evaluated.

INDIAN BRONZEBACK TREE SNAKE
Dendrelaphis tristis PLATE 44
Measurement TL 1,500mm **Identification** Body slender; head distinct from neck; single preocular; postoculars 2; supralabials 9; Supralabials V–VI contact orbit; eye large; pupil rounded; tail about a third of snout-vent length; dorsals smooth with apical pits; midbody scale rows 15; ventrals 163–197, paired; subcaudals 108–145, paired; anal divided. **Coloration** Dorsum unpatterned purplish- or bronzy-brown; vertebral scales on neck and forebody yellow; buff flank stripe from neck to vent; light blue on neck between scales displayed when excited; venter pale grey, green or yellow; iris golden. **Habitat and Behaviour** Inhabits open locations, such as forest edges and areas around human habitation, from sea level to 2,000m asl. Diurnal and arboreal, and also known to forage on land. Capable of crossing gaps

between trees of up to 25m. Diet consists of frogs, lizards, birds' eggs and insects. Oviparous; clutches comprise 6 eggs, measuring 29–39 x 10–12mm, which are deposited in tree hollows or deserted birds' nests. Incubation period 4–8 weeks. **Distribution** Myanmar. Also India, Pakistan, Bangladesh, Sri Lanka and Nepal. **Status** Not Evaluated.

UNDERWOOD'S BRONZEBACK TREE SNAKE
Dendrelaphis underwoodi   NOT ILLUSTRATED
Measurement TL 900mm **Identification** Body moderate; head distinct from neck; loreal present; single preocular; postoculars 2–3; supralabials 9; Supralabials IV–VI contact orbit; infralabials 10–11; 5 infralabials contact anterior chin shields; eye large; pupil rounded; tail slender; vertebrals enlarged; dorsals smooth; midbody scale rows 15; ventrals 183–189; subcaudals 126–133, paired; anal divided. **Coloration** Dorsum olive or olive-brown; supralabials and chin light yellow; black postocular stripe extends from loreal to neck, where it is fragmented into oblique bars; narrow ventrolateral stripe covers third scale row; pale ventrolateral stripe absent; venter pale green. **Habitat and Behaviour** Inhabits lowland forests and the mid-hills at elevations of 335–900m asl. Presumably diurnal and arboreal. Diet and reproductive habits unstudied. **Distribution** Endemic to western Java (Gunung Simpai, Rajamandala Province, and Cilayang Province). **Status** Not Evaluated.

YELLOW LARGE-TOOTHED SNAKE
Dinodon flavozonatum   PLATE 44
Measurement TL 1,440mm **Identification** Body slender, elongated, slightly compressed; head broad, blunt, indistinct from neck; loreal separate from internasal and orbit; single preocular; supralabials 8; eye moderate; pupil elliptical; 3 enlarged posterior maxillary teeth; tail long; dorsals smooth, except median rows 10–12, which are weakly keeled; midbody scale rows 17; ventrals 212–240; subcaudals 75–88; anal entire. **Coloration** Dorsum black with 85–95 narrow yellow cross-bars that bifurcate on flanks, enclosing dark spots; head black with a pale stripe from corner of eyes to angle of jaws, and a parallel one from the posterior margin of parietals; labials edged with black; venter yellow with squarish black spots in middle of ventrals and rounded spots at their edges. **Habitat and Behaviour** Inhabits montane forests. Nocturnal and terrestrial. Diet and reproductive habits unstudied. **Distribution** Northern Myanmar (Nam Thai Valley) and northern Vietnam (Bac Ky or Dông Kinh). Also eastern China. **Status** Not Evaluated.

VIETNAMESE LARGE-TOOTHED SNAKE
Dinodon meridionale   PLATE 44
Measurement TL 1,950mm **Identification** Body slender, elongated, slightly compressed; head short, depressed, indistinct from neck; loreal present; single preocular; postoculars 2; supralabials 8; Supralabials III–V contact orbit; 5 infralabials contact anterior chin shields; anterior chin shield larger than posterior; eye large; pupil elliptical; 3 enlarged posterior maxillary teeth; tail long; first 4 scale rows smooth, other scale rows keeled, especially posteriorly; midbody scale rows 17; ventrals 234–246; subcaudals 98–106; anal entire. **Coloration** Dorsum yellow with 97 black bands on body and 31 on tail; scales on forehead have large grey areas with black edges; supralabials with large black areas; chin white with black stippling; venter unpatterned pale yellow; edges of ventral with dark streaks, darkening posteriorly, especially on subcaudals. **Habitat and Behaviour** Inhabits monsoon forests and rainforests reaching submontane limits, at elevations of 200–1,700m asl. Nocturnal and terrestrial, active on granite and karst areas. Diet consists of skinks, other snakes and bird fledglings. Oviparous; clutches comprise 15 eggs. Incubation period 40–45 days. **Distribution** Northern Laos (Xieng Khuang) and north-western Vietnam (Sa Pá, Lao Cai Province, Tam Dao, Vinh Phu Province, Cuc Phuong, Thanh Hoa Province and Nguyen Binh, Cao Bang Province). Also southern and eastern China (Yunnan and Guangxi Provinces). **Status** Not Evaluated.

PINK LARGE-TOOTHED SNAKE
Dinodon rosozonatum   NOT ILLUSTRATED
Measurement TL 1,060mm **Identification** Body slender, elongated, slightly compressed; head short, depressed, indistinct from neck; single loreal; single preocular; postoculars 2; supralabials 8; maxillary teeth 12–13; vertebral scale rows distinctly enlarged; eye moderate; pupil elliptical; 3 enlarged posterior maxillary teeth; tail long; dorsals smooth; midbody scale rows 19; ventrals 221–234. **Coloration** Dorsum blackish-brown with 28–35 narrow pink bands on body and 9–13 on tail, bands 1–2 scales wide; V-shaped mark on nape reaches parietals; supralabials pinkish-brown, sutures dark; throat cream with blackish-brown spots; anterior ventrals greyish-cream, those at posterior with blackish-brown blotches. **Habitat and Behaviour** Inhabits the lowlands and mid-hills at elevations of under 850m asl. Nocturnal and crepuscular, associated with forested hills and fields of rice paddies. Diet and reproductive habits unstudied. **Distribution** Central Vietnam (Phong Nha-Ke Bang National Park, Quang Binh Province). Also eastern China (Hainan Province). **Status** Not Evaluated.

RED LARGE-TOOTHED SNAKE
Dinodon rufozonatum   PLATE 44
Measurement TL 1,350mm **Identification** Body slender, elongated, slightly compressed; head short, depressed, indistinct from neck; single small preocular; postoculars 2; supralabials 8; Supralabials III–V contact orbit; loreal contacts orbit; infralabials 10; eye moderate; pupil elliptical; 3 enlarged posterior maxillary teeth; tail long; dorsals smooth, except faint keels on a few middorsals; midbody scale rows 17; ventrals 193–200; subcaudals 55–74, paired. **Coloration** Dorsum coral-red with 61–65 black or dark brown bands on body and 18–21 on tail, each

2–3 scales wide; flanks with dark blotches or indistinct vertical bars, alternating with the cross-bars on dorsum; forehead dark brown, scales with lighter margins; sides of head lighter; postocular stripe extends to Supralabial VII, temporal stripe from parietals to neck; distinct V-shaped mark behind postocular and temporal stripes. **Habitat and Behaviour** Inhabits the forested mid-hills at elevations of up to 700m asl. Nocturnal and terrestrial, sometimes associated with granite-dominated landscape. Threat response includes biting and also rolling itself into a compact spherical mass. Diet consists of rodents, birds, lizards, snakes, frogs and fish. Oviparous; clutches comprise 5–20 eggs, measuring 27.9–40.5 x 14.6–18.6mm. Hatchlings 232–242mm. **Distribution** Vietnam. Also central, eastern and southern China, and South Korea. **Status** Not Evaluated.

NORTHERN LARGE-TOOTHED SNAKE
Dinodon septentrionale   NOT ILLUSTRATED
Measurement TL 1,180mm **Identification** Body slender, elongated, slightly compressed; head short, depressed, indistinct from neck; loreal small; single preocular; postoculars 2; supralabials 8; Supralabials III–V contact orbit; eye moderate; pupil elliptical; 3 enlarged posterior maxillary teeth; tail long; dorsals smooth or median 4–7 rows weakly keeled; midbody scale rows 17; ventrals 202–217; subcaudals 80–92, paired; anal entire. **Coloration** Dorsum purplish-black with 20–35 narrow, transverse white bands on body, each 1–1.5 scale wide, and 10–17 on tail that expand on flanks; venter white, sometimes spotted or barred with black; subcaudals with black speckling. **Habitat and Behaviour** Inhabits the mid-hills of evergreen forests at elevations of 220–500m asl. Nocturnal and terrestrial. Diet consists of small vertebrates. Reproductive habits unstudied. **Distribution** Myanmar (Kayin, Kachin and northern Mon States), northern Thailand (Chiang Mai Province), northern Laos (Kiang Kuoang Province), Cambodia (O'Rang District and Mondolkiri Province) and Vietnam (Lai Chau, Lao Cai, Bac Kan, Cao Bang, Vinh Phuc, Nghe An, Quang Binh, Ha Tinh and Thua Thien-Hue Provinces). Also eastern and north-eastern India, and southern China (Yunnan Province). **Status** Not Evaluated.

DAVISON'S BRIDLED SNAKE
Dryocalamus davisonii   PLATE 45
Measurement TL 920mm **Identification** Body slender, compressed; head depressed, distinct from neck; loreal present; preocular absent; postoculars 1–2; supralabials 7–8; Supralabials III–IV contact orbit; infralabials 8; eye large; pupil vertical; dorsals smooth; midbody scale rows 13; ventrals 233–255; subcaudals 90–112, paired; anal entire. **Coloration** Dorsum black with 29–31 irregular pale green or white cross-bars that expand on flanks and 19–21 on tail; bars widest anteriorly, 2–4 scales wide, and narrower posteriorly, where they are broken up, producing a reticulated design; back of forehead white with dark median stripe; supralabials white; tail with black flecks; venter unpatterned white. **Habitat and Behaviour** Inhabits lowland temperate and subtropical forests, including evergreen and moist deciduous forests, and grasslands, up to submontane limits at an elevation of 1,000m asl. Nocturnal and terrestrial, with some arboreal activity, and can climb trees up to at least 2m above substratum. Diet consists of lizards. Oviparous; clutches comprise 3–4 eggs, measuring 35 x 9mm. Incubation period ca 70 days. Hatchlings 250mm. **Distribution** Southern Myanmar (Tenasserim, Tanintharyi Division), Thailand, Laos, Cambodia and Vietnam (Kien Giang, Quang Binh, Quang Tri, Khanh Hoa, Binh Thuan, Dong Nai and Tay Ninh Provinces, and Ho Chi Minh City). **Status** Not Evaluated.

HALF-BANDED BRIDLED SNAKE
Dryocalamus subannulatus   PLATE 45
Measurement TL 600mm **Identification** Body slender, compressed; head depressed, distinct from neck; loreal present; single preocular; postoculars 2; supralabials 7; Supralabials III–IV contact orbit; eye large; pupil vertical; dorsals smooth; midbody scale rows 15; ventrals 225–244; subcaudals 88–107, paired; anal entire. **Coloration** Dorsum tan or light brown with large, transverse brown spots; on flanks, a smaller spot; 2 dark postocular streaks; venter unpatterned yellow. **Habitat and Behaviour** Inhabits lowland forests and disturbed areas. Nocturnal, and terrestrial and arboreal. Diet consists of small vertebrates. Reproductive habits unstudied. **Distribution** Southern Thailand, Peninsular Malaysia, Singapore, Sumatra, the Mentawai and Riau Archipelagos, and Borneo (Sandakan, Sabah State and Brunei). Also Palawan (the Philippines). **Status** Not Evaluated.

THREE-BANDED BRIDLED SNAKE
Dryocalamus tristrigatus   PLATE 45
Measurement TL 650mm **Identification** Body slender, compressed; head depressed, distinct from neck; preocular absent; postoculars 2; supralabials 7 (rarely 6); Supralabials III–IV contact orbit; eye large; pupil vertical; dorsals smooth; midbody scale rows 15; ventrals 218–231; subcaudals 86–96, paired; anal entire. **Coloration** Dorsum dark brown with 3 yellow stripes; forehead shields with white edge; supralabials white; venter cream. **Habitat and Behaviour** Inhabits lowland forests. Associated with rocky biotopes and trees. Diet includes lizards. Reproductive habits unstudied. **Distribution** Borneo and the Natuna Archipelago. Also Balabac and Palawan (the Philippines). **Status** Not Evaluated.

KEEL-BELLIED WHIP SNAKE
Dryophiops rubescens   PLATE 45
Measurement TL 750mm **Identification** Body slender, compressed; head distinct from neck; loreal present; single preocular; postoculars 2–3; supralabials 9; Supralabials IV–VI contact orbit; 4–5 infralabials contact anterior chin shields; eye large; pupil horizontal; tail long, slender; dorsals smooth;

midbody scale rows 15; ventrals 186–199; subcaudals 111–136, paired; anal divided. **Coloration** Dorsum reddish-brown with small dark and pale spots; forehead with dark streaks; dark postocular streak; labials with dark spots; venter yellow or olive. **Habitat and Behaviour** Inhabits lowland forests and forest edges. Diurnal and arboreal, associated with low vegetation as well as with low branches of trees. Diet consists of lizards. Oviparous; clutches comprise 2–3 eggs. **Distribution** Southern Thailand (Phuket, Surat Thani, Trang and Pattani Provinces), Cambodia, Peninsular Malaysia, Singapore, Sumatra, the Mentawai and Natuna Archipelagos, Borneo and Java. **Status** Not Evaluated.

KEELED RAT SNAKE
Elaphe carinata PLATE 45
Measurement TL 2.4m **Identification** Body robust; head indistinct from neck; snout elongated; prefrontals in broad contact with supraoculars; single preocular; postoculars 2; small subocular; eye large; supraoculars large; supralabials 8–9 (rarely 7); Supralabials IV–V contact orbit; infralabials 10–11; dorsals keeled; midbody scale rows 21–23; ventrals 194–226; subcaudals 73–103, paired; anal divided. **Coloration** Dorsum yellow, olive-clay or brown on anterior half, posterior half with 32 indistinct blackish-grey cross-bars; posterior dorsals edged with black; venter yellowish-grey with scattered small dark spots. **Subspecies** Three subspecies are recognized, of which one (*E. c. carinata*) occurs in the region. **Habitat and Behaviour** Associated with seasonal forests, including open forests, scrub, bamboo groves and the vicinity of human settlements, with an elevational range of 0–3,200m asl. Both diurnal and nocturnal, and mostly terrestrial. Produces a strong odour from its cloacal glands when alarmed. Diet consists of rodents, birds and their eggs, and other snakes. Oviparous; clutches comprise 6–17 eggs, measuring 49.6–68.7 x 26.6–34.6mm. Incubation period 40–60 days. Hatchlings 350–450mm. **Distribution** Northern Vietnam (*E. c. carinata*). Also southern and eastern China, including Taiwan (*E. c. carinata*), the Ryukyu Archipelago (Japan) (*E. c. yonaguniensis*) and north-western Yunnan (China) (*E. c. deqenensis*). **Status** Not Evaluated.

DARK-GREY GROUND SNAKE
Elapoides fuscus PLATE 45
Measurement TL 500mm **Identification** Body slender; head indistinct from neck; snout short; loreal and prefrontal contact orbit; preocular absent; single postocular; supralabials 6; Supralabials III–IV contact orbit; 3–4 infralabials contact anterior chin shields; eye reduced; pupil rounded; tail long; dorsals keeled; midbody scale rows 15; ventrals 146–158; subcaudals 74–91; anal entire. **Coloration** Dorsum iridescent black, dark brown or reddish-brown, unpatterned or with yellow or red spots or flecks; sometimes anterior of body yellow with dark brown vertebral stripe and dark spots on flanks; posterior part black and iridescent; venter yellow or cream. **Habitat and Behaviour** Inhabits forested hills at elevations of more than 1,000m asl. Nocturnal and subfossorial. Diet unstudied. Oviparous; clutches comprise 2–4 eggs, measuring 31.5–33.5 x 9mm. **Distribution** Sumatra, Borneo and Java. **Status** Not Evaluated.

SUMATRAN BURROWING SNAKE
Etheridgeum pulchrum NOT ILLUSTRATED
Measurement SVL 143mm (tail-tip in the only known specimen missing) **Identification** Body slender, cylindrical; supralabials 7; Supralabials III–IV contact orbit; nostril lateral; 2 postoculars; eye small; pupil rounded; infralabials 8; dorsals smooth and imbricating, lacking apical pits; midbody scale rows 15; ventrals 114; anal divided. **Coloration** Dorsum golden-tan; scales light gold with irregular dark brown stippling along posterior edges; forehead gold, many head shields have a black-bordered white ocellus; nuchal region with chevron-like blackish-brown bands; posterior-most pair of complete cross-bands in broad contact middorsally, separated ventrolaterally by white triangles; pair of 71 blackish-brown cross-bands and 4 paired cross-bars, followed by cross-bars that break into 3 blackish-brown spots; venter unpatterned yellow; blackish-brown ventrolateral stripe edged dorsally by golden-yellow stripe. **Habitat and Behaviour** Presumably subfossorial and an inhabitant of the leaf litter in forested habitats. Diet and reproductive habits unstudied. **Distribution** Endemic to western Sumatra (Padang, Sumatera Barat Province). **Status** Not Evaluated.

MANDARIN TRINKET SNAKE
Euprepiophis mandarinus PLATE 45
Measurement TL 1,700mm **Identification** Body robust; head short, slightly distinct from neck; snout obtuse; tail short and stout; eye small; single preocular; postoculars 1–2; supralabials 6–8; Supralabials III–IV or IV–V contact orbit; infralabials 8–10; dorsals smooth; midbody scale rows 21 or 23; ventrals 200–241; subcaudals 59–69; anal divided. **Coloration** Dorsum grey to greyish-brown; dorsal scales with brownish-red centres; dorsum and tail with large, rounded yellow blotches edged with black and yellow; forehead with V-shaped dark pattern; venter cream, sometimes with large black blotches. **Habitat and Behaviour** Inhabits the plains to montane regions, in open forests with rocky substrate, scrubland and agricultural fields, at elevations of 500–3,000m asl. Diurnal and terrestrial, frequenting areas in the vicinity of water. Diet consists of mice and shrews. Oviparous; clutches comprise 2–10 eggs, measuring 50–57 x 23–27mm. Incubation period 42–60 days. Hatchlings 192–250mm. **Distribution** Northern and central Myanmar and northern Vietnam (Lai Chau, Lao Cai, Lang Son, Vinh Phuc, Tuyen Quang, Son La and Ha Tinh Provinces). Also north-eastern India, and central, southern and eastern China. Suspected to occur in Laos. **Status** Not Evaluated.

ORANGE-BELLIED SNAKE
Gongylosoma baliodeirum PLATE 45
Measurement TL 450mm **Identification** Body slender; head slightly wider than neck; 2 scales border orbit; nasal divided; loreal present; preoculars 1–2; postoculars 2; supralabials 7; Supralabials III–IV contact orbit; infralabials 7; 4–5 infralabials contact anterior chin shields; eye small; pupil rounded; spines on proximal part of hemipenes not enlarged; dorsals smooth; midbody scale rows 13; ventrals 115–141; subcaudals 42–75; anal divided. **Coloration** Dorsum dark brown to reddish-brown with paired rows of cream spots; upper labials edged with dark grey; venter yellowish-cream, sometimes with fine dark spots. **Subspecies** Three poorly diagnosed subspecies have been described: *G. b. baliodeirum* (from Java), *G. b. cinctus* (from Pulau Nias) and *G. b. cochranae* (from Khao Soi Dao, Thailand); their systematic status is unclear. **Habitat and Behaviour** Associated with lowland to submontane forests at elevations of up to 1,525m asl. Nocturnal and terrestrial, hiding under fallen logs and stones by day. Diet includes spiders and other arthropods, and lizards. Oviparous; clutches comprise 2–3 eggs, measuring 22–24 x 7.5mm. **Distribution** South-eastern Thailand (Khao Soi Dao, Chanthaburi Province), Peninsular Malaysia, Singapore, Sumatra, Pulau Nias, the Natuna Archipelago, Borneo and Java. **Status** Not Evaluated.

STRIPED GROUND SNAKE
Gongylosoma longicauda PLATE 45
Measurement TL 500mm **Identification** Body slender; head wider than neck; loreal present; preoculars 1–2; postoculars 2; supralabials 8; Supralabials III–V contact orbit; 4 infralabials contact anterior chin shields; eye large; tail long and slender; dorsals smooth; midbody scale rows 13; ventrals 110–138; subcaudals 71–105; anal divided. **Coloration** Dorsum brownish-red or black, with yellow or cream chevron at back of head and 5 orange or yellow stripes on dorsum, most distinct anteriorly; venter unpatterned red or cream, or each ventral with brown spot on outer edge. **Habitat and Behaviour** Inhabits lowland rainforests, and occasionally encountered near human habitations or as road kills. Diet includes spiders and lizards. Reproductive habits unstudied. **Distribution** Peninsular Malaysia, Sumatra, Borneo and Java. **Status** Not Evaluated.

PULAU TIOMAN GROUND SNAKE
Gongylosoma mukutense PLATE 45
Measurement TL 429mm (subadult; adults unknown) **Identification** Body slender; head wider than neck; snout short, rounded; single preocular; postoculars 2; supralabials 7; Supralabials III–IV contact orbit; infralabials 8; eye large; pupil rounded; large square posterior temporal scale; thin chevron-shaped nuchal band; anterior and posterior chin shields subequal; tail long; dorsals smooth; midbody scale rows 13; ventrals 134; subcaudals 99, divided. **Coloration** Dorsum red anteriorly fading to brown-grey posteriorly; forehead brick-red; supralabials white edged with black; nuchal band confluent with vertebral stripe; white postocular patch; remnants of 5 thin white stripes anteriorly; venter cream. **Habitat and Behaviour** Inhabits coastal forests on Pulau Tioman at 10m asl. Diurnal and terrestrial. Diet and reproductive habits unstudied. **Distribution** Endemic to Peninsular Malaysia (Pulau Tioman, Pahang State). **Status** Not Evaluated.

INDO-CHINESE GROUND SNAKE
Gongylosoma scriptum NOT ILLUSTRATED
Measurement TL 465mm **Identification** Body slender; head slightly depressed, distinct from neck; 3 scales border orbit; nasals separate; loreal present, small; single preocular; postoculars 2; supralabials 8; Supralabials III–V contact orbit; infralabials 8; eye large; pupil rounded; dorsals smooth; midbody scale rows 13; ventrals 126–145; subcaudals 87–98; anal divided. **Coloration** Dorsum pale brown or greyish-brown; scales edged with black, forming longitudinal stripes; series of small black spots on paravertebral region of anterior body; broad, dark-edged pale nuchal collar, which is sometimes indistinct; labials yellow with black spots; venter yellow or cream. **Habitat and Behaviour** Inhabits the lowlands and low hills. Diurnal and terrestrial. Diet and reproductive habits unstudied. **Distribution** Southern Myanmar (Mottama, Mon State), and western, north-eastern and southern Thailand (Khao Luang and Ronpibon, Nakhon Sri Thammarat Province, and Pulau Panjang, Phuket Province). **Status** Not Evaluated.

ROYAL TREE SNAKE
Gonyophis margaritatus PLATE 46
Measurement TL 2m **Identification** Body robust, elongated, compressed; head distinct from neck; snout elongated, squarish; loreal present; single preocular; postoculars 2; supralabials 8–9; Supralabials IV–VI or III–V contact orbit; infralabials 12; eye large; pupil rounded; tail long, tapering; dorsals smooth; midbody scale rows 19; ventrals 230–249; subcaudals 108–130, paired; anal divided. **Coloration** Dorsum bright green to greyish-olive, each scale edged with black; several yellow or orange-red bands on posterior of body and tail; black postocular stripe; forehead with black streaks; venter yellowish-pink. **Habitat and Behaviour** Inhabits lowland forests at elevations of up to 700m asl. Diurnal and arboreal, known from forest canopies. Diet and reproductive habits unstudied. **Distribution** Peninsular Malaysia (Johor and Selangor States), Singapore and Borneo (Sungei Purulon, Ranau and Long Pasia, Sabah State, and Kuching, Sungei Mengiong, Gunung Dulit and Gunung Merinjak, Sarawak State). **Status** Not Evaluated.

RED-TAILED RACER
Gonyosoma oxycephalum PLATE 46
Measurement TL 2.4m **Identification** Body slender, elongated, compressed; head distinct from neck; snout elongated, squarish; loreal present; single preocular; postoculars 2; supralabials 9–11; Supralabials V–VI, VI–VII or VI–VIII contact orbit; infralabials 12–14;

eye large; pupil rounded; tail long, tapering; dorsals smooth or weakly keeled; midbody scale rows 23 or 25 (rarely 27); ventrals 230–263; subcaudals 120–157, paired; anal divided. **Coloration** Dorsum emerald-green or light green (rarely yellow), with a light green throat and a black stripe along sides of olive-yellow head, from nostril, across eye, to above level of upper jaw; tail-tip yellowish-brown or reddish-orange; venter yellow; juveniles olive-brown with narrow white bars towards posterior; iris grey or yellowish-green; tongue bluish-black. **Habitat and Behaviour** Inhabits lowland tropical and subtropical forests, including mangrove swamps, and also plantations and gardens, from sea level to 750m asl. Diurnal and arboreal, with some terrestrial activity, especially in juveniles, associated with trees and other foliage. Diet consists of rodents and birds. Oviparous; clutches comprise 5–12 eggs, measuring 65mm, and several clutches are produced annually. Incubation period 100–120 days. Hatchlings 240mm. **Distribution** Southern Myanmar, Thailand (Chiang Mai, Trang and Narathiwat Provinces), Laos, Cambodia, Vietnam, Peninsular Malaysia, Singapore, Sumatra, Pulau Nias, Pulau Bangka, Pulau Belitung, the Riau, Mentawai and Natuna Archipelagos, Borneo and Java. Also India (the Andaman Islands), Lombok and Sulawesi (Indonesia), and Balabac, Bohol, Bongao, Dinagat, Lubang, Luzon, Mindoro, Negros and Palawan (the Philippines). **Status** Not Evaluated.

STRIPE-NECKED SNAKE
Liopeltis frenata PLATE 46
Measurement TL 760mm **Identification** Body slender and cylindrical; head distinct from neck; snout not projecting; single preocular; postoculars 1–2; supralabials 7; Supralabials III–IV contact orbit; infralabials 8; single loreal; eye large; pupil rounded; tail long; dorsals smooth, lacking apical pits; midbody scale rows 15; ventrals 140–174; subcaudals 70–105, paired; anal divided. **Coloration** Dorsum brownish-olive with scales edged with black, and sometimes also with white, forming longitudinal stripes on anterior half of body; broad black postocular stripe on neck; supralabials and venter cream; tongue orange. **Habitat and Behaviour** Inhabits subtropical and montane forests at altitudes of 600–1,830m asl. Diurnal and terrestrial. Diet probably consists of frogs. Oviparous; clutches comprise 4–5 eggs, measuring 25–28 x 7–9mm, which are produced in bamboo internodes. Hatchlings 230–249mm. **Distribution** Myanmar, Laos and Vietnam. Also north-eastern India. **Status** Not Evaluated.

STOLICZKA'S RINGNECK
Liopeltis stoliczkae NOT ILLUSTRATED
Measurement TL 600mm **Identification** Body slender, cylindrical; head distinct from neck, depressed; snout projecting, twice as long as diameter of orbit; small nostril in undivided nasal; squarish loreal; supralabials 8; Supralabials IV–V contact orbit; eye large; tail long, slender; dorsals smooth; midbody scale rows 15; ventrals 148–155; subcaudals 116–134; anal divided. **Coloration** Dorsum brown or greyish-brown with broad black stripe on sides of head, extending to anterior body and fading thereafter; grey stripe on outer margins of ventrals; venter pale grey. **Habitat and Behaviour** Inhabits the evergreen and deciduous forested mid-hills at elevations of up to 700m asl. Diurnal and arboreal, associated with bamboo. Diet and reproductive habits unstudied. **Distribution** Northern Myanmar (Karen Hills), Laos (Bolikhamxay Province) and eastern Cambodia (Pichrada District). Also eastern and north-eastern India. **Status** Not Evaluated.

TRICOLOURED RINGNECK
Liopeltis tricolor PLATE 46
Measurement TL 560mm **Identification** Body slender; head indistinct from neck; snout long; nasal single (rarely divided); loreal present; single (rarely 2) preocular; postoculars 2; supralabials 8–9; Supralabials IV–V or V–VI contact orbit; eye large; pupil rounded; tail long; spines on proximal portion of hemipenes enlarged; dorsals smooth; midbody scale rows 15 or 17; ventrals 140–187; subcaudals 103–137; anal divided. **Coloration** Dorsum yellowish-olive; dark postocular streak to beyond neck; venter yellowish-cream with olive streak on sides of each scale. **Habitat and Behaviour** Inhabits lowland forests. Diurnal and arboreal, associated with short trees and other low vegetation. Diet consists of insects and spiders. Reproductive habits unstudied. **Distribution** Southern Thailand, Vietnam, Peninsular Malaysia, Singapore, Sumatra, the Mentawai Archipelago, Borneo (Sungei Tangap, Sarawak State and Batu Apoi, Brunei) and Java. Also Palawan (the Philippines). **Status** Not Evaluated.

DUSKY WOLF SNAKE
Lycodon albofuscus PLATE 46
Measurement TL 2.7m **Identification** Body slender, subcylindrical; head wider than neck; snout short, blunt and depressed; single preocular; postoculars 2; supralabials 8; Supralabials III–V contact orbit; 2 enlarged posterior maxillary teeth; eye small; pupil vertical; tail long; dorsals keeled; midbody scale rows 17; ventrals 225–259; subcaudals 148–208, paired; anal divided. **Coloration** Dorsum unpatterned dark brown or brownish-black in adults; juveniles with 30–40 narrow white or yellow bands on dorsum; supralabials yellow; venter unpatterned cream or yellow. **Habitat and Behaviour** Inhabits open lowland forests and forest edges, typically with streams, from sea level to 500m asl. Nocturnal and terrestrial, with some arboreal activity. Diet consists of lizards including skinks, as well as frogs. Reproductive habits unstudied. **Distribution** Peninsular Malaysia, Sumatra, Pulau Nias and Borneo. **Status** Not Evaluated.

INDIAN WOLF SNAKE
Lycodon aulicus PLATE 46
Measurement TL 800mm **Identification** Body slender, subcylindrical; head flattened; snout projecting beyond lower jaw; single preocular;

postoculars 2; supralabials 9; Supralabials III–V contact orbit; 2 enlarged posterior maxillary teeth; eye small; pupil vertical; tail long; dorsals smooth; midbody scale rows 17; ventrals 172–214; subcaudals 57–80, paired; anal divided. **Coloration** Dorsum brown or greyish-brown with 12–19 white cross-bars, sometimes speckled with brown, expanding laterally to enclose triangular patches; venter cream or yellowish-white. **Habitat and Behaviour** Inhabits the lowlands. Nocturnal and arboreal, entering human habitations and sometimes occupying roofs and abandoned storerooms. Diet consists of geckos, other snakes and rodents. Oviparous; clutches comprise 3–11 eggs, measuring 25–32mm, and more than a single clutch is produced annually. Hatchlings 140–190mm. **Distribution** Myanmar. Also India, Bangladesh, Nepal and Sri Lanka. Possibly also the Maldives Archipelago. **Status** Not Evaluated.

BUTLER'S WOLF SNAKE
Lycodon butleri   PLATE 46
Measurement TL 1,000mm **Identification** Body slender, subcylindrical; head flattened; single preocular; loreal present; single postocular; supralabials 8; Supralabials III–V contact orbit; 2 enlarged posterior maxillary teeth; eye small; pupil vertical; tail long; dorsals weakly keeled; midbody scale rows 17; ventrals 218–234; subcaudals 82–93, paired; anal entire. **Coloration** Dorsum dark bluish-grey or blackish-brown, with 40–50 irregular cross-bars in juveniles, suffused with dark pigments in adults; venter cream with dark brown cross-bars, most distinct at midbody. **Habitat and Behaviour** Inhabits montane forests at elevations of ca 1,220–2,031m asl. Nocturnal and arboreal, sometimes entering human habitations. Threat response includes rolling itself into a compact spherical mass. Diet consists of geckos. Reproductive habits unstudied. **Distribution** Endemic to Peninsular Malaysia (Bukit Larut, Perak State and Cameron Highlands, Pahang State). **Status** Not Evaluated.

ISLAND WOLF SNAKE
Lycodon capucinus   PLATE 47
Measurement TL 760mm **Identification** Body slender, subcylindrical; snout rounded; head flattened; loreal present; single preocular; postoculars 2 (rarely 3); supralabials 8–9; Supralabials III–IV contact orbit; infralabials 10; 2 enlarged posterior maxillary teeth; eye small; pupil vertical; tail long; dorsals smooth; midbody scale rows 17; ventrals 176–224; subcaudals 53–80, paired; anal entire. **Coloration** Dorsum brown, grey-brown or purple with narrow yellow or cream band at back of head that may be spotted with brown; interstitial skin yellow or grey; scales of body light-edged, forming indistinct cross-bars or reticulated pattern; venter cream or light yellow. **Habitat and Behaviour** Associated with lowland forests and the mid-hills. Nocturnal, and both terrestrial and arboreal, sometimes entering human habitations. Diet consists of geckos and skinks; large prey is constricted. Oviparous; clutches comprise 3–11 eggs, measuring 20–30 x 10mm. Incubation period 33–45 days. **Distribution** Myanmar, Thailand, Vietnam, Peninsular Malaysia, Singapore, Sumatra, Borneo, Java and Bali. Also the Andaman Islands (India), Sulawesi and the Lesser Sundas (Indonesia), eastern China and the Philippines. **Status** Not Evaluated.

CARDAMOM MOUNTAINS WOLF SNAKE
Lycodon cardamomensis   PLATE 47
Measurement TL 316mm **Identification** Body slender, subcylindrical; head flattened; loreal present; single preocular; postoculars 3; supralabials 8; Supralabials IV–V contact orbit; infralabials 10; 5 infralabials contact anterior chin shields; 2 enlarged posterior maxillary teeth; eye small; pupil vertical; tail long; dorsals weakly keeled; midbody scale rows 19; ventrals 215, lateral keels present; subcaudals 93, paired; anal entire. **Coloration** Dorsum black with 12 white bands on body and 6 on tail; body bands wider on flanks (5–9 scales) than on vertebral region, 3–5 scales wide; forehead black except for pale suture between frontal and parietal; venter mostly cream. **Habitat and Behaviour** Inhabits dipterocarp-dominated lowland forests at an elevation of 500m asl. Terrestrial and crepuscular. Diet and reproductive habits unstudied. **Distribution** North-eastern Thailand and Cambodia (Cardamom Mountains). **Status** Not Evaluated.

BROWN WOLF SNAKE
Lycodon effraenis   PLATE 47
Measurement TL 1,000mm **Identification** Body slender, subcylindrical; head flattened; rounded snout; loreal present; single preocular; postoculars 1–3; supralabials 9–10; Supralabials III–V contact orbit; 2 enlarged posterior maxillary teeth; eye small; pupil vertical; tail long; dorsals smooth or weakly keeled; midbody scale rows 17; ventrals 215–233; subcaudals 72–100, paired; anal entire. **Coloration** Dorsum reddish-brown or dark brown; 3 broad yellow or cream rings encircle body of juveniles, which also show streaks of same colour on sides of head, or a distinct canthal stripe; venter unpatterned brown. **Habitat and Behaviour** Inhabits lowland forests and disturbed habitats from sea level to ca 700m asl. Nocturnal and diurnal, and mostly terrestrial, capable of climbing. Diet presumably consists of lizards and small snakes. Reproductive habits unstudied. **Distribution** Peninsular Malaysia, Sumatra and Borneo. **Status** Not Evaluated.

BANDED WOLF SNAKE
Lycodon fasciatus   PLATE 47
Measurement TL 895mm **Identification** Body slender, subcylindrical; head flattened; loreal contacts internasal; single preocular, in contact with frontal; postoculars 2; supralabials 8 (rarely 9); Supralabials III–V (rarely IV–V) contact orbit; infralabials 9; 2 enlarged posterior maxillary teeth; eye small; pupil vertical; tail long; dorsals weakly keeled, keels more pronounced posteriorly; midbody scale rows 17;

ventrals 189–225; subcaudals 66–94, paired; anal entire. **Coloration** Dorsum glossy black with 22–48 irregular cross-bars on body and tail, or a reticulated or spotted pattern; venter blotched; iris flecked with grey. **Habitat and Behaviour** Inhabits the temperate evergreen mid-hills to montane limits at elevations of ca 914–2,300m asl. Nocturnal and arboreal, associated with trees and bushes. Diet consists of other snakes, and lizards such as geckos and skinks. Oviparous; clutches comprise 4–14 eggs. Hatchlings ca 216mm. **Distribution** Northern Myanmar (Shan State), north-eastern Thailand (Tawkawbee, nr Umpang, Tak Province), northern Laos and Vietnam. Also eastern and north-eastern India (northern West Bengal and Assam States), and southern China (Tibet, Sichuan and Yunnan Provinces). **Status** Not Evaluated.

YELLOW-SPECKLED WOLF SNAKE
Lycodon jara NOT ILLUSTRATED
Measurement TL 550mm **Identification** Body slender, subcylindrical; head flattened, not projecting beyond lower jaw; loreal present; single preocular; postoculars 2; supralabials 8–9; Supralabials III–V contact orbit; 2 enlarged posterior maxillary teeth; eye small; pupil vertical; tail long; dorsals smooth; midbody scale rows 17; ventrals 167–188, not angular laterally; subcaudals 52–74, paired; anal divided. **Coloration** Dorsum brown or purplish-black, finely stippled throughout with paired yellowish-white spots or short longitudinal lines on each scale; supralabials and venter unpatterned white; juveniles with white or yellow collar. **Habitat and Behaviour** Inhabits both forests and open areas with bushes and scattered trees, as well as agricultural areas in the lowlands. Diet consists of frogs, lizards and small mammals. Oviparous. **Distribution** Myanmar (northern Kachin State). Also Nepal, Bangladesh and northern and eastern India. **Status** Not Evaluated.

KUNDU'S WOLF SNAKE
Lycodon kundui NOT ILLUSTRATED
Measurement TL 225mm **Identification** Body slender; head flattened; supralabials 7; Supralabials III–IV contact orbit; 4 infralabials contact anterior pair of genials; 2 enlarged posterior maxillary teeth; eye small; pupil vertical; tail long; dorsals smooth; midbody scale rows 15; ventrals 186; subcaudals 70, paired; anal divided. **Coloration** Dorsum bluish-black with narrow white cross-bars; on posterior half of body, cross-bars located close to each other and bifurcate or break up on flanks; white nuchal collar; venter white. **Habitat and Behaviour** Nothing known of its natural history. **Distribution** Endemic to Myanmar (Gyobyu, Taikkyi township, Bago Division). **Status** Not Evaluated.

LAOS WOLF SNAKE
Lycodon laoensis PLATE 47
Measurement TL 500mm **Identification** Body slender, subcylindrical; head flattened; loreal present; single preocular; postoculars 2–3; supralabials 9–10; Supralabials III–V contact orbit; infralabials 10; 2 enlarged posterior maxillary teeth; eye small; pupil vertical; tail long; dorsals smooth; midbody scale rows 17; ventrals 163–192; subcaudals 60–76, paired; anal divided. **Coloration** Dorsum shiny black with white-edged yellow cross-bars, numbering 13–29 on body and 8–18 on tail, and becoming narrower towards posterior; forehead and labials deep blue; venter unpatterned cream. **Habitat and Behaviour** Inhabits evergreen forests in the plains and low hills. Nocturnal and arboreal, climbing trees up to at least 9m above substratum. Diet includes frogs and lizards. Oviparous; clutches comprise 5 eggs. **Distribution** Thailand, Laos, Cambodia, Vietnam and northern Peninsular Malaysia (Kedah State). Also north-eastern India and southern China. **Status** Not Evaluated.

ANNAM WOLF SNAKE
Lycodon paucifasciatus PLATE 47
Measurement TL 763mm **Identification** Body slender; head slightly distinct from neck, flattened; single preocular; postoculars 2; supralabials 8; Supralabials III–V contact orbit; 2 enlarged posterior maxillary teeth; eye small; pupil vertical; tail long; dorsals smooth; midbody scale rows 19; ventrals 219; subcaudals 90, paired; anal divided. **Coloration** Dorsum black with 14 cream bands with irregular outline on body, and 8 on tail; white bar at back of head; venter cream with greyish variegation, especially on posterior and subcaudals. **Habitat and Behaviour** Nothing known of its natural history. **Distribution** Endemic to central Vietnam (Thua Lun, 50km south of Hue, Thua Thien-Hue Province). **Status** Not Evaluated.

RUHSTRAT'S WOLF SNAKE
Lycodon ruhstrati PLATE 47
Measurement TL 940mm **Identification** Body slender, subcylindrical; head flattened; loreal in contact with orbit; single preocular; postoculars 2; supralabials 8 (rarely 7); Supralabials III–V contact orbit (rarely III–IV, additionally III or VI may contact orbit); eye small; pupil vertical; vertebral rows not enlarged; dorsals smooth (in Vietnam and Laos) or keeled (Taiwan, the type locality); midbody scale rows 17; ventrals 211–210; subcaudals 88–89, paired; anal entire. **Coloration** Dorsum black with 20–22 pale grey or white bands on body and 11 or more on tail; head black with pale grey or white areas; venter typically cream, though some individuals have black venter. **Habitat and Behaviour** Inhabits submontane forests at elevations of 762–1,067m asl. Nocturnal and diurnal, and arboreal as well as terrestrial. Diet consists of lizards including skinks and lacertids (*Takydromus* spp.). Oviparous; clutches comprise 4 eggs. **Distribution** Laos (Tran Ninh Plateau) and northern Vietnam (Tam Dao, Vinh Phuc Province, Sa Pa, Lao Cai Province, Ngan Son Bac Kan Province, Chin Xai, Ha Tinh Province, Binh Khe and Nui Yen Tu, Quang Ninh Province, Ngoc Lau, Ha Son Binh Province, Cao Bang, Nguyen Binh Province and Da Nang, Ba Na Province). Also southern and eastern

China. More than a single species is suspected to be involved within the complex from China, and if the Taiwanese population proves distinct, the name *Lycodon futsingensis* is available for the Indo-Chinese population. **Status** Not Evaluated.

BARRED WOLF SNAKE
Lycodon striatus PLATE 47
Measurement TL 430mm **Identification** Body slender, subcylindrical; head flattened; snout obtusely rounded; loreal present; single preocular; postoculars 2; supralabials 8; Supralabials I–II contact nasal; Supralabials III–V contact orbit; 2 enlarged posterior maxillary teeth; eye small; pupil vertical; tail long; dorsals smooth; midbody scale rows 17; ventrals 154–195; subcaudals 35–58, paired; anal divided. **Coloration** Dorsum of body and forehead black to dark brown; series of transverse white or yellow marks, distance between which diminishes towards tail; tail dorsum with irregular longitudinal white streaks; supralabials and venter unpatterned white. **Habitat and Behaviour** Inhabits relatively dry regions such as forest edges from the plains up to 1,830m asl. Nocturnal and terrestrial, hiding under stones during the day. Diet consists of geckos and skinks. Oviparous; clutches comprise 2–4 eggs, measuring 25–30 x 9–12mm. **Distribution** Myanmar. Also central Asia, Pakistan, India, Sri Lanka and Nepal. **Status** Not Evaluated.

WHITE-BANDED WOLF SNAKE
Lycodon subcinctus PLATE 47
Measurement TL 1,020mm **Identification** Body slender; head flattened; snout rounded; preocular absent; loreal present; postoculars 2; supralabials 8; Supralabials III–V or IV–V contact orbit; infralabials 9; 2 enlarged posterior maxillary teeth; eye small; pupil vertical; tail long; dorsals weakly keeled; midbody scale rows 17; ventrals 192–230; subcaudals 60–91, paired; anal divided. **Coloration** Dorsum black or dark brown with 9–15 cream bands, 3–5 scales wide; pattern most distinct in juveniles, fading with growth; in adults, vestiges of bands on venter, which is grey or cream. **Habitat and Behaviour** Inhabits lowland forests from sea level to 1,000m asl. Nocturnal and arboreal, with some terrestrial activity. Threat response includes vibration of its tail. Diet includes geckos and skinks. Oviparous; clutches comprise 5–11 eggs, measuring 32–36 x 12.5–13mm. Incubation period 70–83 days. Hatchlings 238mm. **Distribution** Myanmar (Kachin State and Tanintharyi Division), Thailand, Laos, Cambodia, Vietnam, Peninsular Malaysia, Singapore, Sumatra, Pulau Nias, the Mentawai Archipelago, Borneo and Java. Also Lombok and Sumbawa (Indonesia), the Philippines and eastern China. **Status** Not Evaluated.

ZAW'S WOLF SNAKE
Lycodon zawi PLATE 47
Measurement TL 480mm **Identification** Body slender; head flattened, distinct from neck; snout projecting; single preocular; postoculars 1–2; supralabials 8 (rarely 9); Supralabials III–V contact orbit; 2 enlarged posterior maxillary teeth; eye small; pupil vertical; tail long; dorsals smooth; midbody scale rows 17; ventrals 179–207; subcaudals 45–75, paired; anal divided. **Coloration** Dorsum brownish-black with narrow cream bands that are less distinct posteriorly; labials pale brown; venter cream, ventrals with dark edges; iris black. **Habitat and Behaviour** Inhabits dry and moist deciduous forests, and semi-evergreen and tropical evergreen forests, in the lowlands and mid-hills, especially near streams, at elevations of up to 500m asl. Nocturnal and terrestrial. Diet consists of skinks. Reproductive habits unstudied. **Distribution** Myanmar (Rakhine Yoma, Ponnyadaung Range). Also north-eastern India (Khasi Hills, Meghalaya State and Mizoram State). **Status** Not Evaluated.

CHAN-ARD'S REED SNAKE
Macrocalamus chanardi NOT ILLUSTRATED
Measurement TL 263mm **Identification** Body robust, subcylindrical; head small; loreal present; single preocular; single postocular; suboculars absent; supralabials 8; Supralabials IV–V contact orbit; infralabials 7; 4 infralabials contact anterior chin shields; eye small; tail short, tapering; dorsals smooth; midbody scale rows 15; ventrals 104–127 (+ 1 preventral); subcaudals 18–28; anal entire. **Coloration** Dorsum chestnut-brown or pale brown, anteriorly with row of yellow or ochre stripes composed of dark-edged ocelli, sometimes small black dots; dark ventrolateral stripe dorsally edged with yellow or cream stripe; 2–4 oblique yellowish-ochre bars, the first on temporals, extending from parietals to gular region, the others parallel to temporal streak, extending from neck to ventrals; venter red, pink or orange. **Habitat and Behaviour** Inhabits submontane forests at elevations of 1,110–1,500m asl. Diurnal and subfossorial, concealing itself under fallen logs. Diet consists of earthworms, slugs, insects and insect larvae. Reproductive habits unstudied. **Distribution** Endemic to Peninsular Malaysia (Bukit Larut, Perak State, and Cameron Highlands and Bukit Fraser, Pahang State). **Status** Not Evaluated.

GENTING HIGHLANDS REED SNAKE
Macrocalamus gentingensis PLATE 48
Measurement TL 378mm **Identification** Body robust, subcylindrical; head small, wedge-shaped; loreal present; single preocular; single postocular; supralabials 8; Supralabials IV–V contact orbit; infralabials 7; eye small; tail tapering, tip pointed; dorsals smooth; midbody scale rows 15; ventrals 122–145; subcaudals 25–33; anal entire. **Coloration** Dorsum iridescent black with scattered yellow patches on each side of nape; head black with yellow postocular streak that extends to neck; supralabials and infralabials yellow; tail with lateral spots; venter black with narrow yellow median stripe. **Habitat and Behaviour** Inhabits lower montane oak-laurel forests at elevations of 1,181–1,689m asl. Fossorial, found under fallen logs. Diet and reproductive habits unstudied. **Distribution** Endemic to Peninsular

Malaysia (Genting Highlands, Pahang State). **Status** Not Evaluated.

JASON'S REED SNAKE
Macrocalamus jasoni NOT ILLUSTRATED
Measurement TL 752mm **Identification** Body robust, subcylindrical; head small, wedge-shaped; loreal present; single preocular; single postocular; supralabials 8; Supralabials IV–V contact orbit; maxillary teeth 11; eye small; tail tapering; dorsals smooth; midbody scale rows 15; ventrals 131–133; subcaudals 19–22; anal entire. **Coloration** Dorsum iridescent black with a pair of longitudinal reddish-brown or rusty stripes, 2 scales wide, between temporals and tail-tip; head brownish-yellow, some forehead scales with dark areas; venter bright yellow, ventrals tipped with black flecks medially; subcaudals yellow with some black speckling medially. **Habitat and Behaviour** Inhabits montane forests at elevations of 1,768–1,981m asl. Terrestrial, associated with moist leaf litter. Diet and reproductive habits unstudied. **Distribution** Endemic to Peninsular Malaysia (Gunong Benom, Pahang State). **Status** Not Evaluated.

STRIPED REED SNAKE
Macrocalamus lateralis PLATE 48
Measurement TL 298mm **Identification** Body robust, subcylindrical; head small; loreal absent; single preocular; single postocular; supralabials 8; Supralabials IV or V contact orbit; infralabials 7; eye small; tail tapering; dorsals smooth; midbody scale rows 15; ventrals 114–122; subcaudals 14–20; anal entire. **Coloration** Dorsum pale brown or yellowish-brown; dorsolateral row on anterior body with dark-edged pale brown ocelli; wide, paired dark brown ventrolateral stripes separated by narrow yellow or pale brown line; venter red, pink or orange, ventrals edged with dark brown; tail sometimes with dark median subcaudal line. **Habitat and Behaviour** Inhabits hill dipterocarp forests at an elevation of 400m asl. Diurnal and terrestrial. Diet and reproductive habits unstudied. **Distribution** Southern Thailand (Hala Bala Wildlife Sanctuary, Narathiwat Province) and Peninsular Malaysia (Pinang State, without a precise locality). **Status** Not Evaluated.

SCHULZ'S REED SNAKE
Macrocalamus schulzi PLATE 48
Measurement TL 399mm **Identification** Body robust, subcylindrical; head small; loreal present; single preocular; single postocular; supralabials 8; Supralabials IV–V contact orbit; infralabials 7; eye small; tail short, tapering; dorsals smooth; midbody scale rows 15; ventrals 114–134; subcaudals 17–31; anal entire. **Coloration** Dorsum mid-brown lacking ventrolateral stripes; some dorsal scales paler anteriorly, darker posteriorly; outer dorsal scale rows pale yellow with brown mottling below; forehead brown with pale temporal streak; venter bright yellow. **Habitat and Behaviour** Inhabits submontane and montane forests at elevations of 1,000–1,800m asl. Nocturnal, and terrestrial and subfossorial. Diet and reproductive habits unstudied. **Distribution** Endemic to Peninsular Malaysia (Cameron Highlands, Pahang State). **Status** Not Evaluated.

TWEEDIE'S REED SNAKE
Macrocalamus tweediei PLATE 48
Measurement TL 500mm **Identification** Body robust, subcylindrical; head small; loreal present; single preocular; single postocular; subocular absent; supralabials 7–8; Supralabials IV–V contact orbit; infralabials 7; eye small; tail short, tapering; dorsals smooth; midbody scale rows 15; ventrals 128–147; subcaudals 24–32; anal entire. **Coloration** Dorsum uniformly black; head with yellow lateral marking extending ventrally; supralabials and infralabials yellow; venter chequered black and yellow; tail sometimes with median subcaudal line. **Habitat and Behaviour** Inhabits submontane forests at elevations of 1,500–1,829m asl. Terrestrial and subfossorial, hiding under fallen logs. Diet in the wild unknown; in captivity known to eat geckos. Reproductive habits unstudied. **Distribution** Endemic to Peninsular Malaysia (Cameron Highlands and Genting Highlands, Pahang State). **Status** Not Evaluated.

VOGEL'S REED SNAKE
Macrocalamus vogeli PLATE 48
Measurement TL 192mm **Identification** Body robust, subcylindrical; head small; loreal present; single preocular; single postocular; supralabials 8; Supralabials IV–V contact orbit; infralabials 7; 4 infralabials contact anterior chin shields; eye small; tail long, tapering and terminating in spine; dorsals smooth; midbody scale rows 15; ventrals 125 (plus 1 preventral); subcaudals 29, paired; anal entire. **Coloration** Dorsum dark yellowish-brown; many dorsal scales faintly and narrowly edged with dark brown; broad dark ventrolateral stripe, dorsally edged by narrow yellow stripe; broad oblique yellow stripe behind head, followed by 2 narrow oblique lines; small dark-edged yellow ocelli on anterior body; venter yellowish-brown with brownish-black speckling; subcaudals with median stripe. **Habitat and Behaviour** Inhabits montane forests at elevations of 1,650–1,750m asl. Presumably diurnal and subfossorial. Diet and reproductive habits unstudied. **Distribution** Endemic to Peninsular Malaysia (Gunung Tahan, Pahang State). **Status** Not Evaluated.

DICE-LIKE TRINKET SNAKE
Maculophis bellus PLATE 48
Measurement TL 800 (*M. b. bellus*); 927mm (*M. b. chapaensis*) **Identification** Body slender; head indistinct from neck; snout rounded; loreal present; single preocular; supralabials 6–8; Supralabials III–IV contact orbit; infralabials 7–9; anterior maxillary teeth largest; eye small; tail short; midbody scale rows 19, smooth or weakly keeled; ventrals 201–226; subcaudals 40–60; anal divided. **Coloration** Dorsum pale brown with saddle-shaped yellow blotches or transverse or oblique cross-bars; forehead with black Y-

shaped mark or lighter streak edged with black; labials with dark edges; venter yellow with irregular large black blotches on each ventral. **Subspecies** Two subspecies are recognized. *M. b. bellus*, single anterior temporal; dorsals smooth; dorsum with saddle-shaped greyish-brown to brown blotches, edged with black, covering 3–4 scales; saddle-like pattern less distinct on flanks. *M. b. chapaensis*, 2 anterior temporals; dorsals weakly keeled; dorsum with transverse or oblique cross-bars, widened on flanks, where they enclose a black spot. **Habitat and Behaviour** Inhabits submontane and montane forests at elevations of 1,500–2,000m asl. Nothing known of its natural history. **Distribution** North-eastern Myanmar (Patsarlamdan, Kambaiti and Sinlum Kaba, Kachin Hills) (*M. b. bellus*) and north-western Vietnam (Sa Pá, Lào Cai Province) (*M. b. chapaensis*). Also southern China. Records from north-eastern India require verification. **Status** Not Evaluated.

WHITE-BARRED KUKRI SNAKE
Oligodon albocinctus PLATE 48
Measurement TL 1,015mm **Identification** Body robust, subcylindrical; head short, snout blunt and rounded; loreal present; single preocular; postoculars 2; supralabials 7 (rarely 6); Supralabials III–IV contact orbit; maxillary teeth 10–12; eye moderate; pupil rounded; dorsals smooth, lacking apical pits; midbody scale rows 17 (rarely 19); ventrals 177–207; subcaudals 47–68, paired; anal entire. **Coloration** Dorsum brownish-red, sometimes with black-edged white, yellow or fawn cross-bars, numbering 19–27 on body and 4–8 on tail; forehead with dark stripe from supralabials to orbit, V-shaped yellow or cream mark on forehead, edged with black; venter cream, yellow or coral-red with black areas. **Habitat and Behaviour** Inhabits forests and tea gardens from the mid-hills to montane areas at elevations of ca 1,980m asl. Terrestrial and crepuscular. Diet consists of rodents, frogs, lizards and their eggs, and insects. Oviparous; clutches comprise 3 eggs. Hatchlings 200–275mm. **Distribution** Myanmar. Also eastern and north-eastern India, Bangladesh, Bhutan and Nepal. **Status** Not Evaluated.

ANNAM KUKRI SNAKE
Oligodon annamensis NOT ILLUSTRATED
Measurement TL 248mm **Identification** Body robust, subcylindrical; head short, snout blunt and rounded; loreal absent; single preocular; single postocular; supralabials 6; Supralabials III–IV contact orbit; infralabials 6; maxillary teeth 8; eye moderate; pupil rounded; dorsals smooth, lacking apical pits; midbody scale rows 13; ventrals 159–170; subcaudals 30–44, paired; anal entire. **Coloration** Dorsum brown with 10 short, narrow white bars, thinly edged with black; forehead with several black-edged white blotches, including an interocular band, an interparietal bar and isolated patches on anterior snout; long oblique stripe on sides of neck; narrow light nuchal chevron; venter white, many ventrals and subcaudals covered partially or completely by quadrangular dark brown or black blotches. **Habitat and Behaviour** Nothing known of its natural history. **Distribution** Endemic to central Vietnam (B'Lao, Lam Dong Province). **Status** Not Evaluated.

SPOTTED KUKRI SNAKE
Oligodon annulifer PLATE 48
Measurement TL 450mm **Identification** Body robust, subcylindrical; head short, slightly distinct from neck; snout rounded; loreal present; single preocular; postoculars 2; supralabials 7–8; Supralabials III–IV contact orbit; infralabials 8; maxillary teeth 7; mandibular teeth 14–15; eye moderate; pupil rounded; tail short; dorsals smooth; midbody scale rows 15; ventrals 151–162; subcaudals 44–64, paired; anal divided. **Coloration** Dorsum dark brown, lighter on flanks, with 20–26 orange-yellow blotches edged with greyish-brown; chevron at back of head; venter cream with black spots on each side; subcaudals brick-red. **Habitat and Behaviour** Inhabits hill dipterocarp forests at elevations of ca 120m asl. Nocturnal, and arboreal with some terrestrial activity. Diet consists of lizards. Reproductive habits unstudied. **Distribution** Endemic to Borneo (northern Sabah, Batu Apoi, Brunei and Aya Yayang Concession, Kalimantan Province). **Status** Not Evaluated.

BARRON'S KUKRI SNAKE
Oligodon barroni PLATE 48
Measurement TL 401mm **Identification** Body moderate, subcylindrical; head short, indistinct from neck; nasal divided; loreal present; single preocular; postoculars 2; supralabials 7 (rarely 8); Supralabials IV–V contact orbit; infralabials 8–9 (rarely 7); maxillary teeth 10–13, 2 posterior-most enlarged, blade-like; eye moderate; pupil rounded; dorsals smooth; midbody scale rows 17; ventrals 136–160; subcaudals 28–48, paired; anal entire. **Coloration** Dorsum light brown; series of 10–14 transversely arranged large brown or blackish-brown blotches, edged with black, numbering 10–14 on body and 2–3 on tail; sometimes, 3 indistinct cross-bars between each spot; forehead dark brown with broad crescentic mark across eye to labials; oval spot on forehead sometimes confluent with oblique band crossing to sides of throat; apex of heart-shaped mark on nape; venter coral-red with squarish black spots on each side. **Habitat and Behaviour** Inhabits lowlands, including low-lying offshore islands, at elevations of ca 300m asl. Nocturnal and terrestrial. Diet and reproductive habits unstudied. **Distribution** South-eastern and central Thailand (Bangkok, Phra Nakhon Si Ayutthaya Province, Khao Sabap, nr Chanthaburi, Chanthaburi Province, Sriracha, Koh Lam, Chon Buri Province, Muang, Rayong Province and Muak Lek, Pak Chong, Saraburi Province), Laos (Huay Saoe, nr Taong and Xepian National Biodiversity Conservation Area, Champasak Province), Cambodia (Cardamom Mountains) and central and southern Vietnam (Sông Bé, Bình Duong Province, Cheo Reo, Gia Lai Province and Càu Dá, nr Nha Trang, Khán Hoa Province). **Status** Not Evaluated.

JAVANESE MOUNTAIN KUKRI SNAKE
Oligodon bitorquatus PLATE 48
Measurement TL 370mm **Identification** Body robust, subcylindrical; head short, indistinct from neck; loreal present or absent; single preocular; postoculars 2; supralabials 7; Supralabials III–IV contact orbit; infralabials 7–8; eye small; pupil rounded; dorsals smooth; midbody scale rows 17; ventrals 140–165; subcaudals 30–46, paired; anal entire. **Coloration** Dorsum dark brown, purple or greyish-brown with small red or yellow spots, arranged to form bands; median series of larger spots sometimes present; forehead with dark stripes and yellow or grey bands, including one on occipital region; venter red with black spots or blotches. **Habitat and Behaviour** Inhabits submontane forests at elevations of 1,200–1,524m asl. Nocturnal and terrestrial. Diet unstudied. Oviparous; clutches comprise 3 eggs. **Distribution** Java (Gunung Pengalengan, Tjisurupan, Krawang, Bogor, Gunung Salak, Gunung Gedeh, Tjibodas, Gunung Ungaran, Salatiga, Ambarawa, Gunung Wilis, Kediri and Gunung Tengger) and Sumbawa. **Status** Not Evaluated.

BOO LIAT'S KUKRI SNAKE
Oligodon booliati PLATE 49
Measurement TL 510mm **Identification** Body robust, subcylindrical; head short, indistinct from neck; loreal present; single preocular; postoculars 2; supralabials 6–7; Supralabials II–III or III–IV contact orbit; infralabials 7; eye small; pupil rounded; dorsals smooth; midbody scale rows 17; ventrals 143–153; subcaudals 54–60, paired; anal entire. **Coloration** Dorsum and flanks deep maroon-red; 19–22 indistinct transverse brown bars from nape along body, fading towards tail; postocular stripe absent; narrow dark brown stripes on supralabials 5–6; venter salmon-pink, lacking spots. **Habitat and Behaviour** Inhabits lowland forests on Pulau Tioman. Nocturnal and terrestrial. Diet and reproductive habits unstudied. **Distribution** Endemic to Peninsular Malaysia (Pulau Tioman, Pahang State). **Status** Not Evaluated.

CHAIN-BANDED KUKRI SNAKE
Oligodon catenatus PLATE 49
Measurement TL 640mm **Identification** Body robust, subcylindrical; head short, indistinct from neck; loreal absent; single preocular; postoculars 2; supralabials 6; Supralabials III–IV contact orbit; infralabials 7; maxillary teeth 7; eye small; pupil rounded; dorsals smooth; midbody scale rows 13; ventrals 179–212; subcaudals 34–43, paired; anal divided. **Coloration** Dorsum purplish-brown with longitudinal dark brown stripes; median dark stripes separated by yellowish-brown vertebral stripe; head brown with dark brown markings, including snout-tip spot, crescentic mark between labials, through orbit and across snout, elongated spot on frontal and on suture between parietals, and oblique band from parietals to angle of jaws; venter bright red. **Habitat and Behaviour** Inhabits wet subtropical forests. Nocturnal, associated with dense vegetation. Diet unstudied. Oviparous; clutches comprise 3 eggs. **Distribution** Myanmar, northern Thailand and Vietnam. **Status** Not Evaluated.

CHINESE KUKRI SNAKE
Oligodon chinensis PLATE 49
Measurement TL 496mm **Identification** Body robust, subcylindrical; head short, indistinct from neck; loreal and presubocular present; single preocular; postoculars 2; supralabials 7–8 (rarely 9); Supralabials IV–V contact orbit; infralabials 9; maxillary teeth 9–10; eye moderate; pupil rounded; dorsals smooth; midbody scale rows 17; ventrals 170–192; subcaudals 51–61, paired; anal entire. **Coloration** Dorsum greyish- or reddish-brown with dorsal series of narrow, elongated dark brown or blackish spots, edged with black, numbering 9–18 on body and 2–4 on tail; venter unpatterned cream. **Habitat and Behaviour** Inhabits the plains and hills at elevations of 400–1,829m asl. Nocturnal and terrestrial. Diet probably consists of skink eggs. Reproductive habits unstudied. **Distribution** Vietnam (Cao Bang, Bac Kan, Lang Son, Quang Ninh and Vinh Phú Provinces). Also eastern and southern China. **Status** Not Evaluated.

GREY KUKRI SNAKE
Oligodon cinereus PLATE 49
Measurement TL 730mm **Identification** Body robust, subcylindrical; head short, indistinct from neck; loreal present; single preocular; postoculars 2; supralabials 8; Supralabials IV–V contact orbit; infralabials 8; maxillary teeth 10–12; eye moderate; pupil rounded; tail short and blunt; dorsals smooth; midbody scale rows 17; ventrals 155–186; subcaudals 28–42, paired; anal entire. **Coloration** Dorsum reddish-brown or red, unpatterned, or with black-edged white or grey cross-bars; forehead unpatterned brown; venter cream. **Habitat and Behaviour** Inhabits the lowlands and mid-hills at elevations of 457–700m asl. Nocturnal and terrestrial. Diet includes spiders and insects. Oviparous; clutches comprise 4–5 eggs. **Distribution** Myanmar, Thailand (Muang and Chiang Dao, Chiang Mai Province, and Sakaerat, Nakhon Ratchasima Province), Laos, eastern and central Cambodia, and Vietnam. Also eastern China. **Status** Not Evaluated.

PEGU KUKRI SNAKE
Oligodon cruentatus PLATE 49
Measurement TL 410mm **Identification** Body robust, subcylindrical; head short, indistinct from neck; loreal present or absent; single preocular; postoculars 2; supralabials 8; Supralabials IV–V contact orbit; infralabials 5; maxillary teeth 14–16; eye moderate; pupil rounded; dorsals smooth; midbody scale rows 17; ventrals 148–173; subcaudals 27–40, paired; anal divided. **Coloration** Dorsum olive- or pinkish-brown, sometimes with 4 longitudinal dark stripes, median rows separated by 3 scale rows and extending to tail-tip; lateral stripe 2.5 scale

rows above ventrals, extending to tail-tip; transverse dark bar before orbit extends as a subocular stripe; incomplete dark collar extends obliquely to parietals; venter bright yellow with large, squarish black spots; subcaudals red. **Habitat and Behaviour** Inhabits forested lowlands. Threat response includes hiding its head under its coils. Nothing else known of its natural history. **Distribution** Endemic to south-central Myanmar (Mandalay to Bago). **Status** Not Evaluated.

CANTOR'S KUKRI SNAKE
Oligodon cyclurus PLATE 49
Measurement TL 940mm **Identification** Body robust, subcylindrical; head short, indistinct from neck; loreal and presubocular present, single preocular; postoculars 2; supralabials 8; Supralabials IV–V contact orbit; maxillary teeth 9–10; eye moderate; pupil rounded; tail short; dorsals smooth; midbody scale rows 19; ventrals 161–185, angular laterally; subcaudals 36–58, paired; anal entire. **Coloration** Dorsal coloration and pattern variable: ground colour typically yellowish-brown or dark brown, with dark reticulations and black-edged cross-bars or transverse, oval dark spots; head with V-shaped dark markings; venter unpatterned cream or spotted. **Habitat and Behaviour** Inhabits forests as well as agricultural areas. Crepuscular and terrestrial. Diet includes rodents, other snakes, and lizards and their eggs. Oviparous; clutches comprise 3–26 eggs, measuring 23–31 x 16–18.5mm. Hatchlings 165–200mm. **Distribution** Northern, central and western Myanmar (Kachin State and Yangon, Yangon Division). Also eastern India, Bangladesh and Nepal. **Status** Not Evaluated.

DEUVE'S KUKRI SNAKE
Oligodon deuvei NOT ILLUSTRATED
Measurement TL 530mm **Identification** Body moderate, subcylindrical; head short, indistinct from neck; loreal present; single preocular; postoculars 2; supralabials 7–8; Supralabials III–IV contact orbit; infralabials 8–9; maxillary teeth 12–15; eye moderate; pupil rounded; dorsals smooth; midbody scale rows 17; ventrals 142–163; subcaudals 31–47, paired; anal entire. **Coloration** Dorsum greyish-brown, reddish-brown or reddish-tan; dorsal scales with narrow dark brown edges, producing irregular cross-bars; scattered dark brown dots; broad orange, rusty-red or red-ochre vertebral stripe between neck and tail-tip; vertebral stripe edged with scattered rounded dark brown spots, or faint wide ochre or dark reddish-brown paravertebral stripe with scattered black spots on edges; forehead brownish-grey or brown; supralabials cream edged with brownish-red; narrow, transverse dark brown marking on snout; narrow, longitudinal droplet-shaped streak on frontal; oblique dark brownish-red sagittal mark on frontal; arrow- or heart-shaped dark brown nuchal blotch; venter bright pink anteriorly, red posteriorly, with a few scattered rectangular spots, or with numerous large, subrectangular blackish-brown spots near edges of ventrals. **Habitat and Behaviour** Inhabits lowland forests, and apparently also the vicinity of human dwellings such as gardens. Nocturnal and terrestrial. Diet and reproductive habits unstudied. **Distribution** Laos (Vientiane and its vicinity, Tha Ngon, and Wattaï, Vientiane Prefecture), Cambodia (Che Teal Chrum Village, Pursat Province) and southern Vietnam (Dong Nai, Bien Hoa, Dong Nai Province and Ho Chi Minh, Ho Chi Minh District). **Status** Not Evaluated.

SPOT-TAILED KUKRI SNAKE
Oligodon dorsalis PLATE 49
Measurement TL 628mm **Identification** Body robust, subcylindrical; head short, indistinct from neck; loreal present; single preocular; postoculars 2; supralabials 7; Supralabials III–IV contact orbit; maxillary teeth 6–7; eye moderate; pupil rounded; dorsals smooth; midbody scale rows 15; ventrals 162–188; subcaudals 27–51, paired; anal divided. **Coloration** Dorsum dark brown or purple with a light vertebral stripe, sometimes dark-edged, or containing small black spots; second stripe along the second and third dorsal scale rows; tail dorsum has 2–3 large black spots; forehead dark brown with 2 cross-bars; venter orange; subcaudals crimson. **Habitat and Behaviour** Inhabits evegreen forests in the mid-hills to montane forests at elevations of up to 1,980m asl. Diurnal and terrestrial. Diet unstudied. Oviparous; clutches comprise at least 2 eggs. **Distribution** Myanmar (Chin and Kachin States). Also north-eastern India, Bangladesh and Bhutan. **Status** Not Evaluated.

DURHEIM'S KUKRI SNAKE
Oligodon durheimi NOT ILLUSTRATED
Measurement TL 370mm **Identification** Body moderate, subcylindrical; head short, indistinct from neck; loreal present; single preocular; postoculars 2; supralabials 7; Supralabials III–IV contact orbit; maxillary teeth 7–8; eye moderate; pupil rounded; dorsals smooth; midbody scale rows 17; ventrals 171–174; subcaudals 40–41, paired; anal divided. **Coloration** Dorsum olive or brown; dark, black-edged vertebral stripe, wide on nape, encloses pale, black-edged spots; neck with oblique black spot; forehead with crescentic dark band that passes through orbit; venter yellow with transverse black spots; subcaudals with median red line. **Habitat and Behaviour** Inhabits submontane forests in the Battak Highlands at elevations of 800–1,000m asl. Nothing known of its natural history. **Distribution** Endemic to Sumatra (Toba Massif, Sumatera Utara Province). **Status** Not Evaluated.

EBERHARDT'S KUKRI SNAKE
Oligodon eberhardti NOT ILLUSTRATED
Measurement TL 530mm **Identification** Body robust, subcylindrical; head short, snout blunt and rounded; loreal present; single preocular; postoculars 2; supralabials 6; Supralabials III–IV contact orbit; eye moderate; pupil rounded; dorsals smooth, lacking apical pits; midbody scale rows 13; ventrals 165–191; subcaudals 29–40, paired; anal divided. **Coloration** Dorsum greyish-brown with dark brown dorsolateral

stripe, stippled dorsally with light spots; venter alternately marked with red and black; subcaudals mostly red. **Habitat and Behaviour** Inhabits submontane forests at elevations of 900–1,200m asl. Nothing known of its natural history. **Distribution** Northern Laos, Cambodia and northern Vietnam (Tam Dao, Vinh Phú Province). Also eastern China. **Status** Not Evaluated.

JEWELLED KUKRI SNAKE
Oligodon everetti PLATE 49
Measurement TL 420mm **Identification** Body moderate, subcylindrical; head short, indistinct from neck; loreal present; single preocular; postoculars 2; supralabials 7; Supralabials III–IV contact orbit; infralabials 7–8; eye moderate; pupil rounded; dorsals smooth; midbody scale rows 15; ventrals 132–154; subcaudals 46–72, paired; anal single. **Coloration** Dorsum pinkish-red or greyish-brown with 3 blackish-brown stripes, the broadest one on vertebral region, 3 scales wide, enclosing short white and red or orange bars; dark stripe on lower flanks encloses white spots; forehead with V-shaped dark mark, pointed backwards; second mark over snout crosses orbit; venter unpatterned coral-red. **Habitat and Behaviour** Inhabits hill dipterocarp forests, with one possible record from submontane limits, ca 1,000m asl. Nocturnal and terrestrial. Diet consists of skinks. Reproductive habits unstudied. **Distribution** Endemic to Borneo (foothills of Gunung Kinabalu, Tawau Hills, Danum Valley and Malutut, Sabah State, and Banjaran and Tanjung, Kalimantan Province). **Status** Not Evaluated.

SMALL-BANDED KUKRI SNAKE
Oligodon fasciolatus PLATE 49
Measurement TL 882mm **Identification** Body robust, subcylindrical; head short, indistinct from neck; loreal present; single preocular; postoculars 2; supralabials 8; Supralabials IV–V or only V contact orbit; infralabials 9–10; presubocular present, internasals 2; prefrontals 2; eye moderate; pupil rounded; dorsals smooth; midbody scale rows 21 or 23; ventrals 160–190; subcaudals 40–51, paired; anal entire. **Coloration** Dorsum yellowish-olive with 13–18 transverse blotches, separated by 3–4 wide dark cross-bars, or with reticulated pattern; dark postocular stripe does not meet at back of head; sometimes, a pale dorsolateral stripe along body; tail with median longitudinal cream band; venter unpatterned white. **Habitat and Behaviour** Inhabits lowland evergreen forests, as well as open areas such as cultivated fields, in the plains and mid-hills, at elevations of up to ca 260m asl. Nocturnal and terrestrial. Diet in captivity consists of frogs and small mammals. Reproductive habits unstudied. **Distribution** South-eastern Myanmar, Thailand (Samut Prakan, Ubon Ratchathani, Phetch-aburi and Nakhon Ratchasima Provinces), Laos (Luang Prabang and Savannakhet Provinces), Cambodia (Cardamom Mountains) and Vietnam (Lam Dong, An Giang, Khánh Hoa and Ha Tinh Provinces, as well as Con Dao Island and Ho Chi Minh City). **Status** Not Evaluated.

BEAUTIFUL KUKRI SNAKE
Oligodon formosanus PLATE 49
Measurement TL 750mm **Identification** Body robust, subcylindrical; head short, indistinct from neck; loreal and presubocular present; single preocular; postoculars 2; supralabials 8 (rarely 7); Supralabials IV–V contact orbit; maxillary teeth 10–11; eye moderate; pupil rounded; tail short; dorsals smooth; midbody scale rows 19; ventrals 164–182; subcaudals 42–52, paired; anal entire. **Coloration** Dorsum greyish-brown or reddish-brown; dorsal pattern reticulated with irregular thin black cross-bars; broad tan vertebral stripe; forehead with dark brown chevron mark that extends to neck, another incomplete mark anteriorly; supralabials and infralabials cream with brown mottling; venter cream. **Habitat and Behaviour** Inhabits the lowlands up to submontane limits. Nocturnal and terrestrial. Threat display involves raising its head and compressing its neck, which is bent into an S-shaped curve. Diet consists of reptile eggs. Oviparous; clutch size unknown. Hatchlings 130mm. **Distribution** Vietnam (Ha Giang, Cao Bang, Vinh Phúc and Thanh Hoa Provinces, as well as Hanoi and Thieu Yen). Also southern and eastern China. **Status** Not Evaluated.

HAMPTON'S KUKRI SNAKE
Oligodon hamptoni NOT ILLUSTRATED
Measurement TL 540mm **Identification** Body robust, subcylindrical; head short, indistinct from neck; nasal entire; internasal absent; loreal present or absent; single preocular; postoculars 1–2; supralabials 5; Supralabials II–III contact orbit; maxillary teeth 7; eye moderate; pupil rounded; dorsals smooth; midbody scale rows 15; ventrals 160–175; subcaudals 30–32, paired; anal divided. **Coloration** Dorsum reddish-brown; broad yellow vertebral stripe from nape to tail-tip; black intervening areas between dorsum and stripes; flanks bluish-grey with 2 narrow dark brown bands; head yellow with dark brown markings, including snout-tip spot, crescentic mark between labials, through orbit and across snout; elongated frontal spot and on suture between parietals, connected to large occipital bifid spot and oblique band from parietal to angle of jaws; venter red with wide black bars; subcaudals unpatterned red. **Habitat and Behaviour** Inhabits forested hills. Diet and reproductive habits unstudied. **Distribution** Endemic to north-central Myanmar (Mogok, Mandalay Division). **Status** Not Evaluated.

UNICOLOURED KUKRI SNAKE
Oligodon inornatus PLATE 50
Measurement TL 580mm **Identification** Body robust, subcylindrical; head short, indistinct from neck; loreal present; preocular absent; postoculars 2; supralabials 8; Supralabials IV–V contact orbit; infralabials 8; eye moderate; pupil rounded; tail long and thin; dorsals smooth; midbody scale rows 15; ventrals 171–174; subcaudals 36–42, paired; anal entire. **Coloration** Dorsum unpatterned brown or dull red with indistinct black cross-bars; forehead

brown; dark cross-bar crosses eye and reaches Supralabials IV and V; dark frontal spot; oblique dark bar from parietals to sides of neck; supralabials and infralabials pink with dark sutures; large chevron from frontal to nape; venter yellowish-cream with quadrangular dark spots on sides of ventrals. **Habitat and Behaviour** Inhabits evergreen forests in the mid-hills at elevations of up to 400m asl. Nocturnal and terrestrial. Diet and reproductive habits unstudied. **Distribution** Western, north-eastern and south-eastern Thailand (Tak, Uthai Thani, Loei and Chon Buri Provinces), Laos and Cambodia (Cardamom Mountains). **Status** Not Evaluated.

JINTAKUNE'S KUKRI SNAKE
Oligodon jintakunei NOT ILLUSTRATED
Measurement TL 448mm **Identification** Body slender, subcylindrical; head short, distinct from neck; internasals and prefrontals fused; loreal present; single preocular; single postocular; subocular absent; supralabials 7; Supralabials III–IV contact orbit; infralabials 7; maxillary teeth 6, the last 3 enlarged and laterally compressed; eye moderate; pupil rounded; dorsals smooth; midbody scale rows 15; ventrals 189; subcaudals 46, paired; anal divided. **Coloration** Dorsum dark brown with 11 regular-spaced, narrow yellowish-cream rings contacting ventrals; 3 narrow bands of the same colour on tail, contacting subcaudals; head beige with crescentic mark across eyes and snout; chin cream with small brown spots; venter unpatterned white. **Habitat and Behaviour** Inhabits lowland rainforests. Diet and reproductive habits unstudied. **Distribution** Endemic to southern Thailand (Krabi Province). **Status** Not Evaluated.

JOYNSON'S KUKRI SNAKE
Oligodon joynsoni PLATE 50
Measurement TL 865mm **Identification** Body robust, subcylindrical; head short, indistinct from neck; loreal present; preocular single or absent; postoculars 2; subocular present; supralabials 8; Supralabials IV–V contact orbit; eye moderate; pupil rounded; dorsals smooth; midbody scale rows 17; ventrals 180–195; subcaudals 40–50, paired; anal entire. **Coloration** Dorsum purplish-brown with dark reticulations forming cross-bars, alternating with ca 50 transverse black spots; forehead with crescentic dark brown band over prefrontals and across eyes, oblique temporal streak and narrow chevron mark; venter red, unpatterned or with rectangular black spots. **Habitat and Behaviour** Nothing known of its natural history. **Distribution** Northern Thailand (Me Wang and Muang Ngow, Lampang Province) and Laos (Champasak Province). **Status** Not Evaluated.

LACROIX'S KUKRI SNAKE
Oligodon lacroixi PLATE 50
Measurement TL 636mm **Identification** Body robust, subcylindrical; head short, indistinct from neck; loreal absent; single preocular; postoculars 2; supralabials 5; Supralabials II–III contact orbit; maxillary teeth 8–12; eye small; pupil rounded; dorsals smooth; midbody scale rows 15; ventrals 162–178; subcaudals 29–33+, paired; anal divided. **Coloration** Dorsum dark grey or purplish-brown; vertebral series of rounded or oval, black-edged orange spots, numbering 11–12 on body and 2–3 on tail; 4 indistinct longitudinal dark stripes, the median 2 on vertebral region, the outer 2 on scale rows 3; forehead brown with probably red or pink marks on snout; chevron behind eyes; venter red, each ventral scale with black blotches. **Habitat and Behaviour** Inhabits submontane forests at elevations of ca 1,500m asl. Presumably nocturnal and terrestrial. Diet and reproductive habits unstudied. **Distribution** Endemic to northern Vietnam (Sa Pá, Lao Cai Province). Also southern China (Yunnan Province). **Status** Not Evaluated.

ARAKAN KUKRI SNAKE
Oligodon mcdougalli NOT ILLUSTRATED
Measurement TL 337mm **Identification** Body robust, subcylindrical; head short, indistinct from neck; loreal absent; single preocular; postoculars 2; supralabials 7; Supralabials III–IV contact orbit; Supralabial II contacts prefrontal; infralabials 7; eye moderate; pupil rounded; dorsals smooth; midbody scale rows 13; ventrals 199; subcaudals 40, paired; anal entire. **Coloration** Dorsum dusky-black; reddish vertebral stripe from nape to tail-tip, edged with small black spots, especially anteriorly; head black with yellow markings; black line on scale rows 2–3, up to vent; tail with 2 black bars; venter black mottled with brown. **Habitat and Behaviour** Nothing known of its natural history. **Distribution** Endemic to Myanmar (Sandoway and Gwa region, Rakhine State). **Status** Not Evaluated.

LONG-TAILED KUKRI SNAKE
Oligodon macrurus NOT ILLUSTRATED
Measurement TL 480mm **Identification** Body robust, subcylindrical; head short, indistinct from neck; loreal present or absent; single preocular; postoculars 2; subocular present or absent; supralabials 7–8; Supralabials III–IV contact orbit; eye moderate; pupil rounded; tail long; dorsals smooth; midbody scale rows 17; ventrals 143–152; subcaudals 76–83, paired; anal entire. **Coloration** Dorsum light brown with indistinct darker reticulation; head with dark postocular stripe, dark stripe behind angle of jaws and chevron contacting parietals; venter unpatterned cream. **Habitat and Behaviour** Nothing known of its natural history. **Distribution** Endemic to southern Vietnam (Nha Trang, Khanh Hoa Province and Pointe Lagan or Thanh Pho, north of Saigon). **Status** Not Evaluated.

MEYERINK'S KUKRI SNAKE
Oligodon meyerinkii NOT ILLUSTRATED
Measurement TL 379mm **Identification** Body robust, subcylindrical; head short, indistinct from neck; loreal present; single preocular; postoculars 2; supralabials 6; Supralabials III–IV contact orbit; infralabials 7; maxillary teeth 9; eye moderate; pupil

rounded; dorsals smooth; midbody scale rows 17; ventrals 154–169; subcaudals 38–57, paired; anal entire. **Coloration** Dorsum reddish-brown with 5–7 longitudinal salmon-pink stripes, bordered by grey or dark brown stripes on scale rows 4–5, lacking transverse markings; forehead pale brown with 2 black stripes, including an interocular bar that extends to sides of head to cover parts of Supralabials IV–V, and an oblique stripe from angle of jaws to parietals; dark stripe along median plane from parietals to frontal; venter rosy-pink with fine dark brown dots, more intense posteriorly; subcaudals pink with some brown spots. **Habitat and Behaviour** Nothing known of its natural history. **Distribution** Eastern Borneo (Sabah State). Also the Philippines (Bongao, Jolo, Papahang, Sibutu and Tawi-Tawi). **Status** Not Evaluated.

MORICE'S KUKRI SNAKE
Oligodon moricei NOT ILLUSTRATED
Measurement TL 443mm **Identification** Body moderate, laterally compressed; head short, indistinct from neck; loreal present; presubocular absent; postoculars 2; supralabials 8; Supralabials IV–V contact orbit; infralabials 9; maxillary teeth 12; eye moderate; pupil rounded; dorsals smooth; midbody scale rows 17; ventrals 175; subcaudals 41, paired; anal entire. **Coloration** Dorsum dark brownish-grey with scales, especially those of fourth and fifth rows, strongly edged with black, producing irregular reticulation; broad rusty-brown vertebral stripe, covering inner half of adjacent rows, edged on each side with wide black paravertebral stripe, extends from neck to tail-tip; vertebral row irregularly mottled with darker brown; paravertebral stripes reach tail; forehead dark greyish-brown; supralabials pale with faint mottling of dark brown; narrow transverse black line across anterior frontal; oblique dark brown streak on Supralabials V–VI; short elongated blotch on frontal; subrectangular, black-edged dark brown blotch on posterior parietal; venter pale yellow with 1–2 rectangular greyish-brown blotches near tips of ventral, heavily speckled with dark greyish-brown in between on anterior half of body; greyish-brown blotches on posterior half united. **Habitat and Behaviour** Nothing known of its natural history. **Distribution** Endemic to southern Vietnam (Nha Trang, Khanh Hoa Province). **Status** Not Evaluated.

CAMBODIAN KUKRI SNAKE
Oligodon mouhoti NOT ILLUSTRATED
Measurement TL 339mm **Identification** Body moderate, subcylindrical; head short, indistinct from neck; presubocular and loreal present; single preocular; postoculars 2; supralabials 7–8; Supralabials IV–V contact orbit; infralabials 9–10; maxillary teeth 14–16; eye moderate; pupil rounded; dorsals smooth; midbody scale rows 17; ventrals 145–163; subcaudals 29–43, paired; anal entire. **Coloration** Dorsum brownish-grey or greyish-tan, dorsals edged with dark brown; yellow, tan or light brown vertebral stripe edged on each side with ochre-brown paravertebral stripe; irregular dark brown stripe extends on each side from neck to vent; tail with wide vertebral stripe edged with brownish-grey stripe; 2 large dark or blackish-brown vertebral blotches reach middle of sides of tail; forehead brownish-grey or greyish-brown, darker than body; supralabials cream variegated with dark brown; broad, transverse reddish-brown mark on snout extends obliquely across eye down to Supralabials 6–7; short, arrow-shaped longitudinal streak on frontal; large, oblique dark brown or blackish-brown mark extends to frontal, anterior part of parietals and posterior temporals, side of neck behind angle of jaws, then downwards; heart-shaped dark brown nuchal blotch, pointing forwards with 2 short branches; venter cream anteriorly, bright pink-red or coral-red posteriorly; on most ventral scales, irregular subrectangular black blotch near one or both tips. **Habitat and Behaviour** Inhabits lowland forests and the vicinity of rice fields. Nocturnal as well as diurnal, and terrestrial. Diet unstudied. Oviparous; clutches comprise 2 eggs that may be stuck together, measuring 26–28 x 9mm. **Distribution** Eastern and central Thailand (Bangkok, Phra Nakhon Si Ayutthaya, Nakhon Ratchasima, Nakhon Sawan, Phetchaburi and Prachuap Khiri Khan Provinces) and Cambodia. **Status** Not Evaluated.

EYED KUKRI SNAKE
Oligodon ocellatus PLATE 50
Measurement TL 852mm **Identification** Body robust, subcylindrical; head short, indistinct from neck; loreal and presubocular present; single preocular; postoculars 2; supralabials 7–8; Supralabials IV–V contact orbit; anterior temporals 2; eye large; pupil rounded; maxillary teeth 9–11, last 2–3 enlarged; tail robust, tapering; long and deeply forked hemipenes, thin, smooth and not spinose; dorsals smooth; midbody scale rows 19; ventrals 161–173; subcaudals 26–44, paired; anal entire. **Coloration** Dorsum orange, brown to yellow-ochre, scales edged with dark brown-grey, typically forming dark bands; dorsal pattern comprises 11–14 rhomboid orange or brown vertebral blotches; 3 oblique bands, comprising dark anterior edges of scales between vertebral blotches, form zigzag pattern on flanks; forehead ochre-brown or greyish-brown with minute scattered dark dots; large, arrow-shaped dark maroon mark on forehead; venter yellowish-cream anteriorly, salmon-pink posteriorly; subcaudals white. **Habitat and Behaviour** Inhabits hilly evergreen forests at an elevation of 200m asl. Nocturnal and arboreal. One individual was found in a tree-hole. Diet and reproductive habits unstudied. **Distribution** Laos (Champasak Province), Cambodia (Ta Veng, Kirirom, Koh Song and Ratanakiri Provinces) and Vietnam (Quang Nam-Danang, Binh Dinh, Lam Dong, Phu Yen, Dac Lac, Dong Nai and Tay Ninh Provinces). **Status** Not Evaluated.

EIGHT-LINED KUKRI SNAKE
Oligodon octolineatus PLATE 50
Measurement TL 700mm **Identification** Body slender, subcylindrical; head short, indistinct from neck; loreal present; single preocular; postoculars 2;

supralabials 6; Supralabials III–IV contact orbit; infralabials 7–8; 2 anterior temporals; eye moderate; pupil rounded; dorsals smooth; midbody scale rows 17; ventrals 155–197, paired; subcaudals 43–61, paired; anal entire. **Coloration** Dorsum black or dark brown with a red vertebral stripe and 5–7 longitudinal light stripes bordered by 6–8 dark stripes; forehead with 2 black stripes; venter yellow. **Habitat and Behaviour** Inhabits lowland and hill dipterocarp forests. Nocturnal and mostly terrestrial, active on the ground. One was seen more than 8m up on a branch of a tall tree. Diet consists of birds' eggs, frogs and their eggs, lizards and other snakes. Oviparous; clutches comprise 4–5 eggs, measuring 16 x 30mm. **Distribution** Peninsular Malaysia, Singapore, Sumatra, Pulau Nias, the Mentawai Archipelago, Pulau Bangka, Pulau Belitung, the Riau Archipelago, Borneo and Java. Also the Sulu Archipelago (the Philippines) and Sulawesi (Indonesia). **Status** Not Evaluated.

PETRONELLA'S KUKRI SNAKE
Oligodon petronellae NOT ILLUSTRATED
Measurement TL 450+mm **Identification** Body moderate, subcylindrical; head short, indistinct from neck; loreal present; single preocular; postoculars 2; supralabials 7; Supralabials III–IV contact orbit; single anterior temporals; eye moderate; pupil rounded; dorsals smooth; midbody scale rows 15; ventrals 152, paired; subcaudals 19+, paired; anal entire. **Coloration** Dorsum greyish-brown with dark brown blotches bordered by pale spots; brown patch on posterior parietals; paired parallel bands on nape and along dorsum; ca 21 lozenge-shaped marks, some bordered with white spots; transverse dark brown band on snout; supralabials spotted with black; venter brick-red, ventrals showing alternate series of dark brown spots on outer edges; subcaudals red. **Habitat and Behaviour** Nothing known of its natural history. **Distribution** Endemic to Sumatra. **Status** Not Evaluated.

FLAT-HEADED KUKRI SNAKE
Oligodon planiceps NOT ILLUSTRATED
Measurement TL 255mm **Identification** Body robust, subcylindrical; head short, indistinct from neck; loreal absent; single preocular; postoculars 2; supralabials 5 (rarely 4); Supralabial III contacts orbit; infralabials 4; eye moderate; pupil rounded; maxillary teeth 10; dorsals smooth; midbody scale rows 13; ventrals 132–145; subcaudals 22–27, paired; anal divided. **Coloration** Dorsum pale brown, some scales with dark lower borders; black bar behind parietals; oblique black subocular streak; venter cream or pink, ventrals with squarish black blotches laterally, typically paired. **Habitat and Behaviour** Nothing known of its natural history. **Distribution** Endemic to southern Myanmar (Yangon, Yangon Division and Minhla, Bago Province). **Status** Not Evaluated.

PULAU WEH KUKRI SNAKE
Oligodon praefrontalis NOT ILLUSTRATED
Measurement TL 250mm **Identification** Body robust, subcylindrical; head short, indistinct from neck; loreal absent; single preocular; single postocular; supralabials 7; Supralabials III–IV contact orbit; eye small; pupil rounded; dorsals smooth; midbody scale rows 15; ventrals 193; subcaudals 37, paired; anal divided. **Coloration** Dorsum greyish-brown with a yellowish-brown stripe along vertebral region, covering vertebral scales and half of adjacent scales; venter yellowish-white; outer edges of ventrals alternately light and dark. **Habitat and Behaviour** Nothing known of its natural history. **Distribution** Endemic to Pulau Weh. **Status** Not Evaluated.

JAVANESE KUKRI SNAKE
Oligodon propinquus NOT ILLUSTRATED
Measurement TL 290mm **Identification** Body moderate, subcylindrical; head short, indistinct from neck; loreal present; single preocular; postoculars 2; supralabials 7; Supralabials III–IV contact orbit; 4 infrabials contact anterior chin shields; eye moderate; pupil rounded; dorsals smooth; midbody scale rows 15; ventrals 140; subcaudals 27, paired; anal entire. **Coloration** Dorsum black with a series of yellow spots; venter cream. **Habitat and Behaviour** Nothing known of its natural history. **Distribution** Endemic to Java. **Status** Not Evaluated.

FALSE STRIPED KUKRI SNAKE
Oligodon pseudotaeniatus NOT ILLUSTRATED
Measurement TL 258mm **Identification** Body moderate, subcylindrical; head short, indistinct from neck; loreal present; presubocular present; single preocular; postoculars 2; supralabials 8; Supralabials IV–V contact orbit; infrabials 8–9; maxillary teeth 15; eye moderate; pupil rounded; dorsals smooth; midbody scale rows 17; ventrals 137–156; subcaudals 34–46, paired; anal entire. **Coloration** Dorsum brownish-grey, scales edged with dark brown; narrow greyish-brown vertebral stripe, edged with dark brown paravertebral stripe, extends from nuchal region to tail; indistinct greyish-brown dorsolateral stripe with scattered black dots extends to vent; forehead brownish-grey with scattered dark dots; supralabials yellowish-brown; faint, irregular transverse marking on snout extends across orbit; elongated streak, constricted on frontal; large dark marks across neck, of which two upper ones indistinct, reaching mid-parietals and lateral ones directed downwards on sides of neck; venter pale yellow with irregular blackish-brown blotches on sides; these marks are small, rounded and/or absent in first quarter of body, progressively larger, subrectangular and more conspicuous posteriorly. **Habitat and Behaviour** Nothing known of its natural history. **Distribution** Endemic to central Thailand (Korat, Nakhon Ratchasima Province, Bangkok, Phra Nakhon Si Ayutthaya Province and Muak Lek, Saraburi Province). **Status** Not Evaluated.

PADANG KUKRI SNAKE
Oligodon pulcherrimus NOT ILLUSTRATED
Measurement TL 367mm **Identification** Body robust, subcylindrical; head short, slightly distinct from neck; snout rounded; loreal present; single preocular; postoculars 2; supralabials 7; Supralabials III–IV contact orbit; infralabials 8; maxillary teeth 9; mandibular teeth 7–9; eye moderate; pupil rounded; tail short; dorsals smooth; midbody scale rows 15; ventrals 179; subcaudals 30, paired; anal divided. **Coloration** Dorsum brown with a longitudinal dark band, commencing from frontal and widening at nape, enclosing 36 hexagonal black-edged pale brown marks; flanks bluish-grey with black and white spots; head brown with crescentic dark brown bands; venter yellow, outer edges of ventrals spotted alternately with black and white. **Habitat and Behaviour** Nothing known of its natural history. **Distribution** Endemic to western Sumatra (Padang, Sumatera Barat Province). **Status** Not Evaluated.

PURPLE KUKRI SNAKE
Oligodon purpurascens PLATE 50
Measurement TL 950mm **Identification** Body robust, subcylindrical; head short, indistinct from neck; loreal present; single presubocular; preoculars 1–2; postoculars 2–3; supralabials 8; Supralabials IV–V or V contact orbit; infralabials 8–9; eye small; pupil rounded; dorsals smooth; midbody scale rows 19 or 21; ventrals 160–210; subcaudals 40–60, paired; anal entire. **Coloration** Dorsum brownish-purple with wavy dark bands or transverse dark-edged yellow or cream bands or blotches, sometimes separated by narrow dark cross-bars; dark chevron on forehead; venter pink or red with squarish dark spots. **Habitat and Behaviour** Inhabits hill dipterocarp and peat swamp forests at sea level to ca 1,200m asl. Nocturnal and terrestrial. Diet includes frogs' and lizards' eggs, and frogs and tadpoles. Oviparous; clutches comprise 8–13 eggs, measuring 18–22 x 27–33mm. Hatchlings 210mm. **Distribution** Southern Thailand (Na Pradoo, Pattani Province), Peninsular Malaysia, Singapore, Sumatra, the Mentawai Archipelago, Borneo and Java. **Status** Not Evaluated.

FOUR-LINED KUKRI SNAKE
Oligodon quadrilineatus PLATE 50
Measurement TL 402mm **Identification** Body robust, subcylindrical; head short; indistinct from neck; loreal present; single preocular; postoculars 2; supralabials 8; Supralabials IV–V contact orbit; infralabials 9; maxillary teeth 14; eye moderate; pupil rounded; dorsals smooth; midbody scale rows 19; ventrals 157–159; subcaudals 41–44, paired; anal entire. **Coloration** Dorsum greyish-brown, scales narrowly edged with dark brown; narrow greyish-yellow vertebral stripe between nuchal marking to tail base, edged with wide brown paravertebral stripe, sometimes intersected with irregular small, subrectangular dark brown blotches; forehead greyish-brown; supralabials lighter, variegated with brown; transverse dark brown mark on snout extends across eye to Supralabials 5–6; arrow- or heart-shaped dark brown nuchal mark; venter cream-yellow with 2 irregular subrectangular blackish-brown blotches near edges of ventrals. **Habitat and Behaviour** Nothing known of its natural history. **Distribution** Myanmar (Sagaing Division and Shan State), Thailand, Cambodia and Vietnam. **Status** Not Evaluated.

SAINT GIRON'S KUKRI SNAKE
Oligodon saintgironsi NOT ILLUSTRATED
Measurement TL 676mm **Identification** Body robust, subcylindrical; head short, indistinct from neck; snout elongated; loreal present; single presubocular; single preocular; postoculars 2; supralabials 8; anterior temporals 2; maxillary teeth 10–12, last 3 enlarged; eye moderate; pupil rounded; tail long, robust, tapering; long, deeply forked hemipenes, smooth and not spinose throughout; dorsals smooth; midbody scale rows 17–18; ventrals 170; subcaudals 28–29, paired; anal entire. **Coloration** Dorsum brownish-ochre, flanks greyish-tan, scales with dense dark brown dots, and strongly edged with dark brown; pale yellowish-tan vertebral stripe; 13 butterfly-shaped maroon vertebral blotches; 3 irregular oblique bands across vertebral region; forehead greyish-brown with numerous dark dots; large, arrow-shaped maroon cephalic marking; venter pale yellowish-cream, most ventrals with small brown blotch near both tips. **Habitat and Behaviour** Nothing known of its natural history. **Distribution** Cambodia (no specific locality) and southern Vietnam (Dong Nai Province). **Status** Not Evaluated.

SPLENDID KUKRI SNAKE
Oligodon splendidus PLATE 50
Measurement TL 830mm **Identification** Body robust, subcylindrical; head short, indistinct from neck; loreal absent; single preocular; postoculars 2; supralabials 8; Supralabials IV–V contact orbit; infralabials 5; maxillary teeth 10–11; eye moderate; pupil rounded; dorsals smooth; midbody scale rows 21; ventrals 169–193; subcaudals 35–47, paired; anal entire. **Coloration** Dorsum pale brown, each scale with a dark centre; series of dark median blotches edged with dark brown line, with a buff-coloured outer edge, numbering 14–17 on body and 3–5 on tail; flanks with smaller dark spots; venter yellow or cream; forehead with indistinct dark bars, including an oblique one between parietals and angle of jaws; large dark chevron on nuchal region; ventrals with alternating series of dark brown spots laterally. **Habitat and Behaviour** Inhabits lowland forests. Diet unstudied. Oviparous; clutches comprise 3–6 eggs, measuring 457 x 152mm. Hatchlings 155mm. **Distribution** Endemic to Myanmar (Valleys of Ayeyarwaddy and Chindwin, in Sagaing, Mandalay and Shan States). **Status** Not Evaluated.

HALF-KEELED KUKRI SNAKE
Oligodon subcarinatus PLATE 50
Measurement TL 390mm **Identification** Body slender, subcylindrical; head short, indistinct from

neck; loreal present; single preocular; single postocular; supralabials 7; Supralabials III–IV contact orbit; eye small; pupil rounded; dorsals weakly keeled; midbody scale rows 17; ventrals 154–166; subcaudals 50–57, paired; anal entire. **Coloration** Dorsum greyish-brown with 20–30 dark-edged pale brown, reddish-brown or cream cross-bars; chevron pattern on forehead; venter orange-red. **Habitat and Behaviour** Inhabits lowland forests at sea level to 200m asl. Nocturnal and terrestrial, found in buttresses of large trees and at edges of peat swamp forests. Diet and reproductive habits unstudied. **Distribution** Endemic to western Borneo (Mendolong, Sabah State, and Kuching, Bako, Matang, Sungei Mengiong and Sungei Nyabau, Sarawak State). **Status** Not Evaluated.

STRIPED KUKRI SNAKE
Oligodon taeniatus PLATE 50
Measurement TL 447mm **Identification** Body moderate, subcylindrical; head short, indistinct from neck; loreal present; single presubocular; single preocular; postoculars 2; supralabials 8; Supralabials IV–V contact orbit; infralabials 8–9; maxillary teeth 14–17, the last 2 strongly enlarged; eye small; pupil rounded; dorsals smooth; midbody scale rows 19; ventrals 146–165; subcaudals 31–48, paired; anal entire. **Coloration** Dorsum brownish-grey or greyish-tan, dorsals finely edged with dark brown posteriorly; 2 longitudinal dark paravertebral stripes edging a yellow vertebral stripe, and 2 narrower dorsolateral stripes; 5 blotches on forehead: 1 anterior transverse marking across snout; 1 longitudinal blotch on frontal; 2 oblique temporal bars, directed posteriorly downwards; 1 large, arrow-shaped nuchal blotch; bases of oblique central streaks may reach ventral scales at least on one side; venter pink or coral-red, cream on edges, with irregular subrectangular black blotch near one or both tips. **Habitat and Behaviour** Inhabits lowland forests. Nocturnal and terrestrial, hiding under stones, wood and leaves by day. Diet consists of frogs and lizards and their eggs. Threat response consists of raising and curling its tail to reveal the red venter. Oviparous; clutch size unknown. **Distribution** Central, eastern and south-eastern Thailand, southern Laos (Paksé, Champasak Province), southern and central Cambodia, and Vietnam. **Status** Not Evaluated.

MANDALAY KUKRI SNAKE
Oligodon theobaldi PLATE 50
Measurement TL 437mm **Identification** Body slender, subcylindrical; head short, indistinct from neck; loreal present; single preocular; postoculars 2; supralabials 8; Supralabials IV–V contact orbit; maxillary teeth 15–16; eye moderate; pupil rounded; tail short; dorsals smooth; midbody scale rows 17; ventrals 164–180; subcaudals 30–42, paired; anal divided. **Coloration** Dorsum light brown with narrow, closely set transverse or angular dark cross-bars; 4 longitudinal dark stripes along dorsum; venter yellow with or without squarish black spots at outer margins of ventrals. **Habitat and Behaviour** Inhabits the forested mid-hills and plains. Nocturnal and terrestrial. Diet unstudied. Oviparous; clutches comprise 3 eggs, measuring 18mm in length. **Distribution** Myanmar. Also north-eastern India (Meghalaya State). **Status** Not Evaluated.

GARLANDED KUKRI SNAKE
Oligodon torquatus NOT ILLUSTRATED
Measurement TL 300mm **Identification** Body robust, subcylindrical; head short, indistinct from neck; loreal present; single preocular; postoculars 2; supralabials 7; Supralabials III–IV contact orbit; infralabials 4; maxillary teeth 15–16; eye moderate; pupil rounded; dorsals smooth; midbody scale rows 15; ventrals 144–159; subcaudals 27–34, paired; anal divided. **Coloration** Dorsum brown with or without 4 indistinct longitudinal stripes; head with obscure black marks; black subocular streak; flanks with white spots; venter white, a few ventrals with scattered black spots; subcaudals unpatterned white. **Habitat and Behaviour** Inhabits hilly areas, though little else is known about it. Diet unstudied. Oviparous; clutches comprise 2–3 eggs. **Distribution** Endemic to Myanmar (Valley of the Ayeyarwaddy, between Myitkyina and Bhamo). **Status** Not Evaluated.

THREE-LINED KUKRI SNAKE
Oligodon trilineatus NOT ILLUSTRATED
Measurement TL 330mm **Identification** Body robust, subcylindrical; head short, indistinct from neck; loreal present; single presubocular; single preocular; single postocular; supralabials 7; Supralabials III–IV contact orbit; eye small; pupil rounded; dorsals smooth; midbody scale rows 17; ventrals 145–157; subcaudals 52–62, paired; anal entire. **Coloration** Dorsum dark brown or black; red vertebral stripe; narrow dorsolateral stripe; forehead yellowish-brown with oblique dark bands; venter dark grey with a pair of white stripes. **Habitat and Behaviour** Nothing known of its natural history. **Distribution** Pulau Nias and Sumatra (Maninjau, Sumatera Barat Province). **Status** Not Evaluated.

DARK-SPINED KUKRI SNAKE
Oligodon vertebralis NOT ILLUSTRATED
Measurement TL 350mm **Identification** Body moderate, subcylindrical; head short, indistinct from neck; loreal present; single preocular; postoculars 2; supralabials 7; Supralabials III–IV contact orbit; eye moderate; pupil rounded; dorsals smooth; midbody scale rows 15; ventrals 136–154; subcaudals 35–54, paired; anal divided. **Coloration** Dorsum brown with scattered, small yellow spots with black edges; spots largest on vertebral region; head yellow with 2 black-edged brown bands, the anterior one passing across orbit, the posterior across frontal; gular region black spotted; venter yellow. **Habitat and Behaviour** Nothing known of its natural history. **Distribution** Endemic to Borneo (Banjarmasin, Kalimantan Province and Gunung Kinabalu, Sabah State). **Status** Not Evaluated.

MOUNTAIN REED SNAKE
Oreocalamus hanitschi　　　　　　PLATE 51
Measurement TL 570mm **Identification** Body robust, subcylindrical; head short, indistinct from neck; loreal present; single preocular; single postocular; supralabials 7–8; Supralabial I contacts nostril; Supralabials IV–V contact orbit; infralabials 7; 4 infralabials contact anterior chin shields; tail short with a sharp tip; dorsals smooth; midbody scale rows 17; ventrals 125–132; subcaudals 20–32, paired; anal entire. **Coloration** Dorsum brownish-tan with dark brown scales forming zigzag pattern; on lower flanks, dark scales join to form continuous line; dark stripe along eyes; dark inverted chevron on neck; venter yellowish-brown, each ventral with black spots centrally and on edges, forming dark line. **Habitat and Behaviour** Inhabits oak forests at elevations of 1,120–1,700m asl. Terrestrial and/or subfossorial, retreat sites being logs and forest litter. Diet consists of earthworms. Reproductive habits unstudied. **Distribution** Peninsular Malaysia (Gunung Brinchang, Cameron Highlands, Pahang State) and Borneo (Gunung Kinabalu and Mendolong, Sabah State, and Gunung Murud, Sarawak State). **Status** Not Evaluated.

RED BAMBOO TRINKET SNAKE
Oreocryptophis porphyraceus　　　　PLATE 51
Measurement TL 1,250mm **Identification** Body slender; head elongated, slightly distinct from neck; snout rounded; loreal present; single preocular; postoculars 2; supralabials 7–8 (rarely 9–10); Supralabials IV–V (rarely III–IV, IV–VI or V–VI) contact orbit; infralabials 8–11; eye small; pupil rounded; ventral keels absent; midbody scale rows 19; ventrals 175–217; subcaudals 48–80; anal divided. **Coloration** Dorsum deep reddish-brown or reddish-orange, body and tail dark-banded (except in *O. p. coxi*); dark flank stripe along posterior half of body; dark postocular stripe; venter unpatterned cream or yellow; iris yellowish-brown; tongue reddish-brown (except in *O. p. coxi*; unknown in *O. p. nigrofasciatus*). **Subspecies** Seven subspecies are recognized, five of which occur in the region. *O. p. porphyraceus*, ventrals 179–217; subcaudals 52–80; tongue reddish-brown. *O. p. coxi*, ventrals 213; subcaudals 62; tongue rosy-pink. *O. p. laticinctus*, ventrals 191–201; subcaudals 62–67; tongue reddish-brown. *O. p. nigrofasciatus*, ventrals 208; subcaudals 78; tongue colour unknown. *O. p. vaillanti*, ventrals 187–209; subcaudals 55–75; tongue reddish-brown. **Habitat and Behaviour** Inhabits the mid-hills to montane forests at elevations of 116–2,600m. Diurnal and arboreal, associated with low vegetation. Diet consists of small mammals such as voles and shrews. Oviparous; clutches comprise 2–5 eggs, measuring 48mm. Incubation period 50–60 days. **Distribution** Myanmar and northern Thailand (*O. p. porphyraceus*), Laos, Cambodia and Vietnam (*O. p. vaillanti*), north-eastern Thailand (*O. p. coxi*), Peninsular Malaysia (Cameron Highlands, Pahang State), Sumatra (*O. p. laticinctus*) and Singapore (*O. p. nigrofasciatus*). Also eastern and north-eastern India, and possibly Nepal (*O. p. porphyraceus*), southern China (*O. p. pulchra*), eastern China (*O. p. vaillanti*) and Taiwan (*O. p. kawakamii*). **Status** Not Evaluated.

EASTERN TRINKET SNAKE
Orthriophis cantoris　　　　　　　PLATE 51
Measurement TL 2m **Identification** Body slender; head elongated, slightly distinct from neck; single preocular; postoculars 2; supralabials 8–9; Supralabials IV–V or III–V contact orbit; infralabials 8–11; ventral keel distinct; dorsals smooth or weakly keeled; midbody scale rows 19 or 21; ventrals 220–235; subcaudals 62–88; anal entire. **Coloration** Dorsum greyish-brown to yellowish-brown; dark or light scale borders produce reticulated pattern; on the top dark brown or reddish-brown blotches form transverse bands towards posterior of body; smaller blotches on flanks; gular region yellow or orange; venter yellow turning pink towards tail; iris red. **Habitat and Behaviour** Montane species inhabiting forests at elevations of 1,000–2,300m asl. Diet is likely to consist of small mammals and birds. Oviparous; clutches comprise 5–10 eggs. Incubation period 84–102 days. **Distribution** Northern Mynamar (Kachin State). Also north-eastern India and Nepal. **Status** Not Evaluated.

MOELLENDORFF'S TRINKET SNAKE
Orthriophis moellendorffi　　　　　PLATE 51
Measurement TL 2.5m **Identification** Body robust; head elongated, distinct from neck; single preocular; postoculars 2–3; supralabials 9–10; Supralabials V–VI or VI–VII contact orbit; infralabials 11–12; tail long and slender; midbody scale rows 25 or 27, dorsals weakly keeled, flank scales smooth; ventrals 265–284; subcaudals 76–100; anal divided. **Coloration** Dorsum greenish-grey or greyish-pink, with saddle-like rusty-brown or pale grey blotches edged with black; flanks with smaller, rounded blotches of same colour; forehead brown or reddish-brown; supralabials greyish- or yellowish-brown; venter yellowish-white with black blotches; iris red. **Habitat and Behaviour** Inhabits deciduous forests on karst limestone at elevations of 50–300m asl. Diurnal and terrestrial, frequently associated with caves. Diet includes rodents and birds. Oviparous; clutches comprise 6–12 eggs. **Distribution** Northern Vietnam. Also southern and eastern China. **Status** Not Evaluated.

CAVE RACER
Orthriophis taeniurus　　　　　　　PLATE 51
Measurement TL 2m **Identification** Body slender, elongated; head long, distinct from neck; loreal present; single preocular; postoculars 2–3; supralabials 7–10; Supralabials V–VI or IV–V contact orbit; infralabials 10–13; dorsals weakly keeled; scales on flanks smooth; midbody scales 25; ventrals 271–305; subcaudals 86–112; anal divided. **Coloration** Dorsum greyish-brown, blue-grey or greyish-black with cream or tan stripe along middle of back, especially on posterior; forehead olive or blue-grey; sides of head with dark stripe; supralabials and chin

cream; venter yellow or cream. **Subspecies** Seven subspecies are recognized, of which four occur in the region. *O. t. grabowski*, supralabials 9; Supralabials V–VI contact orbit; midbody scale rows 25 or 27; ventrals 275–285; subcaudals 102–114; dorsum lead-grey to greyish-brown. *O. t. mocquardi*, supralabials 9; Supralabials V–VI contact orbit; midbody scale rows 23; ventrals 251–264; subcaudals 90–125; dorsum bright yellow or yellowish-brown. *O. t. ridleyi*, supralabials 9; Supralabials V–VI contact orbit; midbody scale rows 25; ventrals 285–305; subcaudals 105–122; dorsum beige to ochre. *O. t. yunnanensis*, supralabials 7–9; Supralabials IV–V or V–VI contact orbit; midbody scale rows 23; ventrals 236–260; subcaudals 89–120; dorsum yellowish-brown. The nominotypic *O. t. taeniurus* occurs peripherally, to the north of Vietnam, in eastern China: Supralabials 8–9; Supralabials V–VI, IV–V or IV–VI contact orbit; midbody scale rows 23–25; ventrals 225–255; subcaudals 84–112; dorsum yellowish-brown with some orange areas. **Habitat and Behaviour** Inhabits lowland and submontane forests from sea level up to 2,000m asl. Diurnal and nocturnal, and arboreal as well as terrestrial, often found in limestone and other caves, as well as on the forest floor and on trees. Diet consists of bats and swiftlets. Oviparous; clutches comprise 5–14 eggs, measuring 45.7–72 x 20.8–31.4mm. Incubation period 70–75 days. Hatchlings 40cm. **Distribution** Northern Myanmar, Thailand, north-eastern Laos, Cambodia, north-western Vietnam (*O. t. yunnanensis*), southern Thailand, Peninsular Malaysia, Singapore (*O. t. ridleyi*), Borneo, Sumatra (*O. t. grabowski*) and northern Vietnam (*O. t. mocquardi*). Also north-eastern India, southern China (*O. t. yunnanensis*), Taiwan (*O. t. friesi*), eastern China (*O. t. mocquardi*), the Ryukyu Archipelago (Japan) (*O. t. schmackeri*), eastern China and south-eastern Russia (*O. t. taeniurus*). **Status** Not Evaluated.

ANNAM KEELBACK
Parahelicops annamensis PLATE 51
Measurement TL 558mm **Identification** Body slender; head distinct from neck; supralabials 8; 1 or 2 prefrontals; Supralabials IV and V contact orbit; supralabials much longer than high; infralabials 8–10; 2 preoculars; anterior chin shield equal to or smaller than posterior one; enlarged posterior maxillary teeth; eye small; median rows feebly keeled anteriorly, becoming strongly keeled posteriorly; tail scales strongly keeled, midbody scale rows 15; ventrals 167–169; subcaudals 117–123; anal entire. **Coloration** Dorsum iridescent purplish-brown; forehead with light markings around eyes; flanks brown; orange speckles on forehead; orange stripe from posterior margin of eye to body, broken posteriorly as dorsolateral stripe, indistinct posteriorly; indistinct longitudinal rows of dark brown spots; venter cream, gular region with dark speckles; outer margin of ventrals dark brown. **Habitat and Behaviour** Inhabits evergreen forests at elevations of 1,280–1,500m asl. Nocturnal, and aquatic and terrestrial. Diet and reproductive habits unstudied. **Distribution** Central Vietnam and southern Laos. **Status** Not Evaluated.

BOONSONG'S KEELBACK
Parahelicops boonsongi NOT ILLUSTRATED
Measurement TL 640mm **Identification** Body moderate; head indistinct from neck; prefrontals fused; loreal present; single preocular; postoculars 2–3; suboculars 2; supralabials 9; Supralabials IV–V contact orbit; infralabials 10; posterior 2–3 maxillary teeth enlarged; eye large; pupil rounded; nostril dorsolaterally oriented; tail long; dorsals keeled, except anterior part of scales on first row; midbody scale rows 19; ventrals 141; subcaudals >33, paired; anal entire. **Coloration** Dorsum greyish-olive, paler on flanks; supralabials and suboculars cream; venter cream. **Habitat and Behaviour** Nothing known of its natural history. Presumably a diurnal and aquatic species inhabiting hill streams. **Distribution** Endemic to north-eastern Thailand (Loei Province). **Status** Not Evaluated.

CHAPA FLAT-NOSED SNAKE
Pararhabdophis chapaensis NOT ILLUSTRATED
Measurement TL >790mm **Identification** Body slender, cylindrical; head distinct from neck; loreal present; preoculars 2; postoculars 2; supralabials 9; Supralabials IV–VI contact orbit; anterior chin shield shorter than posterior one; maxillary teeth 32, followed by 3 enlarged ones; eye moderate; pupil vertical; dorsals weakly keeled; midbody scale rows 17; ventrals 177; subcaudals 73+, paired. **Coloration** Dorsum dark brown, scales on row 5 with light centres, forming light dorsolateral stripes; labials cream with brown sutures; venter pale brown, outer margins of ventrals paler. **Habitat and Behaviour** Inhabits submontane forests at elevations of ca 1,500m asl. Presumably nocturnal and aquatic. Diet and reproductive habits unstudied. **Distribution** Endemic to northern Vietnam (Sa Pá, Lào Cai Province). **Status** Not Evaluated.

BLAKEWAY'S BLOTCH-NECKED SNAKE
Plagiopholis blakewayi NOT ILLUSTRATED
Measurement TL 500mm **Identification** Body robust, rounded; head distinct from neck; eye small; loreal absent; posterior temporals 0 or 1; supralabials 5; Supralabial III contacts orbit; Supralabial IV largest; tail short, slightly flattened below; midbody scale rows 15; ventrals 110–130. **Coloration** Dorsum greyish-brown with an iridescent sheen; forehead greyish-brown; indistinct, narrow blackish-brown band on nape; chevron-shaped vertebral projection, tip contacting posterior margins of parietals; longitudinal dorsolateral row of small blackish-brown blotches on body and tail, separated by 3 scales; scales of flanks with black lower margins, producing reticulated pattern; venter greyish-cream, dark speckled; dark outer margins. **Habitat and Behaviour** Known from montane forests at elevations of 1,300–2,200m asl. Fossorial and presumably nocturnal. Diet and

reproductive habits unstudied. **Distribution** Myanmar (Kachin and Shan States) and northern Thailand. Also southern China (Yunnan Province). **Status** Not Evaluated.

DELACOUR'S BLOTCH-NECKED SNAKE
Plagiopholis delacouri NOT ILLUSTRATED
Measurement TL 189mm **Identification** Body robust, rounded; head distinct from neck; single preocular; postoculars 2; supralabials 6; Supralabials V–VI contact orbit; eye small; tail short, slightly flattened below; midbody scale rows 15; ventrals 108–129; subcaudals 20–28, divided; anal entire. **Coloration** Dorsum yellowish- or greyish-brown with a series of black dorsolateral spots connected to each other by transverse pale bars or chevrons; large black chevron on nuchal region, edged anteriorly and posteriorly by white lines; labials with black bars; venter yellow with heavy suffusion of dark brown. **Habitat and Behaviour** Inhabits montane forests at elevations of 1,400–2,500m asl. Diet and reproductive habits unstudied. **Distribution** Northern Laos (Xieng-Khouang Province) and north-western Vietnam. **Status** Not Evaluated.

COMMON BLOTCH-NECKED SNAKE
Plagiopholis nuchalis PLATE 51
Measurement TL 450mm **Identification** Body robust, rounded; head indistinct from neck; loreal present; posterior temporals 2; single preocular; supralabials 6; Supralabials III–IV contact orbit; Supralabial V largest; infralabials 6; eye small; tail short, slightly flattened below; dorsals smooth or keeled; midbody scale rows 15; ventrals 122–142; subcaudals 20–30, paired; anal entire. **Coloration** Dorsum blackish-brown, mid-brown or reddish-brown, scales edged with black; dorsolateral series of rounded black spots, connected by pale cross-bars or a series of oblique light brown cross-bars or elongated spots; black chevron on nape; venter pink with dense black speckles, typically with large squarish spots on either side of ventrals. **Habitat and Behaviour** Inhabits the mid-hills up to submontane forests at elevations of 600–1,620m asl. Diet and reproductive habits unstudied. **Distribution** Northern Myanmar (Kachin and Shan States) and northern Thailand (Chiang Mai, Mae Hon Son and Lampang Provinces). Records from Laos and north-western Vietnam require verification. **Status** Not Evaluated.

FUJIAN BLOTCH-NECKED SNAKE
Plagiopholis styani PLATE 51
Measurement TL 396mm **Identification** Body robust, rounded; head short, distinct from neck; loreal absent; single preocular; postoculars 2; supralabials 6–7; Supralabials III–IV contact orbit; eye small; tail short, slightly flattened below; dorsals smooth; midbody scale rows 15; ventrals 109–122; subcaudals 23–34; anal entire. **Coloration** Dorsum mid-brown dotted with black; black nuchal blotch or cross-bars, edged with yellow; labials yellow with black edges; rostral yellow with large black spot; venter and subcaudals yellow dotted and speckled with black on edges. **Habitat and Behaviour** Inhabits submontane forests at elevations of 1,200–1,230m asl. Subfossorial, associated with moss and grass. Threat response consists of dorsoventrally compressing its body. Diet consists of earthworms. Oviparous; clutches comprise 2–8 eggs, measuring 18 x 9mm. **Distribution** Vietnam. Also southern China (Fujian and Sichuan Provinces). **Status** Not Evaluated.

INDO-CHINESE SAND SNAKE
Psammophis indochinensis PLATE 51
Measurement TL 1,075mm **Identification** Body slender; head oval, distinct from neck; loreal present; single preocular; postoculars 2; supralabials 8; Supralabials IV–V contact orbit; infralabials 10; nasal incompletely divided; frontal longer than distance from snout-tip, anterior end less than twice as broad as middle; eye large; pupil rounded; dorsals smooth; midbody scale rows 17 or 19; ventrals 150–173; subcaudals 75–85, paired; anal divided. **Coloration** Dorsum olive-green or buff; interstitial region black with 4 dark brown stripes, 2 scales broad, extending from brown forehead; venter bright yellow or yellowish-cream with black line at outer margins of ventrals. **Habitat and Behaviour** Inhabits grasslands, open forests and agricultural areas at elevations of 100–2,000m asl. Diurnal and terrestrial, with some arboreal activity, occasionally associated with bushes and short trees. Diet consists of rodents, frogs, lizards and snakes. Oviparous; clutch size unknown. **Distribution** Myanmar (Sagaing, Mandalay and Magwe Divisions), Thailand (Bangkok, Phra Nakhon Si Ayutthaya Province, Doi Suthep, Chiang Mai Province and Phu Khieo, Chaiyaphum Province), Laos (Champasak Province), Vietnam (Ninh Thuan Province), eastern Java and Bali. **Status** Not Evaluated.

WHITE-COLLARED REED SNAKE
Pseudorabdion albonuchalis PLATE 52
Measurement TL 270mm **Identification** Body slender; head indistinct from neck; snout pointed; nostril between 2 nasals; loreal and preocular absent; supralabials 5–6; Supralabials III–IV contact orbit; infralabials 5, loreal present; frontal borders eye; eye small; tail short; dorsals smooth; midbody scale rows 15; ventrals 127–144; anal entire; subcaudals 43–64. **Coloration** Dorsum iridescent black or brown; broad yellow or red collar over half of parietals and band of 4 scales behind parietals; venter dark brown. **Habitat and Behaviour** Inhabits lowland forests. Terrestrial and fossorial, associated with leaf litter of moist forests. Diet presumably includes small arthropods. Reproductive habits unstudied. **Distribution** Endemic to Borneo (Sungei Pesu, Sungei Baram and Kubah National Park, Sarawak State, Batu Apoi, Brunei and Marak Parak, Sabah State). **Status** Not Evaluated.

MOCQUARD'S REED SNAKE
Pseudorabdion collaris PLATE 52
Measurement TL 250mm **Identification** Body slender; head indistinct from neck; snout pointed; nostril between 2 nasals; loreal and preocular absent; supralabials 5–6; prefrontal in contact with supralabials; Supralabials III–IV contact orbit; infralabials 5; preoculars not fused; maxillary teeth 20; eye small; tail short; dorsals smooth; midbody scale rows 15; ventrals 110–134; subcaudals 20–41; anal entire. **Coloration** Dorsum shiny black; red or yellow collar sometimes present; venter slightly lighter than dorsum. **Habitat and Behaviour** Inhabits lowland forests at elevations of up to 650m asl. Subfossorial, digging into the soil and leaf litter, and under rocks and logs. Diet consists of earthworms. Reproductive habits unstudied. **Distribution** Endemic to Borneo. **Status** Not Evaluated.

EISELT'S DWARF REED SNAKE
Pseudorabdion eiselti NOT ILLUSTRATED
Measurement TL 200mm **Identification** Body slender; head indistinct from neck; snout pointed; loreal and preocular absent; supralabials 5–6; Supralabials I–II contact internasal; infralabials 5; postocular and supraocular not fused to each other or to ocular; supraocular separates frontal from eye; maxillary teeth 10; eye small; tail short; dorsals smooth; midbody scale rows 15; ventrals 130; subcaudals 12, paired; anal entire. **Coloration** Dorsum dark brown; scales with pale network or light apical spot; head dark brown; supralabials paler; indistinct yellow lateral margin of parietals; venter dark brown. **Habitat and Behaviour** Inhabits lowland forests. Diet and reproductive habits unstudied. **Distribution** The Mentawai Archipelago and western Sumatra (Padang, Sumatera Barat Province). **Status** Not Evaluated.

DWARF REED SNAKE
Pseudorabdion longiceps PLATE 52
Measurement TL 230mm **Identification** Body slender; head indistinct from neck; snout pointed; nostril in single nasal; loreal absent; single preocular; single postocular; supralabials 5–6; Supralabials III–IV contact orbit; infralabials 5; single nasal; internasal separate from supralabials; eye small; tail short; dorsals smooth; midbody scale rows 15; ventrals 129–148; subcaudals 10–31, paired; anal entire. **Coloration** Dorsum iridescent black or brown; white or yellow collar and yellow spot above angle of mouth; venter dark brown. **Habitat and Behaviour** Inhabits lowland rainforests; records from plantations and rice fields may be of individuals deposited there during floods. Nocturnal and subfossorial. Diet consists of litter invertebrates, earthworms, and small insects and larvae. Oviparous; clutches comprise 2–3 narrow and elongated eggs. **Distribution** Southern Thailand (Narathiwat and Yala Provinces), Peninsular Malaysia, Singapore, west coast of Sumatra, Pulau Nias, the Mentawai and Riau Archipelagos, and Borneo. Also Sulawesi (Indonesia). **Status** Not Evaluated.

SARAWAK REED SNAKE
Pseudorabdion saravacensis NOT ILLUSTRATED
Measurement TL 142mm **Identification** Body slender; head indistinct from neck; snout pointed; nostril between 2 nasals; loreal present; preocular absent; single postocular; supralabials 5–6; Supralabials III–IV contact orbit; prefrontal not in contact with supralabials; infralabials 5; nasals enlarged, not divided; anterior genials not in contact with mentals; maxillary teeth 10–21; eye small; tail short, pointed; dorsals smooth; midbody scale rows 15; ventrals 113; subcaudals 26; anal entire. **Coloration** Dorsum iridescent dark brown; red blotch on sides of head above angle of jaws; irregular red band on neck; venter unpatterned brown. **Habitat and Behaviour** Inhabits lowland forests. Diet and reproductive habits unstudied. **Distribution** Endemic to Borneo (Kuching, Sungei Mengiong and Niah, Sarawak State). **Status** Not Evaluated.

MALAYAN KEELED RAT SNAKE
Ptyas carinata PLATE 52
Measurement TL 4m **Identification** Body slender; head distinct from neck; loreals 2–3; single preocular; postoculars 2; single subocular; supralabials 9 (rarely 8 or 10); Supralabials V–VI contact orbit; infralabials 11; 5 infralabials contact anterior chin shields; eye large; pupil rounded; tail long; dorsals smooth, except 2–4 median rows that are keeled; midbody scale rows 16 or 18; ventrals 208–215; subcaudals 110–118, paired; anal divided. **Coloration** Dorsum olive-brown to nearly black anteriorly, sometimes with indistinct yellow cross-bars; posterior dorsum yellow with distinct chequered black pattern, terminating in black tail with yellow spots; venter cream turning grey or black posteriorly. **Habitat and Behaviour** Inhabits lowland forests and agricultural fields. Diet consists of amphibians and rodents. Oviparous; clutches comprise 10 eggs. Incubation period ca 60 days. **Distribution** Myanmar, Thailand, Vietnam, Peninsular Malaysia, Singapore, Sumatra, Borneo and Java. Also southern China and Palawan (the Philippines). **Status** Not Evaluated.

BLACK-STRIPED RAT SNAKE
Ptyas dhumnades PLATE 52
Measurement TL 2.56m **Identification** Body robust; head distinct from neck; preoculars 2; postoculars 2; supralabials 8 (rarely 9); Supralabials IV–V contact orbit; infralabials 10; vertebral keel sharply defined; tail long; 2–6 middorsal rows keeled, other dorsals smooth; midbody scale rows 14; ventrals 186–197; subcaudals 105–124, paired; anal divided. **Coloration** Dorsum with longitudinal black stripes, 2 scales wide; anterior half of body grey, scales black edged with blue centres; posterior half of body slate-black; forehead slate-grey; chin and gular region cream; rest of venter dark grey, lateral tips with black edges. **Subspecies** Three poorly diagnosed subspecies have been described, of which one (*P. d. montana*) occurs in the region. **Habitat and Behaviour** Inhabits the lowlands to submontane limits at elevations of ca 1,500m asl.

Diurnal and terrestrial, associated with open plains, especially near water, and may enter human habitations. Diet consists of frogs. Oviparous; clutches comprise 13–17 eggs, measuring 35.9–65 x 17.9–30mm. **Distribution** Northern Vietnam (Sa Pá, Lao Cai Province and Tam Dao, Vinh Phúc Province). Also eastern China (*P. d. dhumnades* and *P. d. montana*) and Taiwan (*P. d. oshimai*). **Status** Not Evaluated.

WHITE-BELLIED RAT SNAKE
Ptyas fusca PLATE 52
Measurement TL 3m **Identification** Body slender; head distinct from neck; single preocular; postoculars 2; supralabials 9; Supralabials V–VI contact orbit; 5 infralabials contact anterior chin shields; eye large; pupil rounded; tail long; dorsals smooth; midbody scale rows 16; ventrals 183–198; subcaudals 160–179, paired; anal divided. **Coloration** Dorsum mid-brown, pinkish-brown or brownish-grey, typically nearly black anteriorly, lightening posteriorly; red vertebral stripe sometimes present; venter cream. **Habitat and Behaviour** Inhabits lowland forests to the mid-hills at elevations of up to 1,330m asl. Diet includes birds and probably small mammals. Oviparous. **Distribution** Thailand (Phangnga Province), Peninsular Malaysia, Singapore, Sumatra, Pulau Nias, Pulau Bangka, Pulau Belitung, the Natuna Archipelago and Borneo. **Status** Not Evaluated.

JAVAN RAT SNAKE
Ptyas korros PLATE 52
Measurement TL 2.68m **Identification** Body robust; head elongated, distinct from neck; loreals 2–3; single preocular; postoculars 2; supralabials 8; Supralabials IV–V contact orbit; 5 infralabials contact anterior chin shields; eye large; pupil rounded; tail long; dorsals smooth anteriorly, weakly keeled posteriorly; midbody scale rows 15; ventrals 160–187; subcaudals 120–151, paired; anal divided. **Coloration** Forehead and anterior of dorsum grey, reddish-brown or olive-brown; posterior of body darkening to nearly black; scales edged with white, becoming more distinct posteriorly, where they appear as white bands on black background; venter, chin and labials brownish-cream. **Habitat and Behaviour** Inhabits lowland and montane forests from sea level to 3,000m asl. Diurnal and terrestrial, but known to ascend trees to rest and to mate. Diet includes rodents, birds, lizards and frogs. Oviparous; clutches comprise 4–14 eggs, measuring 33.5–48 x 17.1–20.5mm. Incubation period 40–101 days. Hatchlings 290–367mm. **Distribution** Myanmar, Thailand, Laos, Cambodia, Vietnam, Peninsular Malaysia, Singapore, Sumatra, Borneo and Java. Also north-eastern India, Bangladesh and eastern China. **Status** Not Evaluated.

INDIAN RAT SNAKE
Ptyas mucosa PLATE 52
Measurement TL 3.7m **Identification** Body slender; head elongated, distinct from neck; loreals 2–3; single preocular; postoculars 2–3; supralabials 8; Supralabials IV–V contact orbit; infralabials 9–10; eye large; pupil rounded; tail long; dorsals smooth or weakly keeled; midbody scale rows 17; ventrals 187–213; subcaudals 95–146, paired; anal divided. **Coloration** Dorsum yellowish-, reddish- or olivaceous-brown to black; posterior of body with dark bands or reticulated pattern; hatchlings greenish-brown, typically with light bluish-grey cross-bands on anterior of body; venter greyish-white or yellow. **Habitat and Behaviour** Inhabits forested areas as well as agricultural fields, and even parks and gardens in cities, from sea level to 4,000m asl. Combat dances are often sighted, in which rival males are partially entwined with their forebodies raised. Diet consists of rats, frogs, bats, birds, lizards, turtles and other snakes. Oviparous; clutches comprise 5–25 eggs, measuring 42–69 x 22.5–28.5mm. Female guards eggs during incubation period, which lasts 60–95 days. Hatchlings 300–472mm. **Distribution** Myanmar, Thailand, Laos, Cambodia, Vietnam, Peninsular Malaysia, Singapore and Sumatra. Also Turkmenistan, Iran, Pakistan, India (including the Andamans Archipelago), Nepal, Sri Lanka, Bhutan, Bangladesh and southern China. **Status** Not Evaluated.

GREEN RAT SNAKE
Ptyas nigromarginata PLATE 52
Measurement TL 1,915mm **Identification** Body slender, slightly compressed; head elongated; single preocular; postoculars 2; supralabials 8; Supralabials IV–V contact orbit; infralabials 10; eye large; pupil rounded; tail long; dorsals smooth, 4–6 median scale rows distinctly keeled; midbody scale rows 17; ventrals 190–205; subcaudals 119–131, paired; anal divided. **Coloration** Dorsum soft velvety green or olive-green; dorsal scales black-edged; 4 longitudinal black stripes along body and tail in juveniles, which are confined in adults to posterior third of body; forehead reddish-brown with bright yellow patch on temporal region; venter with greenish-yellow tinge. **Habitat and Behaviour** Inhabits disturbed areas and open forests in the plains and mid-hills at elevations of up to 720m asl. Diurnal, and terrestrial and arboreal. Diet consists of rodents, birds, lizards and other snakes. Oviparous; clutches comprise 8–9 eggs. **Distribution** Myanmar and Vietnam. Also north-eastern India, Nepal and southern China (Sichuan Province). **Status** Not Evaluated.

TWO-COLOURED FOREST SNAKE
Rhabdops bicolor NOT ILLUSTRATED
Measurement TL 600mm **Identification** Body robust, cylindrical, elongated; head indistinct from neck, depressed; snout rounded; nostrils dorsolateral; eye small; loreal present; preoculars 1–2; postoculars 2–3; supralabials 5; Supralabial III contacts orbit, or supralabials separated from orbit by suboculars; dorsals smooth; midbody scale rows 17; ventrals 187–214; subcaudals 63–77; anal divided. **Coloration** Dorsum dark brown or black; venter yellowish-cream, the two colours forming a line of demarcation; subcaudals cream, unpatterned or with

black spots. **Habitat and Behaviour** Inhabits the forested mid-hills. Terrestrial or subfossorial. Diet includes earthworms and slugs. Reproductive habits unstudied. **Distribution** Myanmar. Also north-eastern India and southern China (Yunnan Province). **Status** Not Evaluated.

KHASI HILLS TRINKET SNAKE
Rhadinophis frenatus PLATE 53
Measurement TL 1,500mm **Identification** Body slender and compressed; snout elongated, tip pointed and slightly arched forwards; loreal fused with prefrontal; single preocular; postoculars 2; supralabials 8–9; Supralabials III–V or IV–VI contact orbit; infralabials 9–11; eye large; pupil rounded; tail long and prehensile, tapering to a point; ventral keels developed; dorsals weakly keeled, except outer 2–3 rows; midbody scale rows 17 or 19; ventrals 198–235; subcaudals 119–149; anal divided. **Coloration** Dorsum grass-green to olive; supralabials and infralabials light green; black postocular stripe to angle of jaws; venter pale green to white, darkening posteriorly; iris golden-yellow. **Habitat and Behaviour** Inhabits evergreen and submontane forests at elevations of 550–2,000m asl. Diurnal and arboreal, inhabiting shrubs and low trees, in addition to bamboo groves. Diet consists of lizards, rats and birds, and prey is killed by constriction. Oviparous; clutches comprise 9 eggs, measuring 28 x 15mm. **Distribution** Northern Vietnam (Lao Cai, Lang Son, Vinh Phúc and Quang Binh Provinces). Also north-eastern India and eastern China. **Status** Not Evaluated.

GREEN TRINKET SNAKE
Rhadinophis prasinus PLATE 53
Measurement TL 1,200mm **Identification** Body slender; snout long; head slightly distinct from neck; snout rounded; single loreal; single preocular; postoculars 2; supralabials 8–9; Supralabials IV–VI contact orbit; infralabials 7–11; dorsals weakly keeled, except outer 2–3 rows; ventral keels well developed; eye large; pupil rounded; tail long and slender; midbody scale rows 19 (rarely 17); ventrals 186–209; subcaudals 91–111, single; anal entire or divided. **Coloration** Dorsum uniformly green or turquoise, tail-tip brown; labials green or yellowish-green; venter pale green; faint dark postocular stripe; iris yellow; tongue reddish-brown. **Habitat and Behaviour** Inhabits submontane and montane forested hills, typically at elevations of 850–2,650m; lowland records (80m asl) also known. Diurnal and arboreal, associated with bamboo nodes, as well as with roofs of thatched houses. Diet includes lizards, mammals and birds, which are killed by constriction. Oviparous; clutches comprise 5–9 eggs, measuring 31–39 x 16–19mm. Incubation period 56–62 days. Hatchlings 230mm. **Distribution** Myanmar, northern Thailand, Laos, northern Vietnam and Peninsular Malaysia. Also eastern and north-eastern India, and southern and eastern China. **Status** Not Evaluated.

VIETNAMESE HORNED SNAKE
Rhynchophis boulengeri PLATE 53
Measurement TL 1,380mm **Identification** Body slender, slightly compressed; head distinct from neck; loreal present; single preocular; postoculars 2–3; supralabials 8–10; Supralabials IV–VI or V–VII contact orbit; infralabials 10–11; eye large; pupil rounded; fleshy, scale-covered appendage on snout-tip; dorsals smooth; midbody scale rows 19; ventrals 207–216; subcaudals 123–138, paired; anal divided. **Coloration** Dorsum pale green or greyish-cream; interstitial skin on sides of body dark blue and white; white line on lower flanks; forehead and rostral appendage dark green or greyish-cream; supralabials yellow or pale green; narrow stripe between nostril and orbit; narrow dark postocular stripe to level of last supralabial; venter pale green; juveniles pale brown; iris brown, pupil with golden ring. **Habitat and Behaviour** Inhabits primary forests. Diurnal and arboreal. Diet unstudied. Oviparous; clutches comprise 6 eggs, measuring 8 x 35mm. **Distribution** Vietnam. Also eastern China. **Status** Not Evaluated.

TWIN-STREAKED BLACK-HEADED SNAKE
Sibynophis bistrigatus PLATE 53
Measurement TL 300mm **Identification** Body slender, cylindrical; head relatively short, slightly distinct from neck and flattened; loreal small; supralabials 9; Supralabials IV–VI contact orbit; eye large; pupil rounded; dorsals smooth; midbody scale rows 17; ventrals 184–186; subcaudals 73–75, paired; anal divided. **Coloration** Dorsum pale reddish-brown or tan; series of black spots on vertebral region; black lateral stripe on scale rows 4–5; forehead and nape black, edged on head by yellow; labials yellow; paired yellow spots on nuchal region; venter unpatterned yellow. **Habitat and Behaviour** Nothing known of its natural history. **Distribution** Endemic to southern Myanmar (Pyay or Pyè, Bago Division). Record from Kamorta, in the Nicobar Archipelago, India, requires verification. **Status** Not Evaluated.

CHINESE BLACK-HEADED SNAKE
Sibynophis chinensis PLATE 53
Measurement TL 694mm **Identification** Body slender, cylindrical; head relatively short, slightly distinct from neck and flattened; nasal divided; postoculars 2; supralabials 9–10; Supralabials II–V contact orbit; infralabials 9; eye large; pupil rounded; dorsals smooth; midbody scale rows 17; ventrals 168–183; subcaudals 98–122, paired; anal entire or divided. **Coloration** Dorsum pale tan, greyish-brown or dark brown; forehead black; yellow or cream nuchal collar; venter cream with dark spot at posterior edge of ventrals. **Habitat and Behaviour** Inhabits submontane forests at elevations of ca 1,219m asl. Nocturnal and terrestrial, hiding in tall grass and under rocks and logs by day. Diet consists of skinks and lacertids (*Takydromus* spp.). Oviparous; clutches comprise 2–4 eggs. **Distribution** Northern Vietnam (Dong Tam-ve, at present, Quang Tri, Quang Tri Province). Also eastern China. **Status** Not Evaluated.

COLLARED BLACK-HEADED SNAKE
Sibynophis collaris PLATE 53
Measurement TL 760mm **Identification** Body slender, cylindrical; head relatively short, slightly distinct from neck and flattened; loreal present; single preocular; postoculars 2; supralabials 10 (rarely 9 or 11); Supralabials IV–VI contact orbit; infralabials 9; eye large; pupil rounded; dorsals smooth; midbody scale rows 17; ventrals 155–188; subcaudals 92–125, paired; anal divided. **Coloration** Dorsum brown or greyish-brown; black vertebral stripe comprising black spots; occasionally a light, dotted dorsolateral line; dark cross-bar behind eye and one on forehead; transverse black band on nape bordered posteriorly with yellow; venter yellow, ventral with outer black spot, forming dotted dark line on either side. **Habitat and Behaviour** Inhabits forests from the plains to montane limits at an elevation of 3,050m asl. Diurnal and terrestrial. Diet consists of skinks, snakes, frogs and insects. Oviparous; clutches comprise 4–6 eggs, measuring 30 x 10mm. **Distribution** Myanmar, northern and western Thailand, Laos, Cambodia, Vietnam and Peninsular Malaysia (Cameron Highlands and Gunung Tahan, Pahang State). Also northern and north-eastern India, Bangladesh and Nepal. **Status** Not Evaluated.

STRIPED BLACK-HEADED SNAKE
Sibynophis geminatus PLATE 53
Measurement TL 1,000mm **Identification** Body slender, cylindrical; head relatively short, slightly distinct from neck and flattened; single preocular; postoculars 2; supralabials 7–9; Supralabials IV–VI, III–IV or III–V contact orbit; maxillary teeth 35–48; eye large; pupil rounded; dorsals smooth; midbody scale rows 17; ventrals 140–183; subcaudals 73–145, paired; anal divided. **Coloration** Dorsum orange-yellow, darker caudally; bright orange nuchal collar present or absent; black, bluish-grey or dark brown vertebral stripe, sometimes edged on either side by red or reddish-brown stripe, 2.5 scales wide; forehead rusty-brown with small, dark-edged yellow spots; supralabials yellow, sutures black; venter greenish-yellow anteriorly, turning pale green caudally; outer edges of ventrals with black spot, forming narrow stripe along lower flanks. **Subspecies** Two subspecies are recognized. *S. g. geminatus*, maxillary teeth 35–48; supralabials 9 (rarely 8); ventrals 144–183 and light nuchal bar present. *S. g. insularis*, maxillary teeth 33; supralabials 7–8; ventrals 140 and light nuchal bar absent. **Habitat and Behaviour** Inhabits lowland forests and offshore islands at elevations of under 400m asl. Diurnal and terrestrial. Diet consists of skinks. Oviparous; clutches comprise 1 egg, measuring 26.5 x 9mm. **Distribution** Southern Thailand, Peninsular Malaysia, Singapore, Sumatra, Pulau Siberut, Pulau Nias, Pulau Bangka, Pulau Belitung, Borneo, Java and Bali. Also Tawi-Tawi, the Sulu Archipelago (the Philippines) (*S. g. geminatus*) and Pulau Weh, off northern Sumatra (*S. g. insularis*). **Status** Not Evaluated.

WHITE-LIPPED BLACK-HEADED SNAKE
Sibynophis melanocephalus PLATE 53
Measurement TL 600mm **Identification** Body slender, cylindrical; head relatively short, slightly distinct from neck and flattened; loreal present; single preocular; postoculars 2; supralabials 9; Supralabials IV–VI contact orbit; infralabials 9; eye large; pupil rounded; dorsals smooth; midbody scale rows 17; ventrals 145–177; subcaudals 132–136, paired; anal divided. **Coloration** Dorsum reddish-brown, grey or brown with short black cross-bars over lighter band; lower flanks sometimes with small yellow spots; forehead dark olive with olive-yellow spots; supralabials white, sutures with black stripes; venter yellow with orange tinge on sides posteriorly; ventrals with rounded black spot laterally. **Habitat and Behaviour** Inhabits lowland forests. Diurnal and terrestrial, associated with leaf litter on the forest floor and with streams. Caudal autotomy known. Diet consists of skinks. Reproductive habits unstudied. **Distribution** Southern Thailand (Nakhon Si Thammarat, Pattani and Narathiwat Provinces), Vietnam, Peninsular Malaysia, Singapore, Sumatra and Borneo. **Status** Not Evaluated.

TRIANGLED BLACK-HEADED SNAKE
Sibynophis triangularis PLATE 53
Measurement TL 700mm **Identification** Body slender, cylindrical; head relatively short, slightly distinct from neck, and flattened; loreal present; single preocular; postoculars 2; supralabials 10; Supralabials IV–VI contact orbit; infralabials 9; eye large; pupil rounded; dorsals smooth; midbody scale rows 17; ventrals 160–189; subcaudals 113–124, paired; anal divided. **Coloration** Dorsum greyish-brown anteriorly, darker posteriorly, with a series of pale brown spots forming a dorsolateral stripe, most distinct anteriorly; forehead greyish-brown with 2 blackish-brown cross-bars, one across eyes, the second over temporals; yellowish-cream stripe across snout and supralabials widens posteriorly to form inverted chevron at back of neck; venter white, each ventral and subcaudal with a series of black dots on posterior edges. **Habitat and Behaviour** Inhabits hilly evergreen forests at elevations of up to 500m asl. Diurnal and terrestrial. Diet and reproductive habits unstudied. **Distribution** North-eastern and south-eastern Thailand (Chaiya-pum and Chon Buri Provinces), and eastern Cambodia (O'Rang District). Suspected to occur in southern Vietnam. **Status** Not Evaluated.

BORNEAN BLACK SNAKE
Stegonotus borneensis PLATE 53
Measurement TL 1,370mm **Identification** Body robust, cylindrical; head distinct from neck; distinct vertebral ridge; supralabials 9; Supralabials IV–V contact orbit; eye small; pupil vertical; dorsals smooth; vertebral scales enlarged; midbody scale rows 17; ventrals 194–233; subcaudals 60–79, paired. **Coloration** Dorsum grey-black, dark brown or black without markings; supralabials grey with pink tinge; each scale on venter alternately banded with dark and

light grey. **Habitat and Behaviour** Reported from the low hills up to submontane limits at an elevation of 1,800m asl. Active both by day and by night, and terrestial. When threatened, it exudes a secretion with an unpleasant odour from its cloacal glands. Diet unstudied. Oviparous. **Distribution** Endemic to north-western and northern Borneo (Kapit, Sarawak State and Gunung Kinabalu, Sabah State). An individual belonging to this genus was found in southern Thailand. **Status** Not Evaluated.

FRUHSTORFER'S MOUNTAIN SNAKE
Tetralepis fruhstorferi PLATE 53
Measurement TL 500mm **Identification** Body robust, cylindrical; head small, indistinct from neck; single preocular; postoculars 2; supralabials 4; Supralabial III contacts orbit; eye small; pupil rounded; tail short; dorsals smooth; midbody scale rows 15; ventrals 187–201; subcaudals 43–56, paired; anal divided. **Coloration** Dorsum dark reddish-brown; black vertebral stripe; venter bluish- or reddish-grey, each ventral with 2 dark brown spots forming 2 longitudinal rows of stripes. **Habitat and Behaviour** Inhabits submontane forests at an elevation of 1,200m asl. Nothing known of its natural history. **Distribution** Endemic to Java (Gunung Tengger). **Status** Not Evaluated.

ORNATE BROWN SNAKE
Xenelaphis ellipsifer PLATE 53
Measurement TL 2.51m **Identification** Body robust; head distinct from neck; snout rounded; suboculars 2; single preocular; postoculars 2; supralabials 8–9; Supralabial IV or V contacts orbit; infralabials 10; 5 infralabials contact anterior chin shields; eye large; pupil rounded; tail long; dorsals smooth; midbody scale rows 17; ventrals 186–203; subcaudals 124–134, paired; anal divided. **Coloration** Dorsum orange-red with 18–20 large elliptical or squarish brown blotches, black-edged (1–2 scale rows wide), separated by narrow cream interspaces on body; forehead unpatterned reddish-brown; supralabials yellow with black markings; infralabials cream, lacking markings; longitudinal dark streaks on neck; flanks with inverted Y- or V-shaped marking; venter pink or white with dark spots on outer edges of ventrals. **Habitat and Behaviour** Inhabits primary forests at elevations of 800–1,000m asl. Nocturnal and aquatic. Diet presumably consists of fish. Reproductive habits unstudied. **Distribution** Peninsular Malaysia (Sungai Gombak, Selangor State), Sumatra and Borneo (Gunung Kinabalu, Sabah State and Gunung Penrissen, Sarawak State). **Status** Not Evaluated.

MALAYAN BROWN SNAKE
Xenelaphis hexagonotus PLATE 53
Measurement TL 2m **Identification** Body robust; head distinct from neck; snout rounded; loreal present; single preocular; postoculars 2; supralabials 7–8; Supralabial IV contacts orbit, or separated by a subocular; infralabials 10; 4–5 infralabials contact anterior chin shields; eye large; pupil rounded; tail long; vertebral scales enlarged, hexagonal in shape; dorsals smooth; midbody scale rows 17; ventrals 185–198; subcaudals 140–179, paired; anal divided. **Coloration** Dorsum brown, dark green or greenish-olive; narrow vertical black bars from neck and along body, the apices reaching lower flanks; juveniles uniformly light brown; dark forehead contrasts strongly with pale sides of head and throat; venter pale or deep yellow, ventrals with dark spot on margins. **Habitat and Behaviour** Inhabits lowland forests, especially coastal peat swamps, and also mangrove swamps. Nocturnal and terrestrial, associated with waterlogged forests. Diet consists of rodents, frogs and fish. Reproductive habits unstudied. **Distribution** Southern Myanmar, southern Thailand (Surat Thani, Phattalung and Narathiwat Provinces), southern Vietnam, Peninsular Malaysia, Singapore, Sumatra, Pulau Bangka, Pulau Belitung, Borneo and Java. **Status** Not Evaluated.

Family VIPERIDAE VIPERS AND PIT VIPERS

Species with haemotoxic venom, a broad, triangular head, and hinged fangs that are folded back against the palate when not in use and adapted for deep penetration of prey. The elongated ectopterygoid serves as a lever to erect or depress the fangs. The body is mostly heavy and robust, and the dorsal scales are heavily keeled. A deep pocket between the nostril and eye has a highly vascular membrane stretched across it, and is supplied with numerous heat-sensitive receptors. These permit detection of prey via thermal cues, thus permitting hunting of warm-blooded prey in darkness. The Viperidae live in a variety of habitats, typically tropical and temperate forests. Their diet consists of warm-blooded prey, though arthropods are occasionally consumed. Many are of medicinal importance and responsible for a number of bites annually in the region. They are generally ovoviviparous, though oviparity is shown in several lineages. They are cosmopolitan, and are absent only in Australia.

In some works the pit vipers are placed in a separate family (the Crotalidae) from the so-called 'typical' vipers. Here pit vipers are treated as a subfamily within the Viperidae. In recent years a number of lineages have been recognized within the green pit vipers that were referred to the genus *Trimeresurus*. This taxonomy has been followed in the present work. A taxonomic rearrangement – the recognition of the South-East Asian *Daboia* (*D. siamensis*) as distinct from that on the Indian subcontinent (*D. russelii*) – came too late to be included in this work.

MALAYAN PIT VIPER
Calloselasma rhodostoma PLATE 54
Measurement TL 1,450mm **Identification** Body robust; head distinct from neck; snout acute and slightly upturned; preoculars 2; postoculars 2–3; supralabials 6–10; Supralabials III–IV contact large subocular; infralabials 9–14; vertebral ridge distinct; eye small; pupil vertical; tail thin and short; dorsals smooth; midbody scale rows 21 or 23; ventrals 142–163; subcaudals 34–55; anal entire. **Coloration** Dorsum reddish-brown or purplish-brown; flanks paler and speckled with dark brown; series of 19–31 subtriangular dark brown marks on each side, their apices meeting or alternating; head dark brown; dark postocular patch, scalloped ventrally, extends beyond angle of jaws; venter pinkish-cream mottled with brown; neonates dark with white tail. **Habitat and Behaviour** Inhabits dry lowland forests and plantations up to submontane limits at an elevation of 1,524m asl. Nocturnal and terrestrial, associated with dense undergrowth as well as rocky biotope. Threat display consists of dorsoventral flattening of the body followed by striking. Juveniles have white tail-tips, which are used as a lure to attract prey. Diet consists of small rodents and birds. Oviparous; clutches comprise 13–30 eggs. Females coil around the eggs during the duration of incubation, which lasts 30–47 days. Hatchlings 148–222mm. **Distribution** Thailand, Laos, eastern and south-western Cambodia (Pichrada District and Cardamom Mountains), Vietnam, northern Peninsular Malaysia (Kedah and Perlis States), Java, Pulau Karimundjawa and Pulau Kangean. **Status** Not Evaluated.

WHITE-LIPPED PIT VIPER
Cryptelytrops albolabris PLATE 54
Measurement TL 1,040mm **Identification** Body robust; head relatively long, distinct from neck; forehead scales small; supraoculars narrow; supralabials 7–12; Supralabial I partially or completely fused to nasal; temporals smooth; eye large; pupil vertical; tail short, prehensile; dorsals smooth or weakly keeled; midbody scale rows 21; ventrals 149–186; subcaudals 48–78, paired; anal typically entire. **Coloration** Dorsum and forehead bright green or lime-green; sides of head below eye, including labials, yellow, white or pale green, lighter than rest of head; tail reddish-brown; iris yellow; males with white stripe on first row of dorsal scales; females unstriped; venter green, cream or yellowish-cream. **Habitat and Behaviour** Inhabits deciduous, subtropical and temperate forests near streams, including disturbed habitats such as the vicinity of agricultural fields, at elevations of 60–3,050m asl. Nocturnal and arboreal, associated with trees and saplings, and occasionally seen on the ground. Diet consists of rodents, birds, lizards and frogs. Ovoviviparous; clutches comprise 7–17 neonates, measuring 120–195mm. **Distribution** Myanmar, Thailand, Laos, eastern and south-western Cambodia (O'Rang District and Cardamom Mountains), Vietnam and south-eastern Sumatra. Also Bangladesh, eastern and north-eastern India, the Nicobar Islands, southern and eastern China, and Sulawesi (Indonesia). **Status** Not Evaluated.

SPOT-TAILED PIT VIPER
Cryptelytrops erythrurus PLATE 54
Measurement TL 1,050mm **Identification** Body slender; head relatively long, narrower than neck; small scales on forehead; supralabials 9–13; Supralabial I partially or completely fused to nasal; 1–2 rows of scales separate supralabials from subocular; temporal scales small, strongly keeled; eye large; pupil vertical; tail short, prehensile; dorsals strongly keeled; midbody scale rows 23 or 25; ventrals 151–174; subcaudals 49–79, typically paired; anal entire. **Coloration** Dorsum and head dark green; labials yellow; iris yellow; males with white dorsolateral stripe from posterior edge of eye to tail; stripe indistinct or absent in females; venter pale green or yellow; tip of tail usually spotted or mottled with brown. **Habitat and Behaviour** Inhabits lowland rainforests and moist deciduous forests at elevations of under 200m asl. Nocturnal and arboreal, associated with trees and the ground near streams. Diet consists of rodents, birds and lizards. Ovoviviparous; clutches comprise 10 neonates, measuring 231–260mm. **Distribution** Myanmar. Also Bangladesh, Nepal, and eastern and north-eastern India. **Status** Not Evaluated.

HON SON PIT VIPER
Cryptelytrops honsonensis NOT ILLUSTRATED
Measurement TL 648mm **Identification** Body slender; head distinct from neck; supralabials 10–11; infralabials 12; scales across forehead 9–12; smooth occipital and temporal scales; internasals in contact; Supralabials III and IV do not contact subocular scale; eye large; pupil vertical; tail short, prehensile; dorsals keeled; midbody scale rows 21; ventrals 183–186; subcaudals 54–74. **Coloration** Dorsum straw-yellow, with ca 92 irregular dark brown bands; reddish-brown markings ventrolaterally; caudal bands irregular; forehead brown with irregular darker markings; iris orange centrally, brown peripherally; no ventrolateral stripe; postorbital stripes absent; tail orange; venter dull white anteriorly, turning darker posteriorly and dark grey with dense stippling beneath tail. **Habitat and Behaviour** Restricted to a single granitic island, ca 22 sq km, with little or no primary vegetation, at an elevation of 100m asl. Arboreal and apparently crepuscular. Associated with rocks and bamboo. Diet and reproductive habits unstudied. **Distribution** Endemic to southern Vietnam (Hon Son, Kien Giang Province). **Status** Not Evaluated.

LESSER SUNDA WHITE-LIPPED PIT VIPER
Cryptelytrops insularis PLATE 54
Measurement TL 930mm **Identification** Body slender; head distinct from neck; forehead scales granular; postoculars 2–3; supralabials 8–12; subocular sometimes in contact with Supralabial III; temporals with short keel; eye large; pupil vertical; tail short, prehensile; dorsals keeled; midbody scale rows

21 (rarely 19); ventrals 153–168; subcaudals 53–81; anal entire. **Coloration** Dorsum typically bright green, sometimes olive or blue, with transverse dark bands; supralabials yellow to greenish-white; tail with rust-coloured streak; venter greenish-yellow, greenish-white or light blue; iris brown or red. **Habitat and Behaviour** Inhabits forests at elevations from sea level to 880m asl. Nocturnal and arboreal, associated with trees up to 15m above substratum. Diet consists of skinks, geckos, rodents and frogs. Ovoviviparous; clutches comprise 11–17 neonates, measuring 140–195mm. **Distribution** Bali. Also Komodo, Rinca, Adonara, Lembata, Pantar, Alor, Roti, Semau, Wetar, Romang, Flores, Kisar, Lombok, Sumba, Sumbawa and Timor. **Status** Not Evaluated.

KANBURI PIT VIPER
Cryptelytrops kanburiensis PLATE 54
Measurement TL 667mm **Identification** Body slender; head distinct from neck; supraoculars irregular and indented; temporals keeled; single preocular; postoculars 2–3; supralabials 9–11; infralabials 11–13; eye large; pupil vertical; tail short, prehensile; dorsals keeled, except for lowest row; midbody scale rows 19; ventrals 155–178; subcaudals 42–74, paired; anal entire. **Coloration** Dorsum olive or greyish-green; head with dull brown or orange-brown blotches; white or blue spots on labials and on first dorsal scale row; broad dark postocular stripe; venter cream; some ventrals with brown edges, especially in males; iris light orange with brown speckles. **Habitat and Behaviour** Inhabits evergreen forests in mid-elevation limestone hills, especially those dominated by bamboo. Diurnal and possibly also nocturnal, and terrestrial. Diet and reproductive habits unstudied. **Distribution** Endemic to western Thailand (Kanchanaburi Province). **Status** Not Evaluated.

LARGE-EYED PIT VIPER
Cryptelytrops macrops PLATE 54
Measurement TL 710mm **Identification** Body slender; head distinct from neck; internasals typically in contact (rarely separated by 1–3 scales); supraoculars wide; supralabials 9–12; supralabials separated from orbit by row of small scales; infralabials 10–13; temporals keeled; eye large; pupil vertical; tail short, prehensile; dorsals keeled, except first row; midbody scale rows 21 (rarely 19); ventrals 160–177; subcaudals 49–74, paired; anal entire. **Coloration** Dorsum pale green, sometimes with pale blue lateral stripe along first dorsal scale row; labials pale bluish-green; forehead pale green, juveniles and males with white postocular stripe; tail reddish-brown; chin and gular region bluish-white; venter pale blue; subcaudals pale blue; iris golden-yellow in females, reddish-orange in males. **Habitat and Behaviour** Inhabits evergreen hill forests at elevations of 200–500m asl. Nocturnal and crepuscular, and arboreal, associated with trees. Diet consists of birds, small mammals, lizards and tree frogs. Ovoviviparous; clutches comprise 6–12 neonates. **Distribution** Central and south-eastern Thailand, Laos, eastern and south-western Cambodia (O'Rang and Siem Pang Districts and Cardamom Mountains) and southern Vietnam (Dong Nai, Kien Giang and Ca Mau Provinces). **Status** Not Evaluated.

MANGROVE PIT VIPER
Cryptelytrops purpureomaculatus PLATE 54
Measurement TL 1,040mm **Identification** Body slender in males, stout in females; head distinct from neck; supralabials 11–13; supraoculars narrow, sometimes fragmented, separated by 12–15 scales; forehead scales small, tuberculate or granular; temporal scales keeled; eye large; pupil vertical; tail short, prehensile; dorsals keeled; midbody scale rows 25–27; ventrals 160–183; subcaudals 56–76, paired. **Coloration** Dorsum variable, olive, grey or brownish-purple, with darker mottling; forehead olive with intense brown mottling; venter cream-green or brown, unpatterned or spotted with brown; pale line on scale row bordering ventrals sometimes present. **Habitat and Behaviour** Inhabits mangrove forests and coastal swamp forests. Crepuscular and arboreal, associated with shrubs and trees. Diet includes small mammals. Ovoviviparous; clutches comprise up to 15 neonates. **Distribution** Myanmar, peninsular and central Thailand, Peninsular Malaysia, Singapore and western Sumatra (Indragiri, Riau Province). **Status** Not Evaluated.

BEAUTIFUL PIT VIPER
Cryptelytrops venustus PLATE 54
Measurement TL 580mm **Identification** Body slender; head distinct from neck; supraoculars less irregular and indented than in Kanburi Pit Viper (*C. kanburiensis*); temporals and supralabials keeled; supralabials 9–11; infralabials 10–13; postoculars 1–4; eye large; pupil vertical; tail short, prehensile; dorsals keeled; midbody scale rows 21; ventrals 166–183; subcaudals 50–72. **Coloration** Dorsum dull olive or bluish-green in males, grass-green in females; head with dull brown blotches; body with rusty or dull brown bands; labials green with at least 1 brown patch; irregular, broad dark postorbital stripe; pale spot on first dorsal scale row; venter pale brown; some ventrals with dark brown edges; iris light orange with brown speckling. **Habitat and Behaviour** Inhabits limestone hills at elevations of up to ca 700m asl. Diurnal and possibly also nocturnal, and terrestrial, associated with shaded humid valleys. Diet and reproductive habits unstudied. **Distribution** Southern Thailand (Nakhon Si Thammarat, Surat Thani and Krabi Provinces) and northern Peninsular Malaysia (Pulau Langkawi, Kedah State). **Status** Not Evaluated.

SHARP-NOSED VIPER
Deinagkistrodon acutus PLATE 54
Measurement TL 1,490mm **Identification** Body robust; head distinct from neck; snout upturned with rostral raised upwards; preoculars 2; postoculars 2; supralabials 7; infralabials 11; 3 enlarged lower temporals; eye large; pupil vertical; tail-tip acute; dorsals keeled; scale tips bituberculate; midbody scale

rows 21; ventrals 157–174; subcaudals 51–61, paired or undivided; anal entire. **Coloration** Dorsum grey or brown with series of 15–23 pairs of large, subtriangular dark marks on each side that meet or alternate; narrow dark brown or black postocular stripe; venter cream mottled with grey and with distinct black spots. **Habitat and Behaviour** Inhabits forests, agricultural areas and the vicinity of human settlements, from the lowlands to submontane limits, at elevations of 100–1,500m asl. Nocturnal and terrestrial, associated with rocky biotope. Diet consists of small mammals, birds and frogs. Oviparous; clutches comprise 15–35 eggs, measuring 45–55 x 25–30mm, which are guarded by the female. Hatchlings ca 214–300mm. **Distribution** Northern Vietnam (Sa Pá, Lao Cai Province, Lai Chau Province, Cao Bang Province, Mau Son, Lang Son Province and Tham Dao Vinh Phúc Province). Also eastern China. **Status** Not Evaluated.

KINABALU BROWN PIT VIPER ☠
Garthius chaseni PLATE 55
Measurement TL 690mm **Identification** Body robust; head triangular, broad and flattened, distinct from neck; snout blunt; supralabials 6; supralabials separated from orbit by 2 rows of suboculars; eye small; pupil vertical; tail not prehensile; dorsals strongly keeled anteriorly, weakly keeled posteriorly; midbody scale rows 17 or 19; ventrals 131–143; subcaudals 20–30, paired; anal entire. **Coloration** Dorsum dark or reddish-brown with irregular dark brown blotches in paired rows anteriorly, joining to form bands towards posterior; smaller dark blotches on lower flanks; dark postocular stripe, edged ventrally by line of cream or yellow scales, extends to neck; venter yellow with large areas of grey. **Habitat and Behaviour** Inhabits submontane forests at elevations of 915–1,550m asl. Nocturnal and terrestrial, associated with leaf litter. Diet and reproductive habits unstudied, although it is known to be ovoviviparous. **Distribution** Endemic to northern Borneo (Gunung Kinabalu and Crocker Range, Sabah State). **Status** Not Evaluated.

MALAYAN BROWN PIT VIPER ☠
Ovophis convictus PLATE 55
Measurement TL 710mm **Identification** Body robust; head triangular, broad and flattened, distinct from neck; supralabials 8–9; supralabials separated from orbit by 2 rows of suboculars; eye small; pupil vertical; tail not prehensile; dorsals keeled; midbody scale rows 21–25; ventrals 133–138; subcaudals 22–34, paired; anal entire. **Coloration** Dorsum mid-brown or brownish-yellow; subrectangular dark brown or black blotches from nape to mid-tail; forehead greyish-black or dark brown; yellow-brown or pale tan postocular stripe to neck; venter cream with brown mottling. **Habitat and Behaviour** Inhabits submontane and montane forests at elevations of ca 1,737m asl. Nocturnal and terrestrial. Diet consists of small mammals. Ovoviviparous; clutches comprise 21 neonates. **Distribution** Southern Thailand, Cambodia, southern Laos, Vietnam, Peninsular Malaysia (Pulau Pinang, Pinang State, and Cameron Highlands and Genting Highlands, Pahang State), Singapore and Sumatra. **Status** Not Evaluated.

MONTANE PIT VIPER ☠
Ovophis monticola PLATE 55
Measurement TL 1,250mm **Identification** Body robust; head triangular, broad and flattened, distinct from neck; supraoculars large, entire, separated by 5–9 scales; supralabials 7–10; Supralabial I separated from orbit; large internasals separated by single scale; eye small; pupil vertical; tail not prehensile; dorsals smooth or weakly keeled; midbody scale rows 21–25; ventrals 127–157; subcaudals 32–64, single or paired; anal entire. **Coloration** Dorsum brownish-grey or yellowish-pink with squarish dark brown blotches; lateral series of smaller dark brown spots; forehead darker, with yellow, brown or pale tan postocular stripe reaching neck; venter dusted with dark brown. **Subspecies** Three subspecies are known, of which one (*O. m. monticola*) occurs in the region. **Habitat and Behaviour** Inhabits montane and submontane evergreen forests, including coniferous forests, at elevations of 700–1,200m asl. Nocturnal and crepuscular, and terrestrial. Diet includes rats and mice, as well as birds and their eggs, frogs and lizards. Oviparous; clutches comprise 5–18 eggs, measuring 18–20mm, which are guarded by the female. Hatchlings 180–200mm. **Distribution** Myanmar (*O. m. monticola*). Also eastern and north-eastern India, Nepal, Bangladesh and southern China (Yunnan Province) (*O. m. monticola*), eastern China, Taiwan (*O. m. makazayazaya*) and Tibet (China) (*O. m. zayuensis*). **Status** Not Evaluated.

TONKIN PIT VIPER ☠
Ovophis tonkinensis PLATE 55
Measurement TL 561mm **Identification** Body robust; head triangular, broad and flattened, distinct from neck; preoculars 3; postoculars 2; supralabials 10; supralabials separated from orbit by 2 rows of suboculars; infralabials 13; eye small; pupil vertical; tail not prehensile; dorsals smooth; midbody scale rows 21; ventrals 127–144; subcaudals 36–55, paired; anal entire. **Coloration** Dorsum grey, tan or pale brown with indistinct brown mottling and large, black-edged dark blotches; on flanks, smaller brown blotches; white stripe from rostral, across orbit, to neck; labials edged with black; yellow, brown or pale tan postocular stripe to neck; venter cream with brown marbling; tail reddish-brown. **Habitat and Behaviour** Inhabits submontane forests at elevations of 900–1,200m asl. Nocturnal and terrestrial, commonly encountered on stream banks. Diet consists of frogs and small mammals (including rodents and insectivores). Oviparous; clutches comprise 5–7 eggs. Incubation period 38–42 days. Hatchlings 195–210mm. **Distribution** Endemic to northern Vietnam (Bac Ky or Dông Kinh and Tam Dao, Vinh Phú Province). **Status** Not Evaluated.

HAGEN'S GREEN PIT VIPER ☠
Parias hageni   PLATE 55
Measurement TL 1,160mm **Identification** Body robust; head distinct from neck; forehead scales smooth; supralabials 9–12; subocular separates Supralabial V from orbit; infralabials 10–15; internasal typically absent; tail prehensile; dorsals weakly keeled; midbody scale rows 21; ventrals 176–198; subcaudals 63–89; anal entire. **Coloration** Dorsum bright or pale green; forehead and body scales not edged with black; pale line, bordered ventrally by dark line or series of dark spots, extends along dorsal rows 1–2; labials, chin and gular regions pale green; venter pale green, ventrals of some individuals with dark edges. **Habitat and Behaviour** Inhabits tropical forests in the lowlands and mid-hills at an elevation of up to 600m asl. Nocturnal and arboreal, associated with trees and stout saplings. Diet consists of birds and small mammals. Reproductive habits unstudied. **Distribution** Southern Thailand (Surat Thani, Trang and Songkhla Provinces), Peninsular Malaysia, Singapore, Sumatra, Pulau Bangka, Pulau Nias, Pulau Simeulue and possibly the Mentawai Archipelago. **Status** Not Evaluated.

KINABALU GREEN PIT VIPER ☠
Parias malcolmi   PLATE 55
Measurement TL 1,330mm **Identification** Body robust; head narrow, distinct from neck; forehead scales smooth; preoculars 3; postoculars 3; supralabials 8–9; infralabials 8–19; internasal present; eye small; pupil vertical; tail prehensile; dorsals weakly keeled; midbody scale rows 19; ventrals 168–174; subcaudals 61–81. **Coloration** Dorsum green with indistinct black bands on posterior part of scale, creating indistinct bands; forehead black with green-centred scales; juveniles with white stripe along flanks; in adults, flank stripe absent; venter pale green, each ventral green with a dark posterior edge and a pale green anterior edge; tail with parallel red dots within dark green scales; subcaudals light green, each scale edged with black. **Habitat and Behaviour** Inhabits submontane and montane forests at elevations of 1,000–1,700m asl. At the upper limits of its distribution, associated with oak forests; lower limit of distribution coincides with largely cleared and secondary vegetation. Nocturnal and arboreal, associated with trees, but also showing some terrestrial activity. Diet in the wild unknown; in captivity accepts rodents, including rats and ground squirrels. Reproductive habits unstudied. **Distribution** Endemic to northern Borneo (Gunung Kinabalu, Sabah State). **Status** Not Evaluated.

SUMATRAN PIT VIPER ☠
Parias sumatranus   PLATE 55
Measurement TL 1,355mm **Identification** Body robust; head distinct from neck; forehead scales smooth; preoculars 3; postoculars 3; supralabials 7–11; Supralabial V contacts subocular; infralabials 8–11; internasal present; tail prehensile; dorsals weakly keeled; midbody scales 19 or 21 (rarely 23); ventrals 182–191; subcaudals 55–66; anal entire. **Coloration** Dorsum green with black cross-bars especially in adults; forehead scales edged with black; supralabials light blue; tail reddish-brown, being brighter in juveniles; venter yellowish-green. **Habitat and Behaviour** Inhabits tropical forests in the lowlands and mid-hills at elevations of up to 1,000m asl. Nocturnal and arboreal, associated with low vegetation. Diet consists of small mammals, birds and frogs. Reproductive habits unstudied. **Distribution** Southern Thailand (Pattani and Narathiwat Provinces), Peninsular Malaysia, Singapore, Sumatra, the Mentawai Archipelago, Pulau Belitung and Borneo. Possibly also Palawan (the Philippines). **Status** Not Evaluated.

SUMATRAN GREEN PIT VIPER ☠
Popeia barati   PLATE 56
Measurement TL 729mm **Identification** Body slender in males, robust in females; head distinct from neck; Supralabial I distinct from nasal; hemipenes long, lacking spines; midbody scale rows 19 (rarely 17); ventrals 142–158; subcaudals 55–73. **Coloration** Dorsum green, lacking cross-bars; postocular streak absent; ventrolateral stripe present in males, reddish-brown below, white above, absent in females; tail green, rusty-red dorsally, slightly mottled with green laterally, with distinct border in between. **Habitat and Behaviour** Inhabits submontane forests. Diet and reproductive habits unstudied. **Distribution** Western Sumatra and the Mentawai Archipelago (Pagai, and probably Sipura and Siberut). **Status** Not Evaluated.

PULAU TIOMAN PIT VIPER ☠
Popeia buniana   PLATE 56
Measurement TL 783mm **Identification** Body long, slender; head triangular and elongated, distinct from neck; snout pointed; rostral large and triangular, followed posteriorly by large, circular azygous scale bordered laterally by internasals; supralabials 9; infralabials 12; midbody scale rows 21; ventrals 170–174; male subcaudals 76–78. **Coloration** Dorsum pale turquoise or green, sometimes with spots arranged to form 81 transverse maroon bands on body and 39 brownish bands on tail; iris centre copper, outer edge turquoise; maroon postorbital stripe; males with postorbital stripe, lacking white vertebral spots and with bicolored ventrolateral stripe; females with ventrolateral white stripe on flanks and tail, bordered ventrally by red stripe, fading caudally; tail banded with brown. **Habitat and Behaviour** Inhabits lowland and hill dipterocarp forest transitions to hill dipterocarp forests at elevations of 240–810m asl. Nocturnal and arboreal, found 2–10m above ground. Diet includes frogs and possibly lizards. Reproductive habits unstudied. **Distribution** Endemic to Peninsular Malaysia (Pulau Tioman, Pahang State). **Status** Not Evaluated.

THAI PENINSULA PIT VIPER ☠
Popeia fucata   PLATE 56
Measurement TL 868mm **Identification** Body robust, cylindrical; head triangular, distinct from

neck; tail long; occipital and temporals keeled; snout elongated and obliquely truncated; Supralabial I distinct from nasal; hemipenes long, reaching at least Subcaudal XXV, spineless; midbody scale rows 21; ventrals 156–171; subcaudals 59–84; anal entire. **Coloration** Dorsum green in both sexes, with irregular rusty or reddish-brown cross-bands; white or white and reddish-brown postocular streak, and white dots on vertebral region of males; pupil yellowish-green, greenish-gold or yellow-copper in adults of both sexes; bright ventrolateral stripe, orange or red ventrally, white dorsally in males; thin white ventrolateral stripe in females; tail rusty- or reddish-brown, sometimes mottled. **Habitat and Behaviour** Inhabits the mid-hills, extending into submontane and montane forests, at elevations of 400–1,280m asl. Nocturnal and arboreal. Diet includes rodents and possibly birds. Reproductive habits unstudied. **Distribution** Myanmar (Tenasserim, nr Myeik and Kanmaw Kyun, Taninthayi Division), Thailand (Prachuap Khiri Khan, Chumphon, Phang-Nga, Krabi, Nakhon Si Thammarat, Surat Thani and Trang Provinces) and Peninsular Malaysia (Pulau Langkawi, Kedah State, Cameron Highlands, Bukit Fraser, Genting Highlands and Gunung Tahan, Pahang State, Gunung Lawit, Terengganu State, Pulau Pinang, Pinang State and Bukit Larut, Perak State). **Status** Not Evaluated.

CAMERON HIGHLANDS PIT VIPER
Popeia nebularis PLATE 56
Measurement TL 1,002mm **Identification** Body robust, cylindrical; head distinct from neck; scales between supraoculars 9–10; Supralabial I distinct from nasal; hemipenes long, lacking spines; dorsals keeled; midbody scale rows 21; ventrals 147–153; subcaudals 50–65. **Coloration** Dorsum bright green with hint of blue; supralabials bluish-green; chin and throat yellowish-green; postocular streak absent; ventrolateral stripe typically absent; tail rusty-brown vertebrally, green laterally with sharp border; iris green or yellowish-green. **Habitat and Behaviour** Inhabits montane forests. Diet presumably consists of rodents and birds. Reproductive habits unstudied. **Distribution** Endemic to Peninsular Malaysia (Cameron Highlands, Pahang State). A member of this group is also known from the Toba Massif, northern Sumatra. **Status** Not Evaluated.

POPES'S PIT VIPER
Popeia popeiorum PLATE 56
Measurement TL 1,050mm **Identification** Body slender in males, robust in females; head distinct from neck; supralabials 9–11; Supralabial I distinct from nasal; occipital scales distinctly keeled in males; hemipenes long, reaching at least Subcaudal XXV; tail prehensile; forehead scales smooth; dorsals smooth or weakly keeled; midbody scale rows 21 (rarely 23); ventrals 165–170; subcaudals 56–64, paired; anal entire. **Coloration** Dorsum bright green or bluish-green, lacking dark cross-bars; postocular streak narrow and white ventrally, broad and red dorsally in males; postocular streak narrow and white or absent in females; iris red in adults, yellow in juveniles; bicoloured ventrolateral stripe, bright red ventrally, white dorsally, in males; stripe white in females; venter light green; tail brown, mottled with green laterally; tail-tip reddish-brown (except in populations from southern Thailand). **Habitat and Behaviour** Inhabits montane and submontane forests. Nocturnal and arboreal, associated with low vegetation of shrubs and dense undergrowth. Threat display consists of vibration of the tail-tip. Venom neurotoxic and causes human mortality. Diet consists of birds, frogs, lizards and small mammals. Ovoviviparous; clutches comprise 10–15 neonates, measuring 120–200mm. **Distribution** Myanmar (except Taninthayi State), northern Laos, and northern and western Thailand. Also north-eastern India. **Status** Not Evaluated.

SABAH GREEN PIT VIPER
Popeia sabahi PLATE 56
Measurement TL 810mm **Identification** Body slender in males, robust in females; head triangular, distinct from neck; Supralabial I distinct from nasal; occipitals and temporals smooth or weakly keeled; hemipenes long, spineless; dorsals keeled; midbody scale rows 21, ventrals 147–157; subcaudals 59–76; keeled. **Coloration** Dorsum bright green, lacking cross-bars or postocular streak; ventrolateral stripe red or rusty-red in males, white or yellow in females; iris red or orange in males and females, orange or yellowish-green in juveniles. **Habitat and Behaviour** Inhabits submontane tropical forests at elevations of 1,000–1,150m asl. Probably nocturnal, and arboreal, found on low vegetation of shrubs and branches. Diet and reproductive habits unstudied. **Distribution** Endemic to Borneo (Gunung Kinabalu, Crocker Range, Gunung Lumaku, Mendolong, Muruk Miau, Gunung Dulit, Gunung Gading, Gunung Penrissen and Gunung Semedoem). **Status** Not Evaluated.

FAN-SI-PAN HORNED PIT VIPER
Protobothrops cornutus PLATE 56
Measurement TL 696mm **Identification** Body moderate; head distinct from neck; horn-like enlarged supraocular scales; loreals 3; dorsals with a single unserrated keel; supralabials 9; Supralabial II contacts loreal pit; midbody scale rows 21; ventrals 189–193; subcaudals 71–78; tail prehensile, with spine apically. **Coloration** Dorsum pale greyish-brown with a series of 48–51 brown dorsal blotches partly alternating along vertebral region, sometimes forming cross-bars; lateral series of greyish-brown spots without dark borders on pale background; forehead with dark triangle; sides with dark subocular and postocular stripes; iris ochre with horizontal dark brown stripe; chin and throat yellowish-cream with brown stippling; venter light brown powdered with beige; tail-tip reddish-brown or yellow. **Habitat and Behaviour** Inhabits evergreen forests in limestone areas at elevations of 150–2,000m asl. Diurnal, and terrestrial and semi-arboreal, associated with leaf litter. Diet and reproductive habits unstudied. **Distribution**

Northern and central Vietnam (Ha Giang, Lai Chau/Lao Cai, Nghe An, Ha Tinh and Quang Binh Provinces). Records from southern Vietnam (Bach Ma, Thua Thien-Hue Province) require confirmation. Also China (Nanling Mountains). **Status** Not Evaluated.

JERDON'S PIT VIPER ☠
Protobothrops jerdonii   PLATE 56
Measurement TL 990mm (*P. j. jerdoni*); 1,090mm (*P. j. bourreti*) **Identification** Body slender; head distinct from neck; snout elongated; forehead scales reduced; forehead, apart for internasals and supraoculars, covered with small smooth scales; scale rows between supraoculars 6–9; supralabials 7–8; Supralabial I separated from nasal by suture; temporals smooth; dorsals strongly keeled; midbody scale rows 21 or 23; ventrals 164–193; subcaudals 44–78, paired. **Coloration** Dorsum greenish-yellow or olive; series of reddish-brown blotches, edged with black, on dorsum; lower flanks with smaller blotches of the same colour; forehead yellow with symmetrical reddish-brown marks, sometimes nearly black; supralabials yellow; venter yellow or green spotted with black. **Subspecies** Three subspecies are recognized, of which two occur in the region. *P. j. jerdoni*, dorsum greenish-yellow or olive; a series of reddish-brown blotches, edged with black, on dorsum; forehead black with fine yellow lines; ventrals 164–193; subcaudals 44–78. *P. j. bourreti*, dorsum yellowish-green or olive; numerous black scales form bands; forehead black with large, A-shaped yellow mark; ventrals 189–192; subcaudals 65–72. **Habitat and Behaviour** Inhabits montane and temperate forests, including secondary forests, at elevations of 1,200–2,300m asl. Nocturnal, and arboreal and terrestrial, the subspecies *P. j. bourreti* being known from near rocky streams. Diet presumably consists of rodents and birds. Ovoviviparous; clutches comprise 4–8 neonates, measuring 180–230mm. **Distribution** Northern Myanmar (Chin and Kachin States) (*P. j. jerdoni*), Cambodia and northern Vietnam (*P. j. bourreti*). Also north-eastern India (Khasi Hills and Arunachal Pradesh), southern China (Yunnan Province and Tibet) (*P. j. jerdoni*), and eastern and south-eastern China (*P. j. xanthomelas*). **Status** Not Evaluated.

KAULBACK'S PIT VIPER ☠
Protobothrops kaulbacki   PLATE 56
Measurement TL 1,340mm **Identification** Body slender, elongated; head elongated, massive, distinct from neck; snout narrow; canthus rostralis sharp; single large square loreal; preoculars 2–3; postoculars 2; supralabials 7–8; Supralabial I separated from nasal by suture; infralabials 12–14; supraoculars usually single, flat, without vertical projection; subcaudals paired; midbody scale rows 23–25; ventrals 193–217; subcaudals 66–82. **Coloration** Dorsum yellowish-grey or olive-green with large rhombohedral dark blotches that may be fused; small spots on flanks; yellow lines on head. **Habitat and Behaviour** Inhabits wet subtropical forests at elevations of 1,015–1,606m asl. Nocturnal, and terrestrial and arboreal. Diet includes small mammals. Oviparous; clutches comprise 6–32 eggs, which are laid in holes underground and guarded by the female. Hatchlings 260–270mm. **Distribution** Northern Myanmar (Kachin State). Also north-eastern India (Arunachal Pradesh) and Tibet (China). **Status** Not Evaluated.

BROWN-SPOTTED PIT VIPER ☠
Protobothrops mucrosquamatus   PLATE 56
Measurement TL 1,174mm **Identification** Body slender; head distinct from neck; snout elongated; forehead scales reduced; scales strongly keeled posteriorly; supralabials 9–12; Supralabial I separated from nasal; small internasals separated by 4–6 small scales; long, narrow non-erect supraoculars separated by 15–17 scales; midbody scale rows 25; ventrals 198–222; subcaudals 76–100, paired. **Coloration** Dorsum greyish-olive or brown with a series of irregular large brown spots with dark edges; series of smaller dark brown spots on flanks; dark postocular streak; tail light brown with black spots; venter cream with light brown speckles. **Habitat and Behaviour** Inhabits temperate and subtropical evergreen forests at elevations of 250–1,088m asl. Nocturnal, and terrestrial and arboreal. Sedentary and associated with water such as forest streams; microhabitats of concealment include areas under rocks or leaf litter. Diet consists of frogs, rodents, insectivores, snakes and birds, and it may enter chicken coops and human habitations in search of food. Oviparous; clutches comprise 5–13 eggs. Incubation period 40 days. Hatchlings 215–220mm. **Distribution** Myanmar (Kachin State), Laos and northern Vietnam. Also Bangladesh, north-eastern India and southern China. **Status** Not Evaluated.

THREE-HORNED PIT VIPER ☠
Triceratolepidophis sieversorum   PLATE 56
Measurement TL 1,255mm **Identification** Body slender, subcylindrical and laterally compressed; head large and subtriangular, distinct from neck; snout elongated; multiple raised, horn-like supraoculars; internasals 2–4; supralabials 8–10; infralabials 13–14; dorsal scales with 3 keels; midbody scale rows 23; ventrals 228–235; subcaudals 79–82; anal entire. **Coloration** Dorsum light greyish-brown with irregular or rhomboidal, dark greyish-brown blotches, edged finely with yellow, which are sometimes confluent with adjacent blotches; flanks with smaller blotches of the same colour; broad dark brown postocular stripe extends to angle of jaws, edged dorsally by beige or light brown; venter yellowish-brown to beige, partially spotted with grey-brown; outer edges of ventrals with large dark brown blotches. **Habitat and Behaviour** Inhabits tropical semi-evergreen lowland forests dominated by karst landscape within the Annamite Range, at elevations of 100–210m asl. Nocturnal, associated with limestone rocks. Diet unstudied. Oviparous; clutches comprise 10–19 eggs. **Distribution** Eastern Laos (Hin Namno NBCA, Khammouan Province) and central Vietnam

(Phong Nha-Ke Bang National Park, Quang Binh Province). **Status** Not Evaluated.

SUMATRAN PALM PIT VIPER 💀
Trimeresurus andalasensis   PLATE 57

Measurement TL 809mm **Identification** Body robust; head triangular, distinct from neck; snout projecting; Supralabial I distinct from nasal; occipital and temporal scales smooth or weakly keeled; Supralabial II borders entire anterior margin of loreal pit; 1–3 narrow supraoculars; dorsals smooth or weakly keeled; midbody scale rows 19 or 21; ventrals 144–156; subcaudals 46–50. **Coloration** Dorsum dark purplish-brown with 17–25 pale grey cross-bars, best marked in males; pattern in females shows less contrast; venter reddish-brown or dark ochre-brown with brown mottling. **Habitat and Behaviour** Inhabits submontane forests at elevations of 500–1,200m asl. Possibly nocturnal, and terrestrial and semi-arboreal. Diet and reproductive habits unstudied. **Distribution** Endemic to northern and western Sumatra (Blangkejeren, Gunung Leuser National Park, Ketembe, Takengon, Toba Massif, Gunung Gadut and Padang Highlands). **Status** Not Evaluated.

BORNEAN PALM PIT VIPER 💀
Trimeresurus borneensis   PLATE 57

Measurement TL 830mm **Identification** Body robust; head triangular, distinct from neck; snout projecting; forehead scales smooth; Supralabial I separated from nasal; dorsals smooth or weakly keeled; midbody scale rows 19 or 21; ventrals 149–166; subcaudals 41–67; tail prehensile. **Coloration** Dorsum mottled light brown or medium brown, with saddle-like dark brown pattern comprising 20–30 blotches or cross-bars, or may be bright yellow with darker mottling; oblique pale postocular stripe to neck; venter paler. **Habitat and Behaviour** Inhabits lowland swamps, forests and the mid-hills at elevations of up to ca 1,130m asl. Nocturnal as well as diurnal, and terrestrial as juveniles, semi-arboreal as adults, encountered on the forest floor, buttresses or edges of forest trails; juveniles climb low vegetation. Diet consists of small mammals. Oviparous; clutches comprise 7–14 eggs. **Distribution** Endemic to Borneo. **Status** Not Evaluated.

BRONGERSMA'S PALM PIT VIPER 💀
Trimeresurus brongersmai   PLATE 57

Measurement TL 410mm **Identification** Body slender; head triangular, distinct from neck; snout projecting; Supralabial I distinct from nasal; Supralabial II borders entire anterior margin of loreal pit; 4–5 narrow supraoculars, strongly erect and divergent; occipital and temporal scales strongly keeled; midbody scale rows 19 or 21; ventrals 129–149; subcaudals 39–49. **Coloration** Dorsum dark greyish-brown with 25–30 dark cross-bars, light centrally and darker on edges; smaller subrectangular or subtriangular dark greyish-brown blotches on flanks; intervening areas spotted with dark dots; venter greyish-brown or dark ochre-brown mottled with brown and stippled with black. **Habitat and Behaviour** Probably inhabits lowland rainforests. Diet unstudied. Ovoviviparous; clutches comprise 3 neonates, measuring ca 130mm. **Distribution** Endemic to the Mentawai Archipelago (Pulau Simeuleu and Pulau Siberut). **Status** Not Evaluated.

JAVANESE PALM PIT VIPER 💀
Trimeresurus puniceus   PLATE 57

Measurement TL 920mm **Identification** Body slender in males, stout in females; head distinct from neck; distinctly projected and raised snout, strongly obliquely truncated; internasals normal or weakly bilobate, either slightly raised or upturned, or flat; occipitals and temporals keeled in both sexes; Supralabial I distinct from nasal; Supralabial II does not contact loreal pit; dorsals smooth or weakly keeled; midbody scale rows 21 or 23 (rarely 19); ventrals 153–167; subcaudals 41–59. **Coloration** Dorsum grey or yellowish-brown with 20–30 dark cross-bands; males with irregular, constricted dark dorsolateral blotches, often heavily powdered with cream and dark dots; in females, pattern shows less contrast; venter similar to dorsal surface. **Habitat and Behaviour** Inhabits lowland forests to montane forests at elevations of up to 1,600m asl, in addition to tea and coffee plantations. Terrestrial (males) or semi-arboreal (females) known to ascend 1–2m above the ground in coffee trees. Diet includes small vertebrates. Ovoviviparous; clutches comprise 7–33 neonates, measuring 180mm. **Distribution** Southern Sumatra, the Mentawai Archipelago, Java and Pulau Tinjil. **Status** Not Evaluated.

TRUONG SON PIT VIPER 💀
Trimeresurus truongsonensis   PLATE 57

Measurement TL 642mm **Identification** Body slender, cylindrical; head subtriangular, distinct from neck; forehead covered by granulated scales; preoculars 3; postoculars 3; supralabials 9–10; infralabials 11–12; Supralabial I separated from large nostril; single enlarged supraocular; single subocular; anterior chin shields enlarged; posterior chin shields small; hemipenes relatively short, reaching Supracaudal VII, forked and with large and small spines; dorsals rhomboid, strongly keeled, first row with weak keel; midbody scale rows 21; ventrals 166–175; subcaudals 65–69, paired; anal entire. **Coloration** Dorsum greenish-blue in males, light brown in females, with 66–70 reddish-brown cross-bars, 1–3 scales wide; pale ventrolateral stripe along first scale row, with a reddish-brown ventral edge; reddish-brown spots on head; reddish-brown postocular stripe; tail-tip yellowish-orange; venter greyish-blue; iris greenish-yellow. **Habitat and Behaviour** Inhabits evergreen monsoon limestone forests within the Annamite chain (Truong Son) at elevations of 500–600m asl. Nocturnal, with daytime basking after heavy showers. Associated with karst outcrops under branches of trees and bushes. Diet includes rodents. Reproductive habits unstudied.

Distribution Endemic to central Vietnam (Phong Nha-Ke Bang National Park, Quang Binh Province). **Status** Not Evaluated.

WIROT'S PALM PIT VIPER ☠
Trimeresurus wiroti PLATE 57
Measurement TL 889mm **Identification** Body slender in males, stout in females; head distinct from neck; distinctly projected and raised snout, strongly obliquely truncated; supralabials 9–11; Supralabial I distinct from nasal; Supralabial II borders entire anterior margin of loreal pit; infralabials 11–14; 2–4 small and narrow supraoculars, convex or granular and raised; occipital and temporal scales as well as dorsals moderately keeled or smooth; midbody scale rows 21 (rarely 19, 20 or 23); ventrals 148–176; subcaudals 43–58, paired; anal entire. **Coloration** Dorsum dark greyish-brown (in males) or dark brown (in females) with 22–35 darker cross-bands; dark brown dorsolateral blotches, areas between blotches darker than on flanks, suffused with dark and light dots or blotches; venter darker. **Habitat and Behaviour** Inhabits lowland and hill forests. Nocturnal and arboreal, known to ascend canopy at more than 20m above ground. Diet consists of small mammals, birds and frogs. Oviparous; clutches comprise 7–12 eggs. Incubation period ca 6–8 weeks. **Distribution** Endemic to peninsular Thailand (Krabi, Nakhon Si Thammarat, Narathiwat and Trang Provinces) and Peninsular Malaysia (Johore, Pahang, Perak and Selangor States). **Status** Not Evaluated.

BORNEAN KEELED GREEN PIT VIPER ☠
Tropidolaemus subannulatus PLATE 57
Measurement TL 963mm **Identification** Body slender in juveniles, relatively thick in adults; head distinct from neck; tail prehensile; Supralabial III separated from subocular by 1–2 scales; occipital scales distinctly keeled in males; 4–7 scales on snout of males, 5–8 in females; internasals separated by 2 (rarely 1) scales; midbody scale rows 21–23 in males, 21–29 in females; ventrals 127–148; subcaudals 40–54, paired; anal divided. **Coloration** Dorsum green or greenish-blue with blue, white and red spots or stripes in males; dorsum green or bluish-green with blue and red cross-bars in females; cream or yellow area below dark postocular stripe in adult females, white and red stripe in juveniles and males; venter uniform, or blotched or spotted with blue or red. **Habitat and Behaviour** Inhabits lowland forests. Arboreal and found on low vegetation, particularly in riparian forests, but can sometimes be found on trees more than 2m above substrate. Diet consists of birds and rodents. Ovoviviparous. **Distribution** Borneo and Pulau Belitung. Also Buton, the Sangihe Archipelago and Sulawesi (Indonesia), and Balabac, Basilan, Bohol, Dinagat, Jolo, Leyte, Luzon, Mindanao, Negros, Palawan, Panay, Samar, Sibutu and Tumindao (the Philippines). **Status** Not Evaluated.

WAGLER'S KEELED GREEN PIT VIPER ☠
Tropidolaemus wagleri PLATE 57
Measurement TL 920mm **Identification** Body slender, laterally compressed, in juveniles and males, robust in adult females; head triangular, distinct from neck; supralabials 8–10; infralabials 11–12; forehead scales small, distinctly keeled; internasals in contact; scales on throat keeled; tail short, prehensile; midbody scale rows 21–23 in males, 23–27 in females, feebly keeled in males, distinctly keeled in females; ventrals 134–152; subcaudals 45–55, paired; anal divided. **Coloration** Dorsum black with yellow cross-bars in adult females, yellow with white cross-bars in juvenile females, and green with white spots in adult and juvenile males; black postocular stripe in adult females, white and red postocular stripe in males and juveniles; venter banded in adult females, pale in males and juveniles. **Habitat and Behaviour** Inhabits lowland forests. Arboreal and found on low vegetation. Diet consists of birds and rodents. Ovoviviparous; clutches comprise 15–41 neonates, measuring 150mm. **Distribution** Southern Thailand (Phang Nga, Pattani, Surat Thani, Nakhon Si Tammarat, Narathiwat and Yala Provinces), southern Vietnam (Minh Hai and Song Be Provinces), Peninsular Malaysia, Singapore, Sumatra, Pulau Nias, the Mentawai Archipelago, Pulau Bangka, Pulau Natuna and the Riau Archipelago. **Status** Not Evaluated.

GUMPRECHT'S PIT VIPER ☠
Viridovipera gumprechti PLATE 58
Measurement TL 1,280mm **Identification** Body slender, elongated, triangular in cross-section; head distinct from neck; loreal present; preoculars 2; postoculars 2; supralabials 10; Supralabial I separated from nasal; infralabials 12–13; internasals in contact; single narrow supraocular; hemipenes short and spinose; tail long, prehensile; dorsals moderately keeled; midbody scale rows 21; ventrals 162–168; subcaudals 51–71, paired; anal entire. **Coloration** Dorsum bright pale green in males, dark green in females; 3 small white vertebral spots at posterior of body; interstitial skin black; bicolored postocular streak in males, red dorsally, white ventrally; postocular streak white in females; ventrolateral stripe white dorsally, red ventrally in males, white or blue in females; tail rusty or reddish-brown; chin and gular region emerald-green; venter anteriorly emerald-green, turning dark bluish-green, and reverting to emerald-green posteriorly; iris red in males, yellow in females. **Habitat and Behaviour** Inhabits seasonal and dry subtropical forests, near bamboo thickets and hill streams, at elevations of 800–1,200m asl. Nocturnal and arboreal, associated with trees, with some terrestrial activity. Diet consists of skinks and rodents. Ovoviviparous; clutches comprise 9–15 neonates. **Distribution** Myanmar, Laos and southern Vietnam (Annamite Mountains) and north-eastern Thailand. Also north-eastern India and southern China. **Status** Not Evaluated.

MEDO PIT VIPER ☠

Viridovipera medoensis   PLATE 58

Measurement TL 677mm **Identification** Body slender, elongated, cylindrical, slightly compressed laterally; head wide, flattened, triangular, distinct from neck; scales on upper snout enlarged; supralabials 8–9 (rarely 9); Supralabial I separated from nasal; temporals smooth; tail short, prehensile; dorsals with obtuse keels on vertebral region and flanks; midbody scale rows 17; ventrals 138–149; subcaudals 52–65, mostly paired; anal entire. **Coloration** Dorsum dark green, sometimes edged with turquoise-blue, with bicoloured white/red ventrolateral stripe in both sexes; venter lighter green or yellowish-cream; iris green or yellowish-green in both sexes. **Habitat and Behaviour** Inhabits steep slopes of tropical and subtropical montane forests at elevations of 1,000–1,400m asl. Nocturnal and arboreal, associated with bamboo thickets. Hibernates between November and April. Diet consists of frogs and rodents. Reproductive habits unstudied. **Distribution** Northern Myanmar (Kachin State). Also north-eastern India (Arunachal Pradesh) and south-western China. **Status** Not Evaluated.

STEJNEGER'S PIT VIPER ☠

Viridovipera stejnegeri   PLATE 58

Measurement TL 1,120mm **Identification** Body slender, elongated, triangular in cross-section; head distinct from neck; loreal present; preoculars 2; postoculars 2; supralabials 10; Supralabial I separated from nasal; infralabials 12–13; internasals not in contact; single narrow supraocular; hemipenes short, spinose beyond bifurcation; tail long, prehensile; dorsals weakly keeled; midbody scale rows 21; ventrals 150–174; subcaudals 54–77, paired; anal entire. **Coloration** Dorsum and head bright green; interstitial skin dark grey or greyish-brown; venter pale or greenish-cream; males with white postocular stripe; ventrolateral stripe orange, brown or red ventrally, white dorsally in males, white in females; venter pale green; iris red or amber in males, yellow or amber in females. **Habitat and Behaviour** Inhabits hill forests near fast-flowing streams at elevations of 500–2,845m asl. Nocturnal and arboreal, associated with bushes and low trees. Diet consists of rodents, birds, lizards and frogs. Ovoviviparous; clutches comprise 3–10 neonates, measuring 155–175mm. **Distribution** Laos and northern Vietnam. Also central and south-eastern China, including Taiwan. **Status** Not Evaluated.

VOGEL'S PIT VIPER ☠

Viridovipera vogeli   PLATE 58

Measurement TL 1,120mm **Identification** Body slender, elongated; head large, distinct from neck; dorsals smooth or strongly keeled; temporal and rear head scales weakly keeled or smooth; supralabials 8–12; infralabials 10–14; ventrals 157–173; subcaudals 52–72; hemipenes with 10–20 stout spines, largest basally, tips calyculate. **Coloration** Dorsum dark or grass-green; ventrolateral line of adult males with lower red component (Laos and Vietnam), or mostly white (Thailand); interstitial skin bright blue; males with white vertebral flecks, which are lacking in females; venter yellowish-green; tail-tip green, dark grey or brown; yellow or yellowish-green iris in both sexes. **Habitat and Behaviour** Inhabits evergreen, semi-evergreen and dry evergreen forests and grasslands, and plantations, at elevations of 200–1,200m asl. Nocturnal and arboreal, sometimes associated with forest streams. Diet includes small mammals, skinks and insects. Reproductive habits unstudied. **Distribution** Eastern Thailand (western Dongraek Mountains, western Khorat Plateau and south-eastern mountains of Khao Sai Dao Wildlife Sanctuary), Laos (Champasak Province), eastern and south-western Cambodia (Mondolkiri Province, Cardamom Mountains and O'Rang District) and central Vietnam (Kontum Plateau). **Status** Not Evaluated.

YUNNAN PIT VIPER ☠

Viridovipera yunnanensis   PLATE 58

Measurement TL 750mm **Identification** Body robust; head distinct from neck; supralabials 9–11; Supralabial I separated from nasal by suture; single narrow supraocular, sometimes divided by transverse suture; 11–16 scales between supraoculars; midbody scale rows 19 (rarely 21); ventrals 155–170; subcaudals 58–68, paired. **Coloration** Dorsum dark green; bicoloured ventrolateral stripe, ventrally orange or brown, and dorsally white, in males, and white only or absent in females, across outermost scale row and portion of second row; venter pale green to cream; iris red in males, golden-yellow in females. **Habitat and Behaviour** Inhabits montane forests of alpine conifer, as well as subtropical forests, at elevations of 1,206–2,845m asl. Nocturnal and arboreal, associated with trees and shrubs. Diet consists of small vertebrates. Reproductive habits unstudied. **Distribution** Myanmar (Chin State and Kachin State), Laos and Vietnam (Lao Cai, Vinh Phú, and possibly Bac Thai, Da Nang, Gia Lai and Hoa Binh Provinces). Also south-western China. **Status** Not Evaluated.

FEA'S VIPER ☠

Azemiops feae   PLATE 58

Measurement TL 925mm **Identification** Body robust, cylindrical; sensory pit absent; head flattened, distinct from neck; forehead covered with large symmetrical shields; nostril large; loreal shield present, small; preoculars 2; postoculars 2; supralabials 6; Supralabial III contacts orbit; infralabials 8; dorsals smooth; midbody scale rows 17; ventrals 180–189; subcaudals 42–53, mostly paired; anal entire. **Coloration** Dorsum blackish-brown, scales edged with grey, with 14–15 narrow white, red or pink crossbars that may be interrupted middorsally, or alternating with one another laterally; forehead yellowish-orange with a pair of dark brown to black stripes from prefrontals to end of neck. **Habitat and Behaviour** Inhabits submontane and montane forests at elevations of 700–2,000m asl; also paddy fields, grassy fields and the vicinity of villages. Nocturnal and terrestrial, associated with forests of bamboo and tree ferns with high humidity. Threat display entails flattening of its body and tail vibration. Diet comprises small vertebrates such as rodents and shrews. Oviparous; clutches comprise 5 eggs. **Distribution** Northern Myanmar (Kachin State) and northern Vietnam (Ngan Son, Cao Bang Province). Also southern and central China. **Status** Not Evaluated.

RUSSELL'S VIPER ☠
Daboia russelii PLATE 58
Measurement TL 1,850mm **Identification** Body robust; head distinct from neck; forehead with small scales; nostril enlarged, in large nasal shield; supralabials 10–12; eye small; pupil vertical; tail short; midbody scale rows 27–33, all except outermost row strongly keeled; ventrals 153–180; subcaudals 41–64, paired. **Coloration** Dorsum light brown with 5–7 rows of large, black-margined orange-brown blotches along body, a dark brown one along midline and a blackish-brown or black one on flanks; venter yellowish-cream with numerous small crescentic marks on ventrals. **Subspecies** Two subspecies are recognized, of which one (*D. r. siamensis*) occurs in the region. Some recent works give species status to these populations. **Habitat and Behaviour** Inhabits the lowlands in grasslands, scrub forests and other more open forest types, in addition to agricultural fields, from sea level to 2,756m asl. Crepuscular and nocturnal, and terrestial; generally sluggish. Diet consists of rodents, crabs, frogs, lizards and birds. Ovoviviparous; clutches comprise 6–40 (exceptionally 63) neonates, measuring 240–273mm. **Distribution** Myanmar, northern and central Thailand, Cambodia and eastern Java (*D. r. siamensis*). Also India, Pakistan, Bangladesh, Sri Lanka (*D. r. russelii*), eastern China, including Taiwan, and the Lesser Sundas of Indonesia (*D. r. siamensis*). **Status** Not Evaluated.

Family CYLINDROPHIIDAE PIPE SNAKES

Pipe snakes are characterized by a blunt head, a subcylindrical body capable of being dorsoventrally depressed, a short tail, and a venter with a black and white chequered pattern. They are terrestrial and subfossorial, and are specialist feeders on elongated prey, including snakes and eels. Their distribution shows a disjunction, with Sri Lanka in the west and South-East Asia to the east.

ENGKARI PIPE SNAKE
Cylindrophis engkariensis PLATE 59
Measurement TL 485mm **Identification** Body robust, elongated, flattened when displaying; head long, blunt, indistinct from neck; supralabials 6; Supralabial III contacts orbit; mental groove present; tail tapers to narrow point; dorsals smooth; midbody scale rows 17; ventrals 234; subcaudals 6. **Coloration** Dorsum black; short light postocular streak; irregular rows of paravertebral spots; tail black with black and white mottling; venter with black and white crossbars, divided in midline. **Habitat and Behaviour** Inhabits low hills within dipterocarp forests. Subfossorial, associated with leaf litter. Diet and reproductive habits unstudied. **Distribution** Endemic to north-eastern Borneo (Lanjak Entimau, Sarawak State). **Status** Not Evaluated.

LINED PIPE SNAKE
Cylindrophis lineatus PLATE 59
Measurement TL 982mm **Identification** Body robust, elongated, presumably flattened when displaying; mental groove present; head long, blunt, indistinct from neck; supralabials 6; Supralabials III–IV contact orbit; interorbital distance equidistant to eye-snout distance; eye reduced; tail tapers to narrow point; dorsals smooth; midbody scale rows 21; ventrals 210–215; subcaudals 9–10; anal divided. **Coloration** Dorsum yellowish-white; series of longitudinal red or yellow stripes from back of head to base of tail; forehead sometimes with scattered dark spots; venter with black cross-bars. **Habitat and Behaviour** Nothing known of its natural history. **Distribution** Endemic to north-eastern Borneo (Gunung Matang and Gunung Penrissen, Sarawak State). **Status** Not Evaluated.

COMMON PIPE SNAKE
Cylindrophis ruffus PLATE 59
Measurement TL 900mm **Identification** Body robust, elongated, flattened when displaying; mental groove present; head short, blunt, indistinct from neck; supralabials 6; Supralabials III–IV contact orbit; interorbital distance greater than eye-snout distance; eye reduced; tail tapers to narrow point; dorsals smooth; midbody scale rows 19 or 21; ventrals 185–245; subcaudals 5–10; anal divided. **Coloration** Dorsum iridescent dark brown, grey or black, typically with a pale yellow or cream collar; cream bands on dorsum in some populations; venter with black cross-bars. **Habitat and Behaviour** Inhabits low and swampy locations within forested areas, saltwater lagoons and agricultural fields up to the mid-hills at an elevation of 1,676m. Nocturnal and subfossorial. Threat display entails raising and waving its brightly patterned tail. Diet consists of snakes and eels. Ovoviviparous; clutches comprise 5–13 neonates, measuring 205mm. **Distribution** Myanmar, Thailand, Laos, Cambodia, Vietnam, Peninsular Malaysia, Sumatra, Pulau Bangka, Pulau Belitung, the Riau Archipelago, Borneo and Java. Also southern and eastern China, the Sangihe and Sula Archipelagos, and Sulawesi (Indonesia). **Status** Not Evaluated.

Family ELAPIDAE COBRAS, KRAITS, CORAL SNAKES AND SEA SNAKES

Members of this family are allocated to three subfamilies, all with neurotoxic venom. The Elapinae are mostly terrestrial, with permanently erect fangs on the anterior end of immovable maxillae and fitting into grooved slots in the buccal floor, edentulous premaxillaries, longitudinally oriented, shortened maxillaries, a mandible lacking a coronoid bone and dentary with teeth. Members of this subfamily are oviparous, and include cobras and kraits. They are widespread in Asia, New Guinea, Australia and Africa. The Hydrophiinae are restricted to marine habitats, and have laterally compressed bodies, oar-like tails, lingual salt glands and greatly reduced ventrals. They are ovoviviparous, and cosmopolitan in warmer seas. The Laticaudinae are also restricted to marine environments, but are the least aquatic of all sea snakes and are capable of coming ashore. They are morphologically characterized by their oak-like tails, relatively wide ventrals and lateral nostrils. All are dangerously venomous, and in the region have been responsible for a number of bites, leading to mortality when left untreated. They are oviparous, and are found in the Indo-Pacific Oceans.

HIMALAYAN KRAIT ☠
Bungarus bungaroides PLATE 59
Measurement TL 1,400mm **Identification** Body robust; head indistinct from neck; eye rather small; rounded pupil; dorsals smooth; midbody scale rows 15; vertebral scales enlarged; ventrals 220–237; subcaudals 44–51; last few subcaudal scales paired. **Coloration** Dorsum black, blue-black or dark brown with transverse yellowish-cream lines or narrow bars, forming broad bands across venter; white stripe across snout, one across nape and a third one from eye to end of jaws. **Habitat and Behaviour** Inhabits evergreen and subtropical forests in the mid-hills and montane regions, reaching an elevation of 2,040m asl. Nocturnal and terrestrial. Diet and reproductive habits unstudied. **Distribution** Northern Myanmar. Also eastern and north-eastern India, and Nepal (Ilam District). **Status** Not Evaluated.

MALAYAN KRAIT ☠
Bungarus candidus PLATE 59
Measurement TL 1,600mm **Identification** Body robust; head indistinct from neck; single preocular; postoculars 2; supralabials 7; Supralabials III–IV contact orbit; infralabials 7; 3–4 infralabials contact anterior chin shields; eye small; pupil rounded; tail short, with acute tip; dorsals smooth; midbody scale rows 15 (rarely 17); ventrals 194–237; subcaudals 37–56, single; anal entire. **Coloration** Dorsum black or bluish-black with 20–35 broad white cross-bars as wide as or wider than dark interspaces on body, and 7–10 on tail; sometimes an indistinct light chevron on nuchal region; pale bands sometimes absent, especially in individuals from Java and Bali; supralabials and venter unpatterned white; subcaudals with dark brown spots. **Habitat and Behaviour** Inhabits forested lowlands and submontane regions, including the vicinity of villages and agricultural areas, at elevations of up to 1,525m asl. Nocturnal and terrestrial. Diet includes other snakes, and also lizards and toads. Oviparous; clutches comprise 6–10 eggs. **Distribution** Southern Thailand, Laos, eastern and south-western Cambodia (O'Rang District and Bokor, Elephant Mountains), Vietnam, Peninsular Malaysia, Singapore, Sumatra, Pulau Nias, Java, Bawean, Karimunjawa and Bali. Records from Sulawesi suspected to be the result of inadvertent introduction. **Status** Not Evaluated.

BANDED KRAIT ☠
Bungarus fasciatus PLATE 59
Measurement TL 2.25m **Identification** Body robust, triangular in cross-section, with raised vertebral region; head distinct from neck; loreal absent; single preocular; postoculars 2; supralabials 7; Supralabials III–IV contact orbit; infralabials 7; eye small; vertebrals strongly enlarged, as broad as or broader than long; tail blunt-tipped; dorsals smooth; midbody scale rows 15; ventrals 200–236; subcaudals 23–39, single; anal entire. **Coloration** Dorsum yellow or pale brown with black bands that are approximately equal to the pale interspaces; forehead with pale V-shaped marking; venter pale yellow or brown with bands; iris black. **Habitat and Behaviour** Inhabits lightly forested areas, swamps and the vicinity of villages; more abundant in the lowlands, and occasionally recorded from montane forests at elevations of up to 2,500m asl. Nocturnal and terrestrial, and daytime retreat may consist of termite mounds and the burrows of rodents. Non-aggressive by day, preferring to hide its head under the coils of its body. Diet consists of other snakes such as water snakes, rat snakes, pythons and vine snakes; lizards, frogs, fish and reptile eggs are also eaten. Oviparous; clutches comprise 3–14 eggs, which are guarded by the female during incubation. Incubation period 61 days. Hatchlings 250–400mm. **Distribution** Myanmar, Thailand, Cambodia, Laos, Vietnam, Peninsular Malaysia, Singapore, Sumatra, Borneo and Java. Also eastern and north-eastern India, Nepal, Bangladesh, and southern and eastern China. **Status** Not Evaluated.

RED-HEADED KRAIT ☠
Bungarus flaviceps PLATE 59
Measurement TL 2.07m **Identification** Body robust; triangular in cross-section; head large, distinct from neck; snout blunt; single preocular; postoculars 2; supralabials 7; Supralabials III–IV contact orbit;

infralabials 6; eye small; tail relatively short; dorsals smooth; midbody scale rows 13; ventrals 193–236; subcaudals 42–54, anterior ones entire, posterior ones divided; anal entire. **Coloration** Dorsum blue-black with yellow or tan vertebral stripe; red, orange or yellow forehead and tail; venter pink or yellow. **Subspecies** Two subspecies are recognized. *B. f. flaviceps*, head reddish-orange; body without black rings. *B. f. baluensis*, head red (rarely yellow); posterior half of body and tail with 5–8 pairs of thick black rings separated by narrow white ring. **Habitat and Behaviour** Inhabits the forested lowlands and mid-hills at elevations of up to 914m asl (*B. f. flaviceps*) and submontane forests at 550–900m asl (*B. f. baluensis*). Nocturnal and terrestrial, and daytime retreats include leaf litter and areas under logs. Most individuals try to conceal their heads under the coils of their bodies when threatened. Diet consists of other snakes, and lizards such as skinks. Reproductive habits unstudied. **Distribution** Southern Myanmar (Tanintharyi Division), southern Thailand, Cambodia, Vietnam, Peninsular Malaysia, Sumatra, Pulau Belitung, Borneo, Java (*B. f. flaviceps*), and Gunung Kinabalu and Trus Madi, Sabah, Borneo (*B. f. baluensis*). **Status** Not Evaluated.

SPLENDID KRAIT
Bungarus magnimaculatus   PLATE 59

Measurement TL 1,300mm **Identification** Body robust; head indistinct from neck; vertebrals strongly enlarged, as broad as or broader than long; tail tapering, terminating in point; dorsals smooth; midbody scale rows 15; ventrals 214–235; subcaudals 40–48, undivided. **Coloration** Dorsum black or bluish-black with 11–14 broad white cross-bars as wide as or wider than dark interspaces; scale centres spotted with black; white preocular spot present; venter unpatterned white. **Habitat and Behaviour** Inhabits moist dipterocarp forests, including seasonal dry forests, and also known from villages and agricultural areas. Nocturnal and terrestrial. Diet and reproductive habits unstudied. **Distribution** Endemic to Myanmar (Magway, Mandalay and Sagaing Divisions). **Status** Not Evaluated.

MANY-BANDED KRAIT
Bungarus multicinctus   PLATE 59

Measurement TL 1,354mm **Identification** Body robust; head indistinct from neck; single preocular; postoculars 2; supralabials 7; infralabials 7; eye small; pupil rounded; dorsals smooth; tail long, thin and tapering; midbody scale rows 15; ventrals 200–228; subcaudals 44–54, divided; anal entire. **Coloration** Dorsum jet-black, dark brown or bluish-black with 27–44 light cross-bars on body and 7–17 on tail; pale bands expand on flanks; supralabials and venter yellowish-cream; subcaudals mottled with dark brown. **Habitat and Behaviour** Inhabits the lowlands, especially in the vicinity of wetlands, including agricultural areas, up to submontane limits. Nocturnal and terrestrial, inhabiting mesic habitats. Diet includes other snakes; rodents, frogs, lizards and eels are also consumed. Oviparous; clutches comprise 3–12 eggs. Incubation period 48 days. Hatchlings 250–300mm. **Distribution** Myanmar, Laos and Vietnam. Also eastern China. **Status** Not Evaluated.

RED RIVER KRAIT
Bungarus slowinskii   PLATE 59

Measurement TL ca 1,350mm **Identification** Body robust; head indistinct from neck; single preocular; postoculars 2; supralabials 7; Supralabials III–IV contact orbit; infralabials 7; eye small; pupil rounded; dorsals smooth, enlarged and hexagonal; midbody scale rows 15; ventrals 220–228; subcaudals 30–37; last few subcaudal scales paired; anal entire. **Coloration** Dorsum iridescent blue-black; incomplete V-shaped mark from frontal to beyond angle of jaws; incomplete narrow light stripe across snout; dorsum with pale rings, composed of white scales with black bases and margins, giving the impression of 20–33 black rings; venter also with dark and pale bands. **Habitat and Behaviour** Inhabits the low hills at an elevation of 750m asl. Nocturnal and terrestrial, and also known from streams. Diet and reproductive habits unstudied. **Distribution** Endemic to northern Vietnam (Quang Nam, Lao Cai, Quang Tri and Yen Bai Provinces). **Status** Not Evaluated.

WANGHAOTING'S KRAIT
Bungarus wanghaotingi   NOT ILLUSTRATED

Measurement TL 1,100mm **Identification** Body robust; head indistinct from neck; postoculars 2; supralabials 7; Supralabials III–IV contact orbit; infralabials 7; tail tapering to a point; middorsal row of vertebrals strongly enlarged, as broad as or broader than long; dorsals smooth; midbody scale rows 15; ventrals 209–228; subcaudals 44–54, divided; anal entire. **Coloration** Dorsum dark brown-black with 20–31 light cross-bands on body, narrower than black interspaces; 7–11 white cross-bars on tail; venter unpatterned cream. **Habitat and Behaviour** Inhabits wet forests (including bamboo stands in coastal rainforests and the vicinity of streams in moist deciduous forests) and subtropical forests, as well as the vicinity of villages, from the plains to 900m asl. Nocturnal and terrestrial. Diet and reproductive habits unstudied. **Distribution** Myanmar (Kachin and Rakhine State, Sagaing Division, and perhaps the Yangon area) and northern Laos. Also southern China (Yunnan Province). **Status** Not Evaluated.

BLUE CORAL SNAKE
Calliophis bivirgatus   PLATE 60

Measurement TL 1,850mm **Identification** Body slender; head short, indistinct from neck; single preocular; postoculars 2; supralabials 6; Supralabials III–IV contact orbit; 3–4 infralabials contact anterior chin shields; eye small; tail short, terminating in sharp point; dorsals smooth; midbody scale rows 13; ventrals 243–304; subcaudals 34–53, paired; anal divided. **Coloration** Dorsum dark blue to blue-black; some populations with distinct stripe along flanks; head, tail and venter coral-red. **Subspecies** Three subspecies are recognized. *C. b. bivirgatus*, no pale flank stripe; sometimes, a narrow white paravertebral stripe. *C. b. flaviceps*, pale blue flank stripe; no white paravertebral stripe. *C. b. tetrataenia*, cream paravertebral stripes, outer one being the broadest. **Habitat and Behaviour** Inhabits lowland forests up to submontane limits at an elevation of 1,200m asl, and also forest fringes such as those in agricultural areas. Display consists of concealment of its head under the coils of its body, exposing its brightly coloured tail. Diet includes other snakes. Oviparous; clutches comprise 1–3 eggs, measuring 35–35 x 9mm. **Distribution** Myanmar, southern Thailand (Songkhla, Pattani, Yala, Trang and Narathiwat Provinces), Laos, Cambodia, Peninsular Malaysia, Singapore, Sumatra, Pulau Nias, Pulau Bangka, the Mentawai and Riau Archipelagos (*C. b. flaviceps*), Borneo (*C. b. tetrataenia*) and Java (*C. b. bivirgatus*). **Status** Not Evaluated.

SPOTTED CORAL SNAKE ☠

Calliophis gracilis   PLATE 60

Measurement TL 740mm **Identification** Body slender; head small, indistinct from neck; single preocular; postoculars 2; supralabials 6; Supralabials III–IV contact orbit; infralabials 7; 4 infralabials contact anterior chin shields; eye small; pupil rounded; tail short; dorsals smooth; midbody scale rows 13; ventrals 303–324; subcaudals 21–30, paired; anal divided. **Coloration** Dorsum pale brown, reddish-brown or greyish-brown with narrow black vertebral stripe; paired oval black marks that are fused on nape and tail; series of black spots on flanks; venter white with black bars; subcaudals pink. **Habitat and Behaviour** Inhabits lowland forests at elevations of up to 915m asl. Nocturnal and terrestrial, associated with tree roots and rock cracks. Diet consists of other snakes. Reproductive habits unstudied. **Distribution** Southern Thailand (Na Pradoo, Pattani Province), Peninsular Malaysia, Singapore and Sumatra. **Status** Not Evaluated.

MALAYAN STRIPED CORAL SNAKE ☠

Calliophis intestinalis   PLATE 60

Measurement TL 710mm **Identification** Body slender; head small, indistinct from neck; loreal absent; single preocular; postoculars 2; supralabials 6; Supralabials III–IV contact orbit; 3–4 infralabials contact anterior chin shields; eye small; dorsals smooth; midbody scale rows 13 or 15; ventrals 197–273; subcaudals 15–33, divided; anal entire. **Coloration** Dorsum reddish-brown, dark brown or black; flanks with white, tan or red stripes; venter yellow, cream or red with black bands; subcaudals orange. **Subspecies** Six subspecies are recognized, of which three occur in the region. *C. i. intestinalis*, dorsum black with narrow yellow or white lines on vertebral region and lower flanks; dark brown stripe on mid-flanks; venter with broad transverse black and white bands, the black cross-bands in contact with black on sides; forehead black with Y-shaped cream mark. *C. i. lineata*, dorsum greyish-brown with narrow black-edged white lines on vertebral region and on lower flanks extending to tail-tip; lower flanks with narrow white line edged with black; venter with broad transverse black and white bands, the black cross-bands not in contact with black on sides; forehead brown mottled with black; subcaudals orange with 2 narrow bars. *C. i. thepassi*, dorsum black with broad black vertebral stripe extending to tail; dorsolateral stripes 2 scales wide; venter with broad transverse black and white bands, the black cross-bands not in contact with black on sides; forehead rufous brown. **Habitat and Behaviour** Inhabits primary lowland forests and more open areas, such as parks and gardens, ascending to submontane limits at an elevation of 1,100m asl. Nocturnal, and terrestrial as well as semi-fossorial, concealing itself under leaves, fallen logs and other debris on the forest floor by day. Defence display consists of raising its tail, exposing its red or distinctly dark-pale banded underside; or turning over backwards and revealing its bright venter; or dorsolateral flattening of its body, which may be turned into an S-shape, with its head concealed under its body. Diet consists of other snakes. Oviparous; clutches comprise 1–3 eggs, measuring 27–35 x 8–9mm. Incubation period 84 days. **Distribution** Southern Thailand (Nakhon Si Thammarat, Surat Thani, Pattani and Yala Provinces), Cambodia, Vietnam, Peninsular Malaysia, Singapore, Sumatra, Pulau Nias, Pulau Bangka, Pulau Belitung, the Mentawai and Riau Archipelagos (*C. i. lineata*), Borneo (*C. i. thepassi*), eastern Sumatra and Java (*C. i. intestinalis*). Also Palawan, Balabac and Busuanga (the Philippines) (*C. i. bilineata*), Luzon, Mindanao and Samar (the Philippines) (*C. i. philippina*), and the Sulu Archipelago (the Philippines) (*C. i. suluensis*). **Status** Not Evaluated.

SPECKLED CORAL SNAKE ☠

Calliophis maculiceps   PLATE 60

Measurement TL 480mm **Identification** Body slender; head small, indistinct from neck; loreal absent; single preocular, in contact with nasal; postoculars 2; supralabials 7; Supralabials III–IV contact orbit; infralabials 7; eye small; tail short; dorsals smooth; middorsal scales not enlarged; midbody scale rows 13; ventrals 193–285; subcaudals 21–32, paired; anal divided. **Coloration** Dorsum brownish-yellow or reddish-brown with or without black spots in longitudinal series along each side; head and nape black with yellow occipital spot; pale band across forehead present or absent; supralabials yellow; venter pale blue or grey. **Subspecies** Two subspecies are recognized. *C. m. maculiceps*, ventrals 174–203. *C. m. hughi*, ventrals 285. **Habitat and Behaviour** Inhabits tropical forests and plantations in the lowlands to submontane limits at an elevation of 1,330m asl. Nocturnal and terrestrial, active near streams on the forest floor. Daytime retreat consists of vegetation, rocks or logs. Diet consists of other snakes such as blind snakes (*Typhlops* spp). Oviparous; clutches comprise 2 eggs. **Distribution** Southern Myanmar, Thailand, Laos, Cambodia, southern Vietnam, Peninsular Malaysia (*C. m. maculiceps*), and Koh Tao, Surat Thani Province, in the Gulf of Siam, Thailand (*C. m. hughi*). **Status** Not Evaluated.

CHINESE COBRA ☠
Naja atra PLATE 61
Measurement TL 1,650mm **Identification** Body moderate; head large, distinct from neck; hood short; single preocular; postoculars 2–3; supralabials 7; Supralabials III–IV contact orbit; 4 infralabials contact anterior chin shields; eye moderate; pupil rounded; tail short; dorsals smooth; midbody scale rows 19 or 21; ventrals 161–180; subcaudals 37–51. **Coloration** Dorsum black, grey or brown, sometimes with narrow white cross-bars that are frequently split into double or quadruple bands); head uniform dark brown; supralabials and infralabials pale brown; hood mark is a mask, a monocle, a horseshoe or an O-shape; gular region pinkish-buff; venter yellow or cream. **Habitat and Behaviour** Inhabits the lowlands to the mid-hills, in habitats as diverse as wet rice fields, coastal lowlands and the proximity of human habitations, reaching montane limits at an elevation of 2,000m asl. Diurnal and crepuscular, and terrestrial. Rarely spits venom. Diet consists of fish, frogs, birds, rodents, other snakes and lizards. Oviparous; clutches comprise 6–23 eggs. Hatchlings 270mm. **Distribution** Northern Laos, northern Vietnam. Also southern and eastern China, including Taiwan. **Status** Not Evaluated.

MONOCLED COBRA ☠
Naja kaouthia PLATE 61
Measurement TL 2.3m **Identification** Body robust; head large, distinct from neck; frontal short, squarish; a cuneate usually present; single preocular; postoculars 2–3; supralabials 7; Supralabials III–IV contact orbit; 4 infralabials contact anterior chin shields; hood rounded; eye moderate; pupil rounded; tail short; dorsals smooth, glossy; midbody scale rows 19–21 (rarely 23); ventrals 164–197; subcaudals 43–61, paired. **Coloration** Dorsum brown, greyish-brown, blackish-brown or pale yellow; some with darker bands; hood marking typically a light circle, or mask-shaped with a dark centre; 1–2 dark spots sometimes present in pale oval portion; light throat colour with paired lateral spots; rest of venter similar to dorsum or with dark pigmentation towards tail; subcaudals dark-edged. **Habitat and Behaviour** Inhabits more mesic regions than related species, and associated with forests, agricultural fields and plantations, from the plains to 820m asl. Crepuscular and terrestrial; also known to swim in lakes and rivers. Rarely spits venom. Diet consists of rodents, frogs, fish and other snakes. Oviparous; clutches comprise 15–30 eggs. Incubation period 50 days. Hatchlings 200–350mm. **Distribution** Myanmar, central, south-western and peninsular Thailand, southern Laos, Cambodia, southern Vietnam and northern Peninsular Malaysia. Also northern, eastern and north-eastern India, Nepal, Bangladesh and southern China (Sichuan and Yunnan Provinces). **Status** Not Evaluated.

MANDALAY COBRA ☠
Naja mandalayensis PLATE 61
Measurement TL 1,400mm **Identification** Body robust; head large, distinct from neck; single preocular; postoculars 3; supralabials 7; Supralabials III–IV contact orbit; infralabials 8; 4 infralabials contact anterior chin shields; fangs modified for spitting, venom discharge orifice small; hood oval-elongated; eye moderate; pupil rounded; tail short; dorsals smooth; midbody scale rows 19 or 21 (rarely 22); ventrals 173–185; subcaudals 50–58, paired; anal entire. **Coloration** Dorsum mid-brown to dark brown with pale interstitial skin; dorsals with light cross-bars; hood marking comprises an indistinct spectacle, especially in juveniles, or unpatterned; chin and throat dark, separated from first dark band by 2–4 ventrals, either pale or densely mottled, followed by 2–3 broad dark brown bands; remainder of venter cream with occasional dark mottling. **Habitat and Behaviour** Inhabits dry deciduous forests, and also the vicinity of villages and agricultural areas. Nocturnal and terrestrial. Diet and reproductive habits unstudied. **Distribution** Endemic to central Myanmar (dry zone, including parts of Sagaing, Mandalay and Magway Divisions). **Status** Not Evaluated.

INDO-CHINESE SPITTING COBRA ☠
Naja siamensis PLATE 61
Measurement TL 1,600mm **Identification** Body robust; head large, distinct from neck; snout rounded; single cuneate; single preocular; postoculars 2–3; supralabials 7; Supralabials III–IV contact orbit; infralabials 8; 4 infralabials contact anterior chin shields; hood oval; eye moderate; pupil rounded; tail short; dorsals smooth; midbody scale rows 19 or 21 (rarely 22); ventrals 153–174; subcaudals 41–54, basal ones undivided. **Coloration** Dorsal colour variable, ranging from contrasting black and white pattern, venter white with or without black or brown bars, to black or grey with white speckling; the dark dorsum may be interrupted by light cross-bars; dorsum occasionally unpatterned light or dark brown; hood marking absent, or U-, V- or H-shaped. **Habitat and Behaviour** Inhabits dry plains and low hills. Nocturnal and terrestrial. Capable of spitting venom. Diet consists of small mammals. Reproductive habits unstudied. **Distribution** Thailand (throughout except the southern peninsula), southern and western Laos, Cambodia (Trapeang Chan) and southern Vietnam (Kon Tum, Gia Lai, Dong Nai, Binh Phuoc, Tay Ninh, Ho Chi Minh City and Kien Giang Provinces). **Status** Not Evaluated.

EQUITORIAL SPITTING COBRA ☠
Naja sputatrix PLATE 61
Measurement TL 1,500mm **Identification** Body robust; head large, distinct from neck; snout rounded; rostral visible from above; single preocular; postoculars 2–3; supralabials 7; Supralabials III–IV contact orbit; infralabials 9; 4 infralabials contact anterior chin shields; hood elongated; eye moderate; pupil rounded; tail short; dorsals smooth; midbody scale rows 19 or 21 (rarely 18); ventrals 160–187; subcaudals 42–56, divided. **Coloration** Dorsum blackish-grey (west Java), or silvery or brown (east Java and the Lesser Sundas); hood pattern typically chevron-shaped, or

occasionally mask-, horseshoe- or spectacle-shaped, or unpatterned; forehead similar to body; throat pattern not clearly defined; venter yellowish-cream, sometimes with faint scattered spots. **Habitat and Behaviour** Inhabits lowland dry deciduous forests within rocky biotope. Nocturnal as well as diurnal, and terrestrial as well as arboreal, found on trees up to 11m above substratum. Diet consists of toads, rodents, lizards and other snakes. Oviparous; clutches comprise 16–26 eggs, measuring 40–51 x 23–31mm. Incubation period 88 days. Hatchlings 240–284mm. **Distribution** Java and Bali. Also Lombok, Alor, Sumbawa, Komodo, Flores, Lomblen and Alor. Occurence on Sulawesi and Timor requires verification. **Status** Not Evaluated.

SUMATRAN COBRA
Naja sumatrana PLATE 61
Measurement TL 1,500mm **Identification** Body robust; head large, distinct from neck; snout rounded; hood rounded in adults, more elongated in juveniles; single preocular; postoculars 2–3; Supralabials 7; Supralabials III–IV contact orbit; 4 infralabials contact anterior chin shields; eye moderate; pupil rounded; tail short; dorsals smooth; midbody scale rows 15–19; ventrals 179–206; subcaudals 40–57; basal ones typically entire. **Coloration** Dorsal coloration variable, related to geographical origin and size. Northern population from Peninsular Malaysia bluish-black; hood unpatterned; juveniles have pale throat with lateral spots, and often a median spot; dorsum with pale narrow cross-bars. Sumatran population pale to dark brown; hood unpatterned; dorsum with ca 12 narrow pale cross-bars in juveniles. Bornean population metallic bluish-black, hood unpatterned; ca 12 narrow, chevron-shaped white or yellowish-cream cross-bars; throat pale, anterior ventrals bright yellow; rest of venter dark brown or brownish-grey. **Habitat and Behaviour** Inhabits lightly forested areas in the lowlands and mid-hills. Nocturnal and terrestrial, and known to enter human habitations in towns and villages. Sprays a fine jet of venom to a distance of up to 1m. Diet includes rats and other small vertebrates. Reproductive habits unstudied. **Distribution** Southern Peninsular Thailand, Peninsular Malaysia, Singapore, Sumatra, Pulau Bangka, Pulau Belitung, the Riau Archipelago and Borneo. Also Palawan and the Calamianes Archipelago in the south-eastern Philippines. **Status** Not Evaluated.

KING COBRA
Ophiophagus hannah PLATE 61
Measurement TL 5.85m **Identification** Body robust in adults, slender in juveniles; head large in adults, distinct from neck; paired postoccipital scales present; single preocular; postoculars 2; supralabials 7; Supralabials III–IV contact orbit; infralabials 8; eye large; pupil rounded; tail short; dorsals smooth; midbody scale rows 15; ventrals 215–270; subcaudals 74–125, paired or single, or a combination of the two; anal entire. **Coloration** Dorsum brownish-black, nearly unpatterned to distinctly banded in adults, depending on population, with 27–84 yellow bands; scales on posterior and tail slightly lighter centrally; chin and throat yellow, rest of venter dark grey; juveniles dark brown or black with pale yellow or orange bands. **Habitat and Behaviour** Inhabits primary forests, including semi-evergreen, evergreen, moist deciduous and tropical dipterocarp forests and mangrove swamps, and sometimes observed in more open areas such as alluvial grasslands, from sea level to montane forests at an elevation of 2,181m asl. Diurnal and terrestrial, with some populations more arboreal than others, and juveniles more arboreal than adults. Dangerously toxic, with several authenticated cases of bites on humans. Diet includes snakes and monitor lizards. Oviparous; clutches comprise 20–51 eggs, measuring 57.6–64.3 x 32.2–36mm, which are laid in a mound nest of fallen leaves that the female constructs with the aid of her body and tail. Incubation period 63–77 days. Hatchlings 288–640mm. **Distribution** Myanmar, Thailand, Cambodia, Laos, Vietnam, Peninsular Malaysia, Singapore, Sumatra, Pulau Simeulue, Pulau Galang, Pulau Bangka, Pulau Belitung, Borneo, Java and Bali. Also eastern Pakistan, India, Bhutan, Nepal, southern and eastern China, and the Philippines. **Status** Not Evaluated.

KELLOG'S CORAL SNAKE
Sinomicrurus kelloggi PLATE 61
Measurement TL 800mm **Identification** Body slender, cylindrical; head short and rounded; diameter of eye subequal to its distance from mouth; single preocular; postoculars 2; supralabials 7; Supralabials III–IV contact orbit; temporals 1 + 2; eye large; pupil rounded; tail short, pointed; vertebral scales not enlarged; dorsals smooth; midbody scale rows 13; ventrals 184; subcaudals 31, divided. **Coloration** Dorsum reddish-brown with 17 narrow dark cross-bars on body and 8 on tail, faintly edged with white; forehead black with crescentic pale mark on snout; chevron at back of head; venter pale orange; each ventral with dark blotch on sides. **Habitat and Behaviour** Inhabits hill forests, ascending to submontane limits, at elevations of 600–1,500m asl. Nocturnal and terrestrial, associated with leaf litter. Daytime retreat consists of logs and other vegetation. Diet consists of other snakes. Oviparous; clutches comprise 5–8 eggs. **Distribution** Northern Laos and Vietnam. Also eastern China. **Status** Not Evaluated.

MacCLELLAND'S CORAL SNAKE
Sinomicrurus macclellandi PLATE 61
Measurement TL 840mm **Identification** Body slender, cylindrical; head short and rounded; single preocular; postoculars 2; supralabials 7; Supralabials III–IV contact orbit; infralabials 6; temporals 1 + 1; eye small, its diameter less than its distance from mouth; pupil rounded; tail short, pointed; dorsals smooth; vertebral scales not enlarged; midbody scale rows 13; ventrals 182–244; subcaudals 25–36, paired; anal divided. **Coloration** Dorsum reddish-brown with 23–40 narrow yellow or pale brown-edged, black

stripes; head black with cream band behind eyes; tail with 2–6 black bands; transverse black bars reduced to transverse vertebral spots; venter yellowish-cream with black bands or squarish marks. Several poorly diagnosed subspecies have been described. **Habitat and Behaviour** Inhabits the lowlands and mid-hills in temperate evergreen forests at elevations of 55–2,500m asl. Nocturnal, and terrestrial and subfossorial, sheltering under loose soil or in vegetation. Sluggish and shy in habits, flattening itself as a threat response. Diet consists of other snakes, and lizards such as skinks. Oviparous; clutches comprise 6–14 eggs, measuring 20–33.3 x 10.9–12mm. **Distribution** Northern Myanmar (Kachin State), northern and north-eastern Thailand (Doi Suthep, Chiang Mai Province, Phu Luang, Loei Province, Sakaerat nr Khao Yai, Nakornratchasima Province and Khao Ang Ru Nai, Chachaengsao Province), Laos and Vietnam. Also northern to north-eastern India, Nepal, Bangladesh, and southern and eastern China. **Status** Not Evaluated.

HORNED SEA SNAKE
Acalyptophis peronii PLATE 62
Measurement TL 1,250mm **Identification** Body robust; head small, shields symmetrical, some with spines on posterior edges; prefrontals absent; preoculars 1–2; postoculars 1–2; supralabials 6–7; Supralabials III–IV contact orbit; infralabials 5–6; maxillary teeth behind front fangs 5–8; tail flattened; dorsals keeled; midbody scale rows 23–31 (rarely 21 or 32); ventrals 142–222; preanal entire. **Coloration** Dorsum light brown with dark bands encircling body, widest on vertebral region and narrowing ventrally; forehead pale brown; venter paler. **Habitat and Behaviour** Inhabits shallow seas, including reefs at medium depth, and estuaries. Diet consists of burrowing gobies. Ovoviviparous; clutches comprise 10 neonates. **Distribution** Thailand, Vietnam, Peninsular Malaysia and Singapore. Range extends to southern China, northern Australia and New Caledonia. **Status** Not Evaluated.

BEADED SEA SNAKE
Aipysurus eydouxii PLATE 62
Measurement TL 1,500mm **Identification** Body slender; head shields symmetrical, some with spines on posterior edges; supralabials 6, not divided horizontally; venom gland and fangs reduced; maxillary teeth behind front fangs 8–12; pupil rounded; tail flattened; dorsals smooth; midbody scale rows 17; ventrals distinct throughout, 124–155; subcaudals 27–30; preanal divided. **Coloration** Dorsum brownish-olive with 44–55 dark-edged, tan or yellowish-cream bands, which may be broken up on vertebral region; bands widen on flanks; forehead dark brown or olive, nearly black in juveniles. **Habitat and Behaviour** Inhabits shallow coastal waters. Diet consists of benthic fish eggs; to ingest these, strong throat musculature to improve suction, consolidation of lip scales to increase rigidity, reduction and loss of teeth, and decreased body size appear to have evolved.

Reproductive habits unstudied. **Distribution** Thailand, Vietnam, east coast of Peninsular Malaysia, Singapore, Sumatra, Borneo and Java. Also the Philippines, New Guinea, Australia and New Caledonia. **Status** Not Evaluated.

STOKES'S SEA SNAKE
Astrotia stokesii PLATE 62
Measurement TL 1,800mm **Identification** Body short, robust; head enlarged; prefrontal not in contact with supralabial; single preocular; postoculars 2; supralabials 8–10; Supralabials IV–VI contact orbit; infralabials 11–12; maxillary teeth behind front fangs 6–7; tail flattened; dorsals imbricate and keeled; midbody scale rows 46–59; ventrals 226–286, divided into pairs of overlapping scales, except on throat; preanals strongly enlarged. **Coloration** Dorsum yellowish-grey or pale brown with 24–37 black annuli on body and 4–7 on tail, the dark areas slightly wider than the pale ones; head black, dark olivaceous or yellow; ca 20 black spots on midventral line; on tail, last 3 black annuli fuse to form paired yellow spots on each side. **Habitat and Behaviour** Inhabits waters off the coast at depths of 60–70m, and may be associated with coral reefs. Diet consists of benthic fish. Ovoviviparous; clutches comprise 12 neonates, measuring 300mm. **Distribution** Thailand, Vietnam, Peninsular Malaysia, Singapore, Sumatra and Borneo. Also coastal Pakistan, India and Sri Lanka, and in the east, eastern China to Australia. **Status** Not Evaluated.

BEAKED SEA SNAKE
Enhydrina schistosa PLATE 62
Measurement TL 1,580mm **Identification** Body robust; head nearly indistinct from neck; rostral scale extends ventrally; mentals elongated; single preocular; postoculars 1–2; supralabials 7–8; Supralabials III–IV or IV contact orbit; maxillary teeth behind front fangs 3–4; tail flattened; dorsals keeled; midbody scale rows 49–66; ventrals small, numbering 239–322; preanal scales slightly enlarged; . **Coloration** Dorsum greyish-olive or silvery-grey; forehead darker; body with indistinct darker markings, sometimes forming 40–60 transverse dark bands; suborbital stripe absent; venter cream anteriorly, darkening to greenish-yellow caudally; juveniles dark grey dorsally; venter cream with dark grey or black annuli. **Habitat and Behaviour** Inhabits shallow (under 5m depth) sea coasts and mangrove swamps, and also known to travel upriver into freshwater habitats. Active by day and night. Diet consists of marine catfish. Ovoviviparous; clutches comprise 4–33 neonates, measuring 150–280mm. **Distribution** Myanmar, Thailand, Cambodia, Vietnam, Peninsular Malaysia, Singapore, Sumatra and Borneo. Range extends from the Persian Gulf, through southern Asia, east to New Guinea and northern Australia. **Status** Not Evaluated.

AAGAARD'S SEA SNAKE
Hydrophis aagaardi PLATE 62
Measurement TL 1,030mm **Identification** Body moderately robust; head elongated; single preocular

and postocular; supralabials 8; tail flattened; dorsals keeled, tuberculate; midbody scale rows 39–47; ventrals 276–348. **Coloration** Dorsum greyish-yellow or greenish-grey with 48–76 dark olive or black bands; venter yellowish-white. **Habitat and Behaviour** Inhabits deep clear waters up to 32km from the coast. Diet consists of fish and shrimps. Reproductive habits unstudied. **Distribution** Thailand, Peninsular Malaysia, Sumatra and Borneo. **Status** Not Evaluated.

BLACK-HEADED SEA SNAKE ☠
Hydrophis atriceps PLATE 62
Measurement TL 1,200mm **Identification** Body robust posteriorly; head very small; slender neck and forebody; dorsals with central keel; scales juxtaposed or slightly imbricate; single preocular and postocular; supralabials 6–7; Supralabials III–IV contact orbit; infralabials 7–8; maxillary teeth behind front fangs 5–6; anterior temporals 2; tail flattened; midbody scale rows 39–49; ventrals 320–455. **Coloration** Dorsum including forehead and neck dark olive to black with oval pale yellow spots on sides, sometimes connected to form 50–75 cross-bars; posteriorly grey; yellow spot between nostril and eye or behind eye; venter cream; rhomboidal dark spots sometime extend down flanks to form annuli in juveniles. **Habitat and Behaviour** Inhabits river mouths. Diet includes eels and marine invertebrates. Ovoviviparous; clutches comprise 2–3 neonates. **Distribution** Thailand, Cambodia, Vietnam, Singapore and presumably Peninsular Malaysia. Range extends to the Philippines, New Guinea and northern Australia. **Status** Not Evaluated.

CAPTAIN BELCHER'S SEA SNAKE ☠
Hydrophis belcheri PLATE 62
Measurement TL 932mm **Identification** Body robust to moderate; single preocular; postoculars 1–3 (typically 2); cuneate absent; supralabials 7 (rarely 6 or 8); Supralabial IV contacts orbit; infralabials 8–9 (rarely 10); maxillary teeth behind front fangs 6–9; tail flattened; dorsals imbricate and hexagonal or rounded; midbody scale rows 32–36; ventrals 278–313, widened; subcaudals 28–42. **Coloration** Dorsum olive-green with 48–64 dark bands, broad dorsally and narrow on flanks; head black flecked with olive, and with traces of horseshoe-shaped yellow mark on prefrontals and eye. **Habitat and Behaviour** Nothing known of its natural history. **Distribution** Thailand, Vietnam, Borneo and Java. Also New Guinea and Australia. **Status** Not Evaluated.

TWO-WATTLED SEA SNAKE ☠
Hydrophis bituberculatus NOT ILLUSTRATED
Measurement TL 1,050mm **Identification** Body robust; maxillary teeth behind front fangs 7–8; supralabials 7–8; infralabials 8–9; tail flattened; dorsal scales with double-crested keel; midbody scale rows 43–50; ventrals 247–290. **Coloration** Dorsum of body and tail grey, with dark bands that encircle body, numbering 37–51 on body and 6–10 on tail; head blackish-grey with pale ring around eye and light supralabials; venter whitish-grey, greyish-black in anterior part of ventral. **Habitat and Behaviour** Habitat unknown: Thai individuals obtained at Phuket Harbour. Diet includes eels. Ovoviviparous; clutches comprise 3 neonates. **Distribution** West coast of Peninsular Thailand. Also west coast of Sri Lanka. **Status** Not Evaluated.

RAJAH BROOK'S SEA SNAKE ☠
Hydrophis brookii PLATE 62
Measurement TL 1,035mm **Identification** Body robust, elongated; head small; single preocular; postoculars 1–2; supralabials 5–6; Supralabials III–IV contact orbit; infralabials 7–8; maxillary teeth behind front fangs 5; tail flattened; dorsals keeled; midbody scale rows 37–45; ventrals 328–453; preanals enlarged. **Coloration** Dorsum greyish-black with 60–80 dark grey cross-bars or bands, twice broader than their pale interspaces; bands encircle body anteriorly; head black with curved yellow mark across snout along sides of head; venter yellowish-white. **Habitat and Behaviour** Inhabits shallow coastal waters. Diet includes anguiliform eels, and also gobies. Ovoviviparous; clutches comprise 7 neonates. **Distribution** Southern Thailand, Vietnam, Peninsular Malaysia, Singapore, Sumatra, Borneo and Java. **Status** Not Evaluated.

BLUE-GREY SEA SNAKE ☠
Hydrophis caerulescens PLATE 62
Measurement TL 1,090mm **Identification** Body moderately robust; head small, upper jaw projecting; scales quadrangular or hexagonal, weakly imbricate or juxtaposed; single preocular; postoculars 1–2; supralabials 7–8; Supralabials III–IV contact orbit; maxillary teeth behind front fangs 12–18; tail flattened; dorsals keeled; midbody scale rows 38–54; ventrals 253–334, less than twice as large as adjacent scales; preanals divided. **Coloration** Dorsum bluish-white or bluish-grey with 40–60 dark bands that narrow on lower flanks; head dark; juveniles with U-shaped yellow or cream mark on head; venter cream with bands twice as wide as interspaces, tapering ventrally; pattern distinct in juveniles, indistinct in old adults. **Habitat and Behaviour** Inhabits shallow seas, estuaries and at least one inland lake. Diet consists of burrowing gobies. Ovoviviparous; clutches comprise 2–6 neonates. **Distribution** Myanmar, Thailand, Vietnam, Peninsular Malaysia, Singapore, Sumatra and Borneo. Range extends from Pakistan and India, to south-eastern China and western Indonesia, and east to northern Australia. **Status** Not Evaluated.

CANTOR'S SEA SNAKE ☠
Hydrophis cantoris NOT ILLUSTRATED
Measurement TL 1,450mm **Identification** Body slender and elongated; head small; snout projecting beyond lower jaw; single preocular and postocular; supralabials 6; Supralabials III–IV contact orbit; Supralabial III contacts prefrontal; infralabials 7–8; maxillary teeth behind front fangs 5–6; tail flattened; dorsals keeled, juxtaposed; midbody scale rows 41–48; ventrals 404–468, divided posteriorly by longitudinal

fissure; preanals feebly enlarged. **Coloration** Dorsum dark olive-grey with yellow cross-bars or lateral spots on anterior of body; posterior, grey; head black in juveniles, grey or yellowish-green in adults; dark stripe along ventrals sometimes present. **Habitat and Behaviour** Inhabits shallow coastal waters. Diet includes eels and marine invertebrates. Oviviviparous; clutches comprise 10 neonates. **Distribution** Myanmar, Thailand and Peninsular Malaysia. Range extends from Pakistan to South-East Asia. **Status** Not Evaluated.

ANNULATED SEA SNAKE
Hydrophis cyanocinctus PLATE 62
Measurement TL 1,885mm **Identification** Body elongated, moderately robust anteriorly, thickening posteriorly; head small, almost indistinct from neck; single preocular; postoculars 2; supralabials 7–8; Supralabials III–V contact orbit; maxillary teeth behind front fangs 5–8; tail flattened; dorsals strongly keeled; midbody scale rows 37–47; ventrals 290–390; preanals enlarged. **Coloration** Dorsum typically olive or yellow with numerous transverse bluish-black bands that may encircle body; head yellowish-green; venter yellowish-cream. **Habitat and Behaviour** Inhabits shallow coastal waters. Often stranded on beaches, but capable of some movement on land, unlike other sea snakes of the genus. Nocturnal. Diet consists of gobies, eels, other elongated fish and marine invertebrates. Ovoviviparous; clutches comprise 3–16 neonates, measuring 381mm. **Distribution** Thailand, Vietnam, Peninsular Malaysia, Singapore and Borneo. Range extends from the Persian Gulf to eastern China, the Philippines and Japan. **Status** Not Evaluated.

BANDED SEA SNAKE
Hydrophis fasciatus PLATE 63
Measurement TL 1,100mm **Identification** Body slender anteriorly, thickening posteriorly; head small; scales juxtaposed or slightly imbricate; loreal absent; single preocular; postoculars 1–2; supralabials 6–7 (rarely 5); Supralabials III–IV contact orbit; infralabials 6–7; maxillary teeth behind front fangs 5–6; anterior temporals 2; tail flattened; midbody scale rows 47–58; ventrals 410–514; preanal divided. **Coloration** Dorsum, including head and neck, beige, dark olive or black with oval pale yellow spots on flanks, sometimes connected as cross-bars; posterior grey in some individuals; venter cream; rhomboidal dark spots along flanks may form annuli in juveniles. **Habitat and Behaviour** Inhabits shallow coastal waters. Nocturnal. Diet includes anguiliform eels. Reproductive habits unstudied. **Distribution** Myanmar, Thailand, Vietnam, Peninsular Malaysia, Singapore, Sumatra, Borneo and Java. Also India and Pakistan. **Status** Not Evaluated.

NARROW-HEADED SEA SNAKE
Hydrophis gracilis PLATE 63
Measurement TL 950mm **Identification** Body slender anteriorly, thickening posteriorly; head small, upper jaw projecting; neck slender; juxtaposed scales on thickest part of body; single preocular and postocular; supralabials 6; Supralabials III–IV contact orbit; Supralabial III not in contact with prefrontal; infralabials 7–8; maxillary teeth behind front fangs 5–6; tail flattened; dorsals keeled; midbody scale rows 29–43; ventrals 212–298, divided by longitudinal fissure; preanals weakly enlarged. **Coloration** Dorsum bluish-grey with 40–60 dark blue bands or lateral blotches; head blue-black; supralabials cream; neck with dark bands with narrow light grey interspaces; venter cream. **Habitat and Behaviour** Inhabits deep and turbid coastal waters, and may be a bottom-dweller. Diet includes anguiliform eels. Ovoviviparous; clutches comprise up to 6 neonates. **Distribution** Myanmar, Thailand, Vietnam, Peninsular Malaysia, Singapore, Sumatra, Borneo and Java. Range extends from the Persian Gulf to eastern Asia, Australia and Melanesia. **Status** Not Evaluated.

PLAIN SEA SNAKE
Hydrophis inornatus NOT ILLUSTRATED
Measurement TL 1,089mm **Identification** Body moderate; head narrow; single preocular; postoculars 2; supralabials 7; infralabials 9; maxillary teeth behind front fangs 9–12; tail flattened; midbody scale rows 43–44; ventrals 233–253; subcaudals 39–43. **Coloration** Dorsum bluish-grey; juveniles with 50–65 dark blotches or cross-bars; tail with greyish-black bands with narrow cream interspaces that disappear posteriorly; venter white, the colours of the dorsum and venter meeting on flanks. **Habitat and Behaviour** Inhabits seas off the mainland and also off major islands. Nocturnal. Diet and reproductive habits unstudied. **Distribution** Peninsular Malaysia, Singapore and Java. Also the Philippines, eastern China, New Guinea, Australia and elsewhere in the eastern Indian Ocean. **Status** Not Evaluated.

KLOSS'S SEA SNAKE
Hydrophis klossi PLATE 63
Measurement TL 1,190mm **Identification** Body robust, elongated; head small; snout projecting; single preocular and postocular; supralabials 5–6; Supralabials III–IV contact orbit; infralabials 6–7; maxillary teeth behind front fangs 5–6; tail flattened; dorsals smooth or weakly keeled; midbody scale rows 31–39; ventrals 360–430, with 2 tubercles; preanals enlarged. **Coloration** Dorsum blackish-grey to olivaceus-yellow with 50–75 dark cross-bars; bars broadest dorsally and broader than their interspaces; black vertebral stripe sometimes present; forehead black or olivaceous, sometimes with indistinct horseshoe-shaped mark. **Habitat and Behaviour** Inhabits shallow coastal waters. Nothing else known of its natural history. **Distribution** Thailand, Cambodia, southern Vietnam, Peninsular Malaysia, Singapore, Sumatra and Borneo. **Status** Not Evaluated.

LAMBERT'S SEA SNAKE
Hydrophis lamberti NOT ILLUSTRATED
Measurement TL 1,250mm **Identification** Body robust; head large; single preocular; postoculars 2;

supralabials 6–8; infralabials 8–11; maxillary teeth behind front fangs 9–13; tail flattened; midbody scale rows 45–56; ventrals 258–306; subcaudals 34–50. **Coloration** Dorsum greyish-yellow or brown with 26–36 large rounded bands anteriorly, which are replaced by cross-bars with broad light grey interspaces posteriorly; tail with 4–6 bands; venter cream. **Habitat and Behaviour** Inhabits seas off the mainland and also off major islands. Diet consists of marine catfish. Reproductive habits unstudied. **Distribution** Thailand, Vietnam, Peninsular Malaysia and Singapore. Also the Philippines. **Status** Not Evaluated.

PERSIAN GULF SEA SNAKE

Hydrophis lapemoides   PLATE 63

Measurement TL 960mm **Identification** Body robust; head moderate; single preocular; postoculars 2–3; supralabials 8; Supralabials III–IV or III–V contact orbit; infralabials 8; maxillary teeth behind front fangs 8–13; tail flattened; dorsals weakly tuberculate or with short keel; midbody scale rows 40–57; ventrals 288–404; subcaudals 36–56. **Coloration** Dorsum grey (yellowish-green in juveniles) with 29–52 black bands in the shape of rhombic spots, wider on vertebral region and narrower on flanks; head dark dorsally with curved yellow or white mark from forehead to back, which disappears with growth; tail with 5–8 bands that disappear in adults. **Habitat and Behaviour** Inhabits seas with gravel bottom at depths of 27–30m. Diet includes fish such as gobies and mullets. Ovoviviparous; clutches comprise 1–4 neonates, measuring 260mm. **Distribution** Thailand, Malaysia and Singapore. Range extends from the Persian Gulf and the Gulf of Oman, to the Malacca Straits. **Status** Not Evaluated.

LESSER DUSKY SEA SNAKE

Hydrophis melanosoma   PLATE 63

Measurement TL 1,390mm **Identification** Body robust; single preocular and postocular; supralabials 6–7; Supralabials III–IV or III–V contact orbit; infralabials 7–8; maxillary teeth behind front fangs 5–8; tail flattened; dorsals strongly keeled; midbody scale rows 35–45; ventrals 266–368, with 2 tubercles; preanal divided. **Coloration** Dorsum blackish-grey with 50–70 wide white bands, subequal to their interspaces; head and neck black with subovoid yellow marks; forehead with yellow speckles; venter cream or yellow. **Habitat and Behaviour** Inhabits sea coasts off the mainland and large islands, and also known to enter rivers. Diet includes anguiliform eels. Reproductive habits unstudied. **Distribution** Southern Peninsular Malaysia, Singapore, Sumatra and Borneo. Range extends to northern Australia. **Status** Not Evaluated.

BLACK-RINGED SEA SNAKE

Hydrophis nigrocinctus   NOT ILLUSTRATED

Measurement TL 1,080mm **Identification** Body robust; mental scale large, not concealed in mental groove; preoculars 1–2; postoculars 1–2; supralabials 7–9; Supralabials III–V contact orbit; maxillary teeth behind front fangs 1–3; palatine teeth 7–9, subequal to pterygoid teeth; tail flattened; midbody scale rows 39–45, imbricate, keeled; ventrals 296–330, distinct, not twice as large as adjacent scales; preanal scales enlarged. **Coloration** Dorsum olive to brown with 40–60 narrow dark cross-bars; head yellow with supraorbital stripe surrounding black patch on crown that extends forwards to prefrontals; venter yellow. **Habitat and Behaviour** Suspected to be associated with deep and turbid waters with sandy bottoms. Diet and reproductive habits unstudied. **Distribution** Myanmar. Also Sunderbans off eastern India and Bangladesh. Restricted to the Bay of Bengal. **Status** Not Evaluated.

OBSCURE-PATTERNED SEA SNAKE

Hydrophis obscurus   NOT ILLUSTRATED

Measurement TL 1,190mm **Identification** Body robust and elongated; scales rounded or with obtuse tips, imbricate; single preocular; postoculars 1–2; supralabials 6–7; Supralabials III–IV contact orbit; single anterior temporal; maxillary teeth behind front fangs 5–7; tail flattened; midbody scale rows 29–37; ventrals 298–346; preanals moderately enlarged. **Coloration** Dorsum of adults unpatterned grey; juveniles black or bluish-black with 30–60 yellow or cream cross-bars that encircle body and tail; head with curved yellow marking between snout and sides of parietals; venter yellow. **Habitat and Behaviour** Inhabits brackish waters at mouths of rivers, and large coastal lagoons and saltwater lakes. Diet and reproductive habits unstudied. **Distribution** Myanmar. Also the east coast of India. **Status** Not Evaluated.

ORNATE SEA SNAKE

Hydrophis ornatus   PLATE 63

Measurement TL 1,150mm **Identification** Body robust; head large; scales hexagonal in shape, feebly imbricate or juxtaposed; single preocular; postoculars 2–3; supralabials 6–8; infralabials 9–12; maxillary teeth behind front fangs 9–13; tail flattened; dorsals tuberculate or with a short keel; midbody scale rows 42–54; ventrals 235–298, anteriorly ventrals twice as large as adjacent scales; subcaudals 38–50; preanals weakly enlarged. **Coloration** Dorsum greyish-olive or light olive to cream with 30–60 broad dark bars or rhomboidal spots separated by narrow interspaces on body, 6–11 on tail; venter yellow or cream. **Subspecies** Two poorly diagnosed subspecies have been described, *H. o. ornatus* (the Persian Gulf, the Indian Ocean, Java and the Gulf of Siam) and *H. o. maresinensis* (the Ryukyu Archipelago of Japan, southern Vietnam to eastern China). Their systematic status is unclear. **Habitat and Behaviour** Inhabits shallow waters with coral reefs, as well as turbid waters near rivers and estuaries. Active by day and night. Diet includes fish of several families that are free swimming and associated with coral reefs and sandy-bottomed seas; specializes in marine catfish. Ovoviviparous; clutches comprise 1–4 neonates, measuring ca 34mm.

Distribution Myanmar, Thailand, Vietnam, Peninsular Malaysia, Singapore, Sumatra and Borneo. Range extends from the Persian Gulf east to the Philippines, New Guinea and Australia. **Status** Not Evaluated.

BROAD-HEADED SEA SNAKE
Hydrophis pachycercos PLATE 63
Measurement TL 1,110mm **Identification** Body robust; head shields large and regular; single preocular; postoculars 3; supralabials 7; Supralabials I–II contact nasals; Supralabials III–IV contact eye; infralabials 10–11; maxillary teeth 7–8; tail flattened; dorsals keeled; midbody scale rows 39–45; ventrals 247–297, distinct throughout, twice the size of adjacent scales; subcaudals 37–44. **Coloration** Dorsum light yellow with light brown bands; head black or dark grey; nasal region green; supralabials cream; venter white. **Habitat and Behaviour** Known from coastal waters of the South China Sea. Diet unstudied. Ovoviviparous; clutches comprise 5–6 neonates. **Distribution** South China Sea, off Vietnam. **Status** Not Evaluated.

SHORT-HEADED SEA SNAKE
Hydrophis parviceps NOT ILLUSTRATED
Measurement TL 890mm **Identification** Body robust; dorsals with greatly developed keels; head long and narrow; single preocular and postocular; supralabials 5; Supralabial II largest and contacts prefrontal; Supralabials III–IV contact orbit; infralabials 7; maxillary teeth behind front fang 6; tail flattened; midbody scale rows 19; ventrals 340–348, distinct throughout, with 1 or more tubercles; preanals enlarged. **Coloration** Dorsum olivaceous anteriorly, greyish-olive posteriorly, with 69 dark bands; bands indistinct ventrally on posterior of body; head and lower neck jet-black; venter pale. **Habitat and Behaviour** Inhabits coastal waters. Nothing else known of its natural history. **Distribution** Endemic to southern Vietnam. **Status** Not Evaluated.

SIBAU RIVER SEA SNAKE
Hydrophis sibauensis PLATE 63
Measurement TL 735mm **Identification** Body slender; head narrower than body; nostrils situated on top of head; forehead shields large and regular; single preocular; single postocular; supralabials 7; Supralabials III–IV contact orbit; infralabials 8; maxillary teeth behind front fangs 1–18; tail flattened; dorsal scales with median keel; ventrals small and distinct throughout; midbody scale rows 35–37; ventrals 257–264. **Coloration** Dorsum grey-brown, darkening posteriorly; 49–58 yellow to light orange bands on body, bands incomplete; forehead black with small yellow spots and arrow-shaped markings; venter including throat black anteriorly, greyish-yellow from midbody to posterior of body. **Habitat and Behaviour** Inhabits a large west-flowing river in Borneo, and known from freshwater section lying more than 1,000km upriver. Diet unstudied. Ovoviviparous; clutches comprise 7 neonates. **Distribution** Endemic to Borneo (Sungei Sibau, a branch of Sungei Kapuas, Kalimantan Province). **Status** Not Evaluated.

SPIRAL SEA SNAKE
Hydrophis spiralis PLATE 63
Measurement TL 2.75m **Identification** Body elongated, moderately robust anteriorly, thickening posteriorly; head and neck slender; single preocular; postoculars 1–2; supralabials 6–8; Supralabials III–V contact orbit; maxillary teeth behind front fangs 6–7; tail flattened; dorsals smooth or keeled; midbody scale rows 33–38; ventrals 295–362, distinct throughout, about twice as broad as adjacent body scales. **Coloration** Dorsum and forehead olive-yellow to olive-brown with 35–50 encircling dark bands, narrower than interspaces; black dorsal spot between bands sometimes present; in juveniles, head nearly black with horseshoe-shaped yellow mark; in adults, head entirely yellow; flanks yellowish-cream; venter yellow or cream; tip of tail dark brown or black. **Habitat and Behaviour** Inhabits deep-water habitats of more than 10m in depth. Diet unstudied. Ovoviviparous; clutches comprise 5–14 neonates, measuring 200–405mm. **Distribution** Myanmar, Thailand, Peninsular Malaysia, Singapore, Sumatra and Borneo. Range extends from the Persian Gulf to Sulawesi (Indonesia) and the Philippines. **Status** Not Evaluated.

NARROW-NECKED SEA SNAKE
Hydrophis stricticollis PLATE 64
Measurement TL 1,050mm **Identification** Body slender; head small; scales on thickest part of body subquadrangular or hexagonal in shape, feebly imbricate or juxtaposed; single preocular; postoculars 1–2; supralabials 7–8; Supralabials III–IV contact orbit; maxillary teeth behind front fangs 8–11; tail flattened; midbody scale rows 45–55; ventrals 374–452, less than twice as large as adjacent scales. **Coloration** Dorsum greyish-olive with 45–65 dark bands (which are absent in adults), broadest dorsally and narrow on flanks; head black or olive; snout and sides of head with yellow markings; venter yellow. **Habitat and Behaviour** Inhabits shallow coastal waters. Diet and reproductive habits unstudied. **Distribution** Myanmar. Also east coast of India. **Status** Not Evaluated.

GARLANDED SEA SNAKE
Hydrophis torquatus PLATE 64
Measurement TL 895mm **Identification** Body elongated, moderately robust; head moderate; single preocular; postoculars 1–2; supralabials 7–8; Supralabials III–IV contact orbit; maxillary teeth behind front fangs 16–19; tail flattened; midbody scale rows 35–42; ventrals 242–343; preanals enlarged. **Coloration** Dorsum greyish-tan or greenish-grey with more than 50 dark brown bands on body and tail; forehead dark greyish-tan or dark olive, sometimes with horseshoe-shaped yellow mark; venter yellowish-cream. **Habitat and Behaviour** Inhabits brackish waters off river mouths. Nothing else known

of its natural history. **Distribution** Thailand, Cambodia, Vietnam, Peninsular Malaysia, Singapore, Sumatra and Borneo. **Status** Not Evaluated.

SADDLE-BACKED SEA SNAKE ☠
Kerilia jerdoni   PLATE 64

Measurement TL 1,000mm **Identification** Body robust, subcylindrical, nearly uniform in diameter; head short, nearly indistinct from neck; single preocular and postocular; supralabials 6; Supralabials III–IV contact orbit; prefrontals usually not in contact with supralabials; maxillary teeth behind front fangs 7–9; tail flattened; dorsals keeled and imbricate; midbody scale rows 21–23 (rarely 19); ventrals 200–278, small, distinct; preanal divided. **Coloration** Dorsum yellowish-olive with black dorsal spots or rhomboidal marks, forming bands, especially in juveniles, adults showing less contrast; forehead dark grey; sometimes, a black band across neck; venter yellow or cream; tail black, some scales with light centres. **Subspecies** Two poorly diagnosed subspecies have been described: *K. j. jerdoni* and *K. j. siamensis*. Their systematic status remains unclear. **Habitat and Behaviour** Inhabits shallow coastal waters. Diet consists of eels. Ovoviviparous; clutches comprise 4 neonates. **Distribution** Myanmar, Thailand, Vietnam, Peninsular Malaysia, Singapore, Sumatra and Borneo. Range extends from west coast of India and Sri Lanka to South-East Asia and eastern China. **Status** Not Evaluated.

ANNANDALE'S SEA SNAKE ☠
Kolpophis annandalei   PLATE 64

Measurement TL 520mm **Identification** Body short, robust; head small, narrower than widest part of body; fragmented head scales; supralabials 9–11; large scale separates Supralabials V and VI from orbit; tail flattened; midbody scale rows 70–100; ventrals 320–368. **Coloration** Dorsum grey-blue with 35–46 dark bands, the bands being broader than their pale interspaces; head oliveaceous; venter pale yellow or cream. **Habitat and Behaviour** Inhabits coasts, and probably associated with shallow waters with sandy bottoms; also recorded from freshwaters. Diet consists of fish. Reproductive habits unstudied. **Distribution** East coast of peninsular Thailand, Cambodia, Vietnam, Peninsular Malaysia, Singapore, Borneo and possibly Sumatra. **Status** Not Evaluated.

SHORT SEA SNAKE ☠
Lapemis curtus   PLATE 64

Measurement TL 972mm **Identification** Body robust, short; head broad and short; scales squarish or hexagonal, lowermost rows especially in males with short keel; head shields entire; nostrils dorsally situated; single preocular; postoculars 1–2; supralabials 7–8; Supralabials III–IV contact orbit; maxillary teeth behind front fangs 3–6; tail flattened; midbody scale rows 25–43; ventrals 114–274; preanals weakly enlarged. **Coloration** Dorsum in juveniles brownish-grey to olive with 35–55 olive to dark grey bands that taper to a point on flanks; adults typically unpatterned olive to dark grey; venter unpatterned yellow or cream, or with narrow dark ventral stripe or irregular band. **Habitat and Behaviour** Inhabits coasts with muddy bottoms, and also known from coral reefs, at depths of 6–15m. Active by day and night. Diet consists of fish, especially eels, gobies and catfish; squids and other marine invertebrates are also eaten. Ovoviviparous; clutches comprise 1–6 neonates, measuring 250–373mm. **Distribution** Myanmar, Thailand, Vietnam, Peninsular Malaysia, Singapore, Sumatra and Borneo. Range extends from the Persian Gulf to the Philippines, Japan and Australia. Two subspecies are recognized by some authorities: *L. c. curtus*, with a distribution from the Persian Gulf to the west coast of India, and *L. c. hardwickii*, with a distribution along the coasts of Sri Lanka and eastern India, China, the Philippines, Japan, New Guinea, Australia and New Caledonia. **Status** Not Evaluated.

YELLOW-LIPPED SEA KRAIT ☠
Laticauda colubrina   PLATE 64

Measurement TL 1,710mm **Identification** Body robust, especially in adult females, cylindrical; nostrils lateral; rostral undivided; nasals separated by internasals; single preocular; postoculars 2; supralabials 7–8; Supralabials III–IV contact orbit; tail flattened; dorsals smooth; midbody scale rows 21–25; ventrals 213–245, broad; subcaudals 29–47; anal divided. **Coloration** Dorsum blue-grey with black annuli numbering 24–64; supralabials yellow; venter cream. **Habitat and Behaviour** Inhabits shallow seas around rocky islets to depths of ca 60m. Comes ashore to lay eggs, bask, rest and digest food. Active by day and night. Non-aggressive species, but with highly toxic venom. Diet includes anguiliform eels. Oviparous; clutches comprise 3–13 eggs, measuring 44.6–92.2 x 20.3–31.1mm. Incubation period ca 90 days. Hatchlings 210mm. **Distribution** Myanmar, Thailand, Vietnam, Peninsular Malaysia, Singapore, Sumatra, Pulau Belitung, Borneo, Java and Bali. Widespread in the Indian and Pacific Oceans, extending eastwards to Polynesia. **Status** Not Evaluated.

LARGE-SCALED SEA KRAIT ☠
Laticauda laticaudata   PLATE 64

Measurement TL 1,100mm **Identification** Body robust in adult females, moderate in adult males; ventrals large, one-third to over half body width; nostrils lateral; nasals separated by internasals; tail flattened; dorsals smooth; midbody scale rows 19; no azygous prefrontal shield; rostral undivided; ventrals 225–243; subcaudals 30–47. **Coloration** Dorsum bright blue with 20–70 black bands; supralabials dark brown or black; venter cream. **Habitat and Behaviour** Inhabits shallow seas and may rest on small rocky islands; occasionally found near fresh water. Active by day and night. Diet includes eels. Reproductive habits unstudied. **Distribution** Myanmar, Thailand, Peninsular Malaysia, Sumatra, Borneo, Java and Bali. Widespread in the Indian and Pacific Oceans, extending eastwards to Polynesia. **Status** Not Evaluated.

YELLOW-BELLIED SEA SNAKE ☠
Pelamis platura PLATE 64
Measurement TL 1,000mm **Identification** Body slender, compressed; head elongated, bill-like and slightly flattened, distinct from neck; preoculars 1–2; postoculars 1–3; supralabials 7–8; Supralabials III–IV, IV–V or V contact orbit; maxillary teeth behind front fangs 7–11; tail flattened; midbody scale rows 49–67; ventrals 264–406, irregular in shape and indistinct after anterior of body; preanals enlarged. **Coloration** Top half of dorsum black or dark brown, bottom half bright yellow or cream; venter light brown or yellow with black spots or bars; tail with diamond-shaped cream or yellow pattern. **Habitat and Behaviour** A species of the open ocean, mostly known from stranded individuals on sea coasts. Diet consists of fish, and most feeding occurs on the surface or close to the surface. Ovoviviparous; clutches comprise 2–6 neonates, measuring 220–260mm. **Distribution** Myanmar, Thailand, Vietnam, Peninsular Malaysia, Singapore, Sumatra, Borneo and Java. Distribution encompasses the Pacific and Atlantic Oceans, including east Africa, east Asia to southern Siberia, to Australia, and on the west, the Gulf of Panama and western North America, in addition to the west coast of Africa in the Atlantic Ocean, as well as Hawaii and the Galapagos Archipelago. **Status** Not Evaluated.

ANOMALOUS SEA SNAKE ☠
Thalassophis anomalus NOT ILLUSTRATED
Measurement TL 810mm **Identification** Body robust; head short, depressed; rostral fragmented into 4–5 parts; single preocular; postoculars 1–2; supralabials 7–9; Supralabials III–V contact orbit; Supralabial II contacts prefrontal; 4 infralabials contact genials; maxillary teeth behind front fangs 5; tail flattened; midbody scale rows 33 (rarely 35), scales hexagonal with strong keel; ventrals 210–256; subequal to dorsals. **Coloration** Dorsum pale grey with 30–36 dark cross-bars, including 3–4 on tail, which are broader than interspaces, narrowing to a point on flanks; venter unpatterned white. **Habitat and Behaviour** Inhabits river mouths and presumably sandy-bottomed seas. Nothing else known of its natural history. **Distribution** Thailand (mouth of Chanthabun River), Cambodia, Vietnam, Peninsular Malaysia, Singapore, Sumatra, Borneo and Java. Also the Sulu Sea (the Philippines), Flores Sea and waters off Maluku (Indonesia). **Status** Not Evaluated.

GREY SEA SNAKE ☠
Thalassophis viperina PLATE 64
Measurement TL 925mm **Identification** Body robust; head short, depressed; forehead scales entire; single preocular; postoculars 1–2; nostril on dorsal surface; supralabials 7–9; supralabials separated from orbit by row of subocular scales; maxillary teeth behind front fangs 5; tail flattened; midbody scale rows 37–50; scales on body hexagonal and keeled; ventrals as broad as dorsal scales in anterior of body; ventrals 226–274, at anterior covering half body width, narrowing at posterior to twice width of adjacent scales or less. **Coloration** Dorsum unicoloured grey or with lighter mottling, or 25–35 dark cross-bars or spots; forehead grey or black, sometimes with dark mottling; venter cream or white; juveniles with large dark vertebral spots. **Habitat and Behaviour** Inhabits seas up to 32km from the coast, and also known from river mouths, where the water is presumably brackish. Diet consists of marine invertebrates and eels. Ovoviviparous; clutches comprise 3–5 neonates. **Distribution** Myanmar, west coast of Thailand, Vietnam, Peninsular Malaysia, Singapore and Borneo. Range extends from the Persian Gulf to eastern China and South-East Asia. **Status** Not Evaluated.

Family HOMALOPSIDAE PUFF-FACED WATER SNAKES

Snakes in this family possess reduced eyes and valvular nostrils, coinciding with a lifestyle in muddy and turbid habitats. Most inhabit freshwaters, including wet agricultural fields, marshes and ponds; a few are found in brackish waters and even in coastal regions. Their diet consists of fish, although frogs and crustaceans are also consumed by some species that are specialists of these prey items. The group is currently considered basal to the more advanced families of snakes. All snakes from this family are ovoviviparous, producing live young underwater. Their distribution ranges from Pakistan, via South-East Asia, to northern Australia, with one species extending to Micronesia.

KEEL-BELLIED WATER SNAKE
Bitia hydroides PLATE 65
Measurement TL 800mm **Identification** Body slender anteriorly, robust on posterior two-thirds; head narrow in juveniles, indistinct from neck, and moderate in adults, distinct from neck; loreal present; single preocular; postoculars 2; supralabials 7–9; Supralabial IV contacts orbit; infralabials 10–12; parietals fragmented; maxillary teeth 11–14, followed by 2 enlarged grooved teeth; enlarged palatine teeth; eye small, dorsally oriented; nostril valvular; tail compressed; dorsals smooth; midbody scale rows 37–43; ventrals 157–165, narrow; subcaudals 24–34, paired; anal divided. **Coloration** Dorsum greyish-blue, pale brown or greyish-yellow with 40–42 dark grey or black cross-bars, 3 scales wide, on body, and 8–9 on tail; head dark brown or grey with 1–2 small dark brown spots on each scale; supralabials and occipital scales dark spotted; chin, gular region and rest of venter pale yellow. **Habitat and Behaviour** Inhabits coastal areas such as river mouths with mud flats. Aquatic. Diet consists of gobies. Clutches

comprise 1–10 neonates, measuring 155–168mm. **Distribution** Southern Myanmar (Bago Division), southern Thailand (Bangkok, Phra Nakhon Si Ayutthaya Province), coastal Peninsular Malaysia (Pulau Pinang, Pinang State and Sungai Muar, Johor State) and Singapore (Nee Soon Swamps). **Status** Not Evaluated.

YELLOW-BANDED MANGROVE SNAKE
Cantoria violacea PLATE 65
Measurement TL 1,200mm **Identification** Body slender, elongated, cylindrical; head indistinct from neck; prefrontals in broad contact; loreal present; single preocular and postocular; supralabials 5; Supralabials III–IV contact orbit; infrabials 8; eye small; pupil rounded; tail slightly compressed, short and blunt; dorsals smooth; midbody scale rows 19 (rarely 21); ventrals 243–291; subcaudals 52–69, paired; anal divided. **Coloration** Dorsum dark blackish-grey or black with transverse dull yellow bars, 2–3 scales wide and narrower than their interspaces; head white-spotted or with 2 yellow cross-bars; venter unpatterned cream or with grey markings. **Habitat and Behaviour** Inhabits tidal rivers including estuaries, as well as mud flats during low tide and mangrove swamps. Nocturnal and aquatic. Diet consists of snapping shrimps and fish. Reproductive habits unstudied. **Distribution** Myanmar, southern Thailand (Phuket Island, Phuket Province), Peninsular Malaysia, Singapore, Sumatra and Borneo. Also the Andaman Islands (India) and Timor (Indonesia/Timor Leste). **Status** Not Evaluated.

DOG-FACED WATER SNAKE
Cerberus rynchops PLATE 65
Measurement TL 1,270mm **Identification** Body moderately robust; head long and distinct from neck; loreal present; single preocular and postocular; supralabials 9–10; Supralabials V–VI contact orbit; parietals fragmented; eye small and beady; pupil rounded; tail short, slender; dorsals strongly keeled; midbody scale rows 23–25 (rarely 21 or 27); ventrals 122–160; subcaudals 49–72, paired; anal divided. **Coloration** Dorsum dark grey or greyish-green with faint dark blotches; dark postocular stripe reaches sides of neck; venter yellowish-cream with dark grey areas. **Habitat and Behaviour** Inhabits coastal low-lying areas such as mangrove mudflats and rice fields. Daytime is spent concealed in crab holes. Diet consists of fish such as mudskippers and gobies, as well as crabs and frogs. Clutches comprise 6–30 neonates, measuring 150–250mm. **Distribution** Myanmar, Thailand, Vietnam, Peninsular Malaysia, Singapore, Sumatra, the Mentawai and Natuna Archipelagos, Borneo, Java and Bali. Also Pakistan, India, Bangladesh, Sri Lanka, New Guinea and northern Australia. **Status** Not Evaluated.

WHITE-SPOTTED WATER SNAKE
Enhydris albomaculata NOT ILLUSTRATED
Measurement TL 650mm **Identification** Body moderate, subcylindrical; nostrils dorsal; loreal present; single preocular; postoculars 1–2; supralabials 8–9; Supralabials IV–V contact orbit; infrabials 12–14; first 5 (rarely 6) infrabials contact anterior chin shields; maxillary teeth 13; 2 posterior teeth enlarged; eye small; tail short; dorsals smooth; midbody scale rows 27; ventrals 140–151; subcaudals 36–50, paired; anal divided. **Coloration** Dorsum olive-brown or black with small yellow or orange spots; yellow nuchal collar and 1–2 indistinct yellow cross-bars on occipital region; venter with yellow, olive-brown or black variegation. **Habitat and Behaviour** Nothing known of its natural history. **Distribution** Sumatra, Pulau Nias, Pulau Simeulue and Pulau Sibigo. **Status** Not Evaluated.

YELLOW-BANDED WATER SNAKE
Enhydris alternans NOT ILLUSTRATED
Measurement TL 610mm **Identification** Body moderate, subcylindrical; nostrils dorsal; loreal present; single preocular; postoculars 2; supralabials 8; Supralabial IV contacts orbit; infrabials 10–11; 4 (rarely 5) infrabials contact anterior chin shields; paired internasals; 2 posterior teeth enlarged; eye small; tail short; dorsals smooth; midbody scale rows 19; ventrals 125–164; subcaudals 23–36, paired; anal divided. **Coloration** Dorsum purplish-brown with indistinct yellow cross-bars, numbering 40–45 on body and 8–9 on tail, which are sometimes broken up into spots; venter purplish-brown with 39–61 transverse yellow bars, 2 ventrals long; subcaudals with yellowish-brown and yellow variegation. **Habitat and Behaviour** Nothing known of its natural history. **Distribution** Sumatra, Pulau Bangka, Pulau Belitung, Borneo (Kuching, Sarawak State) and Java. **Status** Not Evaluated.

BENNETT'S WATER SNAKE
Enhydris bennetti PLATE 65
Measurement TL 610mm **Identification** Body moderate, subcylindrical; head large; nostrils dorsal; loreal present; single preocular; postoculars 2; supralabials 8; Supralabial IV contacts orbit; 3 infrabials contact anterior chin shields; 2 posterior teeth enlarged; eye small; tail short; dorsals smooth; midbody scale rows 21; ventrals 158–164; subcaudals 64, paired; anal divided. **Coloration** Dorsum greyish-brown with irregular black or greyish-black cross-bars, sometimes forming zigzag pattern; white stripe comprising 4 lowest dorsals adjacent to ventrals; supralabials and infrabials cream; venter cream. **Habitat and Behaviour** Inhabits tidal marshes and mangrove swamps in both brackish and sea water. Diurnal and nocturnal, and aquatic, also capable of ascending trees to rest. Diet presumably consists of fish such as mudskippers. Reproductive habits unstudied. **Distribution** Vietnam (Quang Ninh Province). Also eastern China. **Status** Not Evaluated.

BOCOURT'S WATER SNAKE
Enhydris bocourti PLATE 65
Measurement TL 1,100mm **Identification** Body robust, subcylindrical; head large, blunt, indistinct from neck; loreal present; single preocular; postoculars 2; supralabials 9; Supralabial IV contacts orbit;

infralabials 14–15; 5 infralabials contact anterior chin shield; 2 posterior-most teeth enlarged; dorsals smooth; midbody scale rows 27; ventrals 123–132; subcaudals 32–47, paired; anal divided. **Coloration** Dorsum reddish-brown or dark brown with transverse, black-edged reddish-brown bars, which narrow on flanks to meet ventrals; head reddish-brown; supralabials cream with black bars on sutures; venter yellow. **Habitat and Behaviour** Inhabits sluggish freshwaters in the lowlands and low hills at elevations of up to ca 200m asl. Highly aquatic, concealing itself on logs near water bodies. Diet consists of fish. Clutches comprise 15–20 neonates, measuring 220mm. **Distribution** Thailand (Beung Boraphet, Nakhon Sawan Province, Bangkok, Phra Nakhon Si Ayutthaya Province, Nong Thung Thong, Surat Thani Province and Thale Noi, Phattalung Province), Cambodia, Vietnam and northern Peninsular Malaysia. **Status** Not Evaluated.

CHAN-ARD'S WATER SNAKE
Enhydris chanardi   NOT ILLUSTRATED
Measurement TL 628mm **Identification** Body robust, subcylindrical; head indistinct from neck; loreal present; single preocular; postoculars 2; supralabials 8; Supralabial IV contacts orbit; infralabials 10 (rarely 11 or 12); first 5 infralabials contact anterior chin shields; posterior chin shields longer than anterior; 2 posterior-most teeth enlarged; dorsals smooth; midbody scale rows 21; ventrals 110–122; subcaudals 38–60, paired; anal divided. **Coloration** Dorsum greyish-brown; yellow dorsolateral stripe, edged ventrally by grey stripe; dorsal rows 5–7 with black spots; chin yellow. **Habitat and Behaviour** Inhabits freshwaters in low-lying areas. Aquatic and presumably piscivorous. Dietary habits unstudied. Clutches comprise 14 neonates. **Distribution** Endemic to south-central Thailand (Bangkok, Phra Nakhon Si Ayutthaya Province and the Inner Gulf eastwards to Chanthaburi Province). **Status** Not Evaluated.

CHINESE WATER SNAKE
Enhydris chinensis   PLATE 65
Measurement TL 610mm **Identification** Body moderate, subcylindrical; head rounded; single preocular; postoculars 2; supralabials 8; Supralabial IV contacts orbit; 3 infralabials contact anterior chin shields; nostrils dorsal; 2 posterior-most teeth enlarged; dorsals smooth; midbody scale rows 23; ventrals 134–147; subcaudals 35–43, paired; anal divided. **Coloration** Dorsum olive-brown or grey with irregular scattered black spots; lowest dorsals cream or yellow, sometimes forming zigzag pattern on sides; venter yellow or cream, each ventral with dark edge. **Habitat and Behaviour** Inhabits the lowlands in locations such as agricultural fields, ponds and slow-moving streams at elevations of up to 200m asl. Aquatic, and abundant in fish ponds. Diet includes frogs and fish. Clutches comprise 6–13 neonates, measuring 100–157mm. **Distribution** Vietnam. Also eastern China. **Status** Not Evaluated.

MARQUIS DORIA'S WATER SNAKE
Enhydris doriae   PLATE 65
Measurement TL 700mm **Identification** Body robust, subcylindrical; head small, depressed, slightly distinct from neck; forehead scales fragmented; single preocular; postoculars 2; supraoculars 2–4; supralabials 11–16; suboculars 9–10, separating supralabials from orbit; infralabials 12–16; eye small; pupil rounded or slightly oval; tail short; dorsals smooth; midbody scale rows 29–31; ventrals 137–152; subcaudals 39–60, paired; anal divided. **Coloration** Dorsum reddish-brown to greyish-brown; labials blotched with grey or cream; venter yellow, orange, cream or red, some individuals with dark blotches, darker posteriorly. **Habitat and Behaviour** Inhabits small freshwater streams and swamps at elevations of up to 500m asl. Nocturnal and aquatic. Diet consists of fish. Reproductive habits unstudied. **Distribution** Endemic to Borneo (Sandakan Bay, Sabah State, Gunung Pueh, Baram, Kuching, Sungei Labang and Sungei Mengiong, Sarawak State and Sungei Kapuas, Kalimantan Province). **Status** Not Evaluated.

RAINBOW WATER SNAKE
Enhydris enhydris   PLATE 65
Measurement TL 810mm **Identification** Body robust, subcylindrical; head small, depressed and slightly distinct from neck; snout rounded; single preocular; postoculars 2 (rarely 1); supralabial 8; Supralabial IV contacts orbit; infralabials 9; posterior chin shield larger than anterior; nostrils situated on upper surface of head; eye small; pupil vertical; tail short, tapering; dorsals smooth, lacking apical pits; midbody scale rows 21 or 23 (rarely 19); ventrals 134–177; subcaudals 46–83, paired; anal divided. **Coloration** Dorsum dark brown, greyish-brown or olive-green with dark vertebral and 2 pale brown paravertebral stripes from upper surface of head to tail; venter yellowish-cream, ventrals with dark spot on edges creating dark line on flanks. **Habitat and Behaviour** Inhabits freshwaters, and sometimes brackish water habitats, including wet rice fields. Diet consists of fish, frogs, tadpoles and lizards. Clutches produced several times a year, and comprise 4–20 neonates, measuring 126–180mm. **Distribution** Myanmar, Thailand, Vietnam, Peninsular Malaysia, Singapore, Sumatra, Borneo and Java. Also eastern India, Nepal, Bangladesh and southern China. **Status** Not Evaluated.

GYI'S WATER SNAKE
Enhydris gyii   PLATE 65
Measurement TL 766mm **Identification** Body robust, subcylindrical; head distinct from neck; loreal present; preoculars 2; postoculars 2; supraoculars 2–3; supralabials 15–17; suboculars separate supralabials from orbit; infralabials 16–20; dorsals smooth; midbody scale rows 25 or 27; ventrals 155–159; subcaudals 44–46, paired; anal divided. **Coloration** Dorsum iridescent dark brown, except for scale rows 1–4, which are reddish-brown, each dorsal with red spot; colour changeable to nearly white when kept in

the dark; supralabials anterior to orbit greyish-black; posterior supralabials reddish-brown; dark stripe from nape to angle of jaws; venter reddish-brown. **Habitat and Behaviour** Inhabits rivers within swamp forests at an elevation of ca 50m asl. Nothing else known of its natural history. **Distribution** Endemic to western Borneo (Putussibau, nr the Sungei Kapuas drainage, Kalimantan Province). **Status** Not Evaluated.

INDIAN WATER SNAKE
Enhydris indica NOT ILLUSTRATED
Measurement TL 470mm **Identification** Body robust, subcylindrical; head distinct from neck; paired internasals; single preocular; single postocular; supralabials 8; Supralabials IV contacts orbit; infralabials 10; dorsals smooth; midbody scale rows 19; ventrals 152; subcaudals 33, paired; anal divided. **Coloration** Dorsum glossy dark brown, lowest 3 rows on flanks with yellow patches that are more prominent anteriorly; venter light brown. **Habitat and Behaviour** Inhabits lowland forest streams. Nocturnal and aquatic. Diet consists of fish. Reproductive habits unstudied. **Distribution** Peninsular Malaysia (Selangor State) and probably Singapore. **Status** Not Evaluated.

BLACK-SPOTTED WATER SNAKE
Enhydris innominata PLATE 65
Measurement TL 175mm **Identification** Body robust, subcylindrical; head short, rounded, depressed, indistinct from neck; prefrontals in contact; loreal present; single preocular; postoculars 2; supralabials 8; Supralabial IV contacts orbit; infralabials 10; maxillary teeth 10–12 + 2; tail short and compressed; dorsals smooth; midbody scale rows 21 or 23; ventrals 108–117; subcaudals 40–56, paired; anal divided. **Coloration** Dorsum brownish-grey, sometimes with small black spots forming 3 longitudinal rows; flanks and venter yellowish-white or tan with 38–39 vertical dark brown blotches, 5 scales wide on flanks, extending to ventrals; forehead with black spots; tail with black and yellow rings. **Habitat and Behaviour** Inhabits artificial canals, flooded grasslands and swamp forests. Aquatic. Diet presumably consists of fish. Clutches comprise 13–32 neonates. **Distribution** Endemic to south-western Vietnam (Tay Ninh and Ba Den Mountain, Tay Ninh Province, and Vinh Thuan, Kien Giang Province). **Status** Not Evaluated.

JAGOR'S WATER SNAKE
Enhydris jagorii PLATE 65
Measurement TL 560mm **Identification** Body robust, subcylindrical; head short, rounded, indistinct from neck; loreal present; single preocular; postoculars 2; supralabials 8; Supralabial IV contacts orbit; infralabials 10; posterior chin shields larger than anterior; tail short and compressed; dorsals smooth; midbody scale rows 21; ventrals 117–127; subcaudals 50–68, paired; anal divided. **Coloration** Dorsum greyish-brown with black occelate spots arranged in a linear series between neck and tail; dorsum occasionally cream or pink in region below spots; supralabials yellowish-cream; gular region lavender-brown; venter pale with edges marked with a grey zigzag line and an indistinct series of dots medially. **Habitat and Behaviour** Inhabits freshwater swamps in the Chao Phraya Drainage. Nocturnal and aquatic. Diet and reproductive habits unstudied. **Distribution** Endemic to south-central Thailand (vicinity of Bangkok, Phra Nakhon Si Ayutthaya Province and possibly also Hua Bin Beach, Prachuap Khiri Khan Province). **Status** Not Evaluated.

LONG-TAILED WATER SNAKE
Enhydris longicauda NOT ILLUSTRATED
Measurement TL 682mm **Identification** Body robust, subcylindrical; head short, rounded, indistinct from neck; loreal present; single preocular; postoculars 2–3; supralabials 8–9; Supralabials IV or IV–V contact orbit; infralabials 12; 4 or 5 infralabials contact anterior chin shields; maxillary teeth 12 + 2; tail short and compressed; dorsals smooth; midbody scale rows 21 or 23; ventrals 124–135; subcaudals 53–74, paired; anal divided. **Coloration** Dorsum greyish-brown; vertebral series of large, elongated dark brown or black spots, ca 9 scales wide; paired dark dorsolateral stripes, sometimes broken up into dark spots; head dark brown; chin and gular region white; venter pale brown with small white spots, one series forming a midventral line. **Habitat and Behaviour** Inhabits large freshwater lakes and associated water bodies, including irrigation canals. Diet consists of fish. Clutch size unknown. **Distribution** Endemic to Cambodia (Tonle Sap, Phnom Penh, Trapeang Chang and Snoc Trou). **Status** Not Evaluated.

BROWN-SPOTTED WATER SNAKE
Enhydris maculosa NOT ILLUSTRATED
Measurement TL 308mm **Identification** Body robust, subcylindrical, becoming laterally compressed towards vent; head distinct from neck; single preocular; postoculars 2; supralabials 8; Supralabial VIII contacts orbit; anterior chin shield larger than posterior; tail short; dorsals smooth, lacking apical pits; midbody scale rows 25; ventrals 122; subcaudals 32–33, paired; anal divided. **Coloration** Dorsum olive with 6 irregular rows of spots, each a cluster of 3–5 scales with dark brown pigments; no cream stripe on scale rows 2–4; outer edges of ventrals and first row of dorsal scales with zigzag stripe running the length of body; ventrals with dark spot in middle. **Habitat and Behaviour** Presumably inhabits wetlands in lowland habitats. Diet and reproductive habits unstudied. **Distribution** Endemic to southern Myanmar (Bago Division). **Status** Not Evaluated.

PAHANG WATER SNAKE
Enhydris pahangensis NOT ILLUSTRATED
Measurement TL 265mm (in a juvenile; adults unknown) **Identification** Body robust, subcylindrical; head distinct from neck; single preocular; postoculars 2; supralabials 8; Supralabial IV contacts orbit; infralabials 9–11; 5 infralabials contact anterior chin shields; dorsals smooth; midbody scale rows 25;

ventrals 126–130; subcaudals 52–55, paired; anal divided. **Coloration** Dorsum grey-brown with small dark spots; pale yellow stripe on flanks, covering first 4 lateral dorsal scale rows anteriorly and 3 rows posteriorly; wide white stripe on sides of head, supralabials and rostral; lateral stripe bordered by dark zigzag line; venter white. **Habitat and Behaviour** Inhabits primary and undisturbed hill dipterocarp forests at elevations of 152–305m asl. Nocturnal and aquatic, associated with streams. Diet and reproductive habits unstudied. **Distribution** Peninsular Malaysia (Pahang and Terengganu, on the east coast). **Status** Not Evaluated.

GREY WATER SNAKE

Enhydris plumbea PLATE 66

Measurement TL 480mm **Identification** Body robust, subcylindrical; head short, rounded, indistinct from neck; loreal present; single preocular; postoculars 2; supralabials 8; Supralabials IV or IV–V contact orbit; 5 infralabials contact anterior chin shields; single internasal; eye small; pupil rounded; tail short and compressed; dorsals smooth; midbody scale rows 19; ventrals 112–139; subcaudals 22–46, paired; anal divided. **Coloration** Dorsum grey or greyish-olive, each scale dark brown or black edged; supralabials and venter cream or yellow, the latter with black spots; anal and subcaudals with dark grey median line. **Habitat and Behaviour** Inhabits wetlands such as swamps, marshes, streams, ditches and paddies; occasionally reported from brackish waters. Nocturnal and aquatic. Diet consists of crustaceans, frogs and their eggs and tadpoles, and fish. Clutches comprise 5–30 neonates, measuring 75–150mm. **Distribution** Myanmar, Thailand, Laos, Cambodia, Vietnam, Peninsular Malaysia, Sumatra, Pulau Belitung, Borneo, Java and Bali. Also southern and eastern China, Sulawesi (Indonesia) and the Nicobar Archipelago (India). **Status** Not Evaluated.

SPOTTED WATER SNAKE

Enhydris punctata PLATE 66

Measurement TL 730mm **Identification** Body robust, subcylindrical; head short, rounded, depressed; indistinct from neck; single preocular; postoculars 2; single supraocular; suboculars absent; supralabials 10–14; Supralabials VI–VII contact orbit; infralabials 14–16; 3 to 4 infralabials contact anterior chin shields; anterior chin shields larger than posterior; tail short and compressed; dorsals smooth; midbody scale rows 25 (rarely 23 or 27); ventrals 137–160; subcaudals 27–48, paired; anal divided. **Coloration** Dorsum greyish-black to dark brown; yellow cross-bar, 1 scale wide, on occipital region; sometimes, transverse rows of yellow spots, of which 6–7 anterior rows form transverse bands; caudal scales with a yellow spot; venter and adjacent 3 rows of dorsals yellow; subcaudals yellow edged with brown. **Habitat and Behaviour** Inhabits the lowlands. Aquatic and associated with rivers. Diet and reproductive habits unstudied, but presumed to be piscivorous and ovoviviparous. **Distribution** Peninsular Malaysia (Gunung Pulai, Johor State), Sumatra, Pulau Bangka, Pulau Belitung and Borneo (Sinkawang, Kalimantan Province). **Status** Not Evaluated.

SIEBOLD'S WATER SNAKE

Enhydris sieboldii NOT ILLUSTRATED

Measurement TL 780mm **Identification** Body robust, subcylindrical; head short, rounded, indistinct from neck; single preocular; postoculars 1–2; supralabials 7–8; Supralabial IV contacts orbit; 4 infralabials contact anterior chin shields; tail short and compressed; dorsals smooth; midbody scale rows 27 or 29 (rarely 31 or 33); ventrals 143–158; subcaudals 43–56, paired; anal divided. **Coloration** Dorsum brownish-grey or olive-brown, with cream or pale brown cross-bars, sometimes fused on vertebral region, widening on lower flanks; forehead with lanceolate area; venter cream or grey mottled with dark green. **Habitat and Behaviour** Inhabits rivers and water bodies within river drainages. Diurnal and aquatic, burrowing in mud. Diet consists of frogs and fish. Clutches comprise 7 neonates. **Distribution** Unsubstantiated record from Myanmar (Bago Division). Also Bangladesh, Nepal and India. **Status** Not Evaluated.

SMITH'S WATER SNAKE

Enhydris smithi NOT ILLUSTRATED

Measurement TL 680mm **Identification** Body robust, subcylindrical; head indistinct from neck; loreal present; single preocular (rarely 2); postoculars 2 (rarely 3); supralabials 8; Supralabial IV contacts orbit; infralabials 10–11; 2 posterior-most teeth enlarged; dorsals smooth; midbody scale rows 21; ventrals 118–127; subcaudals 49–56, paired; anal divided. **Coloration** Dorsum black with 38–39 narrow pink bands on vertebral region; bands yellow on flanks, meeting ventrally; head black with pink or grey markings; tail encircled by narrow yellow rings; venter grey. **Habitat and Behaviour** Nothing known of its natural history. **Distribution** Endemic to central and western Thailand (Bangkok, Phra Nakhon Si Ayutthaya Province and Prachuap Khiri Khan Province). **Status** Not Evaluated. **Note** Considered synonymous with Jagor's Water Snake (*E. jagorii*) by recent authors.

INDO-CHINESE WATER SNAKE

Enhydris subtaeniata PLATE 66

Measurement TL 870mm **Identification** Body moderate, subcylindrical; head indistinct from neck; nostrils dorsal; loreal present; single preocular; postoculars 2; supralabials 8; Supralabial IV contacts orbit; infralabials 10; 4 infralabials contact anterior chin shields; 2 posterior-most teeth enlarged; eye small; tail short; dorsals smooth; midbody scale rows 21; ventrals 136–153; subcaudals 46–69, paired; anal divided. **Coloration** Dorsum mid-brown with black ventrolateral stripe covering scale rows 1–3; dark lateral spots; indistinct postocular stripe; labials pale; chin pale, dotted with grey, and sutures darker; venter pale with edges marked with zigzag grey line and an indistinct series of dots medially. **Habitat and Behaviour** Inhabits low-lying drainage of the Mekong River. Diurnal and aquatic, living in streams, ditches, ponds and wet rice fields. Diet consists of fish and frogs. Clutches comprise 7–20 neonates, measuring ca 159mm. **Distribution** South-eastern Thailand (Khorat Plateau), Laos, Cambodia and Vietnam. Isolated population in central Thailand (Chao Phraya Drainage) suspected to be introduced through human agencies. **Status** Not Evaluated.

VORIS'S WATER SNAKE

Enhydris vorisi NOT ILLUSTRATED

Measurement TL 586mm **Identification** Body robust, subcylindrical; head depressed, distinct from neck; loreal present; single preocular; postoculars 2; supralabials 8; Supralabials I–III contact loreal; Supralabial IV contacts orbit; infralabials 10; 4 infralabials contact anterior chin shields; 2 posterior-most teeth enlarged; eye small; tail short; dorsals smooth; dorsal scale rows on neck 26–28; midbody scale rows 25; ventrals 142–152; subcaudals 41–58, paired; anal divided. **Coloration** Dorsum olive; cream stripe on scale rows 2–4, dorsally bordered by dark spots on dorsal edge; uniform coloration in centre of each ventral scale. **Habitat and Behaviour** Known from the freshwater swamp region of the Ayeyarwady Drainage. Diet unstudied. Clutches comprise 5 neonates. **Distribution** Endemic to Myanmar (Maubin, within the Ayeyarwady Drainage). **Status** Not Evaluated.

TENTACLED SNAKE

Erpeton tentaculatus PLATE 66

Measurement TL 770mm **Identification** Body slender; head small, distinct from neck; long, scaly and flexible rostral appendages, as long as snout length; loreal present; single preocular; postoculars 1–2; supralabials 13–15; suboculars separate supralabials from orbit; infralabials 13–15; maxillary teeth 12–14, followed by pair of enlarged grooved teeth; eye small, protruding; pupil vertical; dorsals strongly keeled; midbody scale rows 35–39; ventrals 103–136, narrow; subcaudals 87–126, paired; anal entire. **Coloration** Dorsum olive, grey or brown with 2 indistinct longitudinal dark paravertebral stripes, or variegated black; vertebral region with dark spots or cross-bars; broad dark lateral stripe from snout, across orbit, to along flanks; venter yellowish-brown. **Habitat and Behaviour** Inhabits the lowlands, occurring in ponds and slow-moving water bodies. Exclusively aquatic. Diet consists of fish; macrophytes found in its stomach may be the result of accidental ingestion with animal prey. Clutches comprise 9–13 neonates, measuring 143–168mm. **Distribution** Central Thailand (Bangkok, Phra Nakhon Si Ayutthaya Province, Thale Noi, Phattalung Province and Songkhla Lake, Nakhorn Si Thammarat Province), Cambodia and south-western Vietnam (Long Xuyen, An Gian Province and Tay Ninh, Tay Ninh Province). **Status** Not Evaluated.

CRAB-EATING MANGROVE SNAKE

Fordonia leucobalia PLATE 66

Measurement TL 950mm **Identification** Body robust, cylindrical; head short, wide, indistinct from neck; forehead scales large and distinct; snout rounded; loreal absent; single preocular; postoculars 1–2; supralabials 5; Supralabial III (rarely II–III) contacts orbit; 3 infralabials contact anterior chin shields; lower jaw short; rear fangs enlarged; tail short, with acute tip; dorsals smooth; midbody scales 25–29; ventrals 137–159; subcaudals 27–43, paired; anal divided. **Coloration** Dorsum dark grey or brown with light spots, or light grey, yellow or orange with dark or light spots; labials yellowish-cream; venter pale cream, sometimes with small dark spots. **Habitat and Behaviour** Inhabits tidal rivers and sometimes associated with mangrove forests. Nocturnal and aquatic, sheltering in crab burrows. Diet consists of crabs; small fish and mud lobsters are also consumed. Clutches comprise 6–15 neonates, measuring 180mm. **Distribution** Myanmar, southern Thailand (Phuket and Satun Provinces), Vietnam, Peninsular Malaysia, Singapore, Sumatra, Borneo and Java. Also Sunderbans, and the Andaman and Nicobar Islands (India), the Lesser Sundas (Indonesia), the Philippines, New Guinea and northern Australia. **Status** Not Evaluated.

GLOSSY MARSH SNAKE

Gerarda prevostiana PLATE 66

Measurement TL 530mm **Identification** Body slender, cylindrical; head slightly distinct from neck; loreal present; single preocular; postoculars 2; supralabials 7 (rarely 8); Supralabial IV contacts orbit; infralabials 7; parietals well developed; anterior chin shields larger than posterior; eye small, situated dorsally; pupil vertical; tail short, with acute tip; dorsals smooth; midbody scale rows 17; ventrals 144–157; subcaudals 29–36, paired; anal divided. **Coloration** Dorsum unpatterned grey, greyish-green or brown; sometimes, cream or yellow stripe from labials to tail-tip; labials edged with dark grey or olive; venter grey or brownish-cream with white edges, or white with grey edges; subcaudals dark grey. **Habitat and Behaviour** Inhabits coastal areas such as mangrove swamps and estuaries. Nocturnal, with some diurnal activity, and aquatic, hiding in the earth mounds of Mud Lobster (*Thalassina anomala*). Diet consists of soft-shelled crabs; fish and shrimps are also eaten. Reproductive habits unstudied. **Distribution** Myanmar, Thailand (Ang Sila, Chonburi Province), Peninsular Malaysia and Singapore. Also India, Bangladesh and Sri Lanka. **Status** Not Evaluated.

PUFF-FACED WATER SNAKE

Homalopsis buccata PLATE 66

Measurement TL 1,400mm **Identification** Body robust, flattened dorsoventrally; head large and distinct from neck; snout squarish; loreal present; preoculars 1–2; postoculars 2; supralabials 11–14, not in contact with orbit; 3–4 suboculars; infralabials 13–20; parietals entire; eye and valvular nostrils directed upwards; eye small; pupil vertically oval; tail short, with acute tip; dorsals keeled; ventrals 155–180; subcaudals 68–101, paired; anal divided. **Coloration** Dorsum greyish, dark brown or black (depending on population and growth stage), with 19–51 narrow, black-edged yellow cross-bars; venter cream with black spots. **Habitat and Behaviour** Inhabits slow-moving and stagnant waterways such as peat swamps, ponds, marshes and rice fields, in addition to coastal areas. Active by both day and night, and aquatic. Diet consists of fish, crustaceans and frogs. Clutches comprise 2–22 neonates, measuring 115–230mm. **Distribution** Myanmar, Thailand, Laos, Cambodia, Vietnam, Peninsular Malaysia, Singapore, Sumatra, Pulau Belitung, Borneo and Java. Also north-eastern India. **Status** Not Evaluated.

CAMBODIAN PUFF-FACED WATER SNAKE
Homalopsis nigroventralis PLATE 66
Measurement TL ca 1,400mm **Identification** Body robust, flattened dorsoventrally; supralabials 9–13, not in contact with orbit; 1–4 suboculars; infralabials 14–17; parietals entire; frontal divided; head large and distinct from neck; eye and valvular nostrils directed upwards, snout squarish; eye small; pupil vertically oval; tail short, with acute tip; dorsals keeled; midbody scale rows 35–39; ventrals 154–165; subcaudals 75–94, paired; anal divided. **Coloration** Dorsum black or grey with yellowish-cream cross-bars in adults, pale orange in juveniles; interrupted cream ventrolateral stripe connects dorsal body bands; forehead black or grey, juveniles with yellowish-tan bars or patch across head, and with spots and blotches of the same colour; an X-shaped white mark on chin; venter black with scattered cream or yellow spots in juveniles, yellowish-olive or olive-brown with cream spots in adults. **Habitat and Behaviour** Inhabits bamboo-dominated evergreen and deciduous forests in the low hills at elevations of 100–175m asl. Both diurnal and nocturnal, and aquatic, associated with rocky hill streams, and also ponds and marshes, and water bodies in cities. Diet consists of frogs, fish and crustaceans. Reproductive habits unstudied. **Distribution** Thailand (Sakon Nakhon Province), eastern Cambodia (Ta Veng, Siem Pang District and Stung Treng Province) and Laos (Mekong Valley, at Pakse and Luang Prabang). **Status** Not Evaluated.

Family NATRICIDAE WATER SNAKES

Water snakes lack enlarged grooved rear fangs on the maxillary bone (although some have enlarged teeth at the rear of the maxillary bone), and have backwards-projecting bony processes (hypapophyses) of the vertebrae (adaptive for swimming). They are highly aquatic, and associated with nearly all types of freshwater habitat apart from high montane regions. Some species are able to survive long winters in low-oxygen environments. Their diet includes fish and other aquatic organisms, and their distribution is cosmopolitan in both tropical and temperate parts of Africa and Australia.

ANDREA'S KEELBACK
Amphiesma andreae PLATE 67
Measurement TL 608mm **Identification** Body slender; head distinct from neck; single loreal and preocular; 3 postoculars; supralabials 9; Supralabials IV–VI contact orbit; infralabials 9; posterior chin-shields longer than anterior; tail tapering; dorsals keeled; midbody scale rows 19; ventrals 179; subcaudals 99. **Coloration** Dorsum mid-brown; black-edged pale bar before and after orbit of eye; head and neck with dark-edged pale blotches, turning into black-edged pale bars on anterior body that become indistinct at midbody and turn into small pale blotches posteriorly, becoming a dorsolateral stripe that ends at tail base; venter light, laterally with dark spots. **Habitat and Behaviour** Restricted to limestone forests at an elevation of 450m asl. Nothing else known of its natural history. **Distribution** Endemic to central Vietnam (Phong Nha-Ke Bang National Park, Quang Binh Province). **Status** Not Evaluated.

MOUNTAIN KEELBACK
Amphiesma atemporale PLATE 67
Measurement TL 500mm **Identification** Body slender; head distinct from neck; single preocular; postoculars 2–probably 3; supralabials 5–6; Supralabials III–IV contact orbit; infralabials 6–7; temporals absent, or small one present between Supralabial V and parietal; maxillary teeth 28–30; eye large; pupil rounded; dorsals keeled; midbody scale rows 17; ventrals 129–146; subcaudals 54–79, paired; anal divided. **Coloration** Dorsum reddish-brown to grey; scales edged with black; 2 pale dorsolateral stripes, sometimes broken up into spots; pale collar; venter cream, ventrals with black spot on outer margins. **Habitat and Behaviour** Inhabits submontane forests at elevations of 800–914m asl. Diurnal and terrestrial. Diet and reproductive habits unstudied. **Distribution** Vietnam (Vinh Phu Province). Also eastern China. **Status** Not Evaluated.

TWO-STRIPED KEELBACK
Amphiesma bitaeniatum PLATE 67
Measurement TL 708mm **Identification** Body moderate, elongated; head distinct from neck; nostrils lateral; loreal present; single preocular; postoculars 3; supralabials 7–8; Supralabials III–V contact orbit; infralabials 10; maxillary teeth 19–23; eye large; pupil rounded; tail long, tapering; dorsals keeled, scales notched posteriorly; midbody scale rows 19; ventrals 160–175; subcaudals 78–95, paired; anal divided. **Coloration** Dorsum dark brownish-grey or ochre-brown with broad, continuous beige-yellow dorsolateral stripe, edged by 2 narrow black lines,

from neck to tail-tip; forehead greyish-brown with indistinct vermiculation; supralabials ivory-yellow; first 5 supralabials with dark brown spots; dark brown postocular stripe extends across neck to join dorsolateral stripe; venter yellowish-cream, small dark brown spot on outer edges of ventrals. **Habitat and Behaviour** Inhabits montane forests at elevations of 1,829–2,377m asl. Diurnal and terrestrial. Diet and reproductive habits unstudied. **Distribution** Northern Myanmar (Kachin and Shan States), northern Thailand (Chiang Mai Province), Laos and northern Vietnam (Mount Phang Si Pang, Bình Thuân and Sa Pa, Lào Cai Province). Also southern China (Yunnan Province). **Status** Not Evaluated.

BOULENGER'S KEELBACK

Amphiesma boulengeri PLATE 67
Measurement TL 877mm **Identification** Body slender; head distinct from neck; nostrils lateral; internasals truncated; loreal present; single preocular; postoculars 2; supralabials 9; Supralabials IV–VI contact orbit; infralabials 9–10; 5 infralabials contact anterior chin shields; maxillary teeth 28, in continuous series, gradually becoming larger posteriorly; eye large; pupil rounded; dorsals weakly keeled, except outermost rows, which are smooth; midbody scale rows 19; ventrals 139–156; subcaudals 85–102, paired; anal divided. **Coloration** Dorsum bluish- or greyish-black, or brown, with a pair of longitudinal white dorsal stripes on head, turning pinkish-brown on body; forehead brownish-grey with fine grey vermiculation; distinct narrow white postocular streak from lower margin of orbit to nape, not forming chevron; venter cream with some blotches on edges of ventrals. **Habitat and Behaviour** Inhabits temperate forests in the mid-hills at elevations of 540–900m asl. Diurnal and terrestrial, associated with damp forest floor and rocky streams. Diet unstudied. Oviparous; clutches comprise 3 eggs, measuring 22–29 x 7–11mm. **Distribution** Eastern Thailand (Nakhon Ratchasima and Nakon Si Thammarat Provinces), Laos (Xiengkhuang Province), Cambodia (Kampong Speu and Kampot Provinces) and Vietnam (Vinh Phu, Lào Cai, Gia Lai and Lam Dong Provinces). Also eastern China. **Status** Not Evaluated.

KUATUN KEELBACK

Amphiesma craspedogaster PLATE 67
Measurement TL 635mm **Identification** Body slender; head distinct from neck; nasal divided; loreal 1 (rarely 2); single preocular; postoculars 3 (rarely 4); supralabials 8 (rarely 9); Supralabials III–V enter orbit; infralabials 10 (rarely 11); anterior temporals 2–3; eye large; pupil rounded; dorsals strongly keeled; midbody scale rows 17 or 19; ventrals 138–159; subcaudals 78–96, paired; anal divided. **Coloration** Dorsum dark brown; rusty-red streak with yellow spots along flanks; flanks with indistinct black spots; labials cream or yellow with black bars on sutures; short oblique yellow streak between supralabials to nape; venter yellow, ventrals with elongated black spot at outer extremities, forming paired line on each side of venter and tail. **Habitat and Behaviour** Inhabits submontane and montane forests. Nothing else known of its natural history. **Distribution** Northern Vietnam. Also southern and eastern China (Sichuan, Fujian and Guizhou Provinces). **Status** Not Evaluated.

DESCHAUENSEE'S KEELBACK

Amphiesma deschauenseei NOT ILLUSTRATED
Measurement TL 480mm **Identification** Body slender; head distinct from neck; loreal present; preoculars 2; postoculars 2; posterior temporals 2; supralabials 9; Supralabials V–VI contact orbit; infralabials 9–10; nostrils dorsolateral; eye small; pupil rounded; dorsals keeled; midbody scale rows 19; ventrals 159; subcaudals 137–140, paired; anal divided. **Coloration** Dorsum brownish-yellow; flanks with irregular black blotches; forehead and neck black, extending dorsally as a vertebral stripe; venter pale with a series of 3 grey or brown spots, turning darker posteriorly. **Habitat and Behaviour** Inhabits lowland forests at elevations of ca 300m asl. Nocturnal and terrestrial, associated with forest streams. Diet and reproductive habits unstudied. **Distribution** Northern and western Thailand (Chiang Rai, Chiang Mai and Uthai Thani Provinces), Vietnam (Cao Bang and Tuyen Quang Provinces, and Na Hang District). Also southern and eastern China (Yunnan and Guizhou Provinces). **Status** Not Evaluated.

WHITE-FRONTED KEELBACK

Amphiesma flavifrons PLATE 67
Measurement TL 750mm **Identification** Body slender; head distinct from neck; supralabials 7–9, Supralabials IV–V or V–VI contact orbit; anterior temporals 2; eye large; pupil rounded; dorsals keeled; midbody scale rows 19; ventrals 146–157; subcaudals 92–101, paired; anal divided. **Coloration** Dorsum greyish-brown or olive-grey with darker markings; distinctive white to yellowish-cream spot on snout; juveniles with paired white spots along midback; venter cream with an alternating series of black spots. **Habitat and Behaviour** Inhabits lowlands at the bases of hills, and frequently encountered swimming along rivers with its head held out of the water. Diet includes frogs and their eggs and tadpoles. Reproductive habits unstudied. **Distribution** Endemic to Borneo. **Status** Not Evaluated.

BRIDLED KEELBACK

Amphiesma frenatum NOT ILLUSTRATED
Measurement TL 610mm **Identification** Body slender; head distinct from neck; supralabials 8; Supralabials III–V contact orbit; anterior temporals 2; eye large; pupil rounded; tail long; dorsals keeled; midbody scale rows 17; ventrals 164–166; subcaudals 112–116, paired; anal divided. **Coloration** Dorsum brown with paired black cross-bars, enclosing cream bar on anterior of body, turning darker posteriorly; narrow white chevron on neck extends forwards to contact supralabials; short black stripe covers lower parts of last 3 supralabials; forehead dark with small

black spots; venter grey with large brown and black spots. **Habitat and Behaviour** Inhabits submontane forests. Diurnal and terrestrial. Hides under fallen logs and other debris on the forest floor. Diet and reproductive habits unstudied. **Distribution** Endemic to north-western Borneo (Gunung Murud and Sungei Labang, Sarawak State). **Status** Not Evaluated.

GROUNDWATER'S KEELBACK

Amphiesma groundwateri   PLATE 67

Measurement TL 450mm **Identification** Body slender; head distinct from neck; loreal present; preoculars 2; postoculars 2–3; supralabials 9; Supralabials IV–VI contact orbit; infralabials 9–10; eye large; pupil rounded; dorsals smooth anteriorly, weakly keeled posteriorly; midbody scale rows 17; ventrals 147–154; subcaudals 120–134, paired; anal entire. **Coloration** Dorsum black or dark brown with undulating dorsolateral stripe comprising yellow spots; supralabials and infralabials yellow with black edges; forehead dark; yellow streak from angle of jaws to neck, forming V-shaped mark; venter yellow, ventrals and subcaudals black-spotted posteriorly. **Habitat and Behaviour** Nothing known of its natural history. **Distribution** Endemic to southern Thailand (Tasan, Chumphon Province). **Status** Not Evaluated.

GUNUNG INAS KEELBACK

Amphiesma inas   PLATE 67

Measurement TL 615mm **Identification** Body slender; head distinct from neck; nasals truncated; loreal present; preoculars 2; postoculars 3; posterior temporals 2; supralabials 9; Supralabials II–VI contact orbit; infralabials 10; maxillary teeth 30, the 2 posterior-most enlarged; eye large; pupil rounded; dorsals distinctly keeled; midbody scale rows 19; ventrals 141–151; subcaudals 93–109, paired; anal divided. **Coloration** Dorsum dark olive-brown with rows of indistinct dorsolateral spots; flanks with faint yellow spots; forehead brown variegated with black; posterior supralabials with rounded dark blotches; spots fuse to form pale stripe that extends posteriorly to meet transverse bars on dorsum; chin and throat with black spots; venter white with squarish black spot at outer margin of ventrals. **Habitat and Behaviour** Inhabits the mid-hills. Nocturnal and terrestrial, active on the litter and buttresses. Diet and reproductive habits unstudied. **Distribution** Southern Thailand (Phu Luang Wildlife Sanctuary, Loei Province, Khao Wang Hip, Nakon Si Thammarat Province and Khao Yai National Park, Nakhon Ratchasima Province) and Peninsular Malaysia (Gunung Inas, Perak State, and Cameron Highlands, Bukit Fraser and Genting Highlands, Pahang State). **Status** Not Evaluated.

GUNUNG KERINCHI KEELBACK

Amphiesma kerinciense   NOT ILLUSTRATED

Measurement TL 516mm **Identification** Body robust; head distinct from neck; loreal present; single preocular; postoculars 3; supralabials 9; Supralabials IV–VI contact orbit; infralabials 11; eye large; pupil rounded; dorsals strongly keeled on upper rows, many of which are distinctly notched posteriorly; scales of first dorsal row enlarged, feebly keeled; midbody scale rows 19; ventrals 140; subcaudals 89, paired; anal divided. **Coloration** Dorsum reddish grey-brown; broad dark vertebral band with discontinuous crossbars; pale ochre-brown dorsolateral stripe bordering vertebral band laterally, widening to produce 3–4 irregular blotches on neck; 2 postocular streaks; venter purple greyish-brown at outer border, with a row of well-defined dark brown blotches. **Habitat and Behaviour** Inhabits hill streams in disturbed forests. Diurnal and aquatic, found in shallow (<30cm-deep) streams. Diet includes tadpoles. Reproductive habits unstudied. **Distribution** Endemic to Sumatra (Gunung Kerinchi, Sumatera Barat Province). **Status** Not Evaluated.

KHASI HILLS KEELBACK

Amphiesma khasiense   PLATE 67

Measurement TL 600mm **Identification** Body slender; head distinct from neck; internasals truncated; preoculars 1–2; postoculars 3; supralabials 9 (rarely 8); Supralabials IV–VI contact orbit; eye large; pupil rounded; dorsals keeled; midbody scale rows 19; ventrals 141–153; subcaudals 105–108, paired; anal divided. **Coloration** Dorsum reddish-brown with greyish-black vertebral stripe and dark reddish-brown lower flanks; forehead greyish-brown with short pale brown marks; labials white with dark edges, continuing to sides of neck; posterior supralabials with rounded blotches; venter white, ventrals with reddish-brown outer edges. **Habitat and Behaviour** Inhabits submontane forests at elevations of 900–1,000m asl. Diurnal and terrestrial, associated with the forest floor, especially in the vicinity of streams. Diet consists of insects, frogs and tadpoles. Oviparous; other details of reproductive habits unknown. **Distribution** Myanmar (Chin and Kachin States), Thailand (Chiang Mai and Loei Provinces), Laos (Phongsaly Province) and northern Vietnam (Tam Dao, Vinh Phú Province). Also southern China (Yunnan Province) and north-eastern India (Meghalaya State). **Status** Not Evaluated.

WHITE-LIPPED KEELBACK

Amphiesma leucomystax   PLATE 67

Measurement TL 772mm **Identification** Body slender in males, robust in females; head distinct from neck; single anterior temporal; internasals narrowed anteriorly; single preocular; postoculars 3; eye moderate; pupil rounded; first 2 rows of dorsal scale rows smooth, rest keeled; midbody scale rows 19; ventrals 154–166; subcaudals 95–109, paired; anal divided. **Coloration** Dorsum dark brownish-grey, flanks lighter; dense speckling of minute dark grey spots; scales edged with black, producing loose network on dorsum and flanks; sides with irregular faint black blotches, alternating in 2 rows; pale beige dorsolateral stripe from sides of neck to vent; head dark brown with irregular pale vermiculations and scattered beige dots; short cream sagittal line behind parietal suture; broad white stripe from snout-tip, across supralabials to neck, nearly connecting at back of head; venter cream, tips of ventrals greyish-brown; small, pale greyish-brown blotch on inner side. **Habitat and Behaviour** Inhabits lowland and montane monsoon evergreen forests of the Annamite Chain (Truong Son) and Tay-Nguyen Plateau (central Vietnam) at elevations of 100–1,300m asl. Nocturnal, and aquatic and semi-terrestrial, associated with forest streams; some arboreal habits known, including the fact that it ascends to 2.5m on vegetation. Diet includes amphibians. Oviparous; clutches comprise 3–7 eggs, measuring 25 x 7mm. Hatchlings ca 163–205mm. **Distribution** Vietnam (Nghe An, Ha Tinh, Quang Binh, Quang Tri, Thua Thien-Hue, Quang Nam, Kon Tum and Gia Lai Provinces) and Laos (Khammouan and Xe Kong Provinces). Possibly also Thailand. **Status** Not Evaluated.

GÜNTHER'S KEELBACK

Amphiesma modestum   PLATE 67

Measurement TL 600mm **Identification** Body slender; head distinct from neck; nostrils dorsolateral; preoculars 2; postoculars 3; supralabials 8 (rarely 9); Supralabials IV–VI contact orbit; infralabials 10; maxillary teeth 28–32; eye small; pupil rounded; dorsals keeled; midbody scale rows 19; ventrals 143–168; subcaudals 83–132; anal divided. **Coloration** Dorsum mid-brown with small, regularly arranged black and yellow spots on flanks that may form stripes; postocular streaks absent; labials cream with dark margins or entirely dark; series of yellow spots on flanks that may be fused; venter yellow with black spots that may be fused. **Habitat and Behaviour** Inhabits primary forests at elevations of 600–1,500m asl. Nothing else known of its natural history. **Distribution** Myanmar (Kachin and Kayah States), northern Thailand, northern Laos (Xieng Khouang), south-central Cambodia (Kamchay Mountains) and central Vietnam (Langbian Plateau, Lam Dong Province). Also north-eastern India (Meghalaya) and southern China. **Status** Not Evaluated.

MOUNT EMEI KEELBACK

Amphiesma optatum   PLATE 67

Measurement TL ca 650mm **Identification** Body slender, head distinct from neck; subcylindrical; snout short; maxillary teeth 18–23 + 2; single preocular; postoculars 2–4; supralabials 7–8; Supralabials III–IV, III–V or IV–V contact orbit; infralabials 8–11; eye large; pupil rounded; dorsals weakly keeled; midbody scale rows 19; ventrals 152–169; subcaudals 87–112, paired; anal divided. **Coloration** Dorsum reddish-brown or bluish-black with 18–30 narrow or wide white cross-bars; flanks with vertical yellow bars; forehead dark brown with spots on parietals; narrow, V-shaped white postocular stripe meeting on neck; venter coral-red or yellow. **Habitat and Behaviour** Inhabits primary forests in the mid-hills at elevations of 400–1,000m asl. Nocturnal and terrestrial. Diet consists of fish. Oviparous; clutches comprise 12–19 eggs. **Distribution** Northern Vietnam (Tam Dao, Vinh Phú Province). Also southern China (Sichuan Province). **Status** Not Evaluated.

STRIPED KEELBACK

Amphiesma parallelum   PLATE 67

Measurement TL 635mm **Identification** Body robust; head distinct from neck; maxillary teeth 20–24; supralabials 8; Supralabials III–V contact orbit; enlarged posterior maxillary teeth separated by diastema; eye large; pupil rounded; dorsals strongly keeled except for outer row; midbody scale rows 19; ventrals 163–172; subcaudals 73–108, paired; anal divided. **Coloration** Dorsum reddish-, olive- or greyish-brown with 2 light dorsolateral stripes or series of spots from neck to tail-tip; forehead brown; short yellow vertebral streak behind occiput, and black streak from eye to angle of mouth; venter unpatterned yellow with row of black dots on each side. **Habitat and Behaviour** Inhabits submontane forests of the Himalayas at elevations of ca 1,000m asl. Diet consists of fish. Oviparous; other details of reproductive habits unknown. **Distribution** Northern Myanmar (Kachin State). Also Nepal, eastern and north-eastern India, and southern China. **Status** Not Evaluated.

PETERS'S KEELBACK

Amphiesma petersii   PLATE 68

Measurement TL 600mm **Identification** Body slender; head distinct from neck; anterior temporals 2; single preocular; postoculars 3–4; supralabials 9; Supralabials IV–VI contact orbit; 5 infralabials contact anterior chin shields; eye large; pupil rounded; dorsals keeled; midbody scale rows 19; ventrals 134–150; subcaudals 65–93, paired; anal divided. **Coloration** Dorsum yellowish-pink or dark brown with rounded black spots; forehead dark olive with black vermiculation; labials yellow with black sutures; venter yellow, scutes edged with black. **Habitat and Behaviour** Inhabits lowland forests, especially near streams. Diurnal, and terrestrial and aquatic. Diet includes agamid lizards. Oviparous; other details of reproductive habits unknown. **Distribution** Peninsular Malaysia, Singapore, northern Sumatra (Medan and Tanjungbalai, Sumatera Utara Province) and western Borneo (Kuching, Gunung Gading and Saribas, Sarawak State). **Status** Not Evaluated.

POPE'S MOUNTAIN KEELBACK

Amphiesma popei   NOT ILLUSTRATED

Measurement TL 490mm **Identification** Body slender; head distinct from neck; loreal present; single preocular; postoculars 3; supralabials 8; Supralabials IV–V contact orbit; infralabials 9; 4 infralabials contact anterior chin shields; maxillary teeth 20, the posterior ones gradually enlarged; eye large; pupil rounded; tail long; dorsals keeled, except smooth outer rows; faint apical pits in a few scales; midbody scale rows 17; ventrals 130–142, paired; subcaudals 78–86; anal divided. **Coloration** Dorsum dark brownish-grey with light cross-bars on pale dorsolateral stripes that extend to midbody; forehead and flank scales with fine white spots; supralabials with broad black stripe on posterior sutures; infralabials with narrow black stripe along sutures; large white nuchal blotch; venter white. **Habitat and Behaviour** Nothing known of its natural history. **Distribution** Vietnam (Vinh Phuc Province). Also southern China (Guangxi Zhuang Autonomous Province). **Status** Not Evaluated.

RED MOUNTAIN KEELBACK 💡
Amphiesma sanguineum PLATE 68
Measurement TL 600mm **Identification** Body slender; head distinct from neck; preoculars 2; postoculars 3; supralabials 9; Supralabials II–VI contact orbit; infralabials 5; maxillary teeth 26, the posterior-most 2 longest; eye large; pupil rounded; dorsals keeled; midbody scale rows 19; ventrals 140–155; subcaudals 99–104, paired; anal divided. **Coloration** Dorsum crimson with vertebral band comprising 4–5 rows of olive- and diamond-shaped black marks; flanks with 2 alternating rows of black spots; forehead dark olive; black-edged white postocular stripe, interrupted above angle of jaws, continues to nape; labials pale pink or cream with black sutures; venter crimson with indistinct black spot on outer edges of ventrals; chin and gular region unpatterned cream. **Habitat and Behaviour** Inhabits montane forests. Possibly diurnal and terrestrial. Diet and reproductive habits unstudied. **Distribution** Endemic to Peninsular Malaysia (Cameron Highlands, Pahang State and Gombak Valley, Selangor State). **Status** Not Evaluated.

SARAWAK KEELBACK 💡
Amphiesma saravacense PLATE 68
Measurement TL 780mm **Identification** Body slender; head distinct from neck; loreal present; single preocular; postoculars 3; supralabials 8; Supralabials III–V contact orbit; 5–6 infralabials contact anterior chin shields; anterior temporals 2; eye large; pupil rounded; tail long; dorsals keeled; midbody scale rows 17; ventrals 134–154; subcaudals 52–112, paired; anal divided. **Coloration** Dorsum olive to reddish-brown, the back with squarish black markings; row of light spots on flanks; supralabials yellow or cream; venter yellow and black. **Habitat and Behaviour** Inhabits mid-elevation and submontane forests at elevations of 640–1,700m asl. Diurnal and terrestrial, frequenting the vicinity of streams. Conceals itself under fallen logs and vegetation. Diet consists of frogs and their eggs. Oviparous; clutches comprise 4–5 eggs. **Distribution** Peninsular Malaysia (Cameron Highlands, Pahang State and Gunung Menuang Gasing, Pahang-Selangor State) and Borneo. **Status** Not Evaluated.

SAUTER'S KEELBACK 💡
Amphiesma sauteri PLATE 68
Measurement TL 401mm **Identification** Body slender; head distinct from neck; loreal present; single preocular; postoculars 2–3; supralabials 7; Supralabials IV–V contact orbit; maxillary teeth 22–23; eye large; pupil rounded; dorsals keeled; midbody scale rows 17; ventrals 118–135; subcaudals 61–83, paired; anal divided. **Coloration** Dorsum dark brown with small black and pink spots; indistinct reddish-brown streak on flanks, with small pale spots on anterior flanks; venter yellowish-white, ventrals with black spots on outer ends. **Habitat and Behaviour** Inhabits submontane coniferous forests, and also forest clearings, at an elevation of 1,200m asl. Diurnal and terrestrial, associated with rocks and leaves. Diet unstudied. Oviparous; clutches comprise 1 egg. **Distribution** Northern Vietnam (Tam Dao, Vinh Phú Province). Also southern and eastern China (Sichuan, Guizhou and Hainan Provinces). **Status** Not Evaluated.

SIEBOLD'S KEELBACK 💡
Amphiesma sieboldii PLATE 68
Measurement TL 943mm **Identification** Body slender; head distinct from neck; single preocular; postoculars 2–3; supralabials 8; Supralabials III–V contact orbit; infralabials 9–11; maxillary teeth 17–21 + 2; eye large; pupil rounded; dorsals keeled; midbody scale rows 19; ventrals 168–207; subcaudals 81–111, paired; anal divided. **Coloration** Dorsum unpatterned olive-green or brown; dorsolateral series of small white spots; forehead brown, lighter on sides; paired pale occipital spots and postparietal streak present; supralabials bordered dorsally by dark brown or black stripe, extending to nuchal as crescent; venter light, typically speckled with dark greyish-brown posteriorly. **Habitat and Behaviour** Inhabits submontane and montane forests at elevations of 1,219–3,658m asl. Diurnal and terrestrial. Diet consists of frogs and their eggs and tadpoles, and skinks. Oviparous; clutches comprise 5 eggs. **Distribution** Northern Myanmar (Taungyi). Also India, Nepal and Pakistan. **Status** Not Evaluated.

BUFF-STRIPED KEELBACK 💡
Amphiesma stolatum PLATE 68
Measurement TL 800mm **Identification** Body robust; head distinct from neck; internasals truncated, narrowed anteriorly; loreal present; single preocular; postoculars 3; supralabials 7–8; Supralabials III–V contact orbit; maxillary teeth 21–24; eye large; pupil rounded; tail short, showing autotomy; dorsals keeled; midbody scale rows 19; ventrals 118–161; subcaudals 50–89, paired; anal divided. **Coloration** Dorsum reddish-brown, or olive- to greenish-grey; pattern variable, depending on locality, and typically with buff, orange, pale yellow or orange-yellow dorsolateral stripes on fifth to seventh scale rows; venter pale. **Habitat and Behaviour** Inhabits grasslands and lowland forests at elevations from sea level to 2,000m asl. Diurnal and crepuscular, and terrestrial, active on open grassy areas, generally in the vicinity of water bodies such as lakes and ponds, as well as rice fields. Shelters in rodent burrows and termite mounds. Diet consists of insects, frogs, scorpions, fish and lizards. Oviparous; clutches comprise 3–15 eggs, measuring 22–35 x 12–18mm, which are laid twice a year. Incubation period 36–62 days. **Distribution** Myanmar, Thailand, Laos, Cambodia and Vietnam. Also India, Bhutan, Pakistan, Nepal, Sri Lanka, Bangladesh, and southern and eastern China. **Status** Not Evaluated.

TARON KEELBACK
Amphiesma taronense NOT ILLUSTRATED
Measurement TL ca 780mm **Identification** Body slender; head distinct from neck; maxillary teeth 27–32; eye large; pupil rounded; scales feebly keeled, outer scale rows smooth; midbody scale rows 17; upper scale rows feebly keeled; strong keels on dorsals near vent and on tail; internasals truncated but distinctly narrowed anteriorly, almost as long as prefrontals; ventrals 166–176; subcaudals 84–106, paired; anal divided. **Coloration** Dorsum dark greyish-brown indistinctly chequered with black; venter mottled with black and yellow anteriorly, entirely black posteriorly. **Habitat and Behaviour** Inhabits wet subtropical forests. Terrestrial. Diet consists of frogs and tadpoles. Reproductive habits unstudied. **Distribution** Northern Myanmar (Chin and Kachin States). **Status** Not Evaluated.

VENNING'S KEELBACK
Amphiesma venningi PLATE 68
Measurement TL 780mm **Identification** Body slender; head distinct from neck; internasals truncated but distinctly narrowed anteriorly, almost as long as prefrontals; preoculars 1–2; postoculars 2–3; supralabials 9; Supralabials IV–VI contact orbit; maxillary teeth 27–32; eye large; pupil rounded; dorsals weakly keeled; outer scale rows smooth, upper rows weakly keeled; midbody scale rows 17; ventrals 155–176; subcaudals 117–167, paired; anal divided. **Coloration** Dorsum olive-brown, indistinctly chequered with black; anteriorly with dorsolateral ochre spots; scales flecked with black; head with lighter vermicular marks; supralabials pale ochre with black posterior margins; narrow ochre line from middle to posterior edge of interparietal suture; venter coral-red with outer edges fawn smudged with black. **Habitat and Behaviour** Inhabits evergreen forests at elevations of 1,040–1,400m asl. Diurnal and terrestrial, associated with the forest floor near hill streams. Diet consists of tadpoles and frogs. Reproductive habits unstudied. **Distribution** Northern Myanmar (Chin and Kachin States). Also north-eastern India (Arunachal Pradesh) and southern China (Yunnan Province). **Status** Not Evaluated.

VIPER-LIKE KEELBACK
Amphiesma viperinum NOT ILLUSTRATED
Measurement TL 250mm **Identification** Body slender; head distinct from neck; loreal present; preoculars 2; postoculars 3–4; supralabials 7; Supralabials III–IV contact orbit; 5 infralabials contact anterior chin shields; anterior temporals 2; eye large; pupil rounded; dorsals keeled; midbody scale rows 19; ventrals 101; subcaudals 59, paired; anal divided. **Coloration** Dorsum grey with yellow dorsolateral stripe, extending from supralabials, and ca 30 transverse black spots; forehead reddish-brown with small yellow dots; supralabials pale brown with black spots on posterior supralabials; gular region with pale spots; venter black. **Habitat and Behaviour** Nothing known of its natural history. **Distribution** Endemic to northern Sumatra (Indragiri, Riau Province). **Status** Not Evaluated.

STRANGE-TAILED KEELBACK
Amphiesma xenura PLATE 68
Measurement TL 660mm **Identification** Body slender; head distinct from neck; single preocular; postoculars 3; supralabials 9; Supralabials III–IV contact orbit; anterior and posterior temporals 2; maxillary teeth 22–23; eye large; pupil rounded; dorsals keeled; midbody scale rows 19; ventrals 158–165; subcaudals 81–107, entire; anal entire or divided. **Coloration** Dorsum olive-brown to nearly black with paired series of reddish-orange, pale brown, yellow or white spots on flanks; adjacent spots may be connected by faint black cross-lines; labials white with dark lines on sutures; venter white or yellow, outer edges of ventrals with dark brown spots; subcaudals dark grey. **Habitat and Behaviour** Inhabits submontane forests. Diurnal and terrestrial, associated with the forest floor near streams. Diet and reproductive habits unstudied. **Distribution** Myanmar (Sagaing and Rakhine States). Also north-eastern India (Khasi Hills, Naga Hills and Mizoram). **Status** Not Evaluated.

WHITE-EYED KEELBACK
Amphiesmoides ornaticeps NOT ILLUSTRATED
Measurement TL 840mm **Identification** Body slender; head distinct from neck; anteriorly broad internasals; temporal scales 1–2 + 2–4; single preocular; postoculars 3–4; supralabials 9; Supralabials IV–VI contact orbit; maxillary teeth 42–46, arranged in continuous series, gradually becoming larger posteriorly; eye large; pupil rounded; dorsals keeled; midbody scale rows 19; ventrals 157–168; subcaudals 116–128, paired; anal divided. **Coloration** Dorsum brown; vertical, black-edged pale bar extends upwards from mouth behind orbit, another one anterior to orbit, over preocular and Supralabial IV; large, square dark spots on anterior body, separated by narrow cream interspaces, and greyish-black dorsal stripe on posterior two-thirds of body; venter cream with irregular dark markings at outer edges of ventrals, becoming finer posteriorly. **Habitat and Behaviour** Inhabits primary forests in the mid-hills. Diurnal, and terrestrial and semi-aquatic. Diet includes frogs. Reproductive habits unstudied. **Distribution** Vietnam. Also southern China. **Status** Not Evaluated.

SUMATRAN KEELBACK
Anoplohydrus aemulans NOT ILLUSTRATED
Measurement TL 430mm **Identification** Body robust, rounded, elongated; head short, indistinct from neck; internasals absent; single preocular; postoculars 2; supralabials 7; Supralabials III–IV contact orbit; 4 infralabials contact chin shields; jaws with 5–6 teeth, largest posteriorly; eye small; pupil rounded; tail short, tapering; dorsals smooth, shiny; midbody scale rows 19; ventrals 159; subcaudals 35, some paired; anal divided. **Coloration** Dorsum iridescent blackish-brown with paired rows of

alternating narrow yellow cross-bars; flanks with vertical broad yellow blotches that may alternate with the cross-bars or fuse with them; forehead blackish-brown, scales with yellow centres; chin and gular region dark with yellow spots on scales. **Habitat and Behaviour** Nothing known of its natural history, although presumed to be a water snake associated with freshwaters. **Distribution** Endemic to northern Sumatra (possibly Ujungpadan, Babongan District, Aceh Province). **Status** Not Evaluated.

YELLOW-SPOTTED WATER SNAKE
Hydrablabes periops PLATE 68
Measurement TL 530mm **Identification** Body slender; head small, distinct from neck; paired prefrontals; loreal present; preoculars 2; postoculars 2; supralabials 8–9; eye small; pupil rounded; orbit separated from supralabials by series of suboculars; tail short; dorsals smooth; midbody scale rows 15 or 17; ventrals 179–209; subcaudals 56–76; anal divided. **Coloration** Dorsum olive-brown, sometimes with a pale yellow lateral stripe and a dark grey-black lateroventral one, other individuals unpatterned; venter yellow or grey. **Habitat and Behaviour** Inhabits lowland and mid-hill forests at elevations of 150–600m asl. Diurnal and aquatic, associated with streams. Diet and reproductive habits unstudied. **Distribution** Endemic to Borneo (Sungei Kallang, Sungei Malutut and Sungei Purulon, Sabah State, Batu Apoi, Brunei, Matang, Sungei Mengiong and Sungei Pesu, Sarawak State, and Landak, Kalimantan Province). **Status** Not Evaluated.

KINABALU WATER SNAKE
Hydrablabes praefrontalis NOT ILLUSTRATED
Measurement TL 346mm **Identification** Body slender; head small, distinct from neck; prefrontals fused; preoculars 2; postoculars 2; supralabials 8–9; eye small; pupil rounded; orbit separated from supralabials by series of suboculars; tail short; dorsals smooth; midbody scale rows 15; ventrals 178–202; subcaudals 72; anal divided. **Coloration** Dorsum olive-brown with dark vertebral stripe; paired dark dorsolateral stripe, broken into spots posteriorly; venter yellow, edges of ventrals with dark brown spots; subcaudals edged with brown. **Habitat and Behaviour** Inhabits submontane and montane forests. Nothing known of its natural history, although presumed to be aquatic and associated with hill streams. **Distribution** Endemic to north-western Borneo (Gunung Kinabalu, Sabah State). **Status** Not Evaluated.

IGUANA-JAWED SNAKE
Iguanognathus werneri NOT ILLUSTRATED
Measurement TL 350mm **Identification** Body slender, cylindrical; head small, slightly distinct from neck; snout short, rounded; loreal absent; preoculars 3; postoculars 3; supralabials 8; Supralabial IV contacts orbit; teeth with spatulate crowns ribbed along outer edge; eye small; pupil rounded; nostrils directed dorsally; dorsals smooth; midbody scales 19;

ventrals 136; subcaudals 53; anal divided. **Coloration** Dorsum dark brown; flanks paler with vertical black bars spotted with cream; cream line across snout; cream postocular streak; pair of cream dots on parietals and on first 5 anterior supralabials; venter cream with black cross-bars that are mostly interrupted; subcaudals black, each scale with cream spot. **Habitat and Behaviour** Nothing known of its natural history. **Distribution** Endemic to Sumatra. **Status** Vulnerable.

ORANGE-LIPPED KEELBACK
Macropisthodon flaviceps PLATE 69
Measurement TL 850mm **Identification** Body robust; head distinct from neck; loreal present; single preocular; postoculars 3–4; supralabials 8 (rarely 7); Supralabials IV–V (rarely III–IV) contact orbit; infralabials 5–6; tail short; dorsals keeled; midbody scale rows 19; ventrals 120–138; subcaudals 49–60, paired; anal divided. **Coloration** Dorsum greyish-black with faint light cross-bars, narrow at vertebral region and wide on flanks; forehead light brown, yellowish-brown or olive; rusty-orange supralabials; black-edged orange nuchal loop, especially in juveniles; venter black or dark green with black bands. **Habitat and Behaviour** Inhabits wetlands in tropical forests from sea level to 1,300m asl. Diurnal and semiaquatic. Threat display consists of raising its head and flattening its neck. Diet includes frogs, toads, tadpoles and lizards. Reproductive habits unstudied. **Distribution** Southern Thailand, Peninsular Malaysia, Sumatra, Pulau Bangka, Pulau Nias and Borneo. **Status** Not Evaluated.

OLIVE KEELBACK
Macropisthodon plumbicolor PLATE 69
Measurement TL 485mm **Identification** Body robust; head short and broad, distinct from neck; preoculars 2; postoculars 3–4; supralabials 7; Supralabials III–IV contact orbit; eye large; pupil rounded; tail short; dorsals strongly keeled; nuchal gland on nape; midbody scale rows 17; ventrals 144–162; subcaudals 34–48, paired; anal divided. **Coloration** Dorsum grass-green; juveniles with large V-shaped mark on neck, followed by similar smaller one, the intervening areas yellow or orange; black postocular stripe to angle of jaws; black spots or cross-bars on dorsum and tail; venter white, cream or grey. **Habitat and Behaviour** Inhabits the low hills at elevations of up to 2,134m asl. Diurnal and terrestrial, known from habitat with grass and may even enter houses. Threat display consists of raising its forebody and flattening its neck. Diet consists of toads. Reproductive habits unstudied. **Distribution** Myanmar. Also India and north-western Sri Lanka. **Status** Not Evaluated.

BLUE-NECKED KEELBACK
Macropisthodon rhodomelas PLATE 69
Measurement TL 750mm **Identification** Body robust; head distinct from neck; loreal present; preoculars 1–2; postoculars 3–4; supralabials 7 (rarely

8); Supralabials III–IV (rarely IV–V) contact orbit; infralabials 10; tail short; dorsals keeled; midbody scale rows 19; ventrals 124–143; subcaudals 42–58, paired; anal divided. **Coloration** Dorsum reddish-brown; black vertebral stripe enters nape as inverted chevron; posteriorly, light blue; venter pink, each ventral scale with small dark spot. **Habitat and Behaviour** Inhabits streams and other wetlands in the lowlands at elevations of up to ca 1,000m asl. Threat display consists of raising its head and flattening its neck, and discharging a white secretion from its nuchal gland. Diet consists of frogs and toads; juveniles consume tadpoles. Oviparous; clutches comprise 25 eggs. **Distribution** Southern Thailand, Peninsular Malaysia, Singapore, Sumatra, Borneo and Java. **Status** Not Evaluated.

ANDERSON'S STREAM SNAKE

Opisthotropis andersonii   PLATE 69

Measurement TL 500mm **Identification** Body slender; head small, depressed, indistinct from neck; nostrils dorsal; single preocular and postocular; Supralabial IV contacts orbit; internasals twice as long as broad, not touching loreals; temporals 1–2; dorsals weakly keeled; dorsal scale rows at anterior body 17; midbody scale rows 17; ventrals 168; subcaudals 53–66; anal divided. **Coloration** Dorsum greyish-black, green or olive-brown with indistinct fine black lines crossing scales; lowest row of dorsals, gular region and venter bright yellow. **Habitat and Behaviour** Inhabits the mid-hills. Nocturnal and aquatic, associated with rocky streams. Diet includes earthworms, freshwater shrimps, tadpoles and fish. Reproductive habits unstudied. **Distribution** Vietnam. Also eastern China. **Status** Not Evaluated.

HAINANESE STREAM SNAKE

Opisthotropis balteata   PLATE 69

Measurement TL 1,050mm **Identification** Body slender; head small, depressed, indistinct from neck; internasals as broad as long; single preocular; postoculars 2–3; supralabials 8–9; Supralabials IV, V or IV–V contact orbit; nostrils located dorsally; dorsals smooth; midbody scale rows 19; ventrals 194–205; subcaudals 69–99; anal divided. **Coloration** Dorsum yellowish-orange with numerous paired black bands that extend along flanks and meet on venter; head mottled with black, with vertical yellow markings in front of and behind orbit, and one at angle of jaws. **Habitat and Behaviour** Inhabits primary forests from sea level up to the mid-hills at elevations of ca 200m asl. Diurnal and aquatic, associated with clear rocky streams, hiding under rocks. Tail autotomy known. Diet includes fish. Reproductive habits unstudied. **Distribution** Cambodia and Vietnam. Also eastern China. **Status** Not Evaluated.

DAO VAN TIEN'S STREAM SNAKE

Opisthotropis daovantieni   PLATE 69

Measurement TL 578mm **Identification** Body slender; head small, depressed, indistinct from neck; loreal present; single preocular; postoculars 2; single subocular; supralabials 8–9; Supralabial V contacts orbit; infralabials 10; 5 infralabials contact anterior chin shields; single prefrontal; internasals narrowed and contact loreals; maxillary teeth 20, the last 2 enlarged; dorsals smooth; midbody scale rows 17; ventrals 189–194; subcaudals 39–47, paired; anal divided. **Coloration** Dorsum pale brown or olive-grey; faint flank stripes located within dark dorsal coloration, not sharply separating dark dorsum from pale venter; venter yellow; subcaudals with small dark marks. **Habitat and Behaviour** Inhabits rainforests at an elevation of 750m asl. Nocturnal and aquatic, associated with rocky streams. Diet and reproductive habits unstudied. **Distribution** Endemic to southern Vietnam (Buonloy, Gia Lai-Contum Province). **Status** Not Evaluated.

JACOB'S STREAM SNAKE

Opisthotropis jacobi   NOT ILLUSTRATED

Measurement TL 540mm **Identification** Body slender; head small, depressed, indistinct from neck; posterior dorsals smooth; single preocular and postocular; single temporal; supralabials 7–8; Supralabials IV–V contact orbit; infralabials 8; maxillary teeth 20, the last 2 enlarged; dorsals smooth; midbody scale rows 15; ventrals 159–179; subcaudals 60–90, paired; anal divided. **Coloration** Dorsum iridescent black; ventrals and subcaudals brown with narrow white edges, enlarging slightly on flanks and belly; forehead black; throat scales with narrow white edges. **Habitat and Behaviour** Inhabits submontane forests at elevations of ca 1,500m asl. Diet and reproductive habits unstudied. **Distribution** Endemic to northern Vietnam (Sa Pá, Lao Cai Province and Ba Vi, Ha Tay Province). **Status** Not Evaluated.

MAN-SON MOUNTAIN STREAM SNAKE

Opisthotropis lateralis   PLATE 69

Measurement TL 500mm **Identification** Body moderate; head small, indistinct from neck; nostrils dorsal; preoculars 2, rarely confluent; postoculars 2; subocular absent; supralabials 10 (rarely 9 or 11); Supralabials V–VII contact orbit; 4–5 infralabials contact anterior chin shields; temporals 1–2; upper dorsals typically keeled, lower dorsals smooth; midbody scale rows 17; ventrals 172; subcaudals 45; anal divided. **Coloration** Dorsum dark grey or greyish-brown; supralabials and infralabials yellow; venter pale yellow; faint dark lateral stripe on lower flanks sharply separates dark dorsum from pale venter. **Habitat and Behaviour** Inhabits the mid-hills at elevations of 170–900m asl. Nocturnal and aquatic, associated with rocky streams. Diet consists of freshwater shrimps, and perhaps also fish and tadpoles. Oviparous; clutches comprise 4–5 eggs. **Distribution** Northern Vietnam (Man Son Mountains, Lang Son Province). Also eastern China. **Status** Not Evaluated.

SPOTTED MOUNTAIN STREAM SNAKE
Opisthotropis maculosus PLATE 69
Measurement TL 520mm **Identification** Body slender; head small, depressed, indistinct from neck; loreal present; single preocular; postoculars 2; single supraocular; supralabials 8; Supralabial IV contacts orbit; infralabials 8–9; 5 infralabials contact anterior chin shields; single temporal; anterior pair of chin shields shortest; dorsals smooth; midbody scale rows 15; ventrals 182; subcaudals 67, paired; anal divided. **Coloration** Dorsum glossy black, each scale with yellow spot; yellow spots larger on flank scales; forehead glossy black with scattered yellow flecks near sutures; labials yellow with black sutures; chin yellow; venter including subcaudals yellow with black anterior and lateral margins to scales; iris black. **Habitat and Behaviour** Inhabits the forested mid-hills at an elevation of 190m asl. Nocturnal and aquatic. The only individual known was found swimming in a stream within disturbed bamboo in a mixed evergreen forest. Diet and reproductive habits unstudied. **Distribution** Endemic to north-eastern Thailand (Phu Wua Wildlife Sanctuary, Nong Khai Province). **Status** Not Evaluated.

JAVANESE STREAM SNAKE
Opisthotropis rugosa NOT ILLUSTRATED
Measurement TL 473mm **Identification** Body slender; head small, depressed, indistinct from neck; prefrontal divided; temporals 1–2; single postocular; supralabials 12; Supralabials VI–VIII contact orbit; nasal divided; internasal small, subtriangular; loreal large; single preocular; postoculars 1–2; midbody scale rows 17, strongly keeled and striated; ventrals 170; subcaudals 95, paired; anal divided. **Coloration** Dorsum purplish-brown fading to dark brown; scales edged with white; dark speckling on labials, especially along scale borders; yellow anterior supralabials and outer row of dorsal scales; venter unpatterned; iris black. **Habitat and Behaviour** Inhabits primary forests in the mid-hills at elevations of 300–600m asl. Nocturnal and aquatic, associated with hill streams. Diet and reproductive habits unstudied. **Distribution** Endemic to northern and western Sumatra (Aek Nangali, Sumatera Utara Province and Kayu Tanam, Sumatera Barat Province). **Status** Not Evaluated.

SPENCER'S STREAM SNAKE
Opisthotropis spenceri NOT ILLUSTRATED
Measurement TL 600mm **Identification** Body slender; head small, depressed, indistinct from neck; internasals broader than long, contacting loreal; single preocular; postoculars 2; supralabials 8; Supralabials IV–V contact orbit; eye small; dorsals smooth; posterior temporals 2; midbody scale rows 17; ventrals 183; subcaudals 33, paired; anal divided. **Coloration** Dorsum olive; venter pale yellow, the two meeting on the 3 outer rows of scales; subcaudals grey-mottled. **Habitat and Behaviour** Inhabits forested streams at elevations of ca 300m asl. Diet and reproductive habits unstudied. **Distribution** Endemic to northern Thailand (Muang Ngow, Lampang Province). **Status** Not Evaluated.

TAM DAO STREAM SNAKE
Opisthotropis tamdaoensis NOT ILLUSTRATED
Measurement TL 555mm **Identification** Body robust, cylindrical; head small, depressed, indistinct from neck; nasal incompletely divided below nostril; loreals 1–2, not in contact with internasals; single preocular; postoculars 2; anterior temporals 2; posterior temporals 3–4; supralabials 8–9; Supralabial V or VI contacts orbit; infralabials 9–10; first pair of chin shields longer than second pair; dorsals smooth anteriorly, keeled posteriorly; dorsal scale rows at anterior body 19; midbody scale rows 17; ventrals 171; subcaudals 49, paired; anal divided. **Coloration** Dorsum unpatterned dark olive-grey with longitudinal dark stripe on flanks; venter pale, tip of subcaudal region with dark mottling. **Habitat and Behaviour** Inhabits mountain streams within primary forests at elevations of ca 750m asl. Diet and reproductive habits unstudied. **Distribution** Endemic to northern Vietnam (Tam Dao, Vinh Phú Province). **Status** Not Evaluated.

CORRUGATED WATER SNAKE
Opisthotropis typica PLATE 69
Measurement TL 502mm **Identification** Body slender; head small, depressed, indistinct from neck; prefrontal divided; forehead scales finely striated; loreal large; preoculars 2; postoculars 2; suboculars 3; temporals 1–2; supralabials 11–12; Supralabials VI–VIII contact orbit; infralabials 10; eye small, separated from labials by small scales; dorsals strongly keeled; midbody scale rows 19; ventrals 160–176; subcaudals 82–96; anal divided. **Coloration** Dorsum unpatterned brownish- or blackish-grey; venter unpatterned cream. **Habitat and Behaviour** Inhabits the forested plains and mid-hills at elevations of 75–500m asl. Nocturnal and aquatic, associated with shallow rocky streams, and swamp pools and streams. Diet consists of tadpoles. Reproductive habits unstudied. **Distribution** Borneo (Gunung Kinabalu, Sabah State, Batu Apoi, Brunei and Pelagus Rapids, Sarawak State). Also Palawan, the south-eastern Philippines. **Status** Not Evaluated.

BROWN STREAM SNAKE
Paratapinophis praemaxillaris NOT ILLUSTRATED
Measurement TL 980mm **Identification** Body robust, quadrangular in cross-section in females, relatively slender in males; head distinct from neck; eye lateral; single prefrontal; single loreal; single preocular; postoculars 2; supralabials 8–9; Supralabials IV–V or V contact orbit; infralabials 9–11; posterior temporals 2; enlarged posterior chin shields; dorsals smooth, scales with middle row of tubercles; midbody scale rows 19; ventrals 145–154; subcaudals 64–67, paired; anal divided. **Coloration** Dorsum uniform brownish-grey; edges of dorsals and ventrals greyish-brown; anterior dorsum with indistinct inverted V-shaped mark of pale yellow scales, turning into blue-grey bands, extending across dorsum in females; in males, bluish-grey colours absent; in females, dark blotches between V-shaped mark; forehead

unpatterned brownish-grey; supralabials brownish-grey; infralabials with cream spots; newborns unpatterned dark brown dorsally; venter unpatterned cream. **Habitat and Behaviour** Inhabits rocky streams in subtropical alpine forests at elevations of 475–1,400m asl. Nocturnal and aquatic, living in fast-flowing waters. Diet consists of fish. Oviparous (based on presence of egg tooth in a presumed hatchling); clutch size unknown. Newborns ca 210–214mm. **Distribution** Northern Thailand (Doi Saket, Chang Mai Province and Wang Pian Waterfalls, Nan Province) and northern Laos (Xieng-Khouang). Also southern China (Yunnan Province). **Status** Not Evaluated.

PAINTED MOCK VIPER
Psammodynastes pictus PLATE 69
Measurement TL 550mm **Identification** Body slender; head flattened, distinct from neck; third lower labial large and borders mental groove; loreal present; preoculars 1–3; postoculars 3–4; supralabials 8; Supralabials III–V contact orbit; infralabials 8–9; 2 posterior-most teeth enlarged; eye large; pupil vertical; dorsals smooth, lacking apical pits; midbody scale rows 17; ventrals 152–171; subcaudals 60–80, paired; anal entire. **Coloration** Dorsum brown or tan, sometimes black, with transverse dark-edged light bands; dark streak along eye; venter cream with brown speckles. **Habitat and Behaviour** Inhabits lowland forests. Diet consists of fish, frogs, lizards and prawns. Ambushes fish while suspended from vegetation overhanging water bodies. Ovoviviparous; clutch size unknown. **Distribution** Peninsular Malaysia, Singapore, Sumatra, the Riau and Mentawai Archipelagos, Pulau Belitung and Borneo. Records from Sulawesi (Indonesia) may be based on human introductions. **Status** Not Evaluated.

MOCK VIPER
Psammodynastes pulverulentus PLATE 69
Measurement TL 770mm **Identification** Body slender; head flattened, distinct from neck; snout short, truncated in profile; loreal present; preoculars 1–2; postoculars 2–4; supralabials 8–9; Supralabials III–V contact orbit; infralabials 8–9; 2 posterior-most teeth enlarged; eye large; pupil vertical; dorsals smooth, lacking apical pits; midbody scale rows 17 or 19; ventrals 146–175; subcaudals 44–70, paired; anal divided. **Coloration** Dorsum reddish-brown, to yellowish-grey, to nearly black, with small streaks; typically, longitudinal stripe along middorsal region and 3 longitudinal stripes along flanks; venter spotted with brown or grey, and with dark spots or longitudinal lines. **Habitat and Behaviour** Inhabits wet evergreen and tropical forests from sea level to 2,000m asl. Diurnal and nocturnal, and terrestrial, but can climb low bushes. Diet consists of heavily scaled vertebrates such as skinks and other snakes; frogs and geckos are also consumed. Ovoviviparous; clutches comprise 3–10 neonates, measuring 148–178mm, and are produced several times a year. **Distribution** Myanmar, Thailand, Laos, Cambodia, Vietnam, Peninsular Malaysia, Sumatra, Pulau Bangka, the Mentawai, Natuna and Riau Archipelagos, Borneo and Java. Also northern and north-eastern India, Nepal, southern China, Sulawesi, the Lesser Sundas, the Philippines (*P. p. pulverulentus* or an unassigned subspecies) and Taiwan (*P. p. papenfussi*). **Status** Not Evaluated.

ANGEL'S KEELBACK
Rhabdophis angelii NOT ILLUSTRATED
Measurement TL 429mm **Identification** Body moderate, cylindrical; frontal wider anteriorly; nasal divided; single preocular; postoculars 3; supralabials 6; Supralabials III–IV contact orbit; infralabials 7; maxillary teeth 23, 2 posterior-most enlarged; eye large; pupil rounded; dorsals smooth; midbody scale rows 15; ventrals 117–126; subcaudals 39–46, paired; anal divided. **Coloration** Dorsum dark brown with double longitudinal series of 42 pairs of small orange marks, less distinct posteriorly, each 1 scale wide; V-shaped orange mark on nape; supralabials and throat unpatterned cream; large oblique mark under eye and another on last supralabial; throat orange with dark brown stippling that increases posteriorly, creating a dark venter. **Habitat and Behaviour** Inhabits the mid-hills at elevations of 450–900m asl. Diet and reproductive habits unstudied. **Distribution** Endemic to northern Vietnam (Tam Dao, Vinh Phuc Province). **Status** Not Evaluated.

BLACK-BARRED KEELBACK
Rhabdophis callichromus NOT ILLUSTRATED
Measurement TL 714mm **Identification** Body moderate, cylindrical; head distinct from neck; single preocular; postoculars 3; supralabials 8; Supralabials III–V contact orbit; infralabials 10; 29 maxillary teeth; eye large; pupil rounded; dorsals smooth; midbody scale rows 19; ventrals 152–159; subcaudals 79–86, paired; anal divided. **Coloration** Dorsum greyish-olive with indistinct, narrow transverse dark bars, intersected on dorsolateral line by short cream bars; head and nape with white patch behind parietals; venter cream with fine grey dots. **Habitat and Behaviour** Inhabits lowland forests at elevations of ca 400m asl. Diet and reproductive habits unstudied. **Distribution** Northern Vietnam (Mount Bavi, Son Tay Province). Also eastern China. **Status** Not Evaluated.

SAHUL KEELBACK
Rhabdophis chrysargoides PLATE 70
Measurement TL 870mm **Identification** Body moderate, cylindrical; head distinct from neck; loreal present; single preocular; postoculars 3–4; supralabials 8–9; Supralabials IV–VI, IV–V or V–VI contact orbit; infralabials 10–11; 5–6 infralabials contact anterior chin shields; eye large; pupil rounded; tail relatively short; dorsals keeled; midbody scale rows 19 or 21; ventrals 148–164; subcaudals 53–85, paired; anal divided. **Coloration** Dorsum dark olive, black or blackish-brown; juveniles with brownish-yellow vertebral stripe, scalloped edged anteriorly; white spot on sides of black head and on occiput; white band

across frontal, supraoculars and preoculars; supralabials yellow or pinkish-white, sutures black; venter yellow, pink, greyish-white or slate-blue, ventrals edged with olive; iris orange. **Habitat and Behaviour** Inhabits the forested mid-hills. Terrestrial, associated with streams. Diet and reproductive habits unstudied. **Distribution** Java. Also Sulawesi, Pulau Buton, Pulau Muna and the Sangihe Archipelago. **Status** Not Evaluated.

SPECKLE-BELLIED KEELBACK
Rhabdophis chrysargos   PLATE 70
Measurement TL 980mm **Identification** Body moderate, cylindrical; head distinct from neck; loreal present; preoculars 1–2; postoculars 3–4; supralabials 9; Supralabials III–V or IV–VI contact orbit; infralabials 10; eye large; pupil rounded; tail relatively short; dorsals keeled; midbody scale rows 19; ventrals 139–176; subcaudals 56–94, paired; anal divided. **Coloration** Dorsum olive-grey or olive-brown; supralabials yellow or cream with darker smudges; narrow cream, reddish-brown or orange nuchal chevron, edged with black; reddish-brown or orange band behind neck; rest of back with oblong yellow and brown marks, within darker bands; venter yellow with brown mottling. **Habitat and Behaviour** Inhabits hilly evergreen forests up to submontane limits at elevations of 100–1,676m asl. Terrestrial, associated with streams. Diet includes frogs. Oviparous; clutches comprise 3–10 eggs, measuring 12–21 x 19.5–34mm. Incubation period 51–61 days. Hatchlings 148–220mm. **Distribution** Southern Myanmar, Thailand, Vietnam, Laos, eastern and south-western Cambodia (Siem Pang District and Kirirom), Peninsular Malaysia, Singapore, Sumatra, Pulau Nias, Pulau Simeuleu, the Mentawai and Anamba Archipelagos, Borneo, Java and Bali. Also Sulawesi, Flores, Ternate (Indonesia), Palawan and Balabac (the Philippines). **Status** Not Evaluated.

RED-BELLIED KEELBACK
Rhabdophis conspicillatus   PLATE 70
Measurement TL 550mm **Identification** Body moderate, cylindrical; head distinct from neck; loreal present; single preocular; postoculars 3; supralabials 8; Supralabials III–V contact orbit; 5 infralabials contact chin shields; eye large; pupil rounded; tail short; dorsals keeled, except the smooth outer rows; midbody scale rows 19; ventrals 141–152; subcaudals 51–60, paired; anal divided. **Coloration** Dorsum brown to reddish-brown; sides of head with cream postocular stripe that curves downwards; supralabials cream; nape and neck with 2 narrow cream collars; venter yellow, each ventral scale dark edged. **Habitat and Behaviour** Inhabits lowland forests, reaching submontane limits at elevations of 100–1,000m asl. Nocturnal and terrestrial, associated with water bodies, leaf litter, tree buttresses, and rotting logs and stones. Diet and reproductive habits unstudied. **Distribution** Peninsular Malaysia, Sumatra, Pulau Singkep, the Natuna Archipelago and Borneo. **Status** Not Evaluated.

HIMALAYAN KEELBACK
Rhabdophis himalayanus   PLATE 70
Measurement TL 1,250mm **Identification** Body moderate, cylindrical; head distinct from neck; single preocular; postoculars 3 (rarely 4); supralabials 8; Supralabials IV–V contact orbit; last 2 maxillary teeth abruptly enlarged; eye large; pupil rounded; dorsals keeled; midbody scale rows 19; ventrals 151–177; subcaudals 78–98, paired; anal divided. **Coloration** Dorsum olive to olive-brown or dark brown; 2 dorsolateral rows of widely separated, orange-yellow spots, more numerous anteriorly; anterior of body chequered with vermilion spots covering 2–3 scales per spot; neck with bright yellow collar edged with black; labials yellow or cream edged with black; black subocular stripe; venter yellowish-white, becoming darker apically. **Habitat and Behaviour** Inhabits forests particularly on rocky biotopes, and also agricultural fields in the vicinity of streams, at elevations of up to 1,100m asl. Diet consists of frogs, lizards and fish. Oviparous; clutches comprise 5–7 eggs. **Distribution** Northern Myanmar. Also eastern and north-eastern India, Nepal, Bangladesh and southern China. **Status** Not Evaluated.

LEONARD'S KEELBACK
Rhabdophis leonardi   PLATE 70
Measurement TL 1,060mm **Identification** Body moderate, cylindrical; head distinct from neck; prefrontals paired; loreal present; single preocular; postoculars 2–3; supralabials 7; Supralabials III–IV contact orbit; infralabials 6; Infralabial VI contacts 3 scales posteriorly; 3 posterior-most maxillary teeth enlarged; eye large; pupil rounded; dorsals keeled, except 2–3 lower rows; midbody scale rows 15 or 17; ventrals 145–155; subcaudals 46–64, paired; anal divided. **Coloration** Dorsum olive-brown, edges of scales of anterior body black and cream; head olive-brown turning greyish-cream near labials; narrow black subocular stripe to Supralabial IV; sometimes, a narrow reddish-orange cross-bar on nuchal region; indistinct dark vertebral stripe that becomes darker posteriorly; venter greyish-cream with tiny grey dots, the intensity of which increases posteriorly. **Habitat and Behaviour** Inhabits wet subtropical forests at an elevation of 1,829m asl. Diurnal and terrestrial. Diet and reproductive habits unstudied. **Distribution** North-eastern Myanmar (Sinlum Kaba, Kachin Hills). Also southern China. **Status** Not Evaluated.

GUNUNG MURUD KEELBACK
Rhabdophis murudensis   PLATE 70
Measurement TL 873mm **Identification** Body moderate, cylindrical; head distinct from neck; single anterior temporal; postoculars 3; supralabials 9; Supralabials IV–VI contact orbit; infralabials 11; 6 infralabials contact anterior chin shields; eye large; pupil rounded; dorsals keeled, except the smooth outer rows; midbody scale rows 19; ventrals 176–185; subcaudals 63–97, paired; anal divided. **Coloration** Dorsum brownish-grey with indistinct dark cross-bars; row of light spots on the edges of the dark cross-

bars; supralabials yellow, bright red or brown; venter greyish-yellow with small black spots; iris brown with pale green portion dorsally. **Habitat and Behaviour** Known from submontane and montane regions at elevations of 915–2,500m asl. Diet includes frogs. Reproductive habits unstudied. **Distribution** Endemic to north-western and northern Borneo (Gunung Murud and Gunung Mulu, Sarawak State, and Gunung Kinabalu and Trus Madi, Sabah State). **Status** Not Evaluated.

BLACK-BANDED KEELBACK
Rhabdophis nigrocinctus   PLATE 70

Measurement TL 950mm **Identification** Body moderate, cylindrical; head distinct from neck; nostrils lateral; loreal present; single preocular; postoculars 2–4; temporals 2 + 2; supralabials 8–9; Supralabials IV–VI or III–V contact orbit; infralabials 10; eye large; pupil rounded; dorsals keeled; midbody scale rows 17 or 19; ventrals 150–168; subcaudals 72–96, paired; anal divided. **Coloration** Dorsum olive-green, turning more brown posteriorly, with indistinct, narrow black cross-bars; 2 oblique black stripes on flanks; forehead copper-brown, lighter on sides; oblique black subocular and postocular stripes, and another on nape; venter white; subcaudals pale pink mottled with dark grey; iris dark brown. **Habitat and Behaviour** Inhabits evergreen forests. Terrestrial and aquatic. Diet consists of fish and frogs. Reproductive habits unstudied. **Distribution** Southern Myanmar (Bago Division), Thailand (Me Wang, Me Nga, M. Fang, Tasan, Tapli, Chumphon, Klong Bang Lai, Hup Bon, Klong Yai and Khao Sebab), Laos, Cambodia (Cardamom Mountains) and Vietnam. Also southern China (Yunnan Province). **Status** Not Evaluated.

COLLARED KEELBACK
Rhabdophis nuchalis   PLATE 70

Measurement TL 620mm **Identification** Body moderate, cylindrical; head flat, distinct from neck; nuchal groove distinct; supralabials 6; Supralabials III–IV contact orbit; infralabials 8; eye large; pupil rounded; dorsals keeled, except smooth outer row; midbody scale rows 15 or 17; ventrals 152–156; subcaudals 49–64, paired; anal divided. **Coloration** Dorsum light brown chequered with pale reddish-brown spots; forehead brown speckled with red; 2 oblique black stripes between posterior supralabials; alternate rows of body scales red and brown; interstitial skin bluish-black; fine oblique black lines on dorsolateral scales at midbody; neck reddish-brown from angle of jaws, posteriorly for 11–12 scale rows; juveniles with bright reddish-yellow collar, posteriorly with reddish tinge. **Subspecies** Two subspecies are recognized, of which one (*R. n. pentasupralabialis*) occurs in the region. **Habitat and Behaviour** Inhabits secondary subtropical broadleaved forests, including bamboo brakes, as well as areas of terrace cultivation, at elevations of 1,500–1,750m asl, the Sichuan subspecies ascending to 2,750m asl. Diurnal and terrestrial. Diet unstudied. Oviparous; clutches comprise 8–19 elongated eggs, measuring 26 x 13mm, which are laid communally. **Distribution** Northern Myanmar and northern Vietnam (presumably the subspecies *R. n. pentasupralabialis*). Also north-eastern India (Nagaland), and eastern and southern China (*R. n. nuchalis* and *R. n. pentasupralabialis*). **Status** Not Evaluated.

RED-NECKED KEELBACK
Rhabdophis subminiatus   PLATE 70

Measurement TL 1,300mm **Identification** Body moderate; head distinct from neck; loreal present; single preocular; postoculars 2–4; supralabials 8; Supralabials III–V or III–IV contact orbit; infralabials 10; last 2 maxillary teeth abruptly enlarged; eye large; pupil rounded; dorsals keeled, except the smooth outer rows; midbody scale rows 19; ventrals 132–175; subcaudals 64–96, paired; anal divided. **Coloration** Dorsum olive-brown or green, unpatterned or with black and yellow reticulation; nape with yellow and red band; oblique dark subocular bar; venter yellow, sometimes with a black dot on outer edges of ventrals; juveniles with black cross-bar or triangular mark on nape, bordered with yellow posteriorly; dorsum with oval black spots. **Subspecies** Two subspecies are recognized. *R. s. subminiatus*, ventrals 132–157; nuchal scales not enlarged; nuchal groove indistinct or absent. *R. s. helleri*, ventrals 160–172; nuchal scales enlarged; nuchal groove distinct. **Habitat and Behaviour** Inhabits the forested lowlands and mid-hills, near ponds and streams, at elevations of up to 1,780m asl. Diurnal and nocturnal, and terrestrial and semi-arboreal. Diet consists of frogs and toads; juveniles eat tadpoles and fish. Oviparous; clutches comprise 5–17 eggs, measuring 17.5–27 x 11–15mm, which are guarded by the female. Incubation period 50–70 days. Hatchlings 130–190mm. **Distribution** Myanmar, Thailand, Laos, Cambodia, Vietnam (*R. s. helleri*), Peninsular Malaysia, Singapore, Sumatra, Borneo and Java (*R. s. subminiatus*). Also north-eastern India, Nepal, Bangladesh, southern and eastern China (*R. s. helleri*), Sulawesi and Ternate (*R. s. subminiatus*). **Status** Not Evaluated.

JAPANESE KEELBACK
Rhabdophis tigrinus   PLATE 70

Measurement TL 1,013mm **Identification** Body robust; head large, distinct from neck; loreal present; preoculars 2; postoculars 3; supralabials 7; Supralabials IV–V contact orbit; 5 infralabials contact chin shields; enlarged posterior maxillary teeth; eye large; pupil rounded; tail short; dorsals keeled; midbody scale rows 19; ventrals 144–224; subcaudals 50–69, paired; anal divided. **Coloration** Dorsum greenish-olive anteriorly, orange-red elsewhere, with series of large rectangular black marks on anterior of body, which turn indistinct at midbody; curved black spot on sides of neck; venter dark grey. **Subspecies** Two subspecies are recognized, of which one (*R. t. lateralis*) occurs in the region. **Habitat and Behaviour** Inhabits forested hills and those covered with bushes. Winter hibernation sites may be 0.3m below ground. Diurnal and terrestrial. Diet consists of beetles, frogs, toads and other snakes. Oviparous; clutches comprise 9–27 eggs. Incubation period 29–47 days. Hatchlings 170mm (subspecies *R. t. lateralis*). **Distribution** Vietnam (*R. t. lateralis*). Also eastern Russia, eastern China, Korea (*R. t. lateralis*) and Japan (*R. t. tigrinus*). **Status** Not Evaluated.

CHINESE SPOTTED KEELBACK WATER SNAKE
Sinonatrix aequifasciata PLATE 71
Measurement TL 1,420mm **Identification** Body robust; head elongated; single supralabial enters orbit of eye; supralabials 9, none or 1 contact orbit; dorsals keeled; midbody scale rows 19; ventrals 142–153; subcaudals 67–76; males with tubercles on chin shields and on Infralabial I. **Coloration** Dorsum with 16–21 bands encircling body, constricted on flanks, and 7–12 on tail; interspaces cream with brownish tinge, narrower than the dark bars; venter cream with black markings. **Habitat and Behaviour** Inhabits the lowlands to submontane forests at elevations of 40–1,250m asl. Associated with rocky streams, and frequently encountered basking on vegetation near or above water, into which it retreats. Diet consists of fish such as carp, and also frogs. Oviparous; clutch size unknown. **Distribution** Myanmar, Laos and Vietnam (Lang Son, Lao Cai, Ha Giang, Cao Bang, Vinh Phuc, Bac Giang, Nghe An and Ha Tinh Provinces). Also eastern and southern China. **Status** Not Evaluated.

RINGED KEELBACK WATER SNAKE
Sinonatrix annularis PLATE 71
Measurement TL 941mm **Identification** Body robust; head not elongated, distinct from neck; loreal present; single preocular; postoculars 3; supralabials 9; Supralabial V contacts orbit; infralabials 10; eye large; dorsals strongly keeled, except outer 2–3 scale rows; midbody scale rows 19; ventrals 145–163; subcaudals 53–69; anal divided. **Coloration** Dorsum dark grey with dark cross-bars, numbering 34–46 on body and 16–27 on tail; indistinct dark reticulations on back; supralabials and infralabials greyish-cream, sutures black; indistinct dark postocular streak; venter red between the dark cross-bars. **Habitat and Behaviour** Inhabits the lowlands and mid-hills. Associated with open streams; also recorded from flooded rice fields. Diet consists of fish including eels, and also insects and frogs. Ovoviviparous; clutches comprise 4–18 neonates, measuring 196–239mm. **Distribution** Myanmar. Also southern and eastern China. **Status** Not Evaluated.

CHINESE KEELBACK WATER SNAKE
Sinonatrix percarinata PLATE 71
Measurement TL 1,100mm **Identification** Body robust; head large, distinct from neck; nasals divided; loreal present; single preocular; postoculars 3; supralabials 8–9; Supralabials IV–V contact orbit; dorsals keeled; midbody scale rows 19; ventrals 133–157; subcaudals 68–85, paired; anal divided. **Coloration** Dorsum and forehead olive-grey, dark brown or black, with 28–36 light-edged black bars on flanks, broad dorsally, becoming narrow laterally; rest of head yellowish-cream; venter yellowish-cream anteriorly, with small black speckles towards posterior; subcaudals dark grey with black spots. **Habitat and Behaviour** Inhabits evergreen hill forests near water bodies, reaching montane limits, at elevations of 300–2,000m asl. Diurnal, and aquatic and terrestrial, foraging in water and resting on riverbanks and on bank vegetation. Diet consists of fish and frogs. Oviparous; clutches comprise 4–12 eggs. **Distribution** Myanmar, northern Thailand (Chiang Mai and Phu Luang Wildlife Sanctuary, Loei Province), Laos and Vietnam. Also southern and eastern China, and north-eastern India (Arunachal Pradesh). **Status** Not Evaluated.

YUNNAN KEELBACK WATER SNAKE
Sinonatrix yunnanensis PLATE 71
Measurement TL 498mm **Identification** Body robust; head slightly distinct from neck; loreal present; single preocular; postoculars 3–4; supralabials 9–10; Supralabial IV (rarely IV–VI or V) contacts orbit; infralabials 10–11; dorsals keeled; 1–3 rows of lateral scales smooth; midbody scale rows 19; ventrals 156–165; subcaudals 61–83. **Coloration** Dorsum brown or brownish-black with transverse black lines that form X-pattern on flanks; venter cream. **Habitat and Behaviour** Inhabits submontane to montane forests at elevations of 900–2,000m asl. Diurnal and aquatic. Diet and reproductive habits unstudied. **Distribution** Northern Myanmar (Kachin State). Also southern China (Yunnan Province). **Status** Not Evaluated.

BURMESE KEELBACK WATER SNAKE
Xenochrophis bellula NOT ILLUSTRATED
Measurement TL 500mm **Identification** Body robust, cylindrical; head slightly distinct from neck; single anterior temporal; 3 supralabials contact orbit; midbody scale rows 19; ventrals 139–144; subcaudals 78–83; hemipenes long, deeply forked, extending to Subcaudal X. **Coloration** Dorsum dark olive-green with indistinct black spots; dorsolateral series of light spots or cross-bars; labials cream, edged with black, the pale stripe extending as a vertical bar anterior and posterior to orbit of eye; sides of neck and anterior flanks with vertical white bars; venter cream, scutes dark-edged. **Habitat and Behaviour** Nothing known of its natural history. **Distribution** Endemic to Myanmar. **Status** Not Evaluated.

YELLOW-SPOTTED KEELBACK WATER SNAKE
Xenochrophis flavipunctatus PLATE 71
Measurement TL 974mm **Identification** Body robust, cylindrical; head slightly distinct from neck; loreal present; single preocular; postoculars 3; supralabials 8; Supralabials IV–V contact orbit; infralabials 10; eye large; pupil rounded; tail short; dorsals weakly keeled anteriorly, keels becoming more distinct posteriorly, lacking apical pits; midbody scale rows 19; ventrals 122–143; subcaudals 70–91, paired; anal divided. **Coloration** Dorsum olivaceous or

greenish-grey, sometimes with reddish tinge, with black spots that develop into reticulated pattern posteriorly; head dark olive or grey with 2 dark lines from eye, and an interrupted band from base of jaw, across neck, forming V-shape, and 2 small median bands on forehead; venter green, cream or yellow, each scale with a black line on posterior edge. **Habitat and Behaviour** Inhabits aquatic areas in the lowlands, including ponds, swamps and flooded rice fields. Diurnal. Diet consists of fish and frogs. Oviparous; clutches comprise 20–60 eggs. Incubation period 43 days. Hatchlings 120–150mm. **Distribution** Myanmar, Thailand, Laos, Cambodia, Vietnam and northern Peninsular Malaysia. Also eastern India, and southern and eastern China. **Status** Not Evaluated.

MALAYAN SPOTTED KEELBACK WATER SNAKE
Xenochrophis maculatus PLATE 71
Measurement TL 1,000mm **Identification** Body slender, cylindrical; head distinct from neck; 2 anterior temporals; loreal present; single preocular; postoculars 3; supralabials 9; Supralabials IV–VI contact orbit; 5 infralabials contact anterior chin shields; eye large; pupil rounded; dorsals keeled; midbody scale rows 19; ventrals 140–156; subcaudals 95–117; anal divided. **Coloration** Dorsum brownish-orange with 4 longitudinal series of small squarish dark marks, and a paired row of yellow spots; head dark brown or black; supralabials yellow with darker smudges; venter yellow, ventrals edged with black. **Habitat and Behaviour** Inhabits open forests in the lowlands. Diurnal, and more terrestrial than congeneric species, although associated with streams, ditches and other wetlands. Diet consists of frogs and toads. Reproductive habits unstudied. **Distribution** Peninsular Malaysia, Singapore, Sumatra, Pulau Bangka, Pulau Belitung, the Natuna and Riau Archipelagos, and Borneo. **Status** Not Evaluated.

JAVAN KEELBACK WATER SNAKE
Xenochrophis melanzostus PLATE 71
Measurement TL 975mm **Identification** Body robust, cylindrical; head slightly distinct from neck; tail shorter in females than in related species; dorsals keeled; midbody scale rows 19; ventrals 128–142; subcaudals 66–83. **Coloration** Two colour morphs known: a blotched form with elongated blotches on dorsum, and a striped form with longitudinal broad dark stripes on dorsum and a wide U- or V-shaped mark on nuchal region. **Habitat and Behaviour** Inhabits the lowlands, and associated with freshwaters. Diet and reproductive habits unstudied. **Distribution** Endemic to Java. **Status** Not Evaluated.

CHEQUERED KEELBACK WATER SNAKE
Xenochrophis piscator PLATE 71
Measurement TL 1,020mm **Identification** Body robust, cylindrical; head distinct from neck; single preocular; postoculars 3; supralabials 9; Supralabials IV–V contact orbit; nostrils directed slightly upwards; eye large; pupil round; dorsal scales strongly keeled; midbody scales 19; ventrals 122–158; subcaudals 60–99, paired; anal usually divided. **Coloration** Dorsum olive-brown, black spots arranged in 5–6 rows; head brown with black subocular stripe from eye to supralabial and from postoculars to angle of jaws; inverted V-shape mark on nuchal region. **Habitat and Behaviour** Inhabits freshwaters, including flooded rice fields, ponds, lakes, marshes and rivers, in the plains and mid-hills. Active by both day and night. Diet consists of fish and frogs. Oviparous; clutches comprise 4–100 eggs, measuring 15–40mm, which are guarded by the female. Incubation period 37–90 days. Hatchlings 110mm. **Distribution** Myanmar, northern and north-western Thailand, and north-western Laos. Also eastern Afghanistan, Pakistan, India, northern Sri Lanka, Bangladesh, Nepal and southern China. **Status** Not Evaluated.

SPOTTED KEELBACK WATER SNAKE
Xenochrophis punctulatus NOT ILLUSTRATED
Measurement TL 860mm **Identification** Body robust, cylindrical, slightly depressed; snout short; eye small; head slightly distinct from neck; dorsals keeled; midbody scale rows 17; ventrals 131–151; subcaudals 62–84. **Coloration** Dorsum dark brown with cream-yellow spots, mainly on scale rows 4–5 on forebody; spots give the appearance of a pale line along flanks from behind neck to level of vent and along tail; venter cream, scales edged with dark brown. **Habitat and Behaviour** Inhabits the lowlands, and associated with freshwater habitats. Diet and reproductive habits unstudied. **Distribution** Southern Myanmar and north-western Thailand. Record from Manipur, north-eastern India, requires verification. **Status** Not Evaluated.

ST JOHN'S KEELBACK WATER SNAKE
Xenochrophis sanctijohannis PLATE 71
Measurement TL 710mm **Identification** Body robust, cylindrical; head distinct from neck; single preocular; postoculars 2–3 (rarely 4); supralabials 9; Supralabials IV or IV–V contact eye; infralabials 10; nostrils directed slightly upwards; eye large; pupil rounded; dorsals feebly keeled or nearly smooth; midbody scale rows 19; ventrals 139–154; subcaudals 84–94; anal divided. **Coloration** Pale olive or silvery-grey dorsally, uniform or with indistinct dark spots, sometimes with double series of cream spots along body; no transverse dark neck stripes; venter unpatterned cream or yellow. **Habitat and Behaviour** Inhabits agricultural fields, open water bodies and marshes at elevations of 407–1,400m asl. Probably nocturnal and diurnal, and more terrestrial than Chequered Keelback Water Snake (*X. piscator*). Diet consists of small mammals, frogs and fish. Reproductive habits unstudied. **Distribution** Myanmar. Also Pakistan, India and Nepal. **Status** Not Evaluated.

RED-SIDED KEELBACK WATER SNAKE
Xenochrophis trianguligerus PLATE 71
Measurement TL 1,350mm **Identification** Body slender, cylindrical; head large, distinct from neck;

loreal present; single preocular; postoculars 3; supralabials 9; Supralabials IV–VI contact orbit; infralabials 10; eye large; pupil rounded; tail short; dorsals keeled, except for outer 1–2 smooth rows; midbody scale rows 19; ventrals 132–150; subcaudals 62–105; anal divided. **Coloration** Dorsum blackish-brown with orange-red triangles on sides of neck and front portion of body; bright colours turn olive-brown or grey with age; triangle-shaped dark marks on top of body; some scales on lips black-edged; venter cream. **Habitat and Behaviour** Inhabits streams and standing water bodies, including ditches and puddles, in lowland forests, and rice paddies, ditches and village areas, to 1,400m asl. Nocturnal, and aquatic or active on edges of water bodies. Diet consists of frogs. Oviparous; clutches comprise 5–15 eggs, measuring 15–17 x 29–34mm. Incubation period 59–60 days. **Distribution** Myanmar, Thailand, south-western Cambodia (Cardamom Mountains), Vietnam, Peninsular Malaysia, Singapore, Sumatra, Pulau Nias, Pulau Bangka, Pulau Belitung, the Mentawai and Riau Archipelagos, Borneo and Java. Also Sulawesi, Indonesia and the Nicobar Archipelago, India. **Status** Not Evaluated.

STRIPED KEELBACK WATER SNAKE
Xenochrophis vittatus PLATE 71
Measurement TL 700mm **Identification** Body robust, cylindrical; head slightly distinct from neck; loreal present; single preocular; postoculars 3; supralabials 9; Supralabials IV–VI contact orbit of eye; infralabials 5; 5 infralabials contact anterior chin shields; eye large; pupil rounded; tail short; dorsals keeled, except for smooth outer rows; midbody scale rows 19; ventrals 140–151; subcaudals 53–84; anal divided. **Coloration** Dorsum mid-brown with 3 black stripes; head with dense mottling or dark areas; supralabials cream with dark edges; venter white, ventrals edged with black. **Habitat and Behaviour** Inhabits wetlands at elevations of ca 500m asl. Diurnal and aquatic. Diet consists of fish. Oviparous; clutches comprise 3–11 eggs, measuring 19–28 x 9.5–12mm. Incubation period 40–60 days. Hatchlings 130–180mm. **Distribution** Sumatra, Pulau Weh, Pulau Bangka and Java; introduced into Singapore. Also possibly Sulawesi. **Status** Not Evaluated.

Family PAREATIDAE SLUG-EATING SNAKES

Species in this family have a short skull, large eyes, large nasal glands, chin shields that lack a midline groove and extend across the chin, and a low midbody scale count (just 13–15 rows). Slug-eating snakes inhabit moist forests, and are nocturnal, arboreal, oviparous and specialized feeders on snails and slugs. They are found only in South-East Asia.

BLUNT-HEADED SLUG-EATING SNAKE
Aplopeltura boa PLATE 72
Measurement TL 850mm **Identification** Body slender, laterally compressed; head short, rounded, distinct from neck; snout short; loreals 2–3; preoculars 2; postoculars 2; supralabials 8–10; row of scales separates orbit from supralabials; infralabials 11; 2 pairs of infralabials contact mental; no central groove under chin; eye large, orbit diameter greater than snout length; pupil vertical; tail a third of body length; dorsals smooth; midbody scale rows 13; ventrals 148–191; subcaudals 88–131; anal entire. **Coloration** Dorsum brown to greyish-brown, typically with saddle-like dark-edged markings; flanks sometimes with large cream spots; forehead dark brown; labials cream; cream patch under eye, with dark subtriangular area; venter brown to dark grey. **Habitat and Behaviour** Inhabits lowland forests at elevations of up to 1,500m asl. Nocturnal and arboreal, associated with low vegetation consisting of bushes and thick undergrowth. Diet includes slugs, snails and lizards. Clutches comprise 4–8 eggs, measuring 18–23 x 10–13.5mm. Incubation period 63–69 days. Hatchlings 207–227mm. **Distribution** Myanmar, southern Thailand (Chumphon, Nakhon Si Thammarat and Pattani Provinces), Peninsular Malaysia, Sumatra, Pulau Nias, the Natuna Archipelago, Borneo and Java. Also the Philippines. **Status** Not Evaluated.

SMOOTH SLUG-EATING SNAKE
Asthenodipsas laevis PLATE 72
Measurement TL 600mm **Identification** Body slender, laterally compressed; head short, rounded, distinct from neck; snout short; loreal present; preoculars absent; postoculars 1–2; supralabials 5–6; Supralabials III–IV contact orbit; infralabials 6; eye large; pupil vertical; tail short; distinct vertebral keel present; scales on vertebral region slightly enlarged; dorsals smooth; midbody scale rows 15; ventrals 148–178; subcaudals 34–69, paired; anal entire. **Coloration** Dorsum mid-brown to dark brown with numerous vertical dark bars that extend to venter; forehead darker than dorsum, lacking lines; throat brown; venter cream or pale yellow, edges of each scale with dark spot. **Habitat and Behaviour** Inhabits lowland forests from sea level to 1,150m asl. Nocturnal, and semi-arboreal and terrestrial, associated with low vegetation or forest floor. Diet consists of slugs and snails. Reproductive habits unstudied. **Distribution** Southern Thailand (Nakhon Si Thammarat Province), Peninsular Malaysia, Sumatra, Pulau Bangka, the Natuna and Mentawai Archipelagos, Borneo and Java. **Status** Not Evaluated.

MALAYAN SLUG-EATING SNAKE
Asthenodipsas malaccanus PLATE 72
Measurement TL 450mm **Identification** Body robust, laterally compressed; head short, rounded, distinct from neck; snout short; loreal present;

preoculars absent; postoculars 2; supralabials 7–8; Supralabials III or III–IV contact orbit; eye small; pupil vertical; tail short; distinct vertebral keel; vertebral scale row enlarged; dorsals with weak keels; midbody scale rows 15; ventrals 154–181; subcaudals 26–58, paired; anal entire. **Coloration** Dorsum pale brown, mid-brown to nearly black with irregular brownish-grey cross-bars, 2–4 scales wide, edged with white; forehead darker than dorsum; dark brown or black band on neck and forebody, encircling body except for narrow space on vertebral region; venter yellowish-cream suffused with brown laterally. **Habitat and Behaviour** Inhabits primary forests in the lowlands to submontane limits, from sea level to 1,000m asl. Nocturnal, and semi-arboreal and terrestrial, associated with low vegetation as well as the forest floor. Diet consists of snails and slugs. Reproductive habits unstudied. **Distribution** Southern Thailand (Yala Province), Peninsular Malaysia, Sumatra, the Mentawai Archipelago and Borneo (Bongon, Sabah State, Gunung Dulit, Sarawak State and Samarinda, Kalimantan Timur Province). **Status** Not Evaluated.

MOUNTAIN SLUG-EATING SNAKE
Asthenodipsas vertebralis PLATE 72
Measurement TL 771mm **Identification** Body slender, laterally compressed; head short, rounded, distinct from neck; snout short; preoculars absent; postoculars 2; supralabials 7; Supralabials III–IV contact orbit; eye large; pupil vertical; tail short; dorsals smooth; midbody scale rows 15; ventrals 180–195; subcaudals 56–78; anal entire. **Coloration** Dorsum greyish-brown or dark brown with small dark brown spots and sometimes, indistinct dark cross-bars; interrupted yellow vertebral stripe occasionally present; venter yellow with brown spots laterally. **Habitat and Behaviour** Inhabits submontane and montane forests at elevations of 1,585–2,012m asl. Nocturnal and arboreal, associated with low vegetation. Diet consists of snails and slugs. Reproductive habits unstudied. **Distribution** Endemic to Peninsular Malaysia (Bukit Larut, Perak State and Cameron Highlands, Pahang State). **Status** Not Evaluated.

KEELED SLUG-EATING SNAKE
Pareas carinatus PLATE 72
Measurement TL 600mm **Identification** Body slender, laterally compressed; head short, rounded, distinct from neck; snout short; loreal present; preoculars 1–2; postoculars 1–2; supralabials 7–9; series of 2–4 suboculars separate supralabials from orbit; eye large; pupil vertical; tail short; dorsals enlarged and weakly keeled on 2 median rows; midbody scale rows 15; ventrals 161–207; subcaudals 53–111, paired; anal entire. **Coloration** Dorsum olive-brown, yellow or reddish-brown with indistinct transverse black bars on anterior body; dark streak along each eye; venter pale brown to yellow. **Habitat and Behaviour** Inhabits lowland forests and reaches submontane forests at elevations of 550–1,300m asl. Nocturnal and arboreal, associated with low vegetation. Diet consists of molluscs. Clutches comprise 3–8 eggs, measuring 19–25 x 9–12mm. Incubation period 53–71 days. Hatchlings 150–185mm. **Distribution** Myanmar, Thailand, Laos, south-western and eastern Cambodia (Cardamom Mountains and O'Rang District), Vietnam, Peninsular Malaysia, Sumatra, Borneo, Java and Bali. Also southern China. **Status** Not Evaluated.

HAMPTON'S SLUG-EATING SNAKE
Pareas hamptoni PLATE 72
Measurement TL 705mm **Identification** Body slender, laterally compressed; head short, rounded, distinct from neck; snout short; loreal present; single preocular and postocular; supralabials 7–9; Supralabials IV–V contact orbit, or separated from orbit by suboculars; 7–9 maxillary teeth; eye large; pupil vertical; tail short; dorsals smooth; 1–3 vertebral rows enlarged; midbody scale rows 15; ventrals 180–197; subcaudals 73–98, paired; anal entire. **Coloration** Dorsum light brown with vertical black cross-bars; forehead with dense black spots; venter yellow with brown spots. **Habitat and Behaviour** Inhabits the mid-hills to submontane forests. Nocturnal and arboreal, associated with vegetation near forest streams. Diet and reproductive habits unstudied. **Distribution** Northern Myanmar (Shan State), north-eastern Thailand (Narathiwat Province), Laos (Pak Loi Province) and Vietnam. Also eastern China (Hainan and Guizhou Provinces). **Status** Not Evaluated.

SPOTTED SLUG-EATING SNAKE
Pareas macularius PLATE 72
Measurement TL 450mm **Identification** Body slender, laterally compressed; head short, rounded, distinct from neck; snout short; loreal present; single preocular and postocular; supralabials 6–7; Supralabial IV contacts orbit; eye large; pupil vertical; tail short; vertebrals not enlarged; median 3–7 dorsals keeled; midbody scale rows 15; ventrals 148–166; subcaudals 39–55, paired; anal entire. **Coloration** Dorsum light or dark grey with dark spots or irregular black cross-bars; supralabials and infralabials pale; pale nuchal collar occasionally present; venter cream with irregular black speckles. **Habitat and Behaviour** Inhabits wet subtropical forests. Nocturnal and arboreal, associated with low vegetation. Diet consists of slugs and snails. Clutches comprise up to 6 eggs. **Distribution** Myanmar (Shan State and Tenasserim, Tanintharyi Division), northern Thailand (Doi Inthanon, Chiang Mai Province), Laos and Vietnam. Also eastern India. **Status** Not Evaluated.

WHITE-SPOTTED SLUG-EATING SNAKE
Pareas margaritophorus PLATE 72
Measurement TL 450mm **Identification** Body slender, not laterally compressed; head short, rounded, distinct from neck; snout short; preoculars 1–2; postoculars 1–2; supralabials 6–7; Supralabial IV contacts orbit or separated from orbit by suboculars; infralabials 7; eye large; pupil vertical; tail short; vertebral scale rows enlarged; dorsals smooth;

midbody scale rows 15; ventrals 136–159; subcaudals 32–58, paired; anal entire. **Coloration** Dorsum light or dark grey with irregular black cross-bars bordered with white; supralabials and infralabials cream with black mottling; white or yellow nuchal collar sometimes present; venter cream with irregular black speckles. **Habitat and Behaviour** Inhabits lowland forests, including evergreen forests of bamboo, up to submontane limits, at elevations of 340–1,524m asl. Nocturnal and arboreal, associated with shrubs. Threat response entails curling itself into a tight ball and remaining immobile. Diet consists of snails and slugs. Clutches comprise 6 eggs. Hatchlings ca 70–100mm. **Distribution** Myanmar, Thailand, Laos, south-western and eastern Cambodia (Cardamom Mountains and Siem Pang District), Vietnam and Peninsular Malaysia. Also eastern China. **Status** Not Evaluated.

MONTANE SLUG-EATING SNAKE
Pareas monticola   PLATE 72
Measurement TL 610mm **Identification** Body slender, laterally compressed; head short, rounded, distinct from neck; snout short; loreal present; preoculars absent; postoculars 2; supralabials 6–7; Supralabials III–IV or IV contact orbit; eye large; pupil vertical; tail short; dorsals weakly keeled; vertebral rows enlarged; midbody scale rows 15; ventrals 177–196; subcaudals 69–87, paired. **Coloration** Dorsum mid-brown with vertical blackish-brown bars on flanks; black postocular stripe to nape; black streak from eye to angle of jaws; forehead brown with dense black spots; venter yellow with brown spots. **Habitat and Behaviour** Inhabits temperate evergreen hill forests. Nocturnal and arboreal, associated with low vegetation. Diet includes snails and slugs. Clutches comprise up to 8 eggs. Hatchlings 168–178mm. **Distribution** Northern Myanmar (Chin State) and Vietnam. Also eastern and north-eastern India. **Status** Not Evaluated.

BARRED SLUG-EATING SNAKE
Pareas nuchalis   NOT ILLUSTRATED
Measurement TL 715mm **Identification** Body slender, laterally compressed; head short, rounded, distinct from neck; snout short; preoculars 2; postoculars 2; loreals 1–3; supralabials 8–9; orbit separated from supralabials by small scales; eye large; pupil vertical; tail short; dorsals weakly keeled; vertebrals enlarged; midbody scale rows 15; ventrals 195–218; subcaudals 105–118, paired; anal entire. **Coloration** Dorsum mid-brown to tan, anterior of body with narrow bands, fused vertebrally, and separated by 2–5 scales; dark postocular stripe to angle of jaws, rising to nuchal region, fusing with irregular, V-shaped dark nuchal patch; venter yellowish-cream. **Habitat and Behaviour** Inhabits lowland forests at elevations of up to 850m asl. Nocturnal and arboreal, associated with shrubs and other low vegetation up to 2m above substratum, sites of concealment being bamboo internodes. Diet consists of snails and slugs. Reproductive habits unstudied. **Distribution** Endemic to Borneo. **Status** Not Evaluated.

TAMDAO SLUG-EATING SNAKE
Pareas tamdaoensis   NOT ILLUSTRATED
Measurement TL 563mm **Identification** Body slender, laterally compressed; head short, rounded, distinct from neck; snout short; loreal present, in contact with Supralabial II; single preocular; enlarged subocular; single postocular; supralabials 7; Supralabials III–V contact orbit; Supralabial VII elongated; infralabials 7; Infralabial V enlarged; eye large; pupil vertical; tail short; dorsals keeled, except a few vertebral rows; midbody scale rows 15; ventrals 152–157; subcaudals 43–45; anal entire. **Coloration** Dorsum brown, scales with narrow pale border; other scales black, generally with white posterior edge, arranged in irregular transverse rows of 1–2 scales; forehead scales black with some brown speckles; labials speckled with black; black subocular stripe reaches Supralabials V–VI, another stripe reaches parietals; chin greyish-white with irregular black lines; venter speckled brown. **Habitat and Behaviour** Inhabits submontane forests at elevations of 900–1,200m asl. Nothing else known of its natural history. **Distribution** Endemic to northern Vietnam (Tam Dao, Vinh Phú Province). **Status** Not Evaluated.

Family PSEUDOXENODONTIDAE FALSE COBRAS

False cobras have large eyes, their two posterior-most maxillary teeth are enlarged, and the dorsals on their anterior body are oriented obliquely. A major behavioural characteristic of this group is its threat display, which is somewhat similar to that of elapids such as cobras, entailing raising the head and spreading a small hood. These species inhabit temperate and tropical forests, and feed on small vertebrates. They are found only in eastern and south-eastern Asia.

BAMBOO FALSE COBRA
Pseudoxenodon bambusicola   PLATE 72
Measurement TL 530mm **Identification** Body robust; head distinct from neck; supralabials 8; Supralabials IV–V contact orbit; infralabials 9; eye large; pupil rounded; tail long; dorsals keeled; midbody scale rows 17; ventrals 131–142; subcaudals 42–59; anal divided. **Coloration** Dorsum light brownish-grey or yellow-brown with 15–24 distinct broad black cross-bars, separated by wide interspaces; black or dark brown chevron on forehead; tail with light vertebral line; venter yellowish-cream with isolated quadrangular, transverse brown spots in anterior. **Habitat and Behaviour** Inhabits temperate as well as wet evergreen forests, especially in bamboo groves, at elevations of 600–900m asl. Threat display

entails flattening its neck and part of its body, gaping, drawing up or curling its lips, vibrating its tail and feigning death. Diet includes frogs and lizards. Reproductive habits unstudied. **Distribution** Laos and Vietnam. Also southern China. **Status** Not Evaluated.

BARAM FALSE COBRA
Pseudoxenodon baramensis NOT ILLUSTRATED
Measurement TL 825mm **Identification** Body robust; head short, distinct from neck; single preocular; postoculars 3; supralabials 8; Supralabials IV–V contact orbit; nostrils large; maxillary teeth 21, 2 posterior-most enlarged; eye large; pupil rounded; tail long; dorsals keeled; midbody scale rows 19; ventrals 134; subcaudals 47, paired; anal divided. **Coloration** Dorsum greyish-olive with indistinct dark network; venter yellow, especially posteriorly; outer margins of subcaudals with pale line. **Habitat and Behaviour** Inhabits submontane forests at elevations of ca 1,000m asl. Nothing else known of its natural history. **Distribution** Endemic to Borneo (Gunung Dulit, Gunung Murud and Sungei Mengiong, Sarawak State). **Status** Not Evaluated.

JAVANESE FALSE COBRA
Pseudoxenodon inornatus NOT ILLUSTRATED
Measurement TL 741mm **Identification** Body robust; head short, distinct from neck; loreal present; preoculars 2–3; postoculars 3; supralabials 8; Supralabials IV–V contact orbit; infralabials 9–10; 4 infralabials contact anterior chin shields; nostrils large; maxillary teeth 16–19 + 2; eye large; pupil rounded; tail long; dorsals keeled; midbody scale rows 19; ventrals 118–121; subcaudals 36–41, paired; anal divided. **Coloration** Dorsum olive-brown or pale brown; juveniles with black nuchal chevron and rhomboidal dark marks with a pale centre; cream ventrolateral stripe extends along posterior part of body and tail; venter yellow or mid-brown with dark brown speckles. **Habitat and Behaviour** Inhabits the low hills. Diurnal and terrestrial, known from grassy areas near tea plantations. Diet and reproductive habits unstudied. **Distribution** Endemic to western Java (foot of Gunung Pangrango and Sumadra Estate, nr Garut). **Status** Not Evaluated.

JACOBSON'S FALSE COBRA
Pseudoxenodon jacobsonii NOT ILLUSTRATED
Measurement TL 1,082mm **Identification** Body robust; head short, distinct from neck; loreals 2; single preocular; postoculars 3; supralabials 7; Supralabials III–IV contact orbit; infralabials 9–10; 4 infralabials contact anterior chin shields; 2 pairs of chin shields subequal; maxillary teeth 16 + 2; eye large; pupil rounded; tail long; dorsals smooth, those on the 3 middle rows keeled; midbody scale rows 19; ventrals 145; subcaudals 37, paired; anal divided. **Coloration** Dorsum grey with black chevron and rhomboidal black markings with pale centres; anterior portion of venter yellow with irregular dark spots, which increase posteriorly; subcaudals dark; pale ventrolateral stripe on tail. **Habitat and Behaviour** Inhabits montane forests. Diet consists of frogs; remains of insects in one individual are suspected to derive from a frog that was also consumed. Reproductive habits unstudied. **Distribution** Endemic to Sumatra (Serapai on Gunung Kerinchi, Sumatera Barat Province). **Status** Not Evaluated.

KARL SCHMIDT'S FALSE COBRA
Pseudoxenodon karlschmidtii PLATE 72
Measurement TL 1,730mm **Identification** Body robust; head distinct from neck; loreal present; single preocular; postoculars 3; supralabials 8; Supralabials IV–V contact orbit; infralabials 9–10; 4 infralabials contact anterior chin shields; eye large; pupil rounded; tail long; dorsals keeled; midbody scale rows 19; ventrals 144–157; subcaudals 53–60, paired; anal divided. **Coloration** Dorsum greyish-black; forehead similarly coloured in adults, reddish-brown in juveniles; dark, anteriorly pointed chevron present only in juveniles; vertebral region with 24 light grey or yellow spots, surrounded by part black and part yellow scales; anterior flanks with indistinct rows of spots composed of black-bordered scales; beyond ventral 20, venter with black speckling. **Habitat and Behaviour** Inhabits temperate forests in the highlands. Diet consists of frogs. Reproductive habits unstudied. **Distribution** Northern Vietnam (Tam Dao, Vinh Phú Province). Also southern China. **Status** Not Evaluated.

LARGE-EYED FALSE COBRA
Pseudoxenodon macrops PLATE 72
Measurement TL 1,400mm **Identification** Body robust; head distinct from neck; loreal large; single preocular; postoculars 3–4; supralabials 8; Supralabials IV–V contact orbit; infralabials 9–10; eye large; pupil rounded; tail long; dorsals keeled, except for smooth lower rows; keels on anal region in mature males tuberculate; midbody scale rows 17 or 19; ventrals 151–180; subcaudals 55–80, paired; anal divided. **Coloration** Dorsum brownish-grey, red or olivaceous; series of yellow, reddish-brown or orange, dark-edged cross-bars or spots, and dorsolateral series of dark spots; nape with chevron-shaped dark marking; venter yellow with quadrangular black or dark brown spots, or cross-bars. **Subspecies** Two subspecies are recognized, of which one (*P. m. macrops*) occurs in the region. **Habitat and Behaviour** Inhabits the mid-hills of wet evergreen to montane forests dominated by oaks, at elevations of 1,500–2,020m asl. Threat display entails raising its forebody and flattening its neck. Diurnal and terrestrial. Diet consists of frogs and lizards. Oviparous; clutches comprise 6–10 eggs. **Distribution** Myanmar, northern Thailand (Chiang Mai, Loei and Mae Hong Son Provinces), Laos, Vietnam, Peninsular Malaysia (Cameron Highlands, Pahang State) and Borneo. Also eastern and north-eastern India, Nepal (*P. m. macrops*) and southern China (*P. m. sinensis*). **Status** Not Evaluated.

Family TYPHLOPIDAE BLIND SNAKES

Blind snakes are characterized by a blunt head, a short tail, eyes concealed under scales, ventrals that are not enlarged, a toothed, moveable maxilla and a toothless premaxilla articulated with the snout. They are subfossorial, and may also be active on the surface, in some cases even showing some arboreality. Their diet consists of small arthropods, including termites, ants and their eggs, larvae and pupae, and also earthworms. Most species are probably oviparous, and parthenogenesis is also known in some species. They are found worldwide.

WHITE-HEADED BLIND SNAKE
Ramphotyphlops albiceps PLATE 73
Measurement TL 302mm **Identification** Body slender; head indistinct from neck, spatulate; nasal divided below nostril; single preocular; supralabials 4; midbody scale rows 18; caudal spine present. **Coloration** Dorsum dark brown; forehead red or buff; chin, gular region and tail cream; venter paler than dorsum. **Habitat and Behaviour** Inhabits primary forests in the lowlands, including offshore islands. Subfossorial, associated with loose soil. Diet unstudied. Oviparous; clutches comprise 2–8 eggs. **Distribution** Myanmar, Thailand, Peninsular Malaysia (Sungai Buloh and Bukit Kanching, Selangor State, Bukit Larut, Perak State, Kedah State and Pulau Tioman, Pahang State) and Vietnam. Also eastern China. **Status** Not Evaluated.

BRAHMINY BLIND SNAKE
Ramphotyphlops braminus PLATE 73
Measurement TL 180mm **Identification** Body slender; head indistinct from neck; snout rounded; nostrils placed laterally; supralabials 4; eye distinct; head scales larger than dorsal scales; diameter of body 30–45 times total length; caudal spine present. **Coloration** Dorsum black or brown, venter lighter; snout and tip of tail paler. **Habitat and Behaviour** Encountered in human habitations as well as in lightly forested areas, from sea level to at least 2,000m asl. Nocturnal and subfossorial, or active on the surface, especially after rains. Arboreal behaviour also known. Diet includes termites, ants and their larvae, and earthworms. Parthenogenetic; clutches comprise 1–8 eggs, measuring 11.2–20 x 3.3–5.1mm. **Distribution** Myanmar, Laos, Cambodia, Vietnam, Thailand, Peninsular Malaysia, Singapore, Sumatra, Pulau Belitung, Borneo, Java and Bali. Cosmopolitan: found in tropical, subtropical and even temperate regions of the world. **Status** Not Evaluated.

LINED BLIND SNAKE
Ramphotyphlops lineatus PLATE 73
Measurement TL 480mm **Identification** Body slender; head indistinct from neck; snout rounded, projecting; rostral enlarged, about half head width; nasal incompletely divided; single large ocular; preoculars and suboculars absent; postoculars 2; supralabials 4; eye indistinct; midbody scale rows 22–23; subcaudals 9; caudal spine present. **Coloration** Dorsum cream, pinkish-brown or yellow with 10 longitudinal brown stripes from head to tail-tip; venter cream or yellow. **Habitat and Behaviour** Inhabits tropical forests in the lowlands up to submontane limits, at elevations of 0–1,420m asl. Fossorial. Diet and reproductive habits unstudied. **Distribution** Southern Thailand (Na Pradoo, Pattani Province and Muang District, Narathiwat Province), Laos, Cambodia, Vietnam, Peninsular Malaysia, Singapore, Sumatra, Pulau Bangka, Pulau Nias, Borneo and Java. **Status** Not Evaluated.

LORENZ'S BLIND SNAKE
Ramphotyphlops lorenzi NOT ILLUSTRATED
Measurement TL 337mm **Identification** Body slender; head indistinct from neck; snout with projecting, sharp horizontal edge; rostral less than half head width; nasal incompletely divided; preocular present; supralabials 4; eye distinct; tail longer than broad; midbody scale rows 22; caudal spine present. **Coloration** Dorsum light greyish-brown or greyish-green; rostral brown edged with yellow; venter pale olive-green. **Habitat and Behaviour** Presumed to be a subfossorial species, associated with litter, consuming small arthropods and worms. Reproductive habits unstudied. **Distribution** Endemic to Pulau Miang Besar (Kalimantan Province, off the east coast of Borneo). **Status** Not Evaluated.

OLIVE BLIND SNAKE
Ramphotyphlops olivaceus NOT ILLUSTRATED
Measurement TL 410mm **Identification** Body slender; head indistinct from neck; snout projecting with narrow transverse edge; rostral large, over half head width; nasal incompletely divided; preocular present; supralabials 4; eye distinct; midbody scale rows 20–22; caudal spine present. **Coloration** Dorsum pale brown; venter paler. **Habitat and Behaviour** Presumed to be a subfossorial species, associated with litter, consuming small arthropods and worms. Reproductive habits unstudied. **Distribution** Borneo (Baram, Sarawak State). Also the Sangihe Archipelago, Seram and Mysool (Indonesia), and Samar and Babuan (the Philippines). **Status** Not Evaluated.

BLACK BLIND SNAKE
Typhlops ater PLATE 73
Measurement TL 166mm **Identification** Body slender; head indistinct from neck; snout rounded; nostrils placed laterally; rostral oval, elongated, extending to level of eyes; supralabials 4; eye distinct; midbody scale rows 18; diameter 68 times total length; tail twice as long as wide; caudal spine present. **Coloration** Dorsum black or dark brown; venter

reddish-brown; chin and anal region cream. **Habitat and Behaviour** Presumed to be a subfossorial species, associated with litter, consuming small arthropods and worms. Reproductive habits unstudied. **Distribution** Java. Also Sulawesi, Ternate and Halmahera. **Status** Not Evaluated.

JAVANESE BLIND SNAKE
Typhlops bisbocularis NOT ILLUSTRATED
Measurement TL 131mm **Identification** Body slender; head indistinct from neck; snout rounded, projecting; rostral third head width; nostril laterally situated between 2 nasals; preocular present; supralabials 4; eye indistinct; midbody scale rows 18; caudal spine present. **Coloration** Dorsum dark grey, each scale with narrow pale edges; venter light grey with broad pale edges; white spots on anal region; edges of scales on forehead, chin and tail-tip white. **Habitat and Behaviour** Presumed to be a subfossorial species, associated with litter, consuming small arthropods and worms. Reproductive habits unstudied. **Distribution** Endemic to western Java. **Status** Not Evaluated.

BURMESE BLIND SNAKE
Typhlops bothriorhynchus NOT ILLUSTRATED
Measurement TL 234mm **Identification** Body robust; head indistinct from neck; snout rounded and projecting; shallow groove under nostril on Supralabial III and on each side of rostral; supralabials 4; single preocular; postoculars 2–3; small pits on preocular and Supralabial II; middorsals 283–330; midbody scale rows 22–24; subcaudals 9–11; caudal spine present. **Coloration** Dorsum brown grading into light brown on venter; large dark ocular spot. **Habitat and Behaviour** Presumed to be a subfossorial species, associated with litter, consuming small arthropods and worms. Reproductive habits unstudied. **Distribution** Northern Myanmar (Taunggyi, Shan State) and possibly Peninsular Malaysia (Pinang State). Also north-eastern India (Assam State). **Status** Not Evaluated.

DIARD'S BLIND SNAKE
Typhlops diardii PLATE 73
Measurement TL 430mm **Identification** Body robust; head indistinct from neck; snout rounded, strongly projecting; upper rostral narrow, widening ventrally; eye distinct; midbody scale rows 24–26 (rarely 28); transverse scale rows 260–300; ventrals not enlarged; caudal spine present. **Coloration** Dorsum dark brown, each scale with an indistinct transverse light streak; venter and sides light brown; gradual transition between dark dorsum and pale venter. **Habitat and Behaviour** Inhabits forests in the mid-hills and submontane limits at elevations of at least 1,524m asl. Subfossorial, burrowing in soil, under debris or in rotting logs. Diet consists of insects and their larvae, and earthworms. Oviparous; clutches comprise 4–14 eggs, measuring 4–7mm. **Distribution** Myanmar, Thailand, Laos, Cambodia and southern Vietnam. Also Pakistan, Bangladesh, northern, eastern and north-eastern India, Nepal and eastern China. **Status** Not Evaluated.

FLOWER'S BLIND SNAKE
Typhlops floweri NOT ILLUSTRATED
Measurement TL 228mm **Identification** Body slender; head indistinct from neck, flattened; snout rounded and projecting; rostral over half head width with scattered glands; nasal completely divided; subocular present; preocular small; postoculars 2–3; midbody scale rows 18; middorsal scale rows 478–520; subcaudals 20–23; caudal spine absent. **Coloration** Dorsum black or dark blackish-brown; snout-tip yellow; rostral and nasals light brown with tiny yellow dots; venter pale brown; yellow vent spot. **Habitat and Behaviour** Presumed to be a subfossorial species, associated with litter, consuming small arthropods and worms. Reproductive habits unstudied. **Distribution** Endemic to south-central Thailand (Bangkok, Phra Nakhon Si Ayutthaya Province and Khao Soi Dao Wildlife Sanctuary, Chanthaburi Province). **Status** Not Evaluated.

GIA DINH BLIND SNAKE
Typhlops giadinhensis NOT ILLUSTRATED
Measurement TL 236.5mm **Identification** Body slender; head indistinct from neck; snout rounded, projecting; rostral large, a third head width; prefrontal and frontal smaller than supraocular; supralabials 4; Supralabial IV largest; eye distinct; preocular large, subequal to ocular; tail thicker than long; midbody scale rows 22; subcaudals 10; caudal spine present, backwards pointing. **Coloration** Dorsum bronze-brown; scale borders paler; paired longitudinal stripes on dorsum; gular region with indistinct brown marks; venter unpatterned yellow. **Habitat and Behaviour** Presumed to be a subfossorial species, associated with litter, consuming small arthropods and worms. Reproductive habits unstudied. **Distribution** Endemic to southern Vietnam (Gia Dinh Province). **Status** Not Evaluated.

PITTED SNOUT BLIND SNAKE
Typhlops hypsobothrius NOT ILLUSTRATED
Measurement TL 285mm **Identification** Body robust; head indistinct from neck; snout rounded; rostral half head width; nasals completely divided; conspicuous groove between nostril and rostral; preocular narrower than ocular; supralabials 4; eye distinct; midbody scale rows 20; tail short, as long as broad; caudal spine absent, but tail terminates in pin-like tip, bent downwards. **Coloration** Dorsum light brown; venter white. **Habitat and Behaviour** Presumed to be a subfossorial species, associated with litter, consuming small arthropods and worms. Reproductive habits unstudied. **Distribution** Endemic to Sumatra. **Status** Not Evaluated.

JERDON'S BLIND SNAKE
Typhlops jerdoni PLATE 73
Measurement TL 280mm **Identification** Body slender; head indistinct from neck; snout rounded;

rostral less than half head width; eye distinct; midbody scale rows 22; tail bluntly pointed; caudal spine present. **Coloration** Dorsum dark brown to nearly black; venter light brown; snout and anal region cream. **Habitat and Behaviour** Inhabits forests in the mid-hills at elevations of up to 855m asl. Subfossorial, found under boulders and rocks, and inside dead trees. Diet and reproductive habits unstudied. **Distribution** Northern Myanmar (Lashio, Shan State and Bago Division). Also Bhutan, Nepal, and eastern and north-eastern India. **Status** Not Evaluated.

KLEMMER'S BLIND SNAKE
Typhlops klemmeri   NOT ILLUSTRATED
Measurement TL 151mm **Identification** Body robust; head indistinct from neck; preocular present; subocular absent; nasal partially divided under nostril; gland or pore under nasal absent; midbody scale rows 23; caudal spine present. **Coloration** Dorsum mid-brown; head and nuchals with yellow glands along sutures; venter light yellowish-brown, colours of dorsum and venter not meeting at a distinct line of demarcation; subcaudals mid-brown. **Habitat and Behaviour** Presumed to be a subfossorial species, associated with litter, consuming small arthropods and worms. Reproductive habits unstudied. **Distribution** Endemic to southern Thailand (Koh Phai Archipelago, nr Pattaya). **Status** Not Evaluated.

KOEKKOEK'S BLIND SNAKE
Typhlops koekkoeki   NOT ILLUSTRATED
Measurement TL 445mm **Identification** Body slender; head indistinct from neck; snout rounded and projecting; rostral broad, 1.5 times head width; preocular present; ocular in contact with Supralabials III–IV; eye indistinct; midbody scale rows 26; caudal spine present. **Coloration** In preservative, dorsum greyish-brown; venter paler; colour in life unknown. **Habitat and Behaviour** Presumed to be a subfossorial species, associated with litter, consuming small arthropods and worms. Reproductive habits unstudied. **Distribution** Endemic to Pulau Bunju (Kalimantan Timur, off eastern Borneo). **Status** Not Evaluated.

MÜLLER'S BLIND SNAKE
Typhlops muelleri   PLATE 73
Measurement TL 540mm **Identification** Body robust; head indistinct from neck; snout rounded, strongly projecting; upper rostral narrow, widening ventrally; single preocular; postoculars 2; supralabials 4; eye distinct; midbody scale rows 24–26 (rarely 22, 28 or 30); caudal spine present. **Coloration** Dorsum dark brown, purple or black; venter yellow, cream or gold, the two colours with a clear line of demarcation. **Habitat and Behaviour** Inhabits forests in the lowlands and mid-hills, and also areas of wet agriculture, at elevations of up to 1,676m. Associated with soft soil, including waterlogged areas, as well as areas under rocks and logs. Diet includes larvae of ants and termites; molluscs and other snakes may also be eaten. Ovoviviparous; clutches comprise 5–14 neonates. **Distribution** Southern Myanmar, southern Thailand (Yala Province), Laos, Cambodia, southern Vietnam, Peninsular Malaysia, Singapore, Sumatra, Pulau Bangka, Pulau Nias, Pulau Weh and Borneo. Also New Guinea (Irian Jaya). **Status** Not Evaluated.

OATES'S BLIND SNAKE
Typhlops oatesii   NOT ILLUSTRATED
Measurement TL 200mm **Identification** Body slender; head indistinct from neck; snout rounded, strongly projecting; rostral less than half head width; nasal nearly completely divided; eye distinct; midbody scale rows 24; caudal spine present. **Coloration** Dorsum yellowish-brown with black spots in centre of scales, forming longitudinal stripes; stripes absent on midventer. **Habitat and Behaviour** Inhabits lowlands including agricultural fields that were presumably once forested. Subfossorial. Diet and reproductive habits unstudied. **Distribution** Endemic to the Cocos Islands, Myanmar. Also reported from the adjacent Middle Andaman Island (India). **Status** Not Evaluated.

OZAKI'S BLIND SNAKE
Typhlops ozakiae   NOT ILLUSTRATED
Measurement TL 227mm **Identification** Body slender; head indistinct from neck; snout rounded; rostral less than a third head width; midbody scale rows 20; middorsal scale rows 326; caudal spine present. **Coloration** Dorsum pale brown; venter paler. **Habitat and Behaviour** Presumed to be a subfossorial species, associated with litter, consuming small arthropods and worms. Reproductive habits unstudied. **Distribution** Endemic to south-central Thailand. **Status** Not Evaluated.

LARGE BLIND SNAKE
Typhlops porrectus   NOT ILLUSTRATED
Measurement TL 285mm **Identification** Body slender; head indistinct from neck; snout rounded, strongly projecting; rostral a third to half head width; nasal incompletely divided; eye distinct; postoculars 1–2; midbody scale rows 18; middorsal scale rows 388–468; subcaudals 7; caudal spine present. **Coloration** Dorsum brownish-black or mid-brown; snout, chin and anal region cream; venter pale brown. **Habitat and Behaviour** Inhabits the lowlands and mid-hills. Subfossorial. Diet and reproductive habits unstudied. **Distribution** Northern Myanmar. Also Pakistan, India, Bangladesh and Sri Lanka. **Status** Not Evaluated.

ROXANE'S BLIND SNAKE
Typhlops roxaneae   NOT ILLUSTRATED
Measurement TL 231mm **Identification** Body robust; head indistinct from neck; snout rounded; postnasal border concave; midbody scale rows 20; middorsals 329; tail length less than half tail width; subcaudals 5; caudal spine present. **Coloration** Dorsum golden-brown, covering 5 rows; venter yellowish-tan, covering 13 rows at midbody, the two

colours separated by a row of scales with reduced pigments; snout and cloacal region yellow; yellow ring around tail-tip. **Habitat and Behaviour** Inhabits the lowlands at elevations of 1–2m asl. Presumed to be a subfossorial species, associated with litter, consuming small arthropods and worms. Reproductive habits unstudied. **Distribution** Endemic to south-central Thailand (Bangkok, Phra Nakhon Si Ayutthaya Province). **Status** Not Evaluated.

THAI BLIND SNAKE
Typhlops siamensis NOT ILLUSTRATED
Measurement TL 166mm **Identification** Body robust; head indistinct from neck; snout rounded and moderately projecting; rostral less than a third head width; nasal incompletely divided; eye distinct; midbody scale rows 22; caudal spine present. **Coloration** Dorsum greyish-olive; venter yellow. **Habitat and Behaviour** Presumed to be a subfossorial species, associated with litter, consuming small arthropods and worms. Reproductive habits unstudied. **Distribution** Thailand, Cambodia and Vietnam. **Status** Not Evaluated.

TRANG BLIND SNAKE
Typhlops trangensis NOT ILLUSTRATED
Measurement TL 155mm **Identification** Body short, robust; head indistinct from neck; snout rounded; supralabials 4; nasal incompletely divided, suture reaching Supralabial II; eye not externally visible; middorsal scale rows 370; median rows of scales enlarged; midbody scale rows 24; caudal spine present. **Coloration** Dorsum grey or ultramarine, covering 11 scale rows; rest of body cream-white, covering 13 scale rows; forehead paler than body; pits on large forehead scales darker; chin cream-white. **Habitat and Behaviour** Inhabits lowland forests. Subfossorial, associated with leaf litter and concealing itself under fallen logs. Diet includes small arthropods and worms. Reproductive habits unstudied. **Distribution** Endemic to southern Thailand (Khao Chong, Trang Province) **Status** Not Evaluated.

Family XENODERMATIDAE STRANGE-SKINNED SNAKES

Strange-skinned snakes have scales that are nearly completely fused to their underlying skin (unlike the scales of most other snakes, in which only one scale edge is embedded in the underlying skin). They inhabit lowland forests, and are found only in South-East Asia.

BLACK BURROWING SNAKE
Achalinus ater PLATE 74
Measurement TL 401mm **Identification** Body slender, cylindrical; head indistinct from neck; nostril in concave nasal; loreal large; supralabials 6; Supralabials IV–V contact orbit; 3 infralabials contact anterior genials; eye small; pupil vertical; dorsals strongly keeled; midbody scale rows 23; ventrals 156–170; subcaudals 56–63, single; anal entire. **Coloration** Dorsum iridescent black or dark brown; venter dark, sides of ventrals paler. **Habitat and Behaviour** Inhabits the mid-hills at elevations of 450–900m asl. Probably subfossorial and associated with leaf litter. Diet and reproductive habits unstudied. **Distribution** Endemic to northern Vietnam (Tam Dao, Vinh Phú Province). **Status** Not Evaluated.

RUFOUS BURROWING SNAKE
Achalinus rufescens PLATE 74
Measurement TL 450mm **Identification** Body slender, cylindrical; head indistinct from neck; nostril in concave nasal; loreal large; supralabials 6; Supralabials IV–V contact orbit; 3 infralabials contact anterior genials; eye small; pupil vertical; dorsals strongly keeled; midbody scale rows 23 or 25; ventrals 135–157; subcaudals 54–82, single; anal entire. **Coloration** Dorsum reddish- or greyish-brown; head and body with iridescent sheen; venter unpatterned cream. **Habitat and Behaviour** Inhabits primary forests in the lowlands and mid-hills at elevations of 300–518m asl. Nocturnal and subfossorial, hiding under logs and leaf litter in moist locations. Diet and reproductive habits unstudied. **Distribution** Northern Vietnam (Bac Ky or Dông Kinh). Also eastern China. **Status** Not Evaluated.

GREY BURROWING SNAKE
Achalinus spinalis PLATE 74
Measurement TL 412mm **Identification** Body slender, cylindrical; head indistinct from neck; loreal present; preoculars and postoculars absent; supralabials 6; Supralabials IV–V contact orbit; infralabials 6; 3 infralabials contact anterior chin shields; nostril in concave nasal; eye small; pupil vertical; temporals enter eye; dorsals strongly keeled; midbody scale rows 21 or 23 (rarely 22); ventrals 150–175; subcaudals 39–62, single; anal entire. **Coloration** Dorsum iridescent walnut-brown, more rufous medially; dark vertebral line extends to tail-tip, 1–3 scales wide; nuchals with black central spot; venter yellowish-brown. **Habitat and Behaviour** Inhabits submontane forests at elevations of 1,067–1,828m asl. Subfossorial, and also associated with thick vegetation such as moss in the vicinity of streams. Diet consists of earthworms. Oviparous; clutches comprise 4–7 eggs, measuring 15–19 x 6–7mm. **Distribution** Vietnam. Also southern China (Sichuan Province). **Status** Not Evaluated.

KLOSS'S ROUGH WATER SNAKE
Fimbrios klossi PLATE 74
Measurement TL 395mm **Identification** Body slender, cylindrical; head long; single supraocular; mental and first 7 infralabials bear raised edges; rostral separated from internasals by horizontal ridge of tissue; nostril in anterior concave nasal; suture between internasals shorter than between prefrontals; loreal large; preocular absent; postoculars 2; subocular 1; single large pair of chin shields; eye small; midbody scale rows 30–32; ventrals 161–190; subcaudals

43–60; anal entire. **Coloration** Dorsum uniform dark grey, olivaceous or purple-brown, lacking pale blotches and stripes; head pale yellow; venter yellow; posterior ventrals and subcaudals tinged with grey. **Habitat and Behaviour** Inhabits lowland evergreen forests in and near streams at elevations of 1,000–1,500m asl. Terrestrial and aquatic, found in leaf litter, under fallen logs and in water. Diet consists of fish. Reproductive habits unstudied. **Distribution** Laos (Pakxong District) and Vietnam. **Status** Not Evaluated.

SMITH'S ROUGH WATER SNAKE
Fimbrios smithi   NOT ILLUSTRATED
Measurement TL 440mm **Identification** Body slender, cylindrical; head long; rostral, mental and first 3–4 labials with raised everted edges; rostral separated from internasals by horizontal ridge of tissue; suture between internasals longer than between prefrontals; single pair of enlarged chin shields; large loreal extends from nasal to eye; preocular 1; supraocular 1; postoculars 2; suboculars 2; eye small; dorsals keeled, those on outer edges enlarged; midbody scale rows 31; ventrals 193; subcaudals 72. **Coloration** Dorsum greyish-brown; pale flanks; pale blotches and stripes on nuchal region; venter yellowish-grey, edged with dark grey, especially on posterior and below tail. **Habitat and Behaviour** Inhabits limestone forests at elevations of ca 350m asl. Presumed to be aquatic. Diet and reproductive habits unstudied. **Distribution** Endemic to Vietnam (Truong Son Mountain Range). **Status** Not Evaluated.

STOLICZKA'S STREAM SNAKE
Stoliczkia borneensis   PLATE 74
Measurement TL 750mm **Identification** Body slender, laterally compressed, with sharp ridge on vertebral region; head large, distinct from neck; forehead with 2 rows of small scales in front of eyes, separating prefrontals from frontal; preoculars 2–3; supralabials 11; Supralabials VI–VII contact orbit; nostrils large and flaring; eye small and beady; temporal scales small; tail long and slender; midbody scale rows 30–35, strongly keeled, vertebral scales enlarged; ventrals 205–210; subcaudals 117–124, single; anal entire. **Coloration** Dorsum dark bluish-black or dark bluish-brown with short transverse dark bands, as broad as or broader than their interspaces; several dark bars behind head; venter unpatterned brown. **Habitat and Behaviour** Inhabits submontane and montane forests at elevations of 800–1,800m. Nocturnal and arboreal, associated with low vegetation and stream banks. Diet and reproductive habits unstudied. **Distribution** Endemic to Borneo (Gunung Kinabalu, Crocker Range and Trus Madi, Sabah State, and Gunung Murud, Sarawak State). **Status** Not Evaluated.

ROUGH-BACKED LITTER SNAKE
Xenodermus javanicus   PLATE 74
Measurement TL 650mm **Identification** Body slender, compressed; head large, distinct from neck; supralabials 17–20; 3 rows of large keeled scales on dorsum; nasal scales enlarged; nostrils flaring and pointed forwards; eye large; pupil rounded; exposed skin between scales; tail long; midbody scale rows 40–51; ventrals 165–186; subcaudals 129–165, single; anal entire. **Coloration** Dorsum unpatterned grey; ridges on scales cream; snout slightly paler than rest of head; supralabials, infralabials and throat cream; venter cream with black areas. **Habitat and Behaviour** Inhabits lowland forests and agricultural areas at elevations of 500–1,100m asl. Nocturnal and terrestrial, with daytime spent burrowing in earth or under logs. Diet consists of frogs. Oviparous; clutches comprise 2–4 eggs, measuring 23–28 x 9–11mm. Incubation period 61–65 days. Hatchlings 180–202mm. **Distribution** Myanmar, southern Thailand (Yala Province), Peninsular Malaysia, Sumatra, Borneo and Java. **Status** Not Evaluated.

Family XENOPELTIDAE SUNBEAM SNAKES

Sunbeam snakes typically possess highly iridescent dorsal scales, a depressed snout, a subcylindrical body, a short tail, large scales on their forehead and reduced ventrals. They are associated with lowland tropical forests, and are subfossorial or live under litter. Their diet consists of small vertebrates, which are killed by constriction. The family includes two species from South-East Asia and eastern China, the affinities of which lie with the burrowing boas of the neotropics and Africa.

HAINAN SUNBEAM SNAKE
Xenopeltis hainanensis   PLATE 74
Measurement TL 628mm **Identification** Body robust, cylindrical; head slightly distinct from neck; snout rounded; large interparietal in middle of 4 parietals; single postocular; supralabials 7; Supralabials IV–V contact orbit; infralabials 6–7; mental groove present; maxillary teeth 22–24; eye moderately small; pupil vertical; tail short; dorsals smooth; midbody scale rows 15; ventrals 152–157; subcaudals 16–19; anal divided. **Coloration** Dorsum iridescent bluish-brown; 2 series of white spots in longitudinal series; venter greyish-white. **Subspecies** Two subspecies are recognized, of which one (*X. h. hainanensis*) occurs in the region. **Habitat and Behaviour** Inhabits lowland to submontane forests at elevations of 200–2,000m asl. Diet consists of amphibians and reptiles. Oviparous; clutches comprise 5–8 eggs. **Distribution** Northern Vietnam. Also eastern China (*X. h. hainanensis* and *X. h. jidamingae*). **Status** Not Evaluated.

SUNBEAM SNAKE
Xenopeltis unicolor PLATE 74
Measurement TL 1,140mm **Identification** Body robust, cylindrical; head slightly distinct from neck; snout rounded and depressed; large interparietal in middle of 4 parietals; postoculars 2; supralabials 8; Supralabials IV–V contact orbit; infralabials 7–8; mental groove present; maxillary teeth 35–45; eye small; pupil vertical; tail short; dorsals smooth; midbody scale rows 15; ventrals 164–196; subcaudals 24–31; anal divided. **Coloration** Dorsum iridescent brown, each scale light-edged; juveniles with pale yellow, white or cream collar; venter white or cream. **Habitat and Behaviour** Inhabits primary forests from sea level to submontane areas at elevations of 0–1,402m asl. Nocturnal and subfossorial, inhabiting burrows excavated by rodents, and also leaf litter and crevices within limestones. Occasionally enters human habitation. Diet consists of rodents, birds, lizards and frogs. Oviparous; clutches comprise 3–17 eggs, measuring 18 x 58mm. **Distribution** Myanmar, Thailand, Laos, southern and eastern Cambodia, Vietnam, Peninsular Malaysia, Singapore, Sumatra, Pulau Bangka, Pulau Belitung, the Riau Archipelago, Pulau Natuna Besar, Pulau Sipura, Pulau Simeuleu, Pulau Siberut, Borneo and Java. Also south-eastern China, the Nicobar Archipelago (India), and Palawan and the Sulu Archipelago (the Philippines). **Status** Not Evaluated.

Family XENOPHIDIIDAE SPINE-JAWED SNAKES

Characterisics of species in this family include a compressed body, a short tail, rounded pupils, enlarged prefrontals, no loreal, an undivided nasal, a single anal and subcaudals, a right lung that is not vascularized, no pelvis and an ectopterygoid process on the maxilla. Two species from the tropical forests of Sundaland – one each from Peninsular Malaysia and Borneo – are allocated to the Xenophidiidae. They are both poorly known, and are most closely related to species found in South America.

BORNEAN SPINY-JAWED SNAKE
Xenophidion acanthognathus NOT ILLUSTRATED
Measurement TL 335mm **Identification** Body slender, strongly compressed; head slightly flattened, indistinct from neck; loreal absent; supralabials 8; Supralabials III–IV contact orbit; infralabials 8–9; midbody scale rows 23, keeled (except 2 lowest rows); ventrals 181; subcaudals 51. **Coloration** Dorsum brown with longitudinal zigzag pattern; forehead mid-brown with pale areas; broad white postocular stripe extends up to sides of neck; zigzag tan stripes on flanks; venter black with small squarish yellow spots. **Habitat and Behaviour** Inhabits the mid-hills at elevations of ca 600m asl. Possibly subfossorial. Diet includes skinks. Oviparous; other details of its reproductive habits are unknown. **Distribution** Endemic to northern Borneo (Gunung Kinabalu, Sabah State). **Status** Not Evaluated.

SCHÄFER'S SPINY-JAWED SNAKE
Xenophidion schaeferi PLATE 74
Measurement TL 263mm **Identification** Body slender, moderately compressed; head slightly distinct from neck; loreal absent; supralabials 8; Supralabials III–IV contact orbit; infralabials 8; midbody scale rows 23, keeled; ventrals 178; subcaudals 43. **Coloration** Dorsum dark brown, iridescent, with undulating greyish-whitish stripe on paravertebral region, from neck to tail-tip; venter brown, sides of ventrals with greyish-white areas. **Habitat and Behaviour** Inhabits lowland forests at elevations of ca 100m asl. Nocturnal and terrestrial. Diet probably consists of earthworms, and insects and their larvae. Reproductive habits unstudied. **Distribution** Endemic to Peninsular Malaysia (Templer's Park Nature Reserve, nr Kuala Lumpur, Selangor State). **Status** Not Evaluated.

GLOSSARY OF TECHNICAL TERMS

acrodont dentition dentitional type of lizards in which teeth are fused onto dorsal surface of jawbone
alate winged morph (such as that of ants and termites)
anal plate terminal ventral scale or scute
annulated having ringed segments
antehumeral fold dermal fold in front of shoulder
anterior towards front of body
aquatic living in water
arboreal living in trees
areola (plural areoli) a circular area
arthropod invertebrate with segmented body, joined limbs and external skeleton
auricular opening ear opening or position, indicated by a depression, where tympanum is located
autotomy spontaneous or reflexive separation of a body part (typically, a tail)
axilla armpit
calcified consisting of or containing calcium carbonate
canthals scales of canthal ridge
canthus (plural canthi) corner of eye
canthus rostralis ridge from eye to tip of snout
carapace upper shell of a turtle or tortoise
carnivorous flesh-eating
cartilagineous constructed on firm and flexible connective tissue
caudal towards tail
cervical pertaining to neck
chevron V-shaped mark
cloaca chamber into which intestinal, urinary and reproductive ducts discharge contents
clutch entire compliment of eggs or neonates from a single female
concave bent inwards, rounded
conical scale cone-shaped scale
convex bent outwards, rounded
costals paired, enlarged scales on sides of carapace
crenated having a scalloped edge
crenulated minutely crenated
crepuscular active at dawn and dusk
crest ridge on head, composed of highly modified scales or skin
crustacean invertebrate such as a crab or shrimp, which has a hard external shell and multiple legs
cryptic camouflaged or hidden
cuneate triangular scale lying between two labials
cusp tooth-like projection of a jaw
deciduous plant one that sheds its leaves annually
denticulate tooth-like
dipterocarp member of a tropical family of trees, the Dipterocarpaceae
dimorphism difference in morphology between members of same species
distal further from centre of body
diurnal active during day
dorsal towards upper surface of body
dorsal crest ridge of highly modified (often conical) scales along back
dorsum back or dorsal surface of body
egg-tooth embryonic tooth used in breaking egg-shell during emergence
emarginated indented
evergreen plant one that retains its leaves throughout the year

excrescence outgrowth of skin
fang recurved elongated tooth on upper jaw, through which venom passes
femoral pore opening on underside of thigh
fossorial burrowing
frontal large median unpaired scale between eyes in squamates
furrow well-defined groove
geometrid member of the moth family Geometridae
granular scales small, convex and non-overlapping scales, typically with a pebbly appearance
gravid carrying fertilized eggs or young
gular pertaining to or located on throat
gular crest ridge of modified scales along ventral surface of throat and chin
gular fold transverse fold of skin across throat
hemibaculum terminal cartilage or bone support structure in each lobe of hemipenes of male monitors
hemipenal bulge swelling at base of tail, on account of presence of male hemipenes
hemipenes paired sex organs of males (of squamates)
herbivorous plant-eating
heterogeneous scalation covering of different types of scale
homogeneous scalation covering of scales of same type
hood expanded skin behind head, especially in cobras
humeral pertaining to upper part of forelimb
infralabial lower labial
infraorbital below orbital
inguinal pertaining to region around the bases of the hind limbs
insectivorous feeding on insects
interstititial skin between adjacent scales
invertebrate animal lacking a backbone
iridescence in reptiles, rainbow-like sheen on surface of very smooth scales
juvenile a young or sexually immature individual
keel raised ridge on back, tail or scale
keratin hardened sulphur-protein present in reptile scales
keratinous containing keratin
knob rounded protuberance
labial pertaining to lip
lachrymal ridge raised platform between eye and nostril
lamella (plural lamellae) narrow, plate-like tissue
lanciform scale shaped like a lance, with a long and narrow end
lateritic soil red or yellow soil
larynx structure of muscle and cartilage at upper end of trachaea
lateral pertaining to side of body
lobe segment of casque
loreal scale on side of head, between nasals and preoculars
lumbar pertaining to lower back
macrophyte aquatic plant growing in or near water
maxillary teeth teeth on maxillary bones in upper jaws
medial pertaining to midline of body
melanistic showing abundance of dark pigment
mental single scale at anterior border of lower jaw
mental groove deep sulcus on midline between chin shields
mesoplastron paired bones in anterior of plastron
mollusc invertebrate with a soft body, typically enclosed

within a hard shell, such as a snail or a mussel
nare nostril
nasal scale on side of head containing opening of nostril
nasal glands glands in nostrils that secrete excess salt
neonate newborn
neural associated with bones of the spinal column. In turtles, median row of bony plates of carapace
notch sharp, V-shaped indentation
nuchal unpaired, small anterior-most scale of carapace
nuchal venom gland integumental gland in paravertebral region of neck of several species of snake
ocellated eye-like
ocellus (plural ocelli) rounded, eye-like spot
occipital towards back of head
occipital spines spiny projections extending from occipital region of head
omnivorous feeding on many types of food, especially both plants and animals
osteoderm bony deposit within scale or dermal layer
oviparity reproduction through production of eggs provided with membranes and/or shells
ovoviviparity reproduction through production of live young that hatch from eggs within female oviducts
palpebral referring to eyelid
papillae small, nipple-like projections
papillar scale with nipple-like projections
paravertebral stripe stripe on one side of midline of dorsum
parietal head scale behind frontal
parietal crest ridge along midline of casque or head
parietal eye sensory structure opening through skull top
parietal foramen hole in midline of skull
parthenogenetic referring to reproduction without external fertilization
patagium fold of skin on flanks and supported by ribs (in lizards of genus *Draco*)
pectoral third pair of plastral scutes; also chest region of vertebrates
postanal enlarged scale behind cloaca
postfrontal dermal bone of skull roof
postoccipital scute enlarged anterior nuchal
phalange bone of finger or toe
plastral referring to plastron
plastron ventral shell of a turtle or tortoise
posterior towards rear of body; opposite of 'anterior'
postlabial scale behind labial
postmental scale behind mental along line of chin
postnasal scale behind nasals and anterior to loreal
postocular behind eye
preanal scale anterior to cloacal aperture (also precloacal)
precloacal groove longitudinal groove on mid-venter before cloaca
precloacal pore extension of femoral pore series onto body
prefrontal scale anterior to frontal
prehensile being able to grasp objects
premaxillae paired dermal bones in anterior of skull, forming anterior angle of upper jaw
preocular anterior to eye
primary forest unlogged forest
pterygoid paired dermal bones forming the palate
quadrate skull bone articulating with the lower jaw

reticulation colour pattern resembling mesh of a net
rostral scale at tip of snout, bordering mouth
rostral horn annulated horn protruding from anterior tip of snout
rostrum literally, 'beak'; most anterior part of head snout
rugose wrinkly or warty
saxicolous inhabiting rocky areas
scalation pattern of scales on body or on a specific part of body
scale small, thin and horny or bony plates covering skin
scansorial similar to lamella
scute enlarged scale
secondary forest forest that has been logged and replaced by new growth
serrated with a saw-like appearance
sexual dimorphism condition in which males and females have distinctly different forms
snout-vent length measurement between snout-tip and vent
spatulate shaped like a spatula, flat and rounded at tip
species complex two or more species currently assigned to the same biological species
spicule tiny pointed structure
spinose sharp pointed shape like a thorn
spur sharp spinous appendage
squamation scale arrangement
subdigital setae lamellae under fingers and toes of lizards
subfossorial burrowing
subocular beneath eye
subcaudal scale beneath tail
subdigital lamella scale on ventral surface of digits in lizards
superciliary small scale bordering orbit
supralabial upper labial scale
supranasal scale above nasal
supraocular scale above eye
suture seam or boundary between scales or scutes
tarsal spur small projection on posterior part of hind limb above ankle
temporal scale behind postocular
temporal arch bridge extension of jugal and squamosal bones of skull
temporal fenestra (plural fenestrae) bilaterally symmetrical hole in temporal bones of skull
termitaria mound nest of termites
terrestial living on land
total length in snakes, measurement between snout-tip and tail-tip
transverse cross-wise or diagonal
tubercular scale small knob-like scale
tympanum eardrum
venom substance capable of producing toxic reaction when introduced into tissue
ventral towards belly or underside
venter entire undersurface or abdomen
ventrum underside of body
vertebrals middorsal row of scales
vertebrate an animal with a backbone
whorl symmetrical row of enlarged scales circling a tail segment

SELECTED BIBLIOGRAPHY

Anderson, J. 1879. *Anatomical and Zoological Researches: Comprising an Account of the Zoological Results of the Two Expeditions to Western Yunnan in 1868 and 1875; and a Monograph of the Two Cetacean Genera,* Platanista *and* Orcella. Bernard Quaritch, London. Vol. I (text:xxv + 984 + general index) & II (plates: corrigenda + pls I–LXXXI + 1–29 pp).

Annandale, N. 1912. Reptilia. In: Zoological results of the Abor Expedition, 1911–12. *Records of the Indian Museum* 8(1):37–58, pl. V.

Auffenberg, W. 1980. The herpetofauna of Komodo, with notes on adjacent areas. *Bulletin of the Florida State Museum, Biological Sciences* 25:39–156.

Auliya, M. 2006. *Taxonomy, Life History and Conservation of Giant Reptiles in West Kalimantan (Indonesian Borneo).* Natur und Tier Verlag GmbH, Münster. 432 pp.

Auliya, M. 2007. *An Identification Guide to the Freshwater Turtles and Tortoises of Malaysia, Singapore, Indonesia, Brunei, the Philippines, East Timor and Papua New Guinea.* TRAFFIC Southeast Asia, Kuala Lumpur. 98 pp.

Barbour, T. 1912. A contribution to the zoogeography of East Indian Islands. *Bulletin of the Museum of Comparative Zoölogy, Harvard College* 44:1–203 + 8 pl.

Bartlett, E. 1894. Notes on the chelonians; tortoises and turtles, found in Borneo and the adjacent islands. *Sarawak Gazette* 24:187–8.

Bartlett, E. 1895. Notes on tortoises. No. 2. *Sarawak Gazette* 25:29–30.

Bartlett, E. 1895. Notes on tortoises. No. 3. *Sarawak Gazette* 25:83–4.

Bartlett, E. 1895. Notes on the snakes of Borneo and the adjacent islands. *Sarawak Gazette* 25:160–2.

Bartlett, E. 1895. Notes on the snakes of Borneo and the adjacent islands. Part II. *Sarawak Gazette* 25:182–4.

Bartlett, E. 1896. Notes on the snakes of Borneo and the adjacent islands. Part III. *Sarawak Gazette* 26:153–7.

Bartlett, E. 1896. Notes on tortoises. No. 4. *Sarawak Gazette* 26:113.

Bartlett, E. 1896. Notes on the snakes. Part IV. *Sarawak Gazette* 26:241.

Bauer, A.M. 2003. Descriptions of seven new *Cyrtodactylus* (Squamata: Gekkonidae) with a key to the species of Myanmar (Burma). *Proceedings of the California Academy of Sciences* 54(25):461–96.

Bauer, A.M. & K. Henle. 1994. *Family Gekkonidae (Reptilia, Sauria) I. Australia and Oceania.* Das Tierreich 109. Walter de Gruyter, Berlin and New York. xiii + 306 pp.

Bauer, A.M., M. Sumontha & O.S.G. Pauwels. 2003. Two new species of *Cyrtodactylus* (Reptilia: Squamata: Gekkonidae) from Thailand. *Zootaxa* 376:1–18.

Bauer, A.M., M. Sumontha & O.S.G. Pauwels. 2008. A new red-eyed Gekko (Reptilia: Gekkonidae) from Kanchanaburi Province, Thailand. *Zootaxa* 1750:32–42.

Belt, P., A. Malhotra, R.S. Thorpe, D.A. Warrell & W. Wüster. 1996. Russell's viper in Indonesia: snakebite and systematics. In: *Venomous Snakes: Ecology, Evolution and Snakebite.* pp:219–34. R.S. Thorpe, W. Wüster and A. Malhotra (eds). The Zoological Society of London/Clarendon Press, Oxford.

Bennett, D. 1996. *Warane der Welt. Welt der Warane.* Edition Chimaira, Frankfurt am Main. 383 pp. (English edition: *Monitors of the World,* 1997, Edition Chimaira, Frankfurt am Main. 352 pp).

Blanford, W.T. 1881. On a collection of reptiles and frogs chiefly from Singapore. *Proceedings of the Zoological Society of London 1881*:215–26.

Böhme, W. 1982. Über Schmetterlingsagamen, *Leiolepis b. belliana* (Gray, 1827) der Malayischen Halbinsel und ihre parthenogenetischen Linien (Sauria: Uromastycidae). *Zoologische Jahrbucher – Abteilung für Systematik, Okologie und Geographie der Tiere* 109:157–169.

Böhme, W. 1989. Rediscovery of the Sumatran agamid lizard *Harpesaurus beccarii* Doria 1888, with the first note on a live specimen. *Tropical Zoology* 2:31–5.

Boulenger, G.A. 1885. *Catalogue of Lizards in the British Museum (Natural History).* Second edition. Vol. 1. Geckonidae, Eublepharidae, Uroplatidae, Pygopodidae, Agamidae. British Museum (Natural History), London. xii + 436 pp + pls I–XXXII.

Boulenger, G.A. 1885. *Catalogue of Lizards in the British Musum (Natural History).* Second edition. Volume II. Iguanidae, Xenosauridae, Zonuridae, Anguidae, Anniellidae, Helodermatidae, Varanidae, Xantusiidae, Teiidae, Amphiesbaenidae. British Museum (Natural History), London. xiii + 497 pp + pls I–XXIV.

Boulenger, G.A. 1885. A list of reptiles and batrachians from the island of Nias. *Annals & Magazine of Natural History, Series 5,* 16:388–9.

Boulenger, G.A. 1887. *Catalogue of Lizards in the British Museum (Natural History).* Second edition. Volume III. Lacertidae, Gerrhosauridae, Scincidae, Anelytropidae, Dibamidae, Chamaeleontidae. British Museum (Natural History), London. xii + 575 pp + pls I–XL.

Boulenger, G.A. 1887. An account of the reptiles and batrachians obtained in Tenasserim by M.L. Fea, of the Genoa Civic Museum. *Annali Museo Civici Genoa, Series 2,* 5:474–86; pls VI–VIII.

Boulenger, G.A. 1887. An account of the scincoid lizards collected in Burma, for the Genoa Civic Museum, by Messrs. G.B. Comotto and L. Fea. *Annali Museo Civici Genoa, Series 2,* 4:618–24.

Boulenger, G.A. 1887. On new reptiles and batrachians from North Borneo. *Annals & Magazine of Natural History, Series 5,* 20:95–7.

Boulenger, G.A. 1888. An account of the Reptilia obtained in Burma, north of Tenasserim by M.L. Fea, of the Genoa Civic Museum. *Annali Museo Civici Genoa, Series 2,* 6:593–604; pls V–VII.

Boulenger, G.A. 1890. *The Fauna of British India, Including Ceylon and Burma. Reptilia and Batrachia.* Taylor and Francis, London. xviii + 541 pp.

Boulenger, G.A. 1890. List of the reptiles, batrachians and freshwater fishes collected by Prof. Moesch and Mr. Iversen in the district of Deli, Sumatra. *Proceedings of the Zoological Society of London* 1890:31–40.

Boulenger, G.A. 1891. Remarks on the herpetological fauna of Mount Kina Baloo, North Borneo. *Annals & Magazine of Natural History, Series 6,* 7:341–5.

Boulenger, G.A. 1892. An account of the reptiles and batrachians collected by Mr. C. Hose on Mt. Dulit, Borneo. *Proceedings of the Zoological Society of London* 1892:505–8.

Boulenger, G.A. 1893. Concluding report on the reptiles and batrachians obtained in Burma by Signor L. Fea, dealing with the collection made in Pegu and the Karin Hills in 1887–8. *Annali Museo Civici Genoa, Series 2,* 13:304–47.

Boulenger, G.A. 1894. *Catalogue of the Snakes in the British Museum (Natural History).* Volume II, containing the conclusion of the Colubridae Aglyphae. British Museum (Natural History), London. xi + 382 pp + pls I–XX.

Boulenger, G.A. 1894. A list of the reptiles and batrachians collected by Dr. E. Modigliani on Sereinu (Sipora), Mentawei Islands. *Annali del Museo Civico di Storia Naturale Giacomo Doria, Genova Series 2,* 14:613–18.

Boulenger, G.A. 1896. *Catalogue of the Snakes in the British Musum (Natural History).* Volume III, containing the Colubridae (Opisthoglyphae and Proteroglyphae), Amblycephalidae, and Viperidae. British Museum, London. xiv + 727 pp + pls I–XXV.

Boulenger, G.A. 1903. Report on the batrachians and reptiles. In: *Fasiculi Malayensis – Zoology.* Vol. 1. pp:131–76; pls V–X. T.N. Annandale and H.C. Robinson (eds). Taylor and Francis, London.

Boulenger, G.A. 1908. III. Fishes, batrachians and reptiles. In: Report on the Gunong Tahan Expedition. May–September, 1905. *Journal of the Federated Malay States Museum* 3:61–9; pls. IV–V.

Boulenger, G.A. 1912. *A Vertebrate Fauna of the Malay Peninsula from the Isthmus of Kra to Singapore, Including the Adjacent Islands. Reptilia and Batrachia.* Taylor and Francis, London. 294 pp.

Boulenger, G.A. 1917. A revision of the lizards of the genus *Tachydromus. Memoirs of the Asiatic Society of Bengal* 5:207–35; pl. XLVII.

Boulenger, G.A. 1920. Reptiles and batrachians collected in Korinchi, West Sumatra, by Messr. H.C. Robinson & C. Boden Kloss. *Journal of the Federated Malay States Museum* 8:285–306; pl. VIII.

Bourret, R. 1936. *Les Serpents de l'Indochine. II. Catalogue systématique descriptif.* Henri Basuyau et Cie, Toulouse. 505 pp.

Bourret, R. 1941. *Les Tortues de l'Indochine.* Institut Océanographie l'Indochine, Hanoi. 235 pp.

Brongersma, L.D. 1928. Lizards from Pulu Berhala. *Miscellanea Zoologica Sumatrana* 26:1–3.

Brongersma, L.D. 1929. A list of reptiles from Java. In: *On the zoogeography of Java.* K.W. Dammerman (ed.). Treubia 11:64–8.

Brongersma, L.D. 1930. Notes on the list of reptiles of Java. *Treubia* 12(3–4):299–303.

Brongersma, L.D. 1931. Résultats scientifiques du voyage aux Indes Orientales Néerlandaises de LL. AA. RR. le Prince et la Princesse Léopold de Belgique. Reptilia. *Verhandelingen van het Koninklijk Natuurhistorisch Museum van Belgie* 5(2):1–39; 1p (captions entitled 'Planche I'); pls I–IV.

Brongersma, L.D. 1932. Some notes on the genus *Hemiphyllodactylus* Bleeker. *Zoologische Mededeelingen* 14:1–13; 1 map.

Brongersma, L.D. 1933. Additions to the reptile-fauna of the Batu Islands. *Miscellanea Zoologica Sumatrana* 72:1–2.

Brongersma, L.D. 1933. The herpetological fauna of Pulu Weh. Herpetological notes V. *Zoologische Mededeelingen* 16:12–17.

Brongersma, L.D. 1933. Bemerkungen über einige angeblich aus dem Indo-Australischen Archipel stammenden Reptilien. *Mitteilungen aus dem Zoologischen Museum in Berlin* 18(3):319–21.

Brongersma, L.D. 1934. Contributions to Indo-Australian herpetology. *Zoologische Mededeelingen* 17:161–251; pls 1–2.

Brongersma, L.D. 1945. Notes on the list of reptiles of Java. *Treubia* 12:299–303.

Brongersma, L.D. 1945. On the arrangement of the scales on the dorsal surface of the digits in *Lygosoma* and allied genera. *Zoologische Mededeelingen* 24:153–8.

Brongersma, L.D. 1947. On the subspecies of *Python curtus* Schlegel occurring in Sumatra. *Verhandelingen der Koninklijke Akademie van Wetenschappen (C)* 50(6):666–71.

Brongersma, L.D. 1948. Notes on *Maticora bivirgata* (Boie) and *Bungarus flaviceps* Reinh. *Zoologische Mededeelingen* 30:1–29.

Brongersma, L.D. 1958. Note on *Vipera russelii* (Shaw). *Zoologische Mededeelingen* 36:55–76; pls 1–3.

Brongersma, L.D. & W. Helle. 1951. Notes on Indo-Australian snakes, I. *Verhandelingen der Koninklijke Akademie van Wetenschappen C* 54:1–8.

Brown, R.M. 1999. New species of parachute gecko (Squamata: Gekkonidae: genus *Ptychozoon*) from northeastern Thailand and central Vietnam. *Copeia* 1999 (4):990–1001.

Brown, W.C. 1991. Lizards of the genus *Emoia* (Scincidae) with observations on their evolution and biogeography. *Memoirs of the California Academy of Sciences* 15:1–94.

Brygoo, E.R. 1987. Les *Ophisaurus* (Sauria, Anguidae) d'Asia orientale. *Bulletin du Museum National d'Histoire Naturelle, Paris 4e, Serie 9*, Section A, (3):727–52.

Campden-Main, S. 1970. *A Field Guide to the Snakes of South Vietnam.* US National Museum, Washington, DC v + 1 pl. + 114 pp.

Cantor, T. 1847. Catalogue of reptiles inhabiting the Malayan Peninsula and islands, collected or observed by Theodore Cantor, Esq., M.D. Bengal Medical Service. *Journal of the Asiatic Society of Bengal* 16:607–56; 897–952; 1026–78; pls XIL–XL.

Chan, K.O. & L.L. Grismer. 2008. A new species of *Cnemaspis* Strauch 1887 (Squamata: Gekkonidae) from Selangor, Peninsular Malaysia. *Zootaxa* 1877:49–58.

Chan-ard, T., W. Grossmann, A. Gumprecht & K.D. Schulz. 1999. *Amphibians and Reptiles of Peninsular Malaysia and Thailand: an Illustrated Checklist/ Amphibien und Reptilien der Halbinsel Malaysia und Thailands: eine Illustrierte Checkliste.* Bushmaster Publications, Würselen. 240 pp.

Cox, M.J., P.P. Van Dijk, J. Nabhitabhata & K. Thirakhupt. 1998. *A Photographic Guide to Snakes and Other Reptiles of Peninsular Malaysia, Singapore and Thailand.* New Holland Publishers (UK) Ltd., London. 144 pp.

Cundall, D., V. Wallach & D.A. Rossman. 1993. The systematic relationships of the snake genus *Anomochilus. Zoological Journal of the Linnean Society* 109:275–99.

Daniel, J.C. 2002. *The Book of Indian Reptiles and Amphibians.* Bombay Natural History Society/Oxford University Press, Mumbai. viii + 238 pp.

Darevsky, I.S. 1992. Two new species of worm-like lizard *Dibamus* (Sauria, Dibamidae), with remarks on the distribution and ecology of *Dibamus* in Vietnam. *Asiatic Herpetological Research* 4:1–12.

Darevsky, I.S., L.A. Kupriyanova & V.V. Roshchin. 1984. A new all-female triploid species of gecko and karyological data on the bisexual *Hemidactylus frenatus* from Vietnam. *Journal of Herpetology* 18:277–84.

Darevsky, I.S., N.L. Orlov & H.T. Cuc. 2004. Two new lygosomine skinks of the genus *Sphenomorphus* Fitzinger, 1843 (Sauria: Scincidae) from northern Vietnam. *Russian Journal of Herpetology* 11(2):111–20.

Das, I. 1986. The diversity and utilisation of land tortoises in tropical Asia. *Tigerpaper* 13(3):18–21.

Das, I. 1987. Distribution of the keeled box turtle *Pyxidea mouhotii* (Gray). *Journal of the Bombay Natural History Society* 84(1):221–2.

Das, I. 1989. Asian land tortoises. *Sanctuary Asia* 9(4):40–7.

Das, I. 1992. Eggs and hatchlings of some lizards from Borneo. *Hamadryad* 17:42–5.

Das, I. 1993. Annandale's seasnake, *Kolpophis annandalei* (Laidlaw, 1901): a new record for Borneo (Reptilia: Serpentes: Hydrophiidae). *Raffles Bulletin of Zoology* 41(2):359–61.

Das, I. 1995. Amphibians and reptiles recorded at Batu Apoi, a lowland dipterocarp forest in Brunei Darussalam. *Raffles Bulletin of Zoology* 43(1):157–80.

Das, I. 1996. Snakes. In: *Indonesian Heritage.* Volume 5. pp:32–3. Wildlife. T. Whitten & J. Whitten (eds). Editions Didier Millet/Archipelago Press, Singapore.

Das, I. 1996. Lizards. In: *Indonesian Heritage.* Volume 5. pp:34–5. Wildlife. T. Whitten & J. Whitten (eds). Editions Didier Millet/Archipelago Press, Singapore.

Das, I. 2002. *An Introduction to the Amphibians and Reptiles of Tropical Asia.* Natural History Publications (Borneo), Sdn. Bhd., Kota Kinabalu. 207 pp.

Das, I. 2004. Collecting in the 'Land below the Wind': herpetological explorations of Borneo. In: Herpetological expeditions and voyages. A.M. Bauer (ed.). *Bonner Zoologische Beiträge* 52(2):231–43.

Das, I. 2004. A new species of *Dixonius* (Sauria: Gekkonidae) from Vietnam. *Raffles Bulletin of Zoology* 52(2):629–34.

Das, I. 2004. *A Pocket Guide. Lizards of Borneo.* Natural History Publications (Borneo) Sdn Bhd. Kota Kinabalu. 83 pp.

Das, I. 2005. Bornean geckos of the genus *Cyrtodactylus. Gekko* 4(2):11–19.

Das, I. 2005. Revision of the genus *Cnemaspis* Strauch, 1887 (Sauria: Gekkonidae) from the Mentawai and adjacent archipelagos off western Sumatra, Indonesia, with the description of four new species. *Journal of Herpetology* 39(2):233–47.

Das, I. 2006. *A Photographic Guide to the Snakes and Other Reptiles of Borneo.* New Holland Publishers (UK) Ltd., London. US edition, Ralph Curtis Books, Sanibel Island, Florida. 144 pp.

Das, I. 2007. *A Pocket Guide. Amphibians and Reptiles of Brunei.* Natural History Publications (Borneo) Sdn Bhd. Kota Kinabalu. viii + 200 pp.

Das, I. 2008. *Pelochelys cantorii.* In: Conservation biology of freshwater turtles and tortoises. A.G.J. Rhodin, P.C.H. Pritchard, P.P. van Dijk, R.A. Saumure, K.A. Buhlmann & J.B. Iverson (eds). *Chelonian Research Monograph Number 5.* Chelonian Research Foundation, Lunenberg, Massachusetts. pp:1–4.

Das, I. & C.C. Austin. 2007. New species of *Lipinia* (Sauria: Scincidae) from Borneo, revealed by molecular and morphological data. *Journal of Herpetology* 41(1):61–71.

Das, I. & L.L. Grismer. 2003. Two new species of *Cnemaspis* Strauch, 1887 (Sauria: Gekkonidae) from the Seribuat Archipelago, Pahang and Johor States, West Malaysia. *Herpetologica* 59(4):546–54.

Das, I., M. Lakim & P. Kandaung. 2008. New species of *Luperosaurus* (Squamata: Gekkonidae) from the Crocker Range Park, Sabah, Malaysia (Borneo). *Zootaxa* 1719:53–60.

Das, I. & K.K.P. Lim. 2003. Two new species of *Dibamus* (Squamata: Dibamidae) from Borneo. *Raffles Bulletin of Zoology* 51(1):137–41.

Das, I. & N.S. Yaakob. 2003. A new species of *Dibamus* (Squamata: Dibamidae) from Peninsular Malaysia. *Raffles Bulletin of Zoology* 51(1):143–7.

David, P. 1994. Liste des reptiles actuels du monde I. Chelonii. *Dumerilia* 1:7–127.

David, P., R.H. Bain, N.Q. Truong, N.L. Orlov, G. Vogel, V.N. Thanh & T. Ziegler. 2007. A new species of the natricine snake genus *Amphiesma* from the Indochinese Region (Squamata: Colubridae: Natricinae). *Zootaxa* 1462: 41–60.

David, P., N. Vidal & O.S.G. Pauwels. 2001. A morphological study of Stejneger's pitviper *Trimeresurus stejnegeri* (Serpentes, Viperidae, Crotalinae), with the description of a new species from Thailand. *Russian Journal of Herpetology* 8(3):205–22.

David, P. & G. Vogel. 1996. *The Snakes of Sumatra. An Annotated Checklist and Key with Natural History Notes*. Edition Chimaira, Frankfurt am Main. 260 pp.

David, P., G. Vogel & O.S.G. Pauwels. 1998. *Amphiesma optatum* (Hu & Djao, 1966) (Serpentes, Colubridae): an addition to the snake fauna of Vietnam, with a list of the species of the genus *Amphiesma* and a note on its type species. *Journal of the Taiwan Museum* 51(2):83–92.

David, P., G. Vogel & O.S.G. Pauwels. 2008. A new species of the genus *Oligodon* Fitzinger, 1826 (Squamata: Colubridae) from southern Vietnam and Cambodia. *Zootaxa* 1939:19–37.

David, P., G. Vogel & J. van Rooijen. 2008. A revision of the *Oligodon taeniatus* (Günther, 1861) group (Squamata: Colubridae), with the description of three new species from the Indochinese Region. *Zootaxa* 1965:1–49.

David, P., G. Vogel, S.P. Vijayakumar & N. Vidal. 2006. A revision of the *Trimeresurus puniceus*-complex (Serpentes: Viperidae: Crotalinae) based on morphological and molecular data. *Zootaxa* 1293:1–78.

de Haas, C.P.J. 1950. Checklist of the snakes of the Indo-Australian Archipelago (Reptiles, Ophidia). *Treubia* 20:511–625.

de Lang, R. & G. Vogel. 2005. *The Snakes of Sulawesi. A Field Guide to the Land Snakes of Sulawesi with Identification Keys*. Edition Chimaira, Frankfurt am Main. 312 pp.

de Lisle, H.F. 1996. *The Natural History of Monitor Lizards*. Krieger Publishing Company, Malabar, Florida. xiii + 201 pp.

de Rooij, N. 1915. *The Reptiles of the Indo-Australian Archipelago*. Vol. I. Lacertilia, Chelonia, Emydosauria. E.J. Brill, Leiden. xiv + 384 pp.

de Rooij, N. 1917. *The Reptiles of the Indo-Australian Archipelago*. II. Ophidia. E. J. Brill, Leiden. xiv + 334 pp.

de Silva, G.S. 1980. The status of sea turtle populations in East Malaysia and the South China Sea. In: *Biology and Conservation of Sea Turtles*. pp:327–37. K.A. Bjorndal (ed.). Smithsonian Institution Press, Washington, D.C.

Diong, C.H. & S.S.L. Lim. 1998. Taxonomic review and morphological description of *Bronchocela cristatella* (Kuhl, 1820) (Squamata: Agamidae) with notes on other species in the genus. *Raffles Bulletin of Zoology* 46(2):345–59.

Dowling, H.G. 1993. The name of Russel's viper. *Amphibia-Reptilia* 14:320.

Dring, J.C.M. 1979. Amphibians and reptiles from northern Trengganu, Malaysia, with descriptions of two new geckos: *Cnemaspis* and *Cyrtodactylus*. *Bulletin of the British Museum of Natural History (Zoology)* 34:181–241.

Erdelen, W. 1991. Conservation and population ecology of monitor lizards: the water monitor *Varanus salvator* (Laurenti, 1768) in south Sumatra. *Mertensiella* 2:120–35.

Gauthier, J., A.G. Kluge & T. Rowe. 1988. The early evolution of the Amniota. In: *The Phylogeny and Classification of the Tetrapods*. Volume 1: Amphibians, reptiles, birds. pp:103–55. M.J. Benton (ed.). Clarendon Press, Oxford.

Glodek, G.S. & H.K. Voris. 1982. Marine snake diets: prey composition, diversity and overlap. *Copeia* 1982(3):661–6.

Gloyd, H.K. & R. Conant. 1990. *Snakes of the Agkistrodon complex. A Monographic Review*. Society for the Study of Amphibians and Reptiles. Contributions to Herpetology No. 6. SSAR, Oxford, Ohio. vi + 614 pp; 52 pls.

Gopalakrishnakone, P. & L.M. Chou. (eds). 1990. *Snakes of Medical Importance*. National University of Singapore, Singapore. v + 670 + vi pp.

Grandison, A.G.C. 1972. The Gunong Benom Expedition 1967. 5. Reptiles and amphibians of Gunong Benom with a description of a new species of *Macrocalamus*. *Bulletin of the British Museum (Natural History) Zoology* 23(4):45–101.

Greer, A.E. 1970. A subfamilial classification of scincid lizards. *Bulletin of the Museum of Comparative Zoology* 139:151–83.

Greer, A.E. 1970. The relationships of the skinks referred to the genus *Dasia*. *Breviora* 348:1–30.

Greer, A.E. 1974. The generic relationships of the scincid lizard genus *Leiolopisma* and its relatives. *Australian Journal of Zoology* (Supplement Series) 31:1–67.

Greer, A.E. 1977. The systematics and evolutionary relationships of the scincid lizard genus *Lygosoma*. *Journal of Natural History* 11:515–40.

Greer, A.E. 1985. The relationships of the lizard genera *Anelytropsis* and *Dibamus*. *Journal of Herpetology* 19:116–56.

Grismer, J.L., T.M. Leong & N.S.b. Yaakob. 2003. Two new southeast Asian skinks of the genus *Larutia* and intrageneric phylogenetic relationships. *Herpetologica* 59(4):552–64.

Grismer, L.L. 2008. On the distribution and identification of *Cyrtodactylus brevipalmatus* Smith, 1923 and *Cyrtodactylus elok* Dring, 1979. *Raffles Bulletin of Zoology* 56:177–9.

Grismer, L.L. 2008. A new species of insular skink (genus *Sphenomorphus* Fitzinger 1843) from the Langkawi Archipelago, Kedah, West Malaysia with the first report of the herpetofauna of Pulau Singa Besar and an updated checklist of the herpetofauna of Pulau Langkawi. *Zootaxa* 1691:53–66.

Grismer, L.L. 2008. A revised and updated checklist of the lizards of Peninsular Malaysia. *Zootaxa* 1860:28–34.

Grismer, L.L. & N. Ahmad. 2008. A new insular species of *Cyrtodactylus* (Squamata: Gekkonidae) from the Langkawi Archipelago, Kedah, Peninsular Malaysia. *Zootaxa* 1924:53–68.

Grismer, L.L. & K.O. Chan. 2008. A new species of *Cnemaspis* Strauch 1887 (Squamata: Gekkonidae) from Pulau Perhentian Besar, Terengganu, Peninsular Malaysia. *Zootaxa* 1771:1–15.

Grismer, L.L., K. O. Chan, J.L. Grismer, P.L. Wood, Jr & D. Belabut. 2008. Three new species of *Cyrtodactylus* (Squamata: Gekkonidae) from Peninsular Malaysia. *Zootaxa* 1921:1–23.

Grismer, L.L., K.O. Chan, N. Nasir & M. Sumontha. 2008. A new species of karst dwelling gecko (genus Strauch 1887) from the border region of Thailand and Peninsular Malaysia. *Zootaxa* 1875:51–68.

Grismer, L.L., J.L. Grismer, P.L. Wood, Jr & K.O. Chan. 2008. The distribution, taxonomy, and redescription of the geckos *Cnemaspis affinis* (Stoliczka 1887) and *C. flavolineata* (Nicholls 1949) with descriptions of a new montane species and two new lowland, karst-dwelling species from Peninsular Malaysia. *Zootaxa* 1931:1–24

Grismer, J.L., T.M. Leong & N.S.b. Yaakob. 2003. Two new southeast Asian skinks of the genus *Larutia* and intrageneric phylogenetic relationships. *Herpetologica* 59(4):552–64.

Grismer, L.L. & T.M. Leong. 2005. New species of *Cyrtodactylus* (Squamata: Gekkonidae) from southern Peninsular Malaysia. *Journal of Herpetology* 39(4):584–91.

Grismer, L.L., N. Van Tri & J.L. Grismer. 2008. A new species of insular pitviper of the genus *Cryptelytrops* (Squamata: Viperidae) from southern Vietnam. *Zootaxa* 1715:57–68.

Grismer, L.L., T.M. Youmans, P.L. Wood & J.L. Grismer. 2006. Checklist of the herpetofauna of the Seribuat Archipelago, West Malaysia with comments on biogeography, natural history, and adaptive types. *Raffles Bulletin of Zoology* 54(1):157–80.

Grismer, L.L., T.M. Youmans, P.L. Wood, A. Ponce, S.B. Wright, B.S. Jones, R. Johnson, K.L. Sanders, D.J. Gower, N.S.b. Yaakob & K.K.P. Lim. 2006. Checklist of the herpetofauna of Pulau Langkawi, Malaysia, with comments on taxonomy. *Hamadryad* 30(1&2):60–73.

Gritis, P. & H.K. Voris. 1990. Variability and significance of parietal and ventral scales in the marine snakes of the genus *Lapemis* (Serpentes: Hydrophiidae), with comments on the occurrence of spiny scales in the genus. *Fieldiana Zoology* n.s. (56):i–iii + 1–13.

Groombridge, B. & R. Luxmoore. 1991. *Pythons in South-east Asia. A Review of Distribution, Status and Trade in Three Selected Species.* Convention on International Trade in Endangered Species of Wild Fauna and Flora, Lausanne. 127 pp.

Gumprecht, A., F. Tillack, N.L. Orlov, A. Captain & S. Ryabov. 2004. *Asian Pitvipers.* GeitjeBooks, Berlin. 368 pp.

Günther, A.C.L.G. 1872. On the reptiles and amphibians of Borneo. *Proceedings of the Zoological Society of London* 1872:586–600; pls XXXV– XXXX.

Günther, A.C.L.G. 1895. The reptiles and batrachians of the Natuna Islands. *Novitates Zoology* 2:499–502.

Günther, R. & U. Manthey. 1995. *Xenophidion*, a new genus with two new species of snakes from Malaysia (Serpentes, Colubridae). *Amphibia-Reptilia* 16:229 –40.

Gyi, K.K. 1970. A revision of colubrid snakes of the subfamily Homalopsinae. *University of Kansas Publications of the Museum of Natural History* 20:47–223.

Haile, N.S. 1958. The snakes of Borneo, with a key to the species. *Sarawak Museum Journal* 8:743–71.

Hallermann, J. & W. Böhme. 2000. A review of the genus *Pseudocalotes* (Squamata: Agamidae), with description of a new species from West Malaysia. *Amphibia-Reptilia* 21(2):193–210.

Harrisson, B. 1961. *Lanthonotus borneensis* – habits and observations. *Sarawak Museum Journal* 10:17–18.

Harrisson, B. 1962. Beobachtungen am lebenden Taubwaran *Lanthonotus borneensis*. *Natur und Museum* 92(2):38–45.

Helfenberger, N. 2001. Phylogenetic relationships of Old World ratsnakes based on visceral organ topography, osteology, and allozyme variation. *Russian Journal of Herpetology Supplement*:1–62.

Hendrickson, J.R. 1958. The green turtle, *Chelonia mydas* (Linn.) in Malaya and Sarawak. *Proceedings of the Zoological Society of London* 130:455–566.

Hikida, T. 1980. [Lizards of Borneo.] *Acta hytotaxonomica et Geobotanica* 31:(1–3):97–102. [In Japanese.]

Hikida, T. 1982. A new limbless *Brachymeles* (Sauria: Scincidae) from Mt. Kinabalu, North Borneo. *Copeia* 1982(4):840–4.

Hikida, T. 1990. Bornean gekkonid lizards of the genus *Cyrtodactylus* (Lacertilia: Gekkonidae) with descriptions of three new species. *Japanese Journal of Herpetology* 13(3):91–107.

Hikida, T., N.L. Orlov, J. Nabhitabhata & H. Ota. 2002. Three new depressed-bodied water skinks of the genus *Tropidophorus* (Lacertilia: Scincidae) from Thailand and Vietnam. *Current Herpetology* 21(1):9–23.

Hodges, R. 1993. Snakes of Java with special reference to East Java. *British Herpetological Society Bulletin* (43):15–32.

Hubrecht, A.A.W. 1879. Contributions to the herpetology of Sumatra. *Notes from the Leyden Museum* 1(4):243–5.

Hubrecht, A.A.W. 1881. On a new genus and species of Agamidae from Sumatra. *Notes from the Leyden Museum* 3:51–2.

Inger, R.F. 1958. Three new species related to *Sphenomorphus variegatus* (Peters). *Fieldiana Zoology* 39(24):257–68.

Inger, R.F. 1960. A review of the agamid lizards of the genus *Phoxophrys* Hubrecht. *Copeia* 1960(3): 221–5.

Inger, R.F. 1983. Morphological and ecological variation in the flying lizards (genus *Draco*). *Fieldiana Zoology* n.s. (18):i–iv + 1–35.

Inger, R.F. & W.C. Brown. 1980. Species of the scincid genus *Dasia* Gray. *Fieldiana Zoology* n.s. 3:1–11.

Inger, R.F. & A.E. Leviton. 1961. A new colubrid snake of the genus *Pseudorabdion* from Sumatra. *Fieldiana Zoology* n.s. 44:45–57.

Inger, R.F. & A.E. Leviton. 1966. The taxonomic status of Bornean snakes of the genus *Pseudorabdion* Jan and of the nominal genus *Idiopholis* Mocquard. *Proceedings of the California Academy of Sciences* 34(4):307–14.

Inger, R.F. & H. Marx. 1965. The systematics and evolution of the Oriental colubrid snakes of the genus *Calamaria*. *Fieldiana Zoology* 49:1–304.

Iskandar, D.T. 1987. The occurrence of *Enhydris alternans* at Java. *The Snake* 19(1):72–3.

Iskandar, D.T. 2000. *Turtles and Crocodiles of Insular Southeast Asia and New Guinea*. Institute of Technology, Bandung. xix + 191 pp. Bahasa Indonesia edition 2000: Kura-kura & buaya Indonesia & Papua Nugini dengan catatan mengenai jenis-jenis di Asia Tenggara. Institut Teknologi Bandung, Bandung. xix + 191 pp.

Iverson, J.B. 1992. *A Revised Checklist with Distribution Maps of the Turtles of the World*. Privately published, Richmond, Indiana. xiii + 363 pp.

Iverson, J.B., P.Q. Spinks, H.B. Shaffer, W.P. McCord & I. Das. 2001. Phylogenetic relationships among the Asian tortoises of the genus *Indotestudo* (Reptilia: Chelonii: Testudinidae). *Hamadryad* 26(2):283–7.

Keogh, J.S., D. Barker & R. Shine. 2001. Heavily exploited but poorly known: systematics and biogeography of commercially harvested pythons (*Python curtus* group) in Southeast Asia. *Biological Journal of the Linnean Society* 73:113–29.

Kharin, V.E. 1984. [A review of sea snakes of the group *Hydrophis* sensu lato (Serpentes, Hydrophiidae).] 3. The genus *Leioselasma*. *Zoological Zhurnal* 63:1536–46. [In Russian.]

Kharin, V.E. 1984. [Revision of sea snakes of subfamily Laticaudinae Cope, 1879, sensu lato (Serpentes, Hydrophiidae).] *Proceedings of the Zoological Institute, Leningrad* 124:128–39. [In Russian.]

Kluge, A.G. 1967. Higher taxonomic categories of gekkonid lizards and their evolution. *Bulletin of the American Museum of Natural History* 135(1):1–60; pls 1–5.

Kluge, A.G. 1968. Phylogenetic relationships of the gekkonid lizard genera *Lepidodactylus* Fitzinger, *Hemiphyllodactylus* Bleeker, and *Pseudogekko* Taylor. *Philippine Journal of Science* 95(3):331–52.

Kluge, A.G. 1987. Cladistic relationships in the Gekkonoidea (Squamata, Sauria). *Misc. Publications of the Museum of Zoology, University of Michigan* 173:1–53.

Kluge, A.G. 2001. Gekkotan lizard taxonomy. *Hamadryad* 26(1):1–209.

Kopstein, F. 1938. Ein Beitrag zur Eierkunde und zur Fortpflanzung der malaiischen Reptilien. *Bulletin of the Raffles Museum* 14:81–167.

Kuch, U. & D. Mebs. 2007. The identity of the Javan Krait, *Bungarus javanicus* Kopstein, 1932 (Squamata: Elapidae): evidence from mitochondrial and nuclear DNA sequence analyses and morphology. *Zootaxa* 1426:1–26.

Kuchling, G., W. Ko Ko, T. Lwin, S.A. Min, K. Myo Myo, T.T. Khaing, W.W. Mar & T.T. Khaing. 2004. The softshell turtles and their exploitation at the upper Chindwin River, Myanmar: range extensions for *Amyda cartilaginea*, *Chitra vandijki*, and *Nilssonia formosa*. *Salamandra* 40(3/4):281–96.

Lading, E. & R. Stuebing. 1997. Nest of a false gharial from Sarawak. *Crocodile Specialist Group Newsletter* 16(2):12–13.

Leong, T.-M., L.L. Grismer & Mumpuni. 2003. Preliminary checklists of the herpetofauna of the Anambas and Natuna Islands (South China Sea). *Hamadryad* 27(2):165–74.

Leviton, A.E. & W.C. Brown. 1959. A review of the snakes of the genus *Pseudorabdion* with remarks on the status of the genera *Agrophis* and *Typhlogeophis* (Serpentes: Colubridae). *Proceedings of the California Academy of Sciences*, Series 4, 29(14):475–508

Leviton, A.E., G. Wogan, M. Koo, G.R. Zug & J.V. Vindum. 2003. The dangerously venomous snakes of Myanmar. Illustrated checklist with keys. *Proceedings of the California Academy of Sciences* 54(24):407–62.

Lim, B.L. 1964. Notes on the elephant's trunk snake and the puff-faced water snake. *Malayan Nature Journal* 1964:179–83.

Lim, B.L. 1967. Further comments on rare snakes. *Federated Museums Journal, new series* 12:123–6.

Lim, B.L. 1971. Venomous snakes of southeast Asia. *Southeast Asian Journal of Tropical Medicine & Public Health* 2(1):56–64.

Lim, B.L. 1990. Venomous land snakes of Malaysia. In: *Snakes of Medical Importance (Asia-Pacific Region)*. pp:387–417. P. Gopalakrishnakone & L.M. Chou (eds). National University of Singapore, Singapore.

Lim, B.-L. & I. Das. 1999. *Turtles of Borneo and Peninsular Malaysia*. Natural History Publications (Borneo), Sdn. Bhd., Kota Kinabalu. xii + 151 pp.

Lim, F.L.K. & M.T.-M. Lee 1989. *Fascinating Snakes of Southeast Asia – An Introduction*. Tropical Press Sdn. Bhd., Kuala Lumpur. xviii + 124 pp.

Lim, K.K.P. & F.L.K. Lim. 2002. *A Guide to the Amphibians & Reptiles of Singapore*. Revised edition. Singapore Science Centre, Singapore. 160 pp.

Lonnberg, E. & H. Rendahl. 1925. Dr E. Mjöberg's zoological collections from Sumatra. 2. Reptiles and batrachians. *Archiv för Zoologi* 17A(23):1–3.

Losos, J.B. & H.W. Greene. 1988. Ecological and evolutionary implications of diet in monitor lizards. *Biological Journal of the Linnean Society* 35:379–407.

Loveridge, A. 1938. New snakes of the genera *Calamaria*, *Bungarus* and *Trimeresurus* from Mount Kinabalu, North Borneo. *Proceedings of the Biological Society of Washington* 51:43–6.

Loveridge, A. 1944. A new elapid snake of the genus *Maticora* from Sarawak, Borneo. *Proceedings of the Biological Society of Washington* 57:105–6.

Loveridge, A. 1944. Errata (A new elapid snake of the genus *Maticora* from Sarawak, Borneo). *Proceedings of the Biological Society of Washington* 57:VI.

Loveridge, A. 1945. *Reptiles of the Pacific World*. The Macmillan Company, New York. xii + 259 pp.

McCord, W.P. & P.C.H. Pritchard. 2002. A review of the softshell turtles of the genus *Chitra*, with the description of new taxa from Myanmar and Indonesia (Java). *Hamadryad* 27:11–56.

McDiarmid, R., J.A. Campbell & T.'S.A. Touré. 1999. *Snake Species of the World. A Taxonomic and Geographic Reference*. Volume 1. The Herpetologists' League, Washington, D.C. xi + 511 pp.

McGuire, J.A. & B.-H. Kiew. 2001. Phylogenetic systematics of southeast Asian flying lizards (Iguania: Agamidae: *Draco*) as inferred from mitochondrial DNA sequence data. *Biological Journal of the Linnaean Society* 72:203–29.

McKay, L. 2006. *A Field Guide to the Reptiles and Amphibians of Bali*. Krieger Publishing, Malabar, Florida. 148 pp. Bahasa Indonesia edition, 2006, *Reptil dan amphibi di Bali*. J. Lindley McKay, Blackrock, Victoria. 153 pp.

Malhotra, A. & R.S. Thorpe. 2004. A phylogeny of four mitochondrial gene regions suggests a revised taxonomy for Asian pitvipers (*Trimeresurus* and *Ovophis*). *Molecular Phylogenetics and Evolution* 32(1):83–100. (Also, 2005. Erratum. *Molecular Phylogenetics and Evolution* 34:680–91.)

Malkmus, R., U. Manthey, G. Vogel, P. Hoffmann & J. Kosuch. 2002. *Amphibians and Reptiles of Mount Kinabalu (North Borneo)*. Koeltz Scientific Books, Königstein. 424 pp.

Malnate, E.V. 1960. Systematic division and evolution of the colubrid snake genus *Natrix*, with comments on the subfamily Natricinae. *Proceedings of the Academy of Natural Sciences of Philadelphia* 112: 41–71.

Malnate, E.V. 1962. The relationships of five species of the Asiatic natricine snake genus *Amphiesma*. *Proceedings of the Academy of Natural Sciences of Philadelphia* 114(8):251–99.

Malnate, E.V. 1966. *Amphiesma platyceps* (Blyth) and *Amphiesma sieboldii* (Günther): sibling species (Reptilia: Serpentes). *Journal of the Bombay Natural History Society* 63(1):1–17.

Manthey, U. 2008. *Agamid Lizards of Southern Asia. Draconinae 1/Agamen des Südlichen Asien. Draconinae 1*. Edition Chimaira, Frankfurt am Main. 160 pp; 1 folding pl.

Manthey, U. & W. Denzer. 1991. Die Echten Winkelkopfagamen der Gattung *Gonocephalus* Kaup (Sauria: Agamidae). I. Die *megalepis* – Gruppe mit *Gonocephalus lacunosus* sp. n. aus Nord-Sumatra. *Sauria, Berlin* 13(1):3–10.

Manthey, U. & W. Denzer. 1991. Die Echten Winkelkopfagamen der Gattung *Gonocephalus* Kaup (Sauria: Agamidae). II. Allgemeine Angaben zur Biologie und Terraristik. *Sauria, Berlin* 13(2):19–22.

Manthey, U. & W. Denzer. 1991. Die Echten Winkelkopfagamen der Gattung *Gonocephalus* Kaup (Sauria: Agamidae). III. *Gonocephalus grandis* (Gray, 1845). *Sauria, Berlin* 13(3):3–10.

Manthey, U. & W. Denzer. 1992. Die Echten Winkelkopfagamen der Gattung *Gonocephalus* Kaup (Sauria: Agamidae). IV. *Gonocephalus mjoebergi* Smith, 1925 und *Gonocephalus robinsoni* Boulenger, 1908. *Sauria, Berlin* 14(1):15–19.

Manthey, U. & W. Denzer. 1992. Die Echten Winkelkopfagamen der Gattung *Gonocephalus* Kaup (Sauria: Agamidae). V. Die *bellii*-Gruppe. *Sauria, Berlin* 14(3).

Manthey, U. & W. Denzer. 1993. Die Echten Winkelkopfagamen der Gattung *Gonocephalus* Kaup (Sauria: Agamidae). VI. Die *chamaeleontinus*-Gruppe. *Sauria, Berlin* 14(1):23–8.

Manthey, U. & W. Grossmann. 1997. *Amphibien & Reptilien Südostasiens*. Natur und Tier, Münster. 512 pp.

Mausfeld, P. & A. Schmitz. 2003. Molecular phylogeography, intraspecific variation and speciation of the Asian scincid lizard genus *Eutropis* Fitzinger, 1843 (Squamata: Reptilia: Scincidae): taxonomic and biogeographic implications. *Organisms, Diversity and Evolution* 3:161–71.

Mertens, R. 1924. Ueber einige Reptilien aus Borneo. *Zoologischer Anzeiger* 60:155–9.

Mertens, R. 1927. Neue Amphibien und Reptilien aus dem Indo-Australischen Archipel, gesammelt wahrend der Sunda-Expedition Rensch. *Senckenbergiana* 9(6):234–42.

Mertens, R. 1929. Über eine kleine herpetologische Sammlung aus Java. *Senckenbergiana Biologica* 11(1/2):22–33.

Mertens, R. 1929. Herpetologische Mitteilungen, XXIII–XXV.XXIV: Amphibien und Reptilien aus Atjeh (Nordsumatra), gesammelt von Herrn H.R. Rookmaaker. *Zoologischer Anzeiger* 86(3/4):62–6.

Mertens, R. 1931. *Ablepharus boutonii* (Desjordin) und seine geographische Variation. Zoologische Jahrbucher – Abteilung für Systematik, *Okologie und Geographie der Tiere* 61:63–210.

Mertens, R. 1933. Weitere Mitteilungen über die Rassen von *Ablepharis boutonii* (Desjardin). I. *Zoologischer Anzeiger* 105:92–6.

Mertens, R. 1942. Die Familie der Warane (Varanidae). *Abhandlungen hrgs von der Senckenbergischen naturforschenden Gesellschaft* (462):1–116; (465):117–234; (466):235–391.

Mertens, R. 1954. Über die javanische Eidechse *Dendragama fruhstorferi* und die Gattung *Dendragama*. *Senckenbergiana* 34(4–6):185–6.

Mertens, R. 1956. Eidechsen (Reptilia) vom Karimundjawa-Archipel. *Treubia* 23(2):253–7.

Mertens, R. 1957. Zur Herpetofauna von Ostjava und Bali. *Senckenbergiana* 38(1/2):23–31.

Mertens, R. 1959. Eine Panzerschleichse (*Ophisaurus*) aus Sumatra. *Senckenbergiana Biologica* 40(1/2): 109–11.

Mertens, R. 1959. Liste der Warane Asiens und der Indo-australischen Inselwelt mit systematischen Bemerkungen. *Senckenbergiana Biologica* 40(5/6): 112–47.

Mertens, R. 1959. Liste der Warane Asiens und der Indo-australischen Inselwelt mit systematischen Bemerkungen. *Senckenbergiana Biologica* 40(5/6): 221–40; 5 pl.

Mertens, R. 1964. Weitere Mitteilungen über die Rassen von *Ablepharis boutonii* (Desjardin). III. *Zoologischer Anzeiger* 173:99–110.

Mertens, R. 1968. Die Arten und Unterarten der Schmuckbaumschlangen (*Chrysopelea*). *Senckenbergiana Biologica* 49(3/4):191–217.

Mertens, R. 1971. Zur Kenntnis des javanischen Baumskinks, *Dasia leucosticta*. *Senckenbergiana Biologica* 52(3/5):192–6.

Meylan, P.A. 1987. The phylogenetic relationships of soft-shelled turtles (family Trionychidae). *Bulletin of the American Museum of Natural History* 186(1): 1–101.

Minton, S.A., Jr. 1975. Geographic distribution of sea snakes. In: *The Biology of Sea Snakes*. pp:21–31. W.A. Dunson (ed.). University Park Press, Baltimore.

Minton, S.A., Jr. 1990. Venomous bites by non-venomous snakes: an annotated bibliography of colubrid envenomation. *Journal of Wilderness Medicine* 1(2):119–27.

Mocquard, F. 1890. Récherches sur la faune herpétologique des Iles de Bornéo et de Palawan. *Nouvelles Archives du Museum d'Histoire Naturelle de Paris Series 3* 2:115–68.

Mocquard, F. 1892. Voyage de M. Chaper à Bornéo. Nouvelle contribution à la faune herpétologique de Bornéo. *Memoirs de la Société Zoologique de France* 5:140–206, 7 pl.

Mocquard, F. 1892. Nouvelle contribution à la faune herpétologique de Bornéo. *Memoirs de la Société Zoologique de France* 5:190–206.

Modigliani, E. 1889. Materiali per la fauna erpetologica dell isola Nias. *Annali del Museo Civico di Storia Naturale Giacomo Doria, Genova* 7:113–24; 1 pl.

Moll, E.O. 1989. *Manouria emys* Asian brown tortoise. In: The conservation biology of tortoises. pp:119–120. I.R. Swingland & M.W. Klemens (eds). *Occasional Paper, IUCN Species Survival Commission No. 5*. IUCN, Gland.

Monk, A.R. 1991. A case of mild envenomation from a mangrove snake bite. *Litteratura Serpentium* 11(1):21–3.

Moody, S.M. 1980. Phylogenetic and historical biogeographical relationships of the genera in the family Agamidae (Reptilia: Lacertilia). PhD Dissertation, University of Michigan, Ann Arbor. 5 unnumbered pages + xv + 373 pp.

Mori, A., K. Araya & T. Hikida. 1995. Biology of the poorly known Bornean lizard, *Apterygodon vittatus* (Squamata: Scincidae): an arboreal anteater. *Herpetological Natural History* 3(1):1–14.

Mori, A. & T. Hikida. 1993. Natural history observations of the flying lizard, *Draco volans sumatranus* (Agamidae, Squamata) from Sarawak, Malaysia. *Raffles Bulletin of Zoology* 41(1):83–94.

Mori, A. & T. Hikida. 1994. Field observations on the social behavior of the flying lizard, *Draco volans sumatranus*, in Borneo. *Copeia* 1994(1):124–30.

Müller, S. & H. Schlegel. 1845. Over de Krokodillen van den Indischen Archipel. 28 pp., pls 1–3. In: *Verhandelingen over de natuurlijke geschiedenis der Nederlandsche overzeesche bezittingenis, door de leden der Natuurkundige Commissie in Indië en andere Schrijvers*. C. J. Temminck. (ed.). S. & J. Luchtmans & C.C. Van de Hoek, Leiden.

Murphy, J.C. 2007. *Homolopsid Snakes. Evolution in the Mud*. Krieger Publishing, Malabar, Florida. 260 pp.

Murphy, J.C. & H.K. Voris. 1994. A key to the homalopsine snakes. *The Snake* 26(2):123–33.

Musters, C.J.M. 1983. Taxonomy of the genus *Draco* L. (Agamidae, Lacertilia, Reptilia). *Zoologische Verhandelingen, Leiden* (199):1–120, 4 pls.

Ngo, V.T. & A.M. Bauer. 2008. Descriptions of two new species of *Cyrtodactylus* Gray, 1827 (Squamata: Gekkonidae) endemic to southern Vietnam. *Zootaxa* 1715:27–42.

Nguyen, V.S., T.C. Ho & Q.T. Nguyen. 2005. *A Checklist of Amphibians and Reptiles of Vietnam*. Agriculture Publishing House, Hanoi. 180 pp. [In Vietnamese.]

O'Shea, M. 2007. *Boas and Pythons of the World*. New Holland Publishers (UK) Ltd., London. 160 pp.

Orlov, N.L. 2000. Distribution, biology and comparative morphology of the snakes of *Xenopeltis* genus (Serpentes: Macrostoma: Xenopeltidae) in Vietnam. *Russian Journal of Herpetology* 7(2):103–14.

Orlov, N.L., S.A. Ryabov, V.S. Nguyen & Q.T. Nguyen. 2003. New records and data on the poorly known snakes of Vietnam. *Russian Journal of Herpetology* 10:217–40.

Orlov, N.L., S.A. Ryabov, K.A. Shiryaev & N. Van Sang. 2001. On the biology of pit vipers of *Protobothrops* genus (Serpentes: Colubroidea: Viperidae: Crotalinae). *Russian Journal of Herpetology* 8(2):159–64.

Ota, H. & T. Hikida. 1991. Taxonomic review of the lizards of the genus *Calotes* Cuvier 1817 (Agamidae: Squamata) from Sabah, Malaysia. *Tropical Zoology* 4:179–92.

Ota, H. & J. Nabhitabhata. 1991. A new species of *Gekko* (Gekkonidae: Squamata) from Thailand. *Copeia* 1991(2):503–9.

Ota, H., S. Sengoku & T. Hikida. 1996. Two new species of *Luperosaurus* (Reptilia: Gekkonidae) from Borneo. *Copeia* 1996(2):433–9.

Ouboter, P.E. 1986. A revision of the genus *Scincella* (Reptilia: Sauria: Scincidae) of Asia, with some notes on its evolution. *Zoologische Verhandelingen* (229):1–66.

Parker, H.W. 1924. Description of a new agamid lizard from Sumatra. *Annals & Magazine of Natural History, Series 9*, 24:624–5.

Pauwels, O.S.G., A.M. Bauer, M. Sumontha & L. Chanhome. 2004. *Cyrtodactylus thirakhupti* (Squamata: Gekkonidae), a new cave-dwelling gecko from southern Thailand. *Zootaxa* 772:1–11.

Peters, J.A. 1964. *Dictionary of Herpetology*. Hafner Publishing Company, New York. 392 pp.

Peters, W.C.H. 1871. Über neue Reptilien aus Ostafrica und Sarawak (Borneo), vorzüglich aus der Sammlung des Hern. Marquis J. Doria zu Genua. *Monatsberichte der Königlich Preussischen Akademie der Wissenschaften zu Berlin* 1871:566–81.

Pianka, E.R., D.R. King & R.A. King. 2004. *Varanoid Lizards of the World*. Indiana University Press, Bloomington, Indiana. xiii + 588 pp.

Pope, C.H. 1935. *The Reptiles of China. Natural History of Central Asia.* Vol. 10. American Museum of Natural History, New York. li + 604 pp; 27 pl.

Praschag, P., A.K. Hundsdörfer, A.H.M.A. Reza & U. Fritz. 2007. Genetic evidence for wild-living *Aspideretes nigricans* and a molecular phylogeny of South Asian softshell turtles (Reptilia: Trionychidae: *Aspideretes, Nilssonia*). *Zoologica Scripta* 2007:1–10.

Praschag, P., R.S. Sommer, C. McCarthy, R. Gemel & U. Fritz. 2008. Naming one of the world's rarest chelonians, the southern *Batagur*. *Zootaxa* 1758: 61–8.

Rasmussen, A.R. 1989. An analysis of *Hydrophis ornatus* (Gray), *H. lamberti* (Smith), and *H. inornatus* (Gray) (Hydrophiidae, Serpentes) based on samples from various localities, with remarks on feeding and breeding biology of *H. ornatus*. *Amphibia-Reptilia* 10:397–417.

Rasmussen, A.R. 1992. Rediscovery and redescription of *Hydrophis bituberculatus* Peters, 1872 (Serpentes: Hydrophiidae). *Herpetologica* 48(1):85–97.

Rasmussen, A.R. 1994. A cladistic analysis of *Hydrophis* subgenus *Chitulia* (McDowell, 1972) (Serpentes, Hydrophiidae). *Zoological Journal of the Linnaean Society* 111:161–78.

Rasmussen, A.R. 1996. Systematics of sea snakes: a critical review. In: *Venomous Snakes: Ecology, Evolution and Snakebite*. pp:15–30. R.S. Thorpe, W. Wüster and A. Malhotra (eds). The Zoological Society of London/Clarendon Press, Oxford.

Rösler, H. & F. Glaw. 2008. A new species of *Cyrtodactylus* Gray, 1827 (Squamata: Gekkonidae) from Malaysia including a literature survey of mensural and meristic data in the genus. *Zootaxa* 1729:8–22.

Ross, C.A. 1990. *Crocodylus raninus* S. Müller and Schlegel, a valid species of crocodile (Reptilia: Crocodylia) from Borneo. *Proceedings of the Biological Society of Washington* 103:955–61.

Rossman, D.A. & W.G. Eberle. 1977. Partition of the genus *Natrix*, with preliminary observations on evolutionary trends in natricine snakes. *Herpetologica* 33(1):34–43.

Roux, J. 1911. Elbert-Sunda-Expedition des Frankfurter Vereins für Geographie und Statistik. Reptilien und Amphibien. *Zoologische Jahrbücher (Abteilung für Systematik, Ökologie und Geographie der Tiere)* 30(5):495–508.

Roux, J. 1914. Note sur une espèce nouvelle d'*Oligodon* provenant de Sumatra. *Revue Suisse de Zoologie* 22(2):27–9.

Roux, J. 1925. Note sur une collection de reptiles et d'amphibiens de l'Ile Nias. *Revue Suisse de Zoologie* 32:319–21.

Rummler, H.-J. & U. Fritz. 1991. Geographische Variabilität der Amboina-Scharnierschildkröte *Cuora amboinensis* (Daudin, 1802), mit Beschreibung einer neuen Unterart, *C. a. kamaroma* subsp. nov. *Salamandra* 27(1):17–45.

Schulz, K.-D. 1996. *A Monograph of the Colubrid Snakes of the Genus* Elaphe *Fitzinger*. Koeltz Scientific Books, Havlickuv Brod. iii + 439 pp.

Slowinski, J.B., J. Boundy & R. Lawson. 2001. The phylogenetic relationships of Asian coral snakes (Elapidae: *Calliophis* and *Maticora*) based on morphological and molecular characters. *Herpetologica* 57(2):233–45.

Smedley, N. 1928. Some reptiles and Amphibia from the Anamba Islands. *Journal of the Malayan Branch of the Royal Asiatic Society* 6(3):76–7.

Smedley, N. 1931. On some reptiles and a frog from the Natuna Islands. *Bulletin of the Raffles Museum* 5:46–54.

Smedley, N. 1931. Amphibians and reptiles from the South Natuna Islands. *Bulletin of the Raffles Museum* 6:102–4.

Smith, M.A. 1925. On a collection of reptiles and amphibians from Mt. Murud, Borneo. *Sarawak Museum Journal* 3:5–14.

Smith, M.A. 1925. Contribution to the herpetology of Borneo. *Sarawak Museum Journal* 3(8):15–34.

Smith, M.A. 1926. *Monograph of the Sea Snakes (Hydrophiidae)*. British Museum (Natural History), London. xvii + 130 + (1) pp + pls I–II.

Smith, M.A. 1926. Spolia Mentawia: Reptiles and amphibians. *Annals & Magazine of Natural History, Series 9*, 18:76–81.

Smith, M.A. 1930. The Reptilia and Amphibia of the Malay Peninsula from the Isthmus of Kra to Singapore including the adjacent islands. *Bulletin of the Raffles Museum* 3:(2) + xviii + 1–149.

Smith, M.A. 1931. *The Fauna of British India, Including Ceylon and Burma. Vol. I. Loricata, Testudines.* Taylor and Francis, London. xxviii + 185 pp + 2 pls.

Smith, M.A. 1931. The herpetology of Mt. Kinabalu, North Borneo, 13,455 ft. *Bulletin of the Raffles Museum* 5:3–32.

Smith, M.A. 1935. *The Fauna of British India, Including Ceylon and Burma. Reptilia and Amphibia. Vol. II. Sauria.* Taylor and Francis, London. xiii + 440 pp + 1 pl.

Smith, M.A. 1937. A review of the genus *Lygosoma* (Scincidae: Reptilia) and its allies. *Records of the Indian Museum* 39(3):213–34.

Smith, M.A. 1938. The nucho-dorsal glands of snakes. *Proceedings of the Zoological Society of London Ser. B* 107:575–83.

Smith, M.A. 1943. *The Fauna of British India, Ceylon and Burma, Including the Whole of the Indo-Chinese Region. Vol. III. Serpentes.* Taylor and Francis, London. xii + 583 pp. + 1 map.

Spinks, P.Q., H.B. Shaffer, J.B. Iverson & W.P. McCord. 2004. Phylogenetic hypotheses for the turtle family Geoemydidae. *Molecular Phylogenetics and Evolution* 32:164–82.

Stejneger, L.H. 1922. List of snakes collected in Bulungan, northeast Borneo by Cark Lumholtz, 1914. *Nyt Magazin for Naturvidensk* 60(2):77–84.

Stoliczka, F. 1870. Observations on some Indian and Malayan Amphibia and Reptilia. *Proceedings of the Asiatic Society of Bengal 1870*(4):103–8.

Stoliczka, F. 1870. Observations on some Indian and Malayan Amphibia and Reptilia. *Journal of the Asiatic Society of Bengal* 39(2):134–57 + captions, pl. IX.

Stoliczka, F. 1870. Observations on some Indian and Malayan Amphibia and Reptilia. *Journal of the Asiatic Society of Bengal* 39(3):159–228, pls X–XII.

Stoliczka, F. 1871. Notes on some Indian and Burmese ophidians. *Journal of the Asiatic Society of Bengal* 40(2):421–45.

Stuart, B.L. & S.G. Platt. 2004. Recent records of turtles and tortoises from Laos, Cambodia, and Vietnam. *Asiatic Herpetological Research* 10:129–50.

Stuebing, R.B. 1991. A checklist of the snakes of Borneo. *Raffles Bulletin of Zoology* 39:323–62.

Stuebing, R.B. 1994. A checklist of the snakes of Borneo: addenda and corrigenda. *Raffles Bulletin of Zoology* 42(4):931–6.

Stuebing, R.B. 1994. A new species of *Cylindrophis* (Serpentes: Cylindrophiidae) from Sarawak, western Borneo. *Raffles Bulletin of Zoology* 42(4):967–73.

Stuebing, R.B. & R.F. Inger. 1999. *A Field Guide to the Snakes of Borneo*. Natural History Publications (Borneo) Sdn Bhd, Kota Kinabalu. viii + 254 pp.

Taylor, E.H. 1962. New Oriental reptiles. *University of Kansas Science Bulletin* 43(7):209–63.

Taylor, E.H. 1963. The lizards of Thailand. *University of Kansas Science Bulletin* 44(14):687–1077.

Taylor, E.H. 1965. The serpents of Thailand and adjacent waters. *University of Kansas Science Bulletin* 45(9):609–1096.

Taylor, E.H. & R.E. Elbel. 1958. Contribution to the herpetology of Thailand. *University of Kansas Science Bulletin* 38(13):1033–189.

Theobald, W. 1868. Catalogue of the reptiles of British Birma, embracing the Province of Pegu, Martaban, and Tenasserim; with descriptions of new or little-known species. *Journal of the Linnean Society* 10(41):4–67.

Theobald, W. 1876. *Descriptive Catalogue of the Reptiles of British India*. Thacker, Spink and Co., Calcutta. ix + 238 pp + xxxviii + xiii + 6 pls.

Traeholt, C. 1994. The food and feeding behaviour of the water monitor, *Varanus salvator*, in Malaysia. *Malayan Nature Journal* 47:331–43.

Tweedie, M.W.F. 1949. Reptiles of the Kelabit Plateau. *Sarawak Museum Journal* 5:154–5.

Tweedie, M.W.F. 1983. *The Snakes of Malaya*. Third edition. Singapore National Printers, Singapore. 167 pp.

Underwood, G. & M.S.Y. Lee. 2000. The egg teeth of *Dibamus* and their bearing on possible relationships with gekkotan lizards. *Amphibia-Reptilia* 21(4):507–11.

van der Meer Mohr, J.C. 1927. Reptiles from Pulau Berhala. *Miscellanea Zoologica Sumatrana* (16):1–2.

van der Meer Mohr, J.C. 1930. Notes on the fauna of Pulau Berhala. *Treubia* 12(3–4):277–93.

van Hoesel, J.K.P. 1959. *Ophidia Javanica*. Museum Zoologicum Bogoriensis, Bogor. 188 pp.

van Lidth de Jeude, T.W. 1890. On a collection of reptiles from Deli. *Notes from the Leyden Museum* 12(1):17–27.

van Lidth de Jeude, T.W. 1890. On a collection of reptiles from Nias, and on *Calamaria virgulata*, Boie. *Notes from the Leyden Museum* 12(4):253–6.

van Lidth de Jeude, T.W. 1893. On reptiles from North Borneo. *Notes from the Leyden Museum* 15:250–7.

van Lidth de Jeude, T.W. 1895. Reptiles from Timor and the neighbouring islands. *Notes from the Leyden Museum* 16:119–27.

van Lidth de Jeude, T.W. 1904. Reptilia from the Malay Archipelago. In: *Zoologische Ergebnisse einer Reise in Niederländisch ost-Indien* 2:178–92. M. Weber (ed.). E.J. Brill, Leiden.

van Lidth de Jeude, T.W. 1922. Snakes from Sumatra. *Zoologisches Mededilingen* 6:239–53.

Van Tri, N. 2008. Two new cave-dwelling species of *Cyrtodactylus* Gray (Squamata: Gekkonidae) from southwestern Vietnam. *Zootaxa* 1909:37–51.

Van Tri, N., L.L. Grismer & J.L. Grismer. 2008. A new endemic cave dwelling species of *Cyrtodactylus* Gray, 1827 (Squamata: Gekkonidae) in Kien Giang Biosphere Reserve, southwestern Vietnam. *Zootaxa* 1967:53–62.

Vetter, H. & P.P. van Dijk. 2006. *Turtles of the World*. Vol. 4. Asia/Schildkröten der Welt. Band 4. Asia. Edition Chimaira, Frankfurt am Main. 160 pp.

Vogel, G. 1995. *Dendrelaphis striatus* (Cohn), neu für die Fauna Borneos (Serpentes, Colubridae). *Mitteilungen aus dem Zoologischen Museum in Berlin* 71:147–9.

Vogel, G. 2006. *Venomous Snakes of Asia/Giftschlangen Asiens*. Terralog. Edition Chimaira, Frankfurt am Main. 148 pp.

Vogel, G., P. David & O.S.G. Pauwels. 2004. A review of morphological variation in *Trimeresurus popeiorum* (Serpentes: Viperidae: Crotalidae), with the description of two new species. *Zootaxa* 727:1–63.

Vogel, G., P. David, M. Lutz, J. van Rooijen & N. Vidal. 2007. Revision of the *Tropidolaemus wagleri*-complex (Serpentes: Viperidae: Crotalinae). I. Definition of included taxa and redescription of *Tropidolaemus wagleri* (Boie, 1827). *Zootaxa* 1644: 1–40.

Voris, H.K. & D.R. Karns. 1996. Habitat utilization, movements, and activity patterns of *Enhydris plumbea* (Serpentes: Homalopsinae) in a rice paddy wetland in Borneo. *Herpetological Natural History* 4(2):111–26.

Wall, F. 1907. Some new Asian snakes. *Journal of the Bombay Natural History Society* 17(3):612–18; 2 pls.

Wall, F. 1909. A monograph of the sea-snakes (Hydrophiinae). *Memoirs of the Asiatic Society of Bengal* 2(8):169–251.

Wallach, V. 1988. Status and redescription of the genus *Pandangia* Werner, with comparative visceral data on *Collorhabdium* Smedley and other genera (Serpentes: Colubridae). *Amphibia-Reptilia* 9:61–76.

Warrell, D.A. 1999. Guidelines for the clinical management of snakebite in the south-east Asia region. *Asian Journal of Tropical Medicine and Public Health* 30 (Supplement 1):i–vii + 1–67.

Whitaker, R. & A.S. Captain. 2004. *Snakes of India. The Field Guide*. Draco Books, Chennai. xiv + 481 pp.

Whitmore, T.C. 1984. *Tropical rain forests of the Far East* (with a chapter on soils by C.P. Burnham). Clarendon Press, Oxford. xiii + 282 pp.

Williams, K.L. &. V. Wallach. 1989. *Snakes of the World. Volume 1. Synopsis of snake generic names*. Krieger Publishing Company, Malabar, Florida. viii + 234 pp.

Win Maung & W. Ko Ko. 2002. *Turtles and Tortoises of Myanmar*. Wildlife Conservation Society – Myanmar Program, Yangon. 94 pp.

Wüster, W. 1998a. The cobras of the genus *Naja* in India. *Hamadryad* 23(1):15–32.

Wüster, W. 1998b. The genus *Daboia* (Serpentes: Viperidae): Russell's viper. *Hamadryad* 23(1): 33–40.

Wüster, W., S. Otsuka, A. Malhotra and R.S. Thorpe. 1992. Population systematics of Russell's viper: a multivariate study. *Biological Journal of the Linnaean Society* 47:97–113.

Wüster, W. & R.S. Thorpe. 1989. Population affinities of the Asiatic cobra (*Naja naja*) species complex in southeast Asia: reliability and random resampling. *Biological Journal of the Linnaean Society* 36:391– 409.

Wüster, W., R.S. Thorpe, M.J. Cox, P. Jintakune & J. Nabhitabhata. 1995. Population systematics of the snake genus *Naja* (Reptilia: Serpentes: Elapidae) in Indochina: multivariate morphological and comparative mitochondrial DNA sequencing (cytochrome oxidase I). *Journal of Evolutionary Biology* 8:493–510.

Yaakob, N.S. & J. Abdul. 1998. Prey of reptiles. *Journal of Wildlife & Parks, Kuala Lumpur* 16:51–6.

Zhao, E. & K. Adler. 1993. *Herpetology of China*. Society for the Study of Amphibians and Reptiles, Contributions to Herpetology, No. 10, Oxford, Ohio. 522 pp + 48 pls + 1 folding map.

Ziegler, T. 2002. *Die Amphibien und Reptilien eines Tieflandfeuchtwald-Schutzgebietes in Vietnam*. Natur & Tier – Verlag, Münster. 342 pp.

Ziegler, T., P. David, A. Miralles, D. Van Kien & N.Q. Truong. 2008. A new species of the snake genus *Fimbrios* from Phong Nha-Ke Bang National Park, Truong Son, central Vietnam (Squamata: Xenodermatidae). *Zootaxa* 1729:37–48.

Ziegler, T., R. Hendrix, N.T. Vu, M. Vogt, B. Forster & N.K. Dang. 2007. The diversity of a snake community in a karst forest ecosystem in the central Truong Son, Vietnam, with an identification key. *Zootaxa* 1493:1–40.

Ziegler, T., H.X. Quang & W. Böhme. 1998. Beitrag zur Kenntnis der Schnelläufer-Eidechsen Vietnams (Reptilia: Lacertidae: Takydromus). *Herpetofauna* 20(114):24–34.Ziegler, T., V.N. Thanh & B.N. Thanh. 2005. A new water skink of the genus *Tropidophorus* from the Phong Nha-Ke Bang National Park, central Vietnam (Squamata: Sauria: Scincidae). *Sala-mandra* 41(3):137–46.

Zug, G.R., L. Vitt & J. Caldwell. 2001. *Herpetology: an Introductory Biology of Amphibians and Reptiles*. Second edition. Academic Press, London and New York. xiv + 630 pp.

INTERNET RESOURCES

CHECKLISTS & SPECIES DATA

Aquatic Snakes of Southeast Asia
http://www.fieldmuseum.org/aquaticsnakes/
Website dedicated to the Homalopsidae, Acrochordidae and marine members of the Elapidae.

Crocodiles and Turtles of Borneo
http://www.arbec.com.my/crocodilesturtles/
Website dedicated to the freshwater, terrestrial and marine turtles and crocodiles of Borneo.

Crocodilian, Tuatara, and Turtle Species of the World. An Online Taxonomic and Geographic Reference
http://www.flmnh.ufl.edu/natsci/herpetology/turtcroclist/
An online taxonomic checklist of crocodiles, tuatara and turtles of the world.

Herpetology of Indonesia
http://www.geocities.com/rainforest/6785/
Listing of Indonesian herpetological species and data on their status and use.

Lizards of Borneo
http://www.arbec.com.my/lizards/
Website dedicated to the lizards of Borneo.

The TIGR Reptile Database
http://www.reptile-database.org/
A periodically updated database of reptile species of the world.

Toxicity in Colubrid Snakes of the World. An Annotated Bibliography
http://bio.bd.psu.edu/dmm/snake/snake.htm
Extensive annotated bibliography of envenomation by proteroglyphous snakes.

Venomous Snake Systematics Alert
http://biology.bangor.ac.uk/~bss166/update.htm
The stated aim is '…to provide regular updates on developments in the systematics of venomous snakes for the benefit of toxinological and medical researchers, as well as herpetoculturists and other interested individuals'.

BIBLIOGRAPHIC RESOURCES

Bibliomania!
http://www.herplit.com/
Includes listing of the world's herpetological literature (ca 50,000 citations, from 1586), current contents of herpetological journals, extensive inventory of papers and books on sale, cross-links to other herpetological sites, and so on.

Biodiversity Heritage Library
http://www.biodiversitylibrary.org/
Electronic library of the world's early biodiversity literature, including many classical works.

The Center for North American Herpetology
http://www.cnah.org/index.asp
Despite its name, deals with herpetological matter from all over the world, including a growing collection of recent herpetological literature.

Directory of Open-access Journals
http://www.doaj.org/
A directory of open-access scientific journals, including several zoological ones that cover herpetology.

Herpfaun
http://www.wku.edu/%7esmithch/mamm/herpfaun.ht#D
A bibliography of pre-1993 publications on the geographical distribution of reptiles (and amphibians) of the world.

SOCIETIES & SPECIALIST GROUPS

American Society of Ichthyologists and Herpetologists
http://www.asih.org/
Society home page, with information on journal (*Copeia*).

The Asian Turtle Consortium
http://www.asianturtle.org/htm/
A network dedicated to the in situ and ex situ conservation of Asian freshwater turtles and tortoises.

The European Snake Society
http://www.snakesociety.nl/Y0/y0-index.htm
Society home page, with information on journal (*Literatura Serpentium*).

The Herpetologists League
http://www.asih.org/
Society home page, with information on journal (*Herpetologica*).

The IUCN SSC Crocodile Specialist Group
http://www.iucncsg.org/ph1/modules/Home/
News from the Crocodile Specialist Group of the IUCN, including access to its online journal (*Crocodilia*).

Madras Crocodile Bank Trust
http://www.madrascrocodilebank.org/
Institutional home page, with information on journal (*Hamadryad*).

Marine Turtle Newsletter
http://www.seaturtle.org/mtn/
Information on the activities on marine turtle conservation and online version of the *Marine Turtle Newsletter* of the IUCN.

Society for the Study of Amphibians and Reptiles
http://www.ssarherps.org/)
Society home page, with information on journal (*Journal of Herpetology*) and newsletter (*Herpetological Review*).

Conservation

Convention on International Trade in Endangered Species of Wild Fauna and Flora (CITES)
http://www.cites.org/
Official site of CITES, and includes listing of species protected from international trade.

The Extinction Website
http://www.petermaas.nl/extinct/
A website with details on recently extinct animals and plants, animals and plants extinct in the wild, and rediscovered animals and plants.

The IUCN Red List of Threatened Species. 2009.2
http://www.iucnredlist.org/
Listing of threat categories and of globally threatened species of reptile (and other animals and plants).

The Ramsar Convention on Wetlands
http://www.ramsar.org/
Resources on wetlands, including listing of the world's Ramsar sites, publications, etc.

Turtle Survival Alliance
http://turtlesurvival.org/
An international organization dedicated to freshwater turtle and tortoise conservation.

UNESCO World Heritage Centre
http://whc.unesco.org/
Official site of UNESCO, and includes listing of the world's heritage sites.

World Database on Protected Areas
http://www.wdpa.org/
A database on protected areas of the Earth, including information and maps (via Google Earth) of sites of conservation concern.

Products and Consumables

(Note: references to individual companies or products listed here do not indicate endorsement of, or affiliation with, these companies or products.)

Anti-leech Socks
http://www.mosquitohammock.com/AntiLeechSocks.html
Socks specifically for use in leech-inhabiting tropical and subtropical forests.

The Compleat Naturalist Store
http://www.compleatnaturalist.com/mall/field_guide_covers.htm
E-store for the field naturalist, offering field guide covers, hand lenses, microscopes, telescopes, etc.

Forestry Suppliers, Inc.
https://www.forestry-suppliers.com/
Field supplies, including meters, scales.

Hunter Boots
http://www.wellie-boots.com/
Rubber Wellington boots for field work in swamps.

Led Lenser
http://www.ledlenserusa.com/
LED tactical use lights.

Lowepro
http://www.lowepro.com/
Backpacks, shoulder bags and camera bags.

Midwest Tongs
http://tongs.com/
Snake sticks, snake tongs, snake probe sets, bags and restraining tubes.

Petzl
http://en.petzl.com/petzl/Accueil
Head lamps and accessories.

Onset Computer Systems
http://www.onsetcomp.com/
Data loggers, water-level loggers and weather stations

Pelican
http://www.pelican.com/
Waterproof and humidity-tolerant cases, and heavy-duty xenon or LED flashlights for field use.

Rite in the Rain
http://www.riteintherain.com/
All-weather writing paper, journals, sketch books and spiral notebooks.

Rotring
http://www.rotring.com/
Ink pens and allied products for scientific sketches. Ink is fluid-proof and suitable for label-writing.

INDEX

Numbers in **bold** refer to plate numbers. Numbers in roman are page numbers.

Acalyptophis peronii **62**, 318
Acanthosaura armata **10**, 178
Acanthosaura capra **10**, 178
Acanthosaura coronata **10**, 178
Acanthosaura crucigera **10**, 179
Acanthosaura lepidogaster **10**, 179
Acanthosaura nataliae **10**, 179
Achalinus ater **74**, 351
Achalinus rufescens **74**, 351
Achalinus spinalis **74**, 351
Acrochordus granulatus **38**, 256
Acrochordus javanicus **38**, 256–7
Aeluroscalabotes felinus **20**, 198
Ahaetulla fasciolata **39**, 259
Ahaetulla fronticincta **39**, 259–60
Ahaetulla mycterizans **39**, 260
Ahaetulla nasuta **39**, 260
Ahaetulla prasina **39**, 260
  A. p. prasina 260
  A. p. suluensis 260
Aipysurus eydouxii **62**, 318
Amphiesma andreae **67**, 330
Amphiesma atemporale **67**, 330
Amphiesma bitaeniatum **67**, 330–1
Amphiesma boulengeri **67**, 331
Amphiesma craspedogaster **67**, 331
Amphiesma deschauenseei 331
Amphiesma flavifrons **67**, 331
Amphiesma frenatum 331–2
Amphiesma groundwateri **67**, 332
Amphiesma inas **67**, 332
Amphiesma kerinciense 332
Amphiesma khasiense **67**, 332
Amphiesma leucomystax **67**, 332–3
Amphiesma modestum **67**, 333
Amphiesma optatum **67**, 333
Amphiesma parallelum **67**, 333
Amphiesma petersii **68**, 333
Amphiesma popei 333–4
Amphiesma sanguineum **68**, 334
Amphiesma saravacense **68**, 334
Amphiesma sauteri **68**, 334
Amphiesma sieboldii **68**, 334
Amphiesma stolatum **68**, 334
Amphiesma taronense 335
Amphiesma venningi **68**, 335
Amphiesma viperinum 335
Amphiesma xenura **68**, 335
Amphiesmoides ornaticeps 335
Anyda cartilaginea **8**, 176
Angle-headed Lizard
  Abbott's **15**, 186
  Beyschlag's **15**, 186
  Blue-eyed **15**, 187–8
  Blue-necked **15**, 186
  Bornean **15**, 186
  Giant **15**, 187
  Hervey's 187
  Kloss's **15**, 187
  Kuhl's **15**, 187
  Large-scaled **15**, 188
  Marquis Doria's **15**, 186–7
  Mjöberg's **16**, 188
  Robinson's **16**, 188
  Sikulikap **15**, 187
  Tioman **15**, 186
Anomochilus leonardi **38**, 257
Anomochilus monticola **38**, 257
Anomochilus weberi **38**, 257
Anoplohydrus aemulans 335–6
Aphaniotis acutirostris **10**, 179
Aphaniotis fusca **10**, 179–80
Aphaniotis ornata **10**, 180
Apterygodon vittatum **29**, 228
Arakan Forest Turtle **4**, 170–1
Asthenodipsas boa **72**, 344
Asthenodipsas laevis **72**, 344
Asthenodipsas malaccanus **72**, 344–5
Asthenodipsas vertebralis **72**, 345
Astrotia stokesii **62**, 318
Ateuchosaurus chinensis **29**, 228–9
Azemiops feae **58**, 311–12

Batagur affinis **2**, 167–8
Batagur baska 168
Batagur borneoensis **2**, 168
Batagur trivittata **2**, 168
Bent-toed Gecko
  Angled **21**, 205
  Annandale's 205
  Ayeyarwady **21**, 205
  Baden **21**, 206
  Beautiful **23**, 213
  Black-eyed 212
  Buchard's 206–7
  Butterfly **23**, 213
  Cao Van Sung's **21**, 207
  Cardamom Mountains **22**, 210
  Chanhome's **22**, 207
  Chanquang **22**, 207
  Chua Chan Mountain 210
  Clouded **23**, 211
  Con Dao 207–8
  Dark-collared **22**, 208
  Dawei 208
  Eisenman's 208
  False Lined **23**, 213
  Feae's 208
  Four-striped **23**, 214
  Gans's **22**, 209
  Grismer's 209
  Grooved **23**, 213
  Gunung Lawit **22**, 208–9
  Gunung Panti 212
  Gunung Raya 211
  Hon Tre 209
  Inger's **22**, 210
  Jarujin's **22**, 211
  Khasi Hills **23**, 211
  Kinabalu **21**, 206
  Large-spotted **23**, 210
  Malayan **23**, 211
  Matsui's **23**, 212
  Mon State 205
  Niah **22**, 207
  Paradoxical 212–13
  Pegu **23**, 213
  Peninsular Malaysian **24**, 214
  Peters's **22**, 208
  Phong Nha Ke Bang **23**, 213
  Pulau Aur 205
  Pulau Besar 206
  Pulau Jarak 210–11
  Pulau Tioman **24**, 216
  Russell's **23**, 214
  Seribuat **24**, 214
  Seven-spotted **26**, 220
  Shan State **22**, 207
  Short-fingered **21**, 206
  Short-toed **21**, 206
  Slowinski's **24**, 214
  Stresemann's 214–15
  Sumatran 211
  Sumontha's **24**, 215
  Sworder's **24**, 215
  Ta Kou 215
  Tamarind **22**, 209
  Thirakhupt's **24**, 215
  Tiger **24**, 215
  Variegated **24**, 216
  Wakes's **24**, 216
  Yoshi's **24**, 216
  Ziegler's 216
Bitia hydroides **65**, 324–5
Black Turtle, Indian **5**, 172
Black-headed Snake
  Bornean **53**, 301–2
  Chinese **53**, 300–1
  Collared **53**, 301
  Striped **53**, 301
  Triangled **53**, 301
  Twin-streaked **53**, 300
  White-lipped **53**, 301
Blind Snake, Black **73**, 348–9
  Brahminy **73**, 348
  Burmese 349
  Diard's **73**, 349
  Flower's 349
  Gia Diah 349
  Javanese 349
  Jerdon's **73**, 349–50
  Klemmer's 350
  Koekkoek's 350
  Large 350
  Lined **73**, 348
  Lorenz's 348
  Müller's **73**, 350
  Oates's 350
  Olive 348
  Ozaki's 350
  Pitted Snout 349
  Roxane's 350–1
  Thai 351
  Trang 351
  White-headed **73**, 348
Blotch-necked Snake
  Blakeway's 296–7
  Common **51**, 297
  Delacour's 297
  Fujian **51**, 297
  Blue-tailed Skink **29**, 231
  Chinese **33**, 240–1
  Common **29**, 230–1
  Elegant 241
  Four-lined **33**, 241
  Tamdao **33**, 241
Blythia reticulata **39**, 260
Boiga bengkuluensis **39**, 260–1
Boiga bourreti **39**, 261
Boiga cyanea **39**, 261
Boiga cynodon **39**, 261
Boiga dendrophila 261–2
  B. d. annectens 261–2
  B. d. dendrophila 261–2
  B. d. divergens 262
  B. d. gemmicincta 262
  B. d. latifasciata 262
  B. d. melanota 261–2
  B. d. multicincta 262
  B. d. occidentalis 261–2
Boiga drapiezii **40**, 262
Boiga gokool **40**, 262
Boiga guangxiensis **40**, 262
Boiga jaspidea **40**, 262
Boiga kraepelini **40**, 262–3
Boiga multomaculata **40**, 263
Boiga nigriceps **40**, 263
  B. n. brevicauda 263
  B. n. nigriceps 263
Boiga ochracea **40**, 263
Boiga quincunciata 263–4
Boiga saengsomi **40**, 263–4
Boiga siamensis **40**, 264
Boiga walli 264
Box Turtle,
  Indo-Chinese **2**, 169
  Keeled **2**, 169
  Malayan **2**, 168
  Three-striped **2**, 169
  Vietnamese 168–9
Brachymeles apus **29**, 229
Bridled Snake
  Davison's **45**, 278
  Half-banded **45**, 278
  Three-banded **45**, 278
Broghammerus reticulatus **38**, 258
  B. r. jampeanu 258
  B. r. reticulatus 258
  B. r. saputrai 258
Bronchocela cristatella **11**, 180
Bronchocela hayeki **11**, 180
Bronchocela jubata **11**, 180
Bronchocela orlovi 180–1
Bronchocela smaragdina **11**, 181
Bronchocela vietnamensis **11**, 181
Bronzeback Tree Snake
  Beautiful **44**, 275
  Blue **44**, 275
  Gore's 275
  Haas's 275
  Indian **44**, 276–7
  Kopstein's **44**, 275–6
  Mountain **44**, 276
  Nganson **44**, 276
  Painted **44**, 276
  Striated **44**, 276
  Stripe-tailed **44**, 274–5
  Underwood's 277
Brown Snake
  Malayan **53**, 302
  Ornate **53**, 302
Bungarus bungaroides **59**, 313
Bungarus candidus **59**, 313
Bungarus fasciatus **59**, 313
Bungarus flaviceps **59**, 268, 313–14
  B. f. baluensis **59**, 313–14
  B. f. flaviceps **59**, 313–14
Bungarus magnimaculatus **59**, 314
Bungarus multicinctus **59**, 314
Bungarus slowinskii **59**, 314
Bungarus wanghaotingi 314
Burrowing Snake
  Annam 274
  Black **74**, 351
  Grey **74**, 351
  Rufous **74**, 351

369

Sumatran 279
Butterfly Lizard
   Bell's **28**, 226
   Böhme's **28**, 227
   Eyed **28**, 227
   Pegu 227
   Peters's 227
   Red-banded **28**, 228
   Reeves's **28**, 227–8
   Spotted **28**, 227
   Triploid **28**, 228

Calamaria abstrusa 264
Calamaria albiventer 264
Calamaria alidae 264–5
Calamaria battersbyi **41**, 265
Calamaria bicolor **41**, 265
Calamaria borneensis **41**, 265
Calamaria buchi 265
Calamaria crassa 265
Calamaria eiselti 266
Calamaria everetti **41**, 266
Calamaria forcarti 266
Calamaria gimletti 266
Calamaria grabowskyi **41**, 266
Calamaria gracillima 267
Calamaria griswoldi **41**, 267
Calamaria hilleniusi **41**, 267
Calamaria ingeri **41**, 267
Calamaria javanica 267
Calamaria lateralis 267–8
Calamaria leucogaster **42**, 268
Calamaria linnaei **42**, 268
Calamaria lovii 268
   C. l. ingermarxorum 268
   C. l. lovii 268
   C. l. wermuthi 268
Calamaria lumbricoidea **42**, 268–9
Calamaria lumholzi 269
Calamaria margaritophora 269
Calamaria mecheli 269
Calamaria melanota 269
Calamaria modesta 269
Calamaria pavimentata **42**, 270
Calamaria prakkei 270
Calamaria rebentischi 270
Calamaria schlegeli **42**, 270
Calamaria schmidti **42**, 270
Calamaria septentrionalis **42**, 270–1
Calamaria suluensis 271
Calamaria sumatrana 271
Calamaria thanhi 271
Calamaria ulmeri 271
Calamaria virgulata **42**, 271
Calamaria yunnanensis 272
Calliophis bivirgatus **60**, 268, 314–15
   C. b. bivirgatus **60**, 315
   C. b. flaviceps **60**, 315
   C. b. tetrataenia **60**, 315
Calliophis gracilis **60**, 315
Calliophis intestinalis **60**, 315
   C. i. intestinalis **60**, 315
   C. i. lineata **60**, 315
   C. i. philippina 315
   C. i. suluensis 315
   C.i. bilineata 315
   C.i. thepassi **60**, 315
Calliophis maculiceps **60**, 215
   C. m. hughi 315
   C. m. maculiceps 315
Calloselasma rhodostoma **54**, 303
Calotes chincollium **11**, 181

Calotes emma **11**, 181
   C. e. alticristatus 181
   C. e. emma 181
Calotes htunwini **12**, 181–2
Calotes irawadi **12**, 182
Calotes jerdoni **12**, 182
Calotes kingdonwardi 182
Calotes maria 182
Calotes mystaceus **12**, 182–3
Calotes versicolor **12**, 183
   C. v. farooqi 183
   C. v. versicolor 183
Camouflage Gecko
   Brooks's 223
   Brown's **27**, 223–4
   Crocker Range **27**, 224
   Yasuma's **27**, 224
Cantoria violacea **65**, 325
Caretta caretta **6**, 173
Cat Gecko **20**, 198
Cat Snake
   Assamese 263
   Banded Green **40**, 263–4
   Bengkulu **39**, 260–1
   Black-headed **40**, 263
   Bourret's **39**, 261
   Dog-toothed **39**, 261
   Eastern **40**, 262
   Green **39**, 261
   Guangxi **40**, 262
   Jasper **40**, 262
   Mangrove **40**, 261–2
   Many-spotted **40**, 263
   Square-headed **40**, 262–3
   Tawny **40**, 263
   Thai **40**, 264
   Wall's 264
   White-spotted **40**, 262
Cerberus rynchops **65**, 325
Chelonia mydas **6**, 173
Chitra chitra **8**, 176
   C. c. chitra **8**, 176
   C. c. javanensis 176
Chitra vandijki **8**, 176
Chrysopelea ornata **43**, 272
Chrysopelea paradisi **43**, 272
Chrysopelea pelias **43**, 272–3
Cnemaspis affinis **20**, 200
Cnemaspis argus 200
Cnemaspis aurantiacopes 200
Cnemaspis baueri **20**, 200
Cnemaspis bayuensis 200
Cnemaspis biocellata 200–1
Cnemaspis boulengerii 201
Cnemaspis caudanivea 201
Cnemaspis chanthaburiensis **20**, 201
Cnemaspis dezwaani 201
Cnemaspis dringi 201
Cnemaspis flavigaster 201–2
Cnemaspis flavolineatus **20**, 202
Cnemaspis jacobsoni 202
Cnemaspis karsticola 202
Cnemaspis kendallii **20**, 202
Cnemaspis kumpoli **20**, 202
Cnemaspis limi **20**, 202–3
Cnemaspis mcguirei 203
Cnemaspis modiglianii 203
Cnemaspis nigridia **21**, 203
Cnemaspis nuicamensis 203
Cnemaspis pemanggilensis **21**, 203–4
Cnemaspis perhentiensis 204
Cnemaspis phuketensis **21**, 204
Cnemaspis siamensis 204

Cnemaspis tucdupensis 204
Cnemaspis whittenorum 204–5
Cobra
   Chinese **61**, 316
   Equitorial **61**, 316–17
   Indo-Chinese Spitting **61**, 316
   King **61**, 317
   Mandalay **61**, 316
   Monocled **61**, 316
   Sumatran **61**, 317
Coelognathus enganensis **43**, 272–3
Coelognathus erythrurus **43**, 273
   C. e. celebensis 273
   C. e. philippina 273
   C. e. psephenoura 273
Coelognathus flavolineatus **43**, 273
Coelognathus radiatus **43**, 273
Coelognathus subradiatus **43**, 273
Collorhabdium williamsoni 273–4
Complicitus nigrigularis **12**, 183
Coral Snake
   Blue **60**, 314–15
   Kellog's **61**, 317
   MacClelland's **61**, 317–18
   Malayan Striped **60**, 315
   Speckled **60**, 315
   Spotted **60**, 315
   Striped 268
Crested Dragon, Kinabalu **16**, 189
Crested Lizard, Forest **11**, 181
Ludeking's **17**, 190–1
Crocodile, Bornean Swamp 166
   Marsh **1**, 166
   Saltwater **1**, 166
   Siamese **1**, 166
Crocodile Lizard, Chinese **37**, 255
Crocodylus palustris **1**, 166
Crocodylus porosus **1**, 166
Crocodylus raninus 166
Crocodylus siamensis **1**, 166
Cryptelytrops albolabris **54**, 303
Cryptelytrops erythrurus **54**, 303
Cryptelytrops honsonensis 303
Cryptelytrops insularis **54**, 303–4
Cryptelytrops kanburiensis **54**, 304
Cryptelytrops macrops **54**, 304
Cryptelytrops purpureomaculatus **54**, 304
Cryptelytrops venustus **54**, 304
Cryptoblepharus balinensis **29**, 229
Cryptoblepharus boutonii 229
   C. b. ahli 229
   C. b. ater 229
   C. b. boutonii 229
   C. b. cognatus 229
   C. b. degrijsi 229
   C. b. mayottensis 229
   C. b. nigropunctatus 229
   C. b. quinquetaeniatus 229
   C. b. voeltzkowi 229
Cryptoblepharus cursor **29**, 229
Cryptoblepharus renschi **29**, 229
Cryptophidion annamense 274
Cuora amboinensis **2**, 168
   C. a. amboinensis 168
   C. a. cuoro 168
   C. a. kamaroma 168
   C. a. lineata 168

Cuora cyclornata 168–9
   C. c. cyclornata 169
   C. c. meieri 169
Cuora galbinifrons **2**, 169
Cuora mouhotii **2**, 169
Cuora trifasciata **2**, 169
Cyclemys atripons **3**, 169
Cyclemys dentata **3**, 169
Cyclemys enigmatica 169–70
Cyclemys fusca **3**, 170
Cyclemys oldhamii **3**, 170
Cyclemys pulchristriata **3**, 170
Cyclophiops doriae 274
Cyclophiops hamptoni 274
Cyclophiops major **43**, 274
Cyclophiops multicinctus **43**, 274
Cylindrophis engkariensis **59**, 312
Cylindrophis lineatus **59**, 312
Cylindrophis ruffus **59**, 312
Cyrtodactylus aequalis 205
Cyrtodactylus angularis **21**, 205
Cyrtodactylus annandalei 205
Cyrtodactylus aurensis 205
Cyrtodactylus ayeyarwadyensis **21**, 205
Cyrtodactylus badenensis **21**, 206
Cyrtodactylus baluensis **21**, 206
Cyrtodactylus batucolus 206
Cyrtodactylus brevidactylus **21**, 206
Cyrtodactylus brevipalmatus **21**, 206
Cyrtodactylus buchardi 206–7
Cyrtodactylus caovansungi **21**, 207
Cyrtodactylus cavernicolus **22**, 207
Cyrtodactylus chanhomeae **22**, 207
Cyrtodactylus chauquangensis **22**, 207
Cyrtodactylus chrysopylos **22**, 207
Cyrtodactylus condorensis 207–8
Cyrtodactylus consobrinoides **22**, 208
Cyrtodactylus consobrinus **22**, 208
Cyrtodactylus cryptus **22**, 208
Cyrtodactylus eisenmani 208
Cyrtodactylus elok **22**, 208–9
Cyrtodactylus feae 209
Cyrtodactylus fumosus **22**, 209
Cyrtodactylus gansi **22**, 209
Cyrtodactylus grismeri 209
Cyrtodactylus hontreensis 209
Cyrtodactylus huynhi **22**, 210
Cyrtodactylus ingeri **22**, 210
Cyrtodactylus interdigitalis **22**, 210
Cyrtodactylus intermedius **22**, 210
Cyrtodactylus irregularis 210
Cyrtodactylus jarakensis 210–11
Cyrtodactylus jarujini **22**, 211
Cyrtodactylus khasiensis **23**, 211
Cyrtodactylus lateralis 211
Cyrtodactylus macrotuberculatus 211
Cyrtodactylus malayanus **23**, 211
Cyrtodactylus marmoratus **23**, 211
Cyrtodactylus matsuii **23**, 212
Cyrtodactylus nigriocularis **23**, 212
Cyrtodactylus oldhami **23**, 212
Cyrtodactylus pantiensis 212
Cyrtodactylus paradoxus 212–13
Cyrtodactylus peguensis **23**, 213

C. p. peguensis **23**, 213
C. p. zebraicus **23**, 213
Cyrtodactylus phongnhakebangensis **23**, 213
Cyrtodactylus pseudoquadrivirgatus **23**, 213
Cyrtodactylus pubisulcus **23**, 213
Cyrtodactylus pulchellus **23**, 213
Cyrtodactylus quadrivirgatus **23**, 214
Cyrtodactylus russelli **23**, 214
Cyrtodactylus semenanjungensis **24**, 214
Cyrtodactylus seribuatensis **24**, 214
Cyrtodactylus slowinskii **24**, 214
Cyrtodactylus stresemanni 214–15
Cyrtodactylus sumonthai **24**, 215
Cyrtodactylus sworderi **24**, 215
Cyrtodactylus takouensis 215
Cyrtodactylus thirakhupti **24**, 215
Cyrtodactylus tigroides **24**, 215–16
Cyrtodactylus tiomanensis **24**, 216
Cyrtodactylus variegatus **24**, 216
Cyrtodactylus wakeorum **24**, 216
Cyrtodactylus yoshii **24**, 216
Cyrtodactylus ziegleri 216

Daboia russelii **58**, 302, 312
   D. russelii 302
   D. r. russelii 312
   D. siamensis 302
   D. r. siamensis 312
Dasia grisea **29**, 230
Dasia olivacea **29**, 230
Dasia semicincta **29**, 230
Davewakeum miriamae 230
Day Gecko
   Bauer's **20**, 200
   Boulenger's **20**, 201
   Chanthaburi **20**, 201
   De Zwaan's 201
   Dring's 201
   Gading **21**, 203
   Gua Bayu 200
   Gunung Lawit 200
   Hon Dat Hill 200
   Hon Tre Island 201
   Jacobson's 202
   Karst-dwelling 202
   Kendall's **20**, 202
   Kumpol's **20**, 202
   Lim's **20**, 202–3
   McGuire's 203
   Modigliani's 203
   Nui Cam Hill 203
   Pemanggil **21**, 203–4
   Penang **20**, 200
   Phuket **21**, 204
   Pulau Perhantian 204
   Thai **21**, 204
   Titi Wangsa **20**, 202
   Two-eyed 200–1
   Whittens's 204–5
   Yellow-bellied 201–2
Deinagkistrodon acutus **54**, 304–5
Dendragama boulengeri **12**, 183
Dendrelaphis caudolineatus **44**, 274–5
   D. c. caudolineatus 274–5
   D. c. flavescens 275
   D. c. luzonensis 275
Dendrelaphis cyanochloris **44**, 275

Dendrelaphis formosus **44**, 275
Dendrelaphis gorei 275
Dendrelaphis haasi 275
Dendrelaphis kopsteini **44**, 275–6
Dendrelaphis ngansonensis **44**, 276
Dendrelaphis pictus **44**, 276
Dendrelaphis striatus **44**, 276
Dendrelaphis subocularis **44**, 276
Dendrelaphis tristis **44**, 276–7
Dendrelaphis underwoodi 277
Dermochelys coriacea **6**, 174
Dibamus alfredi 196
Dibamus booliati **19**, 196
Dibamus bourreti 196
Dibamus deharvengi 196
Dibamus dezwaani **19**, 196–7
Dibamus greeri **19**, 197
Dibamus ingeri **19**, 197
Dibamus kondaoensis 197
Dibamus montanus 197
Dibamus smithi 197
Dibamus somsaki 198
Dibamus tiomanensis **19**, 198
Dibamus vorisi **19**, 198
Dinodon flavozonatum **44**, 277
Dinodon meridionale **44**, 277
Dinodon rosozonatum 277
Dinodon rufozonatum **44**, 277–8
Dinodon septentrionale 278
Dixonius hangseesom **25**, 216–17
Dixonius melanostictus **25**, 217
Dixonius siamensis **25**, 217
Dixonius vietnamensis **25**, 217
Dogania subplana **8**, 176
Draco cornutus **13**, 183
Draco cristatellus **13**, 183–4
Draco fimbriatus **13**, 184
Draco formosus **13**, 184
Draco indochinensis **13**, 184
Draco maculatus **14**, 184
   D. m. haasei 184
   D. m. maculatus 184
   D. m. whiteheadi 184
Draco maximus **14**, 185
Draco melanopogon **14**, 185
Draco obscurus **14**, 185
Draco quinquefasciatus **14**, 185
Draco sumatranus **14**, 185
Draco taeniopterus **14**, 185
Draco volans **14**, 186
Dryocalamus davisonii **45**, 278
Dryocalamus subannulatus **45**, 278
Dryocalamus tristrigatus **45**, 278
Dryophiops rubescens **45**, 278–9
Dwarf Snake, Mountain 273–4

Earless Monitor, Bornean **28**, 226
Elaphe carinata **45**, 279
Elapoides fuscus **45**, 279
Emoia atrocostata **29**, 230
Emoia caeruleocauda **29**, 230–1
Emoia cyanura **29**, 231
Emoia laobaoensis 231
Enhydrina schistosa **62**, 318
Enhydris albomaculata 325
Enhydris alternans 325
Enhydris bennetti **65**, 325
Enhydris bocourti **65**, 325–6
Enhydris chanardi 326
Enhydris chinensis **65**, 326
Enhydris doriae **65**, 326
Enhydris enhydris **65**, 326

Enhydris gyii **65**, 326–7
Enhydris indica 327
Enhydris innominata **65**, 327
Enhydris jagorii **65**, 327
Enhydris longicauda 327
Enhydris maculosa 327
Enhydris pahangensis 327–8
Enhydris plumbea **66**, 328
Enhydris punctata **66**, 328
Enhydris sieboldii 328
Enhydris smithi 328
Enhydris subtaeniata **66**, 328–9
Enhydris vorisi 329
Eretmochelys imbricata **6**, 173
Erpeton tentaculatus **66**, 329
Etheridgeum pulchrum 279
Eugongylus rufescens **29**, 231
Euprepiophis mandarinus **45**, 279
Eutropis chapaensis 231
Eutropis darevskii 231
Eutropis dissimilis **30**, 231–2
Eutropis doriae 232
Eutropis indeprensa 232
Eutropis longicaudata **30**, 232
Eutropis macrophthalma **30**, 232
Eutropis macularia **30**, 232
Eutropis multifasciata **30**, 233
   E. m. balinensis 233
   E. m. multifasciata 233
Eutropis novemcarinata 233
Eutropis quinquecarinata 233
Eutropis rudis **30**, 233
Eutropis rugifera **30**, 233–4
Eyed Turtle, Burmese **5**, 172

False Cobra
   Bamboo **72**, 346–7
   Baram 347
   Jacobson's 347
   Javanese 347
   Karl Schmidt's **72**, 347
   Large-eyed **72**, 347
Fan-throated Lizard
   Green **18**, 194
   Mount Victoria **18**, 194
Fimbrios klossi **74**, 351–2
Fimbrios smithi 352
Flapshell Turtle, Burmese **8**, 177
Flat-nosed Snake, Chapa 296
Flat-shelled Turtle, Malayan **5**, 172
Flying Lizard
   Beautiful **13**, 184
   Black-bearded **14**, 185
   Blanford's **13**, 183
   Common **14**, 185
   Crested **13**, 183–4
   Five-banded **14**, 185
   Fringed **13**, 184
   Horned **13**, 183
   Indo-Chinese **13**, 184
   Javanese **14**, 186
   Large **14**, 185
   Narrow-lined **14**, 185
   Obscure **14**, 185
   Orange-Winged **14**, 184
   Red-bearded **13**, 184
Flying Snake
   Garden **43**, 272
   Ornate 272
   Twin-barred **43**, 272
Fordonia leucobalia **66**, 329
Forest Gecko, Frilly **26**, 221
Forest Lizard
   Ayeyarwady **12**, 182
   Blue **12**, 182–3

   Chanthabun **11**, 181
   Hayek's **11**, 180
   Htunwin's **12**, 181–2
   Jerdon's **12**, 182
   Kinabalu **12**, 183
   Kingdon-Ward's 182
   Maned **11**, 180
   Maria's 182
   Orlov's 180–1
   Vietnamese **11**, 181
Forest Skink
   Alfred's 243
   Black-spotted 248
   Blue-throated **34**, 244–5
   Bogor 248
   Bukit Fraser 243–4
   Bukit Larut **35**, 248
   Buon Loi 243
   Butler's 244
   Büttikofer's 243
   Cameron Highlands 244
   Dwarf **35**, 247
   Earless **34**, 244
   Flores 245
   Grandison's 245
   Grasshopper-eating **34**, 245
   Gunung Kinabalu **34**, 246
   Gunung Murud **35**, 247–8
   Gunung Tahan 244
   Haas's 245
   Hallier's 245
   Helen's 246
   Indian **34**, 246
   Ishak's **34**, 246
   Lao Bao 231
   Malayan 247
   Many-scaled **35**, 247
   Modigliani's 247
   Montane 250
   Narrow-necked 250
   Oak **34**, 243
   Pale **36**, 250
   Pulau Langkawi 246
   Pulau Sibu 249
   Red-tailed 248
   Sabah **35**, 248–9
   Selangor **35**, 249
   Shelford's 249
   Spotted **35**, 247
   Spotted-lined 246
   Starred **36**, 249–50
   Sumatran 243
   Temminck's **36**, 250
   Thick 244
   Three-banded 250–1
   Three-toed 250
   Van Heurn's 251
   White-throated 247
   Yellow-lined 249
Forest Snake, Two-coloured 299–300
Four-Clawed Gecko
   Butler's 217
   Common **25**, 218
   Fehlmann's **25**, 217
   Kanchanaburi **25**, 218
   Slender-tailed 217
Four-eyed Turtle **5**, 172–3
   Beale's **5**, 172

Garden Lizard **12**, 183
Garthius chaseni **55**, 305
Gavialis gangeticus **1**, 167
Gecko
   Baden's **25**, 218

Chinese 218
Grossmann's **25**, 218–19
Nutaphandi's 219
Palmated **25**, 219
Sandstone **26**, 219–20
Siamese **26**, 220
Tokay **25**, 218
Ulikovski's **26**, 220
Gehyra angusticaudata 217
Gehyra butleri 217
Gehyra fehlmanni **25**, 217
Gehyra mutilata **25**, 218
Gekko badenii **25**, 218
Gekko chinensis 218
Gekko gecko **25**, 218
 G. g. azhari 218
 G. g. gecko 218
Gekko grossmanni **25**, 218–19
Gekko japonicus **25**, 219
Gekko monarchus **25**, 219
Gekko nutaphandi 219
Gekko palmatus **25**, 219
Gekko petricolus **26**, 219–20
Gekko scientiadventura **26**, 220
Gekko siamensis **26**, 220
Gekko smithii **26**, 220
Gekko ulikovskii **26**, 220
Geochelone elegans 175
Geoemyda spengleri **3**, 170
Gerarda prevostiana **66**, 329
Gharial
 Ganges **1**, 167
 Malayan False **1**, 167
Giant Blind Snake
 Kinabalu **38**, 257
 Malayan **38**, 257
 Sumatran **38**, 257
Giant Gecko
 Japanese **25**, 219
 Smith's **26**, 220
Giant Skink, Bampfylde's **32**, 238
Giant Turtle, Malayan **5**, 172
Glass Snake
 Bornean **19**, 195
 Hart's **19**, 195
 Indian **19**, 195
 Ludovic's 195
 Sokolov's **19**, 195
 Wegner's 196
Gongylosoma baliodeirum **46**, 280
 G. b. baliodeirum 280
 G. b. cinctus 280
 G. b. cochranae 280
Gongylosoma longicauda **46**, 280
Gongylosoma mukutense **46**, 280
Gongylosoma nicobariensis 280
Goniurosaurus araneus **20**, 198–9
Goniurosaurus catbaensis 199
Goniurosaurus huuliensis 199
Goniurosaurus lichtenfelderi **20**, 199
Goniurosaurus luii **20**, 199
Gonocephalus abbotti **15**, 186
Gonocephalus bellii **15**, 186
Gonocephalus beyschlagi **15**, 186
Gonocephalus bornensis **15**, 186
Gonocephalus chamaeleontinus **15**, 186
Gonocephalus doriae **15**, 186–7
Gonocephalus grandis **15**, 187
Gonocephalus herveyi 187
Gonocephalus klossi **15**, 187
Gonocephalus kuhlii **15**, 187

Gonocephalus lacunosus **15**, 187
Gonocephalus liogaster **15**, 187–8
Gonocephalus megalepis **15**, 188
Gonocephalus mjobergi **15**, 188
Gonocephalus robinsoni **15**, 188
Gonyophis margaritatus **46**, 280
Gonyosoma oxycephalum **46**, 280–1
Grass Lizard
 Han's **28**, 225
 Kühne's **28**, 225
 Long-tailed **28**, 225–6
 Wolter's **28**, 226
Green Lizard, Crested **11**, 180
Green Snake
 Doria's 274
 Greater **43**, 274
 Hampton's 274
 Many-banded **43**, 274
Ground Gecko
 Dark-sided **25**, 217
 Orange-tailed **25**, 216–17
 Spotted **25**, 217
 Vietnamese **25**, 217
Ground Skink
 Black-banded **30**, 233
 Black-spotted **33**, 242
 Chapa 231
 Darevski's 231
 Doria's 232
 Five-keeled 233
 Kohtao 242
 Large-eyed **30**, 232
 Little **30**, 232
 Long-tailed **30**, 232
 Marquis Doria's 241
 Mount Victoria **33**, 243
 Nine-keeled 233
 Philippine 232
 Red-throated **30**, 233–4
 Reeves's **33**, 242
 Rock-dwelling 242
 Spot-lined 242
 Striped **30**, 231–2
 Tawny 242
Ground Snake
 Dark-grey **45**, 279
 Indo-Chinese 280
 Pulau Tioman **45**, 280
 Striped **45**, 280

Harpesaurus beccarii **15**, 188
Harpesaurus borneensis **15**, 188
Harpesaurus ensicauda **15**, 188
Harpesaurus modiglianii **16**, 189
Harpesaurus tricinctus **16**, 189
Hemidactylus bowringii **26**, 220
Hemidactylus brookii **26**, 220–1
Hemidactylus craspedotus **26**, 221
Hemidactylus frenatus **26**, 221
Hemidactylus garnotii **26**, 221
Hemidactylus karenorum 221
Hemidactylus platyurus **26**, 221–2
Hemidactylus stejnegeri **26**, 222
Hemidactylus vandermeermohri 222
Hemidactylus vietnamensis 222
Hemiphyllodactylus chapaensis 222
Hemiphyllodactylus harterti **27**, 222
Hemiphyllodactylus typus 223
Hemiphyllodactylus yunnanensis 223

Heosemys annandalii **4**, 170
Heosemys depressa **4**, 170–1
Heosemys spinosa **4**, 171
Hill Turtle, Spiny **4**, 171
Homalopsis buccata **66**, 329–30
Homalopsis nigroventralis **66**, 330
Horned Lizard
 Bornean **16**, 188
 Modigliani's **16**, 189
 Sumatran **18**, 194–5
 Three-banded **16**, 189
 Werner's **18**, 188–9
Horned Snake, Vietnamese **53**, 300
House Gecko
 Asian **26**, 221
 Bowring's **26**, 220
 Brooke's **26**, 220–1
 Frilly **26**, 221–2
 Garnot's **26**, 221
 Stejneger's **26**, 222
 Vandeermeermohr's 222
Hydrablabes periops **68**, 336
Hydrablabes praefrontalis 336
Hydrophis aagaardi **62**, 318–19
Hydrophis atriceps **62**, 319
Hydrophis belcheri **62**, 319
Hydrophis bituberculatus 319
Hydrophis brookii **62**, 319
Hydrophis caerulescens **62**, 319
Hydrophis cantoris 319–20
Hydrophis cyanocinctus **62**, 320
Hydrophis fasciatus **63**, 320
Hydrophis gracilis **63**, 320
Hydrophis inornatus 320
Hydrophis klossi **63**, 320
Hydrophis lamberti 320–1
Hydrophis lapemoides **63**, 321
Hydrophis melanosoma **63**, 321
Hydrophis nigrocinctus 321
Hydrophis obscurus 321
Hydrophis ornatus **63**, 321–2
Hydrophis pachycercos **63**, 322
Hydrophis parviceps 322
Hydrophis sibauensis **64**, 322
Hydrophis spiralis **64**, 322
Hydrophis stricticollis **64**, 322
Hydrophis torquatus **64**, 322–3
Hypsicalotes kinabaluensis **16**, 189

Iguanognathus werneri 336
Indotestudo elongata **7**, 175
Isopachys anguinoides 234
Isopachys borealis 234
Isopachys gyldenstolpei **31**, 234
Isopachys roulei 234

Japalura chapaensis **16**, 189
Japalura fasciata 189
Japalura hamptoni 189
Japalura kaulbacki 189
Japalura planidorsa **16**, 190
Japalura sagittifera 190
Japalura swinhonis **16**, 190
Japalura yunnanensis **16**, 190
 J. y. pogei 190
 J. y. yunnanensis 190
Keelback
 Andrea's **67**, 330
 Angel's 339
 Annam **51**, 296
 Black-banded **70**, 341
 Black-barred 339

Blue-necked **69**, 336–7
Boonsong's 296
Boulenger's **67**, 331
Bridled 331–2
Buff-striped **68**, 334
Collared **70**, 341
Deschauensee's 331
Groundwater's **67**, 332
Günther's **67**, 333
Gunung Inas **67**, 332
Gunung Kerinchi 332
Gunung Murud **70**, 340–1
Himalayan **70**, 340
Japanese **70**, 341–2
Khasi Hills **67**, 332
Kuatun **67**, 331
Leonard's **70**, 340
Mount Emei **67**, 333
Mountain **67**, 330
Olive **69**, 336
Orange-lipped **69**, 336
Peter's **68**, 333
Pope's Mountain 333–4
Red Mountain **68**, 334
Red-bellied **70**, 340
Red-necked **70**, 341
Sahul **70**, 339–40
Sarawak **68**, 334
Sauter's **68**, 334
Siebold's **68**, 334
Speckle-bellied **70**, 340
Strange-tailed **68**, 335
Striped **67**, 333
Sumatran 335–6
Taron 335
Two-striped **67**, 330–1
Venning's **68**, 335
Viper-like 335
White-eyed 335
White-fronted **67**, 331
White-lipped **67**, 332–3
Keelback Water Snake
 Burmese 342
 Chequered **71**, 343
 Chinese **71**, 342
 Chinese Spotted **71**, 342
 Javan **71**, 343
 Malayan Spotted **71**, 343
 Red-sided **71**, 343–4
 Ringed **71**, 342
 Spotted 343
 St John's **71**, 343
 Striped **71**, 344
 Yellow-spotted **71**, 342–3
 Yunnan **71**, 342
Kerilia jerdoni **64**, 323
Kolpophis annandalei **64**, 323
Krait
 Banded **59**, 313
 Himalayan **59**, 313
 Malayan **59**, 313
 Many-banded **59**, 314
 Red River **59**, 314
 Red-headed **59**, 268, 313–14
 Splendid **59**, 314
 Wanghaoting's 314
Kukri Snake
 Annam 286
 Arakan 290
 Barron's **48**, 286
 Beautiful **49**, 289
 Boo Liat's **49**, 287
 Cambodian 291
 Cantor's **49**, 288
 Chain-banded **49**, 287

Chinese **49**, 287
Dark-spined 294
Deuve's 288
Durheim's 288
Eberhardt's 288–9
Eight-lined **50**, 291–2
Eyed **50**, 291
False Striped 292
Flat-headed 292
Four-lined **50**, 293
Garlanded 294
Grey **49**, 287
Half-keeled **50**, 293–4
Hampton's 289
Javanese 292
Javanese Mountain **48**, 287
Jewelled **49**, 289
Jintakune's 290
Joynson's **50**, 290
Lacroix's **50**, 290
Long-tailed 290
Mandalay **50**, 294
Meyerink's 290–1
Morice's 291
Padang 293
Pegu **49**, 287–8
Petronella's 292
Pulau Weh 292
Purple **50**, 293
Saint Giron's 293
Small-banded **49**, 289
Splendid **50**, 293
Spot-tailed **49**, 288
Spotted **48**, 286
Striped **50**, 294
Three-lined 294
Unicoloured **50**, 289–90
White-barred **48**, 286

Lamprolepis leucosticta 234
Lamprolepis nieuwenhuisii 234–5
Lamprolepis vyneri **31**, 235
Lanthanotus borneensis **28**, 226
Lapemis curtus **64**, 323
 L. c. curtus 323
 L. c. hardwickii 323
Large-toothed Snake
 Northern 278
 Pink **44**, 277
 Red **44**, 277–8
 Vietnamese **44**, 277
 Yellow **44**, 277
Larutia larutensis **31**, 235
Larutia miodactyla 235
Larutia puehensis 235
Larutia seribuatensis **31**, 235–6
Larutia sumatrensis **31**, 236
Larutia trifasciata **31**, 236
Laticauda colubrina **64**, 323
Laticauda laticaudata **64**, 323
Leaf Turtle
 Black-breasted **3**, 170
 Black-bridged **3**, 169
 Common **3**, 169
 Gray 170
 Oldham's **3**, 170
 Sunda 169–70
 Vietnamese **3**, 170
Leiolepis belliana **28**, 226
Leiolepis boehmei **28**, 227
Leiolepis guentherpetersi 227
Leiolepis guttata **28**, 227
Leiolepis ocellata **28**, 227
Leiolepis peguensis 227
Leiolepis reevesii **28**, 227–8

Leiolepis rubritaeniata **28**, 228
Leiolepis triploida **28**, 228
Leopard Gecko
 Cat Ba 199
 Chinese **20**, 199
 Huu Lian 199
 Lichtenfelder's **20**, 199
 Vietnam **20**, 198–9
Lepidochelys olivacea **6**, 173–4
Lepidodactylus lugubris **27**, 223
Lepidodactylus ranauensis **27**, 223
Leptoseps osellai 236
Leptoseps poilani 236
Leptoseps tetradactylus 236
Limbless Skink
 Balinese **29**, 229
 Bornean **29**, 229
 Bukit Larut **31**, 235
 Count Gyldenstolpe's **31**, 234
 Eel-like 234
 Gunung Pueh 235
 Miriam's 230
 Northern 234
 Osella's 236
 Poilapoilane's 236
 Roule's 234
 Seribuat **31**, 235–6
 Sumatran **31**, 236
 Three-lined **31**, 236
Liopeltis frenata **46**, 281
Liopeltis stoliczkae 281
Liopeltis tricolor **46**, 281
Lipinia inexpectata **31**, 236–7
Lipinia miangensis 237
Lipinia nitens **31**, 237
Lipinia relicta 237
Lipinia surda **31**, 237
Lipinia vittigera **31**, 237
 L. v. kronfanum 237
 L. v. vittigera 237
Lissemys punctata **8**, 176–7
 L. p. andersoni 177
 L. p. punctata 177
Lissemys scutata **8**, 177
Litter Snake, Rough-backed **74**, 352
Livorimica bacboensis 237
Long-headed Lizard
 Bornean **18**, 193–4
 Bukit Larut **18**, 193
 Dring's 192
 Flower's **17**, 192
 Javanese **18**, 194
 Kakhien Hills **18**, 193
 Khao Nan 193
 Kon Tum **18**, 194
 Poilane's 193
 Short-footed **17**, 192
 Small-scaled 193
 Sumatran 194
 Yellow-throated **17**, 192
Lophocalotes ludekingi **17**, 190–1
Luperosaurus brooksii 223
Luperosaurus browni **27**, 223–4
Luperosaurus sorok **27**, 224
Luperosaurus yasumai **27**, 224
Lycodon albofuscus **46**, 281
Lycodon aulicus **46**, 281–2
Lycodon butleri **46**, 282
Lycodon capucinus **47**, 282
Lycodon cardamomensis **47**, 282
Lycodon effraenis **47**, 282
Lycodon fasciatus **47**, 282–3
Lycodon futsingensis 284
Lycodon jara 283

Lycodon kundui 283
Lycodon laoensis **47**, 283
Lycodon paucifasciatus **47**, 283
Lycodon ruhstrati **47**, 283–4
Lycodon striatus **47**, 284
Lycodon subcinctus **47**, 284
Lycodon zawi **47**, 284
Lygosoma albopunctatum **32**, 238
Lygosoma angeli 238
Lygosoma anguineum 238
Lygosoma bampfyldei **32**, 238
Lygosoma boehmei **32**, 238
Lygosoma bowringii **32**, 238–9
Lygosoma carinatum 239
Lygosoma corpulentum **32**, 239
Lygosoma frontoparietale 239
Lygosoma haroldyoungi **32**, 239
Lygosoma herberti **32**, 239
Lygosoma isodactylum 239
Lygosoma koratense **32**, 240
Lygosoma lineolatum **32**, 240
Lygosoma opisthorhodum 240
Lygosoma punctatum **33**, 240
Lygosoma quadrupes **33**, 240

Macrocalamus chanardi 284
Macrocalamus gentingensis **48**, 284–5
Macrocalamus jasoni 286
Macrocalamus lateralis **48**, 286
Macrocalamus schulzi **48**, 286
Macrocalamus tweediei **48**, 286
Macrocalamus vogeli **48**, 286
Macropisthodon flaviceps **69**, 336
Macropisthodon plumbicolor **69**, 336
Macropisthodon rhodomelas **69**, 336–7
Maculophis bellus **48**, 285–6
 M. b. bellus 285–6
 M. b. chapaensis 286
Malayemys macrocephala **4**, 171
Malayemys subtrijuga **4**, 171
Mangrove Snake
 Crab-eating **66**, 329
 Yellow-Banded **65**, 325
Manouria emys **7**, 175
 M. e. emys **7**, 175
 M. e. phayrei **7**, 175
Manouria impressa **7**, 175
Mantheyus phuwuanensis **17**, 191
Marsh Snake, Glossy **66**, 329
Marsh Turtle, Black **5**, 173
Mauremys annamensis **4**, 171
Mauremys mutica **4**, 171
 M. m. kami 171
 M. m. mutica 171
Mauremys sinensis **4**, 171
Melanochelys trijuga **5**, 172
 M. t. coronata 172
 M. t. edeniana 172
 M. t. indopeninsularis 172
 M. t. parkeri 172
 M. t. thermalis 172
 M. t. trijuga 172
Mock Viper **69**, 339
 Painted **69**, 339
Mole Skink, Indo-Pacific **29**, 231
Monitor Lizard
 Bengal **37**, 255
 Duméril's **37**, 255
 Rough-necked **37**, 256
 South-east Asian **37**, 255
 Water **37**, 256

Morenia ocellata **5**, 172
Mountain Lizard
 Arrow-headed 190
 Banded 189
 Beccari's Horned **16**, 188
 Chapa **16**, 189
 Crestless **16**, 190
 Hampton's 189
 Kaulback's 189–90
 Swinhoe's **16**, 190
 Yunnan **16**, 190
Mountain Snake, Fruhstorfer's **53**, 302
Mourning Gecko
 Common **27**, 223
 Kinabalu **27**, 223

Naja atra **61**, 316
Naja kaouthia **61**, 316
Naja mandalayensis **61**, 316
Naja siamensis **61**, 316
Naja sputatrix **61**, 316–17
Naja sumatrana **61**, 317
Nilssonia formosus **9**, 177

Oligodon albocinctus **48**, 286
Oligodon annamensis 286
Oligodon annulifer **48**, 286
Oligodon barroni **48**, 286
Oligodon bitorquatus **48**, 287
Oligodon booliati **49**, 287
Oligodon catenatus **49**, 287
Oligodon chinensis **49**, 287
Oligodon cinereus **49**, 287
Oligodon cruentatus **49**, 287–8
Oligodon cyclurus **49**, 288
Oligodon deuvei 288
Oligodon dorsalis **49**, 288
Oligodon durheimi **49**, 288
Oligodon eberhardti 288–9
Oligodon everetti **49**, 289
Oligodon fasciolatus **49**, 289
Oligodon formosanus 289
Oligodon hamptoni 289
Oligodon inornatus **50**, 289–90
Oligodon jintakunei 290
Oligodon joynsoni **50**, 290
Oligodon lacroixi **50**, 290
Oligodon macrurus 290
Oligodon mcdougalli 290
Oligodon meyerinkii 290–1
Oligodon moricei 291
Oligodon mouhoti 291
Oligodon ocellatus **50**, 291
Oligodon octolineatus **50**, 291–2
Oligodon petronellae 292
Oligodon planiceps 292
Oligodon praefrontalis 292
Oligodon propinquus 292
Oligodon pseudotaeniatus 292
Oligodon pulcherrimus 293
Oligodon purpurascens **50**, 293
Oligodon quadrilineatus **50**, 293
Oligodon saintgironsi 293
Oligodon splendidus **50**, 293
Oligodon subcarinatus **50**, 293–4
Oligodon taeniatus **50**, 294
Oligodon theobaldi **50**, 294
Oligodon torquatus 294
Oligodon trilineatus 294
Oligodon vertebralis 294
Ophiophagus hannah **61**, 317
Ophisaurus buettikoferi **19**, 195
Ophisaurus gracilis **19**, 195
Ophisaurus harti **19**, 195

Ophisaurus ludovici 195
Ophisaurus sokolovi **19**, 195
Ophisaurus wegneri 196
Opisthotropis andersonii **69**, 337
Opisthotropis balteata **69**, 337
Opisthotropis daovantieni **69**, 337
Opisthotropis jacobi 337
Opisthotropis lateralis **69**, 337
Opisthotropis maculosus **69**, 338
Opisthotropis rugosa 338
Opisthotropis spenceri 338
Opisthotropis tamdaoensis 338
Opisthotropis typica **69**, 338
Oreocalamus hanitschi 295
Oreocryptophis porphyraceus **51**, 295
  O. p. coxi **51**, 295
  O. p. kawakamii 295
  O. p. laticinctus **51**, 295
  O. p. nigrofasciatus 295
  O. p. porphyraceus 295
  O. p. pulchra 295
  O. p. vaillanti **51**, 295
Orlitia borneensis **5**, 172
Orthriophis cantoris **51**, 295
Orthriophis moellendorffi **51**, 295
Orthriophis taeniurus **51**, 295–6
  O. t. friesi 296
  O. t. grabowski **51**, 296
  O. t. mocquardi 296
  O. t. ridleyi **51**, 296
  O. t. schmackeri 296
  O. t. taeniurus **51**, 296
  O. t. yunnanensis **51**, 296
Ovophis convictus **55**, 305
Ovophis monticola **55**, 305
  O. m. makazayazaya 305
  O. m. monticola 305
  O. m. zayuensis 305
Ovophis tonkinensis **55**, 305

Palea steindachneri **9**, 177
Parachute Gecko
  Horsfield's **27**, 224
  Kinabalu **27**, 224–5
  Kuhl's **27**, 224
  Smooth **27**, 224
  Three-banded **27**, 225
Parahelicops annamensis **51**, 296
Parahelicops boonsongi 296
Paralipinia rara **33**, 240
Pararhabdophis chapaensis 296
Paratapinophis praemaxillaris 338–9
Pareas carinatus **64**, 323
Pareas hamptoni **72**, 345
Pareas macularius **72**, 345
Pareas margaritophorus **72**, 345–6
Pareas monticola **72**, 346
Pareas nuchalis 346
Pareas tamdaoensis 346
Parias hageni **55**, 306
Parias malcolmi **55**, 306
Parias sumatranus **55**, 306
Pelamis platura **64**, 324
Pelochelys cantorii **9**, 177
Pelodiscus sinensis **9**, 177
Phoxophrys borneensis **17**, 191
Phoxophrys cephalus **17**, 191
Phoxophrys nigrilabris **17**, 191
Phoxophrys tuberculatus 191–2
Phu Wua Lizard **17**, 191
Physignathus cocincinus **17**, 192
Pipe Snake
  Common **59**, 312

  Engkari **59**, 312
  Lined **59**, 312
Pit Viper
  Beautiful **54**, 304
  Bornean Keeled Green **57**, 310
  Bornean Palm **57**, 309
  Brongersma's Palm **57**, 309
  Brown-spotted **56**, 308
  Cameron Highlands **56**, 307
  Fan-Si-Pan Horned **56**, 307–8
  Fea's **58**, 311–12
  Gumprecht's **58**, 310
  Hagen's Green **55**, 306
  Hon Son 303
  Javanese Palm **57**, 309
  Jerdon's **56**, 308
  Kanburi **54**, 304
  Kaulback's **56**, 308
  Kinabalu Brown **55**, 305
  Kinabalu Green **55**, 305
  Large-eyed **54**, 304
  Lesser Sunda White-lipped **54**, 303–4
  Malayan **54**, 303
  Malayan Brown **55**, 305
  Mangrove **54**, 304
  Medo **58**, 311
  Montane **55**, 305
  Popes's **56**, 307
  Pulau Tioman **55**, 306
  Sabah Green **56**, 307
  Spot-tailed **54**, 303
  Stejneger's **58**, 311
  Sumatran **55**, 306
  Sumatran Green **56**, 306
  Sumatran Palm **57**, 309
  Thai Peninsula **55**, 306–7
  Three-horned **56**, 308–9
  Tonkin **55**, 305
  Truong Son **57**, 309–10
  Vogel's **58**, 311
  Wagler's Keeled Green **57**, 310
  White-lipped **54**, 303
  Wirot's Palm **57**, 310
  Yunnan **58**, 311
Plagiopholis blakewayi 296–7
Plagiopholis delacouri 297
Plagiopholis nuchalis **51**, 297
Plagiopholis styani **51**, 297
Platysternon megacephalum **7**, 174–5
  P. m. megacephalum **7**, 175
  P. m. peguense **7**, 175
  P. m. shiui **7**, 175
Plestiodon chinensis **33**, 240–1
  P. c. chinensis 241
  P. c. daishanensis 241
  P. c. formosensis 241
  P. c. leucostictus 241
  P. c. pulcher 241
Plestiodon elegans 241
Plestiodon quadrilineatus **33**, 241
Plestiodon tamdaoensis 241
Pond Turtle
  Asian Yellow **4**, 171
  Giant Asian **4**, 171
  Vietnamese **4**, 171
Popeia barati **56**, 306
Popeia buniana **56**, 306
Popeia fucata **55**, 306–7
Popeia nebularis **56**, 307
Popeia popeiorum **56**, 307
Popeia sabahi **56**, 307
Protobothrops cornutus **56**, 307–8
Protobothrops jerdonii **56**, 308

  P. j. bourreti **56**, 308
  P. j. jerdoni **56**, 308
  P. j. xanthomelas 308
Protobothrops kaulbacki **56**, 308
Protobothrops mucrosquamatus **56**, 308
Psammodynastes pictus **69**, 339
Psammodynastes pulverulentus **69**, 339
  P. p. papenfussi 339
  P. p. pulverulentus 339
Psammophis indochinensis **51**, 297
Pseudocalotes brevipes **17**, 192
Pseudocalotes dringi 192
Pseudocalotes flavigula **17**, 192
Pseudocalotes floweri **17**, 192
Pseudocalotes kakhienensis **18**, 193
Pseudocalotes khaonanensis 193
Pseudocalotes larutensis **18**, 193
Pseudocalotes microlepis 193
Pseudocalotes poilani 193
Pseudocalotes sarawacensis **18**, 193–4
Pseudocalotes tympanistriga **18**, 194
Pseudocophotis kontumensis **18**, 194
Pseudocophotis sumatrana 194
Pseudorabdion albonuchalis **52**, 297
Pseudorabdion collaris **52**, 298
Pseudorabdion eiseleti 298
Pseudorabdion longiceps **52**, 298
Pseudorabdion saravacensis 298
Pseudoxenodon bambusicola **72**, 346–7
Pseudoxenodon baramensis 347
Pseudoxenodon inornatus 347
Pseudoxenodon jacobsonii 347
Pseudoxenodon karlschmidtii **72**, 347
Pseudoxenodon macrops **72**, 347
  P. m. macrops 347
  P. m. sinensis 347
Ptyas carinata **52**, 298
Ptyas dhumnades **52**, 298–9
  P. d. dhumnades 299
  P. d. montana 298–9
  P. d. oshimai 299
Ptyas fusca **52**, 299
Ptyas korros **52**, 299
Ptyas mucosa **52**, 299
Ptyas nigromarginata **52**, 299
Ptychozoon horsfieldii **27**, 224
Ptychozoon kuhli **27**, 224
Ptychozoon lionotum **27**, 224
Ptychozoon rhacophorus **27**, 224–5
Ptychozoon trinotaterra **27**, 225
Ptyctolaemus collicristatus **18**, 194
Ptyctolaemus gularis **18**, 194
Python
  Bornean Short **38**, 258
  Brongersma's Short **38**, 258
  Reticulated **38**, 258
  Sumatran Short **38**, 259
Python breitensteini **38**, 258
Python brongersmai **38**, 258
Python curtus **38**, 259
Python molurus **38**, 259
  P. m. bivittatus 259
  P. m. molurus 259

Racer
  Cave **51**, 295–6
  Red-tailed **46**, 280–1
Rafetus swinhoei **9**, 177–8
Ramphotyphlops albiceps **73**, 348
Ramphotyphlops braminus **73**, 348
Ramphotyphlops lineatus **73**, 348
Ramphotyphlops lorenzi 348
Ramphotyphlops olivaceus 348
Rat Snake
  Black-striped **52**, 298–9
  Enggano **43**, 272–3
  Green **52**, 299
  Indian **52**, 299
  Javan **52**, 299
  Keeled **45**, 279
  Malayan Keeled **52**, 298
  White-bellied **52**, 299
Reed Snake
  Battersby's 265
  Bengkulu 264–5
  Bicoloured **41**, 265
  Bornean **41**, 265–6
  Brown **42**, 270
  Buch's 265
  Chan-Ard's 284
  Collared **42**, 268
  Döderlein's 265–6
  Dwarf **52**, 298
  Eiselt's 266
  Eiselt's Dwarf 298
  Everett's **41**, 266
  Forcart's 266
  Genting Highlands **48**, 284–5
  Gimlett's 266
  Grabowsky's **41**, 266
  Hillenius's **41**, 267
  Inger's **41**, 267
  Jason's 285
  Javanese 267
  Kapuas 269
  Lined **41**, 267
  Linnaeus's **42**, 268
  Low's 268
  Lumholz's 269
  Mechel's 269
  Mocquard's **52**, 298
  Mountain **51**, 295
  Northern **42**, 270–1
  Padang 264
  Prakke's 270
  Rebentisch's 270
  Red-headed **42**, 270
  Sarawak 298
  Schmidt's **42**, 270
  Schulz's **48**, 285
  Short-tailed **42**, 271
  Slender 267
  Striped **48**, 285
  Stripe-necked 269
  Sumatran 271
  Thanh's 271
  Thick 265
  Tweedie's **48**, 285
  Ulmer's 271
  Variable **42**, 268–9
  Vogel's **48**, 285
  White-bellied 264
  White-collared **52**, 297
  White-striped 267–8
  Yellow-bellied 271
  Yellow-spotted 269
  Yunnan 272
Rhabdophis angelii 339

Rhabdophis callichromus 339
Rhabdophis chrysargoides **70**, 339–40
Rhabdophis chrysargos **70**, 340
Rhabdophis conspicillatus **70**, 340
Rhabdophis himalayanus **70**, 340
Rhabdophis leonardi **70**, 340
Rhabdophis murudensis **70**, 340–1
Rhabdophis nigrocinctus **70**, 341
Rhabdophis nuchalis **70**, 341
  R. n. nuchalis 341
  R. n. pentasupralabialis 341
Rhabdophis subminiatus **70**, 341
  R. s. helleri 341
  R. s. subminiatus 341
Rhabdophis tigrinus **70**, 341–2
  R. t. lateralis 341–2
  R. t. tigrinus 342
Rhabdops bicolor 299–300
Rhadinophis frenatus **53**, 300
Rhadinophis prasinus **53**, 300
Rhynchophis boulengeri **53**, 300
Ringneck
  Stoliczka's 281
  Tricoloured **46**, 281
Rock Python, Indian **38**, 259
Rough Water Snake
  Kloss's **74**, 351–2
  Smith's 352

Sacalia bealei **5**, 172
Sacalia quadriocellata **5**, 172–3
Sand Snake, Indo-Chinese **51**, 297
Scincella doriae 241
Scincella kohtaoensis 242
Scincella melanosticta **33**, 242
Scincella ochracea 242
Scincella punctatolineata 242
Scincella reevesii **33**, 242
Scincella rupicola 242
Scincella victoriana **33**, 243
Sea Krait
  Large-scaled **64**, 323
  Yellow-lipped **64**, 323
Sea Snake
  Aagaard's **62**, 318–19
  Annandale's **64**, 323
  Annulated **62**, 320
  Anomalous 324
  Banded **63**, 320
  Beaded **62**, 318
  Beaked **62**, 318
  Black-headed **62**, 319
  Black-ringed 321
  Blue-grey **62**, 319
  Broad-headed **63**, 322
  Cantor's 319–20
  Captain Belcher's **62**, 319
  Garlanded **64**, 322–3
  Grey **64**, 324
  Horned **62**, 318
  Kloss's **63**, 320
  Lambert's 320–1
  Lesser Dusky **63**, 321
  Narrow-headed **63**, 320
  Narrow-necked **64**, 322
  Obscure-patterned 321
  Ornate **63**, 321–2
  Persian Gulf **63**, 321
  Plain 320
  Rajah Brook's **62**, 319
  Saddle-backed **64**, 323
  Short **64**, 323

Short-headed 322
Sibau River **63**, 322
Spiral **63**, 322
Stoke's **62**, 318
Two-wattled 319
Yellow-bellied **64**, 324
Sea Turtle
  Hawksbill **6**, 173
  Leatherback **6**, 174
  Loggerhead **6**, 173
  Olive Ridley **6**, 173–4
Shinisaurus crocodilurus **37**, 255
Shrub Lizard
  Black-lipped **17**, 191
  Bornean **17**, 191
  Brown **10**, 179–80
  Large-headed **17**, 191
  Long-snouted **10**, 179
  Ornate **10**, 180
  Spiny-headed **17**, 191
  Sumatran 191–2
Sibynophis bistrigatus **53**, 300
Sibynophis chinensis **53**, 300–1
Sibynophis collaris **53**, 301
Sibynophis geminatus **53**, 301
  S. g. geminatus 301
  S. g. insularis 301
Sibynophis melanocephalus **53**, 301
Sibynophis triangularis **53**, 301
Siebenrockiella crassicollis **5**, 173
Sinomicrurus kelloggi **61**, 317
Sinomicrurus macclellandi **61**, 317–18
Sinonatrix aequifasciata **71**, 342
Sinonatrix annularis **71**, 342
Sinonatrix percarinata **71**, 342
Sinonatrix yunnanensis **71**, 342
Skink
  Bacbo 237
  Four-fingered 236
  Mangrove **29**, 230
  One-fingered 235
Slender Skink, Chinese **29**, 228–9
Slider, Red-eared **7**, 174
Slug-eating Snake,
  Barred 346
  Blunt-headed **72**, 344
  Hampton's **72**, 345
  Keeled **72**, 345
  Malayan **72**, 344–5
  Montane **72**, 346
  Mountain **72**, 345
  Smooth **72**, 344
  Spotted **72**, 345
  Tamdao 346
  White-spotted **72**, 345–6
Snail-eating Turtle
  Malayan **4**, 171
  Mekong **4**, 171
Snake, Elephant Trunk **38**, 256–7
  Iguana-jawed 336
  Iridescent **39**, 260
  Orange-bellied **45**, 280
  Stripe-necked **46**, 281
  Tentacled **66**, 329
  Wart **38**, 256
Snake-eyed Skink
  Balinese **29**, 229
  Beach **29**, 229
  Blue-tailed **29**, 229
  Bouton 229
Softshell Turtle

  Asian **8**, 176
  Asian Giant **9**, 177
  Burmese Narrow-headed **8**,176
  Chinese **9**, 177
  Indo-Chinese Giant **9**, 177–8
  Malayan **8**, 176
  Narrow-headed **8**, 176
  Wattle-necked **9**, 177
Sphenomorphus aesculeticola **34**, 243
Sphenomorphus alfredi 243
Sphenomorphus anomalopus 243
Sphenomorphus buenloicus 243
Sphenomorphus buettikoferi 243
Sphenomorphus bukitensis 243–4
Sphenomorphus butleri 244
Sphenomorphus cameronicus 244
Sphenomorphus cophias 244
Sphenomorphus crassus 244
Sphenomorphus cryptotis **34**, 244
Sphenomorphus cyanolaemus **34**, 244–5
Sphenomorphus devorator **34**, 245
Sphenomorphus florensis 245
  S. f. barbouri 245
  S. f. florensis 245
  S. f. weberi 245
Sphenomorphus grandisonae 245
Sphenomorphus haasi 245
Sphenomorphus hallieri 245
Sphenomorphus helenae 246
Sphenomorphus indicus **34**, 246
Sphenomorphus ishaki **34**, 246
Sphenomorphus kinabaluensis **34**, 246
Sphenomorphus langkawiensis 246
Sphenomorphus lineopunctulatus 246
Sphenomorphus maculatus **35**, 247
Sphenomorphus maculicollus 247
Sphenomorphus malayanus 247
Sphenomorphus mimicus **35**, 247
Sphenomorphus modiglianii 247
Sphenomorphus multisquamatus **35**, 247
Sphenomorphus murudensis **35**, 247–8
Sphenomorphus necopinatus 248
Sphenomorphus praesignis **35**, 248
Sphenomorphus puncticentralis 248
Sphenomorphus rufocaudatus 248
Sphenomorphus sabanus **35**, 248–9
Sphenomorphus sanctus 249
Sphenomorphus scotophilus **35**, 249
Sphenomorphus shelfordi 249
Sphenomorphus sibuensis 249
Sphenomorphus stellatus **36**, 249–50
Sphenomorphus tanahtinggi 250
Sphenomorphus temminckii **36**, 250
Sphenomorphus tenuiculus 250
Sphenomorphus tersus **36**, 250
Sphenomorphus tridigitus 250
Sphenomorphus tritaeniatus 250–1
Sphenomorphus vanheurni 251
  S. v. balicus 251
  S. v. vanheurni 251
Spiny Lizard
  Crowned **10**, 178–9

  Greater **10**, 178
  Indo-Chinese **10**, 178
  Masked **10**, 179
  Natalia's **10**, 179
  Scale-bellied **10**, 179
Spiny-jawed Snake
  Bornean 353
  Schäfer's **74**, 353
Spotted Gecko, Burmese 221
Stegonotus borneensis **53**, 301–2
Stoliczkia borneensis **74**, 352
Stream Snake
  Anderson's **69**, 337
  Brown 338–9
  Dao Van Tien's **69**, 337
  Hainanese **69**, 337
  Jacob's 337
  Javanese 338
  Man-Son Mountain **69**, 337
  Spencer's 338
  Spotted Mountain **69**, 338
  Stoliczka's **74**, 352
  Tam Dao 338
Striped Skink
  Bornean **31**, 236–7
  Common **31**, 237
  False **33**, 240
  Malayan **31**, 237
  Mentawai 237
  Pulau Miang 237
  Sarawak **31**, 237
Stripe-necked Turtle, Chinese **4**, 171
Sun Skink, Common **30**, 233
Sunbeam Snake **74**, 353
  Hainan **74**, 352
Supple Skink
  Angel's 238
  Annam **32**, 239
  Böhme's **32**, 238
  Bowring's **32**, 238–9
  Burmese 238
  Central 239
  Harold Young's **32**, 239
  Herbert's **32**, 239
  Indragiri 240
  Keeled 239
  Korat **32**, 240
  Lined **32**, 240
  Pygmy 239
  Short-limbed **33**, 240
  Spotted **33**, 240
  White-spotted **32**, 238

Takydromus hani **28**, 225
Takydromus kuehnei **28**, 225
  T. k. kuehnei 225
  T. k. vietnamensis 225
Takydromus sexlineatus **28**, 225–6
  T. s. ocellatus 225
  T. s. sexlineatus 225
Takydromus wolteri **28**, 226
Temple Turtle, Yellow-headed **4**, 170
Terrapin
  Burmese Painted **2**, 168
  Northern River 168
  Painted **2**, 168
  Southern River **2**, 167–8
Tetralepis fruhstorferi **53**, 302
Thalassophis anomalus 324
Thalassophis viperina **64**, 324
Thaumatorhynchus brooksi **18**, 194–5

Tomistoma schlegelii **1**, 167
Tortoise
    Asian Giant **7**, 175
    Burmese Star **7**, 175
    Elongated **7**, 175
    Impressed **7**, 175
    Indian Star 175
Trachemys scripta **7**, 174
Tree Lizard, Boulenger's **12**, 183
Tree Skink
    Grey **29**, 230
    Half-banded **29**, 230
    Nieuwenhuis's 234–5
    Olive **29**, 230
    Striped Bornean **29**, 228
    Vyner's **31**, 235
    White-spotted 234
Tree Snake, Royal **46**, 280
Triceratolepidophis sieversorum **56**, 308–9
Trimeresurus andalasensis **57**, 309
Trimeresurus borneensis **57**, 309
Trimeresurus brongersmai **57**, 309
Trimeresurus puniceus **57,** 309
Trimeresurus truongsonensis **57**, 309–10
Trimeresurus wiroti **58**, 310
Trinket Snake
    Copper-head **43**, 273
    Dice-like **48**, 285–6
    Eastern **51**, 295
    Green **53**, 300
    Indonesian **43**, 273
    Khasi Hills **53**, 300
    Mandarin **45**, 279
    Moellendorff's **51**, 295
    Philippine **43**, 273
    Red Bamboo **51**, 295
    Yellow-striped **43**, 273
Tropidolaemus subannulatus **58**, 310
Tropidolaemus wagleri **58**, 310
Tropidophorus baviensis **36**, 251
Tropidophorus beccarii **36**, 251
Tropidophorus berdmorei **36**, 251
Tropidophorus brookei **36**, 251–2
Tropidophorus cocincinensis **36**, 252
Tropidophorus hainanus 252
Tropidophorus hangnam **36**, 252
Tropidophorus iniquus 252
Tropidophorus laotus **36**, 252
Tropidophorus latiscutatus 252–3
Tropidophorus matsuii 253
Tropidophorus microlepis 253
Tropidophorus micropus 253

Tropidophorus mocquardii **37**, 253
Tropidophorus murphyi **37**, 253
Tropidophorus noggei **37**, 253–4
Tropidophorus perplexus **37**, 254
Tropidophorus robinsoni 254
Tropidophorus sinicus **37**, 254
Tropidophorus thai 254
Turtle
    Big-headed **7**, 174–5
    Green **6**, 173
Typhlops ater **73**, 348–9
Typhlops bisubocularis 349
Typhlops bothriorhynchus 349
Typhlops diardii **73**, 349
Typhlops floweri 349
Typhlops giadinhensis 349
Typhlops hypsobothrius 349
Typhlops jerdoni **73**, 349–50
Typhlops klemmeri 350
Typhlops koekkoeki 350
Typhlops muelleri **73**, 350
Typhlops oatesii 350
Typhlops ozakiae 350
Typhlops porrectus 350
Typhlops roxaneae 350–1
Typhlops siamensis 351
Typhlops trangensis 351

Varanus bengalensis **37**, 255
Varanus dumerilii **37**, 255
Varanus nebulosus **37**, 255
Varanus rudicollis **37**, 256
Varanus salvator **37**, 256
    V. s. bivittatus 256
    V. s. macromaculatus 256
    V. s. salvator 256
Vietnamese Skink, Rough 254
Vietnascincus rugosus 254
Vine Snake
    Indian **39**, 260
    Malayan **39**, 260
    Oriental **39**, 260
    River **39**, 259–60
    Speckle-headed **39**, 259
Viper
    Russell's **58**, 312
    Sharp-nosed **54**, 304–5
Viridovipera gumprechti **58**, 310
Viridovipera medoensis **58**, 311
Viridovipera stejnegeri **58**, 311
Viridovipera vogeli **58**, 311
Viridovipera yunnanensis **58**, 311

Water Dragon, Indo-Chinese **17**, 192

Water Skink
    Bavi **36**, 251
    Beccari's **36**, 251
    Berdmore's **36**, 251
    Broad-scaled 252–3
    Brooke's **36**, 251–2
    Chinese **37**, 254
    Hainan 252
    Indo-Chinese **36**, 252
    Laotian **36**, 252
    Matsui's 253
    Mocquard's **37**, 253
    Murphy's **37**, 253
    Nogge's **37**, 253–4
    Perplexing **37**, 254
    Robinson's 254
    Small-footed 253
    Small-scaled 253
    Spiny-tailed **36**, 252
    Sungei Kajan 252
    Thai 254
Water Snake
    Bennett's **65**, 325
    Black-spotted **65**, 327
    Bocourt's **65**, 325–6
    Brown-spotted 327
    Cambodian Puff-faced **66**, 330
    Chan-Ard's 326
    Chinese **65**, 326
    Corrugated **69**, 338
    Dog-faced **65**, 325
    Grey **66**, 328
    Gyi's **65**, 326–7
    Indian 327
    Indo-Chinese **66**, 328–9
    Jagor's **65**, 327
    Keel-bellied **65**, 324–5
    Kinabalu 336
    Long-tailed 327
    Marquis Doria's **65**, 326
    Pahang 327–8
    Puff-faced **66**, 329–30
    Rainbow **65**, 326
    Siebold's 328
    Smith's 328
    Spotted **66**, 328
    Voris's 329
    White-spotted 325
    Yellow-banded 325
    Yellow-spotted **68**, 336
Whip Snake, Keel-bellied **45**, 278–9
Wolf Snake
    Annam **47**, 283
    Banded **47**, 282–3
    Barred **47**, 284

    Brown **47**, 282
    Butler's **46**, 282
    Cardamom Mountains **47**, 282
    Dusky **46**, 281
    Indian **46**, 281–2
    Island **47**, 282
    Kundu's 283
    Laos **47**, 283
    Ruhstrat's **47**, 283–4
    White-banded **47**, 284
    Yellow-speckled 283
    Zaw's **47**, 284
Worm Gecko
    Chapa 222
    Common **27**, 223
    Hartert's **27**, 222
    Yunnan 223
Worm Lizard
    Alfred's 196
    Boo Liat's **19**, 196
    Bourret's 196
    Condao 197
    De Harveng's 196
    De Zwaan's **19**, 196–7
    Greer's **19**, 197
    Inger's **19**, 197
    Montane 197
    Smith's 197
    Somsak's 198
    Tioman **19**, 198
    Voris's **19**, 198
    White-tailed **19**, 197

Xenelaphis ellipsifer **53**, 302
Xenelaphis hexagonotus **53**, 302
Xenochrophis bellula 342
Xenochrophis flavipunctatus **71**, 342–3
Xenochrophis maculatus **71**, 343
Xenochrophis melanzostus **71**, 343
Xenochrophis piscator **71**, 343
Xenochrophis punctulatus 343
Xenochrophis sanctijohannis **71**, 343
Xenochrophis trianguligerus **71**, 343–4
Xenochrophis vittatus **71**, 344
Xenodermus javanicus **74**, 352
Xenopeltis hainanensis **74**, 352
Xenopeltis unicolor **74**, 353
Xenophidion acanthognathus 353
Xenophidion schaeferi **74**, 353